SELECTED SOLUTIONS MANUAL

Joseph Topich
Virginia Commonwealth University

CHEMISTRY

Seventh Edition

John E. McMurry
Robert C. Fay
Jill K. Robinson

PEARSON

Editor-in-Chief: Jeanne Zalesky
Acquisitions Editor: Chris Hess
Project Manager: Crissy Dudonis
Program Manager: Lisa Pierce
Program Management Team Lead: Kristen Flatham
Project Management Team Lead: David Zielonka
Compositor: Lumina Datamatics Ltd.
Cover Designer: Elise Lansdon
Executive Marketing Manager: Will Moore
Cover Photo Credit: Dr. Keith Wheeler / Science Source

3 16

PEARSON

www.pearsonhighered.com

ISBN 10: 0-133-88879-7
ISBN 13: 978-0-13-388879-9

Contents

Preface

Problem solving is your key to success in chemistry! CHEMISTRY, 7/e by McMurry, Fay, and Robinson contains thousands of questions and problems for you to answer. Develop a problem-solving strategy. Read each problem carefully. List the information contained in the problem. Understand what the problem is asking. Use your knowledge of chemistry principles to identify connections between the information in the problem and the solution that you are seeking. Set up and attempt to solve the problem. Look at your answer. Is your answer reasonable? Are the units correct? Then, and only then, check your answer with the *Selected Solutions Manual*.

The *Selected Solutions Manual* to accompany CHEMISTRY, 7/e by McMurry, Fay, and Robinson contains the solutions to all in-chapter and even-numbered end-of-chapter questions and problems.

I have worked to ensure that the solutions in this manual are as error free as possible. Solutions have been double-checked and in many cases triple-checked. Small differences in numerical answers between student results and those in the *Selected Solutions Manual* may result because of rounding and significant figure differences. It also should be noted that there is, in many instances, more than one acceptable setup for a problem.

I would like to thank John McMurry, Robert Fay, and Jill Robinson for the opportunity to contribute to their CHEMISTRY package. I also want to thank them for their helpful comments as I worked on this solutions manual. I also want to acknowledge and thank Alton Hassell (accuracy checker) and the entire Pearson staff. Finally, I want to thank in a very special way my wife, Ruth, and our daughter, Judy, for their constant encouragement and support as I worked on this project.

Joseph Topich
Department of Chemistry
Virginia Commonwealth University

1 Chemical Tools: Experimentation and Measurement

1.1 (a) 5×10^{-9} m; 5 nm (b) 4.0075017×10^{7} m; 40.075017 Mm

1.2 (a) 7×10^{-5} m (b) 2×10^{13} kg

1.3 $$^{\circ}C = \frac{5}{9} \times (^{\circ}F - 32) = \frac{5}{9} \times (1474 - 32) = 801\ ^{\circ}C$$

$$K = {}^{\circ}C + 273.15 = 801 + 273.15 = 1074.15\ K \text{ or } 1074\ K$$

1.4 The melting point of gallium is converted from 302.91 K to °F for comparison.

$$^{\circ}C = K - 273.15 = 302.91 - 273.15 = 29.76\ ^{\circ}C$$

$$^{\circ}F = (\frac{9}{5} \times {}^{\circ}C) + 32 = (\frac{9}{5} \times 29.76) + 32 = 85.57\ ^{\circ}F$$

The temperature in the compartment (88 °F) is above the melting point, so the liquid state exists.

1.5 $$\text{Volume} = 9.37\ g \times \frac{1\ mL}{1.483\ g} = 6.32\ mL$$

1.6 Bracelet mass = 80.0 g

Bracelet volume = 17.61 mL – 10.0 mL = 7.61 mL

$$\text{Bracelet density} = \frac{80.0\ g}{7.61\ mL} = 10.5\ g/mL$$

The density of the bracelet matches the density of silver. Since density is one way to identify an unknown substance, it is likely that the bracelet is made of pure silver.

1.7 (a) $$E_K = \frac{1}{2}mv^2 = \frac{1}{2}(6.6 \times 10^{-27}\ kg)\left(\frac{1.5 \times 10^{7}\ m}{s}\right)^2 = 7.4 \times 10^{-13}\ \frac{kg{\cdot}m^2}{s^2} = 7.4 \times 10^{-13}\ J$$

(b) 0.74 pJ

1.8 $450\ g = 0.450\ kg$; $E_K = 406\ J = 406\ \frac{kg \cdot m^2}{s^2}$

$$E_K = \frac{1}{2}mv^2$$

$$v = \sqrt{\frac{2 \times E_K}{m}} = \sqrt{\frac{2 \times 406\ kg{\cdot}m^2/s^2}{0.450\ kg}} = 42.5\ m/s$$

1.9 (a) 76.600 kJ has 5 significant figures because zeros at the end of a number and after the decimal point are always significant.
(b) 4.502 00 x 10^3 g has 6 significant figures because zeros in the middle of a number are significant and zeros at the end of a number and after the decimal point are always significant.
(c) 3000 nm has 1, 2, 3, or 4 significant figures because zeros at the end of a number and before the decimal point may or may not be significant.
(d) 0.003 00 mL has 3 significant figures because zeros at the beginning of a number are not significant and zeros at the end of a number and after the decimal point are always significant.
(e) 18 students has an infinite number of significant figures because this is an exact number.
(f) 3 x 10^{-5} g has 1 significant figure.
(g) 47.60 mL has 4 significant figures because a zero at the end of a number and after the decimal point is always significant.
(h) 2070 mi has 3 or 4 significant figures because a zero in the middle of a number is significant and a zero at the end of a number and before the decimal point may or may not be significant.

1.10 To indicate the uncertainty in a measurement, the value you record should use all the digits you are sure of plus one additional digit that you estimate. The volume can be read to the tenths place and therefore the hundredths place should be estimated. The volume reported to the correct number of significant figures is 4.55 mL.

1.11 (a) In figure (c) darts are scattered (low precision) and are away from the bull's-eye (low accuracy).
(b) In figure (b) darts are clustered together (high precision) and hit the bull's-eye (high accuracy).

1.12 The three measurements are 0.7783 g, 0.7780 g, and 0.7786 g. There is little variation between the three measurements so they have fairly high precision. However, the measurements are all lower than the true value and therefore, the accuracy is low.

1.13 (a)

$$\begin{array}{r} 24.567\ \ \ \ \text{g} \\ +\ 0.044\ 78\ \text{g} \\ \hline 24.611\ 78\ \text{g} \end{array}$$

This result should be expressed with 3 decimal places. Because the digit to be dropped (7) is greater than 5, round up. The result is 24.612 g (5 significant figures).

(b) 4.6742 g / 0.003 71 L = 1259.89 g/L
0.003 71 has only 3 significant figures so the result of the division should have only 3 significant figures. Because the digit to be dropped (first 9) is greater than 5, round up. The result is 1260 g/L (3 significant figures), or 1.26 x 10^3 g/L.

(c)

$$\begin{array}{r} 0.378\ \ \ \text{mL} \\ +\ 42.3\ \ \ \ \ \ \ \text{mL} \\ -\ \ 1.5833\ \text{mL} \\ \hline 41.0947\ \text{mL} \end{array}$$

This result should be expressed with 1 decimal place. Because the digit to be dropped (9) is greater than 5, round up. The result is 41.1 mL (3 significant figures).

1.14 NaCl mass = 36.2365 g − 35.6783 g = 0.5582 g
NaCl concentration = 0.5582 g/25.0 mL = 0.0223 g/mL = 2.23×10^{-2} g/mL

1.15 1 carat = 200 mg = 200×10^{-3} g = 0.200 g

Mass of Hope Diamond in grams = $44.4\ \text{carats} \times \dfrac{0.200\ \text{g}}{1\ \text{carat}} = 8.88\ \text{g}$

1 ounce = 28.35 g

Mass of Hope Diamond in ounces = $8.88\ \text{g} \times \dfrac{1\ \text{ounce}}{28.35\ \text{g}} = 0.313\ \text{ounces}$

1.16 Volume of Hope Diamond = $8.88\ \text{g} \times \dfrac{1\ \text{cm}^3}{3.52\ \text{g}} = 2.52\ \text{cm}^3$

1.17 (a) area = $113.112\ \text{in}^2 \times \left(\dfrac{2.54\ \text{cm}}{1\ \text{in}}\right)^2 = 729.753\ \text{cm}^2$

(b) volume = $355\ \text{mL} \times \dfrac{1\ \text{cm}^3}{1\ \text{mL}} \times \left(\dfrac{1 \times 10^{-2}\ \text{m}}{1\ \text{cm}}\right)^3 = 3.55 \times 10^{-4}\ \text{m}^3$

1.18 Volume of a cylinder = $\pi r^2 h$
Cell radius = 6×10^{-6} m/2 = 3×10^{-6} m
Cell volume = $\pi(3 \times 10^{-6}\ \text{m})^2(2 \times 10^{-6}\ \text{m}) = 6 \times 10^{-17}\ \text{m}^3$

Cell volume = $6 \times 10^{-17}\ \text{m}^3 \times \left(\dfrac{1\ \text{cm}}{1 \times 10^{-2}\ \text{m}}\right)^3 = 6 \times 10^{-11}\ \text{cm}^3 = 6 \times 10^{-11}\ \text{mL}$

Cell volume = $6 \times 10^{-11}\ \text{mL} \times \dfrac{1 \times 10^{-3}\ \text{L}}{1\ \text{mL}} = 6 \times 10^{-14}\ \text{L} = 0.06 \times 10^{-12}\ \text{L} = 0.06\ \text{pL}$

1.19 The diameter of a human hair (~1×10^{-5} m) is approximately 1,000 times larger than the diameter of a 10 nm nanoparticle. (b) A red blood cell (~1×10^{-6} m) is approximately 10,000 times larger than a glucose molecule (1×10^{-10} m).

1.20 Assume that individual atoms pack as cubes in the nanoparticle.
5.0 nm = 5.0×10^{-9} m; 10.0 nm = 10.0×10^{-9} m; 250 pm = 250×10^{-12} m
Atom volume = $(250 \times 10^{-12}\ \text{m})^3 = 1.6 \times 10^{-29}\ \text{m}^3$
(a) Particle volume = $(5.0 \times 10^{-9}\ \text{m})^3 = 1.3 \times 10^{-25}\ \text{m}^3$

Atoms/particle = $\dfrac{\text{particle volume}}{\text{atom volume}} = \dfrac{1.3 \times 10^{-25}\ \text{m}^3/\text{particle}}{1.6 \times 10^{-29}\ \text{m}^3/\text{atom}} = 8125\ \text{atoms/particle}$

Atom face area = $(250 \times 10^{-12}\ \text{m})^2 = 6.25 \times 10^{-20}\ \text{m}^2$
Particle face area = $(5.0 \times 10^{-9}\ \text{m})^2 = 2.5 \times 10^{-17}\ \text{m}^2$

Atoms/particle face = $\dfrac{\text{particle face area}}{\text{atom face area}} = \dfrac{2.5 \times 10^{-17}\ \text{m}^2/\text{particle}}{6.25 \times 10^{-20}\ \text{m}^2/\text{atom}} = 400\ \text{atoms/particle face}$

% atoms on surface = $\dfrac{(6\ \text{faces})(400\ \text{atoms/face})}{8125\ \text{atoms}} \times 100 = 30\%$

(b) Particle volume = $(10.0 \times 10^{-9}\ m)^3 = 1.0 \times 10^{-24}\ m^3$

$$\text{Atoms/particle} = \frac{\text{particle volume}}{\text{atom volume}} = \frac{1.0 \times 10^{-24}\ m^3/\text{particle}}{1.6 \times 10^{-29}\ m^3/\text{atom}} = 62{,}500\ \text{atoms/particle}$$

Atom face area = $(250 \times 10^{-12}\ m)^2 = 6.25 \times 10^{-20}\ m^2$
Particle face area = $(10.0 \times 10^{-9}\ m)^2 = 1.0 \times 10^{-16}\ m^2$

$$\text{Atoms/particle face} = \frac{\text{particle face area}}{\text{atom face area}} = \frac{1.0 \times 10^{-16}\ m^2/\text{particle}}{6.25 \times 10^{-20}\ m^2/\text{atom}} = 1{,}600\ \text{atoms/particle face}$$

$$\%\ \text{atoms on surface} = \frac{(6\ \text{faces})(1{,}600\ \text{atoms/face})}{62{,}500\ \text{atoms}} \times 100\% = 15\%$$

1.21 (a) Diameter = 5.0 nm = 5.0×10^{-9} m; radius = 2.5 nm = 2.5×10^{-9} m
$SA = 4\pi r^2 = 4\pi(2.5 \times 10^{-9}\ m)^2 = 7.9 \times 10^{-17}\ m^2$

$$SA = (7.9 \times 10^{-17}\ m^2)\left(\frac{1\ \mu m}{1 \times 10^{-6}\ m}\right)^2 = 7.9 \times 10^{-5}\ \mu m^2$$

Diameter = 5.0 μm = 5.0×10^{-6} m; radius = 2.5 μm = 2.5×10^{-6} m
$SA = 4\pi r^2 = 4\pi(2.5 \times 10^{-6}\ m)^2 = 7.9 \times 10^{-11}\ m^2$

$$SA = (7.9 \times 10^{-11}\ m^2)\left(\frac{1\ \mu m}{1 \times 10^{-6}\ m}\right)^2 = 79\ \mu m^2$$

(b) Diameter = 5.0 nm = 5.0×10^{-9} m; radius = 2.5 nm = 2.5×10^{-9} m

$$\text{Volume} = \frac{4}{3}\pi r^3 = \frac{4}{3}\pi(2.5 \times 10^{-9}\ m)^3 = 6.5 \times 10^{-26}\ m^3$$

$$\text{Volume} = (6.5 \times 10^{-26}\ m^3)\left(\frac{1\ \mu m}{1 \times 10^{-6}\ m}\right)^3 = 6.5 \times 10^{-8}\ \mu m^3$$

Diameter = 5.0 μm = 5.0×10^{-6} m; radius = 2.5 μm = 2.5×10^{-6} m

$$\text{Volume} = \frac{4}{3}\pi r^3 = \frac{4}{3}\pi(2.5 \times 10^{-6}\ m)^3 = 6.5 \times 10^{-17}\ m^3$$

$$\text{Volume} = (6.5 \times 10^{-17}\ m^3)\left(\frac{1\ \mu m}{1 \times 10^{-6}\ m}\right)^3 = 65\ \mu m^3$$

(c) 5.0 nm particle $\dfrac{SA}{Volume} = \dfrac{7.9 \times 10^{-5}\ \mu m^2}{6.5 \times 10^{-8}\ \mu m^3} = 1{,}200\ \mu m^{-1}$

5.0 μm particle $\dfrac{SA}{Volume} = \dfrac{79\ \mu m^2}{65\ \mu m^3} = 1.2\ \mu m^{-1}$

(d) $1{,}200\ \mu m^{-1}/1.2\ \mu m^{-1} = 1{,}000$ times

1.22 (a) As particle size decreases there is a larger fraction of atoms on the surface and surface atoms are more reactive.
(b) Smaller particles maximize the number of reactive atoms while minimizing the total amount of the expensive metal.
(c) Color, electrical conductivity, or melting point.

Conceptual Problems

1.24 The level of the liquid in the thermometer is just past the 32 °C mark on the thermometer. The temperature is 32.2°C (3 significant figures).

1.26

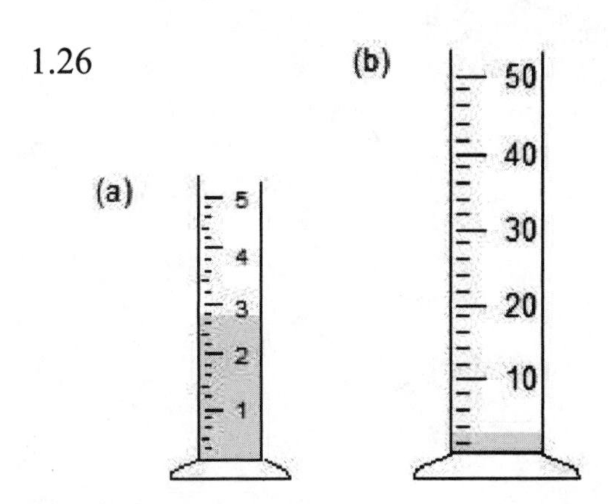

The 5 mL graduated cylinder is marked every 0.2 mL and can be read to ± 0.02 mL. The 50 mL graduated cylinder is marked every 2 mL and can only be read to ± 0.2 mL. The 5 mL graduated cylinder will give more accurate measurements.

Section Problems

Scientific Method (Section 1.1)

1.28 (a) experiment (b) hypothesis (c) observation

1.30 (c) is the correct statement.

1.32 Molecular models are simplified representations of more complex systems. These models can be used to visualize structure-function relationships that help make theories more concrete.

1.34 (a), (b) and (d) are quantitative. (c) and (e) are qualitative.

Units and Significant Figures (Sections 1.2–1.10)

1.36 Mass measures the amount of matter in an object, whereas weight measures the pull of gravity on an object by the earth or other celestial body.

1.38 (a) kilogram, kg (b) meter, m (c) kelvin, K
(d) cubic meter, m^3 (e) joule, $(kg \cdot m^2)/s^2$ (f) kg/m^3 or g/cm^3

1.40 A Celsius degree is larger than a Fahrenheit degree by a factor of $\frac{9}{5}$.

1.42 The volume of a cubic decimeter (dm^3) and a liter (L) are the same.

1.44 (a) and (b) are exact numbers because they are both definitions.
(c) and (d) are not exact numbers because they result from measurements.

1.46 cL is centiliter (10^{-2} L)

1.48 (a) Convert cm to km and compare the two quantities.

$$5.63 \text{ x } 10^6 \text{ cm x } \frac{1 \text{ x } 10^{-2} \text{ m}}{1 \text{ cm}} \text{ x } \frac{1 \text{ km}}{1{,}000 \text{ m}} = 5.63 \text{ x } 10^1 \text{ km}$$

6.02 x 10^1 km is larger.

(b) Convert μs to ms and compare the two quantities.

$$46 \ \mu\text{s x } \frac{1 \text{ x } 10^{-6} \text{ s}}{1 \ \mu\text{s}} \text{ x } \frac{1 \text{ ms}}{1 \text{ x } 10^{-3} \text{ s}} = 4.6 \text{ x } 10^{-2} \text{ ms}$$

46 μs is larger.

(c) Convert g to kg and compare the two quantities.

$$200{,}098 \text{ g x } \frac{1 \text{ kg}}{1000 \text{ g}} = 20.0098 \text{ x } 10^1 \text{ kg}$$

200,098 g is larger.

1.50 1 mg = 1 x 10^{-3} g and 1 pg = 1 x 10^{-12} g

$$\frac{1 \text{ x } 10^{-3} \text{ g}}{1 \text{ mg}} \text{ x } \frac{1 \text{ pg}}{1 \text{ x } 10^{-12} \text{ g}} = 1 \text{ x } 10^9 \text{ pg/mg}$$

35 ng = 35 x 10^{-9} g

$$\frac{35 \text{ x } 10^{-9} \text{ g}}{35 \text{ ng}} \text{ x } \frac{1 \text{ pg}}{1 \text{ x } 10^{-12} \text{ g}} = 3.5 \text{ x } 10^4 \text{ pg/35 ng}$$

1.52 (a) 5 pm = 5 x 10^{-12} m

$$5 \text{ x } 10^{-12} \text{ m x } \frac{1 \text{ cm}}{1 \text{ x } 10^{-2} \text{ m}} = 5 \text{ x } 10^{-10} \text{ cm}$$

$$5 \text{ x } 10^{-12} \text{ m x } \frac{1 \text{ nm}}{1 \text{ x } 10^{-9} \text{ m}} = 5 \text{ x } 10^{-3} \text{ nm}$$

(b)

$$8.5 \text{ cm}^3 \text{ x } \left(\frac{1 \text{ x } 10^{-2} \text{ m}}{1 \text{ cm}}\right)^3 = 8.5 \text{ x } 10^{-6} \text{ m}^3$$

$$8.5 \text{ cm}^3 \text{ x } \left(\frac{10 \text{ mm}}{1 \text{ cm}}\right)^3 = 8.5 \text{ x } 10^3 \text{ mm}^3$$

(c)

$$65.2 \text{ mg x } \frac{1 \text{ x } 10^{-3} \text{ g}}{1 \text{ mg}} = 0.0652 \text{ g}$$

$$65.2 \text{ mg x } \frac{1 \text{ x } 10^{-3} \text{ g}}{1 \text{ mg}} \text{ x } \frac{1 \text{ pg}}{1 \text{ x } 10^{-12} \text{ g}} = 6.52 \text{ x } 10^{10} \text{ pg}$$

1.54 (a) 35.0445 g has 6 significant figures because zeros in the middle of a number are significant.

(b) 59.0001 cm has 6 significant figures because zeros in the middle of a number are significant.

(c) 0.030 03 kg has 4 significant figures because zeros at the beginning of a number are not significant and zeros in the middle of a number are significant.

(d) 0.004 50 m has 3 significant figures because zeros at the beginning of a number are not significant and zeros at the end of a number and after the decimal point are always significant.
(e) 67,000 m^2 has 2, 3, 4, or 5 significant figures because zeros at the end of a number and before the decimal point may or may not be significant.
(f) 3.8200 x 10^3 L has 5 significant figures because zeros at the end of a number and after the decimal point are always significant.

1.56 To convert 3,666,500 m^3 to scientific notation, move the decimal point 6 places to the left and include an exponent of 10^6. The result is 3.6665 x 10^6 m^3. Because the digit to be dropped is 5 with nothing following, round down. The result is 3.666 x 10^6 m^3 (4 significant figures). Because the digit to be dropped (the second 6) is greater than 5, round up. The result is 3.7 x 10^6 m^3 (2 significant figures).

1.58 (a) To convert 453.32 mg to scientific notation, move the decimal point 2 places to the left and include an exponent of 10^2. The result is 4.5332 x 10^2 mg.
(b) To convert 0.000 042 1 mL to scientific notation, move the decimal point 5 places to the right and include an exponent of 10^{-5}. The result is 4.21 x 10^{-5} mL.
(c) To convert 667,000 g to scientific notation, move the decimal point 5 places to the left and include an exponent of 10^5. The result is 6.67 x 10^5 g.

1.60 (a) Because the digit to be dropped (0) is less than 5, round down. The result is 3.567 x 10^4 or 35,670 m (4 significant figures).
Because the digit to be dropped (the second 6) is greater than 5, round up. The result is 35,670.1 m (6 significant figures).
(b) Because the digit to be dropped is 5 with nonzero digits following, round up. The result is 69 g (2 significant figures).
Because the digit to be dropped (0) is less than 5, round down. The result is 68.5 g (3 significant figures).
(c) Because the digit to be dropped is 5 with nothing following, round down. The result is 4.99 x 10^3 cm (3 significant figures).
(d) Because the digit to be dropped is 5 with nothing following, round down. The result is 2.3098 x 10^{-4} kg (5 significant figures).

1.62 (a) 4.884 x 2.05 = 10.012
The result should contain only 3 significant figures because 2.05 contains 3 significant figures (the smaller number of significant figures of the two). Because the digit to be dropped (1) is less than 5, round down. The result is 10.0.
(b) 94.61 / 3.7 = 25.57
The result should contain only 2 significant figures because 3.7 contains 2 significant figures (the smaller number of significant figures of the two). Because the digit to be dropped (second 5) is 5 with nonzero digits following, round up. The result is 26.
(c) 3.7 / 94.61 = 0.0391
The result should contain only 2 significant figures because 3.7 contains 2 significant figures (the smaller number of significant figures of the two). Because the digit to be dropped (1) is less than 5, round down. The result is 0.039.

(d)

$$\begin{array}{r} 5502.3 \\ 24 \\ +\quad 0.01 \\ \hline 5526.31 \end{array}$$

This result should be expressed with no decimal places. Because the digit to be dropped (3) is less than 5, round down. The result is 5526.

(e)

$$\begin{array}{r} 86.3 \\ +\ 1.42 \\ -\ 0.09 \\ \hline 87.63 \end{array}$$

This result should be expressed with only 1 decimal place. Because the digit to be dropped (3) is less than 5, round down. The result is 87.6.

(f) $5.7 \times 2.31 = 13.167$
The result should contain only 2 significant figures because 5.7 contains 2 significant figures (the smaller number of significant figures of the two). Because the digit to be dropped (second 1) is less than 5, round down. The result is 13.

1.64 1 mile = 1.6093 km; The time is 1 h, 5 min, and 26.6 s.
Convert the time to seconds and then hours.

$$\text{time} = \left(1\ \text{hr} \times \frac{60\ \text{min}}{1\ \text{h}} \times \frac{60\ \text{s}}{1\ \text{min}}\right) + \left(5\ \text{min} \times \frac{60\ \text{s}}{1\ \text{min}}\right) + 26.6\ \text{s} = 3926.6\ \text{s}$$

$$\text{time} = 3926.6\ \text{s} \times \frac{1\ \text{min}}{60\ \text{s}} \times \frac{1\ \text{h}}{60\ \text{min}} = 1.0907\ \text{h}$$

Convert meters to miles.

$$20{,}000\ \text{m} \times \frac{1\ \text{km}}{1000\ \text{m}} \times \frac{1\ \text{mi}}{1.6093\ \text{km}} = 12.4278\ \text{mi}$$

$$\text{average speed} = \frac{12.4278\ \text{mi}}{1.0907\ \text{h}} = 11.394\ \text{mi/h}$$

Unit Conversions (Section 1.11)

1.66 (a) $0.25\ \text{lb} \times \frac{453.59\ \text{g}}{1\ \text{lb}} = 113.4\ \text{g} = 110\ \text{g}$

(b) $1454\ \text{ft} \times \frac{12\ \text{in.}}{1\ \text{ft}} \times \frac{2.54\ \text{cm}}{1\ \text{in.}} \times \frac{1 \times 10^{-2}\ \text{m}}{1\ \text{cm}} = 443.2\ \text{m}$

(c) $2{,}941{,}526\ \text{mi}^2 \times \left(\frac{1.6093\ \text{km}}{1\ \text{mi}}\right)^2 \times \left(\frac{1000\ \text{m}}{1\ \text{km}}\right)^2 = 7.6181 \times 10^{12}\ \text{m}^2$

1.68 (a) $1\ \text{acre-ft} \times \frac{1\ \text{mi}^2}{640\ \text{acres}} \times \left(\frac{5280\ \text{ft}}{1\ \text{mi}}\right)^2 = 43{,}560\ \text{ft}^3$

(b) $116\ \text{mi}^3 \times \left(\frac{5280\ \text{ft}}{1\ \text{mi}}\right)^3 \times \frac{1\ \text{acre-ft}}{43{,}560\ \text{ft}^3} = 3.92 \times 10^8\ \text{acre-ft}$

1.70 $8.65 \text{ stones} \times \frac{14 \text{ lb}}{1 \text{ stone}} = 121 \text{ lb}$

1.72 $160 \text{ lb} \times \frac{1 \text{ kg}}{2.2046 \text{ lb}} = 72.6 \text{ kg}$

$$72.6 \text{ kg} \times \frac{20\ \mu\text{g}}{1 \text{ kg}} \times \frac{1 \text{ mg}}{1 \times 10^3\ \mu\text{g}} = 1.452 \text{ mg} = 1.5 \text{ mg}$$

Temperature (Section 1.5)

1.74 $°F = (\frac{9}{5} \times °C) + 32$

$$°F = (\frac{9}{5} \times 39.9°C) + 32 = 103.8\ °F \qquad \text{(goat)}$$

$$°F = (\frac{9}{5} \times 22.2°C) + 32 = 72.0\ °F \qquad \text{(Australian spiny anteater)}$$

1.76 $°F = (\frac{9}{5} \times °C) + 32 = (\frac{9}{5} \times 175) + 32 = 347\ °F$

1.78

Ethanol boiling point	78.5 °C	173.3 °F	200 °E
Ethanol melting point	–117.3 °C	–179.1 °F	0 °E

(a) $\frac{200\ °E}{[78.5\ °C - (-117.3\ °C)]} = \frac{200\ °E}{195.8\ °C} = 1.021\ °E/°C$

(b) $\frac{200\ °E}{[173.3\ °F - (-179.1\ °F)]} = \frac{200\ °E}{352.4\ °F} = 0.5675\ °E/°F$

(c) $°E = \frac{200}{195.8} \times (°C + 117.3)$

H_2O melting point = 0°C; $°E = \frac{200}{195.8} \times (0 + 117.3) = 119.8\ °E$

H_2O boiling point = 100°C; $°E = \frac{200}{195.8} \times (100 + 117.3) = 222.0\ °E$

(d) $°E = \frac{200}{352.4} \times (°F + 179.1) = \frac{200}{352.4} \times (98.6 + 179.1) = 157.6\ °E$

(e) $°F = \left(°E \times \frac{352.4}{200}\right) - 179.1 = \left(130 \times \frac{352.4}{200}\right) - 179.1 = 50.0\ °F$

Because the outside temperature is 50.0°F, I would wear a sweater or light jacket.

Density (Section 1.7)

1.80 $d = \frac{m}{V} = \frac{27.43 \text{ g}}{12.40 \text{ cm}^3} = 2.212 \text{ g/cm}^3$

1.82 $3.10\ g/cm^3 = 3.10\ g/mL$

$$\text{mass} = 3.10\ g/mL \times \frac{1\ kg}{1000\ g} \times \frac{1\ mL}{1 \times 10^{-3}\ L} \times 4.67\ L = 14.5\ kg$$

1.84 For H_2: $V = 1.0078\ g \times \dfrac{1\ L}{0.0899\ g} = 11.2\ L$

For Cl_2: $V = 35.45\ g \times \dfrac{1\ L}{3.214\ g} = 11.03\ L$

1.86 $$d = \frac{m}{V} = \frac{220.9\ g}{(0.50 \times 1.55 \times 25.00)\ cm^3} = 11.4\ \frac{g}{cm^3} = 11\ \frac{g}{cm^3}$$

1.88 Silverware mass = 80.56 g

Silverware volume = 15.90 mL – 10.00 mL = 5.90 mL

Silverware density = $\dfrac{80.56\ g}{5.90\ mL} = 13.7\ g/mL$

The density of the silverware and pure silver are different. The silverware is not pure silver.

Energy (Section 1.8)

1.90 Car: $E_K = \frac{1}{2}(1400\ kg)\left(\dfrac{115 \times 10^3\ m}{3600\ s}\right)^2 = 7.1 \times 10^5\ J$

Truck: $E_K = \frac{1}{2}(12{,}000\ kg)\left(\dfrac{38 \times 10^3\ m}{3600\ s}\right)^2 = 6.7 \times 10^5\ J$

The car has more kinetic energy.

1.92 1 oz = 28.35 g

$$\text{energy} = 0.450\ oz \times \frac{28.35\ g}{1\ oz} \times \frac{2498\ kJ}{45.0\ g} \times \frac{1\ kcal}{4.184\ kJ} = 169\ kcal$$

1.94 (a) $540\ Cal \times \dfrac{1000\ cal}{1\ Cal} \times \dfrac{4.184\ J}{1\ cal} \times \dfrac{1\ kJ}{1000\ J} = 2259\ kJ = 2300\ kJ$

(b) 100 watts = 100 J/s

$$\text{time} = 2259\ kJ \times \frac{1000\ J}{1\ kJ} \times \frac{1\ s}{100\ J} \times \frac{1\ min}{60\ s} \times \frac{1\ h}{60\ min} = 6.275\ h = 6.3\ h$$

Chapter Problems

1.96 $$d = \frac{m}{V} = \frac{8.763\ g}{(28.76 - 25.00)\ mL} = \frac{8.763\ g}{3.76\ mL} = 2.331\ \frac{g}{cm^3} = 2.33\ \frac{g}{cm^3}$$

1.98 NaCl melting point = 1074 K

°C = K – 273.15 = 1074 – 273.15 = 800.85 °C = 801 °C

$$°F = (\frac{9}{5} \text{ x } °C) + 32 = (\frac{9}{5} \text{ x } 800.85) + 32 = 1473.53\ °F = 1474\ °F$$

NaCl boiling point = 1686 K

°C = K – 273.15 = 1686 – 273.15 = 1412.85 °C = 1413 °C

$$°F = (\frac{9}{5} \text{ x } °C) + 32 = (\frac{9}{5} \text{ x } 1412.85) + 32 = 2575.13\ °F = 2575\ °F$$

1.100 $V = 112.5 \text{ g x } \frac{1 \text{ mL}}{1.4832 \text{ g}} = 75.85 \text{ mL}$

1.102 (a) density = $\frac{1 \text{ lb}}{1 \text{ pint}} \text{ x } \frac{8 \text{ pints}}{1 \text{ gal}} \text{ x } \frac{1 \text{ gal}}{3.7854 \text{ L}} \text{ x } \frac{453.59 \text{ g}}{1 \text{ lb}} \text{ x } \frac{1 \text{ L}}{1000 \text{ mL}} = 0.958\ 61 \text{ g/mL}$

(b) area in m^2 =

$$1 \text{ acre x } \frac{1 \text{ mi}^2}{640 \text{ acres}} \text{ x } \left(\frac{5280 \text{ ft}}{1 \text{ mi}}\right)^2 \text{ x } \left(\frac{12 \text{ in.}}{1 \text{ ft}}\right)^2 \text{ x } \left(\frac{2.54 \text{ cm}}{1 \text{ in.}}\right)^2 \text{ x } \left(\frac{1 \text{ x } 10^{-2} \text{ m}}{1 \text{ cm}}\right)^2 = 4047 \text{ m}^2$$

(c) mass of wood =

$$1 \text{ cord x } \frac{128 \text{ ft}^3}{1 \text{ cord}} \text{ x } \left(\frac{12 \text{ in.}}{1 \text{ ft}}\right)^3 \text{ x } \left(\frac{2.54 \text{ cm}}{1 \text{ in.}}\right)^3 \text{ x } \frac{0.40 \text{ g}}{1 \text{ cm}^3} \text{ x } \frac{1 \text{ kg}}{1000 \text{ g}} = 1450 \text{ kg} = 1400 \text{ kg}$$

(d) mass of oil =

$$1 \text{ barrel x } \frac{42 \text{ gal}}{1 \text{ barrel}} \text{ x } \frac{3.7854 \text{ L}}{1 \text{ gal}} \text{ x } \frac{1 \text{ mL}}{1 \text{ x } 10^{-3} \text{ L}} \text{ x } \frac{0.85 \text{ g}}{1 \text{ mL}} \text{ x } \frac{1 \text{ kg}}{1000 \text{ g}} = 135.1 \text{ kg} = 140 \text{ kg}$$

(e) fat Calories =

$$0.5 \text{ gal x } \frac{32 \text{ servings}}{1 \text{ gal}} \text{ x } \frac{165 \text{ Calories}}{1 \text{ serving}} \text{ x } \frac{30.0 \text{ Cal from fat}}{100 \text{ Cal total}} = 792 \text{ Cal from fat}$$

1.104 (a) number of Hershey's Kisses =

$$2.0 \text{ lb x } \frac{453.59 \text{ g}}{1 \text{ lb}} \text{ x } \frac{1 \text{ serving}}{41 \text{ g}} \text{ x } \frac{9 \text{ kisses}}{1 \text{ serving}} = 199 \text{ kisses} = 200 \text{ kisses}$$

(b) Hershey's Kiss volume = $\frac{41 \text{ g}}{1 \text{ serving}} \text{ x } \frac{1 \text{ serving}}{9 \text{ kisses}} \text{ x } \frac{1 \text{ mL}}{1.4 \text{ g}} = 3.254 \text{ mL} = 3.3 \text{ mL}$

(c) Calories/Hershey's Kiss = $\frac{230 \text{ Cal}}{1 \text{ serving}} \text{ x } \frac{1 \text{ serving}}{9 \text{ kisses}} = 25.55 \text{ Cal/kiss} = 26 \text{ Cal/kiss}$

(d) % fat Calories =

$$\frac{13 \text{ g fat}}{1 \text{ serving}} \text{ x } \frac{9 \text{ Cal from fat}}{1 \text{ g fat}} \text{ x } \frac{1 \text{ serving}}{230 \text{ Cal total}} \text{ x } 100\% = 51\% \text{ Calories from fat}$$

1.106 $°C = \frac{5}{9} \text{ x } (°F - 32)$

Set °C = °F: $°C = \frac{5}{9} \text{ x } (°C - 32)$

Solve for °C: $°C \times \frac{9}{5} = °C - 32$

$(°C \times \frac{9}{5}) - °C = -32$

$°C \times \frac{4}{5} = -32$

$°C = \frac{5}{4}(-32) = -40\ °C$

The Celsius and Fahrenheit scales "cross" at −40 °C (−40 °F).

1.108 $d = 0.037\ \frac{lbs}{in^3} \times \frac{453.59\ g}{1\ lb} \times \left(\frac{1\ in}{2.54\ cm}\right)^3 = 1.0\ g/cm^3$

1.110 Convert 8 min, 25 s to s. $8\ min \times \frac{60\ s}{1\ min} + 25\ s = 505\ s$

Convert 293.2 K to °F:

$293.2 - 273.15 = 20.05\ °C$ and $°F = (\frac{9}{5} \times 20.05) + 32 = 68.09\ °F$

Final temperature $= 68.09\ °F + 505\ s \times \frac{3.0\ °F}{60\ s} = 93.34\ °F$

$°C = \frac{5}{9} \times (93.34 - 32) = 34.1\ °C$

1.112 Average brass density = $(0.670)(8.92\ g/cm^3) + (0.330)(7.14\ g/cm^3) = 8.333\ g/cm^3$

length $= 1.62\ in. \times \frac{2.54\ cm}{1\ in.} = 4.115\ cm$

diameter $= 0.514\ in. \times \frac{2.54\ cm}{1\ in.} = 1.306\ cm$

volume $= \pi r^2 h = (3.1416)[(1.306\ cm)/2]^2(4.115\ cm) = 5.512\ cm^3$

mass $= 5.512\ cm^3 \times \frac{8.333\ g}{1\ cm^3} = 45.9\ g$

1.114 (a) Ga density $= \frac{0.2133\ lb}{1\ in.^3} \times \frac{453.59\ g}{1\ lb} \times \frac{1\ in.^3}{(2.54\ cm)^3} = 5.904\ g/cm^3$

(b)

Ga boiling point	2204 °C	1000 °G
Ga melting point	29.78 °C	0 °G

$$\frac{1000\ °G - 0\ °G}{2204\ °C - 29.78\ °C} = \frac{1000\ °G}{2174.22\ °C} = 0.4599\ °G/°C$$

$^{\circ}G = 0.4599 \times (^{\circ}C - 29.78)$
$^{\circ}G = 0.4599 \times (801 - 29.78) = 355\ ^{\circ}G$

The melting point of sodium chloride (NaCl) on the gallium scale is 355 °G.

2 Atoms, Molecules, and Ions

2.1 It is a metal, and most likely near the end of the transition metals because it can be found in nature in its pure form. A likely candidate is silver.

2.2 (a) K, potassium, metal (b) shiny, metallic solid
(c) It would deform and not crack since metals are malleable.
(d) It would conduct electricity, but it is not a good choice for wiring because it is reacts when exposed to oxygen and humidity (water) in the atmosphere. Potassium is also a soft metal which, makes it unsuitable for use as a wire.

2.3 First, find the S:O ratio in each compound.
Substance A: S:O mass ratio = (6.00 g S) / (5.99 g O) = 1.00
Substance B: S:O mass ratio = (8.60 g S) / (12.88 g O) = 0.668

$$\frac{\text{S:O mass ratio in substance A}}{\text{S:O mass ratio in substance B}} = \frac{1.00}{0.668} = 1.50 = \frac{3}{2}$$

2.4 In compound A the O/S mass ratio is 1. In compound B, the O/S mass ratio is 3/2. This means there is 3/2 times more O in compound B. To find the formula of Compound B multiply the subscript on O in Compound A by 3/2. Compound B is SO_3.

2.5 $$0.005\text{ mm} \times \frac{1 \times 10^{-3}\text{ m}}{1\text{ mm}} \times \frac{1\text{ Au atom}}{2.9 \times 10^{-10}\text{ m}} = 2 \times 10^{4}\text{ Au atoms}$$

2.6 $$1 \times 10^{19}\text{ C atoms} \times \frac{1.5 \times 10^{-10}\text{ m}}{\text{C atom}} \times \frac{1\text{ km}}{1000\text{ m}} \times \frac{1\text{ time}}{40{,}075\text{ km}} = 37.4\text{ times} \sim 40\text{ times}$$

2.7 $^{75}_{34}Se$ has 34 protons, 34 electrons, and (75 – 34) = 41 neutrons.

2.8 The element with 24 protons is Cr. The mass number is the sum of the protons and the neutrons, 24 + 28 = 52. The isotope symbol is $^{52}_{24}Cr$.

2.9 atomic mass = (0.6915 x 62.93) + (0.3085 x 64.93) = 63.55
(63.546 from periodic table)

2.10 (a) The atomic weight of Ga is 69.7231. This mass is closer to the mass of gallium-69 than to gallium-71; therefore gallium-69 must be more abundant.
(b) The total abundance of both isotopes must be 100.00%. Let Y be the natural abundance of ^{69}Ga and [1 – Y] the natural abundance of ^{71}Ga.
(Y x 68.9256) + ([1 – Y] x 70.9247) = 69.7231

Solve for Y. $Y = \frac{-1.2016}{-1.9991} = 0.6011$

^{69}Ga natural abundance = 60.11% and ^{71}Ga natural abundance = 100.00 – 60.11 = 39.89%

2.11 Pt, 195.078

$$\text{mol Pt} = 9.50 \text{ g Pt} \times \frac{1 \text{ mol Pt}}{195.078 \text{ g Pt}} = 0.0487 \text{ mol Pt}$$

$$\text{atoms Pt} = 0.0487 \text{ mol Pt} \times \frac{6.022 \times 10^{23} \text{ atoms Pt}}{1 \text{ mol Pt}} = 2.93 \times 10^{22} \text{ atoms Pt}$$

2.12 $\text{atomic mass in g} = \frac{1.50 \text{ g}}{2.26 \times 10^{22} \text{ atoms}} \times 6.022 \times 10^{23} \text{ atoms} = 40.0 \text{ g}$; Y = Ca

2.13 Figure (b) represents a collection of hydrogen peroxide (H_2O_2) molecules.

2.14 (a) Figures (b) and (d) illustrate pure substances.
(b) Figures (a) and (c) illustrate mixtures.
(c) Figures (b) and (d) illustrate the law of multiple proportions.

2.15

```
    H   H
    |   |
H—C—N—H
    |
    H
```

2.16 adrenaline, $C_9H_{13}NO_3$

2.17 (a) LiBr is composed of a metal (Li) and nonmetal (Br) and is ionic.
(b) $SiCl_4$ is composed of only nonmetals and is molecular.
(c) BF_3 is composed of only nonmetals and is molecular.
(d) CaO is composed of a metal (Ca) and nonmetal (O) and is ionic.

2.18 Figure (a) most likely represents an ionic compound because there are no discrete molecules, only a regular array of two different chemical species (ions). Figure (b) most likely represents a molecular compound because discrete molecules are present.

2.19 (a) magnesium fluoride, MgF_2
(b) tin(IV) oxide, SnO_2
(c) iron(III) sulfide, Fe_2S_3

2.20 red – potassium sulfide, K_2S; green – strontium iodide, SrI_2; blue – gallium oxide, Ga_2O_3

2.21 (a) potassium hypochlorite, KClO
(b) silver(I) chromate, Ag_2CrO_4
(c) iron(III) carbonate, $Fe_2(CO_3)_3$

2.22 Drawing 1 represents ionic compounds with one cation and two anions. Only (c) $CaCl_2$ is consistent with drawing 1.
Drawing 2 represents ionic compounds with one cation and one anion. Both (a) LiBr and (b) $NaNO_2$ are consistent with drawing 2.

2.23 (a) disulfur dichloride, S_2Cl_2
(b) iodine monochloride, ICl
(c) nitrogen triiodide, NI_3

2.24 (a) PCl_5, phosphorus pentachloride (b) N_2O, dinitrogen monoxide

2.25 The green chemistry principle of atom economy accounts for the mass of every atom in a chemical reaction.

2.26 $C_3H_8O \rightarrow C_3H_6 + H_2O$

Reactants:
3 C = (3)(12.0) = 36.0
8 H = (8)(1.0) = 8.0
1 O = (1)(16.0) = 16.0
Σ = 60.0

Desired Product:
3 C = (3)(12.0) = 36.0
6 H = (6)(1.0) = 6.0
Σ = 42.0

$$\% \text{ Atom Economy} = \frac{\Sigma \text{ Atomic Weight}_{\text{(atoms in desired product)}}}{\Sigma \text{ Atomic Weight}_{\text{(atoms in all reactants)}}} = \frac{42.0}{60.0} \times 100 = 70\%$$

2.27 (a) Reaction 1 has the higher % atom economy because all reactant atoms are used.
(b) rxn 1: $C_2H_4 + Cl_2 \rightarrow C_2H_4Cl_2$
There are no undesired products, therefore the % Atom Economy = 100%.

rxn 2: $CH_3Cl + Br^- \rightarrow CH_3Br + Cl^-$

Reactants:
1 C = (1)(12.0) = 12.0
3 H = (3)(1.0) = 3.0
1 Cl = (1)(35.5) = 35.5
1 Br = (1)(80.0) = 80.0
Σ = 130.5

Desired Product:
1 C = (1)(12.0) = 12.0
3 H = (3)(1.0) = 3.0
1 Br = (1)(80.0) = 80.0
Σ = 93.0

$$\% \text{ Atom Economy} = \frac{\Sigma \text{ Atomic Weight}_{\text{(atoms in desired product)}}}{\Sigma \text{ Atomic Weight}_{\text{(atoms in all reactants)}}} = \frac{93.0}{130.5} \times 100 = 71\%$$

2.28 Ibuprofen, $C_{13}H_{18}O_2$

2.29 (a) $1 \text{ mol Na} \times \frac{23.0 \text{ g Na}}{1 \text{ mol Na}} = 23.0 \text{ g Na}$

$23 \text{ mol H} \times \frac{1.0 \text{ g H}}{1 \text{ mol H}} = 23.0 \text{ g H}$

$1 \text{ mol N} \times \frac{14.0 \text{ g N}}{1 \text{ mol N}} = 14.0 \text{ g N}$

$7 \text{ mol C} \times \frac{12.0 \text{ g C}}{1 \text{ mol C}} = 84.0 \text{ g C}$

$8 \text{ mol O} \times \frac{16.0 \text{ g O}}{1 \text{ mol O}} = 128.0 \text{ g O}$

$1 \text{ mol Cl} \times \frac{35.5 \text{ g Cl}}{1 \text{ mol Cl}} = 35.5 \text{ g Cl}$

(b) 23.0 g + 23.0 g + 14.0 g + 84.0 g + 128.0 g + 35.5 g = 307.5 g

(c) $6.6 \times 10^7 \text{ mol Ibuprofen} \times \frac{307.5 \text{ g waste}}{1 \text{ mol Ibuprofen}} \times \frac{1 \text{ kg}}{1000 \text{ g}} = 2.0 \times 10^7 \text{ kg waste}$

2.30 (a) $4 \text{ mol H} \times \frac{1.0 \text{ g H}}{1 \text{ mol H}} = 4.0 \text{ g H}$

$2 \text{ mol C} \times \frac{12.0 \text{ g C}}{1 \text{ mol C}} = 24.0 \text{ g C}$

$2 \text{ mol O} \times \frac{16.0 \text{ g O}}{1 \text{ mol O}} = 32.0 \text{ g O}$

(b) 4.0 g + 24.0 g + 32.0 g = 60.0 g

(c) $6.6 \times 10^7 \text{ mol Ibuprofen} \times \frac{60.0 \text{ g waste}}{1 \text{ mol Ibuprofen}} \times \frac{1 \text{ kg}}{1000 \text{ g}} = 4.0 \times 10^6 \text{ kg waste}$

(d) savings = 2.0×10^7 kg – 4.0×10^6 kg = 1.6×10^7 kg

Conceptual Problems

2.32

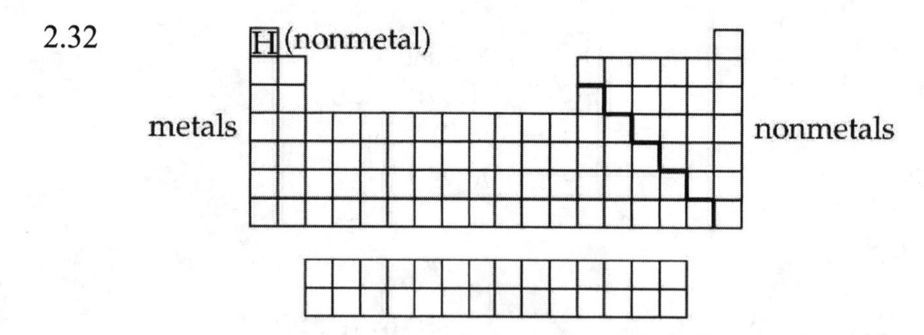

2.34 The element is americium (Am) with atomic number = 95. It is in the actinide series.

2.36 Drawing (a) represents a collection of SO_2 units. Drawing (d) represents a mixture of S atoms and O_2 units.

2.38 Figures (b) and (c) both contain two protons but different numbers of neutrons. They are isotopes of the same element. Figure (a) contains only one proton. It is a different element than (b) and (c).

2.40 thymine, $C_5H_6N_2O_2$

2.42 A Na atom has 11 protons and 11 electrons [drawing (b)].
A Ca^{2+} ion has 20 protons and 18 electrons [drawing (c)].
A F^- ion has 9 protons and 10 electrons [drawing (a)].

Section Problems

Elements and the Periodic Table (Sections 2.1–2.3)

2.44 118 elements are presently known. About 90 elements occur naturally.

2.46 (a) gadolinium, Gd (b) germanium, Ge (c) technetium, Tc (d) arsenic, As

2.48 (a) Te, tellurium (b) Re, rhenium (c) Be, beryllium (d) Ar, argon
(e) Pu, plutonium

2.50 (a) Tin is Sn. Ti is titanium. (b) Manganese is Mn. Mg is magnesium.
(c) Potassium is K. Po is polonium.
(d) The symbol for helium is He. The second letter is lowercase.

2.52 The rows are called periods, and the columns are called groups.

2.54 Elements within a group have similar chemical properties.

2.56

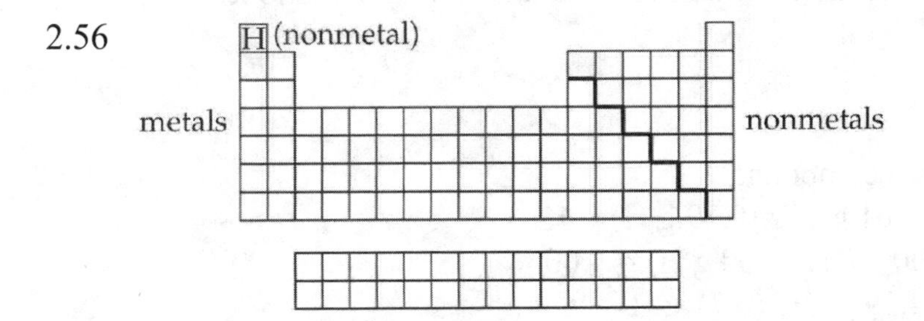

2.58 (a) Ti, metal (b) Te, semimetal (c) Se, nonmetal
(d) Sc, metal (e) Si, semimetal

2.60 (a) The alkali metals are shiny, soft, low-melting metals that react rapidly with water to form products that are alkaline.
(b) The noble gases are gases of very low reactivity.
(c) The halogens are nonmetallic and corrosive. They are found in nature only in combination with other elements.

2.62 F, Cl, Br, and I

2.64 An element that is a soft, silver-colored solid that reacts violently with water and is a good conductor of electricity has the characteristics of a metal.

2.66 An element that is a yellow crystalline solid, does not conduct electricity, and when hit with a hammer it shatters has the characteristics of a nonmetal.

2.68 All match in groups 2A and 7A.

Atomic Theory (Sections 2.4 and 2.5)

2.70 The law of mass conservation in terms of Dalton's atomic theory states that chemical reactions only rearrange the way that atoms are combined; the atoms themselves are not changed.
The law of definite proportions in terms of Dalton's atomic theory states that the chemical combination of elements to make different substances occurs when atoms join together in small, whole-number ratios.

2.72 In any chemical reaction, the combined mass of the final products equals the combined mass of the starting reactants.
mass of reactants = mass of products
mass of reactants = mass of Hg + mass of O_2 = 114.0 g + 12.8 g = 126.8 g
mass of products = mass of HgO + mass of left over O_2 = 123.1 g + mass of left over O_2
126.8 g = 123.1 g + mass of left over O_2
mass of left over O_2 = 126.8 g – 123.1 g = 3.7 g

2.74 For the "other" compound: C:H mass ratio = (32.0 g C) / (8.0 g H) = 4
The "other" compound is not methane because the methane C:H mass ratio is 3.

$$\frac{\text{C:H mass ratio in "other"}}{\text{C:H mass ratio in methane}} = \frac{4}{3}$$

2.76 First, find the C:H ratio in each compound.
Benzene: C:H mass ratio = (4.61 g C) / (0.39 g H) = 12
Ethane: C:H mass ratio (4.00 g C) / (1.00 g H) = 4.00
Ethylene: C:H mass ratio = (4.29 g C) / (0.71 g H) = 6.0

$$\frac{\text{C:H mass ratio in benzene}}{\text{C:H mass ratio in ethane}} = \frac{12}{4.00} = \frac{3}{1}$$

$$\frac{\text{C:H mass ratio in benzene}}{\text{C:H mass ratio in ethylene}} = \frac{12}{6.0} = \frac{2}{1}$$

$$\frac{\text{C:H mass ratio in ethylene}}{\text{C:H mass ratio in ethane}} = \frac{6.0}{4.00} = \frac{3}{2}$$

2.78 Assume a 100.0 g sample for each compound and then find the O:C ratio in each compound.
Compound 1: O:C mass ratio = (57.1 g O)/(42.9 g C) = 1.33
Compound 2: O:C mass ratio = (72.7 g O)/(27.3 g C) = 2.66

$$\frac{\text{O:C mass ratio in compound 2}}{\text{O:C mass ratio in compound 1}} = \frac{2.66}{1.33} = \frac{2}{1}$$

If compound 1 is CO, and the O:C mass ratio is 2 times that of compound 1, then the formula for compound 2 is CO_2.

2.80 Assume a 1.00 g sample of the binary compound of zinc and sulfur.
0.671 x 1.00 g = 0.671 g Zn; 0.329 x 1.00 g = 0.329 g S

$$0.671 \text{ g} \times \frac{1 \text{ u}}{1.6605 \times 10^{-24} \text{ g}} \times \frac{1 \text{ Zn atom}}{65.39} = 6.18 \times 10^{21} \text{ Zn atoms}$$

$$0.329 \text{ g} \times \frac{1}{1.6605 \times 10^{-24} \text{ g}} \times \frac{1 \text{ S atom}}{32.066} = 6.18 \times 10^{21} \text{ S atoms}$$

$$\frac{\text{Zn}}{\text{S}} = \frac{6.18 \times 10^{21} \text{ Zn atoms}}{6.18 \times 10^{21} \text{ S atoms}} = \frac{1}{1}$$

Elements and Atoms (Sections 2.6 –2.8)

2.82 electron

2.84 (d) The mass to charge ratio of the electron.

2.86 (a) -1.010×10^{-18} C because it is not an integer multiple of the electron charge.

2.88 (a) The alpha particles would pass right through the gold foil with little to no deflection.

2.90 350 pm = 350×10^{-12} m

$$\text{Pb atoms} = 0.25 \text{ in} \times \frac{2.54 \text{ cm}}{1 \text{ in}} \times \frac{1 \times 10^{-2} \text{ m}}{1 \text{ cm}} \times \frac{1 \text{ Pb atom}}{350 \times 10^{-12} \text{ m}} = 1.8 \times 10^{7} \text{ Pb atoms thick}$$

2.92 The atomic number is equal to the number of protons.
The mass number is equal to the sum of the number of protons and the number of neutrons.

2.94 The subscript giving the atomic number of an atom is often left off of an isotope symbol because one can readily look up the atomic number in the periodic table.

2.96 (a) carbon, C (b) argon, Ar (c) vanadium, V

2.98 (a) ${}^{220}_{86}Rn$ (b) ${}^{210}_{84}Po$ (c) ${}^{197}_{79}Au$

2.100 (a) ${}^{15}_{7}N$, 7 protons, 7 electrons, (15 – 7) = 8 neutrons
(b) ${}^{60}_{27}Co$, 27 protons, 27 electrons, (60 – 27) = 33 neutrons
(c) ${}^{131}_{53}I$, 53 protons, 53 electrons, (131 – 53) = 78 neutrons
(d) ${}^{142}_{58}Ce$, 58 protons, 58 electrons, (142 – 58) = 84 neutrons

2.102 (a) $^{24}_{12}Mg$, magnesium (b) $^{58}_{28}Ni$, nickel

(c) $^{104}_{46}Pd$, palladium (d) $^{183}_{74}W$, tungsten

2.104 $^{12}_{5}C$, the atomic number for carbon is 6, not 5.

$^{33}_{35}Br$, the mass number must be greater than the atomic number.

$^{11}_{5}Bo$, the element symbol for boron is B.

Atomic Weight and Moles (Section 2.9)

2.106 An element's atomic mass is the weighted average of the isotopic masses of the element's naturally occurring isotopes. The atomic mass for Cu (63.546) must fall between the masses of its two isotopes. If one isotope is ^{65}Cu, the other isotope must be ^{63}Cu, and not ^{66}Cu. If the other isotope was ^{66}Cu, the atomic mass for Cu would be greater than 65.

2.108 (0.199 x 10.0129) + (0.801 x 11.009 31) = 10.8 for B

2.110 24.305 = (0.7899 x 23.985) + (0.1000 x 24.986) + (0.1101 x Z)
Solve for Z. Z = 25.982 for ^{26}Mg.

2.112 atomic mass = (0.6915 x 62.93) + (0.3085 x 64.93) = 63.55

2.114 (a) $\text{g Ti} = 1.505 \text{ mol Ti} \times \frac{47.867 \text{ g Ti}}{1 \text{ mol Ti}} = 72.04 \text{ g Ti}$

(b) $\text{g Na} = 0.337 \text{ mol Na} \times \frac{22.989\ 770 \text{ g Na}}{1 \text{ mol Na}} = 7.75 \text{ g Na}$

(c) $\text{g U} = 2.583 \text{ mol U} \times \frac{238.028\ 91 \text{ g U}}{1 \text{ mol U}} = 614.8 \text{ g U}$

2.116 The mass of 6.02×10^{23} atoms is its atomic mass expressed in grams. If the atomic mass of an element is X, then 6.02×10^{23} atoms of this element weighs X grams.

2.118 The mass of 6.02×10^{23} atoms is its atomic mass expressed in grams. If the mass of 6.02×10^{23} atoms of element Y is 83.80 g, then the atomic mass of Y is 83.80. Y is Kr.

Chemical Compounds (Sections 2.10 and 2.11)

2.120 A covalent bond results when two atoms share several (usually two) of their electrons. An ionic bond results from a complete transfer of one or more electrons from one atom to another. The C–H bonds in methane (CH_4) are covalent bonds. The bond in NaCl (Na^+Cl^-) is an ionic bond.

2.122 Element symbols are composed of one or two letters. If the element symbol is two letters, the first letter is uppercase and the second is lowercase. CO stands for carbon and oxygen in carbon monoxide.

2.124 (a) Be^{2+}, 4 protons and 2 electrons (b) Rb^{+}, 37 protons and 36 electrons
(c) Se^{2-}, 34 protons and 36 electrons (d) Au^{3+}, 79 protons and 76 electrons

2.126 C_3H_8O

2.128

```
    H   H   H   H
    |   |   |   |
H — C — C — C — C — H
    |   |   |   |
    H   H   H   H
```

2.130

```
           H              H
           |              |
       H — C — H      H — C — H
   H       |      H       |       H
   |       |      |       |       |
H— C ———— C ———— C ———— C ———— C — H
   |       |      |       |       |
   H       |      H       H       H
       H — C — H
           |
           H
```

Naming Compounds (Section 2.12)

2.132 (a) CsF, cesium fluoride (b) K_2O, potassium oxide (c) CuO, copper(II) oxide

2.134 (a) potassium chloride, KCl (b) tin(II) bromide, $SnBr_2$ (c) calcium oxide, CaO
(d) barium chloride, $BaCl_2$ (e) aluminum hydride, AlH_3

2.136 (a) calcium acetate, $Ca(CH_3CO_2)_2$ (b) iron(II) cyanide, $Fe(CN)_2$
(c) sodium dichromate, $Na_2Cr_2O_7$ (d) chromium(III) sulfate, $Cr_2(SO_4)_3$
(e) mercury(II) perchlorate, $Hg(ClO_4)_2$

2.138 (a) $Ca(ClO)_2$, calcium hypochlorite
(b) $Ag_2S_2O_3$, silver(I) thiosulfate or silver thiosulfate
(c) NaH_2PO_4, sodium dihydrogen phosphate (d) $Sn(NO_3)_2$, tin(II) nitrate
(e) $Pb(CH_3CO_2)_4$, lead(IV) acetate (f) $(NH_4)_2SO_4$, ammonium sulfate

2.140 (a) $CaBr_2$ (b) $CaSO_4$ (c) $Al_2(SO_4)_3$

2.142 (a) $CaCl_2$ (b) CaO (c) CaS

2.144 (a) sulfite ion, SO_3^{2-} (b) phosphate ion, PO_4^{3-} (c) zirconium(IV) ion, Zr^{4+}
(d) chromate ion, CrO_4^{2-} (e) acetate ion, $CH_3CO_2^{-}$ (f) thiosulfate ion, $S_2O_3^{2-}$

2.146 (a) CCl_4, carbon tetrachloride (b) ClO_2, chlorine dioxide
(c) N_2O, dinitrogen monoxide (d) N_2O_3, dinitrogen trioxide

2.148 (a) NO, nitrogen monoxide (b) N_2O, dinitrogen monoxide (c) NO_2, nitrogen dioxide
(d) N_2O_4, dinitrogen tetroxide (e) N_2O_5, dinitrogen pentoxide

2.150 (a) Na^+ and SO_4^{2-}; therefore the formula is Na_2SO_4
(b) Ba^{2+} and PO_4^{3-}; therefore the formula is $Ba_3(PO_4)_2$
(c) Ga^{3+} and SO_4^{2-}; therefore the formula is $Ga_2(SO_4)_3$

Chapter Problems

2.152 atomic mass = (0.205 x 69.924) + (0.274 x 71.922)
+ (0.078 x 72.923) + (0.365 x 73.921) + (0.078 x 75.921) = 72.6

2.154 For NH_3, $(2.34 \text{ g N})\left(\frac{3 \text{ x } 1.0079 \text{ H}}{14.0067 \text{ N}}\right) = 0.505 \text{ g H}$

For N_2H_4, $(2.34 \text{ g N})\left(\frac{4 \text{ x } 1.0079 \text{ H}}{2 \text{ x } 14.0067 \text{ N}}\right) = 0.337 \text{ g H}$

2.156 (a) I (b) Kr

2.158 $\frac{12.0000}{15.9994} = \frac{X}{16.0000}$; X = 12.0005 for ^{12}C prior to 1961.

2.160 molecular mass = (8 x 12.011) + (9 x 1.0079) + (1 x 14.0067)
+ (2 x 15.9994) = 151.165 g/mol

2.162

```
 H    O    H
  \  / \  /
H—C     C—H
   \   /
    C—C—H
 H / \
   H   H
```


3 Mass Relationships in Chemical Reactions

3.1 $3\ A_2 + 2\ B \rightarrow 2\ BA_3$

3.2

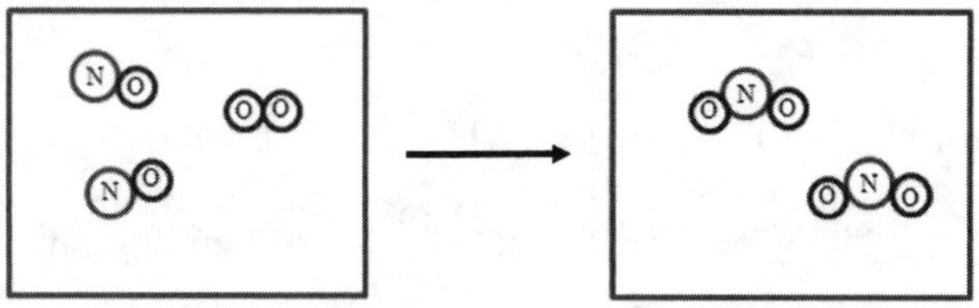

3.3 (a) $C_6H_{12}O_6 \rightarrow 2\ C_2H_6O + 2\ CO_2$
(b) $2\ NaClO_3 \rightarrow 2\ NaCl + O_2$
(c) $4\ NH_3 + Cl_2 \rightarrow N_2H_4 + 2\ NH_4Cl$

3.4 $8\ KClO_3 + C_{12}H_{22}O_{11} \rightarrow 8\ KCl + 12\ CO_2 + 11\ H_2O$

3.5 (a) Fe_2O_3 $2(55.85) + 3(16.00) = 159.7$
(b) H_2SO_4 $2(1.01) + 1(32.07) + 4(16.00) = 98.1$
(c) $C_6H_8O_7$ $6(12.01) + 8(1.01) + 7(16.00) = 192.1$
(d) $C_{16}H_{18}N_2O_4S$ $16(12.01) + 18(1.01) + 2(14.01) + 4(16.00) + 1(32.07) = 334.4$

3.6 sucrose, $C_{12}H_{22}O_{11}$ $12(12.01) + 22(1.01) + 11(16.00) = 342.3$
molar mass = 342.3 g/mol

3.7 $NaHCO_3$, 84.0

$$5.26\text{ g } NaHCO_3 \times \frac{1\text{ mol } NaHCO_3}{84.0\text{ g } NaHCO_3} = 0.0626\text{ mol } NaHCO_3$$

3.8 glucose, $C_6H_{12}O_6$, 180.0

(a) $$0.0833\text{ mol glucose} \times \frac{180.0\text{ g glucose}}{1\text{ mol glucose}} = 15.0\text{ g glucose}$$

(b) $$15.0\text{ g glucose} \times \frac{1\text{ glucose tablet}}{3.75\text{ g glucose}} = 4\text{ glucose tablets}$$

(c) $$0.0833\text{ mol glucose} \times \frac{6.022 \times 10^{23}\text{ glucose molecules}}{1\text{ mol glucose}} = 5.02 \times 10^{22}\text{ glucose molecules}$$

3.9 salicylic acid, $C_7H_6O_3$, 138.1; acetic anhydride, $C_4H_6O_3$, 102.1

(a) $$4.50\text{ g } C_7H_6O_3 \times \frac{1\text{ mol } C_7H_6O_3}{138.1\text{ g } C_7H_6O_3} \times \frac{1\text{ mol } C_4H_6O_3}{1\text{ mol } C_7H_6O_3} \times \frac{102.1\text{ g } C_4H_6O_3}{1\text{ mol } C_4H_6O_3} = 3.33\text{ g } C_4H_6O_3$$

3.10 salicylic acid, $C_7H_6O_3$, 138.1; acetic anhydride, $C_4H_6O_3$, 102.1
aspirin, $C_9H_8O_4$, 180.2; acetic acid, CH_3CO_2H, 60.1

(a) $10.0\ g\ C_9H_8O_4 \times \frac{1\ mol\ C_9H_8O_4}{180.2\ g\ C_9H_8O_4} \times \frac{1\ mol\ C_7H_6O_3}{1\ mol\ C_9H_8O_4} \times \frac{138.1\ g\ C_7H_6O_3}{1\ mol\ C_7H_6O_3} = 7.66\ g\ C_7H_6O_3$

(b) $10.0\ g\ C_9H_8O_4 \times \frac{1\ mol\ C_9H_8O_4}{180.2\ g\ C_9H_8O_4} \times \frac{1\ mol\ CH_3CO_2H}{1\ mol\ C_9H_8O_4} \times \frac{60.1\ g\ CH_3CO_2H}{1\ mol\ CH_3CO_2H} = 3.34\ g\ CH_3CO_2H$

3.11 C_2H_4, 28.1; C_2H_6O, 46.1

$4.6\ g\ C_2H_4 \times \frac{1\ mol\ C_2H_4}{28.1\ g\ C_2H_4} \times \frac{1\ mol\ C_2H_6O}{1\ mol\ C_2H_4} \times \frac{46.1\ g\ C_2H_6O}{1\ mol\ C_2H_6O} = 7.5\ g\ C_2H_6O$ (theoretical yield)

$\text{Percent yield} = \frac{\text{Actual yield}}{\text{Theoretical yield}} \times 100\% = \frac{4.7\ g}{7.5\ g} \times 100\% = 63\%$

3.12 C_2H_6O, 46.1; $C_4H_{10}O$, 74.1

(a) $40.0\ g\ C_2H_6O \times \frac{1\ mol\ C_2H_6O}{46.1\ g\ C_2H_6O} \times \frac{1\ mol\ C_4H_{10}O}{2\ mol\ C_2H_6O} \times \frac{74.1\ g\ C_4H_{10}O}{1\ mol\ C_4H_{10}O} = 32.1\ g\ C_4H_{10}O$ (theoretical yield)

actual yield = (32.1 g)(0.870) = 27.9 g $C_4H_{10}O$

(b) 100.0 g = (theoretical yield)(0.870)
theoretical yield = 100.0 g/0.870 = 115 g $C_4H_{10}O$

$115\ g\ C_4H_{10}O \times \frac{1\ mol\ C_4H_{10}O}{74.1\ g\ C_4H_{10}O} \times \frac{2\ mol\ C_2H_6O}{1\ mol\ C_4H_{10}O} \times \frac{46.1\ g\ C_2H_6O}{1\ mol\ C_2H_6O} = 143\ g\ C_2H_6O$

3.13 (a) $A + B_2 \rightarrow AB_2$
There is a 1:1 stoichiometry between the two reactants. A is the limiting reactant because there are fewer reactant A's than there are reactant B_2's.
(b) 1.0 mol of AB_2 can be made from 1.0 mol of A and 1.0 mol of B_2.

3.14

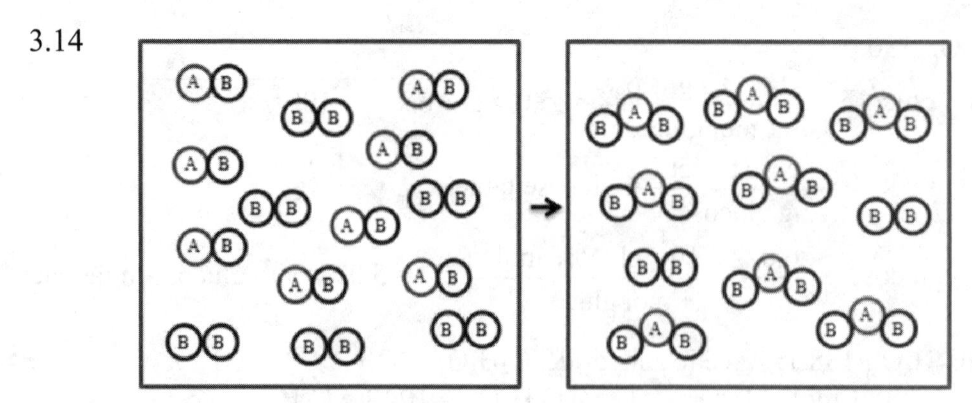

B_2 is in excess, AB is the limiting reactant.

3.15 Li_2O, 29.9: 65 kg = 65,000 g; H_2O, 18.0: 80.0 kg = 80,000 g; LiOH, 23.9

(a) $65{,}000 \text{ g } Li_2O \times \dfrac{1 \text{ mol } Li_2O}{29.9 \text{ g } Li_2O} = 2.17 \times 10^3 \text{ mol } Li_2O$

$80{,}000 \text{ g } H_2O \times \dfrac{1 \text{ mol } H_2O}{18.0 \text{ g } H_2O} = 4.44 \times 10^3 \text{ mol } H_2O$

The reaction stoichiometry between Li_2O and H_2O is one to one. There are twice as many moles of H_2O as there are moles of Li_2O. Therefore, Li_2O is the limiting reactant.

(b) $(4.44 \times 10^3 \text{ mol} - 2.17 \times 10^3 \text{ mol}) = 2.27 \times 10^3 \text{ mol } H_2O$ remaining

$2.27 \times 10^3 \text{ mol } H_2O \times \dfrac{18.0 \text{ g } H_2O}{1 \text{ mol } H_2O} = 40{,}860 \text{ g } H_2O = 40.9 \text{ kg} = 41 \text{ kg } H_2O$

(c) $2.17 \times 10^3 \text{ mol } Li_2O \times \dfrac{2 \text{ mol LiOH}}{1 \text{ mol } Li_2O} \times \dfrac{23.9 \text{ g LiOH}}{1 \text{ mol LiOH}} \times \dfrac{1 \text{ kg}}{1000 \text{ g}} = 104 \text{ kg LiOH}$

3.16 $LiHCO_3$, 68.0; LiOH, 23.9

$500.0 \text{ g } LiHCO_3 \times \dfrac{1 \text{ mol } LiHCO_3}{68.0 \text{ g } LiHCO_3} \times \dfrac{1 \text{ mol LiOH}}{1 \text{ mol } LiHCO_3} \times \dfrac{23.9 \text{ g LiOH}}{1 \text{ mol LiOH}} = 175.7 \text{ g LiOH}$

You start with 400.0 g of LiOH. 500.0 g of $LiHCO_3$ are produced from 175.7 g of LiOH. Over 200 g of LiOH remain. Additional CO_2 can be removed.

3.17 Assume a 100.0 g sample. From the percent composition data, a 100.0 g sample contains 14.25 g C, 56.93 g O, and 28.83 g Mg.

$14.25 \text{ g C} \times \dfrac{1 \text{ mol C}}{12.011 \text{ g C}} = 1.19 \text{ mol C}$

$56.93 \text{ g O} \times \dfrac{1 \text{ mol O}}{15.999 \text{ g O}} = 3.56 \text{ mol O}$

$28.83 \text{ g Mg} \times \dfrac{1 \text{ mol Mg}}{24.305 \text{ g Mg}} = 1.19 \text{ mol Mg}$

$Mg_{1.19}C_{1.19}O_{3.56}$; divide each subscript by the smallest, 1.19.

$Mg_{1.19\,/\,1.19}C_{1.19\,/\,1.19}O_{3.56\,/\,1.19}$

The empirical formula is $MgCO_3$.

3.18 Assume a 100.0 g sample. From the percent composition data, a 100.0 g sample contains 40.0 g C, 13.3 g H, and 46.7 g N.

$40.0 \text{ g C} \times \dfrac{1 \text{ mol C}}{12.01 \text{ g C}} = 3.33 \text{ mol C}$

$13.3 \text{ g H} \times \dfrac{1 \text{ mol H}}{1.01 \text{ g H}} = 13.2 \text{ mol H}$

$46.7 \text{ g N} \times \dfrac{1 \text{ mol N}}{14.01 \text{ g N}} = 3.33 \text{ mol N}$

$C_{3.33}H_{13.2}N_{3.33}$; divide each subscript by the smallest, 3.33.

$C_{3.33\,/\,3.33}H_{13.2\,/\,3.33}N_{3.33\,/\,3.33}$

The empirical formula is CH_4N, 30.0.
60.0 / 30.0 = 2; molecular formula = $C_{(2 \times 1)}H_{(2 \times 4)}N_{(2 \times 1)} = C_2H_8N_2$

3.19 glucose, $C_6H_{12}O_6$, 180.0
Divide each subscript by the smallest, 6, to get the empirical formula.
$C_{6/6}H_{12/6}O_{6/6}$
The empirical formula is CH_2O.
Percent composition:

$$\%C = \frac{6 \times 12.0 \text{ g}}{180.0 \text{ g}} \times 100\% = 40.0\%$$

$$\%H = \frac{12 \times 1.01 \text{ g}}{180.0 \text{ g}} \times 100\% = 6.7\%$$

$$\%O = \frac{6 \times 16.0 \text{ g}}{180.0 \text{ g}} \times 100\% = 53.3\%$$

3.20

$$1.161 \text{ g } H_2O \times \frac{1 \text{ mol } H_2O}{18.0 \text{ g } H_2O} \times \frac{2 \text{ mol H}}{1 \text{ mol } H_2O} = 0.129 \text{ mol H}$$

$$2.818 \text{ g } CO_2 \times \frac{1 \text{ mol } CO_2}{44.0 \text{ g } CO_2} \times \frac{1 \text{ mol C}}{1 \text{ mol } CO_2} = 0.0640 \text{ mol C}$$

$$0.129 \text{ mol H} \times \frac{1.01 \text{ g H}}{1 \text{ mol H}} = 0.130 \text{ g H}$$

$$0.0640 \text{ mol C} \times \frac{12.0 \text{ g C}}{1 \text{ mol C}} = 0.768 \text{ g C}$$

1.00 g total – (0.130 g H + 0.768 g C) = 0.102 g O

$$0.102 \text{ g O} \times \frac{1 \text{ mol O}}{16.0 \text{ g O}} = 0.006\ 38 \text{ mol O}$$

$C_{0.0640}H_{0.129}O_{0.006\ 38}$; divide each subscript by the smallest, 0.006 38.
$C_{0.0640 / 0.006\ 38}H_{0.129 / 0.006\ 38}O_{0.006\ 38 / 0.006\ 38}$
$C_{10.03}H_{20.22}O_1$
The empirical formula is $C_{10}H_{20}O$.

3.21

$$0.697 \text{ g } H_2O \times \frac{1 \text{ mol } H_2O}{18.0 \text{ g } H_2O} \times \frac{2 \text{ mol H}}{1 \text{ mol } H_2O} = 0.0774 \text{ mol H}$$

$$1.55 \text{ g } CO_2 \times \frac{1 \text{ mol } CO_2}{44.0 \text{ g } CO_2} \times \frac{1 \text{ mol C}}{1 \text{ mol } CO_2} = 0.0352 \text{ mol C}$$

$C_{0.0352}H_{0.0774}$; divide each subscript by the smaller, 0.0352.
$C_{0.0352 / 0.0352}H_{0.0774 / 0.0352}$
CH_2
The empirical formula is CH_2, 14.0.
142.0 / 14.0 = 10.1 = 10; molecular formula = $C_{(10 \times 1)}H_{(10 \times 2)} = C_{10}H_{20}$

3.22 (a) $0.138 \text{ g } H_2O \times \frac{1 \text{ mol } H_2O}{18.0 \text{ g } H_2O} \times \frac{2 \text{ mol H}}{1 \text{ mol } H_2O} = 0.0153 \text{ mol H}$

$1.617 \text{ g } CO_2 \times \frac{1 \text{ mol } CO_2}{44.0 \text{ g } CO_2} \times \frac{1 \text{ mol C}}{1 \text{ mol } CO_2} = 0.0368 \text{ mol C}$

$0.0153 \text{ mol H} \times \frac{1.01 \text{ g H}}{1 \text{ mol H}} = 0.016 \text{ g H}$

$0.0368 \text{ mol C} \times \frac{12.0 \text{ g C}}{1 \text{ mol C}} = 0.442 \text{ g C}$

1.0 g total – (0.016 g H + 0.442 g C) = 0.542 g Cl

$0.542 \text{ g Cl} \times \frac{1 \text{ mol Cl}}{35.5 \text{ g Cl}} = 0.0153 \text{ mol Cl}$

$C_{0.0368}H_{0.0153}Cl_{0.0153}$; divide each subscript by the smallest, 0.0153.

$C_{0.0368 / 0.0153}H_{0.0153 / 0.0153}Cl_{0.0153 / 0.0153}$

$C_{2.5}HCl$

Multiply each subscript by 2 to get all integer subscripts.

$C_{(2 \times 2.5)}H_{(2 \times 1)}Cl_{(2 \times 1)}$

The empirical formula is $C_5H_2Cl_2$, 133.0.

(b) 326.26/133.0 = 2.45 = 2.5; molecular formula = $C_{(2.5 \times 5)}H_{(2.5 \times 2)}Cl_{(2.5 \times 2)} = C_{12}H_5Cl_5$

(c) No, because Cl does not form a useful oxide, so both O and Cl would have to be obtained by mass difference but that is impossible. You can only determine one element by mass difference.

3.23 The empirical formula is C_6H_5, 77.0.
The peak with the largest mass in the mass spectrum occurs near a mass of 154, which would be the molecular weight of the compound.
154/77.0 = 2; molecular formula = $C_{(2 \times 6)}H_{(2 \times 5)} = C_{12}H_{10}$

3.24 $0.57 \text{ g } H_2O \times \frac{1 \text{ mol } H_2O}{18.0 \text{ g } H_2O} \times \frac{2 \text{ mol H}}{1 \text{ mol } H_2O} = 0.063 \text{ mol H}$

$2.79 \text{ g } CO_2 \times \frac{1 \text{ mol } CO_2}{44.0 \text{ g } CO_2} \times \frac{1 \text{ mol C}}{1 \text{ mol } CO_2} = 0.063 \text{ mol C}$

$0.063 \text{ mol H} \times \frac{1.01 \text{ g H}}{1 \text{ mol H}} = 0.063 \text{ g H}$

$0.063 \text{ mol C} \times \frac{12.0 \text{ g C}}{1 \text{ mol C}} = 0.76 \text{ g C}$

1.00 g total – (0.063 g H + 0.76 g C) = 0.18 g N

$0.18 \text{ g N} \times \frac{1 \text{ mol N}}{14.0 \text{ g N}} = 0.013 \text{ mol N}$

$C_{0.063}H_{0.063}N_{0.013}$; divide each subscript by the smallest, 0.013.

$C_{0.063 / 0.013}H_{0.063 / 0.013}N_{0.013 / 0.013}$

$C_{4.85}H_{4.85}N$

The empirical formula is C_5H_5N, 79.0.

The peak with the largest mass in the mass spectrum occurs near a mass of 79, which would be the molecular weight of the compound, therefore the empirical formula and molecular formula are the same.

3.25 (a) C_2H_5OH, 46.1; CO_2, 44.0

$C_2H_5OH + 3O_2 \rightarrow 2\,CO_2 + 3\,H_2O$

$$14.7 \text{ gal} \times \frac{1 \text{ L}}{0.2642 \text{ gal}} \times \frac{0.79 \text{ kg}}{1 \text{ L}} = 44 \text{ kg } C_2H_5OH$$

$$44 \text{ kg } C_2H_5OH \times \frac{1000 \text{ g}}{1 \text{ kg}} \times \frac{1 \text{ mol } C_2H_5OH}{46.1 \text{ g } C_2H_5OH} \times \frac{2 \text{ mol } CO_2}{1 \text{ mol } C_2H_5OH}$$

$$\times \frac{44.0 \text{ g } CO_2}{1 \text{ mol } CO_2} \times \frac{1 \text{ kg}}{1000 \text{ g}} = 84 \text{ kg } CO_2$$

(b) C_3H_8, 44.1; CO_2, 44.0

$C_3H_8 + 5O_2 \rightarrow 3\,CO_2 + 4\,H_2O$

$$13.7 \text{ gal} \times \frac{1 \text{ L}}{0.2642 \text{ gal}} \times \frac{0.49 \text{ kg}}{1 \text{ L}} = 25 \text{ kg } C_2H_5OH$$

$$25 \text{ kg } C_2H_5OH \times \frac{1000 \text{ g}}{1 \text{ kg}} \times \frac{1 \text{ mol } C_3H_8}{44.1 \text{ g } C_3H_8} \times \frac{3 \text{ mol } CO_2}{1 \text{ mol } C_3H_8}$$

$$\times \frac{44.0 \text{ g } CO_2}{1 \text{ mol } CO_2} \times \frac{1 \text{ kg}}{1000 \text{ g}} = 75 \text{ kg } CO_2$$

(c) CH_4, 16.0; CO_2, 44.0

$CH_4 + 2O_2 \rightarrow CO_2 + 2\,H_2O$

$$25.7 \text{ kg } C_2H_5OH \times \frac{1000 \text{ g}}{1 \text{ kg}} \times \frac{1 \text{ mol } CH_4}{16.0 \text{ g } CH_4} \times \frac{1 \text{ mol } CO_2}{1 \text{ mol } CH_4}$$

$$\times \frac{44.0 \text{ g } CO_2}{1 \text{ mol } CO_2} \times \frac{1 \text{ kg}}{1000 \text{ g}} = 70.7 \text{ kg } CO_2$$

(d) Electricity from a Coal-Burning Power Plant

Assume an electric car is driven for 250 miles.

$$250 \text{ mi} \times \frac{35 \text{ kWh}}{100 \text{ mi}} \times \frac{0.94 \text{ kg } CO_2}{1 \text{ kWh}} = 82 \text{ kg } CO_2$$

(e) Electricity from a Natural Gas-Burning Power Plant

Assume an electric car is driven for 250 miles.

$$250 \text{ mi} \times \frac{35 \text{ kWh}}{100 \text{ mi}} \times \frac{0.55 \text{ kg } CO_2}{1 \text{ kWh}} = 48 \text{ kg } CO_2$$

(f) Electricity from natural gas produced the least amount of CO_2 when burned to provide energy for a car.

3.26 (a) $6\,CO_2 + 6\,H_2O \rightarrow C_6H_{12}O_6 + 6\,O_2$

(b) CO_2, 44.0; $C_6H_{12}O_6$, 180.0

$$1400 \text{ lb } CO_2 \times \frac{1 \text{ kg}}{2.2 \text{ lb}} \times \frac{1000 \text{ g}}{1 \text{ kg}} \times \frac{1 \text{ mol } CO_2}{44.0 \text{ g } CO_2}$$

$$\times \frac{1 \text{ mol } C_6H_{12}O_6}{6 \text{ mol } CO_2} \times \frac{180.0 \text{ g } C_6H_{12}O_6}{1 \text{ mol } C_6H_{12}O_6} \times \frac{1 \text{ kg}}{1000 \text{ g}} = 434 \text{ kg } C_6H_{12}O_6$$

3.27 (a) CO_2, 44.0; $HOCH_2CH_2NH_2$, 61.1

$CO_2 + 2\ HOCH_2CH_2NH_2 \rightarrow HOCH_2CH_2NH_3^+ + HOCH_2CH_2NHCO_2^-$

$$20 \times 10^6 \text{ tons} \times \frac{907.2 \text{ kg}}{1 \text{ ton}} \times \frac{1000 \text{ g}}{1 \text{ kg}} \times \frac{1 \text{ mol } CO_2}{44.0 \text{ g } CO_2}$$

$$\times \frac{2 \text{ mol } HOCH_2CH_2NH_2}{1 \text{ mol } CO_2} \times \frac{61.1 \text{ g } HOCH_2CH_2NH_2}{1 \text{ mol } HOCH_2CH_2NH_2}$$

$$\times \frac{1 \text{ kg}}{1000 \text{ g}} \times \frac{1 \text{ ton}}{907.2 \text{ kg}} = 55.5 \times 10^6 \text{ tons } HOCH_2CH_2NH_2$$

(b) $$25 \times 10^6 \text{ tons } HOCH_2CH_2NH_2 \times \frac{907.2 \text{ kg}}{1 \text{ ton}} \times \frac{1000 \text{ g}}{1 \text{ kg}} \times \frac{1 \text{ mol } HOCH_2CH_2NH_2}{61.1 \text{ g } HOCH_2CH_2NH_2}$$

$$\times \frac{1 \text{ mol } CO_2}{2 \text{ mol } HOCH_2CH_2NH_2} \times \frac{44.0 \text{ g } CO_2}{1 \text{ mol } CO_2}$$

$$\times \frac{1 \text{ kg}}{1000 \text{ g}} \times \frac{1 \text{ ton}}{907.2 \text{ g}} = 9 \times 10^6 \text{ tons } CO_2$$

(20×10^6 tons – 9×10^6 tons) = 11×10^6 (11 million) tons of CO_2 would escape into the atmosphere if only 25 million tons of $HOCH_2CH_2NH_2$ were used.

3.28 Open response question

Conceptual Problems

3.30 reactants, box (d), and products, box (c)

3.32 $C_{17}H_{18}F_3NO$ 17(12.01) + 18(1.01) + 3(19.00) + 1(14.01) + 1(16.00) = 309.36

3.34 (a) $A_2 + 3\ B_2 \rightarrow 2\ AB_3$; B_2 is the limiting reactant because it is completely consumed.
(b) For 1.0 mol of A_2, 3.0 mol of B_2 are required. Because only 1.0 mol of B_2 is available, B_2 is the limiting reactant.

$$1 \text{ mol } B_2 \times \frac{2 \text{ mol } AB_3}{3 \text{ mol } B_2} = 2/3 \text{ mol } AB_3$$

Section Problems
Balancing Equations (Section 3.2)

3.36 Equation (b) is balanced, (a) is not balanced.

3.38 (a) $Mg + 2\ HNO_3 \rightarrow H_2 + Mg(NO_3)_2$
(b) $CaC_2 + 2\ H_2O \rightarrow Ca(OH)_2 + C_2H_2$
(c) $2\ S + 3\ O_2 \rightarrow 2\ SO_3$
(d) $UO_2 + 4\ HF \rightarrow UF_4 + 2\ H_2O$

3.40 (a) $SiCl_4 + 2\ H_2O \rightarrow SiO_2 + 4\ HCl$
(b) $P_4O_{10} + 6\ H_2O \rightarrow 4\ H_3PO_4$
(c) $CaCN_2 + 3\ H_2O \rightarrow CaCO_3 + 2\ NH_3$
(d) $3\ NO_2 + H_2O \rightarrow 2\ HNO_3 + NO$

Molecular Weights and Stoichiometry (Section 3.3)

3.42 (a) Hg_2Cl_2: $2(200.59) + 2(35.45) = 472.1$
(b) $C_4H_8O_2$: $4(12.01) + 8(1.01) + 2(16.00) = 88.1$
(c) CF_2Cl_2: $1(12.01) + 2(19.00) + 2(35.45) = 120.9$

3.44 (a) $C_{33}H_{35}FN_2O_5$: $33(12.01) + 35(1.01) + 1(19.00) + 2(14.01) + 5(16.00) = 558.7$
(b) $C_{22}H_{27}F_3O_4S$: $22(12.01) + 27(1.01) + 3(19.00) + 4(16.00) + 1(32.06) = 444.5$
(c) $C_{16}H_{16}ClNO_2S$: $16(12.01) + 16(1.01) + 1(35.45) + 1(14.01) + 2(16.00) + 1(32.06) = 321.8$

3.46 One mole equals the atomic weight or molecular weight in grams.
(a) Ti, 47.87 g (b) Br_2, 159.81 g (c) Hg, 200.59 g (d) H_2O, 18.02 g

3.48 There are 3 ions (one Mg^{2+} and 2 Cl^-) per formula unit of $MgCl_2$.
$MgCl_2$, 95.2

$$27.5\text{ g } MgCl_2 \times \frac{1\text{ mol } MgCl_2}{95.2\text{ g } MgCl_2} \times \frac{3\text{ mol ions}}{1\text{ mol } MgCl_2} = 0.867\text{ mol ions}$$

3.50 $$\text{Molar mass} = \frac{3.28\text{ g}}{0.0275\text{ mol}} = 119\text{ g/mol; molecular weight} = 119.$$

3.52 $FeSO_4$, 151.9; 300 mg = 0.300 g

$$0.300\text{ g } FeSO_4 \times \frac{1\text{ mol } FeSO_4}{151.9\text{ g } FeSO_4} = 1.97 \times 10^{-3}\text{ mol } FeSO_4$$

$$1.97 \times 10^{-3}\text{ mol } FeSO_4 \times \frac{6.022 \times 10^{23}\text{ Fe(II) atoms}}{1\text{ mol } FeSO_4} = 1.19 \times 10^{21}\text{ Fe(II) atoms}$$

3.54 $C_8H_{10}N_4O_2$, 194.2; 125 mg = 0.125 g

$$0.125\text{ g caffeine} \times \frac{1\text{ mol caffeine}}{194.2\text{ g caffeine}} = 6.44 \times 10^{-4}\text{ mol caffeine}$$

$$0.125\text{ g caffeine} \times \frac{1\text{ mol caffeine}}{194.2\text{ g caffeine}} \times \frac{6.022 \times 10^{23}\text{ molecules}}{1\text{ mol}} = 3.88 \times 10^{20}\text{ caffeine molecules}$$

3.56 By definition, 6.022×10^{23} particles are 1.000 mole of particles.
mol Ar = 0.2500 mol and mol "other" = 0.7500 mol

$$\text{mass Ar} = 0.2500 \text{ mol Ar} \times \frac{39.95 \text{ g Ar}}{1 \text{ mol Ar}} = 9.99 \text{ g Ar}$$

mass "other" = total mass – mass Ar = 25.12 g – 9.99 g = 15.13 g "other"

$$\text{molar mass "other"} = \frac{15.13 \text{ g "other"}}{0.7500 \text{ mol "other"}} = 20.17 \text{ g/mol}$$

The "other" element is neon (Ne).

3.58 TiO_2, 79.87; $100.0 \text{ kg Ti} \times \frac{79.87 \text{ kg } TiO_2}{47.87 \text{ kg Ti}} = 166.8 \text{ kg } TiO_2$

3.60 (a) $2\ Fe_2O_3 + 3\ C \rightarrow 4\ Fe + 3\ CO_2$

(b) Fe_2O_3, 159.7; $525 \text{ g } Fe_2O_3 \times \frac{1 \text{ mol } Fe_2O_3}{159.7 \text{ g } Fe_2O_3} \times \frac{3 \text{ mol C}}{2 \text{ mol } Fe_2O_3} = 4.93 \text{ mol C}$

(c) $4.93 \text{ mol C} \times \frac{12.01 \text{ g C}}{1 \text{ mol C}} = 59.2 \text{ g C}$

3.62 (a) $2\ Mg + O_2 \rightarrow 2\ MgO$

(b) Mg, 24.30; O_2, 32.00; MgO, 40.30

$$25.0 \text{ g Mg} \times \frac{1 \text{ mol Mg}}{24.30 \text{ g Mg}} \times \frac{1 \text{ mol } O_2}{2 \text{ mol Mg}} \times \frac{32.00 \text{ g } O_2}{1 \text{ mol } O_2} = 16.5 \text{ g } O_2$$

$$25.0 \text{ g Mg} \times \frac{1 \text{ mol Mg}}{24.30 \text{ g Mg}} \times \frac{2 \text{ mol MgO}}{2 \text{ mol Mg}} \times \frac{40.30 \text{ g MgO}}{1 \text{ mol MgO}} = 41.5 \text{ g MgO}$$

(c) $25.0 \text{ g } O_2 \times \frac{1 \text{ mol } O_2}{32.00 \text{ g } O_2} \times \frac{2 \text{ mol Mg}}{1 \text{ mol } O_2} \times \frac{24.30 \text{ g Mg}}{1 \text{ mol Mg}} = 38.0 \text{ g Mg}$

$$25.0 \text{ g } O_2 \times \frac{1 \text{ mol } O_2}{32.00 \text{ g } O_2} \times \frac{2 \text{ mol MgO}}{1 \text{ mol } O_2} \times \frac{40.30 \text{ g MgO}}{1 \text{ mol MgO}} = 63.0 \text{ g MgO}$$

3.64 (a) $2\ HgO \rightarrow 2\ Hg + O_2$

(b) HgO, 216.6; Hg, 200.6; O_2, 32.0

$$45.5 \text{ g HgO} \times \frac{1 \text{ mol HgO}}{216.6 \text{ g HgO}} \times \frac{2 \text{ mol Hg}}{2 \text{ mol HgO}} \times \frac{200.6 \text{ g Hg}}{1 \text{ mol Hg}} = 42.1 \text{ g Hg}$$

$$45.5 \text{ g HgO} \times \frac{1 \text{ mol HgO}}{216.6 \text{ g HgO}} \times \frac{1 \text{ mol } O_2}{2 \text{ mol HgO}} \times \frac{32.00 \text{ g } O_2}{1 \text{ mol } O_2} = 3.36 \text{ g } O_2$$

(c) $33.3 \text{ g } O_2 \times \frac{1 \text{ mol } O_2}{32.00 \text{ g } O_2} \times \frac{2 \text{ mol HgO}}{1 \text{ mol } O_2} \times \frac{216.6 \text{ g HgO}}{1 \text{ mol HgO}} = 451 \text{ g HgO}$

3.66 $2.00\ g\ Ag \times \frac{1\ mol\ Ag}{107.9\ g\ Ag} = 0.0185\ mol\ Ag$; $0.657\ g\ Cl \times \frac{1\ mol\ Cl}{35.45\ g\ Cl} = 0.0185\ mol\ Cl$

$Ag_{0.0185}Cl_{0.0185}$; divide both subscripts by 0.0185. The empirical formula is AgCl.

3.68 N_2H_4, 32.05; I_2, 253.8; HI, 127.9

(a) $36.7\ g\ N_2H_4 \times \frac{1\ mol\ N_2H_4}{32.05\ g\ N_2H_4} \times \frac{2\ mol\ I_2}{1\ mol\ N_2H_4} \times \frac{253.8\ g\ I_2}{1\ mol\ I_2} = 581\ g\ I_2$

(b) $115.7\ g\ N_2H_4 \times \frac{1\ mol\ N_2H_4}{32.05\ g\ N_2H_4} \times \frac{4\ mol\ HI}{1\ mol\ N_2H_4} \times \frac{127.9\ g\ HI}{1\ mol\ HI} = 1847\ g\ HI$

Limiting Reactants and Reaction Yield (Sections 3.4 and 3.5)

3.70 $3.44\ mol\ N_2 \times \frac{3\ mol\ H_2}{1\ mol\ N_2} = 10.3\ mol\ H_2$ required.

Because there is only 1.39 mol H_2, H_2 is the limiting reactant.

$1.39\ mol\ H_2 \times \frac{2\ mol\ NH_3}{3\ mol\ H_2} \times \frac{17.03\ g\ NH_3}{1\ mol\ NH_3} = 15.8\ g\ NH_3$

$1.39\ mol\ H_2 \times \frac{1\ mol\ N_2}{3\ mol\ H_2} \times \frac{28.01\ g\ N_2}{1\ mol\ N_2} = 13.0\ g\ N_2$ reacted

$3.44\ mol\ N_2 \times \frac{28.01\ g\ N_2}{1\ mol\ N_2} = 96.3\ g\ N_2$ initially

(96.3 g – 13.0 g) = 83.3 g N_2 left over

3.72 C_2H_4, 28.05; Cl_2, 70.91; $C_2H_4Cl_2$, 98.96

$15.4\ g\ C_2H_4 \times \frac{1\ mol\ C_2H_4}{28.05\ g\ C_2H_4} = 0.549\ mol\ C_2H_4$

$3.74\ g\ Cl_2 \times \frac{1\ mol\ Cl_2}{70.91\ g\ Cl_2} = 0.0527\ mol\ Cl_2$

Because the reaction stoichiometry between C_2H_4 and Cl_2 is one to one, Cl_2 is the limiting reactant.

$0.0527\ mol\ Cl_2 \times \frac{1\ mol\ C_2H_4Cl_2}{1\ mol\ Cl_2} \times \frac{98.96\ g\ C_2H_4Cl_2}{1\ mol\ C_2H_4Cl_2} = 5.22\ g\ C_2H_4Cl_2$

3.74 H_2SO_4, 98.08; $NiCO_3$, 118.7; $NiSO_4$, 154.8

(a) $14.5\ g\ NiCO_3 \times \frac{1\ mol\ NiCO_3}{118.7\ g\ NiCO_3} \times \frac{1\ mol\ H_2SO_4}{1\ mol\ NiCO_3} \times \frac{98.08\ g\ H_2SO_4}{1\ mol\ H_2SO_4}$

$= 12.0\ g\ H_2SO_4$

(b) $14.5\ g\ NiCO_3 \times \frac{1\ mol\ NiCO_3}{118.7\ g\ NiCO_3} \times \frac{1\ mol\ NiSO_4}{1\ mol\ NiCO_3} \times \frac{154.8\ g\ NiSO_4}{1\ mol\ NiSO_4} \times 0.789$

$= 14.9\ g\ NiSO_4$

3.76 $CaCO_3$, 100.1; HCl, 36.46

$$CaCO_3 + 2\,HCl \rightarrow CaCl_2 + H_2O + CO_2$$

$$2.35\ g\ CaCO_3 \times \frac{1\ mol\ CaCO_3}{100.1\ g\ CaCO_3} = 0.0235\ mol\ CaCO_3$$

$$2.35\ g\ HCl \times \frac{1\ mol\ HCl}{36.46\ g\ HCl} = 0.0645\ mol\ HCl$$

The reaction stoichiometry is 1 mole of $CaCO_3$ for every 2 moles of HCl. For 0.0235 mol $CaCO_3$, we only need 2(0.0235 mol) = 0.0470 mol HCl. We have 0.0645 mol HCl, therefore $CaCO_3$ is the limiting reactant.

$$0.0235\ mol\ CaCO_3 \times \frac{1\ mol\ CO_2}{1\ mol\ CaCO_3} \times \frac{22.4\ L}{1\ mol\ CO_2} = 0.526\ L\ CO_2$$

3.78 $CH_3CO_2H + C_5H_{12}O \rightarrow C_7H_{14}O_2 + H_2O$

CH_3CO_2H, 60.05; $C_5H_{12}O$, 88.15; $C_7H_{14}O_2$, 130.19

$$3.58\ g\ CH_3CO_2H \times \frac{1\ mol\ CH_3CO_2H}{60.05\ g\ CH_3CO_2H} = 0.0596\ mol\ CH_3CO_2H$$

$$4.75\ g\ C_5H_{12}O \times \frac{1\ mol\ C_5H_{12}O}{88.15\ g\ C_5H_{12}O} = 0.0539\ mol\ C_5H_{12}O$$

Because the reaction stoichiometry between CH_3CO_2H and $C_5H_{12}O$ is one to one, isopentyl alcohol ($C_5H_{12}O$) is the limiting reactant.

$$0.0539\ mol\ C_5H_{12}O \times \frac{1\ mol\ C_7H_{14}O_2}{1\ mol\ C_5H_{12}O} \times \frac{130.19\ g\ C_7H_{14}O_2}{1\ mol\ C_7H_{14}O_2} = 7.02\ g\ C_7H_{14}O_2$$

7.02 g $C_7H_{14}O_2$ is the theoretical yield. Actual yield = (7.02 g)(0.45) = 3.2 g.

3.80 $CH_3CO_2H + C_5H_{12}O \rightarrow C_7H_{14}O_2 + H_2O$

CH_3CO_2H, 60.05; $C_5H_{12}O$, 88.15; $C_7H_{14}O_2$, 130.19

$$1.87\ g\ CH_3CO_2H \times \frac{1\ mol\ CH_3CO_2H}{60.05\ g\ CH_3CO_2H} = 0.0311\ mol\ CH_3CO_2H$$

$$2.31\ g\ C_5H_{12}O \times \frac{1\ mol\ C_5H_{12}O}{88.15\ g\ C_5H_{12}O} = 0.0262\ mol\ C_5H_{12}O$$

Because the reaction stoichiometry between CH_3CO_2H and $C_5H_{12}O$ is one to one, isopentyl alcohol ($C_5H_{12}O$) is the limiting reactant.

$$0.0262\ mol\ C_5H_{12}O \times \frac{1\ mol\ C_7H_{14}O_2}{1\ mol\ C_5H_{12}O} \times \frac{130.19\ g\ C_7H_{14}O_2}{1\ mol\ C_7H_{14}O_2} = 3.41\ g\ C_7H_{14}O_2$$

3.41 g $C_7H_{14}O_2$ is the theoretical yield.

$$\%\ Yield = \frac{Actual\ yield}{Theoretical\ yield} \times 100\% = \frac{2.96\ g}{3.41\ g} \times 100\% = 86.8\%$$

Formulas and Elemental Analysis (Sections 3.6 and 3.7)

3.82 CH_4N_2O, 60.1

$$\%C = \frac{12.0 \text{ g C}}{60.1 \text{ g}} \times 100\% = 20.0\%$$

$$\%H = \frac{4 \times 1.01 \text{ g H}}{60.1 \text{ g}} \times 100\% = 6.72\%$$

$$\%N = \frac{2 \times 14.0 \text{ g N}}{60.1 \text{ g}} \times 100\% = 46.6\%$$

$$\%O = \frac{16.0 \text{ g O}}{60.1 \text{ g}} \times 100\% = 26.6\%$$

3.84 Assume a 100.0 g sample of liquid. From the percent composition data, a 100.0 g sample of liquid contains 5.57 g H, 28.01 g Cl, and 66.42 g C.

$$66.42 \text{ g C} \times \frac{1 \text{ mol C}}{12.01 \text{ g C}} = 5.530 \text{ mol C}$$

$$5.57 \text{ g H} \times \frac{1 \text{ mol H}}{1.01 \text{ g H}} = 5.51 \text{ mol H}$$

$$28.01 \text{ g Cl} \times \frac{1 \text{ mol Cl}}{35.45 \text{ g Cl}} = 0.7901 \text{ mol Cl}$$

$C_{5.530}H_{5.51}Cl_{0.7901}$; divide each subscript by the smallest, 0.7901.

$C_{5.530/0.7901}H_{5.51/0.7901}Cl_{0.7901/0.7901}$

C_7H_7Cl

The empirical formula is C_7H_7Cl, 126.59.

Because the molecular weight equals the empirical formula weight, the empirical formula is also the molecular formula.

3.86 Assume a 100.0 g sample. From the percent composition data, a 100.0 g sample contains 24.25 g F and 75.75 g Sn.

$$24.25 \text{ g F} \times \frac{1 \text{ mol F}}{19.00 \text{ g F}} = 1.276 \text{ mol F}$$

$$75.75 \text{ g Sn} \times \frac{1 \text{ mol Sn}}{118.7 \text{ g Sn}} = 0.6382 \text{ mol Sn}$$

$Sn_{0.6382}F_{1.276}$; divide each subscript by the smaller, 0.6382.

$Sn_{0.6382/0.6382}F_{1.276/0.6382}$

The empirical formula is SnF_2.

3.88 Mass of toluene sample = 45.62 mg = 0.045 62 g; mass of CO_2 = 152.5 mg = 0.1525 g; mass of H_2O = 35.67 mg = 0.035 67 g

$$0.1525 \text{ g } CO_2 \times \frac{1 \text{ mol } CO_2}{44.01 \text{ g } CO_2} \times \frac{1 \text{ mol C}}{1 \text{ mol } CO_2} = 0.003\ 465 \text{ mol C}$$

$$\text{mass C} = 0.003\ 465 \text{ mol C} \times \frac{12.011 \text{ g C}}{1 \text{ mol C}} = 0.041\ 62 \text{ g C}$$

$$0.035\ 67 \text{ g } H_2O \times \frac{1 \text{ mol } H_2O}{18.02 \text{ g } H_2O} \times \frac{2 \text{ mol H}}{1 \text{ mol } H_2O} = 0.003\ 959 \text{ mol H}$$

$$\text{mass H} = 0.003\ 959 \text{ mol H} \times \frac{1.008 \text{ g H}}{1 \text{ mol H}} = 0.003\ 991 \text{ g H}$$

The (mass C + mass H) = 0.041 62 g + 0.003 991 g = 0.045 61 g. The calculated mass of (C + H) essentially equals the mass of the toluene sample, this means that toluene contains only C and H and no other elements.
$C_{0.003\ 465}H_{0.003\ 959}$; divide each subscript by the smaller, 0.003 465.
$C_{0.003\ 465\ /\ 0.003\ 465}H_{0.003\ 959\ /\ 0.003\ 465}$
$CH_{1.14}$; multiply each subscript by 7 to obtain integers.
The empirical formula is C_7H_8.

3.90 Let X equal the molecular weight of cytochrome c.

$$0.0043 = \frac{55.847 \text{ u}}{X}; \quad X = \frac{55.847 \text{ u}}{0.0043} = 13{,}000 \text{ u}$$

3.92 Let X equal the molecular weight of disilane.

$$0.9028 = \frac{2 \times 28.09 \text{ u}}{X}; \quad X = \frac{2 \times 28.09 \text{ u}}{0.9028} = 62.23 \text{ u}$$

62.23 – 2(Si atomic weight) = 62.23 – 2(28.09) = 6.05
6.05 is the total mass of H atoms.

$$6.05 \text{ u} \times \frac{1 \text{ H atom}}{1.01 \text{ u}} = 6 \text{ H atoms; Disilane is } Si_2H_6.$$

3.94 $C_{12}Br_{10}$, 943.2; $C_{12}Br_{10}O$, 959.2; 17.33 mg = 0.017 33 g

$$\text{For } C_{12}Br_{10}, \%\text{ C} = \frac{12 \times 12.01 \text{ g C}}{943.2 \text{ g}} \times 100\% = 15.28\%$$

$$\text{For } C_{12}Br_{10}O, \%\text{ C} = \frac{12 \times 12.01 \text{ g C}}{959.2 \text{ g}} \times 100\% = 15.03\%$$

Calculate the mass of C in 17.33 mg of CO_2.

$$0.017\ 33 \text{ g } CO_2 \times \frac{1 \text{ mol } CO_2}{44.01 \text{ g } CO_2} \times \frac{1 \text{ mol C}}{1 \text{ mol } CO_2} \times \frac{12.01 \text{ g C}}{1 \text{ mol C}} = 0.004\ 729 \text{ g} = 4.729 \text{ mg C}$$

Calculate the %C in the 31.472 mg sample.

$$\%\text{C} = \frac{4.729 \text{ mg C}}{31.472 \text{ mg}} \times 100\% = 15.03\%$$

Decabrom is $C_{12}Br_{10}O$.

Mass Spectrometry (Section 3.8)

3.96 A neutral molecule will travel in a straight, undeflected, path in a mass spectrometer. Ionization is necessary as electric and magnetic fields will only exert a force on a charged species, not a neutral molecule. Ions of different masses are then accelerated by an electric field and passed between the poles of a strong magnet, which deflects them through a curved, evacuated pipe. The radius of deflection of a charged ion, M^+, as it passes between the magnet poles depends on its mass, with lighter ions deflected more strongly than heavier ones.

3.98 Mass of the sample is 70.042 11.
For C_5H_{10}, mass = 5(12.000 000) + 10(1.007 825) = 70.078 250
For C_4H_6O, mass = 4(12.000 000) + 6(1.007 825) + 1(15.994 915) = 70.041 865
For $C_3H_6N_2$, mass = 3(12.000 000) + 6(1.007 825) + 2(14.003 074) = 70.053 098
The sample is C_4H_6O.

3.100 mass of CO_2 = 169.2 mg = 0.1692 g; mass of H_2O = 34.6 mg = 0.0346 g

$$0.1692\text{ g } CO_2 \times \frac{1\text{ mol } CO_2}{44.01\text{ g } CO_2} \times \frac{1\text{ mol C}}{1\text{ mol } CO_2} = 0.003\ 845\text{ mol C}$$

$$0.0346\text{ g } H_2O \times \frac{1\text{ mol } H_2O}{18.02\text{ g } H_2O} \times \frac{2\text{ mol H}}{1\text{ mol } H_2O} = 0.003\ 84\text{ mol H}$$

$C_{0.003\ 845}H_{0.003\ 84}$; divide each subscript by the smaller, 0.003 84.
$C_{0.003\ 845\ /\ 0.003\ 84}H_{0.003\ 84\ /\ 0.003\ 84}$
The empirical formula is CH, 13.0.
The peak with the largest mass in the mass spectrum occurs near a mass of 78, which would be the molecular weight of the compound.
78/13.0 = 6; molecular formula = $C_{(6 \times 1)}H_{(6 \times 1)} = C_6H_6$

Chapter Problems

3.102 NaCl, 58.4; KCl, 74.6; $CaCl_2$, 111.0; 500 mL = 0.500 L

$$4.30\text{ g NaCl} \times \frac{1\text{ mol NaCl}}{58.4\text{ g NaCl}} = 0.0736\text{ mol NaCl}$$

$$0.150\text{ g KCl} \times \frac{1\text{ mol KCl}}{74.6\text{ g KCl}} = 0.002\ 01\text{ mol KCl}$$

$$0.165\text{ g } CaCl_2 \times \frac{1\text{ mol } CaCl_2}{111.0\text{ g } CaCl_2} = 0.001\ 49\text{ mol } CaCl_2$$

0.0736 mol + 0.002 01 mol + 2(0.001 49 mol) = 0.0786 mol Cl^-

$$Na^+\text{ molarity} = \frac{0.0736\text{ mol}}{0.500\text{ L}} = 0.147\text{ M}$$

$$Ca^{2+}\text{ molarity} = \frac{0.001\ 49\text{ mol}}{0.500\text{ L}} = 0.002\ 98\text{ M}$$

K^+ molarity $= \dfrac{0.002\ 01\ \text{mol}}{0.500\ \text{L}} = 0.004\ 02$ M

Cl^- molarity $= \dfrac{0.0786\ \text{mol}}{0.500\ \text{L}} = 0.157$ M

3.104 H_2S, 34.08; O_2, 32.00; Ag_2S, 247.8

(a) $4\ Ag + 2\ H_2S + O_2 \rightarrow 2\ Ag_2S + 2\ H_2O$

(b) Compute the theoretical yield for Ag_2S from each reactant to determine the limiting reactant.

$$496\ \text{g Ag} \times \frac{1\ \text{mol Ag}}{107.9\ \text{g Ag}} \times \frac{2\ \text{mol}\ Ag_2S}{4\ \text{mol Ag}} \times \frac{247.8\ \text{g}\ Ag_2S}{1\ \text{mol}\ Ag_2S} = 570\ \text{g}\ Ag_2S$$

$$80.0\ \text{g}\ H_2S \times \frac{1\ \text{mol}\ H_2S}{34.08\ \text{g}\ H_2S} \times \frac{2\ \text{mol}\ Ag_2S}{2\ \text{mol}\ H_2S} \times \frac{247.8\ \text{g}\ Ag_2S}{1\ \text{mol}\ Ag_2S} = 582\ \text{g}\ Ag_2S$$

$$40.0\ \text{g}\ O_2 \times \frac{1\ \text{mol}\ O_2}{32.00\ \text{g}\ O_2} \times \frac{2\ \text{mol}\ Ag_2S}{1\ \text{mol}\ O_2} \times \frac{247.8\ \text{g}\ Ag_2S}{1\ \text{mol}\ Ag_2S} = 619\ \text{g}\ Ag_2S$$

The smallest theoretical yield (570 g) means that Ag is the limiting reactant.
With a 95% yield, the mass of Ag_2S produced = (0.95)(570 g) = 541 g Ag_2S

3.106 (a) Assume a 100.0 g sample of aspirin. From the percent composition data, a 100.0 g sample contains 60.00 g C, 35.52 g O, and 4.48 g H.

$$60.00\ \text{g C} \times \frac{1\ \text{mol C}}{12.01\ \text{g C}} = 4.996\ \text{mol C}$$

$$35.52\ \text{g O} \times \frac{1\ \text{mol O}}{16.00\ \text{g O}} = 2.220\ \text{mol O}$$

$$4.48\ \text{g H} \times \frac{1\ \text{mol H}}{1.01\ \text{g H}} = 4.44\ \text{mol H}$$

$C_{4.996}H_{4.44}O_{2.220}$; divide each subscript by the smallest, 2.220.

$C_{4.996\,/\,2.220}H_{4.44\,/\,2.220}O_{2.220\,/\,2.220}$

$C_{2.25}H_2O_1$; multiply each subscript by 4 to obtain integers.

The empirical formula is $C_9H_8O_4$.

(b) Assume a 100.0 g sample of ilmenite. From the percent composition data, a 100.0 g sample contains 31.63 g O, 31.56 g Ti, and 36.81 g Fe.

$$31.63\ \text{g O} \times \frac{1\ \text{mol O}}{16.00\ \text{g O}} = 1.977\ \text{mol O}$$

$$31.56\ \text{g Ti} \times \frac{1\ \text{mol Ti}}{47.87\ \text{g Ti}} = 0.6593\ \text{mol Ti}$$

$$36.81\ \text{g Fe} \times \frac{1\ \text{mol Fe}}{55.85\ \text{g Fe}} = 0.6591\ \text{mol Fe}$$

$Fe_{0.6591}Ti_{0.6593}O_{1.977}$; divide each subscript by the smallest, 0.6591.

$Fe_{0.6591\,/\,0.6591}Ti_{0.6593\,/\,0.6591}O_{1.977\,/\,0.6591}$

The empirical formula is $FeTiO_3$.

(c) Assume a 100.0 g sample of sodium thiosulfate. From the percent composition data, a 100.0 g sample contains 30.36 g O, 29.08 g Na, and 40.56 g S.

$$30.36 \text{ g O} \times \frac{1 \text{ mol O}}{16.00 \text{ g O}} = 1.897 \text{ mol O}$$

$$29.08 \text{ g Na} \times \frac{1 \text{ mol Na}}{22.99 \text{ g Na}} = 1.265 \text{ mol Na}$$

$$40.56 \text{ g S} \times \frac{1 \text{ mol S}}{32.07 \text{ g S}} = 1.265 \text{ mol S}$$

$Na_{1.265}S_{1.265}O_{1.897}$; divide each subscript by the smallest, 1.265.
$Na_{1.265 / 1.265}S_{1.265 / 1.265}O_{1.897 / 1.265}$
$NaSO_{1.5}$; multiply each subscript by 2 to obtain integers.
The empirical formula is $Na_2S_2O_3$.

3.108 NaH, 24.00; B_2H_6, 27.67; $NaBH_4$, 37.83

(a) $2\ NaH + B_2H_6 \rightarrow 2\ NaBH_4$

$$8.55 \text{ g NaH} \times \frac{1 \text{ mol NaH}}{24.00 \text{ g NaH}} = 0.356 \text{ mol NaH}$$

$$6.75 \text{ g } B_2H_6 \times \frac{1 \text{ mol } B_2H_6}{27.67 \text{ g } B_2H_6} = 0.244 \text{ mol } B_2H_6$$

For 0.244 mol B_2H_6, 2 x (0.244) = 0.488 mol NaH are needed. Because only 0.356 mol of NaH is available, NaH is the limiting reactant.

$$0.356 \text{ mol NaH} \times \frac{2 \text{ mol } NaBH_4}{2 \text{ mol NaH}} \times \frac{37.83 \text{ g } NaBH_4}{1 \text{ mol } NaBH_4} = 13.5 \text{ g } NaBH_4 \text{ produced}$$

(b) $$0.356 \text{ mol NaH} \times \frac{1 \text{ mol } B_2H_6}{2 \text{ mol NaH}} \times \frac{27.67 \text{ g } B_2H_6}{1 \text{ mol } B_2H_6} = 4.93 \text{ g } B_2H_6 \text{ reacted}$$

B_2H_6 left over = 6.75 g – 4.93 g = 1.82 g B_2H_6

3.110 $$\text{Mass of 1 HCl molecule} = (36.5 \frac{\text{u}}{\text{molecule}})(1.6605 \times 10^{-24} \frac{\text{g}}{\text{u}}) = 6.06 \times 10^{-23} \text{ g/molecule}$$

$$\text{Avogadro's number} = \left(\frac{36.5 \text{ g/mol}}{6.06 \times 10^{-23} \text{ g/molecule}}\right) = 6.02 \times 10^{23} \text{ molecules/mol}$$

3.112 High-resolution mass spectrometry is capable of measuring the mass of molecules with a particular isotopic composition.

3.114 The combustion reaction is: $2\ C_8H_{18} + 25\ O_2 \rightarrow 16\ CO_2 + 18\ H_2O$

C_8H_{18}, 114.23; CO_2, 44.01

$$\text{pounds } CO_2 = 1.00 \text{ gal} \times \frac{3.7854 \text{ L}}{1 \text{ gal}} \times \frac{1000 \text{ mL}}{1 \text{ L}} \times \frac{0.703 \text{ g } C_8H_{18}}{1 \text{ mL}} \times \frac{1 \text{ mol } C_8H_{18}}{114.23 \text{ g } C_8H_{18}} \times$$

$$\frac{16 \text{ mol } CO_2}{2 \text{ mol } C_8H_{18}} \times \frac{44.01 \text{ g } CO_2}{1 \text{ mol } CO_2} \times \frac{1 \text{ lb}}{453.59 \text{ g}} = 18.1 \text{ lb } CO_2$$

3.116 $CaCO_3$, 100.09

$$\% \text{ Ca} = \frac{40.08 \text{ g Ca}}{100.09 \text{ g}} \times 100\% = 40.04\%$$

$$\% \text{ C} = \frac{12.01 \text{ g C}}{100.09 \text{ g}} \times 100\% = 12.00\%$$

$$\% \text{ O} = \frac{3 \times 16.00 \text{ g O}}{100.09 \text{ g}} \times 100\% = 47.96\%$$

Because the mass %'s for the pulverized rock are different from the mass %'s for pure $CaCO_3$ calculated here, the pulverized rock cannot be pure $CaCO_3$.

3.118 FeO, 71.85; Fe_2O_3, 159.7

Let X equal the mass of FeO and Y the mass of Fe_2O_3 in the 10.0 g mixture. Therefore, X + Y = 10.0 g.

$$\text{mol Fe} = 7.43 \text{ g} \times \frac{1 \text{ mol Fe}}{55.85 \text{ g Fe}} = 0.133 \text{ mol Fe}$$

$$\text{mol FeO} + 2 \times \text{mol Fe}_2\text{O}_3 = 0.133 \text{ mol Fe}$$

$$X \times \frac{1 \text{ mol FeO}}{71.85 \text{ g FeO}} + 2 \times \left(Y \times \frac{1 \text{ mol Fe}_2\text{O}_3}{159.7 \text{ g Fe}_2\text{O}_3} \right) = 0.133 \text{ mol Fe}$$

Rearrange to get X = 10.0 g – Y and then substitute it into the equation above to solve for Y.

$$(10.0 \text{ g} - Y) \times \frac{1 \text{ mol FeO}}{71.85 \text{ g FeO}} + 2 \times \left(Y \times \frac{1 \text{ mol Fe}_2\text{O}_3}{159.7 \text{ g Fe}_2\text{O}_3} \right) = 0.133 \text{ mol Fe}$$

$$\frac{10.0 \text{ mol}}{71.85} - \frac{Y \text{ mol}}{71.85 \text{ g}} + \frac{2\,Y \text{ mol}}{159.7 \text{ g}} = 0.133 \text{ mol}$$

$$-\frac{Y \text{ mol}}{71.85 \text{ g}} + \frac{2\,Y \text{ mol}}{159.7 \text{ g}} = 0.133 \text{ mol} - \frac{10.0 \text{ mol}}{71.85} = -0.0062 \text{ mol}$$

$$\frac{(-Y \text{ mol})(159.7 \text{ g}) + (2\,Y \text{ mol})(71.85 \text{ g})}{(71.85 \text{ g})(159.7 \text{ g})} = -0.0062 \text{ mol}$$

$$\frac{-16.0\,Y \text{ mol}}{11474 \text{ g}} = -0.0062 \text{ mol}; \quad \frac{16.0\,Y}{11474 \text{ g}} = 0.0062$$

Y = (0.0062)(11474 g)/16.0 = 4.44 g = 4.4 g Fe_2O_3

X = 10.0 g – Y = 10.0 g – 4.4 g = 5.6 g FeO

3.120 $C_6H_{12}O_6 + 6\,O_2 \rightarrow 6\,CO_2 + 6\,H_2O$; $C_6H_{12}O_6$, 180.16; CO_2, 44.01

$$66.3 \text{ g C}_6\text{H}_{12}\text{O}_6 \times \frac{1 \text{ mol C}_6\text{H}_{12}\text{O}_6}{180.16 \text{ g C}_6\text{H}_{12}\text{O}_6} \times \frac{6 \text{ mol CO}_2}{1 \text{ mol C}_6\text{H}_{12}\text{O}_6} \times \frac{44.01 \text{ g CO}_2}{1 \text{ mol CO}_2} = 97.2 \text{ g CO}_2$$

$$66.3 \text{ g C}_6\text{H}_{12}\text{O}_6 \times \frac{1 \text{ mol C}_6\text{H}_{12}\text{O}_6}{180.16 \text{ g C}_6\text{H}_{12}\text{O}_6} \times \frac{6 \text{ mol CO}_2}{1 \text{ mol C}_6\text{H}_{12}\text{O}_6} \times \frac{25.4 \text{ L CO}_2}{1 \text{ mol CO}_2} = 56.1 \text{ L CO}_2$$

3.122 Mass of added Cl = mass of XCl_5 – mass of XCl_3 = 13.233 g – 8.729 g = 4.504 g

mass of Cl in XCl_5 = 5 Cl's x $\frac{4.504\ g}{2\ Cl's}$ = 11.26 g Cl

mass of X in XCl_5 = 13.233 g – 11.26 g = 1.973 g X

$11.26\ g\ Cl \times \frac{1\ mol\ Cl}{35.45\ g\ Cl} = 0.3176\ mol\ Cl$

$0.3176\ mol\ Cl \times \frac{1\ mol\ X}{5\ mol\ Cl} = 0.063\ 52\ mol\ X$

molar mass of X = $\frac{1.973\ g\ X}{0.063\ 52\ mol\ X}$ = 31.1 g/mol; atomic weight =31.1, X = P

3.124 NH_4NO_3, 80.04; $(NH_4)_2HPO_4$, 132.06

Assume you have a 100.0 g sample of the mixture.

Let X = grams of NH_4NO_3 and (100.0 – X) = grams of $(NH_4)_2HPO_4$.

Both compounds contain 2 nitrogen atoms per formula unit.

Because the mass % N in the sample is 30.43%, the 100.0 g sample contains 30.43 g N.

$$\text{mol } NH_4NO_3 = (X) \times \frac{1\ mol\ NH_4NO_3}{80.04\ g}$$

$$\text{mol } (NH_4)_2HPO_4 = (100.0 - X) \times \frac{1\ mol\ (NH_4)_2HPO_4}{132.06\ g}$$

$$\text{mass N} = \left(\left((X) \times \frac{1\ mol\ NH_4NO_3}{80.04\ g}\right) + \left((100.0 - X) \times \frac{1\ mol\ (NH_4)_2HPO_4}{132.06\ g}\right)\right) \times \left(\frac{2\ mol\ N}{1\ mol\ ammonium\ cmpds}\right) \times \left(\frac{14.0067\ g\ N}{1\ mol\ N}\right) = 30.43\ g$$

Solve for X.

$$\left(\frac{X}{80.04} + \frac{100.0 - X}{132.06}\right)(2)(14.0067) = 30.43$$

$$\left(\frac{X}{80.04} + \frac{100.0 - X}{132.06}\right) = 1.08627$$

$$\frac{(132.06)(X) + (100.0 - X)(80.04)}{(80.04)(132.06)} = 1.08627$$

$(132.06)(X) + (100.0 – X)(80.04) = (1.08627)(80.04)(132.06)$

$132.06X + 8004 – 80.04X = 11481.96$

$132.06X – 80.04X = 11481.96 – 8004$

$52.02X = 3477.96$

$X = \frac{3477.96}{52.02} = 66.86\ g\ NH_4NO_3$

$(100.0 – X) = (100.0 – 66.86) = 33.14\ g\ (NH_4)_2HPO_4$

$$\frac{mass_{NH_4NO_3}}{mass_{(NH_4)_2HPO_4}} = \frac{66.86\ g}{33.14\ g} = 2.018$$

The mass ratio of NH_4NO_3 to $(NH_4)_2HPO_4$ in the mixture is 2 to 1.

3.126 (a) 56.0 mL = 0.0560 L

$$\text{mol } X_2 = (0.0560 \text{ L } X_2)\left(\frac{1 \text{ mol}}{22.41 \text{ L}}\right) = 0.00250 \text{ mol } X_2$$

mass X_2 = 1.12 g MX_2 – 0.720 g MX = 0.40 g X_2

$$\text{molar mass } X_2 = \frac{0.40 \text{ g}}{0.00250 \text{ mol}} = 160 \text{ g/mol}$$

atomic weight of X = 160/2 = 80; X is Br.

(b) $$\text{mol MX} = 0.00250 \text{ mol } X_2 \times \frac{2 \text{ mol MX}}{1 \text{ mol } X_2} = 0.00500 \text{ mol MX}$$

$$\text{mass of X in MX} = 0.00500 \text{ mol MX} \times \frac{1 \text{ mol X}}{1 \text{ mol MX}} \times \frac{80 \text{ g X}}{1 \text{ mol X}} = 0.40 \text{ g X}$$

mass of M in MX = 0.720 g MX – 0.40 g X = 0.32 g M

$$\text{molar mass M} = \frac{0.32 \text{ g}}{0.00500 \text{ mol}} = 64 \text{ g/mol}$$

atomic weight of X = 64; M is Cu.

4 Reactions in Aqueous Solution

4.1 $C_{12}H_{22}O_{11}$, 342.3; 355 mL = 0.355 L

$$43.0\text{ g } C_{12}H_{22}O_{11} \times \frac{1\text{ mol } C_{12}H_{22}O_{11}}{342.3\text{ g } C_{12}H_{22}O_{11}} = 0.126\text{ mol } C_6H_{12}O_6$$

$$\text{molarity} = \frac{0.126\text{ mol}}{0.355\text{ L}} = 0.355\text{ M}$$

4.2 The solution is not 1.00 M. The solution was incorrectly prepared by combining 0.500 mol of solute with 500 mL of water; resulting in a volume larger than 500 mL. Instead the solute should be added to a 500 mL volumetric flask and solvent added to bring the final volume to 500 mL.

4.3 (a) 125 mL = 0.125 L; (0.20 mol/L)(0.125 L) = 0.025 mol $NaHCO_3$
(b) 650.0 mL = 0.6500 L; (2.50 mol/L)(0.6500 L) = 1.62 mol H_2SO_4

4.4 (a) NaOH, 40.0; 500.0 mL = 0.5000 L

$$\frac{1.25\text{ mol NaOH}}{\text{L}} \times 0.500\text{ L} \times \frac{40.0\text{ g NaOH}}{1\text{ mol NaOH}} = 25.0\text{ g NaOH}$$

(b) $C_6H_{12}O_6$, 180.2

$$\frac{0.250\text{ mol } C_6H_{12}O_6}{\text{L}} \times 1.50\text{ L} \times \frac{180.2\text{ g } C_6H_{12}O_6}{1\text{ mol } C_6H_{12}O_6} = 67.6\text{ g } C_6H_{12}O_6$$

4.5 (a) $C_{27}H_{46}O$, 386.7; 750 mL = 0.750 L

$$\frac{0.005\text{ mol } C_{27}H_{46}O}{\text{L}} \times 0.750\text{ L} \times \frac{386.7\text{ g } C_{27}H_{46}O}{1\text{ mol } C_{27}H_{46}O} = 1\text{ g } C_{27}H_{46}O$$

(b)
$$25\text{ mg } C_{27}H_{46}O \times \frac{1 \times 10^{-3}\text{ g}}{1\text{ mg}} \times \frac{1\text{ mol } C_{27}H_{46}O}{386.7\text{ g } C_{27}H_{46}O}$$
$$\times \frac{1\text{ L}}{0.005\text{ mol } C_{27}H_{46}O} \times \frac{1\text{ mL}}{1 \times 10^{-3}\text{ L}} = 13\text{ mL blood}$$

4.6 $M_i \times V_i = M_f \times V_f$; $M_f = \dfrac{M_i \times V_i}{V_f} = \dfrac{3.50\text{ M} \times 75.0\text{ mL}}{400.0\text{ mL}} = 0.656\text{ M}$

4.7 $M_i \times V_i = M_f \times V_f$; $V_i = \dfrac{M_f \times V_f}{M_i} = \dfrac{0.500\text{ M} \times 250.0\text{ mL}}{18.0\text{ M}} = 6.94\text{ mL}$

4.8 $FeBr_3$ contains 3 Br^- ions. The molar concentration of Br^- ions = 3 x 0.225 M = 0.675 M.

4.9 (a) A_2Y is the strongest electrolyte because all the molecules have dissociated into ions. A_2X is the weakest because only 1 out of 5 molecules have dissociated into ions.
(b) In a 0.350 M solution of A_2Y, the molar concentration of A ions = 2 x 0.350 = 0.700 M.
In a 0.350 M solution of A_2Y, the molar concentration of Y ions = 1 x 0.350 = 0.350 M.
(c) There are 5 A_2X molecules initially in the solution. Only one of them is ionized. The percent ionization = (1/5) x 100% = 20%

4.10 (a) Ionic equation:
$2\ Ag^+(aq) + 2\ NO_3^-(aq) + 2\ Na^+(aq) + CrO_4^{2-}(aq) \rightarrow Ag_2CrO_4(s) + 2\ Na^+(aq) + 2\ NO_3^-(aq)$
Delete spectator ions from the ionic equation to get the net ionic equation.
Net ionic equation: $2\ Ag^+(aq) + CrO_4^{2-}(aq) \rightarrow Ag_2CrO_4(s)$
(b) Ionic equation:
$2\ H^+(aq) + SO_4^{2-}(aq) + MgCO_3(s) \rightarrow H_2O(l) + CO_2(g) + Mg^{2+}(aq) + SO_4^{2-}(aq)$
Delete spectator ions from the ionic equation to get the net ionic equation.
Net ionic equation: $2\ H^+(aq) + MgCO_3(s) \rightarrow H_2O(l) + CO_2(g) + Mg^{2+}(aq)$
(c) Ionic equation:
$Hg^{2+}(aq) + 2\ NO_3^-(aq) + 2\ NH_4^+(aq) + 2\ I^-(aq) \rightarrow HgI_2(s) + 2\ NH_4^+(aq) + 2\ NO_3^-(aq)$
Delete spectator ions from the ionic equation to get the net ionic equation.
Net ionic equation: $Hg^{2+}(aq) + 2\ I^-(aq) \rightarrow HgI_2(s)$

4.11 Spectator ions: Ca^{2+} and NO_3^-
Net ionic equation: $Ag^+(aq) + Cl^-(aq) \rightarrow AgCl(s)$
Molecular equation: $2\ AgNO_3(aq) + CaCl_2(aq) \rightarrow 2\ AgCl(s) + Ca(NO_3)_2(aq)$

4.12 (a) Ionic equation:
$Ni^{2+}(aq) + 2\ Cl^-(aq) + 2\ NH_4^+(aq) + S^{2-}(aq) \rightarrow NiS(s) + 2\ NH_4^+(aq) + 2\ Cl^-(aq)$
Delete spectator ions from the ionic equation to get the net ionic equation.
Net ionic equation: $Ni^{2+}(aq) + S^{2-}(aq) \rightarrow NiS(s)$
(b) Ionic equation:
$2\ Na^+(aq) + CrO_4^{2-}(aq) + Pb^{2+}(aq) + 2\ NO_3^-(aq) \rightarrow PbCrO_4(s) + 2\ Na^+(aq) + 2\ NO_3^-(aq)$
Delete spectator ions from the ionic equation to get the net ionic equation.
Net ionic equation: $Pb^{2+}(aq) + CrO_4^{2-}(aq) \rightarrow PbCrO_4(s)$
(c) Ionic equation:
$2\ Ag^+(aq) + 2\ ClO_4^-(aq) + Ca^{2+}(aq) + 2\ Br^-(aq) \rightarrow 2\ AgBr(s) + Ca^{2+}(aq) + 2\ ClO_4^-(aq)$
Delete spectator ions from the ionic equation and reduce coefficients to get the net ionic equation.
Net ionic equation: $Ag^+(aq) + Br^-(aq) \rightarrow AgBr(s)$
(d) Ionic equation:
$Zn^{2+}(aq) + 2\ Cl^-(aq) + 2\ K^+(aq) + CO_3^{2-}(aq) \rightarrow ZnCO_3(s) + 2\ K^+(aq) + 2\ Cl^-(aq)$
Delete spectator ions from the ionic equation to get the net ionic equation.
Net ionic equation: $Zn^{2+}(aq) + CO_3^{2-}(aq) \rightarrow ZnCO_3(s)$

4.13 $3\ CaCl_2(aq) + 2\ Na_3PO_4(aq) \rightarrow Ca_3(PO_4)_2(s) + 6\ NaCl(aq)$
Ionic equation:
$3\ Ca^{2+}(aq) + 6\ Cl^-(aq) + 6\ Na^+(aq) + 2\ PO_4^{3-}(aq) \rightarrow Ca_3(PO_4)_2(s) + 6\ Na^+(aq) + 6\ Cl^-(aq)$
Delete spectator ions from the ionic equation to get the net ionic equation.
Net ionic equation: $3\ Ca^{2+}(aq) + 2\ PO_4^{3-}(aq) \rightarrow Ca_3(PO_4)_2(s)$

4.14 A precipitate results from the reaction. The precipitate contains cations and anions in a 3:2 ratio. The precipitate is either $Mg_3(PO_4)_2$ or $Zn_3(PO_4)_2$.

4.15 The precipitate is $PbBr_2$.

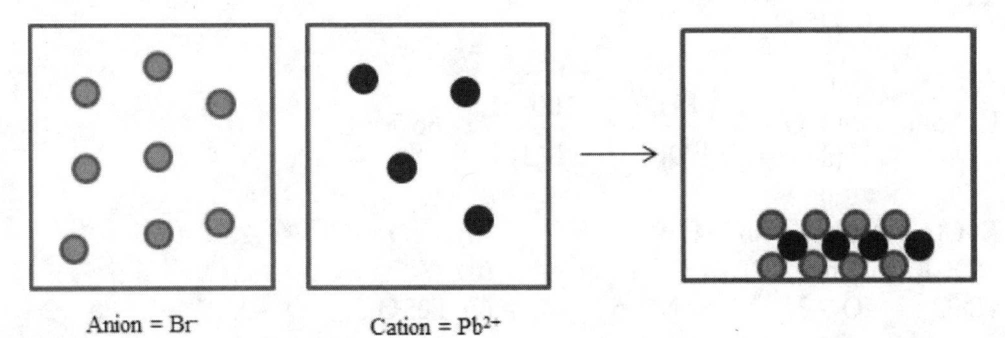

4.16 (a) HI, hydroiodic acid (b) $HBrO_2$, bromous acid (c) H_2CrO_4, chromic acid

4.17 (a) H_3PO_3 (b) H_2Se

4.18 (a) Ionic equation:
$2\ Cs^+(aq) + 2\ OH^-(aq) + 2\ H^+(aq) + SO_4^{2-}(aq) \rightarrow 2\ Cs^+(aq) + SO_4^{2-}(aq) + 2\ H_2O(l)$
Delete spectator ions from the ionic equation and reduce coefficients to get the net ionic equation.
Net ionic equation: $H^+(aq) + OH^-(aq) \rightarrow H_2O(l)$
(b) Ionic equation:
$Ca^{2+}(aq) + 2\ OH^-(aq) + 2\ CH_3CO_2H(aq) \rightarrow Ca^{2+}(aq) + 2\ CH_3CO_2^-(aq) + 2\ H_2O(l)$
Delete spectator ions from the ionic equation and reduce coefficients to get the net ionic equation.
Net ionic equation: $CH_3CO_2H(aq) + OH^-(aq) \rightarrow CH_3CO_2^-(aq) + H_2O(l)$

4.19 $Mg^{2+}(aq) + 2\ OH^-(aq) + 2\ H^+(aq) + 2\ Cl^-(aq) \rightarrow Mg^{2+}(aq) + 2\ Cl^-(aq) + H_2O(l)$
$H^+(aq) + OH^-(aq) \rightarrow H_2O(l)$

4.20 50.0 mL = 0.0500 L
(0.100 mol/L)(0.0500 L) = 5.00×10^{-3} mol NaOH

$$5.00 \times 10^{-3}\ \text{mol NaOH} \times \frac{1\ \text{mol}\ H_2SO_4}{2\ \text{mol NaOH}} = 2.50 \times 10^{-3}\ \text{mol}\ H_2SO_4$$

$$2.50 \times 10^{-3}\ \text{mol}\ H_2SO_4 \times \frac{1\ \text{L}}{0.250\ \text{mol}\ H_2SO_4} \times \frac{1000\ \text{mL}}{1\ \text{L}} = 10.0\ \text{mL}$$

4.21 25.0 mL = 0.0250 L; 68.5 mL = 0.0685 L
From the reaction stoichiometry, moles KOH = moles HNO_3
(0.150 mol/L)(0.0250 L) = 3.75×10^{-3} mol KOH = 3.75×10^{-3} mol HNO_3

$$\text{molarity} = \frac{3.75 \times 10^{-3}\text{ mol}}{0.0685\text{ L}} = 0.0547\text{ M}$$

4.22 25.0 mL = 0.0250 L; 94.7 mL = 0.0947 L
From the reaction stoichiometry, moles NaOH = moles CH_3CO_2H
(0.200 mol/L)(0.0947 L) = 0.018 94 mol NaOH = 0.018 94 mol CH_3CO_2H

$$\text{molarity} = \frac{0.018\ 94\text{ mol}}{0.0250\text{ L}} = 0.758\text{ M}$$

4.23 $$\text{vol } H^+ \text{ solution} = 8\ OH^- \times \frac{1\ H^+}{1\ OH^-} \times \frac{100\text{ mL}}{12\ H^+} = 66.7\text{ mL}$$

4.24 (a) $SnCl_4$: Cl –1, Sn +4
(b) CrO_3: O –2, Cr +6
(c) $VOCl_3$: O –2, Cl –1, V +5
(d) V_2O_3: O –2, V +3
(e) HNO_3: O –2, H +1, N +5
(f) $FeSO_4$: O –2, S +6, Fe +2

4.25 (a) +2, ClO, chlorine monoxide; +3, Cl_2O_3, dichlorine trioxide; +6, ClO_3, chlorine trioxide; +7, Cl_2O_7, dichlorine heptoxide
(b) Cl_2O_7 cannot react with O_2 because Cl has its maximum oxidation number, +7, and cannot be further oxidized.

4.26 (a) $SnO_2(s) + 2\ C(s) \rightarrow Sn(s) + 2\ CO(g)$
C is oxidized (its oxidation number increases from 0 to +2). C is the reducing agent.
The Sn in SnO_2 is reduced (its oxidation number decreases from +4 to 0). SnO_2 is the oxidizing agent.
(b) $Sn^{2+}(aq) + 2\ Fe^{3+}(aq) \rightarrow Sn^{4+}(aq) + 2\ Fe^{2+}(aq)$
Sn^{2+} is oxidized (its oxidation number increases from +2 to +4). Sn^{2+} is the reducing agent.
Fe^{3+} is reduced (its oxidation number decreases from +3 to +2). Fe^{3+} is the oxidizing agent.
(c) $4\ NH_3(g) + 5\ O_2(g) \rightarrow 4\ NO(g) + 6\ H_2O(l)$
The N in NH_3 is oxidized (its oxidation number increases from –3 to +2). NH_3 is the reducing agent.
Each O in O_2 is reduced (its oxidation number decreases from 0 to –2). O_2 is the oxidizing agent.

4.27 (a) C in C_2H_5OH oxidized and Cr in $K_2Cr_2O_7$ gets reduced.
(b) $K_2Cr_2O_7$ is the oxidizing agent; C_2H_5OH is the reducing agent.

4.28 (a) Pt is below H in the activity series; therefore NO REACTION.
(b) Mg is below Ca in the activity series; therefore NO REACTION.

4.29 "Any element higher in the activity series will react with the ion of any element lower in the activity series."
$A + D^+ \rightarrow A^+ + D$; therefore A is higher than D.
$B^+ + D \rightarrow B + D^+$; therefore D is higher than B.
$C^+ + D \rightarrow C + D^+$; therefore D is higher than C.
$B + C^+ \rightarrow B^+ + C$; therefore B is higher than C.
The net result is A > D > B > C.

4.30 31.50 mL = 0.031 50 L; 10.00 mL = 0.010 00 L

$$0.031\ 50\ \text{L} \times \frac{0.105\ \text{mol}\ BrO_3^-}{1\ \text{L}} \times \frac{6\ \text{mol}\ Fe^{2+}}{1\ \text{mol}\ BrO_3^-} = 1.98 \times 10^{-2}\ \text{mol}\ Fe^{2+}$$

$$\text{molarity} = \frac{1.98 \times 10^{-2}\ \text{mol}\ Fe^{2+}}{0.010\ 00\ \text{L}} = 1.98\ \text{M}\ Fe^{2+}\ \text{solution}$$

4.31 14.92 mL = 0.014 92 L

$$0.014\ 92\ \text{L} \times \frac{0.100\ \text{mol}\ Cr_2O_7^{-2}}{1\ \text{L}} \times \frac{6\ \text{mol}\ Fe^{2+}}{1\ \text{mol}\ Cr_2O_7^{-2}} \times \frac{55.85\ \text{g}\ Fe^{2+}}{1\ \text{mol}\ Fe^{2+}} \times \frac{1000\ \text{mg}}{1\ \text{g}} = 50.0\ \text{mg}\ Fe^{2+}$$

4.32 Sodium chloride, sodium citrate, and potassium dihydrogen phosphate are strong electrolytes since they are soluble ionic compounds; citric acid and vitamin B3 are weak electrolytes due the presence of the carboxylic acid (CO_2H) group; and fructose is a nonelectrolyte. (b) Sodium chloride, potassium dihydrogen phosphate, and sodium citrate replenish electrolytes with important biological functions.

4.33 150 mg = 0.15 g; 35 mg = 0.035 g; 360 mL = 0.36 L

$$\text{molarity} = \frac{\left(0.15\ \text{g}\ Na^+ \times \frac{1\ \text{mol}\ Na^+}{23\ \text{g}\ Na^+}\right)}{0.36\ \text{L}} = 0.018\ \text{M}$$

$$\text{molarity} = \frac{\left(0.35\ \text{g}\ K^+ \times \frac{1\ \text{mol}\ K^+}{39\ \text{g}\ K^+}\right)}{0.36\ \text{L}} = 0.0025\ \text{M}$$

4.34 NaCl, 58.4; 0.500 L = 500 mL

$$0.416\ \text{mg}\ Na^+/\text{mL} \times \frac{1 \times 10^{-3}\ \text{g}}{1\ \text{mg}} \times 500\ \text{mL} \times \frac{1\ \text{mol}\ Na^+}{23.0\ \text{g}\ Na^+}$$
$$\times \frac{1\ \text{mol NaCl}}{1\ \text{mol}\ Na^+} \times \frac{58.4\ \text{g NaCl}}{1\ \text{mol NaCl}} = 0.528\ \text{g NaCl}$$

4.35 (a) Ba^{2+} (b) $3\ Ba^{2+}(aq) + 2\ PO_4^{3-}(aq) \rightarrow Ba_3(PO_4)_2(s)$

4.36 AgCl, 143.3; 172 mg = 0.172 g; 100.0 mL = 0.1000 L

$$0.172 \text{ g AgCl} \times \frac{1 \text{ mol AgCl}}{143.3 \text{ g AgCl}} \times \frac{1 \text{ mol Cl}^-}{1 \text{ mol AgCl}} = 0.001\ 20 \text{ mol Cl}^-$$

$$\text{Cl}^- \text{ molarity} = \frac{0.001\ 20 \text{ mol Cl}^-}{0.1000 \text{ L}} = 0.0120 \text{ M}$$

4.37 35.6 mL = 0.0356 L; 25.0 mL = 0.0250 L
(0.0356 L)(0.0400 mol/L) = 0.001 42 mol NaOH

$$0.001\ 42 \text{ mol NaOH} \times \frac{1 \text{ mol } H_3C_6H_5O_7}{3 \text{ mol NaOH}} = 4.75 \times 10^{-4} \text{ mol } H_3C_6H_5O_7$$

$$H_3C_6H_5O_7 \text{ molarity} = \frac{4.75 \times 10^{-4} \text{ mol}}{0.0250 \text{ L}} = 0.0190 \text{ M}$$

Conceptual Problems

4.38 The concentration of a solution is cut in half when the volume is doubled. This is best represented by box (b).

4.40 (a) $2\,Na^+(aq) + CO_3^{2-}(aq)$ does not form a precipitate. This is represented by box (1).
(b) $Ba^{2+}(aq) + CrO_4^{2-}(aq) \rightarrow BaCrO_4(s)$. This is represented by box (2).
(c) $2\,Ag^+(aq) + SO_4^{2-}(aq) \rightarrow Ag_2SO_4(s)$. This is represented by box (3).

4.42 HY is the strongest acid because it is completely dissociated.
HX is the weakest acid because it is the least dissociated.

4.44 The concentration in the buret is three times that in the flask. The NaOCl concentration is 0.040 M. Because the I^- concentration in the buret is three times the OCl^- concentration in the flask and the reaction requires 2 I^- ions per OCl^- ion, 2/3 or 67% of the I^- solution from the buret must be added to the flask to react with all of the OCl^-.

4.46 (a) $Sr^+ + At \rightarrow Sr + At^+$ No reaction.
(b) $Si + At^+ \rightarrow Si^+ + At$ Reaction would occur.
(c) $Sr + Si^+ \rightarrow Sr^+ + Si$ Reaction would occur.

Section Problems

Molarity and Dilution (Sections 4.1 and 4.2)

4.48 (a) 35.0 mL = 0.0350 L; $\frac{1.200 \text{ mol HNO}_3}{\text{L}} \times 0.0350 \text{ L} = 0.0420 \text{ mol HNO}_3$

(b) 175 mL = 0.175 L; $\frac{0.67 \text{ mol } C_6H_{12}O_6}{\text{L}} \times 0.175 \text{ L} = 0.12 \text{ mol } C_6H_{12}O_6$

4.50 $BaCl_2$, 208.2

$$15.0 \text{ g } BaCl_2 \times \frac{1 \text{ mol } BaCl_2}{208.2 \text{ g } BaCl_2} = 0.0720 \text{ mol } BaCl_2$$

$$0.0720 \text{ mol} \times \frac{1.0 \text{ L}}{0.45 \text{ mol}} = 0.16 \text{ L}; \quad 0.16 \text{ L} = 160 \text{ mL}$$

4.52 NaCl, 58.4; 400 mg = 0.400 g; 100 mL = 0.100 L

$$0.400 \text{ g NaCl} \times \frac{1 \text{ mol NaCl}}{58.4 \text{ g NaCl}} = 0.006\ 85 \text{ mol NaCl}$$

$$\text{molarity} = \frac{0.006\ 85 \text{ mol}}{0.100 \text{ L}} = 0.0685 \text{ M}$$

4.54 $$3.045 \text{ g Cu} \times \frac{1 \text{ mol Cu}}{63.546 \text{ g Cu}} = 0.047\ 92 \text{ mol Cu}; \ 50.0 \text{ mL} = 0.0500 \text{ L}$$

$$Cu(NO_3)_2 \text{ molarity} = \frac{0.047\ 92 \text{ mol}}{0.0500 \text{ L}} = 0.958 \text{ M}$$

4.56 $$M_f \times V_f = M_i \times V_i; \quad M_f = \frac{M_i \times V_i}{V_f} = \frac{12.0 \text{ M} \times 35.7 \text{ mL}}{250.0 \text{ mL}} = 1.71 \text{ M HCl}$$

4.58 $CaCl_2$, 111.0; 500 mL = 0.500 L
(0.500 L)(0.33 mol/L) = 0.165 mol $CaCl_2$

$$0.165 \text{ mol } CaCl_2 \times \frac{111.0 \text{ g } CaCl_2}{1 \text{ mol } CaCl_2} = 18.3 \text{ g } CaCl_2$$

Place 18.3 g of $CaCl_2$ in a 500 mL volumetric flask and fill it to the mark with water.

Electrolytes, Net Ionic Equations, and Aqueous Reactions (Sections 4.3–4.5)

4.60 (a) strong electrolyte, bright (b) nonelectrolyte, dark (c) weak electrolyte, dim

4.62 $Ba(OH)_2$ is soluble in aqueous solution, dissociates into Ba^{2+}(aq) and 2 OH^-(aq), and conducts electricity. In aqueous solution, H_2SO_4 dissociates into H^+(aq) and HSO_4^-(aq). H_2SO_4 solutions conduct electricity. When equal molar solutions of $Ba(OH)_2$ and H_2SO_4 are mixed, the insoluble $BaSO_4$ is formed along with two H_2O. In water, $BaSO_4$ does not produce any appreciable amount of ions and the mixture does not conduct electricity.

4.64 (a) HBr, strong electrolyte
(b) HF, weak electrolyte
(c) $NaClO_4$, strong electrolyte
(d) $(NH_4)_2CO_3$, strong electrolyte
(e) NH_3, weak electrolyte
(f) C_2H_5OH, nonelectrolyte

4.66 (a) K_2CO_3 contains 3 ions (2 K^+ and 1 CO_3^{2-}).
The molar concentration of ions = 3 x 0.750 M = 2.25 M.

(b) $AlCl_3$ contains 4 ions (1 Al^{3+} and 3 Cl^-).
The molar concentration of ions = 4 x 0.355 M = 1.42 M.

4.68 (a) precipitation (b) redox (c) acid-base neutralization

4.70 (a) Ionic equation:
$Hg^{2+}(aq) + 2\ NO_3^-(aq) + 2\ Na^+(aq) + 2\ I^-(aq) \rightarrow 2\ Na^+(aq) + 2\ NO_3^-(aq) + HgI_2(s)$
Delete spectator ions from the ionic equation to get the net ionic equation.
Net ionic equation: $Hg^{2+}(aq) + 2\ I^-(aq) \rightarrow HgI_2(s)$

(b) $2\ HgO(s) \xrightarrow{Heat} 2\ Hg(l) + O_2(g)$
(c) Ionic equation:
$H_3PO_4(aq) + 3\ K^+(aq) + 3\ OH^-(aq) \rightarrow 3\ K^+(aq) + PO_4^{3-}(aq) + 3\ H_2O(l)$
Delete spectator ions from the ionic equation to get the net ionic equation.
Net ionic equation: $H_3PO_4(aq) + 3\ OH^-(aq) \rightarrow PO_4^{3-}(aq) + 3\ H_2O(l)$

Precipitation Reactions and Solubility Guidelines (Section 4.6)

4.72 (a) $PbSO_4$, insoluble (b) $Ba(NO_3)_2$, soluble
(c) $SnCO_3$, insoluble (d) $(NH_4)_3PO_4$, soluble

4.74 (a) No precipitate will form. (b) $Fe^{2+}(aq) + 2\ OH^-(aq) \rightarrow Fe(OH)_2(s)$
(c) No precipitate will form. (d) No precipitate will form.

4.76 (a) 0.10 M $LiNO_3$ will not form a precipitate.
(b) $BaSO_4(s)$ will precipitate. (c) $AgCl(s)$ will precipitate.

4.78 (a) $Pb(NO_3)_2(aq) + Na_2SO_4(aq) \rightarrow PbSO_4(s) + 2\ NaNO_3(aq)$
(b) $3\ MgCl_2(aq) + 2\ K_3PO_4(aq) \rightarrow Mg_3(PO_4)_2(s) + 6\ KCl(aq)$
(c) $ZnSO_4(aq) + Na_2CrO_4(aq) \rightarrow ZnCrO_4(s) + Na_2SO_4(aq)$

4.80 $Ag^+(aq) + NO_3^-(aq) + H^+(aq) + Cl^-(aq) \rightarrow AgCl(s) + H^+(aq) + NO_3^-(aq)$

$$\text{mol } Cl^- = 30.0 \text{ mL} \times \frac{1 \times 10^{-3} \text{ L}}{1 \text{ mL}} \times 0.150 \text{ mol/L} = 0.00450 \text{ mol}$$

$$\text{mol } Ag^+ = 25.0 \text{ mL} \times \frac{1 \times 10^{-3} \text{ L}}{1 \text{ mL}} \times 0.200 \text{ mol/L} = 0.00500 \text{ mol}$$

Because the mole ratio between Ag^+ and Cl^- is 1 to 1, Cl^- is the limiting reactant because it is the smaller mol quantity. 0.00450 mol of AgCl is produced.

AgCl, 143.3; $\text{mass AgCl} = 0.00450 \text{ mol AgCl} \times \frac{143.3 \text{ g AgCl}}{1 \text{ mol AgCl}} = 0.645 \text{ g AgCl}$

4.82 Add HCl(aq); it will selectively precipitate AgCl(s).

4.84 Ag^+ is eliminated because it would have precipitated as AgCl(s); Ba^{2+} is eliminated because it would have precipitated as $BaSO_4(s)$. The solution might contain Cs^+ and/or NH_4^+. Neither of these will precipitate with OH^-, SO_4^{2-}, or Cl^-.

Acids, Bases, and Neutralization Reactions (Section 4.7)

4.86 Add the solution to an active metal, such as magnesium. Bubbles of H_2 gas indicate the presence of an acid.

4.88 (a) $2\ H^+(aq) + 2\ ClO_4^-(aq) + Ca^{2+}(aq) + 2\ OH^-(aq) \rightarrow Ca^{2+}(aq) + 2\ ClO_4^-(aq) + 2\ H_2O(l)$
(b) $CH_3CO_2H(aq) + Na^+(aq) + OH^-(aq) \rightarrow CH_3CO_2^-(aq) + Na^+(aq) + H_2O(l)$

4.90 (a) $LiOH(aq) + HI(aq) \rightarrow LiI(aq) + H_2O(l)$
Ionic equation: $Li^+(aq) + OH^-(aq) + H^+(aq) + I^-(aq) \rightarrow Li^+(aq) + I^-(aq) + H_2O(l)$
Delete spectator ions from the ionic equation to get the net ionic equation.
Net ionic equation: $H^+(aq) + OH^-(aq) \rightarrow H_2O(l)$
(b) $2\ HBr(aq) + Ca(OH)_2(aq) \rightarrow CaBr_2(aq) + 2\ H_2O(l)$
Ionic equation:
$2\ H^+(aq) + 2\ Br^-(aq) + Ca^{2+}(aq) + 2\ OH^-(aq) \rightarrow Ca^{2+}(aq) + 2\ Br^-(aq) + 2\ H_2O(l)$
Delete spectator ions from the ionic equation to get the net ionic equation.
Net ionic equation: $H^+(aq) + OH^-(aq) \rightarrow H_2O(l)$

Solution Stoichiometry and Titration (Sections 4.8 and 4.9)

4.92 $2\ HBr(aq) + K_2CO_3(aq) \rightarrow 2\ KBr(aq) + CO_2(g) + H_2O(l)$
K_2CO_3, 138.2; 450 mL = 0.450 L

$$\frac{0.500\ \text{mol HBr}}{\text{L}} \times 0.450\ \text{L} = 0.225\ \text{mol HBr}$$

$$0.225\ \text{mol HBr} \times \frac{1\ \text{mol}\ K_2CO_3}{2\ \text{mol HBr}} \times \frac{138.2\ \text{g}\ K_2CO_3}{1\ \text{mol}\ K_2CO_3} = 15.5\ \text{g}\ K_2CO_3$$

4.94 $H_2C_2O_4$, 90.0

$$3.225\ \text{g}\ H_2C_2O_4 \times \frac{1\ \text{mol}\ H_2C_2O_4}{90.0\ \text{g}\ H_2C_2O_4} \times \frac{2\ \text{mol}\ KMnO_4}{5\ \text{mol}\ H_2C_2O_4} = 0.0143\ \text{mol}\ KMnO_4$$

$$0.0143\ \text{mol} \times \frac{1\ \text{L}}{0.250\ \text{mol}} = 0.0572\ \text{L} = 57.2\ \text{mL}$$

4.96 (a) $LiOH(aq) + HBr(aq) \rightarrow LiBr(aq) + H_2O(l)$
25.0 mL = 0.0250 L and 75.0 mL = 0.0750 L
(0.240 mol/L LiOH)(0.0250 L) = 0.006 00 mol LiOH
(0.200 mol/L HBr)(0.0750 L) = 0.0150 mol HBr
HBr and LiOH react in a one to one mole ratio.
mol HBr left over = 0.0150 mol HBr – 0.006 00 mol LiOH = 0.0090 mol HBr
$KOH(aq) + HBr(aq) \rightarrow KBr(aq) + H_2O(l)$
HBr and KOH react in a one to one mole ratio.
0.0090 mol KOH will neutralize the solution.

$$\text{volume} = \frac{\text{mol}}{\text{M}} = \frac{0.0090\ \text{mol KOH}}{1.00\ \text{mol/L}} = 0.0090\ \text{L} = 9.0\ \text{mL KOH}$$

(b) $HCl(aq) + NaOH(aq) \rightarrow NaCl(aq) + H_2O(l)$
45.0 mL = 0.0450 L and 10.0 mL = 0.0100 L
(0.300 mol/L HCl)(0.0450 L) = 0.0135 mol HCl
(0.250 mol/L NaOH)(0.0100 L) = 0.002 50 mol NaOH
HCl and NaOH react in a one to one mole ratio.
mol HCl left over = 0.0135 mol HCl – 0.002 50 mol NaOH = 0.0110 mol HCl
$KOH(aq) + HCl(aq) \rightarrow KCl(aq) + H_2O(l)$
HCl and KOH react in a one to one mole ratio.
0.0110 mol KOH will neutralize the solution.

$$\text{volume} = \frac{\text{mol}}{\text{M}} = \frac{0.0110 \text{ mol KOH}}{1.00 \text{ mol/L}} = 0.0110 \text{ L} = 11.0 \text{ mL KOH}$$

4.98 (a) $HBr(aq) + KOH(aq) \rightarrow KBr(aq) + H_2O(l)$
50.0 mL = 0.0500 L and 30.0 mL = 0.0300 L
(0.100 mol/L HBr)(0.0500 L) = 0.005 00 mol HBr
(0.200 mol/L KOH)(0.0300 L) = 0.006 00 mol KOH
HBr and KOH react in a one to one mole ratio. The resulting solution is basic because there is an excess of KOH.
(b) $2\ HCl(aq) + Ba(OH)_2(aq) \rightarrow BaCl_2(aq) + 2\ H_2O(l))$
100.0 mL = 0.1000 L and 75 mL = 0.0750 L
(0.0750 mol/L HCl)(0.1000 L) = 0.007 50 mol HCl
(0.100 mol/L $Ba(OH)_2$)(0.0750 L) = 0.007 50 mol $Ba(OH)_2$

$$0.007\ 50 \text{ mol HCl x } \frac{1 \text{ mol Ba(OH)}_2}{2 \text{ mol HCl}} = 0.003\ 75 \text{ mol Ba(OH)}_2 \text{ needed}$$

The resulting solution is basic because there is an excess of $Ba(OH)_2$.

Redox Reactions, Oxidation Numbers, and Activity Series (Sections 4.10–4.12)

4.100 The best reducing agents are at the bottom left of the periodic table. The best oxidizing agents are at the top right of the periodic table (excluding the noble gases).

4.102 (a) An oxidizing agent gains electrons.
(b) A reducing agent loses electrons.
(c) A substance undergoing oxidation loses electrons.
(d) A substance undergoing reduction gains electrons.

4.104 (a) NO_2 O –2, N +4
(b) SO_3 O –2, S +6
(c) $COCl_2$ O –2, Cl –1, C +4
(d) CH_2Cl_2 Cl –1, H +1, C 0
(e) $KClO_3$ O –2, K +1, Cl +5
(f) HNO_3 O –2, H +1, N +5

4.106 (a) ClO_3^- O –2, Cl +5
(b) SO_3^{2-} O –2, S +4
(c) $C_2O_4^{2-}$ O –2, C +3
(d) NO_2^- O –2, N +3
(e) BrO^- O –2, Br +1
(f) AsO_4^{3-} O –2, As +5

4.108 (a) +1, N_2O, nitrous oxide; +2, NO, nitric oxide; +4, N_2O_4, dinitrogen tetroxide; +5, N_2O_5, dinitrogen pentoxide
(b) N_2O_5 cannot react with O_2 because N has its maximum oxidation number, +5, and cannot be further oxidized.

4.110 (a) $Ca(s) + Sn^{2+}(aq) \rightarrow Ca^{2+}(aq) + Sn(s)$
Ca(s) is oxidized (oxidation number increases from 0 to +2).
$Sn^{2+}(aq)$ is reduced (oxidation number decreases from +2 to 0).
(b) $ICl(s) + H_2O(l) \rightarrow HCl(aq) + HOI(aq)$
No oxidation numbers change. The reaction is not a redox reaction.

4.112 (a) Zn is below Na^+; therefore no reaction.
(b) Pt is below H^+; therefore no reaction.
(c) Au is below Ag^+; therefore no reaction.
(d) Ag is above Au^{3+}; the reaction is $Au^{3+}(aq) + 3\,Ag(s) \rightarrow 3\,Ag^+(aq) + Au(s)$.

4.114 (a) "Any element higher in the activity series will react with the ion of any element lower in the activity series."
$A + B^+ \rightarrow A^+ + B$; therefore A is higher than B.
$C^+ + D \rightarrow$ no reaction; therefore C is higher than D.
$B + D^+ \rightarrow B^+ + D$; therefore B is higher than D.
$B + C^+ \rightarrow B^+ + C$; therefore B is higher than C.
The net result is $A > B > C > D$.
(b) (1) C is below A^+; therefore no reaction.
(2) D is below A^+; therefore no reaction.

Redox Titrations (Section 4.13)

4.116 $I_2(aq) + 2\,S_2O_3^{2-}(aq) \rightarrow S_4O_6^{2-}(aq) + 2\,I^-(aq)$; 35.20 mL = 0.032 50 L

$$0.035\ 20\ \text{L} \times \frac{0.150\ \text{mol}\ S_2O_3^{2-}}{\text{L}} \times \frac{1\ \text{mol}\ I_2}{2\ \text{mol}\ S_2O_3^{2-}} \times \frac{253.8\ \text{g}\ I_2}{1\ \text{mol}\ I_2} = 0.670\ \text{g}\ I_2$$

4.118 46.99 mL = 0.046 99 L; 50.00 mL = 0.050 00 L
$(0.2004\ \text{mol/L}\ Cr_2O_7^{2-})(0.046\ 99\ \text{L}) = 0.009\ 417\ \text{mol}\ Cr_2O_7^{2-}$

$$0.009\ 417\ \text{mol}\ Cr_2O_7^{2-} \times \frac{6\ \text{mol}\ Fe^{2+}}{1\ \text{mol}\ Cr_2O_7^{2-}} = 0.056\ 50\ \text{mol}\ Fe^{2+}$$

$$[Fe^{2+}] = \frac{0.056\ 50\ \text{mol}}{0.050\ 00\ \text{L}} = 1.130\ \text{M}$$

4.120 $3\,H_3AsO_3(aq) + BrO_3^-(aq) \rightarrow Br^-(aq) + 3\,H_3AsO_4(aq)$
22.35 mL = 0.022 35 L and 50.00 mL = 0.050 00 L

$$0.022\ 35\ \text{L} \times \frac{0.100\ \text{mol}\ BrO_3^-}{\text{L}} \times \frac{3\ \text{mol}\ H_3AsO_3}{1\ \text{mol}\ BrO_3^-} = 6.70 \times 10^{-3}\ \text{mol}\ H_3AsO_3$$

$$\text{molarity} = \frac{6.70 \times 10^{-3}\ \text{mol}}{0.050\ 00\ \text{L}} = 0.134\ \text{M As(III)}$$

4.122 $2\ Fe^{3+}(aq) + Sn^{2+}(aq) \rightarrow 2\ Fe^{2+}(aq) + Sn^{4+}(aq)$; 13.28 mL = 0.013 28 L

$$0.013\ 28\ \text{L} \times \frac{0.1015\ \text{mol}\ Sn^{2+}}{\text{L}} \times \frac{2\ \text{mol}\ Fe^{3+}}{1\ \text{mol}\ Sn^{2+}} \times \frac{55.845\ \text{g}\ Fe^{3+}}{1\ \text{mol}\ Fe^{3+}} = 0.1506\ \text{g}\ Fe^{3+}$$

$$\text{mass \% Fe} = \frac{0.1506\ \text{g}}{0.1875\ \text{g}} \times 100\% = 80.32\%$$

4.124 $C_2H_5OH(aq) + 2\ Cr_2O_7^{2-}(aq) + 16\ H^+(aq) \rightarrow 2\ CO_2(g) + 4\ Cr^{3+}(aq) + 11\ H_2O(l)$
C_2H_5OH, 46.1; 8.76 mL = 0.008 76 L

$$0.008\ 76\ \text{L} \times \frac{0.049\ 88\ \text{mol}\ Cr_2O_7^{2-}}{\text{L}} \times \frac{1\ \text{mol}\ C_2H_5OH}{2\ \text{mol}\ Cr_2O_7^{2-}} \times \frac{46.1\ \text{g}\ C_2H_5OH}{1\ \text{mol}\ C_2H_5OH}$$

$$= 0.0101\ \text{g}\ C_2H_5OH$$

$$\text{mass \%}\ C_2H_5OH = \frac{0.0101\ \text{g}}{10.002\ \text{g}} \times 100\% = 0.101\%$$

Chapter Problems

4.126 NaCl, 58.4; KCl, 74.6; $CaCl_2$, 111.0; 500 mL = 0.500 L

$$4.30\ \text{g NaCl} \times \frac{1\ \text{mol NaCl}}{58.4\ \text{g NaCl}} = 0.0736\ \text{mol NaCl}$$

$$0.150\ \text{g KCl} \times \frac{1\ \text{mol KCl}}{74.6\ \text{g KCl}} = 0.002\ 01\ \text{mol KCl}$$

$$0.165\ \text{g}\ CaCl_2 \times \frac{1\ \text{mol}\ CaCl_2}{111.0\ \text{g}\ CaCl_2} = 0.001\ 49\ \text{mol}\ CaCl_2$$

$0.0736\ \text{mol} + 0.002\ 01\ \text{mol} + 2(0.001\ 49\ \text{mol}) = 0.0786\ \text{mol}\ Cl^-$

$$Na^+\ \text{molarity} = \frac{0.0736\ \text{mol}}{0.500\ \text{L}} = 0.147\ \text{M}$$

$$Ca^{2+}\ \text{molarity} = \frac{0.001\ 49\ \text{mol}}{0.500\ \text{L}} = 0.002\ 98\ \text{M}$$

$$K^+\ \text{molarity} = \frac{0.002\ 01\ \text{mol}}{0.500\ \text{L}} = 0.004\ 02\ \text{M}$$

$$Cl^-\ \text{molarity} = \frac{0.0786\ \text{mol}}{0.500\ \text{L}} = 0.157\ \text{M}$$

4.128 Na_2SO_4, 142.04; Na_3PO_4, 163.94; Li_2SO_4, 109.95; 100.00 mL = 0.100 00 L

$$0.550\ \text{g}\ Na_2SO_4 \times \frac{1\ \text{mol}\ Na_2SO_4}{142.04\ \text{g}\ Na_2SO_4} = 0.003\ 872\ \text{mol}\ Na_2SO_4$$

$$1.188 \text{ g } Na_3PO_4 \text{ x } \frac{1 \text{ mol } Na_3PO_4}{163.94 \text{ g } Na_3PO_4} = 0.007\ 247 \text{ mol } Na_3PO_4$$

$$0.223 \text{ g } Li_2SO_4 \text{ x } \frac{1 \text{ mol } Li_2SO_4}{109.95 \text{ g } Li_2SO_4} = 0.002\ 028 \text{ mol } Li_2SO_4$$

$$Na^+ \text{ molarity} = \frac{(2 \text{ x } 0.003\ 872 \text{ mol}) + (3 \text{ x } 0.007\ 247 \text{ mol})}{0.100\ 00 \text{ L}} = 0.295 \text{ M}$$

$$Li^+ \text{ molarity} = \frac{2 \text{ x } 0.002\ 028 \text{ mol}}{0.100\ 00 \text{ L}} = 0.0406 \text{ M}$$

$$SO_4^{2-} \text{ molarity} = \frac{(1 \text{ x } 0.003\ 872 \text{ mol}) + (1 \text{ x } 0.002\ 028 \text{ mol})}{0.100\ 00 \text{ L}} = 0.0590 \text{ M}$$

$$PO_4^{3-} \text{ molarity} = \frac{1 \text{ x } 0.007\ 247 \text{ mol}}{0.100\ 00 \text{ L}} = 0.0725 \text{ M}$$

4.130 $H_2C_2O_4$, 90.03; 22.35 mL = 0.022 35 L

$$0.5170 \text{ g } H_2C_2O_4 \text{ x } \frac{1 \text{ mol } H_2C_2O_4}{90.03 \text{ g } H_2C_2O_4} \text{ x } \frac{2 \text{ mol } KMnO_4}{5 \text{ mol } H_2C_2O_4} = 2.297 \text{ x } 10^{-3} \text{ mol } KMnO_4$$

$$KMnO_4 \text{ molarity} = \frac{2.297 \text{ x } 10^{-3} \text{ mol}}{0.022\ 35 \text{ L}} = 0.1028 \text{ M}$$

4.132 57.91 mL = 0.057 91 L

$$0.057\ 91 \text{ L x } \frac{0.1018 \text{ mol } Ce^{4+}}{\text{L}} \text{ x } \frac{1 \text{ mol } Fe^{2+}}{1 \text{ mol } Ce^{4+}} \text{ x } \frac{55.85 \text{ g } Fe^{2+}}{1 \text{ mol } Fe^{2+}} = 0.3292 \text{ g } Fe^{2+}$$

$$\text{mass \% Fe} = \frac{0.3292 \text{ g}}{1.2284 \text{ g}} \text{ x } 100\% = 26.80\%$$

4.134 (a) "Any element higher in the activity series will react with the ion of any element lower in the activity series."
$C + B^+ \rightarrow C^+ + B$; therefore C is higher than B.
$A^+ + D \rightarrow$ no reaction; therefore A is higher than D.
$C^+ + A \rightarrow$ no reaction; therefore C is higher than A.
$D + B^+ \rightarrow D^+ + B$; therefore D is higher than B.
The net result is C > A > D > B.
(b) (1) The reaction, $A^+ + C \rightarrow A + C^+$, will occur because C is above A in the activity series.
(2) The reaction, $A^+ + B \rightarrow A + B^+$, will not occur because B is below A in the activity series.

4.136 10.49 mL = 0.010 49 L
(0.100 mol/L $S_2O_3^{2-}$)(0.010 49 L) = 0.001 049 mol $S_2O_3^{2-}$

$$0.001\ 049 \text{ mol } S_2O_3^{2-} \text{ x } \frac{1 \text{ mol } I_3^-}{2 \text{ mol } S_2O_3^{2-}} \text{ x } \frac{2 \text{ mol } Cu^{2+}}{1 \text{ mol } I_3^-} = 0.001\ 049 \text{ mol } Cu^{2-}$$

$$0.001\ 049 \text{ mol } Cu^{2-} \times \frac{63.55 \text{ g Cu}}{1 \text{ mol } Cu^{2+}} = 0.066\ 66 \text{ g Cu}$$

$$\text{mass \% Cu} = \frac{0.066\ 66 \text{ g Cu}}{14.98 \text{ g sample}} \times 100\% = 0.4450 \text{ \% Cu}$$

4.138 $MgF_2(s) \rightleftarrows Mg^{2+}(aq) + 2\,F^-(aq)$

x 2x

$[Mg^{2+}] = x = 2.6 \times 10^{-4}$ M and $[F^-] = 2x = 2(2.6 \times 10^{-4}\text{ M}) = 5.2 \times 10^{-4}$ M in a saturated solution.

$K_{sp} = [Mg^{2+}][F^-]^2 = (2.6 \times 10^{-4}\text{ M})(5.2 \times 10^{-4}\text{ M})^2 = 7.0 \times 10^{-11}$

4.140 (a) Add HCl to precipitate Hg_2Cl_2. $Hg_2^{2+}(aq) + 2Cl^-(aq) \rightarrow Hg_2Cl_2(s)$

(b) Add H_2SO_4 to precipitate $PbSO_4$. $Pb^{2+}(aq) + SO_4^{2-}(aq) \rightarrow PbSO_4(s)$

(c) Add Na_2CO_3 to precipitate $CaCO_3$. $Ca^{2+}(aq) + CO_3^{2-}(aq) \rightarrow CaCO_3(s)$

(d) Add Na_2SO_4 to precipitate $BaSO_4$. $Ba^{2+}(aq) + SO_4^{2-}(aq) \rightarrow BaSO_4(s)$

4.142 100.0 mL = 0.1000 L; 47.14 mL = 0.047 14 L

$$\text{mol HCl and HBr} = \text{mol } H^+ = 0.1235 \frac{\text{mol NaOH}}{1 \text{ L}} \times 0.047\ 14 \text{ L} = 5.8218 \times 10^{-3} \text{ mol}$$

mass of AgCl and AgBr = 0.9974 g; mol Ag = mol H^+ = 5.8218×10^{-3} mol

$$\text{mass of Ag} = 5.8218 \times 10^{-3} \text{ mol Ag} \times \frac{107.87 \text{ g Ag}}{1 \text{ mol Ag}} = 0.6280 \text{ g Ag}$$

mass of Cl and Br = 0.9974 g – 0.6280 g = 0.3694 g of Cl and Br

Let Y = moles Cl and Z = moles Br in 0.3694 g of Cl and Br.

Let (Y + Z) = moles Ag in 0.6280 g Ag.

For Ag: 0.6280 g = (Y + Z) x 107.87 g

For Cl and Br: 0.3694 g = (Y x 35.453 g) + (Z x 79.904 g)

Solve the simultaneous equations for Y and Z.

$$\text{Rearrange the Ag equation: } \left(\frac{0.6280 \text{ g}}{107.87 \text{ g}} - Z\right) = Y$$

Substitute for Y in the Cl and Br equation above and solve for Z.

$$0.3694 \text{ g} = \left[\left(\frac{0.6280 \text{ g}}{107.87 \text{ g}} - Z\right) \times 35.453 \text{ g}\right] + (Z \times 79.904 \text{ g}); \quad Z = \frac{0.1630}{44.451} = 3.667 \times 10^{-3}$$

$$Y = \left(\frac{0.6280 \text{ g}}{107.87 \text{ g}} - Z\right) = \left(\frac{0.6280 \text{ g}}{107.87 \text{ g}} - 3.667 \times 10^{-3}\right) = 2.155 \times 10^{-3}$$

$$\text{HCl molarity} = \frac{2.155 \times 10^{-3} \text{ mol}}{0.1000 \text{ L}} = 0.021\ 55 \text{ M}$$

$$\text{HBr molarity} = \frac{3.667 \times 10^{-3} \text{ mol}}{0.1000 \text{ L}} = 0.036\ 67 \text{ M}$$

4.144 (a) PbI_2, 461.01

$Pb(NO_3)_2(aq) + 2\,KI(aq) \rightarrow PbI_2(s) + 2\,KNO_3(aq)$

75.0 mL = 0.0750 L and 100.0 mL = 0.1000 L

mol $Pb(NO_3)_2$ = (0.0750 L)(0.100 mol/L) = 7.50×10^{-3} mol $Pb(NO_3)_2$
mol KI = (0.1000 L)(0.190 mol/L) = 1.90×10^{-2} mol KI

$$\text{mols KI needed} = 7.50 \times 10^{-3}\ \text{mol Pb(NO}_3)_2 \times \frac{2\ \text{mol KI}}{1\ \text{mol Pb(NO}_3)_2} = 1.50 \times 10^{-2}\ \text{mol KI}$$

There is an excess of KI, so $Pb(NO_3)_2$ is the limiting reactant.

$$\text{mass PbI}_2 = 7.50 \times 10^{-3}\ \text{mol Pb(NO}_3)_2 \times \frac{1\ \text{mol PbI}_2}{1\ \text{mol Pb(NO}_3)_2} \times \frac{461.01\ \text{g PbI}_2}{1\ \text{mol PbI}_2} = 3.46\ \text{g PbI}_2$$

(b) Because $Pb(NO_3)_2$ is the limiting reactant, Pb^{2+} is totally consumed and $[Pb^{2+}] = 0$.
mol K^+ = mol KI = 1.90×10^{-2} mol

$$\text{mol NO}_3^- = 7.50 \times 10^{-3}\ \text{mol Pb(NO}_3)_2 \times \frac{2\ \text{mol NO}_3^-}{1\ \text{mol Pb(NO}_3)_2} = 0.0150\ \text{mol NO}_3^-$$

mol I^- = (initial mol KI) – (mol KI needed) = 0.0190 mol – 0.0150 mol = 0.0040 mol I^-
total volume = 0.0750 L + 0.1000 L = 0.1750 L

$$[K^+] = \frac{0.0190\ \text{mol}}{0.1750\ \text{L}} = 0.109\ \text{M}$$

$$[NO_3^-] = \frac{0.0150\ \text{mol}}{0.1750\ \text{L}} = 0.0857\ \text{M}$$

$$[I^-] = \frac{0.0040\ \text{mol}}{0.1750\ \text{L}} = 0.023\ \text{M}$$

4.146 Mass S in MS = 1.504 g MS – 1.000 g M = 0.504 g S

$$0.504\ \text{g S} \times \frac{1\ \text{mol S}}{32.06\ \text{g S}} = 0.0157\ \text{mol S}$$

$$0.0157\ \text{mol S} \times \frac{1\ \text{mol M}}{1\ \text{mol S}} = 0.0157\ \text{mol M}$$

$$\text{molar mass M} = \frac{1.000\ \text{g M}}{0.0157\ \text{mol M}} = 63.69\ \text{g/mol; M is Cu}$$

4.148 48.39 mL = 0.048 39 L
(0.1116 mol/L MnO_4^-)(0.048 39 L) = 5.400×10^{-3} mol MnO_4^-

$$5.400 \times 10^{-3}\ \text{mol MnO}_4^- \times \frac{5\ \text{mol Fe}^{2+}}{1\ \text{mol MnO}_4^-} \times \frac{55.84\ \text{g Fe}^{2+}}{1\ \text{mol Fe}^{2+}} = 1.508\ \text{g Fe}^{2+}$$

$$\text{mass \% Fe} = \frac{1.508\ \text{g Fe}^{2+}}{2.368\ \text{g}} \times 100\% = 63.68\%$$

Multiconcept Problems

4.150 Let SA stand for salicylic acid.

$$\text{mol C in 1.00 g of SA} = 2.23\ \text{g CO}_2 \times \frac{1\ \text{mol CO}_2}{44.01\ \text{g CO}_2} \times \frac{1\ \text{mol C}}{1\ \text{mol CO}_2} = 0.0507\ \text{mol C}$$

mass C = 0.0507 mol C x $\frac{12.011 \text{ g C}}{1 \text{ mol C}}$ = 0.609 g C

mol H in 1.00 g of SA = 0.39 g H_2O x $\frac{1 \text{ mol } H_2O}{18.02 \text{ g } H_2O}$ x $\frac{2 \text{ mol H}}{1 \text{ mol } H_2O}$ = 0.043 mol H

mass H = 0.043 mol H x $\frac{1.008 \text{ g H}}{1 \text{ mol H}}$ = 0.043 g H

mass O = 1.00 g – mass C – mass H = 1.00 – 0.609 g – 0.043 g = 0.35 g O

mol O in 1.00 g of = 0.35 g N x $\frac{1 \text{ mol O}}{16.00 \text{ g O}}$ = 0.022 mol O

Determine empirical formula.

$C_{0.0507}H_{0.043}O_{0.022}$; divide each subscript by the smallest, 0.022.

$C_{0.0507 / 0.022}H_{0.043 / 0.022}O_{0.022 / 0.022}$

$C_{2.3}H_2O$, multiply each subscript by 3 to get integers.

The empirical formula is $C_7H_6O_3$. The empirical formula mass = 138.12 g/mol.

Because salicylic acid has only one acidic hydrogen, there is a 1 to 1 mol ratio between salicylic acid and NaOH in the acid-base titration.

mol SA in 1.00 g SA = 72.4 mL x $\frac{1 \times 10^{-3} \text{ L}}{1 \text{ mL}}$ x $\frac{0.100 \text{ mol NaOH}}{1 \text{ L}}$ x $\frac{1 \text{ mol SA}}{1 \text{ mol NaOH}}$ = 0.00724 mol SA

SA molar mass = $\frac{1.00 \text{ g}}{0.00724 \text{ mol}}$ = 138 g/mol

Because the empirical formula mass and the molar mass are the same, the empirical formula is the molecular formula for salicylic acid.

4.152 100.00 mL = 0.100 00 L; 71.02 mL = 0.071 02 L

mol H_2SO_4 = $\frac{0.1083 \text{ mol } H_2SO_4}{\text{L}}$ x 0.100 00 L = 0.010 83 mol H_2SO_4

mol NaOH = $\frac{0.1241 \text{ mol NaOH}}{\text{L}}$ x 0.071 02 L = 0.008 814 mol NaOH

H_2SO_4 + 2 NaOH → Na_2SO_4 + 2 H_2O

mol H_2SO_4 reacted with NaOH = 0.008 814 mol NaOH x $\frac{1 \text{ mol } H_2SO_4}{2 \text{ mol NaOH}}$ = 0.004 407 mol H_2SO_4

mol H_2SO_4 reacted with MCO_3 = 0.010 83 mol – 0.004 407 mol = 0.006 423 mol H_2SO_4

mol H_2SO_4 reacted with MCO_3 = mol CO_3^{2-} in MCO_3 = mol CO_2 produced = 0.006 423 mol CO_2

(a) CO_3^{2-}, 60.01; 0.006 423 mol CO_3^{2-} x $\frac{60.01 \text{ g } CO_3^{2-}}{1 \text{ mol } CO_3^{2-}}$ = 0.3854 g CO_3^{2-}

mass of M = 1.268 g – 0.3854 g = 0.8826 g M

molar mass of M = $\frac{0.8826 \text{ g}}{0.006 423 \text{ mol}}$ = 137.4 g/mol; M is Ba

(b) $0.006\ 423 \text{ mol } CO_2 \text{ x } \dfrac{44.01 \text{ g } CO_2}{1 \text{ mol } CO_2} \text{ x } \dfrac{1 \text{ L}}{1.799 \text{ g}} = 0.1571 \text{ L } CO_2$

4.154 NaOH, 40.00; $Ba(OH)_2$, 171.34

Let X equal the mass of NaOH and Y the mass of $Ba(OH)_2$ in the 10.0 g mixture. Therefore, X + Y = 10.0 g.

$$\text{mol HCl} = 108.9 \text{ mL x } \frac{1 \text{ x } 10^{-3} \text{ L}}{1 \text{ mL}} \text{ x } \frac{1.50 \text{ mol HCl}}{1 \text{ L}} = 0.163 \text{ mol HCl}$$

$$\text{mol NaOH} + 2 \text{ x mol } Ba(OH)_2 = 0.163 \text{ mol HCl}$$

$$\text{X x } \frac{1 \text{ mol NaOH}}{40.00 \text{ g NaOH}} + 2 \text{ x } \left(\text{Y x } \frac{1 \text{ mol } Ba(OH)_2}{171.34 \text{ g } Ba(OH)_2} \right) = 0.163 \text{ mol HCl}$$

Rearrange to get X = 10.0 g – Y and then substitute it into the equation above to solve for Y.

$$(10.0 \text{ g} - \text{Y}) \text{ x } \frac{1 \text{ mol NaOH}}{40.00 \text{ g NaOH}} + 2 \text{ x } \left(\text{Y x } \frac{1 \text{ mol } Ba(OH)_2}{171.34 \text{ g } Ba(OH)_2} \right) = 0.163 \text{ mol HCl}$$

$$\frac{10.00 \text{ mol}}{40.00} - \frac{\text{Y mol}}{40.00 \text{ g}} + \frac{2 \text{ Y mol}}{171.34 \text{ g}} = 0.163 \text{ mol}$$

$$-\frac{\text{Y mol}}{40.00 \text{ g}} + \frac{2 \text{ Y mol}}{171.34 \text{ g}} = 0.163 \text{ mol} - \frac{10.00 \text{ mol}}{40.00} = -0.087 \text{ mol}$$

$$\frac{(-\text{Y mol})(171.34 \text{ g}) + (2 \text{ Y mol})(40.00 \text{ g})}{(40.00 \text{ g})(171.34 \text{ g})} = -0.087 \text{ mol}$$

$$\frac{-91.34 \text{ Y mol}}{6853.6 \text{ g}} = -0.087 \text{ mol}; \quad \frac{91.34 \text{ Y}}{6853.6 \text{ g}} = 0.087$$

$\text{Y} = (0.087)(6853.6 \text{ g})/91.34 = 6.5 \text{ g } Ba(OH)_2$

$\text{X} = 10.0 \text{ g} - \text{Y} = 10.0 \text{ g} - 6.5 \text{ g} = 3.5 \text{ g NaOH}$

4.156 KNO_3, 101.10; $BaCl_2$, 208.24; NaCl, 58.44; $BaSO_4$, 233.40; AgCl, 143.32

(a) The two precipitates are $BaSO_4(s)$ and AgCl(s).

(b) H_2SO_4 only reacts with $BaCl_2$.

$H_2SO_4(aq) + BaCl_2(aq) \rightarrow BaSO_4(s) + 2 HCl(aq)$

Calculate the number of moles of $BaCl_2$ in 100.0 g of the mixture.

$$\text{mol } BaCl_2 = 67.3 \text{ g } BaSO_4 \text{ x } \frac{1 \text{ mol } BaSO_4}{233.40 \text{ g } BaSO_4} \text{ x } \frac{1 \text{ mol } BaCl_2}{1 \text{ mol } BaSO_4} = 0.288 \text{ mol } BaCl_2$$

Calculate mass and moles of $BaCl_2$ in 250.0 g sample.

$$\text{mass } BaCl_2 = 0.288 \text{ mol } BaCl_2 \text{ x } \frac{208.24 \text{ g } BaCl_2}{1 \text{ mol } BaCl_2} \text{ x } \frac{250.0 \text{ g}}{100.0 \text{ g}} = 150. \text{ g } BaCl_2$$

$$\text{mol } BaCl_2 = 150. \text{ g } BaCl_2 \text{ x } \frac{1 \text{ mol } BaCl_2}{208.24 \text{ g } BaCl_2} = 0.720 \text{ mol } BaCl_2$$

$AgNO_3$ reacts with both NaCl and $BaCl_2$ in the remaining 150.0 g of the mixture.

$3 AgNO_3(aq) + NaCl(aq) + BaCl_2(aq) \rightarrow 3 AgCl(s) + NaNO_3(aq) + Ba(NO_3)_2(aq)$

Calculate the moles of AgCl that would have been produced from the 250.0 g mixture.

$$\text{mol AgCl} = 197.6 \text{ g AgCl} \times \frac{1 \text{ mol AgCl}}{143.32 \text{ g AgCl}} \times \frac{250.0 \text{ g}}{150.0 \text{ g}} = 2.30 \text{ mol AgCl}$$

mol AgCl = 2 x (mol $BaCl_2$) + mol NaCl

Calculate the moles and mass of NaCl in the 250.0 g mixture.

2.30 mol AgCl = 2 x 0.720 mol $BaCl_2$ + mol NaCl

mol NaCl = 2.30 mol – 2(0.720 mol) = 0.86 mol NaCl

$$\text{mass NaCl} = 0.86 \text{ mol NaCl} \times \frac{58.44 \text{ g NaCl}}{1 \text{ mol NaCl}} = 50. \text{ g NaCl}$$

Calculate the mass of KNO_3 in the 250.0 g mixture.

total mass = mass $BaCl_2$ + mass NaCl + mass KNO_3

250.0 g = 150. g $BaCl_2$ + 50. g NaCl + mass KNO_3

mass KNO_3 = 250.0 g – 150. g $BaCl_2$ – 50. g NaCl = 50. g KNO_3

4.158 (a) $Cr^{2+}(aq) + Cr_2O_7^{2-}(aq) \rightarrow Cr^{3+}(aq)$

$[Cr^{2+}(aq) \rightarrow Cr^{3+}(aq) + e^-] \times 6$ (oxidation half reaction)

$Cr_2O_7^{2-}(aq) \rightarrow Cr^{3+}(aq)$

$Cr_2O_7^{2-}(aq) \rightarrow 2\,Cr^{3+}(aq)$

$Cr_2O_7^{2-}(aq) \rightarrow 2\,Cr^{3+}(aq) + 7\,H_2O(l)$

$14\,H^+(aq) + Cr_2O_7^{2-}(aq) \rightarrow 2\,Cr^{3+}(aq) + 7\,H_2O(l)$

$6\,e^- + 14\,H^+(aq) + Cr_2O_7^{2-}(aq) \rightarrow 2\,Cr^{3+}(aq) + 7\,H_2O(l)$ (reduction half reaction)

Combine the two half reactions.

$14\,H^+(aq) + Cr_2O_7^{2-}(aq) + 6\,Cr^{2+}(aq) \rightarrow 8\,Cr^{3+}(aq) + 7\,H_2O(l)$

(b) total volume = 100.0 ml + 20.0 mL = 120.0 mL = 0.1200 L

Initial moles:

$$0.120 \frac{\text{mol } Cr(NO_3)_2}{1 \text{ L}} \times 0.1000 \text{ L} = 0.0120 \text{ mol } Cr(NO_3)_2$$

$$0.500 \frac{\text{mol } HNO_3}{1 \text{ L}} \times 0.1000 \text{ L} = 0.0500 \text{ mol } HNO_3$$

$$0.250 \frac{\text{mol } K_2Cr_2O_7}{1 \text{ L}} \times 0.0200 \text{ L} = 0.005\,00 \text{ mol } K_2Cr_2O_7$$

Check for the limiting reactant. 0.0120 mol of Cr^{2+} requires (0.0120)/6 = 0.00200 mol $Cr_2O_7^{2-}$ and (14/6)(0.0120) = 0.0280 mol H^+. Both are in excess of the required amounts, so Cr^{2+} is the limiting reactant.

	$14\,H^+(aq)$ +	$Cr_2O_7^{2-}(aq)$ +	$6\,Cr^{2+}(aq)$ →	$8\,Cr^{3+}(aq)$ + $7\,H_2O(l)$
Initial moles	0.0500	0.00500	0.0120	0
Change	–14x	–x	–6x	+8x

Because Cr^{2+} is the limiting reactant, 6x = 0.0120 and x = 0.00200

Final moles	0.0220	0.00300	0	0.0160

$$\text{mol } K^+ = 0.00500 \text{ mol } K_2Cr_2O_7 \times \frac{2 \text{ mol } K^+}{1 \text{ mol } K_2Cr_2O_7} = 0.0100 \text{ mol } K^+$$

$$\text{mol } NO_3^- = 0.0120 \text{ mol } Cr(NO_3)_2 \times \frac{2 \text{ mol } NO_3^-}{1 \text{ mol } Cr(NO_3)_2}$$

$$+ 0.0500 \text{ mol } HNO_3 \times \frac{1 \text{ mol } NO_3^-}{1 \text{ mol } HNO_3} = 0.0740 \text{ mol } NO_3^-$$

mol H^+ = 0.0220 mol; mol $Cr_2O_7^{2-}$ = 0.00300 mol; mol Cr^{3+} = 0.01600 mol

Check for charge neutrality.
Total moles of +charge = 0.0100 + 0.0220 + 3 x (0.01600) = 0.0800 mol +charge
Total moles of –charge = 0.0740 + 2 x (0.00300) = 0.0800 mol –charge
The charges balance and there is electrical neutrality in the solution after the reaction.

$$K^+ \text{ molarity} = \frac{0.0100 \text{ mol } K^+}{0.1200 \text{ L}} = 0.0833 \text{ M}$$

$$NO_3^- \text{ molarity} = \frac{0.0740 \text{ mol } NO_3^-}{0.1200 \text{ L}} = 0.617 \text{ M}$$

$$H^+ \text{ molarity} = \frac{0.0220 \text{ mol } H^+}{0.1200 \text{ L}} = 0.183 \text{ M}$$

$$Cr_2O_7^{2-} \text{ molarity} = \frac{0.00300 \text{ mol } Cr_2O_7^{2-}}{0.1200 \text{ L}} = 0.0250 \text{ M}$$

$$Cr^{3+} \text{ molarity} = \frac{0.0160 \text{ mol } Cr^{3+}}{0.1200 \text{ L}} = 0.133 \text{ M}$$

4.160 (a) (1) $Cu(s) \rightarrow Cu^{2+}(aq)$
$[Cu(s) \rightarrow Cu^{2+}(aq) + 2\,e^-] \times 3$ (oxidation half reaction)

$NO_3^-(aq) \rightarrow NO(g)$
$NO_3^-(aq) \rightarrow NO(g) + 2\,H_2O(l)$
$4\,H^+(aq) + NO_3^-(aq) \rightarrow NO(g) + 2\,H_2O(l)$
$[3\,e^- + 4\,H^+(aq) + NO_3^-(aq) \rightarrow NO(g) + 2\,H_2O(l)] \times 2$ (reduction half reaction)

Combine the two half reactions.
$3\,Cu(s) + 8\,H^+(aq) + 2\,NO_3^-(aq) \rightarrow 3\,Cu^{2+}(aq) + 2\,NO(g) + 4\,H_2O(l)$

(2) $Cu^{2+}(aq) + SCN^-(aq) \rightarrow CuSCN(s)$
$[e^- + Cu^{2+}(aq) + SCN^-(aq) \rightarrow CuSCN(s)] \times 2$ (reduction half reaction)

$HSO_3^-(aq) \rightarrow HSO_4^-(aq)$
$H_2O(l) + HSO_3^-(aq) \rightarrow HSO_4^-(aq)$
$H_2O(l) + HSO_3^-(aq) \rightarrow HSO_4^-(aq) + 2\,H^+(aq)$
$H_2O(l) + HSO_3^-(aq) \rightarrow HSO_4^-(aq) + 2\,H^+(aq) + 2\,e^-$ (oxidation half reaction)

Combine the two half reactions.
$2\ Cu^{2+}(aq) + 2\ SCN^-(aq) + H_2O(l) + HSO_3^-(aq) \rightarrow$
$2\ CuSCN(s) + HSO_4^-(aq) + 2\ H^+(aq)$

(3) $Cu^+(aq) \rightarrow Cu^{2+}(aq)$
$[Cu^+(aq) \rightarrow Cu^{2+}(aq) + e^-] \times 10$ (oxidation half reaction)

$IO_3^-(aq) \rightarrow I_2(aq)$
$2\ IO_3^-(aq) \rightarrow I_2(aq)$
$2\ IO_3^-(aq) \rightarrow I_2(aq) + 6\ H_2O(l)$
$12\ H^+(aq) + 2\ IO_3^-(aq) \rightarrow I_2(aq) + 6\ H_2O(l)$
$10\ e^- + 12\ H^+(aq) + 2\ IO_3^-(aq) \rightarrow I_2(aq) + 6\ H_2O(l)$ (reduction half reaction)
Combine the two half reactions.
$10\ Cu^+(aq) + 12\ H^+(aq) + 2\ IO_3^-(aq) \rightarrow 10\ Cu^{2+}(aq) + I_2(aq) + 6\ H_2O(l)$

(4) $I_2(aq) \rightarrow I^-(aq)$
$I_2(aq) \rightarrow 2\ I^-(aq)$
$2\ e^- + I_2(aq) \rightarrow 2\ I^-(aq)$ (reduction half reaction)

$S_2O_3^{2-}(aq) \rightarrow S_4O_6^{2-}(aq)$
$2\ S_2O_3^{2-}(aq) \rightarrow S_4O_6^{2-}(aq)$
$2\ S_2O_3^{2-}(aq) \rightarrow S_4O_6^{2-}(aq) + 2\ e^-$ (oxidation half reaction)

Combine the two half reactions.
$I_2(aq) + 2\ S_2O_3^{2-}(aq) \rightarrow 2\ I^-(aq) + S_4O_6^{2-}(aq)$

(5) $2\ ZnNH_4PO_4 \rightarrow Zn_2P_2O_7 + H_2O + 2\ NH_3$

(b) 10.82 mL = 0.01082 L
mol $S_2O_3^{2-}$ = (0.1220 mol/L)(0.01082 L) = 0.00132 mol $S_2O_3^{2-}$

$$\text{mol } I_2 = 0.00132 \text{ mol } S_2O_3^{2-} \times \frac{1 \text{ mol } I_2}{2 \text{ mol } S_2O_3^{2-}} = 6.60 \times 10^{-4} \text{ mol } I_2$$

$$\text{mol } Cu^+ = 6.60 \times 10^{-4} \text{ mol } I_2 \times \frac{10 \text{ mol } Cu^+}{1 \text{ mol } I_2} = 6.60 \times 10^{-3} \text{ mol } Cu^+ \text{ (Cu)}$$

g Cu = (6.60×10^{-3} mol)(63.546 g/mol) = 0.419 g Cu

$$\text{mass \% Cu in brass} = \frac{0.419 \text{ g Cu}}{0.544 \text{ g brass}} \times 100\% = 77.1\% \text{ Cu}$$

(c) $Zn_2P_2O_7$, 304.72

$$\text{mass \% Zn in } Zn_2P_2O_7 = \frac{2 \times 65.39 \text{ g}}{304.72 \text{ g}} \times 100\% = 42.92\%$$

mass of Zn in $Zn_2P_2O_7$ = (0.4292)(0.246 g) = 0.106 g Zn

$$\text{mass \% Zn in brass} = \frac{0.106 \text{ g Zn}}{0.544 \text{ g brass}} \times 100\% = 19.5\% \text{ Zn}$$

4.162 (a) $H_3MO_3(aq) \rightarrow H_3MO_4(aq)$

$H_3MO_3(aq) + H_2O(l) \rightarrow H_3MO_4(aq)$

$H_3MO_3(aq) + H_2O(l) \rightarrow H_3MO_4(aq) + 2\ H^+(aq)$

$[H_3MO_3(aq) + H_2O(l) \rightarrow H_3MO_4(aq) + 2\ H^+(aq) + 2\ e^-] \times 5$ (oxidation half reaction)

$MnO_4^-(aq) \rightarrow Mn^{2+}(aq)$

$MnO_4^-(aq) \rightarrow Mn^{2+}(aq) + 4\ H_2O(l)$

$MnO_4^-(aq) + 8\ H^+(aq) \rightarrow Mn^{2+}(aq) + 4\ H_2O(l)$

$[MnO_4^-(aq) + 8\ H^+(aq) + 5\ e^- \rightarrow Mn^{2+}(aq) + 4\ H_2O(l)] \times 2$ (reduction half reaction)

Combine the two half reactions.

$5\ H_3MO_3(aq) + 5\ H_2O(l) + 2\ MnO_4^-(aq) + 16\ H^+(aq) \rightarrow$
$5\ H_3MO_4(aq) + 10\ H^+(aq) + 2\ Mn^{2+}(aq) + 8\ H_2O(l)$

$5\ H_3MO_3(aq) + 2\ MnO_4^-(aq) + 6\ H^+(aq) \rightarrow 5\ H_3MO_4(aq) + 2\ Mn^{2+}(aq) + 3\ H_2O(l)$

(b) 10.7 mL = 0.0107 L

mol MnO_4^- = (0.0107 L)(0.100 mol/L) = 1.07×10^{-3} mol MnO_4^-

$$\text{mol } H_3MO_3 = 1.07 \times 10^{-3} \text{ mol } MnO_4^- \times \frac{5 \text{ mol } H_3MO_3}{2 \text{ mol } MnO_4^-} = 2.67 \times 10^{-3} \text{ mol } H_3MO_3$$

$$\text{mol } M_2O_3 = 2.67 \times 10^{-3} \text{ mol } H_3MO_3 \times \frac{1 \text{ mol } M_2O_3}{2 \text{ mol } H_3MO_3} = 1.34 \times 10^{-3} \text{ mol } M_2O_3$$

$$\text{mol M in } M_2O_3 = 1.34 \times 10^{-3} \text{ mol } M_2O_3 \times \frac{2 \text{ mol M}}{1 \text{ mol } M_2O_3} = 2.68 \times 10^{-3} \text{ mol M}$$

(c) $$\text{M molar mass} = \frac{0.200 \text{ g}}{2.68 \times 10^{-3} \text{ mol}} = 74.6 \text{ g/mol; M atomic weight} = 74.6$$

M is As.

5 Periodicity and the Electronic Structure of Atoms

5.1 $\nu = 102.5 \text{ MHz} = 102.5 \times 10^6 \text{ Hz} = 102.5 \times 10^6 \text{ s}^{-1}$

$$\lambda = \frac{c}{\nu} = \frac{3.00 \times 10^8 \text{ m/s}}{102.5 \times 10^6 \text{ s}^{-1}} = 2.93 \text{ m}$$

$\nu = 9.55 \times 10^{17} \text{ Hz} = 9.55 \times 10^{17} \text{ s}^{-1}$

$$\lambda = \frac{c}{\nu} = \frac{3.00 \times 10^8 \text{ m/s}}{9.55 \times 10^{17} \text{ s}^{-1}} = 3.14 \times 10^{-10} \text{ m}$$

5.2 The wave with the shorter wavelength (b) has the higher frequency. The wave with the larger amplitude (b) represents the more intense beam of light. The wave with the shorter wavelength (b) represents blue light. The wave with the longer wavelength (a) represents red light.

5.3 IR, $\lambda = 1.55 \times 10^{-6} \text{ m}$

$$E = \frac{hc}{\lambda} = (6.626 \times 10^{-34} \text{ J}\cdot\text{s})\left(\frac{3.00 \times 10^8 \text{ m/s}}{1.55 \times 10^{-6} \text{ m}}\right)(6.022 \times 10^{23}\text{/mol})$$

$E = 7.72 \times 10^4 \text{ J/mol} = 77.2 \text{ kJ/mol}$

UV, $\lambda = 250 \text{ nm} = 250 \times 10^{-9} \text{ m}$

$$E = \frac{hc}{\lambda} = (6.626 \times 10^{-34} \text{ J}\cdot\text{s})\left(\frac{3.00 \times 10^8 \text{ m/s}}{250 \times 10^{-9} \text{ m}}\right)(6.022 \times 10^{23}\text{/mol})$$

$E = 4.79 \times 10^5 \text{ J/mol} = 479 \text{ kJ/mol}$

X ray, $\lambda = 5.49 \text{ nm} = 5.49 \times 10^{-9} \text{ m}$

$$E = \frac{hc}{\lambda} = (6.626 \times 10^{-34} \text{ J}\cdot\text{s})\left(\frac{3.00 \times 10^8 \text{ m/s}}{5.49 \times 10^{-9} \text{ m}}\right)(6.022 \times 10^{23}\text{/mol})$$

$E = 2.18 \times 10^7 \text{ J/mol} = 2.18 \times 10^4 \text{ kJ/mol}$

5.4 $$E = \frac{hc}{\lambda} = (6.626 \times 10^{-34} \text{ J}\cdot\text{s})\left(\frac{3.00 \times 10^8 \text{ m/s}}{2.3 \times 10^{-3} \text{ m}}\right)(6.022 \times 10^{23}\text{/mol})$$

$E = 52 \text{ J/mol} = 0.052 \text{ kJ/mol}$

$$74 \text{ kJ} \times \frac{1 \text{ mol photons}}{0.052 \text{ kJ}} = 1.4 \times 10^3 \text{ mol photons}$$

5.5 $$E = \frac{hc}{\lambda} = (6.626 \times 10^{-34} \text{ J}\cdot\text{s})\left(\frac{3.00 \times 10^8 \text{ m/s}}{390 \times 10^{-9} \text{ m}}\right)(6.022 \times 10^{23}\text{/mol})$$

$E = 3.07 \times 10^5$ J/mol = 307 kJ/mol
The energy is less than the work function. Electrons will not be ejected.

5.6 (a) Ag is predicted to have the higher work function because Rb is further left on the periodic table and holds its electrons less tightly.
(b) Rb because lower energies correspond to longer wavelength.

5.7 $m = 2$; $R_\infty = 1.097 \times 10^{-2}\ \text{nm}^{-1}$

$$\frac{1}{\lambda} = R_\infty\left[\frac{1}{m^2} - \frac{1}{n^2}\right];\ \frac{1}{\lambda} = R_\infty\left[\frac{1}{2^2} - \frac{1}{7^2}\right];\ \frac{1}{\lambda} = 2.519 \times 10^{-3}\ \text{nm}^{-1};\ \lambda = 397.0\ \text{nm}$$

5.8 $m = 3$; $R_\infty = 1.097 \times 10^{-2}\ \text{nm}^{-1}$

(a) $$\frac{1}{\lambda} = R_\infty\left[\frac{1}{m^2} - \frac{1}{n^2}\right];\ \frac{1}{\lambda} = R_\infty\left[\frac{1}{3^2} - \frac{1}{4^2}\right];\ \frac{1}{\lambda} = 5.333 \times 10^{-4}\ \text{nm}^{-1};\ \lambda = 1875\ \text{nm}$$

(b) $$\frac{1}{\lambda} = R_\infty\left[\frac{1}{m^2} - \frac{1}{n^2}\right];\ \frac{1}{\lambda} = R_\infty\left[\frac{1}{3^2} - \frac{1}{\infty^2}\right];\ \frac{1}{\lambda} = 1.219 \times 10^{-3}\ \text{nm}^{-1};\ \lambda = 820.4\ \text{nm}$$

5.9 $$\lambda = \frac{h}{mv} = \frac{6.626 \times 10^{-34}\ \text{kg m}^2\ \text{s}^{-1}}{(1150\ \text{kg})(24.6\ \text{m/s})} = 2.34 \times 10^{-38}\ \text{m}$$
This wavelength is shorter than the diameter of an atom.

5.10 $\lambda = 1\text{nm}/10 = 1 \times 10^{-10}$ m

$$\lambda = \frac{h}{mv};\quad v = \frac{h}{m\lambda} = \frac{6.626 \times 10^{-34}\ \text{kg m}^2\ \text{s}^{-1}}{(9.1 \times 10^{-31}\ \text{kg})(1 \times 10^{-10}\ \text{m})} = 7 \times 10^6\ \text{m/s}$$

5.11 (a) 2p (b) 4f (c) 3d

5.12

n	l	m_l	Orbital	No. of Orbitals
5	0	0	5s	1
	1	−1, 0, +1	5p	3
	2	−2, −1, 0, +1, +2	5d	5
	3	−3, −2, −1, 0, +1, +2, +3	5f	7
	4	−4, −3, −2, −1, 0, +1, +2, +3, +4	5g	9

There are 25 possible orbitals in the fifth shell.

5.13 $n = 4$, $l = 0$, 4s

5.14 The g orbitals have four nodal planes.

5.15 (a) Ti, $1s^2 2s^2 2p^6 3s^2 3p^6 4s^2 3d^2$ or [Ar] $4s^2 3d^2$

[Ar] ↑↓ (4s) ↑ ↑ _ _ _ (3d)

(b) Zn, $1s^2 2s^2 2p^6 3s^2 3p^6 4s^2 3d^{10}$ or [Ar] $4s^2 3d^{10}$

[Ar] ↑↓ (4s) ↑↓ ↑↓ ↑↓ ↑↓ ↑↓ (3d)

(c) Sn, $1s^2 2s^2 2p^6 3s^2 3p^6 4s^2 3d^{10} 4p^6 5s^2 4d^{10} 5p^2$ or [Kr] $5s^2 4d^{10} 5p^2$

[Kr] ↑↓ (5s) ↑↓ ↑↓ ↑↓ ↑↓ ↑↓ (4d) ↑ ↑ _ (5p)

(d) Pb, [Xe] $6s^2 4f^{14} 5d^{10} 6p^2$

[Xe] ↑↓ (6s) ↑↓ ↑↓ ↑↓ ↑↓ ↑↓ ↑↓ ↑↓ (4f) ↑↓ ↑↓ ↑↓ ↑↓ ↑↓ (5d) ↑ ↑ _ (6p)

5.16 (a) 43 electrons = Tc (b) 28 electrons = Ni

5.17 (a) Sn; atoms get larger as you go down a group.
(b) Lu; atoms get smaller as you go across a period.

5.18 Iodine has the largest atomic radius of the three halogens. C–I would be the longest bond length.

5.19 (a) [Xe] $6s^2 4f^{14} 5d^{10}$

(b) [Xe] ↑↓ (6s) ↑↓ ↑↓ ↑↓ ↑↓ ↑↓ ↑↓ ↑↓ (4f) ↑↓ ↑↓ ↑↓ ↑↓ ↑↓ (5d)

(c) There are no unpaired electrons.

5.20 (a) [Xe] $6s^1 4f^{14} 5d^{10} 6p^1$

(b) [Xe] ↑ (6s) ↑↓ ↑↓ ↑↓ ↑↓ ↑↓ ↑↓ ↑↓ (4f) ↑↓ ↑↓ ↑↓ ↑↓ ↑↓ (5d) ↑ _ _ (6p)

(c) There are 2 unpaired electrons.

5.21 (a) 7d, $n = 7$, $l = 2$, $m_l = -2, -1, 0, 1, 2$
(b) 6p, $n = 6$, $l = 1$, $m_l = -1, 0, 1$

(c) $$E = \frac{hc}{\lambda} = (6.626 \times 10^{-34}\ \text{J}\cdot\text{s})\left(\frac{3.00 \times 10^{8}\ \text{m/s}}{434.7 \times 10^{-9}\ \text{m}}\right)(6.022 \times 10^{23}\ /\text{mol})$$

$E = 2.75 \times 10^5$ J/mol = 275 kJ/mol

5.22 The shortest wavelength corresponds to the highest energy, therefore, 126.8 nm corresponds to 8p → 6s; 140.2 nm corresponds to 7p → 6s; and 185.0 nm corresponds to 6p → 6s.

5.23 (a) The fluorescent bulb does not emit all the wavelengths of light that would be emitted from a white light source. Notice that there are dark regions between the colored peaks.
(b) Fluorescent light does appear as "white light" because its line spectrum has contributions from all the colors (blue, green, yellow, orange, and red).

Conceptual Problems

5.24 The wave with the larger amplitude (a) has the greater intensity. The wave with the shorter wavelength (a) has the higher energy radiation. The wave with the shorter wavelength (a) represents yellow light. The wave with the longer wavelength (b) represents infrared radiation.

5.26 (a) $3p_y$ $n = 3, l = 1$ (b) $4d_{z^2}$ $n = 4, l = 2$

5.28 The green element, molybdenum, has an anomalous electron configuration. Its predicted electron configuration is [Ar] $5s^2\ 4d^4$. Its anomalous electron configuration is [Ar] $5s^1\ 4d^5$ because of the resulting half-filled d-orbitals.

5.30 There are 34 total electrons in the atom, so there are also 34 protons in the nucleus. The atom is selenium (Se)

Se, [Ar] ⥮ (4s) ⥮ ⥮ ⥮ ⥮ ⥮ (3d) ⥮ ↑ ↑ (4p)

Section Problems

Electromagnetic Energy and Atomic Spectra (Sections 5.1–5.4)

5.32 Violet has the higher frequency and energy. Red has the higher wavelength.

5.34 $1.15 \times 10^{-7}\text{ m} = 115 \times 10^{-9}\text{ m} = 115\text{ nm} = \text{UV}$

$2.0 \times 10^{-6}\text{ m} = 2000 \times 10^{-9}\text{ m} = 2000\text{ nm} = \text{IR}$

The visible region is (380 to 780 nm) is completely within this range. The ultraviolet and infrared regions are partially in this range.

5.36 $$\lambda = \frac{c}{\nu} = \frac{3.00 \times 10^{8}\text{ m/s}}{5.5 \times 10^{15}\text{ s}^{-1}} = 5.5 \times 10^{-8}\text{ m}$$

5.38 (a) $\nu = 99.5\text{ MHz} = 99.5 \times 10^{6}\text{ s}^{-1}$

$E = h\nu = (6.626 \times 10^{-34}\text{ J}\cdot\text{s})(99.5 \times 10^{6}\text{ s}^{-1})(6.022 \times 10^{23}\text{ /mol})$

$E = 3.97 \times 10^{-2}\text{ J/mol} = 3.97 \times 10^{-5}\text{ kJ/mol}$

$\nu = 1150\text{ kHz} = 1150 \times 10^{3}\text{ s}^{-1}$

$E = h\nu = (6.626 \times 10^{-34}\text{ J}\cdot\text{s})(1150 \times 10^{3}\text{ s}^{-1})(6.022 \times 10^{23}\text{ /mol})$

$E = 4.589 \times 10^{-4}\text{ J/mol} = 4.589 \times 10^{-7}\text{ kJ/mol}$

The FM radio wave (99.5 MHz) has the higher energy.

(b) $\lambda = 3.44 \times 10^{-9}\text{ m}$

$$E = \frac{hc}{\lambda} = (6.626 \times 10^{-34}\text{ J}\cdot\text{s})\left(\frac{3.00 \times 10^{8}\text{ m/s}}{3.44 \times 10^{-9}\text{ m}}\right)(6.022 \times 10^{23}\text{/mol})$$

$E = 3.48 \times 10^{7}\text{ J/mol} = 3.48 \times 10^{4}\text{ kJ/mol}$

$\lambda = 6.71 \times 10^{-2}\text{ m}$

$$E = \frac{hc}{\lambda} = (6.626 \text{ x } 10^{-34} \text{ J·s})\left(\frac{3.00 \text{ x } 10^{8} \text{ m/s}}{6.71 \text{ x } 10^{-2} \text{ m}}\right)(6.022 \text{ x } 10^{23}/\text{mol})$$

E = 1.78 J/mol = 1.78 x 10^{-3} kJ/mol

The X ray (λ = 3.44 x 10^{-9} m) has the higher energy.

5.40 (a) $\lambda = \frac{c}{\nu} = \frac{3.00 \text{ x } 10^{8} \text{ m/s}}{825 \text{ x } 10^{6} \text{ s}^{-1}} \text{ x } \frac{1 \text{ cm}}{1 \text{ x } 10^{-2} \text{ m}} = 36.4 \text{ cm}$

(b) $\lambda = \frac{c}{\nu} = \frac{3.00 \text{ x } 10^{8} \text{ m/s}}{875 \text{ x } 10^{6} \text{ s}^{-1}} \text{ x } \frac{1 \text{ cm}}{1 \text{ x } 10^{-2} \text{ m}} = 34.3 \text{ cm}$

5.42 (a) $E = 90.5 \text{ kJ/mol x } \frac{1000 \text{ J}}{1 \text{ kJ}} \text{ x } \frac{1 \text{ mol}}{6.02 \text{ x } 10^{23}} = 1.50 \text{ x } 10^{-19} \text{ J}$

$\nu = \frac{E}{h} = \frac{1.50 \text{ x } 10^{-19} \text{ J}}{6.626 \text{ x } 10^{-34} \text{ J·s}} = 2.27 \text{ x } 10^{14} \text{ s}^{-1}$

$\lambda = \frac{c}{\nu} = \frac{3.00 \text{ x } 10^{8} \text{ m/s}}{2.27 \text{ x } 10^{14} \text{ s}^{-1}} = 1.32 \text{ x } 10^{-6} \text{ m} = 1320 \text{ x } 10^{-9} \text{ m} = 1320 \text{ nm}$, near IR

(b) $E = 8.05 \text{ x } 10^{-4} \text{ kJ/mol x } \frac{1000 \text{ J}}{1 \text{ kJ}} \text{ x } \frac{1 \text{ mol}}{6.02 \text{ x } 10^{23}} = 1.34 \text{ x } 10^{-24} \text{ J}$

$\nu = \frac{E}{h} = \frac{1.34 \text{ x } 10^{-24} \text{ J}}{6.626 \text{ x } 10^{-34} \text{ J·s}} = 2.02 \text{ x } 10^{9} \text{ s}^{-1}$

$\lambda = \frac{c}{\nu} = \frac{3.00 \text{ x } 10^{8} \text{ m/s}}{2.02 \text{ x } 10^{9} \text{ s}^{-1}} = 0.149 \text{ m}$, radio wave

(c) $E = 1.83 \text{ x } 10^{3} \text{ kJ/mol x } \frac{1000 \text{ J}}{1 \text{ kJ}} \text{ x } \frac{1 \text{ mol}}{6.02 \text{ x } 10^{23}} = 3.04 \text{ x } 10^{-18} \text{ J}$

$\nu = \frac{E}{h} = \frac{3.04 \text{ x } 10^{-18} \text{ J}}{6.626 \text{ x } 10^{-34} \text{ J·s}} = 4.59 \text{ x } 10^{15} \text{ s}^{-1}$

$\lambda = \frac{c}{\nu} = \frac{3.00 \text{ x } 10^{8} \text{ m/s}}{4.59 \text{ x } 10^{15} \text{ s}^{-1}} = 6.54 \text{ x } 10^{-8} \text{ m} = 65.4 \text{ x } 10^{-9} \text{ m} = 65.4 \text{ nm}$, UV

5.44 (a) $\lambda = \frac{c}{\nu} = \frac{3.00 \text{ x } 10^{8} \text{ m/s}}{3.85 \text{ x } 10^{14} \text{ s}^{-1}} = 7.79 \text{ x } 10^{-7} \text{ m} = 779 \text{ x } 10^{-9} \text{ m} = 779 \text{ nm}$

$E = h\nu = (6.626 \text{ x } 10^{-34} \text{ J·s})(3.85 \text{ x } 10^{14} \text{ s}^{-1}) = 2.55 \text{ x } 10^{-19} \text{ J}$

(b) $\lambda = \frac{c}{\nu} = \frac{3.00 \text{ x } 10^{8} \text{ m/s}}{4.62 \text{ x } 10^{14} \text{ s}^{-1}} = 6.49 \text{ x } 10^{-7} \text{ m} = 649 \text{ x } 10^{-9} \text{ m} = 649 \text{ nm}$

$E = h\nu = (6.626 \text{ x } 10^{-34} \text{ J·s})(4.62 \text{ x } 10^{14} \text{ s}^{-1}) = 3.06 \text{ x } 10^{-19} \text{ J}$

(c) $\lambda = \frac{c}{\nu} = \frac{3.00 \text{ x } 10^{8} \text{ m/s}}{7.41 \text{ x } 10^{14} \text{ s}^{-1}} = 4.05 \text{ x } 10^{-7} \text{ m} = 405 \text{ x } 10^{-9} \text{ m} = 405 \text{ nm}$

$E = h\nu = (6.626 \text{ x } 10^{-34} \text{ J·s})(7.41 \text{ x } 10^{14} \text{ s}^{-1}) = 4.91 \text{ x } 10^{-19} \text{ J}$

5.46 For n = 3; $\lambda = 656.3 \text{ nm} = 656.3 \times 10^{-9} \text{ m}$

$$E = \frac{hc}{\lambda} = (6.626 \times 10^{-34} \text{ J}\cdot\text{s})\left(\frac{2.998 \times 10^{8} \text{ m/s}}{656.3 \times 10^{-9} \text{ m}}\right)\left(\frac{1 \text{ kJ}}{1000 \text{ J}}\right)(6.022 \times 10^{23}/\text{mol})$$

$E = 182.3 \text{ kJ/mol}$

For n = 4; $\lambda = 486.1 \text{ nm} = 486.1 \times 10^{-9} \text{ m}$

$$E = \frac{hc}{\lambda} = (6.626 \times 10^{-34} \text{ J}\cdot\text{s})\left(\frac{2.998 \times 10^{8} \text{ m/s}}{486.1 \times 10^{-9} \text{ m}}\right)\left(\frac{1 \text{ kJ}}{1000 \text{ J}}\right)(6.022 \times 10^{23}/\text{mol})$$

$E = 246.1 \text{ kJ/mol}$

For n = 5; $\lambda = 434.0 \text{ nm} = 434.0 \times 10^{-9} \text{ m}$

$$E = \frac{hc}{\lambda} = (6.626 \times 10^{-34} \text{ J}\cdot\text{s})\left(\frac{2.998 \times 10^{8} \text{ m/s}}{434.0 \times 10^{-9} \text{ m}}\right)\left(\frac{1 \text{ kJ}}{1000 \text{ J}}\right)(6.022 \times 10^{23}/\text{mol})$$

$E = 275.6 \text{ kJ/mol}$

5.48 $m = 1, n = \infty;\ R_\infty = 1.097 \times 10^{-2} \text{ nm}^{-1}$

$$\frac{1}{\lambda} = R_\infty\left[\frac{1}{m^2} - \frac{1}{n^2}\right];\ \frac{1}{\lambda} = R_\infty\left[\frac{1}{1^2} - \frac{1}{\infty^2}\right] = R_\infty = 1.097 \times 10^{-2} \text{ nm}^{-1};\ \lambda = 91.16 \text{ nm}$$

$$E = \frac{hc}{\lambda} = (6.626 \times 10^{-34} \text{ J}\cdot\text{s})\left(\frac{2.998 \times 10^{8} \text{ m/s}}{91.16 \times 10^{-9} \text{ m}}\right)\left(\frac{1 \text{ kJ}}{1000 \text{ J}}\right)(6.022 \times 10^{23}/\text{mol})$$

$E = 1312 \text{ kJ/mol}$

5.50 $m = 4, n = 5;\ R_\infty = 1.097 \times 10^{-2} \text{ nm}^{-1}$

$$\frac{1}{\lambda} = R_\infty\left[\frac{1}{m^2} - \frac{1}{n^2}\right];\ \frac{1}{\lambda} = R_\infty\left[\frac{1}{4^2} - \frac{1}{5^2}\right] = 2.468 \times 10^{-4} \text{ nm}^{-1};\ \lambda = 4051 \text{ nm}$$

$$E = \frac{hc}{\lambda} = (6.626 \times 10^{-34} \text{ J}\cdot\text{s})\left(\frac{2.998 \times 10^{8} \text{ m/s}}{4051 \times 10^{-9} \text{ m}}\right)\left(\frac{1 \text{ kJ}}{1000 \text{ J}}\right)(6.022 \times 10^{23}/\text{mol})$$

$E = 29.55 \text{ kJ/mol}$, IR

$m = 4, n = 6;\ R_\infty = 1.097 \times 10^{-2} \text{ nm}^{-1}$

$$\frac{1}{\lambda} = R_\infty\left[\frac{1}{m^2} - \frac{1}{n^2}\right];\ \frac{1}{\lambda} = R_\infty\left[\frac{1}{4^2} - \frac{1}{6^2}\right] = 3.809 \times 10^{-4} \text{ nm}^{-1};\ \lambda = 2625 \text{ nm}$$

$$E = \frac{hc}{\lambda} = (6.626 \times 10^{-34} \text{ J}\cdot\text{s})\left(\frac{2.998 \times 10^{8} \text{ m/s}}{2625 \times 10^{-9} \text{ m}}\right)\left(\frac{1 \text{ kJ}}{1000 \text{ J}}\right)(6.022 \times 10^{23}/\text{mol})$$

$E = 45.60 \text{ kJ/mol}$, IR

5.52 Both (c) & (d) are below the threshold energy and no electrons would be ejected. (b) would eject the least number of electrons.

5.54 $E = (436\text{ kJ/mol})\left(\dfrac{1000\text{ J}}{1\text{ kJ}}\right)\left(\dfrac{1\text{ mol}}{6.022 \times 10^{23}\text{ photon}}\right) = 7.24 \times 10^{-19}\text{ J/photon}$

$$\nu = \frac{E}{h} = \frac{7.24 \times 10^{-19}\text{ J}}{6.626 \times 10^{-34}\text{ J}\cdot\text{s}} = 1.09 \times 10^{15}\text{ s}^{-1} = 1.09 \times 10^{15}\text{ Hz}$$

5.56 The deuterium lamp produces a continuous emission spectrum.

Particles and Waves (Section 5.5)

5.58 $\lambda = \dfrac{h}{mv} = \dfrac{6.626 \times 10^{-34}\text{ kg m}^2\text{ s}^{-1}}{(9.11 \times 10^{-31}\text{ kg})(0.99 \times 3.00 \times 10^8\text{ m/s})} = 2.45 \times 10^{-12}\text{ m}, \gamma\text{ ray}$

5.60 156 km/h = 156×10^3 m/3600 s = 43.3 m/s; 145 g = 0.145 kg

$$\lambda = \frac{h}{mv} = \frac{6.626 \times 10^{-34}\text{ kg m}^2\text{ s}^{-1}}{(0.145\text{ kg})(43.3\text{ m/s})} = 1.06 \times 10^{-34}\text{ m}$$

The wavelength is too small, compared to the object, to observe.

5.62 145 g = 0.145 kg; 0.500 nm = 0.500×10^{-9} m

$$v = \frac{h}{m\lambda} = \frac{6.626 \times 10^{-34}\text{ kg m}^2\text{ s}^{-1}}{(0.145\text{ kg})(0.500 \times 10^{-9}\text{ m})} = 9.14 \times 10^{-24}\text{ m/s}$$

5.64 0.68 g = 0.68×10^{-3} kg

$$(\Delta x)(\Delta mv) \ge \frac{h}{4\pi}; \quad \Delta x \ge \frac{h}{4\pi(\Delta mv)} = \frac{6.626 \times 10^{-34}\text{ kg m}^2\text{ s}^{-1}}{4\pi(0.68 \times 10^{-3}\text{ kg})(0.1\text{ m/s})} = 8 \times 10^{-31}\text{ m}$$

Orbitals and Quantum Mechanics (Sections 5.6–5.9)

5.66 The Heisenberg uncertainty principle states that one can never know both the position and the velocity of an electron beyond a certain level of precision. This means we cannot think of electrons circling the nucleus in specific orbital paths, but we can think of electrons as being found in certain three-dimensional regions of space around the nucleus, called orbitals.

5.68 n is the principal quantum number. The size and energy level of an orbital depends on n. l is the angular-momentum quantum number. l defines the three-dimensional shape of an orbital. m_l is the magnetic quantum number. m_l defines the spatial orientation of an orbital. m_s is the spin quantum number. m_s indicates the spin of the electron and can have either of two values, +½ or –½.

5.70 (a) 4s $n = 4$; $l = 0$; $m_l = 0$; $m_s = \pm\frac{1}{2}$
(b) 3p $n = 3$; $l = 1$; $m_l = -1, 0, +1$; $m_s = \pm\frac{1}{2}$
(c) 5f $n = 5$; $l = 3$; $m_l = -3, -2, -1, 0, +1, +2, +3$; $m_s = \pm\frac{1}{2}$
(d) 5d $n = 5$; $l = 2$; $m_l = -2, -1, 0, +1, +2$; $m_s = \pm\frac{1}{2}$

5.72 A 4s orbital has three nodal surfaces.

4s orbital

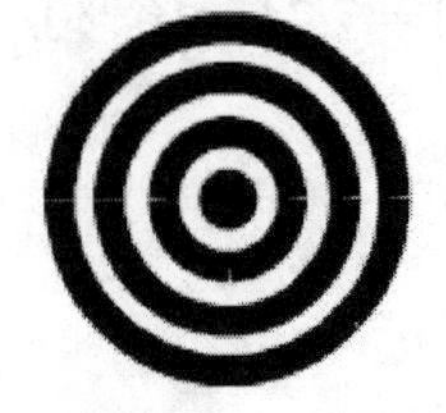

nodes are white

regions of maximum electron probability are black

5.74 Co $1s^2\, 2s^2\, 2p^6\, 3s^2\, 3p^6\, 4s^2\, 3d^7$

(a) is not allowed because for $l = 0$, $m_l = 0$ only.

(b) is not allowed because n = 4 and $l = 2$ is for a 4d orbital.

(c) is allowed because n = 3 and $l = 1$ is for a 3p orbital.

5.76 For n = 5, the maximum number of electrons will occur when the 5g orbital is filled:
[Rn] $7s^2\, 5f^{14}\, 6d^{10}\, 7p^6\, 8s^2\, 5g^{18}$ = 138 electrons

5.78 $\lambda = 330\text{ nm} = 330 \times 10^{-9}\text{ m}$

$$E = \frac{hc}{\lambda} = (6.626 \times 10^{-34}\text{ J}\cdot\text{s})\left(\frac{3.00 \times 10^8\text{ m/s}}{330 \times 10^{-9}\text{ m}}\right)\left(\frac{1\text{ kJ}}{1000\text{ J}}\right)(6.022 \times 10^{23}/\text{mol})$$

$$E = 363\text{ kJ/mol}$$

5.80 C $1s^2\, 2s^2\, 2p^2$

n	l	m_l	m_s
1	0	0	+½
1	0	0	−½
2	0	0	+½
2	0	0	−½
2	1	−1	+½
2	1	0	+½

5.82 Sr [Kr] $5s^2$

n	l	m_l	m_s
5	0	0	+½
5	0	0	−½

Electron Configurations (Sections 5.10–5.13)

5.84 Part of the electron-nucleus attraction is canceled by the electron-electron repulsion, an effect we describe by saying that the electrons are shielded from the nucleus by the other electrons. The net nuclear charge actually felt by an electron is called the effective nuclear charge, Z_{eff}, and is often substantially lower than the actual nuclear charge, Z_{actual}.

$$Z_{eff} = Z_{actual} - \text{electron shielding}$$

5.86 4s > 4d > 4f

5.88 The number of elements in successive periods of the periodic table increases by the progression 2, 8, 18, 32 because the principal quantum number n increases by 1 from one period to the next. As the principal quantum number increases, the number of orbitals in a shell increases. The progression of elements parallels the number of electrons in a particular shell.

5.90 (a) 5d (b) 4s (c) 6s

5.92 (a) 3d after 4s (b) 4p after 3d (c) 6d after 5f (d) 6s after 5p

5.94
(a) Ti, Z = 22 $1s^2 2s^2 2p^6 3s^2 3p^6 4s^2 3d^2$
(b) Ru, Z = 44 $1s^2 2s^2 2p^6 3s^2 3p^6 4s^2 3d^{10} 4p^6 5s^2 4d^6$
(c) Sn, Z = 50 $1s^2 2s^2 2p^6 3s^2 3p^6 4s^2 3d^{10} 4p^6 5s^2 4d^{10} 5p^2$
(d) Sr, Z = 38 $1s^2 2s^2 2p^6 3s^2 3p^6 4s^2 3d^{10} 4p^6 5s^2$
(e) Se, Z = 34 $1s^2 2s^2 2p^6 3s^2 3p^6 4s^2 3d^{10} 4p^4$

5.96
(a) Rb, Z = 37 [Kr] ↑ (5s)
(b) W, Z = 74 [Xe] ↑↓ (6s) ↑↓ ↑↓ ↑↓ ↑↓ ↑↓ ↑↓ ↑↓ (4f) ↑ ↑ ↑ ↑ _ (5d)
(c) Ge, Z = 32 [Ar] ↑↓ (4s) ↑↓ ↑↓ ↑↓ ↑↓ ↑↓ (3d) ↑ ↑ _ (4p)
(d) Zr, Z = 40 [Kr] ↑↓ (5s) ↑ ↑ _ _ _ (4d)

5.98
(a) O $1s^2 2s^2 2p^4$ ↑↓ ↑ ↑ (2p) 2 unpaired e^-
(b) Si $1s^2 2s^2 2p^6 3s^2 3p^2$ ↑ ↑ _ (3p) 2 unpaired e^-
(c) K [Ar] $4s^1$ 1 unpaired e^-
(d) As [Ar] $4s^2 3d^{10} 4p^3$ ↑ ↑ ↑ (4p) 3 unpaired e^-

5.100 Order of orbital filling:
1s→2s→2p→3s→3p→4s→3d→4p→5s→4d→5p→6s→4f→5d→6p→7s→5f→6d→7p→8s→5g
Z = 121

5.102 Na^+ $1s^2 2s^2 2p^6$

Electron Configurations and Periodic Properties (Section 5.14)

5.104 Atomic radii increase down a group because the electron shells are farther away from the nucleus.

5.106 F < O < S

5.108 (a) K, lower in group 1A
(b) Ta, lower in group 5B
(c) V, farther to the left in same period
(d) Ba, four periods lower and only one group to the right

5.110 Z = 116 [Rn] $7s^2 5f^{14} 6d^{10} 7p^4$

Chapter Problems

5.112 m = 2; n = 3; R = 1.097×10^{-2} nm^{-1}

$$\frac{1}{\lambda} = Z^2R\left[\frac{1}{m^2} - \frac{1}{n^2}\right]; \quad \frac{1}{\lambda} = (2^2)R\left[\frac{1}{2^2} - \frac{1}{3^2}\right] = 6.094 \times 10^{-3}\ \text{nm}^{-1}; \quad \lambda = 164\ \text{nm}$$

5.114 m = 2; $R_\infty = 1.097 \times 10^{-2}$ nm^{-1}

$$\frac{1}{\lambda} = R_\infty\left[\frac{1}{m^2} - \frac{1}{n^2}\right]; \quad \frac{1}{\lambda} = R_\infty\left[\frac{1}{2^2} - \frac{1}{6^2}\right] = 2.438 \times 10^{-3}\ \text{nm}^{-1}$$

$\lambda = 410.2\ \text{nm} = 410.2 \times 10^{-9}\ \text{m}$

$$E = \frac{hc}{\lambda} = (6.626 \times 10^{-34}\ \text{J}\cdot\text{s})\left(\frac{2.998 \times 10^8\ \text{m/s}}{410.2 \times 10^{-9}\ \text{m}}\right)\left(\frac{1\ \text{kJ}}{1000\ \text{J}}\right)(6.022 \times 10^{23}/\text{mol})$$

$E = 291.6\ \text{kJ/mol}$

5.116 m = 5, n = ∞; $R_\infty = 1.097 \times 10^{-2}$ nm^{-1}

$$\frac{1}{\lambda} = R_\infty\left[\frac{1}{5^2} - \frac{1}{\infty^2}\right] = R_\infty\left[\frac{1}{25}\right] = 4.388 \times 10^{-4}\ \text{nm}^{-1}; \quad \lambda = 2279\ \text{nm}$$

5.118 (a) $E = h\nu = (6.626 \times 10^{-34}\ \text{J}\cdot\text{s})(3.79 \times 10^{11}\ \text{s}^{-1})\left(\frac{1\ \text{kJ}}{1000\ \text{J}}\right)(6.022 \times 10^{23}/\text{mol}) = 0.151\ \text{kJ/mol}$

(b) $E = h\nu = (6.626 \times 10^{-34}\ \text{J}\cdot\text{s})(5.45 \times 10^{4}\ \text{s}^{-1})\left(\frac{1\ \text{kJ}}{1000\ \text{J}}\right)(6.022 \times 10^{23}/\text{mol}) = 2.17 \times 10^{-8}\ \text{kJ/mol}$

(c) $E = h\nu = (6.626 \times 10^{-34}\ \text{J}\cdot\text{s})\left(\frac{3.00 \times 10^8\ \text{m/s}}{4.11 \times 10^{-5}\ \text{m}}\right)\left(\frac{1\ \text{kJ}}{1000\ \text{J}}\right)(6.022 \times 10^{23}/\text{mol}) = 2.91\ \text{kJ/mol}$

5.120 (a) Ra [Rn] $7s^2$ [Rn] ⇅ (7s)

(b) Sc [Ar] $4s^2 3d^1$ [Ar] ⇅ (4s) ↑ __ __ __ __ (3d)

(c) Lr [Rn] $7s^2 5f^{14} 6d^1$ [Rn] ⇅ (7s) ⇅ ⇅ ⇅ ⇅ ⇅ ⇅ ⇅ (5f) ↑ __ __ __ __ (6d)

(d) B [He] $2s^2 2p^1$ [He] ⇅ (2s) ↑ __ __ (2p)

(e) Te [Kr] $5s^2 4d^{10} 5p^4$ [Kr] ⇅ (5s) ⇅ ⇅ ⇅ ⇅ ⇅ (4d) ⇅ ↑ ↑ (5p)

5.122 206.5 kJ = 206.5 x 10^3 J; $E = \frac{206.5 \times 10^3\ \text{J}}{1\ \text{mol}} \times \frac{1\ \text{mol}}{6.022 \times 10^{23}} = 3.429 \times 10^{-19}\ \text{J}$

$$E = \frac{hc}{\lambda},\quad \lambda = \frac{hc}{E} = \frac{(6.626 \times 10^{-34}\ \text{J}\cdot\text{s})(3.00 \times 10^8\ \text{m/s})}{3.429 \times 10^{-19}\ \text{J}} = 5.797 \times 10^{-7}\ \text{m} = 580.\ \text{nm}$$

5.124 (a) Sr, Z = 38 [Kr] ⇅ (5s)

(b) Cd, Z = 48 [Kr] ⇅ (5s) ⇅ ⇅ ⇅ ⇅ ⇅ (4d)

(c) Z = 22, Ti [Ar] ⇅ (4s) ↑ ↑ _ _ _ (3d)

(d) Z = 34, Se [Ar] ⇅ (4s) ⇅ ⇅ ⇅ ⇅ ⇅ (3d) ⇅ ↑ ↑ (4p)

5.126 (a) $\lambda = \frac{c}{\nu} = \frac{3.00 \times 10^8\ \text{m/s}}{2.9 \times 10^{18}\ \text{s}^{-1}} = 1.0 \times 10^{-10}\ \text{m}$

(b) $E = h\nu = (6.626 \times 10^{-34}\ \text{J}\cdot\text{s})(2.9 \times 10^{18}\ \text{s}^{-1})\left(\frac{1\ \text{kJ}}{1000\ \text{J}}\right)(6.022 \times 10^{23}/\text{mol})$

$E = 1.2 \times 10^6$ kJ/mol

(c) X rays

5.128 For K, $Z_{eff} = \sqrt{\frac{(418.8\ \text{kJ/mol})(4^2)}{1312\ \text{kJ/mol}}} = 2.26$

For Kr, $Z_{eff} = \sqrt{\frac{(1350.7\ \text{kJ/mol})(4^2)}{1312\ \text{kJ/mol}}} = 4.06$

5.130 q = (350 g)(4.184 J/g·°C)(95 °C – 20 °C) = 109,830 J

$\lambda = 15.0\ \text{cm} = 15.0 \times 10^{-2}\ \text{m}$

$E = (6.626 \times 10^{-34}\ \text{J}\cdot\text{s})\left(\frac{3.00 \times 10^8\ \text{m/s}}{15.0 \times 10^{-2}\ \text{m}}\right) = 1.33 \times 10^{-24}\ \text{J/photon}$

number of photons $= \frac{109{,}830\ \text{J}}{1.33 \times 10^{-24}\ \text{J/photon}} = 8.3 \times 10^{28}$ photons

5.132 48.2 nm = 48.2 x 10^{-9} m

$E(\text{photon}) = 6.626 \times 10^{-34}\ \text{J}\cdot\text{s} \times \frac{3.00 \times 10^8\ \text{m/s}}{48.2 \times 10^{-9}\ \text{m}} \times \frac{1\ \text{kJ}}{1000\ \text{J}} \times \frac{6.022 \times 10^{23}}{\text{mol}} = 2.48 \times 10^3\ \text{kJ/mol}$

$E_K = E(\text{electron}) = \tfrac{1}{2}(9.109 \times 10^{-31}\ \text{kg})(2.371 \times 10^6\ \text{m/s})^2\left(\frac{1\ \text{kJ}}{1000\ \text{J}}\right)\left(\frac{6.022 \times 10^{23}}{\text{mol}}\right)$

$E_K = 1.54 \times 10^3$ kJ/mol

$E(\text{photon}) = E_i + E_K$; $E_i = E(\text{photon}) - E_K = (2.48 \times 10^3) - (1.54 \times 10^3) = 940$ kJ/mol

5.134 Substitute the equation for the orbit radius, r, into the equation for the energy level, E, to get $E = \dfrac{-Ze^2}{2\left(\dfrac{n^2 a_o}{Z}\right)} = \dfrac{-Z^2e^2}{2a_o n^2}$

Let E_1 be the energy of an electron in a lower orbit and E_2 the energy of an electron in a higher orbit. The difference between the two energy levels is

$$\Delta E = E_2 - E_1 = \frac{-Z^2e^2}{2a_o n_2^2} - \frac{-Z^2e^2}{2a_o n_1^2} = \frac{-Z^2e^2}{2a_o n_2^2} + \frac{Z^2e^2}{2a_o n_1^2} = \frac{Z^2e^2}{2a_o n_1^2} - \frac{Z^2e^2}{2a_o n_2^2}$$

$$\Delta E = \frac{Z^2e^2}{2a_o}\left[\frac{1}{n_1^2} - \frac{1}{n_2^2}\right]$$

Because Z, e, and a_o are constants, this equation shows that ΔE is proportional to $\left[\dfrac{1}{n_1^2} - \dfrac{1}{n_2^2}\right]$ where n_1 and n_2 are integers with $n_2 > n_1$.

This is similar to the Balmer-Rydberg equation where $1/\lambda$ or ν for the emission spectra of atoms is proportional to $\left[\dfrac{1}{m^2} - \dfrac{1}{n^2}\right]$ where m and n are integers with $n > m$.

5.136 (a) 3d, $n = 3, l = 2$
(b) 2p, $n = 2, l = 1, m_l = -1, 0, +1$
3p, $n = 3, l = 1, m_l = -1, 0, +1$
3d, $n = 3, l = 2, m_l = -2, -1, 0, +1, +2$
(c) N, $1s^2 2s^2 2p^3$ so the 3s, 3p, and 3d orbitals are empty.
(d) C, $1s^2 2s^2 2p^2$ so the 1s and 2s orbitals are filled.
(e) Be, $1s^2 2s^2$ so the 2s orbital contains the outermost electrons.
(f) 2p and 3p (↿ ↿ __) and 3d (↿ ↿ __ __ __).

5.138 (a) $E = h\nu$; $\nu = \dfrac{E}{h} = \dfrac{7.21 \times 10^{-19}\ J}{6.626 \times 10^{-34}\ J \cdot s} = 1.09 \times 10^{15}\ s^{-1}$

(b) E(photon) = $E_i + E_K$; from (a), $E_i = 7.21 \times 10^{-19}$ J

$$E(\text{photon}) = \frac{hc}{\lambda} = (6.626 \times 10^{-34}\ J \cdot s)\left(\frac{3.00 \times 10^8\ m/s}{2.50 \times 10^{-7}\ m}\right) = 7.95 \times 10^{-19}\ J$$

E_K = E(photon) – E_i = $(7.95 \times 10^{-19}\ J) - (7.21 \times 10^{-19}\ J) = 7.4 \times 10^{-20}\ J$

Calculate the electron velocity from the kinetic energy, E_K.

$E_K = 7.4 \times 10^{-20}\ J = 7.4 \times 10^{-20}\ kg \cdot m^2/s^2 = \frac{1}{2}mv^2 = \frac{1}{2}(9.109 \times 10^{-31}\ kg)v^2$

$$v = \sqrt{\frac{2 \times (7.4 \times 10^{-20}\ kg \cdot m^2/s^2)}{9.109 \times 10^{-31}\ kg}} = 4.0 \times 10^5\ m/s$$

$$\text{de Broglie wavelength} = \frac{h}{mv} = \frac{6.626 \times 10^{-34}\ kg \cdot m^2/s}{(9.109 \times 10^{-31}\ kg)(4.0 \times 10^5\ m/s)} = 1.8 \times 10^{-9}\ m = 1.8\ nm$$

Multiconcept Problems

5.140 (a) 5f subshell: $n = 5, l = 3, m_l = -3, -2, -1, 0, +1, +2, +3$
3d subshell: $n = 3, l = 2, m_l = -2, -1, 0, +1, +2$

(b) In the H atom the subshells in a particular energy level are all degenerate, i.e., all have the same energy. Therefore, you only need to consider the principal quantum number, n, to calculate the wavelength emitted for an electron that drops from the 5f to the 3d subshell.

$m = 3, n = 5;\ R_\infty = 1.097 \times 10^{-2}\ nm^{-1}$

$$\frac{1}{\lambda} = R_\infty\left[\frac{1}{m^2} - \frac{1}{n^2}\right];\ \frac{1}{\lambda} = R_\infty\left[\frac{1}{3^2} - \frac{1}{5^2}\right];\ \frac{1}{\lambda} = 7.801 \times 10^{-4}\ nm^{-1};\quad \lambda = 1282\ nm$$

(c) $m = 3, n = \infty;\ R_\infty = 1.097 \times 10^{-2}\ nm^{-1}$

$$\frac{1}{\lambda} = R_\infty\left[\frac{1}{m^2} - \frac{1}{n^2}\right];\ \frac{1}{\lambda} = R_\infty\left[\frac{1}{3^2} - \frac{1}{\infty^2}\right];\ \frac{1}{\lambda} = R_\infty\left[\frac{1}{3^2}\right] = 1.219 \times 10^{-3}\ nm^{-1};\ \lambda = 820.4\ nm$$

$$E = (6.626 \times 10^{-34}\ J\cdot s)\left(\frac{3.00 \times 10^8\ m/s}{820.4 \times 10^{-9}\ m}\right)(6.022 \times 10^{23}/mol) = 1.46 \times 10^5\ J/mol = 146\ kJ/mol$$

6 Ionic Compounds: Periodic Trends and Bonding Theory

6.1 (a) Ra^{2+} [Rn] (b) Ni^{2+} [Ar] $3d^8$ (c) N^{3-} [Ne]

6.2 (b) Ti^{4+}, Ca^{2+}, and Cl^- are isoelectronic. They all have the electron configuration of Ar.
(c) Na^+, Mg^{2+}, and Al^{3+} are isoelectronic. They all have the electron configuration of Ne.

6.3 (a) O^{2-}; decrease in effective nuclear charge and an increase in electron-electron repulsions leads to the anion being larger.
(b) Fe; in Fe^{3+} electrons are removed from a larger valence shell and there is an increase in effective nuclear charge leading to the smaller cation.
(c) H^-; decrease in effective nuclear charge and an increase in electron-electron repulsions leads to the anion being larger.

6.4 K^+, Ca^{2+}, and Cl^- are isoelectronic. The Z_{eff} for $Ca^{2+} > K^+ > Cl^-$.
In terms of size, $Cl^- > K^+ > Ca^{2+}$. Cl^- is yellow, K^+ is green, and Ca^{2+} is red.

6.5 Ionization energy generally increases from left to right across a row of the periodic table and decreases from top to bottom down a group.
(a) Br (b) S (c) Se (d) Ne

6.6 (c) < (b) < (a) < (d)

6.7 (a) Be $1s^2 2s^2$; N $1s^2 2s^2 2p^3$
Be would have the larger third ionization energy because this electron would come from the 1s orbital.
(b) Ga [Ar] $4s^2 3d^{10} 4p^1$; Ge [Ar] $4s^2 3d^{10} 4p^2$
Ga would have the larger fourth ionization energy because this electron would come from the 3d orbitals. Ge would be losing a 4s electron.

6.8 The first 3 electrons are relatively easy to remove compared to the fourth and fifth electrons. The atom is Al.

6.9 Cr [Ar] $4s^1 3d^5$; Mn [Ar] $4s^2 3d^5$; Fe [Ar] $4s^2 3d^6$
Cr can accept an electron into a 4s orbital. The 4s orbital is lower in energy than a 3d orbital. Both Mn and Fe accept the added electron into a 3d orbital that contains an electron, but Mn has a lower value of Z_{eff}. Therefore, Mn has a less negative E_{ea} than either Cr or Fe.

6.10 The least favorable E_{ea} is for Kr (red) because it is a noble gas with a filled set of 4p orbitals. The most favorable E_{ea} is for Ge (blue) because the 4p orbitals would become half filled. In addition, Z_{eff} is larger for Ge than it is for K (green).

6.11 (a) Rb would lose one electron and adopt the Kr noble gas configuration.
(b) Ba would lose two electrons and adopt the Xe noble gas configuration.
(c) Ga would lose three electrons and adopt an Ar-like noble gas configuration (note that Ga^{3+} has ten 3d electrons in addition to the two 3s and six 3p electrons).
(d) F would gain one electron and adopt the Ne noble gas configuration.

6.12 Group 6A elements will gain 2 electrons. The ion charge will be 2–.

6.13

$Mg(s) \rightarrow Mg(g)$	+147.7	kJ/mol
$Mg(g) \rightarrow Mg^{+}(g) + e^{-}$	+737.7	kJ/mol
$Mg^{+}(g) \rightarrow Mg^{2+}(g) + e^{-}$	+1450.7	kJ/mol
$F_2(g) \rightarrow 2\ F(g)$	+158	kJ/mol
$2[F(g) + e^{-} \rightarrow F^{-}(g)]$	2(–328)	kJ/mol
$Mg^{2+}(g) + 2\ F^{-}(g) \rightarrow MgF_2(s)$	–2957	kJ/mol
Sum =	–1119	kJ/mol for $Mg(s) + F_2(g) \rightarrow MgF_2(s)$

6.14

$Li(s) \rightarrow Li(g)$	+161	kJ/mol
$Li(s) \rightarrow Li(g) + e^{-}$	+520	kJ/mol
½ $[Cl_2(g) \rightarrow 2\ Cl(g)]$	+243/2	kJ/mol
$Cl(g) + e^{-} \rightarrow Cl^{-}(g)$	–349	kJ/mol
$Li^{+}(g) + Cl^{-}(g) \rightarrow LiCl(s)$	–U	kJ/mol
Sum =	–409	kJ/mol for $Li(s) + ½\ Cl_2(g) \rightarrow LiCl(s)$

electrostatic attraction = –U = – 409 – 161 – 520 – 243/2 + 349 = –863 kJ/mol

6.15 (a) KCl has the higher lattice energy because of the smaller K^{+}.
(b) CaF_2 has the higher lattice energy because of the smaller Ca^{2+}.
(c) CaO has the higher lattice energy because of the higher charge on both the cation and anion.

6.16 The anions are larger than the cations. Cl^{-} is larger than O^{2-} because it is below it in the periodic table. Therefore, (a) is NaCl and (b) is MgO. Because of the higher ion charge and shorter cation–anion distance, MgO has the larger lattice energy.

6.17 In ionic liquids the cation has an irregular shape and one or both of the ions are large and bulky to disperse charges over a large volume. Both factors minimize the crystal lattice energy, making the solid less stable and favoring the liquid.

6.18 (a) Iodide ions are larger than bromide ions. Tetraheptylammonium bromide corresponds to picture (ii) and tetraheptylammonium iodide corresponds to picture (i).
(b) Tetraheptylammonium bromide has the larger lattice energy because bromide ions are smaller than iodide ions.
(c) Tetraheptylammonium bromide has melting point of 88 °C and tetraheptylammonium iodide has a melting point of 39 °C.

6.19 (a) F^{-}, $1s^2 2s^2 2p^6$
Se^{2-}, $1s^2 2s^2 2p^6 3s^2 3p^6 4s^2 3d^{10} 4p^6$

O^{2-}, $1s^2 2s^2 2p^6$
Br^-, $1s^2 2s^2 2p^6 3s^2 3p^6 4s^2 3d^{10} 4p^6$
(b) F^- and O^{2-} are isoelectronic; Se^{2-} and Br^- are isoelectronic.
(c) The larger Br^-.

Conceptual Problems

6.20

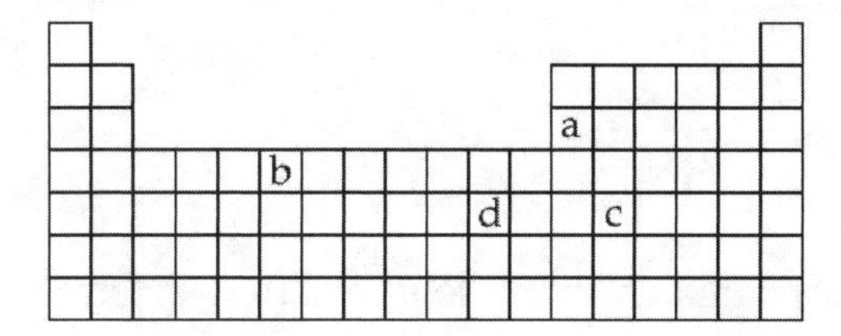

(a) Al^{3+}
(b) Cr^{3+}
(c) Sn^{2+}
(d) Ag^+

6.22

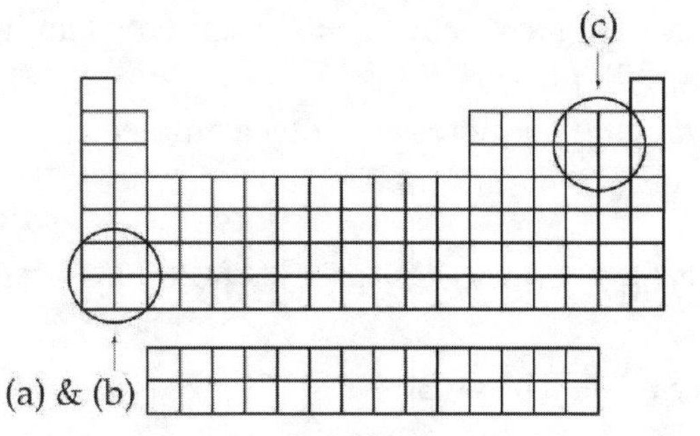

6.24 The first 2 electrons are relatively easy to remove compared to the third and fourth electron. The atom is Mg.

6.26 (a) shows an extended array, which represents an ionic compound.
(b) shows discrete units, which represent a covalent compound.

6.28 (c) has the largest lattice energy because the charges are closest together.
(a) has the smallest lattice energy because the charges are farthest apart.

6.30 Green, CBr_4; Blue, SrF_2; Red, PbS or PbS_2

Section Problems
Ions, Ionization Energy, and Electron Affinity (Sections 6.1–6.5)

6.32 A covalent bond results when two atoms share several (usually two) of their electrons. An ionic bond results from a complete transfer of one or more electrons from one atom to another.

6.34 A molecule is the unit of matter that results when two or more atoms are joined by covalent bonds. An ion results when an atom gains or loses electrons.

6.36 (a) Be^{2+}, 4 protons and 2 electrons
(b) Rb^+, 37 protons and 36 electrons
(c) Se^{2-}, 34 protons and 36 electrons
(d) Au^{3+}, 79 protons and 76 electrons

6.38 (a) La^{3+}, [Xe] (b) Ag^{+}, [Kr] $4d^{10}$ (c) Sn^{2+}, [Kr] $5s^2 4d^{10}$

6.40 Ca^{2+}, [Ar]; Ti^{2+}, [Ar] $3d^2$

6.42 The neutral atom contains 12 e^- and is Mg. The ion is Mg^{2+}.

6.44 Cr^{2+} [Ar] $3d^4$ ↑ ↑ ↑ ↑ __ (3d)

Fe^{2+} [Ar] $3d^6$ ↑↓ ↑ ↑ ↑ ↑ (3d)

6.46 (a) S^{2-}; decrease in effective nuclear charge and an increase in electron-electron repulsions leads to the anion being larger.
(b) Ca; in Ca^{2+} electrons are removed from a larger valence shell and there is an increase in effective nuclear charge leading to the smaller cation.
(c) O^{2-}; decrease in effective nuclear charge and an increase in electron-electron repulsions leads to O^{2-} being larger.

6.48 Sr^{2+}, Se^{2-}, Br^-, and Rb^+ are isoelectronic. The Z_{eff} for $Sr^{2+} > Rb^+ > Br^- > Se^{2-}$. The smallest ion has the largest Z_{eff}.
Ions arranged from smallest to largest are $Sr^{2+} < Rb^+ < Br^- < Se^{2-}$.

6.50 The largest E_{i1} are found in Group 8A because of the largest values of Z_{eff}.
The smallest E_{i1} are found in Group 1A because of the smallest values of Z_{eff}.

6.52 Using Figure 6.3 as a reference:

	Lowest E_{i1}	Highest E_{i1}
(a)	K	Li
(b)	B	Cl
(c)	Ca	Cl

6.54 (a) K [Ar] $4s^1$ Ca [Ar] $4s^2$
Ca has the smaller second ionization energy because it is easier to remove the second 4s valence electron in Ca than it is to remove the second electron in K from the filled 3p orbitals.
(b) Ca [Ar] $4s^2$ Ga [Ar] $4s^2 3d^{10} 4p^1$
Ca has the larger third ionization energy because it is more difficult to remove the third electron in Ca from the filled 3p orbitals than it is to remove the third electron (second 4s valence electron) from Ga.

6.56 (a) $1s^2 2s^2 2p^6 3s^2 3p^3$ is P (b) $1s^2 2s^2 2p^6 3s^2 3p^6$ is Ar (c) $1s^2 2s^2 2p^6 3s^2 3p^6 4s^2$ is Ca
Ar has the highest E_{i2}. Ar has a higher Z_{eff} than P. The 4s electrons in Ca are easier to remove than any 3p electrons.
Ar has the lowest E_{i7}. It is difficult to remove 3p electrons from Ca, and it is difficult to remove 2p electrons from P.

6.58 The likely second row element is boron because it has three valence electrons. The large fourth ionization energy is from an inner shell 1s electron.

6.60 The relationship between the electron affinity of a univalent cation and the ionization energy of the neutral atom is that they have the same magnitude but opposite signs.

6.62 Na^+ has a more negative electron affinity than either Na or Cl because of its positive charge.

6.64 Energy is usually released when an electron is added to a neutral atom but absorbed when an electron is removed from a neutral atom because of the positive Z_{eff}.

6.66 The electron-electron repulsion is large and Z_{eff} is low.

Octet Rule, Ionic Bonds, and Lattice Energy (Sections 6.6–6.8)

6.68 (a) [Ne], N^{3-} (b) [Ar], Ca^{2+} (c) [Ar], S^{2-} (d) [Kr], Br^-

6.70 (a) Because X reacts by losing electrons, it is likely to be a metal.
(b) Because Y reacts by gaining electrons, it is likely to be a nonmetal.
(c) X_2Y_3
(d) X is likely to be in group 3A and Y is likely to be in group 6A.

6.72 $MgCl_2 > LiCl > KCl > KBr$

6.74
$Li \rightarrow Li^+ + e^-$ +520 kJ/mol
$Br + e^- \rightarrow Br^-$ –325 kJ/mol
+195 kJ/mol

6.76
$Li(s) \rightarrow Li(g)$ +159.4 kJ/mol
$Li(s) \rightarrow Li(g) + e^-$ +520 kJ/mol
$\frac{1}{2}[Br_2(l) \rightarrow Br_2(g)]$ +15.4 kJ/mol
$\frac{1}{2}[Br_2(g) \rightarrow 2\,Br(g)]$ +112 kJ/mol
$Br(g) + e^- \rightarrow Br^-(g)$ –325 kJ/mol
$Li^+(g) + Br^-(g) \rightarrow LiBr(s)$ –807 kJ/mol
Sum = –325 kJ/mol for $Li(s) + \frac{1}{2}Br_2(l) \rightarrow LiBr(s)$

6.78
$Na(s) \rightarrow Na(g)$ +107.3 kJ/mol
$Na(g) \rightarrow Na^+(g) + e^-$ +495.8 kJ/mol
$\frac{1}{2}[H_2(g) \rightarrow 2\,H(g)]$ ½(+435.9) kJ/mol
$H(g) + e^- \rightarrow H^-(g)$ –72.8 kJ/mol
$Na^+(g) + H^-(g) \rightarrow NaH(s)$ –U
Sum = –60 kJ/mol for $Na(s) + \frac{1}{2}H_2(g) \rightarrow NaH(s)$

$-U = -60 - 107.3 - 495.8 - 435.9/2 + 72.8 = -808$ kJ/mol; $U = 808$ kJ/mol

6.80

$Cs(s) \rightarrow Cs(g)$	+76.1 kJ/mol
$Cs(g) \rightarrow Cs^+(g) + e^-$	+375.7 kJ/mol
$\frac{1}{2}[F_2(g) \rightarrow 2\ F(g)]$	+79 kJ/mol
$F(g) + e^- \rightarrow F^-(g)$	–328 kJ/mol
$Cs^+(g) + F^-(g) \rightarrow CsF(s)$	–740 kJ/mol
Sum =	–537 kJ/mol for $Cs(s) + \frac{1}{2} F_2(g) \rightarrow CsF(s)$

6.82

$Ca(s) \rightarrow Ca(g)$	+178.2 kJ/mol
$Ca(g) \rightarrow Ca^+(g) + e^-$	+589.8 kJ/mol
$\frac{1}{2}[Cl_2(g) \rightarrow 2\ Cl(g)]$	+121.5 kJ/mol
$Cl(g) + e^- \rightarrow Cl^-(g)$	–348.6 kJ/mol
$Ca^+(g) + Cl^-(g) \rightarrow CaCl(s)$	–717 kJ/mol
Sum =	–176 kJ/mol for $Ca(s) + \frac{1}{2} Cl_2(g) \rightarrow CaCl(s)$

6.84

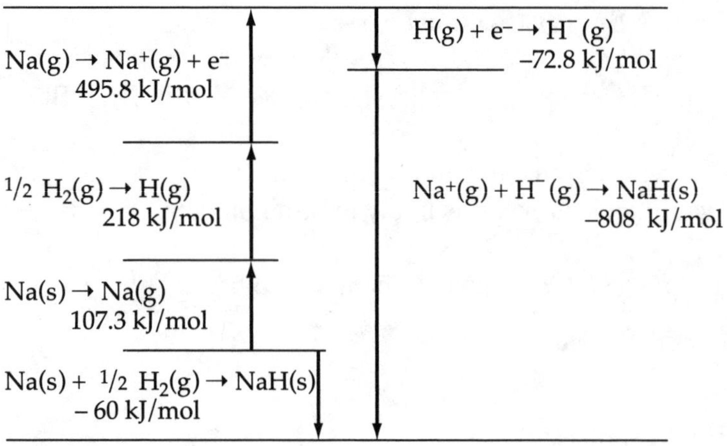

$Na(g) \rightarrow Na^+(g) + e^-$
495.8 kJ/mol

$H(g) + e^- \rightarrow H^-(g)$
–72.8 kJ/mol

$1/2\ H_2(g) \rightarrow H(g)$
218 kJ/mol

$Na^+(g) + H^-(g) \rightarrow NaH(s)$
–808 kJ/mol

$Na(s) \rightarrow Na(g)$
107.3 kJ/mol

$Na(s) + 1/2\ H_2(g) \rightarrow NaH(s)$
– 60 kJ/mol

Chapter Problems

6.86 Cu^{2+} has fewer electrons and a larger effective nuclear charge; therefore it has the smaller ionic radius.

6.88

$Mg(s) \rightarrow Mg(g)$	+147.7 kJ/mol
$Mg(g) \rightarrow Mg^+(g) + e^-$	+737.7 kJ/mol
$\frac{1}{2} F_2(g) \rightarrow F(g)$	+79 kJ/mol
$F(g) + e^- \rightarrow F^-(g)$	–328 kJ/mol
$Mg^+(g) + F^-(g) \rightarrow MgF(s)$	–930 kJ/mol
Sum =	–294 kJ/mol for $Mg(s) + \frac{1}{2} F_2(g) \rightarrow MgF(s)$

$Mg(s) \rightarrow Mg(g)$	+147.7 kJ/mol
$Mg(g) \rightarrow Mg^+(g) + e^-$	+737.7 kJ/mol
$Mg^+(g) \rightarrow Mg^{2+}(g) + e^-$	+1450.7 kJ/mol
$F_2(g) \rightarrow 2\ F(g)$	+158 kJ/mol
$2[F(g) + e^- \rightarrow F^-(g)]$	2(–328) kJ/mol
$Mg^{2+}(g) + 2\ F^-(g) \rightarrow MgF_2(s)$	–2952 kJ/mol
Sum =	–1114 kJ/mol for $Mg(s) + F_2(g) \rightarrow MgF_2(s)$

In the reaction of magnesium with fluorine, MgF_2 will form because the overall energy for the formation of MgF_2 is much more negative than for the formation of MgF.

6.90

$Na(s) \rightarrow Na(g)$	+107.3 kJ/mol
$Na(g) + e^- \rightarrow Na^-(g)$	−52.9 kJ/mol
$\frac{1}{2}[Cl_2(g) \rightarrow 2\,Cl(g)]$	+122 kJ/mol
$Cl(g) \rightarrow Cl^+(g) + e^-$	+1251 kJ/mol
$Na^-(g) + Cl^+(g) \rightarrow ClNa(s)$	−787 kJ/mol
Sum =	+640 kJ/mol for $Na(s) + \frac{1}{2}Cl_2(g) \rightarrow Cl^+Na^-(s)$

The formation of Cl^+Na^- from its elements is not favored because the net energy change is positive whereas it is negative for the formation of Na^+Cl^-.

6.92 When moving diagonally down and right on the periodic table, the increase in atomic radius caused by going to a larger shell is offset by a decrease caused by a higher Z_{eff}. Thus, there is little net change in the charge density.

6.94

$Mg(s) \rightarrow Mg(g)$	+147.7 kJ/mol
$Mg(g) \rightarrow Mg^+(g) + e^-$	+738 kJ/mol
$Mg^+(g) \rightarrow Mg^{2+}(g) + e^-$	+1451 kJ/mol
$\frac{1}{2}[O_2(g) \rightarrow 2\,O(g)]$	+249.2 kJ/mol
$O(g) + e^- \rightarrow O^-(g)$	−141.0 kJ/mol
$O^-(g) + e^- \rightarrow O^{2-}(g)$	E_{ea2}
$Mg^{2+}(g) + O^{2-}(g) \rightarrow MgO(s)$	−3791 kJ/mol
$Mg(s) + \frac{1}{2}O_2(g) \rightarrow MgO(s)$	−601.7 kJ/mol

$$147.7 + 738 + 1451 + 249.2 - 141.0 + E_{ea2} - 3791 = -601.7$$
$$E_{ea2} = -147.7 - 738 - 1451 - 249.2 + 141.0 + 3791 - 601.7 = +744 \text{ kJ/mol}$$

Because E_{ea2} is positive, O^{2-} is not stable in the gas phase. It is stable in MgO because of the large lattice energy that results from the +2 and −2 charge of the ions and their small size.

6.96 (a) The more negative the E_{ea}, the greater the tendency of the atom to accept an electron, and the more stable the anion that results. Be, N, O, and F are all second row elements. F has the most negative E_{ea} of the group because the anion that forms, F^-, has a complete octet of electrons and its nucleus has the highest effective nuclear charge.
(b) Se^{2-} and Rb^+ are below O^{2-} and F^- in the periodic table and are the larger of the four. Se^{2-} and Rb^+ are isoelectronic, but Rb^+ has the higher effective nuclear charge so it is smaller. Therefore Se^{2-} is the largest of the four ions.

6.98

$Cr(s) \rightarrow Cr(g)$	+397 kJ/mol
$Cr(g) \rightarrow Cr^+(g)$	+652 kJ/mol
$Cr^+(g) \rightarrow Cr^{2+}(g)$	+1588 kJ/mol
$Cr^{2+}(g) \rightarrow Cr^{3+}(g)$	+2882 kJ/mol
$\frac{1}{2}(I_2(s) \rightarrow I_2(g))$	+62/2 kJ/mol
$\frac{1}{2}(I_2(g) \rightarrow 2\,I(g))$	+151/2 kJ/mol
$I(g) + e^- \rightarrow I^-(g)$	–295 kJ/mol
$Cl_2(g) \rightarrow 2\,Cl(g)$	+243 kJ/mol
$2(Cl(g) + e^- \rightarrow Cl^-(g))$	2(–349) kJ/mol
$Cr^{3+}(g) + 2\,Cl^-(g) + I^-(g) \rightarrow CrCl_2I(s)$	–U
$Cr(s) + Cl_2(g) + \frac{1}{2}\,I_2(g) \rightarrow CrCl_2I(s)$	– 420 kJ/mol

$-U = -420 - 397 - 652 - 1588 - 2882 - 62/2 - 151/2 + 295 - 243 + 2(349) = -5295.5$ kJ/mol

$U = 5295$ kJ/mol

Multiconcept Problems

6.100 (a)

Fe	$[Ar]\ 4s^2\,3d^6$
Fe^{2+}	$[Ar]\ 3d^6$
Fe^{3+}	$[Ar]\ 3d^5$

(b) A 3d electron is removed on going from Fe^{2+} to Fe^{3+}. For the 3d electron, $n = 3$ and $l = 2$.

(c) $$E(\text{J/photon}) = 2952\ \text{kJ/mol} \times \frac{1\ \text{mol photons}}{6.022 \times 10^{23}\ \text{photons}} \times \frac{1000\ \text{J}}{1\ \text{kJ}} = 4.90 \times 10^{-18}\ \text{J/photon}$$

$$E = \frac{hc}{\lambda}$$

$$\lambda = \frac{hc}{E} = \frac{(6.626 \times 10^{-34}\ \text{J}\cdot\text{s})(3.00 \times 10^{8}\ \text{m/s})}{4.90 \times 10^{-18}\ \text{J}} = 4.06 \times 10^{-8}\ \text{m} = 40.6 \times 10^{-9}\ \text{m} = 40.6\ \text{nm}$$

(d) Ru is directly below Fe in the periodic table and the two metals have similar electron configurations. The electron removed from Ru to go from Ru^{2+} to Ru^{3+} is a 4d electron. The electron with the higher principal quantum number, $n = 4$, is farther from the nucleus, less tightly held, and requires less energy to remove.

7 Covalent Bonding and Electron-Dot Structures

7.1 (a) $SiCl_4$

chlorine	EN = 3.0	
silicon	EN = 1.8	
	ΔEN = 1.2	The Si–Cl bond is polar covalent.

(b) CsBr

bromine	EN = 2.8	
cesium	EN = 0.7	
	ΔEN = 2.1	The Cs^+Br^- bond is ionic.

(c) $FeBr_3$

bromine	EN = 2.8	
iron	EN = 1.8	
	ΔEN = 1.0	The Fe–Br bond is polar covalent.

(d) CH_4

carbon	EN = 2.5	
hydrogen	EN = 2.1	
	ΔEN = 0.4	The C–H bond is polar covalent.

7.2 H is positively polarized (blue). O is negatively polarized (red). This is consistent with the electronegativity values for O (3.5) and H (2.1). The more negatively polarized atom should be the one with the larger electronegativity.

7.3 (a) H : S̈ : H

(b)
```
      H
 :C̈l:C̈:C̈l:
     :C̈l:
```

7.4 (a) OF_2 (b) $SiCl_4$

7.5 (a)
```
    H
    |
 H—C—F̈:
    |
   :F:
```

(b)
```
 :C̈l—Al—C̈l:
      |
    :C̈l:
```

(c)
```
 [     :Ö:       ]⁻
 [      |        ]
 [ :Ö—Cl—Ö:      ]
 [      |        ]
 [     :Ö:       ]
```

(d)
```
      :C̈l:
        |
 :C̈l—P—C̈l:
      /  \
  :C̈l    C̈l:
```

(e)
```
      :Ö:
 :F̈    |    F̈:
    \  |  /
      Xe
    /     \
 :F̈         F̈:
```

(e)
```
 [    H    ]⁺
 [    |    ]
 [ H—N—H   ]
 [    |    ]
 [    H    ]
```

7.6 (a) is incorrect, the least electronegative atom (P) should be placed in the center.
(b) correct
(c) is incorrect, oxygen does not have a complete octet.

7.7 (a) :C≡O: (b) H—C≡N: (c) $[\,:\ddot{O}{=}C(—\ddot{O}:)—\ddot{O}:\,]^{2-}$ (d) :Ö=C=Ö:

7.8 (a) correct
(b) is incorrect, the electron-dot structure has 34 electrons, but there are only 32 valence electrons in phosphate.
(c) correct
(d) is incorrect, the oxygen with a double bond does not have a complete octet and there are only 30 electrons in the electron-dot structure while there are 32 valence electrons.

Electron-dot structures are a model for bonding and it is possible to draw several structures that meet the criteria for a valid structure. If more than one electron-dot structure can be drawn, the molecule exhibits resonance.

7.9 (a) :Ö—Cl—Ö: (b) :Ȯ—H (c) $H_3C—CH_2\cdot$

7.10 O_2^- is a radical and the presence of an unpaired electron leads to a highly reactive species.

7.11 (a) $H_3C—\ddot{N}H_2$ (b) $H_2C{=}CH_2$ (c) $H_3C—CH_2—CH_3$

(d) H—Ö—Ö—H (e) $H_2\ddot{N}—\ddot{N}H_2$

7.12 $H_3C—CH_2—\ddot{O}—H$ and $H_3C—\ddot{O}—CH_3$

7.13 Molecular formula: $C_4H_5N_3O$

(ring structure: $H_2\ddot{N}$—C=$\ddot{N}$—C(=Ö:)—$\ddot{N}$—H, C=C with H, H)

7.14 (a)

```
        H   H   H   H                     H
        |   |   |   |   ..                |
H—C≡C—C=C—C=C—N—H                 :N—H
                        |                 |
                        H            H  C  H
                                      C   C
                                      ‖   |
                                      C   C
                                    H   C   H
                                        |
                                        H
```

(b)

```
  H   H   Ö:                  H   Ö:  H
  |   |   ‖                   |   ‖   |
H—N—C—C—Ö—H             H—Ö—C—C—N—H
  ..  |       ..              ..  |       ..
      H                           H
```

7.15 (a) N=N=O:

(b) N=N=O: ⟷ :N—N≡O:

N=N=O: ⟷ :N≡N—O:

7.16 (a) :O—S=O: ⟷ :O=S—O:

(b) [O=C(—O:)(—O:)]$^{2-}$ ⟷ [:O—C(=O)(—O:)]$^{2-}$ ⟷ [:O—C(—O:)(=O)]$^{2-}$

(c) :F:—B(—F:)(—F:) ⟷ :F:—B(—F:)(=F:) ⟷ :F:—B(=F:)(—F:) ⟷ :F=B(—F:)(—F:)

7.17 (a)

```
        H                 :O:
        |        H   H    ‖
        N         \ /     C     H
H—C   /   \ ..      C   /  \ .. /
    ‖        C   /   C      O
    N—C           / \      ..
   ..   \     :N     H
         H    / \
             H   H
```

This is a valid resonance structure.

(b)

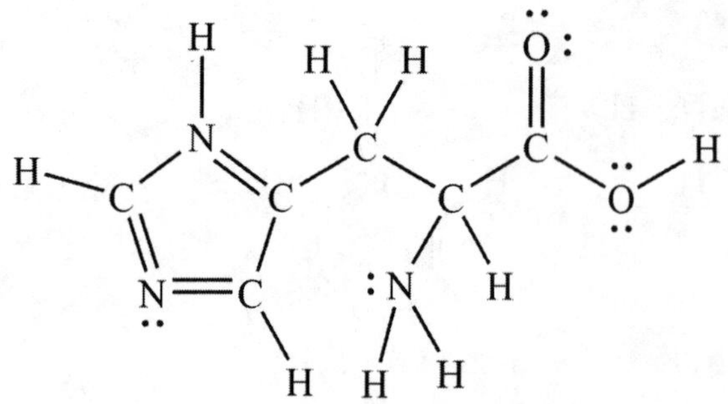

This is not a valid resonance structure. The N with two double bonds has 10 electrons, not 8.

7.18 (a)

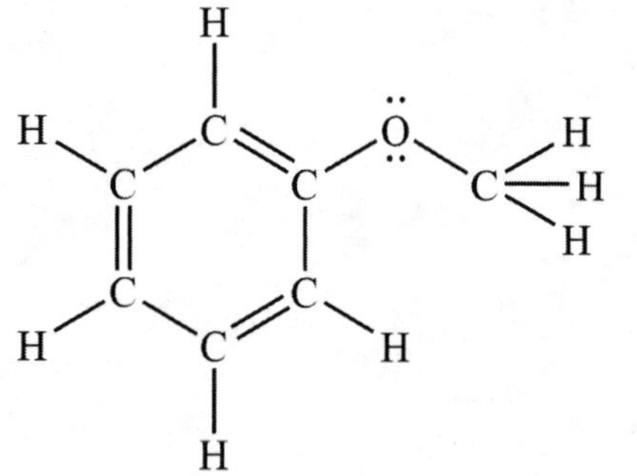

(b)

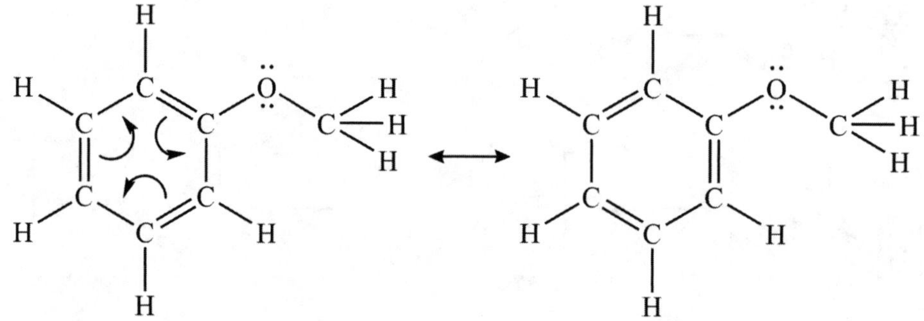

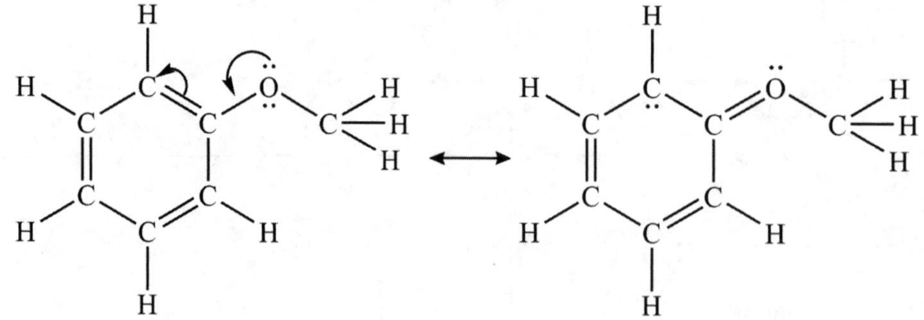

7.19 (a) $[:\ddot{N}=C=\ddot{O}:]^-$

For nitrogen:	Isolated nitrogen valence electrons	5
	Bound nitrogen bonding electrons	4
	Bound nitrogen nonbonding electrons	4
	Formal charge = 5 – ½(4) – 4 = –1	
For carbon:	Isolated carbon valence electrons	4
	Bound carbon bonding electrons	8
	Bound carbon nonbonding electrons	0
	Formal charge = 4 – ½ (8) – 0 = 0	
For oxygen:	Isolated oxygen valence electrons	6
	Bound oxygen bonding electrons	4
	Bound oxygen nonbonding electrons	4
	Formal charge = 6 – ½(4) – 4 = 0	

(b) $:\underset{..}{\ddot{O}}-\ddot{O}=\ddot{O}:$

For left oxygen:	Isolated oxygen valence electrons	6
	Bound oxygen bonding electrons	2
	Bound oxygen nonbonding electrons	6
	Formal charge = 6 – ½(2) – 6 = –1	
For central oxygen:	Isolated oxygen valence electrons	6
	Bound oxygen bonding electrons	6
	Bound oxygen nonbonding electrons	2
	Formal charge = 6 – ½(6) – 2 = +1	
For right oxygen:	Isolated oxygen valence electrons	6
	Bound oxygen bonding electrons	4
	Bound oxygen nonbonding electrons	4
	Formal charge = 6 – ½(4) – 4 = 0	

7.20 (a) $[\ddot{N}=C=\ddot{O}]^- \longleftrightarrow [:N\equiv C-\ddot{O}:]^-$

$[\ddot{N}=C=\ddot{O}]^- \longleftrightarrow [:\ddot{N}-C\equiv O:]^-$

(b) $[\ddot{N}=C=\ddot{O}]^-$

For nitrogen:	Isolated nitrogen valence electrons	5
	Bound nitrogen bonding electrons	4
	Bound nitrogen nonbonding electrons	4
	Formal charge = 5 – ½(4) – 4 = –1	

For carbon:	Isolated carbon valence electrons	4
	Bound carbon bonding electrons	8
	Bound carbon nonbonding electrons	0
	Formal charge = 4 – ½(8) – 0 = 0	
For oxygen:	Isolated oxygen valence electrons	6
	Bound oxygen bonding electrons	4
	Bound oxygen nonbonding electrons	4
	Formal charge = 6 – ½(4) – 4 = 0	

$\left[:N\equiv C-\ddot{\underset{\cdot\cdot}{O}}:\right]^-$

For nitrogen:	Isolated nitrogen valence electrons	5
	Bound nitrogen bonding electrons	6
	Bound nitrogen nonbonding electrons	2
	Formal charge = 5 – ½(6) – 2 = 0	
For carbon:	Isolated carbon valence electrons	4
	Bound carbon bonding electrons	8
	Bound carbon nonbonding electrons	0
	Formal charge = 4 – ½(8) – 0 = 0	
For oxygen:	Isolated oxygen valence electrons	6
	Bound oxygen bonding electrons	2
	Bound oxygen nonbonding electrons	6
	Formal charge = 6 – ½(2) – 6 = –1	

$\left[:\ddot{\underset{\cdot\cdot}{N}}-C\equiv O:\right]^-$

For nitrogen:	Isolated nitrogen valence electrons	5
	Bound nitrogen bonding electrons	2
	Bound nitrogen nonbonding electrons	6
	Formal charge = 5 – ½(2) – 6 = –2	
For carbon:	Isolated carbon valence electrons	4
	Bound carbon bonding electrons	8
	Bound carbon nonbonding electrons	0
	Formal charge = 4 – ½(8) – 0 = 0	
For oxygen:	Isolated oxygen valence electrons	6
	Bound oxygen bonding electrons	6
	Bound oxygen nonbonding electrons	2
	Formal charge = 6 – ½(6) – 2 = +1	

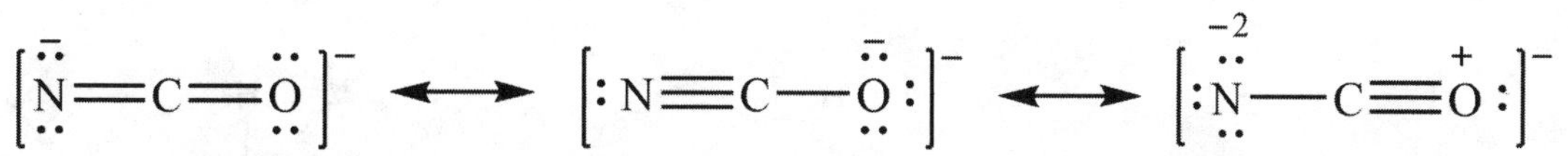

$[\overset{..}{\underset{..}{\bar{N}}}=C=\overset{..}{\underset{..}{O}}]^- \longleftrightarrow [:N\equiv C-\overset{..}{\underset{..}{\bar{O}}}:]^- \longleftrightarrow [:\overset{-2}{\overset{..}{\underset{..}{N}}}-C\equiv\overset{+}{O}:]^-$

The first two structures make the largest contribution to the resonance hybrid because the −1 formal charge is on either of the electronegative N or O. The third structure has a +1 formal charge on O and does not significantly contribute to the resonance hybrid.
(c) Carbon–nitrogen because of the triple bond contribution to the resonance hybrid.

7.21

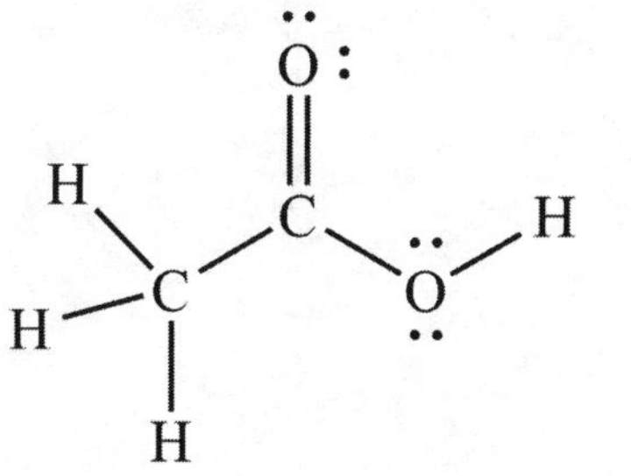

All atoms in this structure have 0 formal charge.

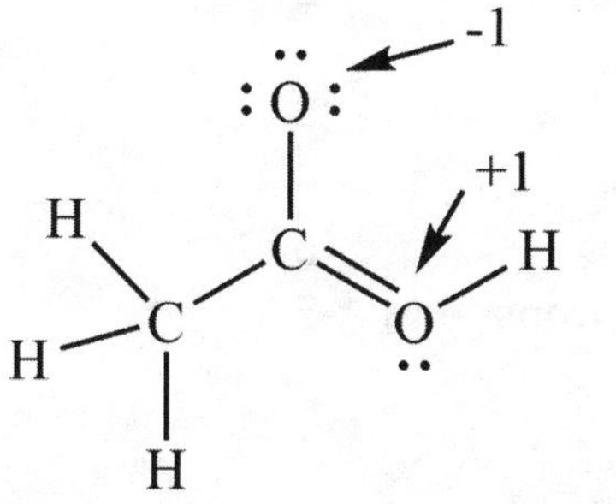

For top oxygen:	Isolated oxygen valence electrons	6
	Bound oxygen bonding electrons	2
	Bound oxygen nonbonding electrons	6
	Formal charge = 6 – ½(2) – 6 = –1	

For right oxygen:	Isolated oxygen valence electrons	6
	Bound oxygen bonding electrons	6
	Bound oxygen nonbonding electrons	2
	Formal charge = 6 – ½(6) – 2 = +1	

All the other atoms in this structure have 0 formal charge.

The structure without formal charges makes a larger contribution to the resonance hybrid because energy is required to separate + and – charges. Thus, the actual electronic structure of acetic acid is closer to that of the more favorable, lower energy structure.

7.22

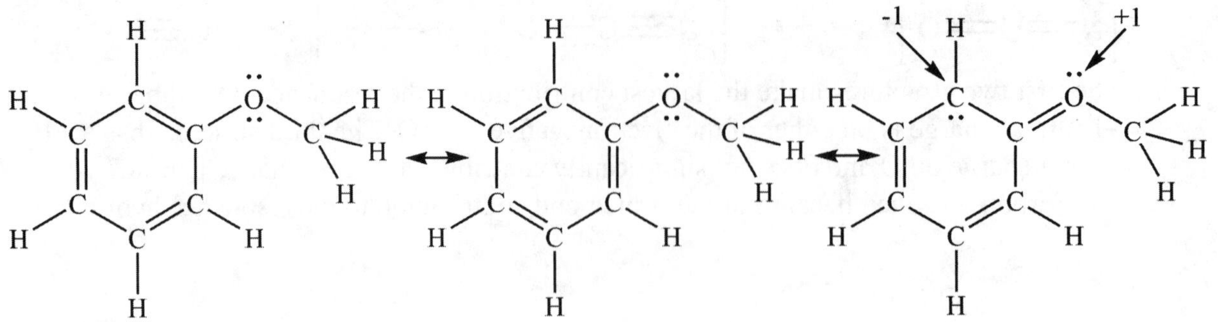

All atoms in structures 1 and 2 have 0 formal charge.

In structure 3:

For carbon:	Isolated carbon valence electrons	4
	Bound carbon bonding electrons	6
	Bound carbon nonbonding electrons	2
	Formal charge = 4 – ½(6) – 2 = –1	
For oxygen:	Isolated oxygen valence electrons	6
	Bound oxygen bonding electrons	6
	Bound oxygen nonbonding electrons	2
	Formal charge = 6 – ½(6) – 2 = +1	

All the other atoms in structure 3 have 0 formal charge.

The structures 1 and 2 without formal charges make the larger contribution to the resonance hybrid because energy is required to separate + and – charges.

7.23 ΔEN(P=O) = 1.4 and ΔEN(P=S) = 0.4. Both bonds are polar covalent, but the phosphorus-oxygen bond is more polar.

7.24 For the reaction between an organophosphate insecticide and the enzyme to occur, the phosphorus atom must bear a positive charge. Greater positive charge leads to increased rate of reaction and increased toxicity of the insecticide. The more electronegative the X group, the more positive the phosphorus.
(a) Cl (b) CF_3

7.25 Phosphorus is in the third row of the periodic table and can utilize d orbitals to hold extra electrons, and therefore form more than four bonds.

7.26 $C_{11}H_{19}N_2PSO_3$

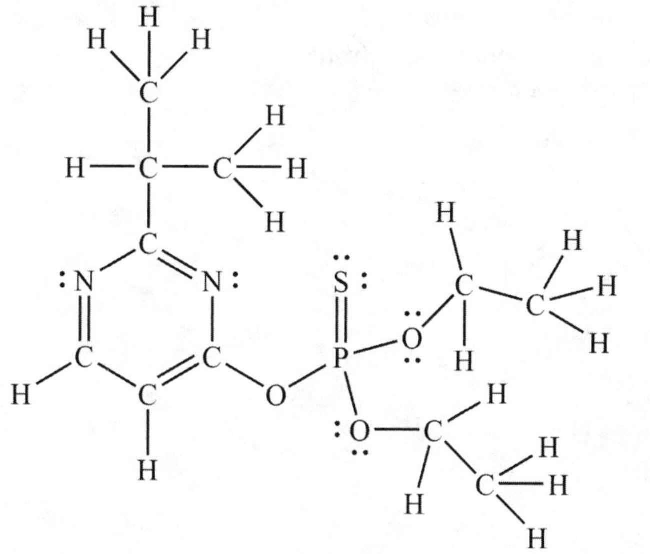

7.27

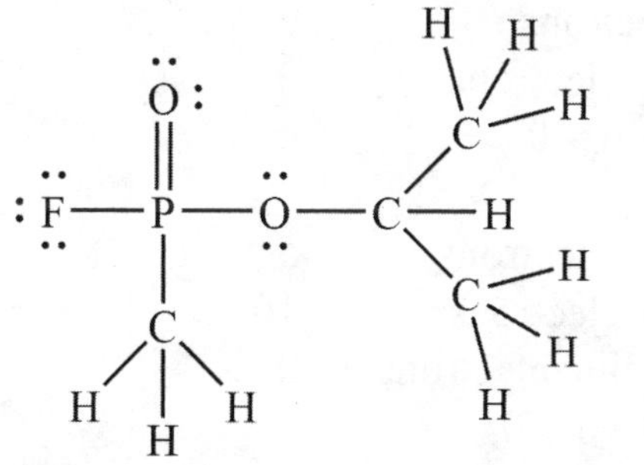

For phosphorus:	Isolated phosphorus valence electrons	5
	Bound phosphorus bonding electrons	10
	Bound phosphorus nonbonding electrons	0
	Formal charge = 5 – ½(10) – 0 = 0	
For top oxygen:	Isolated oxygen valence electrons	6
	Bound oxygen bonding electrons	4
	Bound oxygen nonbonding electrons	4
	Formal charge = 6 – ½(4) – 4 = 0	
For right oxygen:	Isolated oxygen valence electrons	6
	Bound oxygen bonding electrons	4
	Bound oxygen nonbonding electrons	4
	Formal charge = 6 – ½(4) – 4 = 0	
For fluorine:	Isolated fluorine valence electrons	7
	Bound fluorine bonding electrons	2
	Bound fluorine nonbonding electrons	6
	Formal charge = 7 – ½(2) – 6 = 0	

For carbon:	Isolated carbon valence electrons	4
	Bound carbon bonding electrons	8
	Bound carbon nonbonding electrons	0
	Formal charge = 4 – ½(8) – 0 = 0	

7.28 (a)

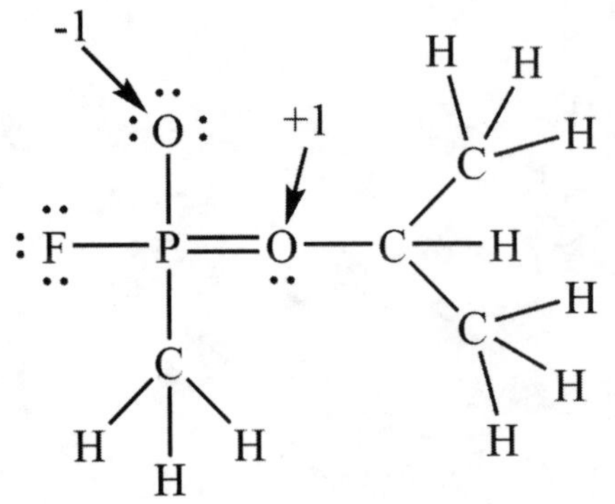

For fluorine:	Isolated fluorine valence electrons	7
	Bound fluorine bonding electrons	2
	Bound fluorine nonbonding electrons	6
	Formal charge = 7 – ½(2) – 6 = 0	

For phosphorus:	Isolated phosphorus valence electrons	5
	Bound phosphorus bonding electrons	10
	Bound phosphorus nonbonding electrons	0
	Formal charge = 5 – ½(10) – 0 = 0	

For top oxygen:	Isolated oxygen valence electrons	6
	Bound oxygen bonding electrons	2
	Bound oxygen nonbonding electrons	6
	Formal charge = 6 – ½(2) – 6 = –1	

For right oxygen:	Isolated oxygen valence electrons	6
	Bound oxygen bonding electrons	6
	Bound oxygen nonbonding electrons	2
	Formal charge = 6 – ½(6) – 2 = +1	

(b)

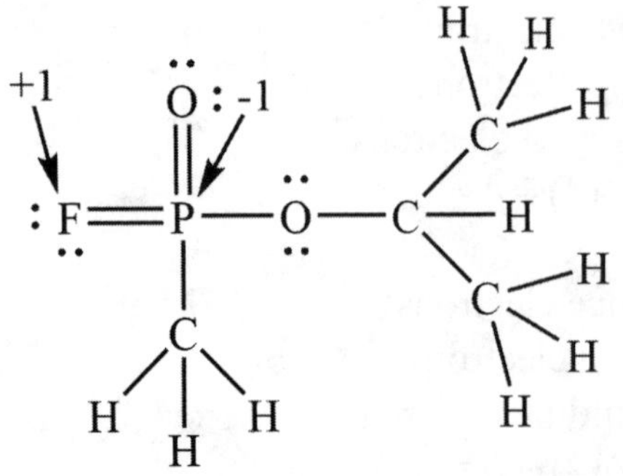

For fluorine:	Isolated fluorine valence electrons	7
	Bound fluorine bonding electrons	4
	Bound fluorine nonbonding electrons	4
	Formal charge = 7 – ½(4) – 4 = +1	
For phosphorus:	Isolated phosphorus valence electrons	5
	Bound phosphorus bonding electrons	12
	Bound phosphorus nonbonding electrons	0
	Formal charge = 5 – ½(12) – 0 = –1	
For top oxygen:	Isolated oxygen valence electrons	6
	Bound oxygen bonding electrons	4
	Bound oxygen nonbonding electrons	4
	Formal charge = 6 – ½(4) – 4 = 0	
For right oxygen:	Isolated oxygen valence electrons	6
	Bound oxygen bonding electrons	4
	Bound oxygen nonbonding electrons	4
	Formal charge = 6 – ½(4) – 4 = 0	

(c)

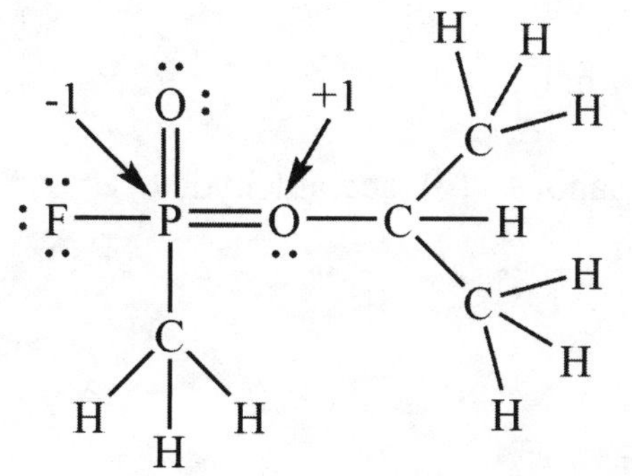

For fluorine:	Isolated fluorine valence electrons	7
	Bound fluorine bonding electrons	2
	Bound fluorine nonbonding electrons	6
	Formal charge = 7 – ½(2) – 6 = 0	
For phosphorus:	Isolated phosphorus valence electrons	5
	Bound phosphorus bonding electrons	12
	Bound phosphorus nonbonding electrons	0
	Formal charge = 5 – ½(12) – 0 = –1	
For top oxygen:	Isolated oxygen valence electrons	6
	Bound oxygen bonding electrons	4
	Bound oxygen nonbonding electrons	4
	Formal charge = 6 – ½(4) – 4 = 0	

For right oxygen:	Isolated oxygen valence electrons	6
	Bound oxygen bonding electrons	6
	Bound oxygen nonbonding electrons	2
	Formal charge = 6 – ½(6) – 2 = +1	

(d)

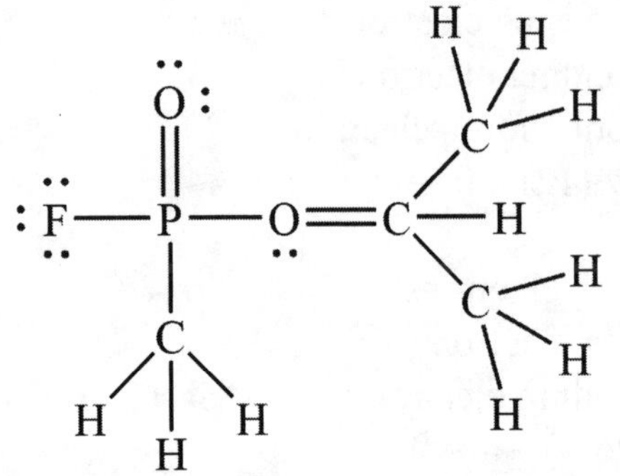

This structure is not valid because one carbon has five bonds.

The resonance structure in problem 7.27 is the major contributor to the resonance hybrid because formal charges on all atoms are zero.

Conceptual Problems

7.30 C–D is the stronger bond. A–B is the longer bond.

7.32 (a) fluoroethane (b) ethane (c) ethanol (d) acetaldehyde

7.34 (a) $C_8H_9NO_2$

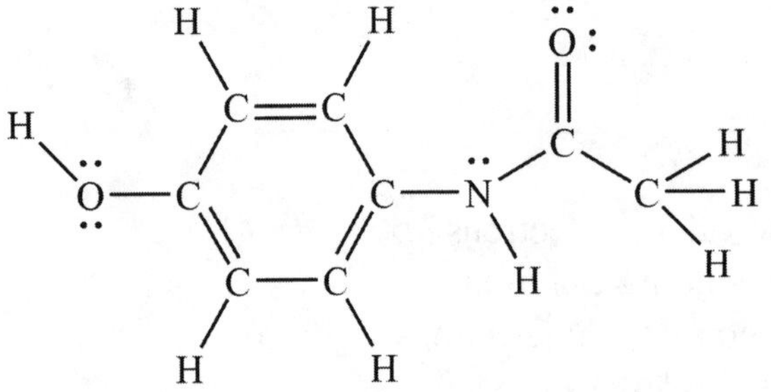

7.36

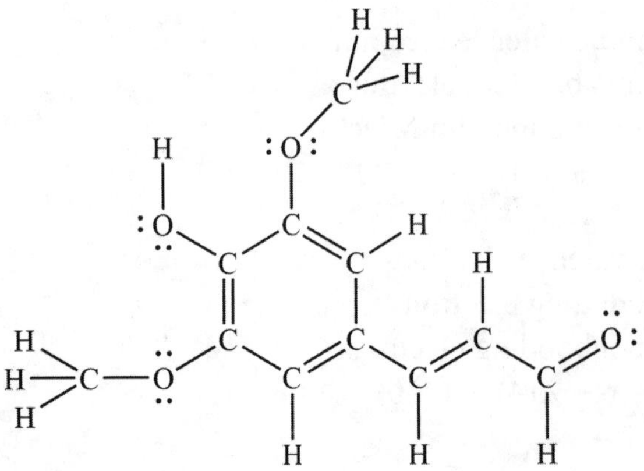

Section Problems

Strengths of Covalent Bonds and Electronegativity (Sections 7.1–7.4)

7.38 (a) ionic (b) nonpolar covalent (c) covalent

7.40 Electronegativity increases from left to right across a period and decreases down a group.

7.42 K < Li < Mg < Pb < C < Br

7.44 (a) HF

fluorine	EN = 4.0	
hydrogen	EN = 2.1	
	ΔEN = 1.9	HF is polar covalent.

(b) HI

iodine	EN = 2.5	
hydrogen	EN = 2.1	
	ΔEN = 0.4	HI is polar covalent.

(c) $PdCl_2$

chlorine	EN = 3.0	
palladium	EN = 2.2	
	ΔEN = 0.8	$PdCl_2$ is polar covalent.

(d) BBr_3

bromine	EN = 2.8	
boron	EN = 2.0	
	ΔEN = 0.8	BBr_3 is polar covalent.

(e) NaOH — $Na^+ – OH^-$ is ionic

OH^-

oxygen	EN = 3.5	
hydrogen	EN = 2.1	
	ΔEN = 1.4	OH^- is polar covalent.

(f) CH_3Li

lithium	EN = 1.0	
carbon	EN = 2.5	
	ΔEN = 1.5	CH_3Li is polar covalent.

7.46 (a) $\overset{\delta-}{C} - \overset{\delta+}{H}$ $\overset{\delta+}{C} - \overset{\delta-}{Cl}$ (b) $\overset{\delta-}{Si} - \overset{\delta+}{Li}$ $\overset{\delta+}{Si} - \overset{\delta-}{Cl}$

(c) N – Cl $\overset{\delta-}{N} - \overset{\delta+}{Mg}$

7.48 (a) MgO, $BaCl_2$, LiBr (b) P_4 (c) $CdBr_2$, BrF_3, NF_3, $POCl_3$

7.50 (a) CCl_4

chlorine	EN = 3.0	
carbon	EN = 2.5	
	ΔEN = 0.5	

(b) $BaCl_2$ chlorine EN = 3.0
barium EN = 0.9
ΔEN = 2.1

(c) $TiCl_3$ chlorine EN = 3.0
titanium EN = 1.5
ΔEN = 1.5

(d) Cl_2O oxygen EN = 3.5
chlorine EN = 3.0
ΔEN = 0.5

Increasing ionic character: $CCl_4 \sim ClO_2 < TiCl_3 < BaCl_2$

7.52 (a) $MgBr_2$ (b) PBr_3

7.54 H—N̈=N̈—H

$H_2\ddot{N}{-}\ddot{N}H_2$ (each N bonded to two H)

N_2H_2 has the stronger N–N bond because of the higher bond order.

7.56 C–F 450 kJ/mol ΔEN = EN(F) – EN(C) = 4.0 – 2.5 = 1.5
N–F 270 kJ/mol ΔEN = EN(F) – EN(N) = 4.0 – 3.0 = 1.0
O–F 180 kJ/mol ΔEN = EN(F) – EN(O) = 4.0 – 3.5 = 0.5
F–F 159 kJ/mol ΔEN = EN(F) – EN(F) = 4.0 – 4.0 = 0
In general, increased bond polarity leads to increased bond strength.

7.58 N–H ΔEN = EN(N) – EN(H) = 3.0 – 2.1 = 0.9
O–H ΔEN = EN(O) – EN(H) = 3.5 – 2.1 = 1.4
S–H ΔEN = EN(S) – EN(H) = 2.5 – 2.1 = 0.4
In general, increased bond polarity leads to increased bond strength. The most polar bond is the O–H bond and should be the strongest of the three.

7.60 (a) Phosphorus trichloride (b) Dinitrogen trioxide (c) Tetraphosphorus heptoxide
(d) Bromine trifluoride (e) Nitrogen trichloride (f) Tetraphosphorus hexoxide;
(g) Disulfur difluoride (h) Selenium dioxide

7.62 (b) and (c) are ionic compounds; (a) is a covalent compound and is most likely a gas at room temperature.

Electron-Dot Structures and Resonance (Sections 7.5–7.9)

7.64 The octet rule states that main-group elements tend to react so that they attain a noble gas electron configuration with filled s and p sublevels (8 electrons) in their valence electron shells. The transition metals are characterized by partially filled d orbitals that can be used to expand their valence shell beyond the normal octet of electrons.

7.66 (a) 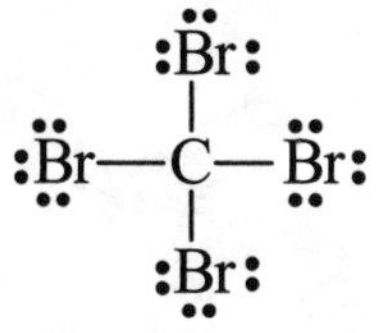 :Br: / :Br—C—Br: / :Br: (b) :Cl—N—Cl: / :Cl: (c) H H / H—C—C—Cl: / H H (d) 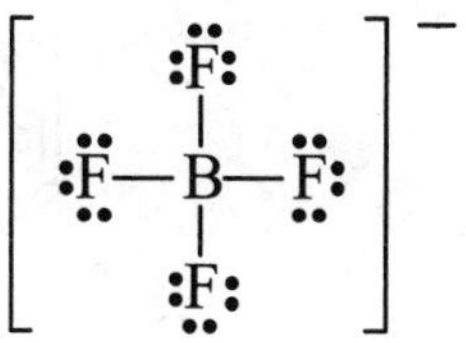 [:F: / :F—B—F: / :F:]$^-$

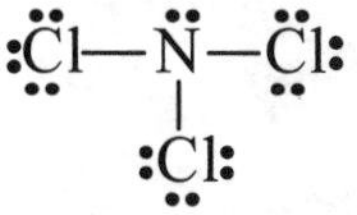

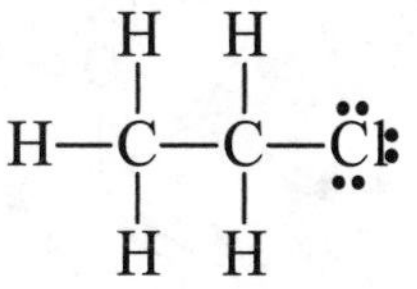

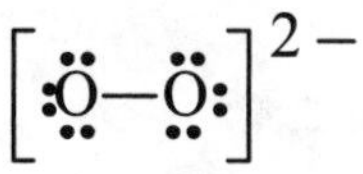

(e) $[:\ddot{O}—\ddot{O}:]^{2-}$ (f) $[:N\equiv O:]^+$

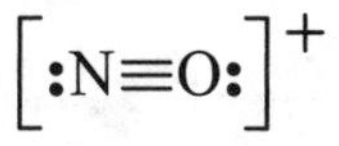

7.68 (c) is the correct electron-dot for XeF_5^+ because it accounts for the required number of bonding (10) and lone pair (32) electrons.

7.70

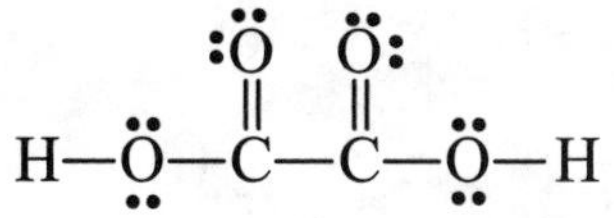

:Ö Ö:
‖ ‖
H—Ö—C—C—Ö—H

7.72 (a) The anion has 32 valence electrons. Each Cl has seven valence electrons (28 total). The minus one charge on the anion accounts for one valence electron. This leaves three valence electrons for X. X is Al.
(b) The cation has eight valence electrons. Each H has one valence electron (4 total). X is left with four valence electrons. Since this is a cation, one valence electron was removed from X. X has five valence electrons. X is P.

7.74 (a)

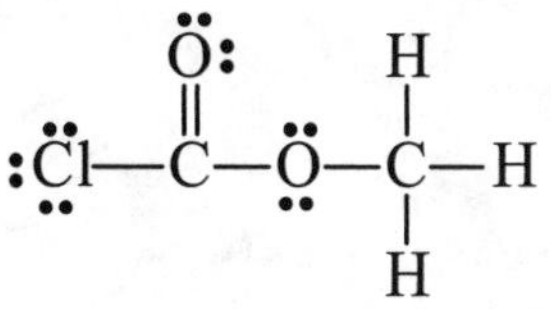

Ö: H
‖ |
:Cl—C—Ö—C—H
|
H

(b)

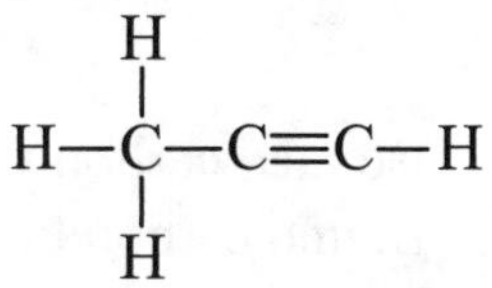

H
|
H—C—C≡C—H
|
H

7.76

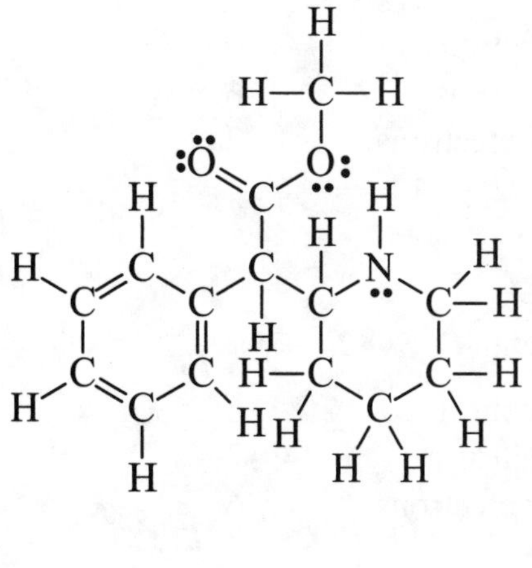

7.78 (a) 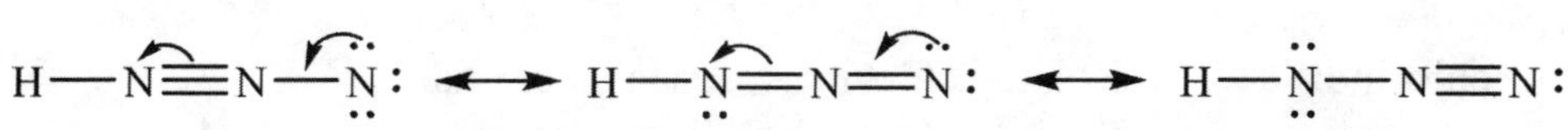 H—N≡N—N: ⟷ H—N=N=N: ⟷ H—N—N≡N:

(b)

$$\begin{matrix} & :\ddot{O} & \\ & \| & \\ & S & \\ :\ddot{O} & & \ddot{O}: \end{matrix} \leftrightarrow \begin{matrix} & :\ddot{O}: & \\ & | & \\ & S & \\ :O & & \ddot{O}: \end{matrix} \leftrightarrow \begin{matrix} & :\ddot{O}: & \\ & | & \\ & S & \\ :\ddot{O} & & O: \end{matrix}$$

(c) $[:N\equiv C-\ddot{S}:]^- \leftrightarrow [:\ddot{N}=C=\ddot{S}:]^- \leftrightarrow [:\ddot{N}-C\equiv S:]^-$

7.80 (a) yes (b) no (c) yes (d) yes

7.82

$$\begin{matrix} & :\ddot{O}-H & \\ & | & \\ H-C & C & C-H \\ H-C & C & C-H \\ & | & \\ & H & \end{matrix} \leftrightarrow \begin{matrix} & :\ddot{O}-H & \\ & | & \\ H-C & C & C-H \\ H-C & C & C-H \\ & | & \\ & H & \end{matrix} \leftrightarrow \begin{matrix} & \ddot{O}-H & \\ & \| & \\ H-\ddot{C} & C & C-H \\ H-C & C & C-H \\ & | & \\ & H & \end{matrix}$$

Formal Charges (Section 7.10)

7.84 $:C\equiv O:$

For carbon:	Isolated carbon valence electrons	4
	Bound carbon bonding electrons	6
	Bound carbon nonbonding electrons	2
	Formal charge = 4 – ½(6) – 2 = –1	
For oxygen:	Isolated oxygen valence electrons	6
	Bound oxygen bonding electrons	6
	Bound oxygen nonbonding electrons	2
	Formal charge = 6 – ½(6) – 2 = +1	

7.86 $[:\ddot{O}-\ddot{Cl}-\ddot{O}:]^-$

For both oxygens:	Isolated oxygen valence electrons	6
	Bound oxygen bonding electrons	2
	Bound oxygen nonbonding electrons	6
	Formal charge = 6 – ½(2) – 6 = –1	
For chlorine:	Isolated chlorine valence electrons	7
	Bound chlorine bonding electrons	4
	Bound chlorine nonbonding electrons	4
	Formal charge = 7 – ½(4) – 4 = +1	

$$\left[:\ddot{\underset{\cdot\cdot}{O}}-\ddot{\underset{\cdot\cdot}{Cl}}=\ddot{\underset{\cdot\cdot}{O}}\right]^-$$

For left oxygen:	Isolated oxygen valence electrons	6
	Bound oxygen bonding electrons	2
	Bound oxygen nonbonding electrons	6
	Formal charge = 6 – ½(2) – 6 = –1	
For right oxygen:	Isolated oxygen valence electrons	6
	Bound oxygen bonding electrons	4
	Bound oxygen nonbonding electrons	4
	Formal charge = 6 – ½(4) – 4 = 0	
For chlorine:	Isolated chlorine valence electrons	7
	Bound chlorine bonding electrons	6
	Bound chlorine nonbonding electrons	4
	Formal charge = 7 – ½(6) – 4 = 0	

7.88 (a)

$$H_2C=N=\ddot{N}:$$

For hydrogen:	Isolated hydrogen valence electrons	1
	Bound hydrogen bonding electrons	2
	Bound hydrogen nonbonding electrons	0
	Formal charge = 1 – ½(2) – 0 = 0	
For nitrogen: (central)	Isolated nitrogen valence electrons	5
	Bound nitrogen bonding electrons	8
	Bound nitrogen nonbonding electrons	0
	Formal charge = 5 – ½(8) – 0 = +1	
For nitrogen: (terminal)	Isolated nitrogen valence electrons	5
	Bound nitrogen bonding electrons	4
	Bound nitrogen nonbonding electrons	4
	Formal charge = 5 – ½(4) – 4 = –1	
For carbon:	Isolated carbon valence electrons	4
	Bound carbon bonding electrons	8
	Bound carbon nonbonding electrons	0
	Formal charge = 4 – ½(8) – 0 = 0	

(b) H₂C–N̈=N̈: (H, H bonded to C; C–N̈=N̈:)

For hydrogen:	Isolated hydrogen valence electrons	1
	Bound hydrogen bonding electrons	2
	Bound hydrogen nonbonding electrons	0
	Formal charge = 1 – ½(2) – 0 = 0	
For nitrogen: (central)	Isolated nitrogen valence electrons	5
	Bound nitrogen bonding electrons	6
	Bound nitrogen nonbonding electrons	2
	Formal charge = 5 – ½(6) – 2 = 0	
For nitrogen: (terminal)	Isolated nitrogen valence electrons	5
	Bound nitrogen bonding electrons	4
	Bound nitrogen nonbonding electrons	4
	Formal charge = 5 – ½(4) – 4 = –1	
For carbon:	Isolated carbon valence electrons	4
	Bound carbon bonding electrons	6
	Bound carbon nonbonding electrons	0
	Formal charge = 4 – ½(6) – 0 = +1	

Structure (a) is more important because of the octet of electrons around carbon.

7.90 :C̈l–C(Cl)(Cl)–N⁺(=Ö:)–Ö:⁻ ⟷ :C̈l–C(Cl)(Cl)–N⁺(–Ö:⁻)=Ö:

For chlorine:	Isolated chlorine valence electrons	7
	Bound chlorine bonding electrons	2
	Bound chlorine nonbonding electrons	6
	Formal charge = 7 – ½(2) – 6 = 0	
For carbon:	Isolated carbon valence electrons	4
	Bound carbon bonding electrons	8
	Bound carbon nonbonding electrons	0
	Formal charge = 4 – ½(8) – 0 = 0	
For nitrogen:	Isolated nitrogen valence electrons	5
	Bound nitrogen bonding electrons	8
	Bound nitrogen nonbonding electrons	0
	Formal charge = 5 – ½(8) – 0 = +1	

For oxygen: (double bonded)	Isolated oxygen valence electrons	6
	Bound oxygen bonding electrons	4
	Bound oxygen nonbonding electrons	4
	Formal charge = 6 – ½(4) – 4 = 0	

For oxygen: (single bonded)	Isolated oxygen valence electrons	6
	Bound oxygen bonding electrons	2
	Bound oxygen nonbonding electrons	6
	Formal charge = 6 – ½(2) – 6 = –1	

7.92

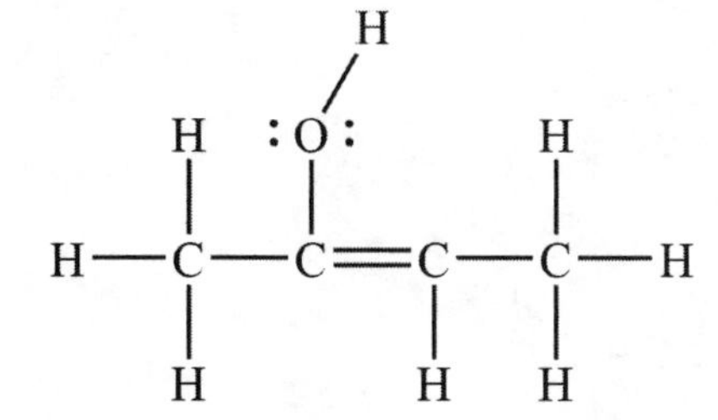

All atoms have 0 formal charge.

```
 +1    H
   \  /
      -1
H  :O     /    H
   ||    ..    |
H—C—C—C—C—H
  |      |  |
  H      H  H
```

For oxygen:	Isolated oxygen valence electrons	6
	Bound oxygen bonding electrons	6
	Bound oxygen nonbonding electrons	2
	Formal charge = 6 – ½(6) – 2 = +1	

For carbon:	Isolated carbon valence electrons	4
	Bound carbon bonding electrons	6
	Bound carbon nonbonding electrons	2
	Formal charge = 4 – ½(6) – 2 = –1	

All other atoms have 0 formal charge.

The original structure is the larger contributor.

Chapter Problems

7.94 (a)

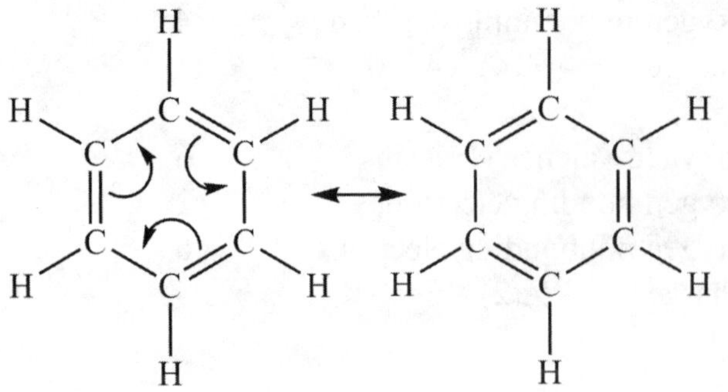

(b) ii)

7.96 (a) (b)

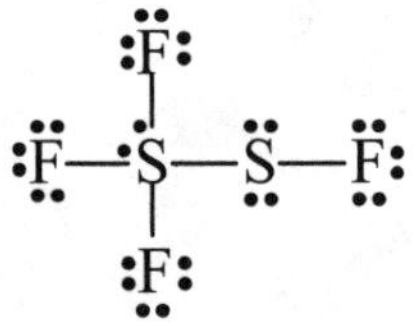

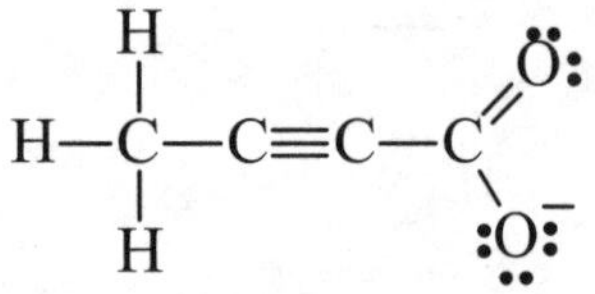

7.98

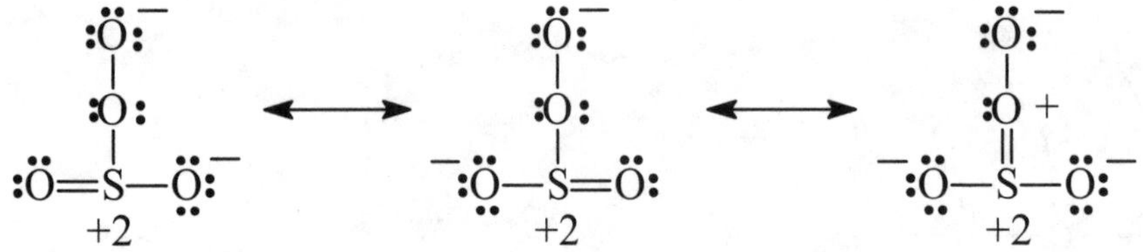

For

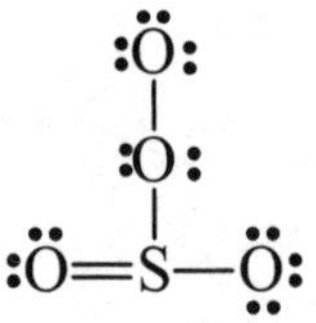

For oxygen: (top)	Isolated oxygen valence electrons	6
	Bound oxygen bonding electrons	2
	Bound oxygen nonbonding electrons	6
	Formal charge = 6 – ½(2) – 6 = –1	

For oxygen: (middle)	Isolated oxygen valence electrons	6
	Bound oxygen bonding electrons	4
	Bound oxygen nonbonding electrons	4
	Formal charge = 6 – ½(4) – 4 = 0	

For oxygen: (left)	Isolated oxygen valence electrons	6
	Bound oxygen bonding electrons	4
	Bound oxygen nonbonding electrons	4
	Formal charge = 6 – ½(4) – 4 = 0	

For oxygen: (right)	Isolated oxygen valence electrons	6
	Bound oxygen bonding electrons	2
	Bound oxygen nonbonding electrons	6
	Formal charge = 6 – ½(2) – 6 = –1	
For sulfur:	Isolated sulfur valence electrons	6
	Bound sulfur bonding electrons	8
	Bound sulfur nonbonding electrons	0
	Formal charge = 6 – ½(8) – 0 = +2	

For

```
      :Ö:
       |
      :O:
       |
:Ö—S=Ö:
  ‥
```

For oxygen: (top)	Isolated oxygen valence electrons	6
	Bound oxygen bonding electrons	2
	Bound oxygen nonbonding electrons	6
	Formal charge = 6 – ½(2) – 6 = –1	
For oxygen: (middle)	Isolated oxygen valence electrons	6
	Bound oxygen bonding electrons	4
	Bound oxygen nonbonding electrons	4
	Formal charge = 6 – ½(4) – 4 = 0	
For oxygen: (left)	Isolated oxygen valence electrons	6
	Bound oxygen bonding electrons	2
	Bound oxygen nonbonding electrons	6
	Formal charge = 6 – ½(2) – 6 = –1	
For oxygen: (right)	Isolated oxygen valence electrons	6
	Bound oxygen bonding electrons	4
	Bound oxygen nonbonding electrons	4
	Formal charge = 6 – ½(4) – 4 = 0	
For sulfur:	Isolated sulfur valence electrons	6
	Bound sulfur bonding electrons	8
	Bound sulfur nonbonding electrons	0
	Formal charge = 6 – ½(8) – 0 = +2	

For

```
      :Ö:
       |
      :O
       ||
:Ö—S—Ö:
  ‥      ‥
```

For oxygen: (top)	Isolated oxygen valence electrons	6
	Bound oxygen bonding electrons	2
	Bound oxygen nonbonding electrons	6
	Formal charge = 6 – ½(2) – 6 = –1	
For oxygen: (middle)	Isolated oxygen valence electrons	6
	Bound oxygen bonding electrons	6
	Bound oxygen nonbonding electrons	2
	Formal charge = 6 – ½(6) – 2 = +1	
For oxygen: (left)	Isolated oxygen valence electrons	6
	Bound oxygen bonding electrons	2
	Bound oxygen nonbonding electrons	6
	Formal charge = 6 – ½(2) – 6 = –1	
For oxygen: (right)	Isolated oxygen valence electrons	6
	Bound oxygen bonding electrons	2
	Bound oxygen nonbonding electrons	6
	Formal charge = 6 – ½(2) – 6 = –1	
For sulfur:	Isolated sulfur valence electrons	6
	Bound sulfur bonding electrons	8
	Bound sulfur nonbonding electrons	0
	Formal charge = 6 – ½(8) – 0 = +2	

7.100

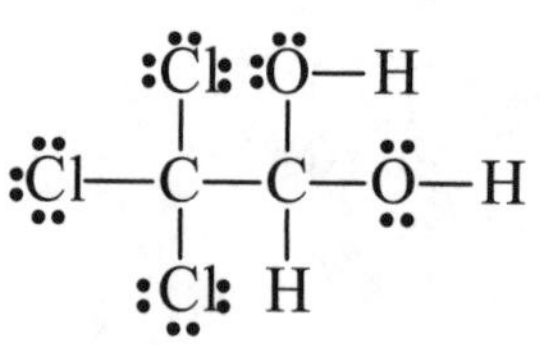

```
      :Cl: :O—H
        |    |
:Cl—C—C—O—H
        |    |
      :Cl:  H
```

7.102

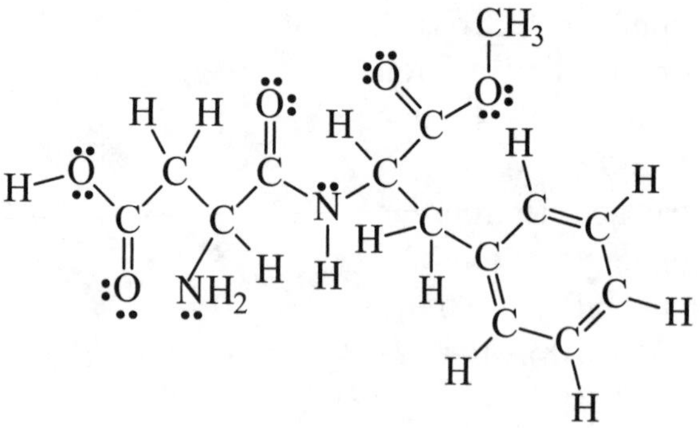

```
                          CH3
                          |
                    :O   :O:
        H  H   O:  H   \\ /
         \ /   ||   \   C
  H—O    C     C     C       H
     \  / \   / \   / \     \      H
      C     C     N    C     C==C
      ||   / \    |  H—C—C        \
     :O  NH2  H   H    |   ||      C—H
                       H   C—C  //
                          /    \
                         H      H
```

7.104 (a)

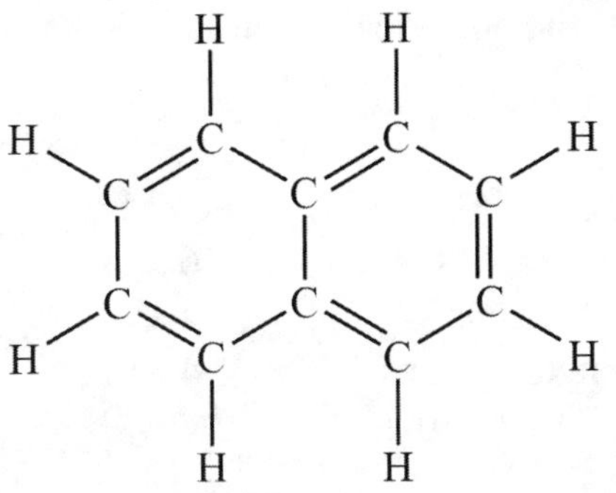

(b)

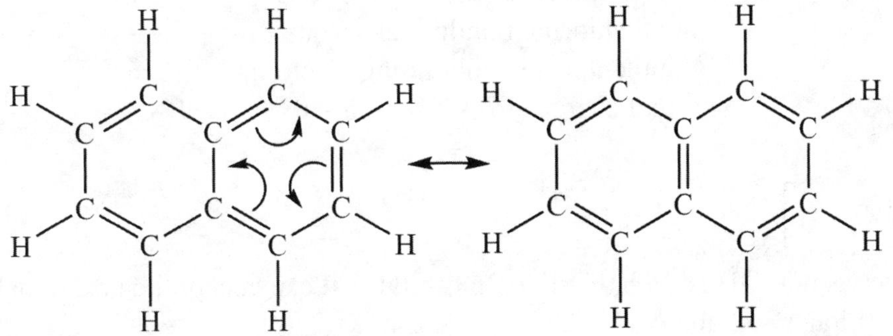

7.106 (a) 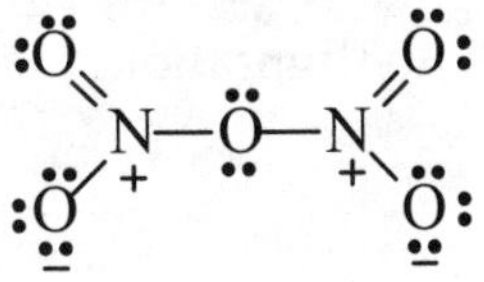 (b) 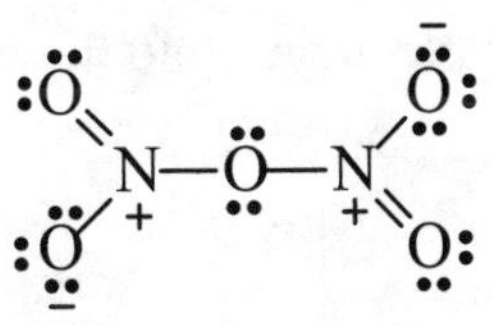 (c)

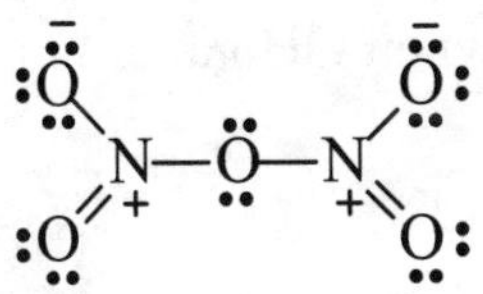

(d) 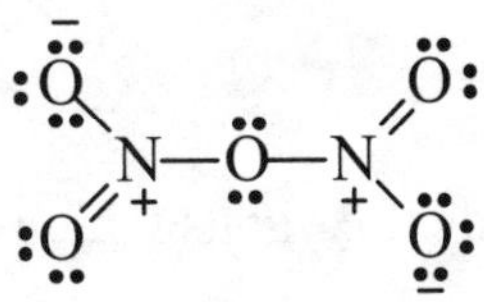 (e) 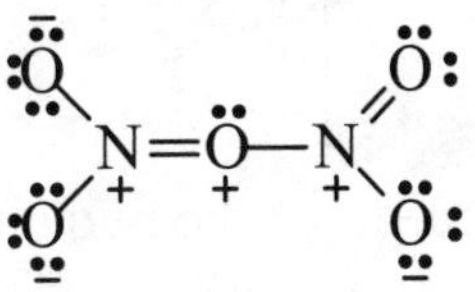 (f)

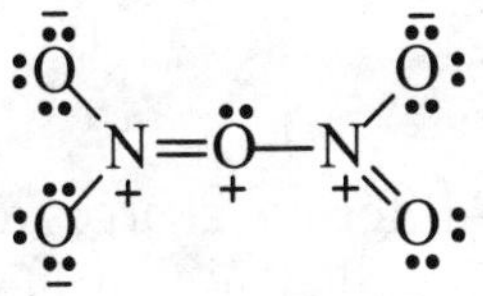

(g) 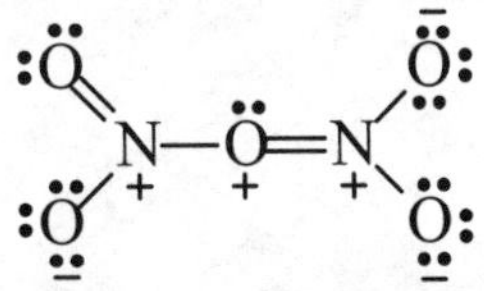 (h)

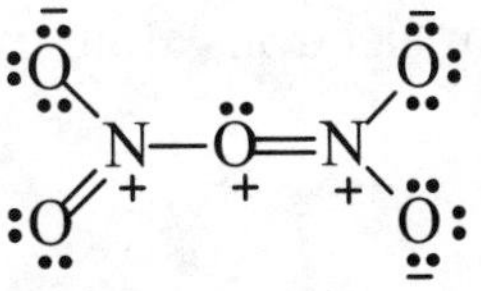

Structures (a) – (d) make more important contributions to the resonance hybrid because of only –1 and 0 formal charges on the oxygens. A +1 formal charge is unlikely.

Multiconcept Problems

7.108 (a) Assume a 100.0 g sample. From the percent composition data, a 100.0 g sample contains 47.5 g S and 52.5 g Cl.

$$47.5 \text{ g S} \times \frac{1 \text{ mol S}}{32.065 \text{ g S}} = 1.48 \text{ mol S}$$

$$52.5 \text{ g Cl} \times \frac{1 \text{ mol Cl}}{35.453 \text{ g Cl}} = 1.48 \text{ mol Cl}$$

Because the two mole quantities are the same, the empirical formula is SCl.

:C̈l—S̈—S̈—C̈l:

For sulfur:	Isolated sulfur valence electrons	6
	Bound sulfur bonding electrons	4
	Bound sulfur nonbonding electrons	4
	Formal charge = 6 – ½(4) – 4 = 0	
For chlorine:	Isolated chlorine valence electrons	7
	Bound chlorine bonding electrons	2
	Bound chlorine nonbonding electrons	6
	Formal charge = 7 – ½(2) – 6 = 0	

7.110 (a) [:Ö—H]⁻   :Ȯ—H

(b) The oxygen in OH has a half-filled 2p orbital that can accept the additional electron. For a 2p orbital, n = 2 and l = 1.
(c) The electron affinity for OH is slightly more negative than for an O atom because when OH gains an additional electron, it achieves an octet configuration.

7.112 (a)

$$\left[\begin{array}{ccccc} & :\ddot{O} & & :\ddot{O} & \\ & \| & & \| & \\ :O= & Cr & -\ddot{O}- & Cr & =O: \\ & | & & | & \\ & :O: & & :O: & \end{array} \right]^{2-}$$

All formal charges are zero except for the two single bonded oxygens which each have a formal charge of –1.

(b) Each Cr atom has 6 pairs of electrons around it.

Covalent Compounds: Bonding Theories and Molecular Structure

8.1

	Number of Bonded Atoms	Number of Lone Pairs	Shape
(a) O_3	2	1	bent
(b) H_3O^+	3	1	trigonal pyramidal
(c) XeF_2	2	3	linear
(d) PF_6^-	6	0	octahedral
(e) $XeOF_4$	5	1	square pyramidal
(f) AlH_4^-	4	0	tetrahedral
(g) BF_4^-	4	0	tetrahedral
(h) $SiCl_4$	4	0	tetrahedral
(i) ICl_4^-	4	2	square planar
(j) $AlCl_3$	3	0	trigonal planar

8.2 (a) 4 charge clouds, tetrahedral charge cloud arrangement, tetrahedral molecular geometry
(b) 5 charge clouds, trigonal bipyramidal charge cloud arrangement, seesaw molecular geometry.

8.3

Tetrahedral (109.5° bond angles)

Trigonal planar (120° bond angles)

8.4 The bond angle around every carbon is 120° (trigonal planar). Benzene is a planar hexagon.

8.5 CH_2Cl_2; The C is sp^3 hybridized. The C–H bonds are formed by the overlap of one singly occupied sp^3 orbital on C with a singly occupied H 1s orbital. The C–Cl bonds are formed by the overlap of one singly occupied sp^3 orbital on C with a singly occupied Cl 2p orbital.

8.6 Each C is sp^3 hybridized. The C–C bonds are formed by the overlap of one singly occupied sp^3 hybrid orbital from each C. The C–H bonds are formed by the overlap of one singly occupied sp^3 orbital on C with a singly occupied H 1s orbital.

8.7

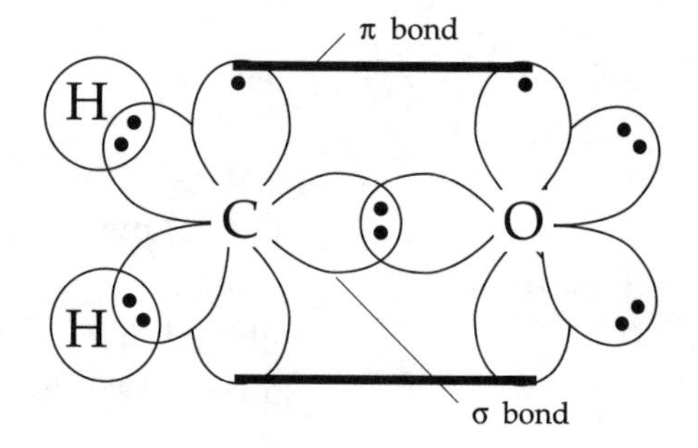

The carbon in formaldehyde is sp^2 hybridized.

8.8 The two Cs with four single bonds are sp^3 hybridized. The C with a double bond is sp^2 hybridized. The C–C bonds are formed by the overlap of one singly occupied sp^3 or sp^2 hybrid orbital from each C. The C–H bonds are formed by the overlap of one singly occupied sp^3 orbital on C with a singly occupied H 1s orbital.

8.9

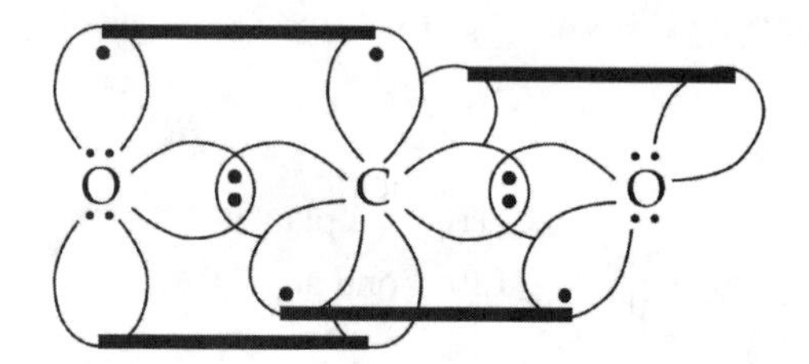

In CO_2 the carbon is sp hybridized.

8.10

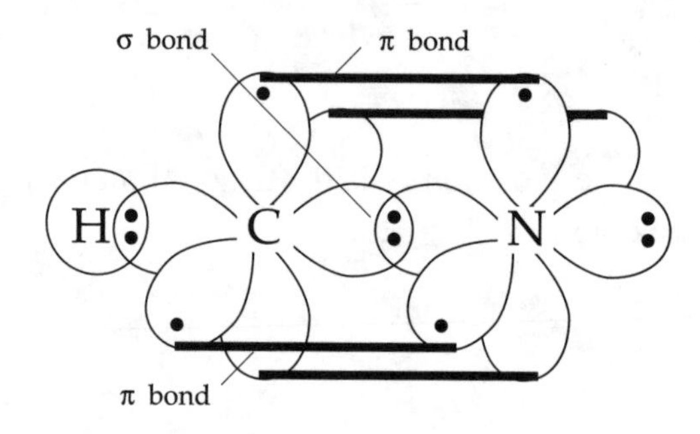

In HCN the carbon is sp hybridized.

8.11 (a)

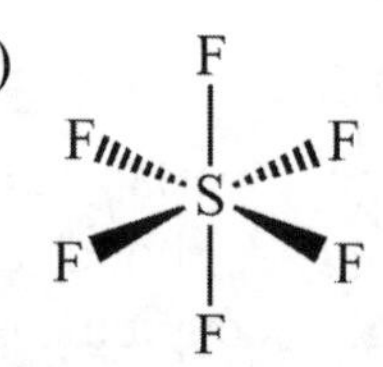

SF_6 has polar covalent bonds but the molecule is symmetrical (octahedral). The individual bond polarities cancel, and the molecule has no dipole moment.

(b)

The C–F bonds in CH_2CF_2 are polar covalent bonds, and the molecule is polar.

(c)

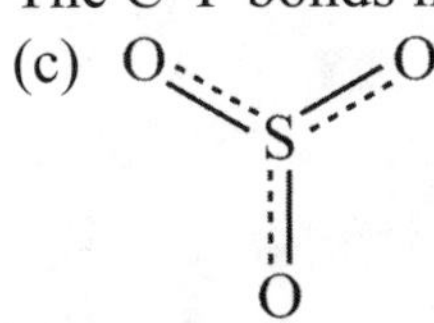

SO_3 has polar covalent bonds but the molecule is symmetrical (trigonal planar). The individual bond polarities cancel, and the molecule has no dipole moment.

(d)

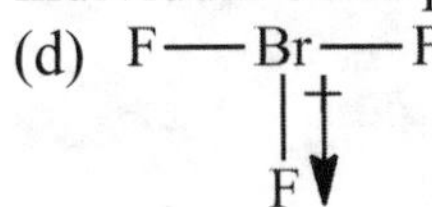

The Br–F bonds in BrF_3 are polar covalent bonds, and the molecule is polar.

8.12 (a) CF_4 (b) CH_2F_2 (c) CHF_3 (d) CH_3F

8.13 $\mu = Q \times r = (1.60 \times 10^{-19}\ C)(92 \times 10^{-12}\ m)\left(\dfrac{1\ D}{3.336 \times 10^{-30}\ C \cdot m}\right) = 4.41\ D$

$\%\ \text{ionic character for HF} = \dfrac{1.83\ D}{4.41\ D} \times 100\% = 41\%$

HF has more ionic character than HCl. HCl has only 18% ionic character.

8.14 HBr is predicted to have greater percent ionic character than HI because the difference in electronegativity between hydrogen and bromine (0.7) is greater than between hydrogen and iodine (0.4).

HBr, $\mu = Q \times r = (1.60 \times 10^{-19}\ C)(142 \times 10^{-12}\ m)\left(\dfrac{1\ D}{3.336 \times 10^{-30}\ C \cdot m}\right) = 6.81\ D$

$\%\ \text{ionic character for HF} = \dfrac{0.82\ D}{6.81\ D} \times 100\% = 12\%$

HI, $\mu = Q \times r = (1.60 \times 10^{-19}\ C)(161 \times 10^{-12}\ m)\left(\dfrac{1\ D}{3.336 \times 10^{-30}\ C \cdot m}\right) = 7.72\ D$

$\%\ \text{ionic character for HF} = \dfrac{0.38\ D}{7.72\ D} \times 100\% = 4.9\%$

The calculation of percent ionic character supports the prediction.

8.15 (b) and (c) are correct depictions of hydrogen bonding. (a) and (d) are incorrect because hydrogen is covalently bonded to C, which is not one of the highly electronegative elements F, O, or N.

8.16 Hydrogen bonds are shown as dashed lines.

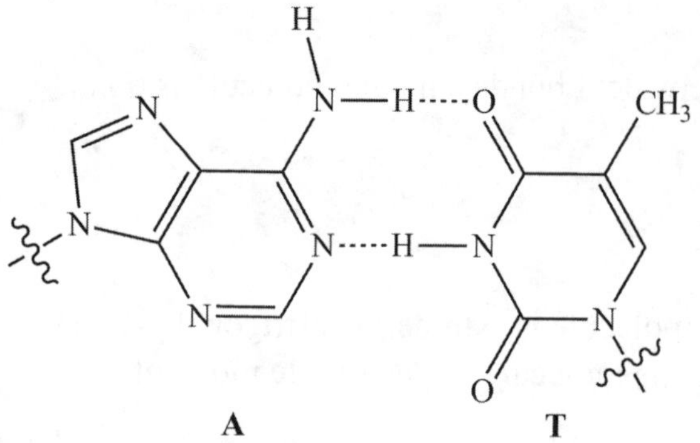

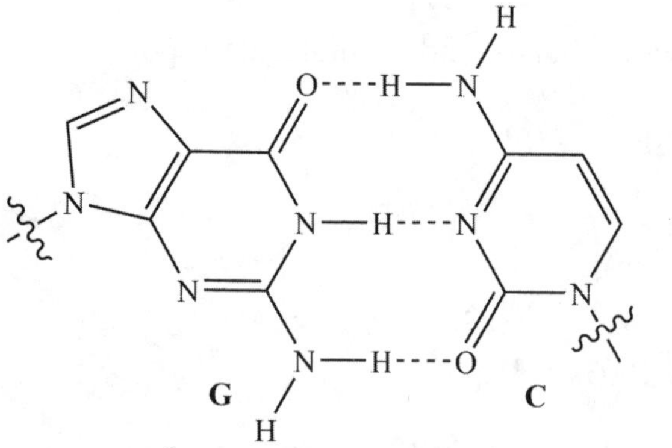

Because of the three hydrogen bonds, DNA regions that are high in G–C pairs would have the higher melting point.

8.17 (a) Both CH_3F and HNO_3 have net dipole moments and dipole–dipole forces.
(b) Only HNO_3 can hydrogen bond.
(c) Ar has fewer electrons than Cl_2 and CCl_4, and has the smallest dispersion forces.
(d) CCl_4 is larger than Ar and Cl_2, has more electrons and the largest dispersion forces.

8.18 H_2S dipole-dipole, dispersion
CH_3OH hydrogen bonding, dipole-dipole, dispersion
C_2H_6 dispersion
Ar dispersion
$Ar < C_2H_6 < H_2S < CH_3OH$

8.19 For He_2^+ σ^*_{1s} $\uparrow$
σ_{1s} $\uparrow\downarrow$

$$He_2^+ \text{ Bond order} = \frac{\left(\begin{array}{c}\text{number of}\\ \text{bonding electrons}\end{array}\right) - \left(\begin{array}{c}\text{number of}\\ \text{antibonding electrons}\end{array}\right)}{2} = \frac{2-1}{2} = 1/2$$

He_2^+ should be stable with a bond order of 1/2.

8.20 The bond order in He_2^{2+} is 1, which is greater than the bond order of 1/2 in He_2^+; therefore He_2^{2+} is predicted to have a stronger bond and be a more stable species

8.21 For B_2

σ^*_{2p} —
π^*_{2p} — —
σ_{2p} —
π_{2p} ↑ ↑
σ^*_{2s} ↑↓
σ_{2s} ↑↓

$$B_2 \text{ Bond order} = \frac{\left(\begin{array}{c}\text{number of}\\ \text{bonding electrons}\end{array}\right) - \left(\begin{array}{c}\text{number of}\\ \text{antibonding electrons}\end{array}\right)}{2} = \frac{4-2}{2} = 1$$

B_2 is paramagnetic because it has two unpaired electrons in the π_{2p} molecular orbitals.

For C_2

σ^*_{2p} —
π^*_{2p} — —
σ_{2p} —
π_{2p} ↑↓ ↑↓
σ^*_{2s} ↑↓
σ_{2s} ↑↓

C_2 Bond order $= \dfrac{6-2}{2} = 2$; C_2 is diamagnetic because all electrons are paired.

8.22 The bond orders are: $O_2^{2-} = 1$, $O_2^- = 1.5$, $O_2 = 2$, $O_2^+ = 2.5$, $O_2^{2+} = 3$.
The order from weakest to strongest bond is: $O_2^{2-} < O_2^- < O_2 < O_2^+ < O_2^{2+}$.
The order from shortest to longest bond is: $O_2^{2+} < O_2^+ < O_2 < O_2^- < O_2^{2-}$.

8.23

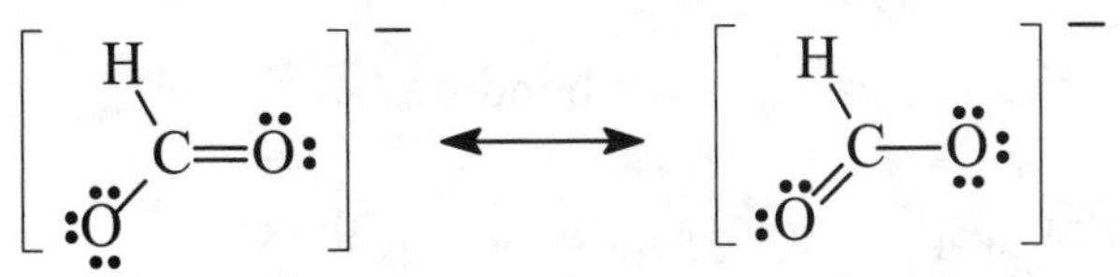

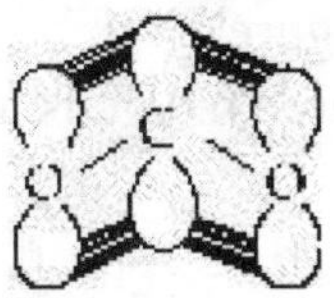

[H–C(=Ö:)–:Ö:]$^-$ ⟷ [H–C(–Ö:)=Ö:]$^-$

H
C
O O

8.24

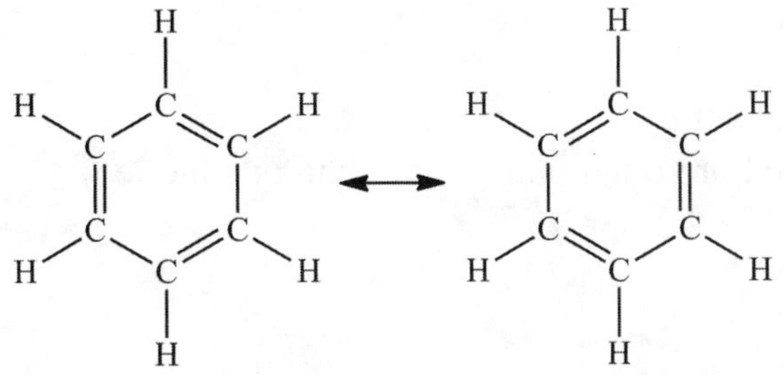

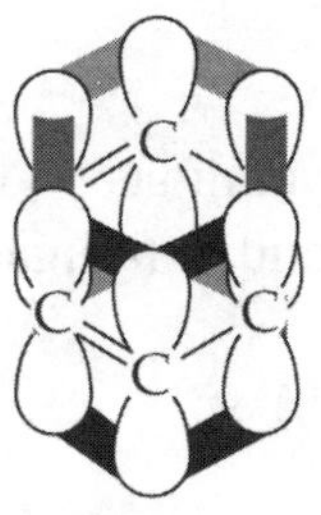

H H H H H H (C–C ring) ⟷ H H H H H H (C–C ring)

C C C C

8.25 (b), (c), and (e) are chiral.

8.26 The mirror image of molecule (a) has the same shape as (a) and is identical to it in all respects, so there is no handedness associated with it. The mirror image of molecule (b) is different than (b) so there is a handedness to this molecule.

8.27 Only (a), lactic acid, is chiral.

8.28

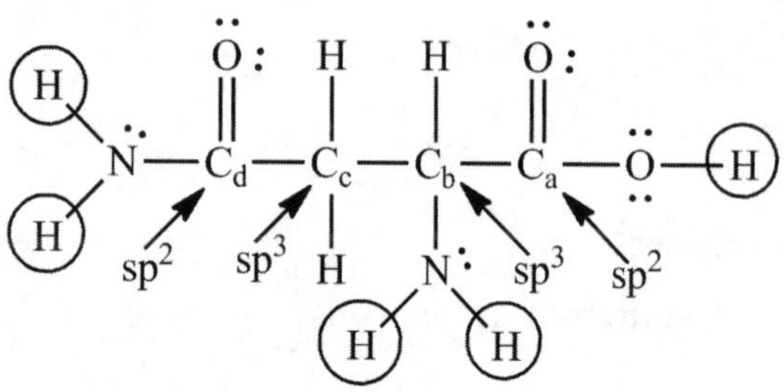

(a) C_b is the chiral center. There are four different groups attached to it.
(b) 16 σ bonds and 2 π bonds.
(c) and (d) see figure

8.29 (a)

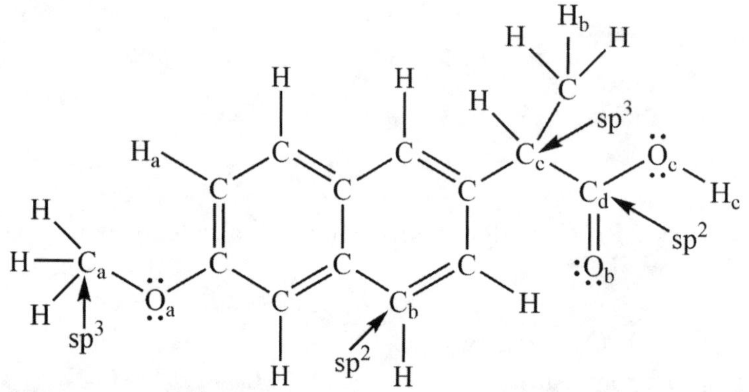

(b) C_c is the chiral center. There are four different groups attached to it.
(c) 32 σ bonds and 6 π bonds.
(d) see figure
(e) All three oxygens and H_c can participate in hydrogen bonding.

8.30 (a) C_c (b) C_a, sp^3; C_b, sp^2; C_c, sp^3; C_d, sp^2 (c) yes, because of the O.

Conceptual Problems

8.32 (a) trigonal bipyramidal (b) tetrahedral
(c) square pyramidal (4 ligands in the horizontal plane, including one hidden)

8.34 (a) sp^2 (b) sp (c) sp^3

8.36 (a) $C_8H_9NO_2$
(b), (c), and (d)

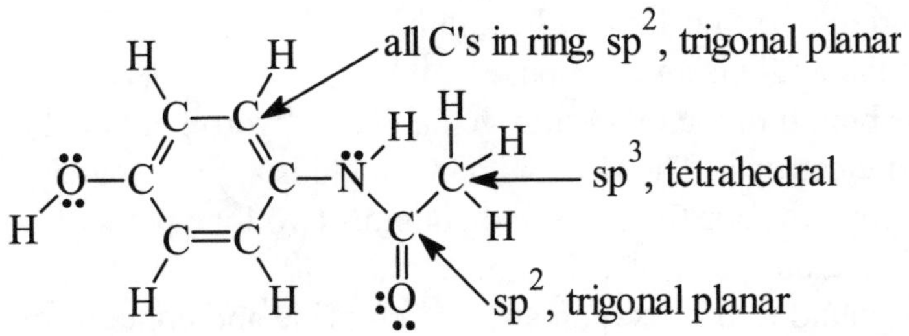

all C's in ring, sp^2, trigonal planar

sp^3, tetrahedral

sp^2, trigonal planar

8.38 The electronegative O atoms are electron rich (red), while the rest of the molecule is electron poor (blue).

8.40 The N atom is electron rich (red) because of its high electronegativity. The C and H atoms are electron poor (blue) because they are less electronegative.

Section Problems
The VSEPR Model (Section 8.1)

8.42 From data in Table 8.1:
(a) trigonal planar (b) trigonal bipyramidal (c) linear (d) octahedral

8.44 From data in Table 8.1:
(a) tetrahedral, 4 (b) octahedral, 6 (c) bent, 3 or 4
(d) linear, 2 or 5 (e) square pyramidal, 6 (f) trigonal pyramidal, 4

8.46

	Number of Bonded Atoms	Number of Lone Pairs	Shape
(a) H_2Se	2	2	bent
(b) $TiCl_4$	4	0	tetrahedral
(c) O_3	2	1	bent
(d) GaH_3	3	0	trigonal planar

8.48

	Number of Bonded Atoms	Number of Lone Pairs	Shape
(a) SbF_5	5	0	trigonal bipyramidal
(b) IF_4^+	4	1	see saw
(c) SeO_3^{2-}	3	1	trigonal pyramidal
(d) CrO_4^{2-}	4	0	tetrahedral

8.50

	Number of Bonded Atoms	Number of Lone Pairs	Shape
(a) PO_4^{3-}	4	0	tetrahedral
(b) MnO_4^-	4	0	tetrahedral
(c) SO_4^{2-}	4	0	tetrahedral
(d) SO_3^{2-}	3	1	trigonal pyramidal
(e) ClO_4^-	4	0	tetrahedral
(f) SCN^- (C is the central atom)	2	0	linear

8.52 (a) In SF_2 the sulfur is bound to two fluorines and contains two lone pairs of electrons. SF_2 is bent and the F–S–F bond angle is approximately 109°.
(b) In N_2H_2 each nitrogen is bound to the other nitrogen and one hydrogen. Each nitrogen has one lone pair of electrons. The H–N–N bond angle is approximately 120°.
(c) In KrF_4 the krypton is bound to four fluorines and contains two lone pairs of electrons. KrF_4 is square planar, and the F–Kr–F bond angle is 90°.
(d) In NOCl the nitrogen is bound to one oxygen and one chlorine and contains one lone pair of electrons. NOCl is bent, and the Cl–N–O bond angle is approximately 120°.

8.54 $H_2C_a{=}C_b(H){-}C_c{\equiv}N$

$H-C_a-H$	~ 120°	C_b-C_c-N	180°
$H-C_a-C_b$	~ 120°	C_a-C_b-H	~ 120°
$C_a-C_b-C_c$	~ 120°	$H-C_b-C_c$	~ 120°

8.56 $F_3C{-}SF_5$

The bond angles are 109.5° around carbon and 90° around S.

8.58 All six carbons in cyclohexane are bonded to two other carbons and two hydrogens (i.e., four charge clouds). The geometry about each carbon is tetrahedral with a C–C–C bond angle of approximately 109°. Because the geometry about each carbon is tetrahedral, the cyclohexane ring cannot be flat.

Valence Bond Theory and Hybridization (Sections 8.2–8.4)

8.60 In a π bond, the shared electrons occupy a region above and below a line connecting the two nuclei. A σ bond has its shared electrons located along the axis between the two nuclei.

8.62 See Table 8.2. (a) sp (b) sp^2 (c) sp^3

8.64 (a) sp^2 (b) sp^2 (c) sp^3 (d) sp^2

8.66 $H{-}O{-}C_a(=O){-}C_b(=O){-}C_cH_2{-}C_d(=O){-}O{-}H$

Carbons a, b, and d are sp^2 hybridized and carbon c is sp^3 hybridized.

The bond angles around carbons a, b, and d are ~120°. The bond angles around carbon c are ~109°. The terminal H–O–C bond angles are ~109°.

8.68 

In Cl_2CO the carbon is sp2 hybridized.

Dipole Moments and Intermolecular Forces (Sections 8.5 and 8.6)

8.70 If a molecule has polar covalent bonds, the molecular shape (and location of lone pairs of electrons) determines whether the bond dipoles cancel and thus whether the molecule has a dipole moment.

8.72 (a) $CHCl_3$ has a permanent dipole moment. Dipole-dipole intermolecular forces are important. London dispersion forces are also present.
(b) O_2 has no dipole moment. London dispersion intermolecular forces are important.
(c) Polyethylene, C_nH_{2n+2}. London dispersion intermolecular forces are important.
(d) CH_3OH has a permanent dipole moment. Dipole-dipole intermolecular forces and hydrogen bonding are important. London dispersion forces are also present.

8.74 For CH_3OH and CH_4, dispersion forces are small. CH_3OH can hydrogen bond; CH_4 cannot. This accounts for the large difference in boiling points.
For 1-decanol and decane, dispersion forces are comparable and relatively large along the C–H chain. 1-decanol can hydrogen bond; decane cannot. This accounts for the 57 °C higher boiling point for 1-decanol.

8.76 (a)

O, Cl, Cl — net

(b) 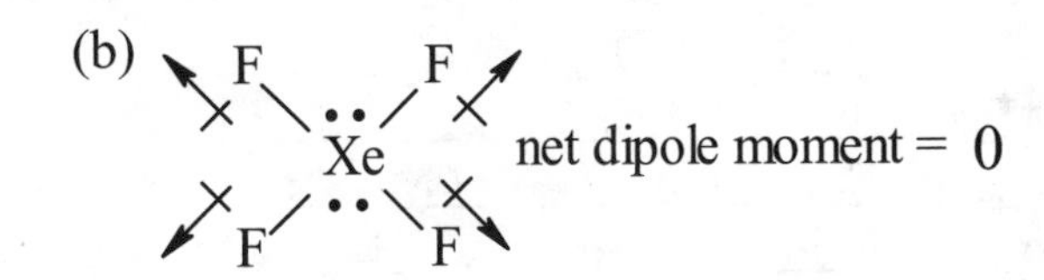

(c) 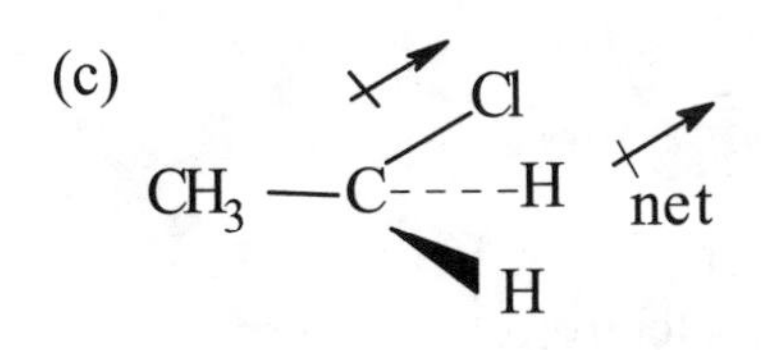

(d)

F, B, F, F

net dipole moment = 0

8.78 $\mu = Q \times r = (1.60 \times 10^{-19}\ C)(213.9 \times 10^{-12}\ m)\left(\dfrac{1\ D}{3.336 \times 10^{-30}\ C \cdot m}\right) = 10.26\ D$

$\%\ \text{ionic character for BrCl} = \dfrac{0.518\ D}{10.26\ D} \times 100\% = 5.05\%$

8.80 S, O, O — net

$O{=}C{=}O$

net dipole moment = 0

SO_2 is bent and the individual bond dipole moments add to give the molecule a net dipole moment.
CO_2 is linear and the individual bond dipole moments point in opposite directions to cancel each other out. CO_2 has no net dipole moment.

8.82 (a) $[PtCl_2Br_2]^{2-}$: Cl—Pt—Cl horizontal, Br above and below Pt (trans). (b) $[PtCl_2Br_2]^{2-}$: Cl—Pt—Br horizontal, Br above and Cl below Pt (cis).

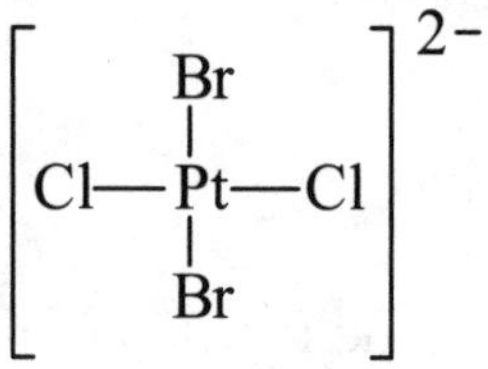

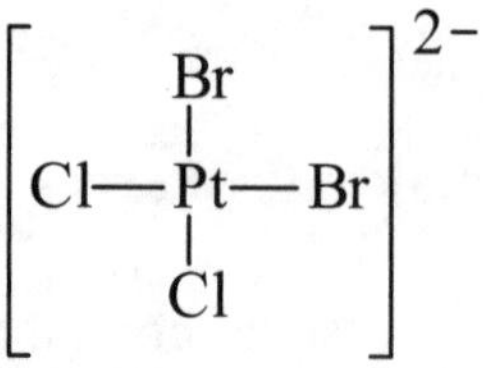

8.84 :NH_2—H- - -:NH_3 (H—N⋯H–N hydrogen bond between two ammonia molecules)

hydrogen bond

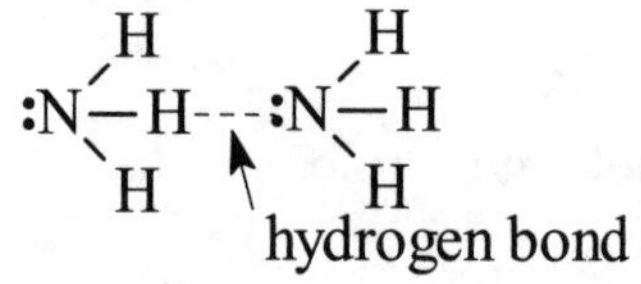

8.86 Illustrations (ii) and (iii) depict the hydrogen bonding that occurs between methylamine and water.

Molecular Orbital Theory (Sections 8.7– 8.9)

8.88 Electrons in a bonding molecular orbital spend most of their time in the region between the two nuclei, helping to bond the atoms together. Electrons in an antibonding molecular orbital cannot occupy the central region between the nuclei and cannot contribute to bonding.

8.90

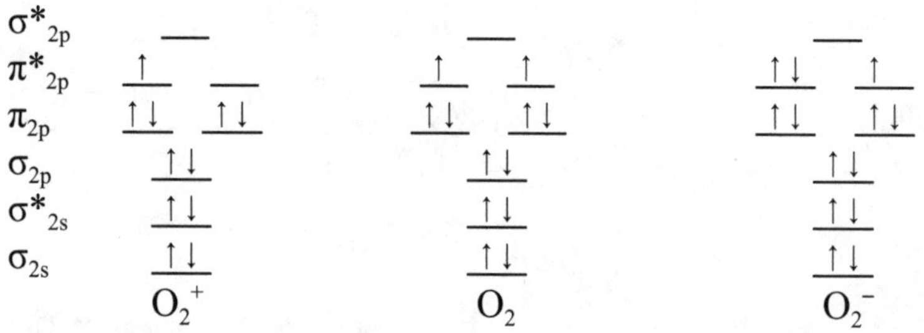

	O_2^+	O_2	O_2^-
σ^*_{2p}	__	__	__
π^*_{2p}	↑ __	↑ ↑	↑↓ ↑
π_{2p}	↑↓ ↑↓	↑↓ ↑↓	↑↓ ↑↓
σ_{2p}	↑↓	↑↓	↑↓
σ^*_{2s}	↑↓	↑↓	↑↓
σ_{2s}	↑↓	↑↓	↑↓

$$\text{Bond order} = \frac{\begin{pmatrix}\text{number of}\\ \text{bonding electrons}\end{pmatrix} - \begin{pmatrix}\text{number of}\\ \text{antibonding electrons}\end{pmatrix}}{2}$$

O_2^+ bond order $= \dfrac{8-3}{2} = 2.5$ O_2 bond order $= \dfrac{8-4}{2} = 2$

O_2^- bond order $= \dfrac{8-5}{2} = 1.5$

All are stable with bond orders between 1.5 and 2.5. All have unpaired electrons.

8.92

	C_2	C_2^-
σ^*_{2p}	—	—
π^*_{2p}	— —	— —
σ_{2p}	—	↑
π_{2p}	↑↓ ↑↓	↑↓ ↑↓
σ^*_{2s}	↑↓	↑↓
σ_{2s}	↑↓	↑↓

$$\text{Bond order} = \frac{\left(\begin{array}{c}\text{number of}\\ \text{bonding electrons}\end{array}\right) - \left(\begin{array}{c}\text{number of}\\ \text{antibonding electrons}\end{array}\right)}{2}$$

(a) C_2 bond order $= \dfrac{6 - 2}{2} = 2$

(b) Add one electron because it will go into a bonding molecular orbital.

(c) C_2^- bond order $= \dfrac{7 - 2}{2} = 2.5$

8.94

(a) C_2^{2-}	(b) C_2^{2+}	(c) F_2^-	(d) Cl_2	(e) Li_2^+
σ^*_{2p}: —	σ^*_{2p}: —	σ^*_{2p}: ↑	σ^*_{2p}: —	σ^*_{2p}: —
π^*_{2p}: — —	π^*_{2p}: — —	π^*_{2p}: ↑↓ ↑↓	π^*_{2p}: ↑↓ ↑↓	π^*_{2p}: — —
σ_{2p}: ↑↓	σ_{2p}: —	π_{2p}: ↑↓ ↑↓	π_{2p}: ↑↓ ↑↓	σ_{2p}: —
π_{2p}: ↑↓ ↑↓	π_{2p}: ↑ ↑	σ_{2p}: ↑↓	σ_{2p}: ↑↓	π_{2p}: ↑ —
σ^*_{2s}: ↑↓	σ^*_{2s}: ↑↓	σ^*_{2s}: ↑↓	σ^*_{2s}: ↑↓	σ^*_{2s}: ↑↓
σ_{2s}: ↑↓	σ_{2s}: ↑↓	σ_{2s}: ↑↓	σ_{2s}: ↑↓	σ_{2s}: ↑↓
diamagnetic	paramagnetic	paramagnetic	diamagnetic	paramagnetic

8.96

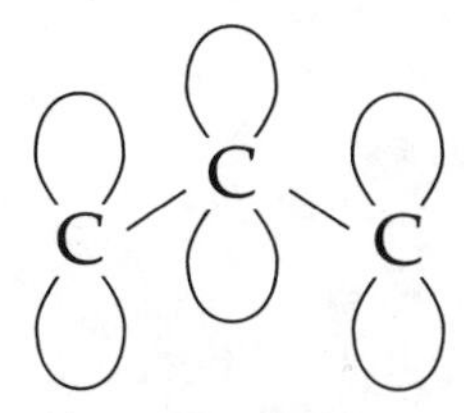

p orbitals in allyl cation

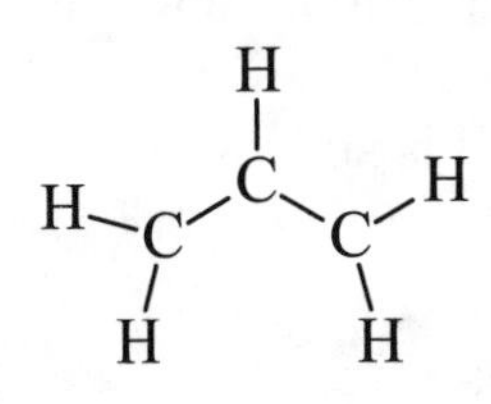

allyl cation showing only the σ bonds (each C is sp^2 hybridized)

delocalized MO model for π bonding in the allyl cation

Chapter Problems

8.98 Because chlorine is larger than fluorine, the charge separation is larger in CH_3Cl compared to CH_3F, resulting in CH_3Cl having a slightly larger dipole moment.

8.100

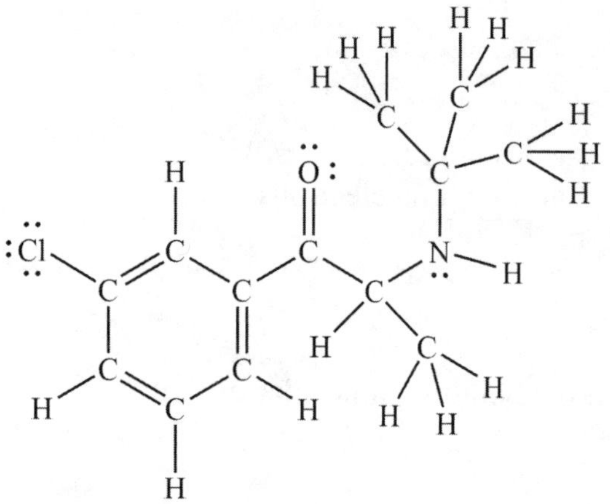

(a) There are 34 σ bonds and 4 π bonds.
(b) and (c) Each C with four single bonds is sp^3 hybridized with bond angles of 109.5°. Each C with a double bond is sp^2 hybridized with bond angles of 120°.
(d) The nitrogen is sp^3 hybridized.

8.102 Both the B and N are sp^2 hybridized. All bond angles are ~120°. The overall geometry of the molecule is planar.

8.104 (a) H—C≡C—H (b) H—N̈=N̈—H (c) :Ö: | :C̈l—S̈—C̈l:

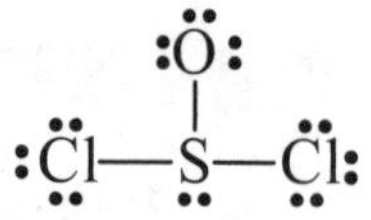

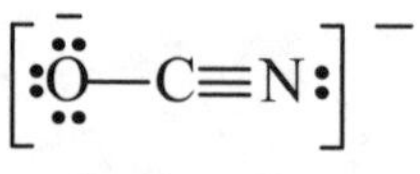

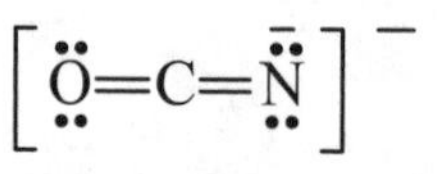

8.106 (a) (1) [:Ö—C≡N:]⁻ (2) [Ö=C=N̈]⁻ (3) [:O≡C—N̈:]⁻

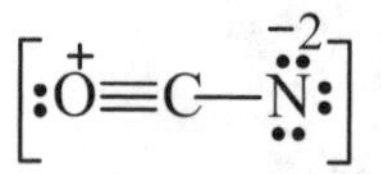

(b) Structure (1) makes the greatest contribution to the resonance hybrid because of the −1 formal charge on the oxygen. Structure (3) makes the least contribution to the resonance hybrid because of the +1 formal charge on the oxygen.
(c) and (d) OCN^- is linear because the C has 2 charge clouds. It is sp hybridized in all three resonance structures. It forms two π bonds.

8.108 $[H—C≡N—Xe—F:]^+$ Both the carbon and nitrogen are sp hybridized.

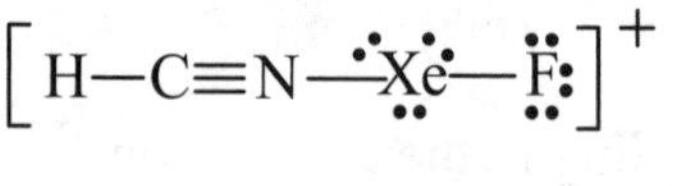

8.110 Every carbon is sp^2 hybridized. There are 18 σ bonds and 5 π bonds.

8.112

Each C with four single bonds is sp^3 hybridized.
Each C with a double bond is sp^2 hybridized.

8.114 C_2^{2-}

σ^*_{2p} —
π^*_{2p} — —
σ_{2p} ↑↓
π_{2p} ↑↓ ↑↓
σ^*_{2s} ↑↓
σ_{2s} ↑↓

$$\text{Bond order} = \frac{\begin{pmatrix}\text{number of}\\ \text{bonding electrons}\end{pmatrix} - \begin{pmatrix}\text{number of}\\ \text{antibonding electrons}\end{pmatrix}}{2}$$

C_2^{2-} bond order $= \frac{8-2}{2} = 3$; there is a triple bond between the two carbons.

8.116 (a) CO

σ^*_{2p} —
π^*_{2p} — —
σ_{2p} ↑↓
π_{2p} ↑↓ ↑↓
σ^*_{2s} ↑↓
σ_{2s} ↑↓

(b) All electrons are paired, CO is diamagnetic.

(c)

$$\text{Bond order} = \frac{\begin{pmatrix}\text{number of}\\ \text{bonding electrons}\end{pmatrix} - \begin{pmatrix}\text{number of}\\ \text{antibonding electrons}\end{pmatrix}}{2}$$

CO bond order $= \frac{8-2}{2} = 3$

The bond order here matches that predicted by the electron-dot structure (:C≡O:).

(d)

8.118

Multiconcept Problems

8.120 (a)

H₂C=CCl₂	*cis*-ClHC=CHCl	*trans*-ClHC=CHCl
polar	polar	nonpolar

(b) All three molecules are planar. The first two structures are polar because they both have an unsymmetrical distribution of atoms about the center of the molecule (the middle of the double bond), and bond polarities do not cancel. Structure 3 is nonpolar because the H's and Cl's, respectively, are symmetrically distributed about the center of the molecule, both being opposite each other. In this arrangement, bond polarities cancel.

(c) 200 nm = 200×10^{-9} m

$$E = \frac{hc}{\lambda} = \frac{(6.626 \times 10^{-34}\ \text{J}\cdot\text{s})(3.00 \times 10^{8}\ \text{m/s})}{200 \times 10^{-9}\ \text{m}}(6.022 \times 10^{23}/\text{mol})$$

$E = 5.99 \times 10^{5}$ J/mol = 599 kJ/mol

(d)

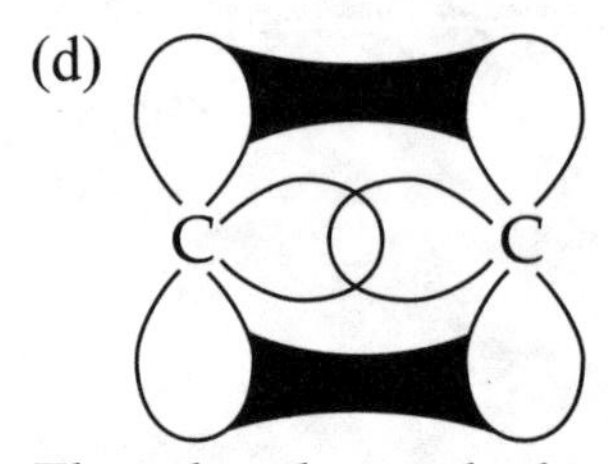

The π bond must be broken before rotation can occur.

9 Thermochemistry: Chemical Energy

9.1 $\Delta V = (4.3\ L - 8.6\ L) = -4.3\ L$

$w = -P\Delta V = -(44\ atm)(-4.3\ L) = +189.2\ L\cdot atm$

$w = (189.2\ L\cdot atm)(101\ \frac{J}{L\cdot atm}) = +1.9 \times 10^4\ J$

The positive sign for the work indicates that the surroundings do work on the system. Energy flows into the system.

9.2 $w = -P\Delta V = -(2.5\ atm)(3\ L - 2\ L) = -2.5\ L\cdot atm$

$w = (-2.5\ L\cdot atm)\left(101\ \frac{J}{L\cdot atm}\right) = -252.5\ J = -250\ J = -0.25\ kJ$

The negative sign indicates that the expanding system loses work energy and does work on the surroundings.

9.3 $\Delta H^\circ = \frac{-484\ kJ}{2\ mol\ H_2}$

$P\Delta V = (1.00\ atm)(-24.4\ L) = -24.4\ L\cdot atm$

$P\Delta V = (-24.4\ L\cdot atm)(101\ \frac{J}{L\cdot atm}) = -2464\ J = -2.46\ kJ$

$w = -P\Delta V = +2.46\ kJ$

$\Delta H^\circ = \frac{-484\ kJ}{2\ mol\ H_2}$

$\Delta E = \Delta H - P\Delta V = -484\ kJ - (-2.46\ kJ) = -481.5\ kJ = -482\ kJ$

9.4 (a) $w = -P\Delta V$ is positive and $P\Delta V$ is negative for this reaction because the system volume is decreased at constant pressure.

(b) $P\Delta V$ is small compared to ΔE.

$\Delta H = \Delta E + P\Delta V$; ΔH is negative. Its value is slightly more negative than ΔE.

9.5 (a) H_2, 2.016

$10.00\ g\ H_2 \times \frac{1\ mol\ H_2}{2.016\ g\ H_2} \times \frac{-571.6\ kJ}{2\ mol\ H_2} = -1418\ kJ$

1418 kJ of heat are evolved.

(b) $5.500\ mol\ H_2O \times \frac{571.6\ kJ}{2\ mol\ H_2O} = 1572\ kJ$

1572 kJ of heat are absorbed.

9.6 C_3H_8, 44.09

$$1.8 \times 10^6 \text{ kJ} \times \frac{1 \text{ mol } C_3H_8}{2044 \text{ kJ}} \times \frac{44.09 \text{ g } C_3H_8}{1 \text{ mol } C_3H_8} \times \frac{1 \text{ kg}}{1000 \text{ g}} = 39 \text{ kg } C_3H_8$$

9.7 (a) heat transfers from system to surroundings, exothermic, ΔH° is negative
(b) heat transfers from system to surroundings, exothermic, ΔH° is negative
(c) heat transfers from surroundings to system, endothermic, ΔH° is positive

9.8 LiF, 25.9; $BaSO_4$, 233.4
Only LiF and $BaSO_4$ will produce endothermic reactions and lower the water temperature.

$$10.00 \text{ g LiF} \times \frac{1 \text{ mol LiF}}{25.9 \text{ g LiF}} \times \frac{5.5 \text{ kJ}}{1 \text{ mol LiF}} = 2.12 \text{ kJ}$$

$$10.00 \text{ g } BaSO_4 \times \frac{1 \text{ mol } BaSO_4}{233.4 \text{ g } BaSO_4} \times \frac{26.3 \text{ kJ}}{1 \text{ mol } BaSO_4} = 1.13 \text{ kJ}$$

10.00 g of LiF would result in the largest temperature decrease.

9.9 $q = (\text{specific heat capacity}) \times m \times \Delta T$

$$\text{specific heat capacity} = \frac{q}{m \times \Delta T} = \frac{97.2 \text{ J}}{(75.0 \text{ g})(10.0\ ^\circ\text{C})} = 0.130 \text{ J/(g} \cdot {}^\circ\text{C)}$$

$$\text{molar heat capacity} = 0.130 \text{ J/(g} \cdot {}^\circ\text{C)} \times \frac{207.2 \text{ g Pb}}{1 \text{ mol Pb}} = 26.9 \text{ J/(mol} \cdot {}^\circ\text{C)}$$

9.10 $$\text{heat capacity} = \frac{1650 \text{ J}}{2.00\ ^\circ\text{C}} = 825 \text{ J/}^\circ\text{C}$$

9.11 25.0 mL = 0.0250 L and 50.0 mL = 0.0500 L
mol H_2SO_4 = (1.00 mol/L)(0.0250 L) = 0.0250 mol H_2SO_4
mol NaOH = (1.00 mol/L)(0.0500 L) = 0.0500 mol NaOH
NaOH and H_2SO_4 are present in a 2:1 mol ratio. This matches the stoichiometric ratio in the balanced equation.
$q = (\text{specific heat}) \times m \times \Delta T$
m = (25.0 mL + 50.0 mL)(1.00 g/mL) = 75.0 g

$$q = (4.18 \frac{\text{J}}{\text{g} \cdot {}^\circ\text{C}})(75.0 \text{ g})(33.9^\circ\text{C} - 25.0\ ^\circ\text{C}) = 2790 \text{ J}$$

$$\text{mol } H_2SO_4 = 0.0250 \text{ L} \times 1.00 \frac{\text{mol}}{\text{L}} H_2SO_4 = 0.0250 \text{ mol } H_2SO_4$$

$$\text{Heat evolved per mole of } H_2SO_4 = \frac{2.79 \times 10^3 \text{ J}}{0.0250 \text{ mol } H_2SO_4} = 1.1 \times 10^5 \text{ J/mol } H_2SO_4$$

Because the reaction evolves heat, the sign for ΔH is negative.

$$\Delta H = -1.1 \times 10^5 \text{ J/mol} \times \frac{1 \text{ kJ}}{1000 \text{ J}} = -1.1 \times 10^2 \text{ kJ/mol}$$

9.12 $q = (4.18 \frac{J}{g\cdot°C})(250.0\ g)(18.6\ °C) + (623\ J/°C)(18.6\ °C) = 31{,}025\ J$

$\Delta H = -q = \Delta E = -31{,}025\ J/2\ g \times \frac{1\ kJ}{1000\ J} = -15.5\ kJ/g$

$\Delta H = -q = \Delta E = -15.5\ kJ/g \times \frac{1\ Cal}{4.184\ kJ} = -3.71\ Cal/g$

9.13

	$C(s) + O_2(g) \rightarrow CO_2(g)$	$\Delta H° = -393.5\ kJ$
	$H_2O(g) \rightarrow 1/2\ O_2(g) + H_2(g)$	$\Delta H° = +483.6/2\ kJ$
	$CO2(g) \rightarrow 1/2\ O_2(g) + CO(g)$	$\Delta H° = +566.0/2\ kJ$
sum	$C(s) + H_2O(g) \rightarrow CO(g) + H_2(g)$	$\Delta H° = +131.3\ kJ$

9.14 (a) $A + 2\ B \rightarrow D$; $\Delta H° = -100\ kJ + (-50\ kJ) = -150\ kJ$
(b) The red arrow corresponds to step 1: $A + B \rightarrow C$
The green arrow corresponds to step 2: $C + B \rightarrow D$
The blue arrow corresponds to the overall reaction.
(c) The top energy level represents A + 2 B.
The middle energy level represents C + B.
The bottom energy level represents D.

9.15 $4\ NH_3(g) + 5\ O_2(g) \rightarrow 4\ NO(g) + 6\ H_2O(g)$
$\Delta H°_{rxn} = [4\ \Delta H°_f(NO) + 6\ \Delta H°_f(H_2O)] - [4\ \Delta H°_f(NH_3)]$
$\Delta H°_{rxn} = [(4\ mol)(91.3\ kJ/mol) + (6\ mol)(-241.8\ kJ/mol)] - [(4\ mol)(-46.1\ kJ/mol)]$
$\Delta H°_{rxn} = -901.2\ kJ$

9.16 $2\ C_8H_{18}(l) + 25\ O_2(g) \rightarrow 8\ CO_2(g) + 9\ H_2O(l)$
$\Delta H°_{rxn} = \Delta H°_c = -5220\ kJ$
$\Delta H°_{rxn} = [8\ \Delta H°_f(CO_2) + 9\ \Delta H°_f(H_2O)] - [2\ \Delta H°_f(C_8H_{18})]$
$-5220\ kJ = [(8\ mol)(-393.5\ kJ/mol) + (9\ mol)(-285.8\ kJ/mol)] - [(2\ mol)(\Delta H°_f(C_8H_{18}))]$
Solve for $\Delta H°_f(C_8H_{18})$
$-5220\ kJ = -5720.2\ kJ - (2\ mol)(\Delta H°_f(C_8H_{18}))$
$+500\ kJ = -(2\ mol)(\Delta H°_f(C_8H_{18}))$
$\Delta H°_f(C_8H_{18}) = \frac{500\ kJ}{-2\ mol} = -250\ kJ/mol$

9.17 $H_2C{=}CH_2(g) + H_2O(g) \rightarrow C_2H_5OH(g)$
$\Delta H°_{rxn} = D\ (\text{Reactant bonds}) - D\ (\text{Product bonds})$
$\Delta H°_{rxn} = (D_{C=C} + 4\ D_{C-H} + 2\ D_{O-H}) - (D_{C-C} + D_{C-O} + 5\ D_{C-H} + D_{O-H})$
$\Delta H°_{rxn} = [(1\ mol)(728\ kJ/mol) + (4\ mol)(410\ kJ/mol) + (2\ mol)(460\ kJ/mol)]$
$- [(1\ mol)(350\ kJ/mol) + (1\ mol)(350\ kJ/mol) + (5\ mol)(410\ kJ/mol) + (1\ mol)(460\ kJ/mol)]$
$\Delta H°_{rxn} = +78\ kJ$

9.18 $\Delta H°_{rxn} = D\ (\text{Reactant bonds}) - D\ (\text{Product bonds})$
(a) $2\ C_6H_6(g) + 15\ O_2(g) \rightarrow 12\ CO_2(g) + 6\ H_2O(g)$
$\Delta H°_{rxn} = (12\ D_{C-C} + 12\ D_{C-H} + 15\ D_{O=O}) - (24\ D_{C=O} + 12\ D_{O-H}) = -6339\ kJ$

$\Delta H^\circ_{rxn} = [(12\ \text{mol})(D_{C\text{–}C}) + (12\ \text{mol})(410\ \text{kJ/mol}) + (15\ \text{mol})(498\ \text{kJ/mol})]$
$\quad - [(24\ \text{mol})(804\ \text{kJ/mol}) + (12\ \text{mol})(460\ \text{kJ/mol})] = -6339\ \text{kJ}$

$(12\ \text{mol})(D_{C\text{–}C}) = -6339\ \text{kJ} + [(24\ \text{mol})(804\ \text{kJ/mol}) + (12\ \text{mol})(460\ \text{kJ/mol})]$
$\quad - [(12\ \text{mol})(410\ \text{kJ/mol}) + (15\ \text{mol})(498\ \text{kJ/mol})]$

$(12\ \text{mol})(D_{C\text{–}C}) = 6087\ \text{kJ}$

$$D_{C\text{–}C} = \frac{6087\ \text{kJ}}{12\ \text{mol}} = 507\ \text{kJ/mol for C– C bond in benzene}$$

9.19 ΔS° is negative because the reaction decreases the number of moles of gaseous molecules.

9.20 The reaction proceeds from a solid and a gas (reactants) to all gas (product). Randomness increases, so ΔS° is positive.

9.21 The reaction involves only bond making, so it is exothermic and ΔH is negative. There are more reactant atoms than product molecules. The randomness of the system decreases on going from reactant to product, therefore ΔS is negative. Because the reaction is spontaneous, ΔG is negative.

9.22 (a) $2\ A_2 + B_2 \rightarrow 2\ A_2B$
(b) Because the reaction is exothermic, ΔH is negative. There are more reactant molecules than product molecules. The randomness of the system decreases on going from reactant to product, therefore ΔS is negative.
(c) Because $\Delta G = \Delta H - T\Delta S$, a reaction with both ΔH and ΔS negative is favored at low temperatures where the negative ΔH term is larger than the positive $- T\Delta S$, and ΔG is negative.

9.23 $\Delta G^\circ = \Delta H^\circ - T\Delta S^\circ = (-92.2\ \text{kJ}) - (298\ \text{K})(-0.199\ \text{kJ/K}) = -32.9\ \text{kJ}$
Because ΔG° is negative, the reaction is spontaneous.
Set $\Delta G^\circ = 0$ and solve for T.

$$\Delta G^\circ = 0 = \Delta H^\circ - T\Delta S^\circ;\quad T = \frac{\Delta H^\circ}{\Delta S^\circ} = \frac{-92.2\ \text{kJ}}{-0.199\ \text{kJ/K}} = 463\ \text{K} = 190\ ^\circ\text{C}$$

9.24 $\Delta G^\circ = \Delta H^\circ - T\Delta S^\circ$
Set $\Delta G^\circ = 0$ and solve for T.

$$\Delta G^\circ = 0 = \Delta H^\circ - T\Delta S^\circ;\quad T = \frac{\Delta H^\circ}{\Delta S^\circ} = \frac{+75.0\ \text{kJ}}{+0.231\ \text{kJ/K}} = 325\ \text{K} = 52\ ^\circ\text{C}$$

The reaction is at equilibrium at 325 K (52 °C). The temperature should be increased to make the reaction spontaneous.

9.25 (a) $C_2H_6O(l) + 3\ O_2(g) \rightarrow 2\ CO_2(g) + 3\ H_2O(l)$
(b) $\Delta H^\circ_c = [2\ \Delta H^\circ_f(CO_2) + 3\ \Delta H^\circ_f(H_2O)] - \Delta H^\circ_f(C_2H_6O)$
$\Delta H^\circ_c = [(2\ \text{mol})(-393.5\ \text{kJ/mol}) + (3\ \text{mol})(-285.8\ \text{kJ/mol})] - [(1\ \text{mol})(-277.7\ \text{kJ/mol})]$
$\Delta H^\circ_c = -1367\ \text{kJ}$

(c) C_2H_6O, 46.1; $\Delta H^\circ_c = -1367\ \text{kJ/mol} \times \dfrac{1\ \text{mol}\ C_2H_6O}{46.1\ \text{g}\ C_2H_6O} = -29.7\ \text{kJ/g}$

(d) $\Delta H^\circ_c = (-29.7\ \text{kJ/g})(0.789\ \text{g/mL}) = -23.4\ \text{kJ/mL}$

9.26 $\Delta H^\circ_{rxn} = D$ (Reactant bonds) – D (Product bonds)

$C_2H_6O(l) + 3\ O_2(g) \rightarrow 2\ CO_2(g) + 3\ H_2O(l)$

$\Delta H^\circ_c = (D_{C\text{–}C} + 5\ D_{C\text{–}H} + D_{C\text{–}O} + D_{O\text{–}H} + 3\ D_{O=O}) - (4\ D_{C=O} + 6\ D_{O\text{–}H})$

ΔH°_c = [(1 mol)(350 kJ/mol) + (5 mol)(410 kJ/mol) + (1 mol)(350 kJ/mol) + (1 mol)(460 kJ/mol) + (3 mol)(498 kJ/mol)] – [(4 mol)(804 kJ/mol) + (6 mol)(460 kJ/mol)] = –1272 kJ/mol

9.27 (a) ΔV = (49.0 L – 73.5 L) = –24.5 L

$P\Delta V$ = (1.00 atm)(–24.5 L) = –24.5 L· atm

$P\Delta V = (-24.5\ \text{L}\cdot\text{atm})(101\ \dfrac{\text{J}}{\text{L}\cdot\text{atm}}) = -2475\ \text{J} = -2.47\ \text{kJ}$

$w = -P\Delta V$ = +2.47 kJ

Work energy is transferred to the system.

(b) $\Delta E = \Delta H - P\Delta V$ = –1367 kJ – (–2.47 kJ) = –1365 kJ/mol

9.28 (a) $q = (4.18\ \dfrac{\text{J}}{\text{g}\cdot{}^\circ\text{C}})(300.0\ \text{g})(6.85\,^\circ\text{C}) + (675\ \text{J}/^\circ\text{C})(6.85\,^\circ\text{C}) = 13{,}214\ \text{J}$

$\Delta H = -q = \Delta E = -13{,}214\ \text{J}/0.350\ \text{g} \times \dfrac{1\ \text{kJ}}{1000\ \text{J}} = -37.8\ \text{kJ/g}$

(b) $C_{19}H_{38}O_2$, 298.5

$\Delta E = -37.8\ \text{kJ/g} \times \dfrac{298.5\ \text{g}\ C_{19}H_{38}O_2}{1\ \text{mol}\ C_{19}H_{38}O_2} = -1.13 \times 10^4\ \text{kJ/mol}$

(c) ΔE = (–37.8 kJ/g)(0.880 g/mL) = –33.3 kJ/mL

9.29 C_2H_6O, 46.1

$\text{heat} = 35.8\ \text{kg}\ C_2H_6O \times \dfrac{1000\ \text{g}}{1\ \text{kg}} \times \dfrac{1\ \text{mol}\ C_2H_6O}{46.1\ \text{g}\ C_2H_6O} \times \dfrac{-1367\ \text{kJ}}{1\ \text{mol}\ C_2H_6O} \times \dfrac{1\ \text{MJ}}{1000\ \text{kJ}} = -1060\ \text{MJ}$

$C_{19}H_{38}O_2$, 298.5

$\text{heat} = 39.9\ \text{kg}\ C_{19}H_{38}O_2 \times \dfrac{1000\ \text{g}}{1\ \text{kg}} \times \dfrac{1\ \text{mol}\ C_{19}H_{38}O_2}{298.5\ \text{g}\ C_{19}H_{38}O_2} \times \dfrac{-11{,}236\ \text{kJ}}{1\ \text{mol}\ C_{19}H_{38}O_2} \times \dfrac{1\ \text{MJ}}{1000\ \text{kJ}} = -1500\ \text{MJ}$

heat (octane) = 1530 MJ

Conceptual Problems

9.30 $\Delta H > 0$ and $w < 0$

9.32 (a) (b)

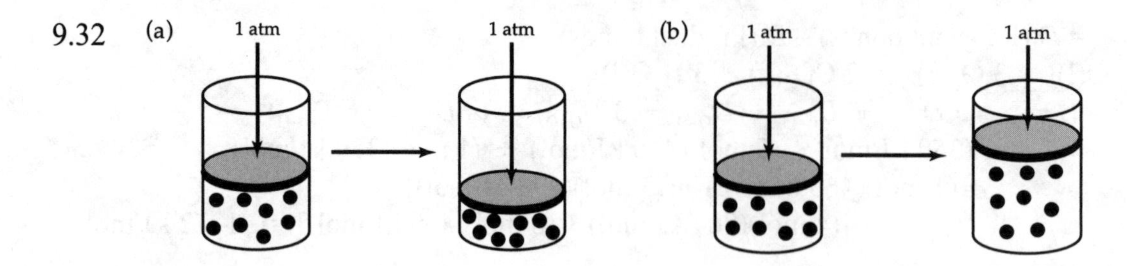

9.34 (a) Diagram 1 (b) $\Delta H^\circ = +30$ kJ

9.36 (a) $2\ AB_2 \rightarrow A_2 + 2\ B_2$
(b) Because the reaction is exothermic, ΔH is negative. Because the number of molecules increases going from reactants to products, ΔS is positive.
(c) When ΔH is negative and ΔS is positive, the reaction is likely to be spontaneous at all temperatures.

9.38 The change is the spontaneous conversion of a liquid to a gas. ΔG is negative because the change is spontaneous. The conversion of a liquid to a gas is endothermic, therefore ΔH is positive. ΔS is positive because randomness increases when a liquid is converted to a gas.

Section Problems
Heat, Work, and Energy (Sections 9.1–9.3)

9.40 Heat is the energy transferred from one object to another as the result of a temperature difference between them. Temperature is a measure of the kinetic energy of molecular motion. Energy is the capacity to do work or supply heat. Work is defined as the distance moved times the force that opposes the motion (w = d x F).
Kinetic energy is the energy of motion. Potential energy is stored energy.

9.42 Car: $E_K = \frac{1}{2}(1400 \text{ kg})\left(\frac{115 \times 10^3 \text{ m}}{3600 \text{ s}}\right)^2 = 7.1 \times 10^5 \text{ J}$

Truck: $E_K = \frac{1}{2}(12{,}000 \text{ kg})\left(\frac{38 \times 10^3 \text{ m}}{3600 \text{ s}}\right)^2 = 6.7 \times 10^5 \text{ J}$

The car has more kinetic energy.

9.44 (a) and (b) are state functions; (c) is not.

9.46 $w = -P\Delta V = -(3.6 \text{ atm})(3.4 \text{ L} - 3.2 \text{ L}) = -0.72 \text{ L} \cdot \text{atm}$

$w = (-0.72 \text{ L} \cdot \text{atm})\left(\frac{101 \text{ J}}{1 \text{ L} \cdot \text{atm}}\right) = -72.7 \text{ J} = -70 \text{ J}$; The energy change is negative.

9.48 (a) 100 W = 100 J/s

$$50 \text{ Cal} \times \frac{1000 \text{ cal}}{1 \text{ Cal}} \times \frac{4.184 \text{ J}}{1 \text{ cal}} \times \frac{1 \text{ s}}{100 \text{ J}} \times \frac{1 \text{ min}}{60 \text{ s}} = 35 \text{ min}$$

(b) 23 W = 23 J/s

$$50 \text{ Cal} \times \frac{1000 \text{ cal}}{1 \text{ Cal}} \times \frac{4.184 \text{ J}}{1 \text{ cal}} \times \frac{1 \text{ s}}{23 \text{ J}} \times \frac{1 \text{ min}}{60 \text{ s}} = 150 \text{ min}$$

9.50 (a) $w = -P\Delta V = -(0.975 \text{ atm})(18.0 \text{ L} - 12.0 \text{ L}) = -5.85 \text{ L} \cdot \text{atm}$

$$w = (-5.85 \text{ L} \cdot \text{atm})\left(\frac{101.325 \text{ J}}{1 \text{ L} \cdot \text{atm}}\right) = -593 \text{ J}$$

(b) $\Delta V = 6.0 \text{ L} = 6000 \text{ mL} = 6000 \text{ cm}^3$; $r = \frac{17.0 \text{ cm}}{2} = 8.50 \text{ cm}$

$$V = \pi r^2 h; \; h = \frac{V}{\pi r^2} = \frac{6000 \text{ cm}^3}{\pi (8.50 \text{ cm})^2} = 26.4 \text{ cm}$$

Energy and Enthalpy (Section 9.4)

9.52 $\Delta E = q_v$ is the heat change associated with a reaction at constant volume. Since $\Delta V = 0$, no PV work is done.
$\Delta H = q_p$ is the heat change associated with a reaction at constant pressure. Since $\Delta V \neq 0$, PV work can also be done.

9.54 $\Delta H = \Delta E + P\Delta V$; ΔH and ΔE are nearly equal when there are no gases involved in a chemical reaction, or, if gases are involved, $\Delta V = 0$ (i.e., there are the same number of reactant and product gas molecules).

9.56 $\Delta V = 448$ L and assume P = 1.00 atm

$$w = -P\Delta V = -(1.00 \text{ atm})(448 \text{ L}) = -448 \text{ L} \cdot \text{atm}$$

$$w = -(448 \text{ L} \cdot \text{atm})\left(101 \frac{\text{J}}{\text{L} \cdot \text{atm}}\right) = -4.52 \times 10^4 \text{ J}$$

$$w = -4.52 \times 10^4 \text{ J} \times \frac{1 \text{ kJ}}{1000 \text{ J}} = -45.2 \text{ kJ}$$

9.58 $P\Delta V = -7.6$ J (from Problem 9.47)

$$\Delta H = \Delta E + P\Delta V$$

$$\Delta E = \Delta H - P\Delta V = -0.31 \text{ kJ} - (-7.6 \times 10^{-3} \text{ kJ}) = -0.30 \text{ kJ}$$

9.60 $$\Delta H = \Delta E + P\Delta V = 44.0 \text{ kJ} + (1.00 \text{ atm})(14.0 \text{ L})\left(\frac{101.325 \text{ J}}{1 \text{ L} \cdot \text{atm}} \times \frac{1 \text{ kJ}}{1000 \text{ J}}\right) = 45.4 \text{ kJ}$$

Thermochemical Equations for Chemical and Physical Changes (Sections 9.5 and 9.6)

9.62 (a) heat is transferred to the system, endothermic, $\Delta H^\circ > 0$
(b) heat is transferred to the surroundings, exothermic, $\Delta H^\circ < 0$
(c) heat is transferred to the system, endothermic, $\Delta H^\circ > 0$

9.64 $C_4H_{10}O$, 74.12; mass of $C_4H_{10}O$ = (0.7138 g/mL)(100 mL) = 71.38 g

$$\text{mol } C_4H_{10}O = 71.38 \text{ g} \times \frac{1 \text{ mol}}{74.12 \text{ g}} = 0.9630 \text{ mol}$$

$$q = n \times \Delta H_{vap} = 0.9630 \text{ mol} \times 26.5 \text{ kJ/mol} = 25.5 \text{ kJ}$$

9.66 Al, 26.98

$$\text{mol Al} = 5.00 \text{ g} \times \frac{1 \text{ mol}}{26.98 \text{ g}} = 0.1853 \text{ mol}$$

$$q = n \times \Delta H^\circ = 0.1853 \text{ mol Al} \times \frac{-1408.4 \text{ kJ}}{2 \text{ mol Al}} = -131 \text{ kJ};\ 131 \text{ kJ is released.}$$

9.68 (a) C_3H_8, 44.10; $\Delta H^\circ = -2220$ kJ/mol C_3H_8

$$15.5 \text{ g} \times \frac{1 \text{ mol } C_3H_8}{44.10 \text{ g } C_3H_8} \times \frac{-2220 \text{ kJ}}{1 \text{ mol } C_3H_8} = -780. \text{ kJ}$$

780. kJ of heat is evolved.

(b) $Ba(OH)_2 \cdot 8\,H_2O$, 315.5; $\Delta H^\circ = +80.3$ kJ/mol $Ba(OH)_2 \cdot 8\,H_2O$

$$4.88 \text{ g} \times \frac{1 \text{ mol } Ba(OH)_2 \cdot 8\,H_2O}{315.5 \text{ g } Ba(OH)_2 \cdot 8\,H_2O} \times \frac{80.3 \text{ kJ}}{1 \text{ mol } Ba(OH)_2 \cdot 8\,H_2O} = +1.24 \text{ kJ}$$

1.24 kJ of heat is absorbed.

9.70 Fe_2O_3, 159.7; mol $Fe_2O_3 = 2.50 \text{ g} \times \frac{1 \text{ mol}}{159.7 \text{ g}} = 0.015\ 65 \text{ mol}$

$$q = n \times \Delta H^\circ = 0.015\ 65 \text{ mol } Fe_2O_3 \times \frac{-24.8 \text{ kJ}}{1 \text{ mol } Fe_2O_3} = -0.388 \text{ kJ};\ 0.388 \text{ kJ is evolved.}$$

Because ΔH is negative, the reaction is exothermic.

Calorimetry and Heat Capacity (Section 9.7)

9.72 Heat capacity is the amount of heat required to raise the temperature of a substance a given amount. Specific heat is the amount of heat necessary to raise the temperature of exactly 1 g of a substance by exactly 1 °C.

9.74 Na, 22.99

$$\text{specific heat} = 28.2 \frac{\text{J}}{\text{mol} \cdot {}^\circ\text{C}} \times \frac{1 \text{ mol}}{22.99 \text{ g}} = 1.23 \text{ J/(g} \cdot {}^\circ\text{C)}$$

9.76 $q = (\text{specific heat}) \times m \times \Delta T = (4.18 \ \frac{J}{g\cdot{}^\circ C})(350 \ g)(3\ {}^\circ C - 25\ {}^\circ C) = -3.2 \times 10^4 \ J$

$q = -3.2 \times 10^4 \ J \times \frac{1 \ kJ}{1000 \ J} = -32 \ kJ$

9.78 NH_4NO_3, 80.04; assume 125 mL = 125 g H_2O

$50.0 \ g \ NH_4NO_3 \times \frac{1 \ mol \ NH_4NO_3}{80.04 \ g \ NH_4NO_3} = 0.625 \ mol \ NH_4NO_3$

$q_p = \Delta H \times n = (+25.7 \ kJ/mol)(0.625 \ mol) = 16.1 \ kJ = 16{,}100 \ J$

$q_{soln} = -q_p = -16{,}100 \ J$

$q_{soln} = (\text{specific heat}) \times m \times \Delta T$

$$\Delta T = \frac{q_{soln}}{(\text{specific heat}) \times m} = \frac{-16{,}100 \ J}{\left(4.18 \ \frac{J}{g\cdot{}^\circ C}\right)(50 \ g + 125 \ g)} = -22.0\ {}^\circ C$$

$\Delta T = -22.0\ {}^\circ C = T_{final} - T_{initial} = T_{final} - 25.0\ {}^\circ C$

$T_{final} = -22.0\ {}^\circ C + 25.0\ {}^\circ C = 3.0\ {}^\circ C$

9.80 Mass of solution = 50.0 g + 1.045 g = 51.0 g

$q = (\text{specific heat}) \times m \times \Delta T$

$$q = \left(4.18 \ \frac{J}{g\cdot{}^\circ C}\right)(51.0 \ g)(32.3\ {}^\circ C - 25.0\ {}^\circ C) = 1.56 \times 10^3 \ J = 1.56 \ kJ$$

CaO, 56.08; $\text{mol CaO} = 1.045 \ g \times \frac{1 \ mol}{56.08 \ g} = 0.018\ 63 \ mol$

$\text{Heat evolved per mole of CaO} = \frac{1.56 \ kJ}{0.018\ 63 \ mol} = 83.7 \ kJ/mol \ CaO$

Because the reaction evolves heat, the sign for ΔH is negative. $\Delta H = -83.7$ kJ

9.82 C_6H_6, 78.11

$$\Delta E = q_V = -q_{H_2O} = -\left(4.18 \ \frac{J}{g\cdot{}^\circ C}\right)(250.0 \ g)(4.53\ {}^\circ C) - (525 \ J/{}^\circ C)(4.53\ {}^\circ C)$$

$\Delta E = -7112 \ J = -7.11 \ kJ$

$0.187 \ g \ C_6H_6 \times \frac{1 \ mol \ C_6H_6}{78.11 \ g \ C_6H_6} = 0.002\ 39 \ mol \ C_6H_6$

$\Delta E(\text{per mole}) = (-7.11 \ kJ)/(0.002\ 39 \ mol) = -2.97 \times 10^3 \ kJ/mol$

$\Delta E \ (\text{per gram } C_6H_6) = (-2.97 \times 10^3 \ kJ/mol)/(78.11 \ g/mol) = -38.1 \ kJ/g$

Hess's Law and Heats of Formation (Sections 9.8 and 9.9)

9.84 The standard state of an element is its most stable form at 1 atm and the specified temperature, usually 25 °C.
Because enthalpy is a state function, ΔH is the same regardless of the path taken between

reactants and products. Thus, the sum of the enthalpy changes for the individual steps in a reaction is equal to the overall enthalpy change for the entire reaction, a relationship known as Hess's law. Hess's law works because of the law of the conservation of energy.

9.86

$CH_4(g) + Cl_2(g) \rightarrow CH_3Cl(g) + HCl(g)$ $\quad \Delta H°_1 = -98.3$ kJ

$\underline{CH_3Cl(g) + Cl_2(g) \rightarrow CH_2Cl_2(g) + HCl(g)}$ $\quad \Delta H°_2 = -104$ kJ

Sum $CH_4(g) + 2\ Cl_2(g) \rightarrow CH_2Cl_2(g) + 2\ HCl(g)$

$\Delta H° = \Delta H°_1 + \Delta H°_2 = -202$ kJ

9.88 A compound's standard heat of formation is the amount of heat associated with the formation of 1 mole of a compound from its elements (in their standard states).

9.90 (a) Cl_2, gas (b) Hg, liquid (c) CO_2, gas (d) Ga, solid

9.92 (a) $2\ Fe(s) + 3/2\ O_2(g) \rightarrow Fe_2O_3(s)$

(b) $12\ C(s) + 11\ H_2(g) + 11/2\ O_2(g) \rightarrow C_{12}H_{22}O_{11}(s)$

(c) $U(s) + 3\ F_2(g) \rightarrow UF_6(s)$

9.94

$S(s) + O_2(g) \rightarrow SO_2(g)$ $\quad \Delta H°_1 = -296.8$ kJ

$\underline{SO_2 + ½\ O_2(g) \rightarrow SO_3(g)}$ $\quad \Delta H°_2 = -98.9$ kJ

Sum $S(s) + 3/2\ O_2(g) \rightarrow SO_3(g)$ $\quad \Delta H°_3 = \Delta H°_1 + \Delta H°_2$

$\Delta H°_f = \Delta H°_3 = -296.8\text{ kJ} + (-98.9\text{ kJ}) = -395.7$ kJ/mol

9.96

$SO_3(g) + H_2O(l) \rightarrow H_2SO_4(aq)$ $\quad \Delta H°_1 = -227.8$ kJ

$H_2(g) + ½\ O_2(g) \rightarrow H_2O(l)$ $\quad \Delta H°_2 = \Delta H°_f = -285.8$ kJ

$\underline{S(s) + 3/2\ O_2(g) \rightarrow SO_3(g)}$ $\quad \Delta H°_3 = \Delta H°_f = -395.7$

Sum $S(s) + H_2(g) + 2\ O_2(g) \rightarrow H_2SO_4(aq)$ $\quad \Delta H°_f\ (H_2SO_4) = ?$

$\Delta H°_f\ (H_2SO_4) = \Delta H°_1 + \Delta H°_2 + \Delta H°_3 = -909.3$ kJ

9.98 $C_8H_8(l) + 10\ O_2(g) \rightarrow 8\ CO_2(g) + 4\ H_2O(l)$

$\Delta H°_{rxn} = \Delta H°_c = -4395$ kJ

$\Delta H°_{rxn} = [8\ \Delta H°_f(CO_2) + 4\ \Delta H°_f(H_2O)] - \Delta H°_f(C_8H_8)$

$-4395\text{ kJ} = [(8\text{ mol})(-393.5\text{ kJ/mol}) + (4\text{ mol})(-285.8\text{ kJ/mol})] - [(1\text{ mol})(\Delta H°_f(C_8H_8))]$

Solve for $\Delta H°_f(C_8H_8)$

$-4395\text{ kJ} = -4291.2\text{ kJ} - (1\text{ mol})(\Delta H°_f(C_8H_8))$

$-103.8\text{ kJ} = -(1\text{ mol})(\Delta H°_f(C_8H_8))$

$$\Delta H°_f(C_8H_8) = \frac{-103.8\text{ kJ}}{-1\text{ mol}} = +103.8\text{ kJ/mol} = +104\text{ kJ/mol}$$

9.100 $\Delta H°_{rxn} = \Delta H°_f(\text{MTBE}) - [\Delta H°_f(\text{2-Methylpropene}) + \Delta H°_f(CH_3OH)]$

$-57.5\text{ kJ} = -313.6\text{ kJ} - [(1\text{ mol})(\Delta H°_f(\text{2-Methylpropene})) + (-239.2\text{ kJ})]$

Solve for $\Delta H°_f$(2-Methylpropene).

$-16.9\text{ kJ} = (1\text{ mol})(\Delta H°_f(\text{2-Methylpropene}))$

$\Delta H°_f(\text{2-Methylpropene}) = -16.9$ kJ/mol

9.102 $CaCO_3(s) \rightarrow CaO(s) + CO_2(g)$

$\Delta H^\circ_{rxn} = [\Delta H^\circ_f(CaO) + \Delta H^\circ_f(CO_2)] - \Delta H^\circ_f(CaCO_3)$

$\Delta H^\circ_{rxn} = [(1\text{ mol})(-634.9\text{ kJ/mol}) + (1\text{ mol})(-393.5\text{ kJ/mol})] - [(1\text{ mol})(-1207.6\text{ kJ/mol})]$

$\Delta H^\circ_{rxn} = +179.2\text{ kJ}$

9.104 $CaCO_3(s) \rightarrow CaO(s) + CO_2(g)$

$\Delta H^\circ_{rxn} = [\Delta H^\circ_f(CaO) + \Delta H^\circ_f(CO_2)] - \Delta H^\circ_f(CaCO_3)$

$\Delta H^\circ_{rxn} = [(1\text{ mol})(-634.9\text{ kJ/mol}) + (1\text{ mol})(-393.5\text{ kJ/mol})] - [(1\text{ mol})(-1207.6\text{ kJ/mol})]$

$\Delta H^\circ_{rxn} = +179.2\text{ kJ}$

Bond Dissociation Energies (Section 9.10)

9.106 $H_2C{=}CH_2(g) + H_2(g) \rightarrow CH_3CH_3(g)$

$\Delta H^\circ_{rxn} = D\text{ (Reactant bonds)} - D\text{ (Product bonds)}$

$\Delta H^\circ_{rxn} = (D_{C=C} + 4\,D_{C\text{–}H} + D_{H\text{–}H}) - (6\,D_{C\text{–}H} + D_{C\text{–}C})$

$\Delta H^\circ_{rxn} = [(1\text{ mol})(728\text{ kJ/mol}) + (4\text{ mol})(410\text{ kJ/mol}) + (1\text{ mol})(436\text{ kJ/mol})]$
$- [(6\text{ mol})(410\text{ kJ/mol}) + (1\text{ mol})(350\text{ kJ/mol})] = -6\text{ kJ}$

9.108 $\Delta H^\circ_{rxn} = D\text{ (Reactant bonds)} - D\text{ (Product bonds)}$

(a) $2\,CH_4(g) \rightarrow C_2H_6(g) + H_2(g)$

$\Delta H^\circ_{rxn} = (8\,D_{C\text{–}H}) - (D_{C\text{–}C} + 6\,D_{C\text{–}H} + D_{H\text{–}H})$

$\Delta H^\circ_{rxn} = [(8\text{ mol})(410\text{ kJ/mol})] - [(1\text{ mol})(350\text{ kJ/mol}) + (6\text{ mol})(410\text{ kJ/mol})$
$+ (1\text{ mol})(436\text{ kJ/mol})] = +34\text{ kJ}$

(b) $C_2H_6(g) + F_2(g) \rightarrow C_2H_5F(g) + HF(g)$

$\Delta H^\circ_{rxn} = (6\,D_{C\text{–}H} + D_{C\text{–}C} + D_{F\text{–}F}) - (5\,D_{C\text{–}H} + D_{C\text{–}C} + D_{C\text{–}F} + D_{H\text{–}F})$

$\Delta H^\circ_{rxn} = [(6\text{ mol})(410\text{ kJ/mol}) + (1\text{ mol})(350\text{ kJ/mol}) + (1\text{ mol})(159\text{ kJ/mol})]$
$- [(5\text{ mol})(410\text{ kJ/mol}) + (1\text{ mol})(350\text{ kJ/mol}) + (1\text{ mol})(450\text{ kJ/mol})$
$+ (1\text{ mol})(570\text{ kJ/mol})] = -451\text{ kJ}$

(c) $N_2(g) + 3\,H_2(g) \rightarrow 2\,NH_3(g)$

The bond dissociation energy for N_2 is 945 kJ/mol.

$\Delta H^\circ_{rxn} = (D_{N_2} + 3\,D_{H\text{–}H}) - (6\,D_{N\text{–}H})$

$\Delta H^\circ_{rxn} = [(1\text{ mol})(945\text{ kJ/mol}) + (3\text{ mol})(436\text{ kJ/mol})] - [(6\text{ mol})(390\text{ kJ/mol})] = -87\text{ kJ}$

Fossil Fuels and Heats of Combustion (Section 9.11)

9.110 $C_4H_{10}(l) + \frac{13}{2}\,O_2(g) \rightarrow 4\,CO_2(g) + 5\,H_2O(g)$

$\Delta H^\circ_{rxn} = [4\,\Delta H^\circ_f(CO_2) + 5\,\Delta H^\circ_f(H_2O)] - \Delta H^\circ_f(C_4H_{10})$

$\Delta H^\circ_{rxn} = [(4\text{ mol})(-393.5\text{ kJ/mol}) + (5\text{ mol})(-241.8\text{ kJ/mol})] - [(1\text{ mol})(-147.5\text{ kJ/mol})]$

$\Delta H^\circ_{rxn} = -2635.5\text{ kJ};\quad \Delta H^\circ_C = -2635.5\text{ kJ/mol}$

C_4H_{10}, 58.12; $\Delta H^\circ_C = \left(-2635.5\,\frac{\text{kJ}}{\text{mol}}\right)\left(\frac{1\text{ mol}}{58.12\text{ g}}\right) = -45.35\text{ kJ/g}$

$\Delta H^\circ_C = \left(-45.35\,\frac{\text{kJ}}{\text{g}}\right)\left(0.579\,\frac{\text{g}}{\text{mL}}\right) = -26.3\text{ kJ/mL}$

Free Energy and Entropy (Sections 9.12 and 9.13)

9.112 Entropy is a measure of molecular randomness.

9.114 A reaction can be spontaneous yet endothermic if ΔS is positive (more randomness) and the $T\Delta S$ term is larger than ΔH.

9.116 (a) positive (more randomness) (b) negative (less randomness)

9.118 (a) zero (equilibrium) (b) zero (equilibrium)
(c) negative (spontaneous)

9.120 ΔS is positive. The reaction increases the total number of molecules.

9.122 $\Delta G = \Delta H - T\Delta S$
(a) $\Delta G = -48\text{ kJ} - (400\text{ K})(135 \times 10^{-3}\text{ kJ/K}) = -102\text{ kJ}$
$\Delta G < 0$, spontaneous; $\Delta H < 0$, exothermic.
(b) $\Delta G = -48\text{ kJ} - (400\text{ K})(-135 \times 10^{-3}\text{ kJ/K}) = +6\text{ kJ}$
$\Delta G > 0$, nonspontaneous; $\Delta H < 0$, exothermic.
(c) $\Delta G = +48\text{ kJ} - (400\text{ K})(135 \times 10^{-3}\text{ kJ/K}) = -6\text{ kJ}$
$\Delta G < 0$, spontaneous; $\Delta H > 0$, endothermic.
(d) $\Delta G = +48\text{ kJ} - (400\text{ K})(-135 \times 10^{-3}\text{ kJ/K}) = +102\text{ kJ}$
$\Delta G > 0$, nonspontaneous; $\Delta H > 0$, endothermic.

9.124 $\Delta G = \Delta H - T\Delta S$; Set $\Delta G = 0$ and solve for T (the crossover temperature).

$$T = \frac{\Delta H}{\Delta S} = \frac{-33\text{ kJ}}{-0.058\text{ kJ/K}} = 570\text{ K}$$

9.126 (a) $\Delta H < 0$ and $\Delta S > 0$; reaction is spontaneous at all temperatures.
(b) $\Delta H < 0$ and $\Delta S < 0$; reaction has a crossover temperature.
(c) $\Delta H > 0$ and $\Delta S > 0$; reaction has a crossover temperature.
(d) $\Delta H > 0$ and $\Delta S < 0$; reaction is nonspontaneous at all temperatures.

9.128 $T = -114.1\text{ °C} = 273.15 + (-114.1) = 159.0\text{ K}$
$\Delta G_{fus} = \Delta H_{fus} - T\Delta S_{fus}$; $\Delta G = 0$ at the melting point temperature.
Set $\Delta G = 0$ and solve for ΔS_{fus}.
$\Delta G = 0 = \Delta H_{fus} - T\Delta S_{fus}$

$$\Delta S_{fus} = \frac{\Delta H_{fus}}{T} = \frac{5.02\text{ kJ/mol}}{159.0\text{ K}} = 0.0316\text{ kJ/(K·mol)} = 31.6\text{ J/(K·mol)}$$

Chapter Problems

9.130 $\Delta H = \Delta E + P\Delta V$
$\Delta H = -5.67\text{ kJ} = -7.20\text{ kJ} + P\Delta V$
$P\Delta V = +1.53\text{ kJ}$

$w = -P\Delta V = -1.53\ kJ$

$$w = -P\Delta V = -1.53\ kJ \times \frac{1000\ J}{1\ kJ} \times \frac{1}{\left(\frac{101\ J}{L\cdot atm}\right)} = -15.1\ L\cdot atm$$

$P\Delta V = 15.1\ L\cdot atm = (1\ atm)(\Delta V)$
$\Delta V = 15.1\ L$

9.132 C_2H_4, 28.05; HCl, 36.46
$w = -P\Delta V = -(1.00\ atm)(-71.5\ L) = 71.5\ L\cdot atm$

$$w = (71.5\ L\cdot atm)\left(\frac{101\ J}{1\ L\cdot atm}\right) = 7222\ J = 7.22\ kJ$$

$$89.5\ g\ C_2H_4 \times \frac{1\ mol\ C_2H_4}{28.05\ g\ C_2H_4} = 3.19\ mol\ C_2H_4; \quad 125\ g\ HCl \times \frac{1\ mol\ HCl}{36.46\ g\ HCl} = 3.43\ mol\ HCl$$

Because the reaction stoichiometry between C_2H_4 and HCl is 1:1, C_2H_4 is the limiting reactant.
$\Delta H^\circ = -72.3\ kJ/mol\ C_2H_4$
$q = (-72.3\ kJ/mol)(3.19\ mol) = -231\ kJ$
$\Delta E = \Delta H - P\Delta V = -231\ kJ - (-7.22\ kJ) = -224\ kJ$

9.134 (a) $C(s) + CO_2(g) \rightarrow 2\ CO(g)$
$\Delta H^\circ_{rxn} = [2\ \Delta H^\circ_f(CO)] - \Delta H^\circ_f(CO_2)$
$\Delta H^\circ_{rxn} = [(2\ mol)(-110.5\ kJ/mol)] - [(1\ mol)(-393.5\ kJ/mol)] = +172.5\ kJ$
(b) $2\ H_2O_2(aq) \rightarrow 2\ H_2O(l) + O_2(g)$
$\Delta H^\circ_{rxn} = [2\ \Delta H^\circ_f(H_2O)] - [2\ \Delta H^\circ_f(H_2O_2)]$
$\Delta H^\circ_{rxn} = [(2\ mol)(-285.8\ kJ/mol)] - [(2\ mol)(-191.2\ kJ/mol)] = -189.2\ kJ$
(c) $Fe_2O_3(s) + 3\ CO(g) \rightarrow 2\ Fe(s) + 3\ CO_2(g)$
$\Delta H^\circ_{rxn} = [3\ \Delta H^\circ_f(CO_2)] - [\Delta H^\circ_f(Fe_2O_3) + 3\ \Delta H^\circ_f(CO)]$
$\Delta H^\circ_{rxn} = [(3\ mol)(-393.5\ kJ/mol)]$
$- [(1\ mol)(-824.2\ kJ/mol) + (3\ mol)(-110.5\ kJ/mol)] = -24.8\ kJ$

9.136 $\Delta G = \Delta H - T\Delta S$; at equilibrium $\Delta G = 0$. Set $\Delta G = 0$ and solve for T.

$$\Delta G = 0 = \Delta H - T\Delta S; \qquad T = \frac{\Delta H}{\Delta S} = \frac{30.91\ kJ/mol}{93.2 \times 10^{-3}\ kJ/(K\cdot mol)} = 332\ K = 59\ ^\circ C$$

9.138 $HgS(s) + O_2(g) \rightarrow Hg(l) + SO_2(g)$
(a) $\Delta H^\circ_{rxn} = \Delta H^\circ_f(SO_2) - \Delta H^\circ_f(HgS)$
$\Delta H^\circ_{rxn} = [(1\ mol)(-296.8\ kJ/mol)] - [(1\ mol)(-58.2\ kJ/mol)] = -238.6\ kJ$
(b) and (c) Because $\Delta H < 0$ and $\Delta S > 0$, the reaction is spontaneous at all temperatures.

9.140 (a) $2\ C_8H_{18}(l) + 25\ O_2(g) \rightarrow 16\ CO_2(g) + 18\ H_2O(l)$
(b) $C_8H_{18}(l) + 25/2\ O_2(g) \rightarrow 8\ CO_2(g) + 9\ H_2O(l)$
$\Delta H^\circ_{rxn} = \Delta H^\circ_c = -5461\ kJ$
$\Delta H^\circ_{rxn} = [8\ \Delta H^\circ_f(CO_2) + 9\ \Delta H^\circ_f(H_2O)] - \Delta H^\circ_f(C_8H_{18})$
$-5461\ kJ = [(8\ mol)(-393.5\ kJ/mol) + (9\ mol)(-285.8\ kJ/mol)] - [(1\ mol)(\Delta H^\circ_f(C_8H_{18}))]$
Solve for $\Delta H^\circ_f(C_8H_{18})$.

$-5461 \text{ kJ} = -5720.2 \text{ kJ} - [(1 \text{ mol})(\Delta H^\circ_f(C_8H_{18}))]$
$259 \text{ kJ} = -(1 \text{ mol})(\Delta H^\circ_f(C_8H_{18}))$
$\Delta H^\circ_f(C_8H_{18}) = -259 \text{ kJ/mol}$

9.142 (a) $\Delta S_{total} = \Delta S_{system} + \Delta S_{surr}$ and $\Delta S_{surr} = -\Delta H/T$
$\Delta S_{total} = \Delta S_{system} + (-\Delta H/T) = \Delta S_{system} - \Delta H/T$
$\Delta S_{system} = \Delta S_{total} + \Delta H/T$
$\Delta G = \Delta H - T\Delta S$ (substitute ΔS_{system} for ΔS in this equation)
$\Delta G = \Delta H - T(\Delta S_{total} + \Delta H/T) = -T\Delta S_{total}$
$\Delta G = -T\Delta S_{total}$ For a spontaneous reaction, if $\Delta S_{total} > 0$ then $\Delta G < 0$.
(b) $\Delta G^\circ = \Delta H^\circ - T\Delta S^\circ$

$\Delta H^\circ = \Delta G^\circ + T\Delta S^\circ$

$$\Delta S_{surr} = -\frac{\Delta H^\circ}{T} = -\frac{[\Delta G^\circ + T\Delta S^\circ]}{T} = -\frac{[2879 \times 10^3 \text{ J/mol} + (298 \text{ K})(-262 \text{ J/(K}\cdot\text{mol}))]}{298 \text{ K}}$$

$\Delta S_{surr} = -9399 \text{ J/(K}\cdot\text{mol)}$

9.144 $q_{Mo} = (110.0 \text{ g})(\text{specific heat Mo})(28.0\ ^\circ\text{C} - 100.0\ ^\circ\text{C})$
$q_{H_2O} = (150.0 \text{ g})[4.184 \text{ J/(g}\cdot{}^\circ\text{C)}](28.0\ ^\circ\text{C} - 24.6\ ^\circ\text{C})$

$q_{Mo} = -q_{H_2O}$

$(110.0 \text{ g})(\text{specific heat Mo})(28.0\ ^\circ\text{C} - 100.0\ ^\circ\text{C}) = -(150.0 \text{ g})[4.184 \text{ J/(g}\cdot{}^\circ\text{C)}](28.0\ ^\circ\text{C} - 24.6\ ^\circ\text{C})$

$$\text{specific heat Mo} = \frac{-(150.0 \text{ g})[4.184 \text{ J/(g}\cdot{}^\circ\text{C)}](28.0\ ^\circ\text{C} - 24.6\ ^\circ\text{C})}{(110.0 \text{ g})(28.0\ ^\circ\text{C} - 100.0\ ^\circ\text{C})} = 0.27 \text{ J/(g}\cdot{}^\circ\text{C)}$$

9.146 There is a large excess of NaOH. 5.00 mL = 0.005 00 L
mol citric acid = (0.005 00 L)(0.64 mol/L) = 0.0032 mol citric acid
$q_{H_2O} = (51.6 \text{ g})[4.0 \text{ J/(g}\cdot{}^\circ\text{C)}](27.9\ ^\circ\text{C} - 26.0\ ^\circ\text{C}) = 392 \text{ J}$

$q_{rxn} = -q_{H_2O} = -392 \text{ J}$

$$\Delta H = -\frac{392 \text{ J}}{0.0032 \text{ mol}} \times \frac{1 \text{ kJ}}{1000 \text{ J}} = -123 \text{ kJ/mol} = -120 \text{ kJ/mol citric acid}$$

9.148 $NaNO_3$, 84.99; KF, 58.10

For $NaNO_3(s) \rightarrow NaNO_3(aq)$, $q = 20.4 \text{ kJ/mol} \times \frac{1 \text{ mol } NaNO_3}{84.99 \text{ g } NaNO_3} = 0.240 \text{ kJ/g}$

For $KF(s) \rightarrow KF(aq)$, $q = -17.7 \text{ kJ/mol} \times \frac{1 \text{ mol KF}}{58.10 \text{ g KF}} = -0.305 \text{ kJ/g}$

$q_{soln} = (110.0 \text{ g})[4.18 \text{ J/(g}\cdot{}^\circ\text{C)}](2.22\ ^\circ\text{C}) = 1021 \text{ J} = 1.02 \text{ kJ}$

$q_{rxn} = -q_{soln} = -1.02 \text{ kJ}$

Let X = mass of $NaNO_3$ and Y = mass of KF
X + Y = 10.0 g, so Y = 10.0 g – X
$q_{rxn} = -1.02 \text{ kJ} = X(0.240 \text{ kJ/g}) + Y(-0.305 \text{ kJ/g})$ (substitute for Y and solve for X)
$-1.02 \text{ kJ} = X(0.240 \text{ kJ/g}) + (10.0 \text{ g} - X)(-0.305 \text{ kJ/g})$

$-1.02 \text{ kJ} = (0.240 \text{ kJ})X - 3.05 \text{ kJ} + (0.305 \text{ kJ})X$
$2.03 \text{ kJ} = (0.545 \text{ kJ})X$
$X = \frac{2.03 \text{ kJ}}{0.545 \text{ kJ}} = 3.72 \text{ g NaNO}_3$
$Y = 10.0 \text{ g} - X = 10.0 \text{ g} - 3.72 \text{ g} = 6.28 \text{ g KF} = 6.3 \text{ g KF}$

Multiconcept Problems

9.150 (a) Each S has 2 bonding pairs and 2 lone pairs of electrons. Each S is sp^3 hybridized and the geometry around each S is bent.
(b) $\Delta H = D\text{ (reactant bonds)} - D\text{ (product bonds)} = (8\ D_{S\text{–}S}) - (4\ D_{S=S}) = +237 \text{ kJ}$
$\Delta H = [(8 \text{ mol})(225 \text{ kJ/mol}) - [(4 \text{ mol})(D_{S=S})] = +237 \text{ kJ}$
$-(4 \text{ mol})(D_{S=S}) = 237 \text{ kJ} - 1800 \text{ kJ} = -1563 \text{ kJ}$
$D_{S=S} = (1563 \text{ kJ})/(4 \text{ mol}) = 391 \text{ kJ/mol}$
(c)
σ^*_{3p} __
π^*_{3p} ↑ ↑
π_{3p} ↑↓ ↑↓
σ_{3p} ↑↓
σ^*_{3s} ↑↓
σ_{3s} ↑↓
S_2
S_2 should be paramagnetic with two unpaired electrons in the π^*_{3p} MOs.

9.152 (a) (1) $2\ CH_3CO_2H(l) + Na_2CO_3(s) \rightarrow 2\ CH_3CO_2Na(aq) + CO_2(g) + H_2O(l)$
(2) $CH_3CO_2H(l) + NaHCO_3(s) \rightarrow CH_3CO_2Na(aq) + CO_2(g) + H_2O(l)$
(b) CH_3CO_2H, 60.05; Na_2CO_3, 105.99; $NaHCO_3$, 84.01

$$1 \text{ gal} \times \frac{3.7854 \text{ L}}{1 \text{ gal}} \times \frac{1 \text{ mL}}{1 \times 10^{-3} \text{ L}} \times \frac{1.049 \text{ g } CH_3CO_2H}{1 \text{ mL}} = 3971 \text{ g } CH_3CO_2H$$

$$3971 \text{ g } CH_3CO_2H \times \frac{1 \text{ mol } CH_3CO_2H}{60.05 \text{ g } CH_3CO_2H} = 66.13 \text{ mol } CH_3CO_2H$$

For reaction (1)

$$66.13 \text{ mol } CH_3CO_2H \times \frac{1 \text{ mol } Na_2CO_3}{2 \text{ mol } CH_3CO_2H} \times \frac{105.99 \text{ g } Na_2CO_3}{1 \text{ mol } Na_2CO_3} \times \frac{1 \text{ kg}}{1000 \text{ g}} = 3.505 \text{ kg } Na_2CO_3$$

For reaction (2)

$$66.13 \text{ mol } CH_3CO_2H \times \frac{1 \text{ mol } NaHCO_3}{1 \text{ mol } CH_3CO_2H} \times \frac{84.01 \text{ g } NaHCO_3}{1 \text{ mol } NaHCO_3} \times \frac{1 \text{ kg}}{1000 \text{ g}} = 5.556 \text{ kg } NaHCO_3$$

(c) $2\ CH_3CO_2H(l) + Na_2CO_3(s) \rightarrow 2\ CH_3CO_2Na(aq) + CO_2(g) + H_2O(l)$
$\Delta H^\circ_{rxn} = [2\ \Delta H^\circ_f(CH_3CO_2Na) + \Delta H^\circ_f(CO_2) + \Delta H^\circ_f(H_2O)] - [2\ \Delta H^\circ_f(CH_3CO_2H) + \Delta H^\circ_f(Na_2CO_3)]$
$\Delta H^\circ_{rxn} = [(2 \text{ mol})(-726.1 \text{ kJ/mol}) + (1 \text{ mol})(-393.5 \text{ kJ/mol}) + (1 \text{ mol})(-285.8 \text{ kJ/mol})] - [(2 \text{ mol})(-484.5 \text{ kJ/mol}) + (1 \text{ mol})(-1130.7 \text{ kJ/mol})]$
$\Delta H^\circ_{rxn} = -31.8 \text{ kJ for } 2 \text{ mol } CH_3CO_2H$

$$\text{Heat} = -\frac{31.8\ \text{kJ}}{2\ \text{mol}\ CH_3CO_2H} \times 66.13\ \text{mol}\ CH_3CO_2H = -1050\ \text{kJ (liberated)}$$

$CH_3CO_2H(l) + NaHCO_3(s) \rightarrow CH_3CO_2Na(aq) + CO_2(g) + H_2O(l)$

$\Delta H^\circ_{rxn} = [\Delta H^\circ_f(CH_3CO_2Na) + \Delta H^\circ_f(CO_2) + \Delta H^\circ_f(H_2O)]$
$\quad - [\Delta H^\circ_f(CH_3CO_2H) + \Delta H^\circ_f(NaHCO_3)]$

$\Delta H^\circ_{rxn} = [(1\ \text{mol})(-726.1\ \text{kJ/mol}) + (1\ \text{mol})(-393.5\ \text{kJ/mol}) + (1\ \text{mol})(-285.8\ \text{kJ/mol})]$
$\quad - [(1\ \text{mol})(-484.5\ \text{kJ/mol}) + (1\ \text{mol})(-950.8\ \text{kJ/mol})]$

$\Delta H^\circ_{rxn} = +29.9\ \text{kJ for 1 mol}\ CH_3CO_2H$

$$q = \frac{29.9\ \text{kJ}}{1\ \text{mol}\ CH_3CO_2H} \times 66.13\ \text{mol}\ CH_3CO_2H = +1980\ \text{kJ (absorbed)}$$

9.154 (a)

```
    H   H
    |   |
H — N — N — H
    ••  ••
```

Each N is sp[3] hybridized and the geometry about each N is trigonal pyramidal.

(b)

Reaction	ΔH°
$2/4\ NH_3(g) + 3/4\ N_2O(g) \rightarrow N_2(g) + 3/4\ H_2O(l)$	$\Delta H^\circ_1 = \frac{-1011.2\ \text{kJ}}{4}$
$1/4\ H_2O(l) + 1/4\ N_2H_4(l) \rightarrow 1/8\ O_2(g) + 1/2\ NH_3(g)$	$\Delta H^\circ_2 = \frac{+286\ \text{kJ}}{8}$
$3/4\ N_2H_4(l) + 3/4\ H_2O(l) \rightarrow 3/4\ N_2O(g) + 9/4\ H_2(g)$	$\Delta H^\circ_3 = \frac{(3)(+317\ \text{kJ})}{4}$
$9/4\ H_2(g) + 9/8\ O_2(g) \rightarrow 9/4\ H_2O(l)$	$\Delta H^\circ_4 = \frac{(9)(-285.8\ \text{kJ})}{4}$
Sum $N_2H_4(l) + O_2(g) \rightarrow N_2(g) + 2\ H_2O(l)$	$\Delta H^\circ = -622\ \text{kJ}$

(c) N_2H_4, 32.045

$$\text{mol}\ N_2H_4 = 100.0\ \text{g}\ N_2H_4 \times \frac{1\ \text{mol}\ N_2H_4}{32.045\ \text{g}\ N_2H_4} = 3.12\ \text{mol}\ N_2H_4$$

$q = (3.12\ \text{mol}\ N_2H_4)(622\ \text{kJ/mol}) = 1940\ \text{kJ}$

10 Gases: Their Properties and Behavior

10.1 $P = 28.48 \text{ in Hg} \times \dfrac{2.54 \text{ cm}}{1 \text{ in}} \times \dfrac{10 \text{ mm}}{1 \text{ cm}} \times \dfrac{1 \text{ atm}}{760 \text{ mm Hg}} = 0.952 \text{ atm}$

$P = 28.48 \text{ in Hg} \times \dfrac{2.54 \text{ cm}}{1 \text{ in}} \times \dfrac{10 \text{ mm}}{1 \text{ cm}} \times \dfrac{101{,}325 \text{ Pa}}{760 \text{ mm Hg}} = 9.64 \times 10^4 \text{ Pa}$

$P = 9.64 \times 10^4 \text{ Pa} \times \dfrac{1 \text{ bar}}{10^5 \text{ Pa}} = 0.964 \text{ bar}$

10.2 (a) $P = 760 \text{ mm Hg} \times \dfrac{13.6 \text{ mm } H_2O}{1 \text{ mm Hg}} \times \dfrac{1 \times 10^{-3} \text{ m}}{1 \text{ mm}} = 10.3 \text{ m } H_2O$

(b) A barometer filled with water would be too tall to be practical.

10.3 The pressure in the flask is less than 0.975 atm because the liquid level is higher on the side connected to the flask. The 24.7 cm of Hg is the difference between the two pressures.

$\text{Pressure difference} = 24.7 \text{ cm Hg} \times \dfrac{1.00 \text{ atm}}{76.0 \text{ cm Hg}} = 0.325 \text{ atm}$

Pressure in flask = 0.975 atm – 0.325 atm = 0.650 atm

10.4 (a)

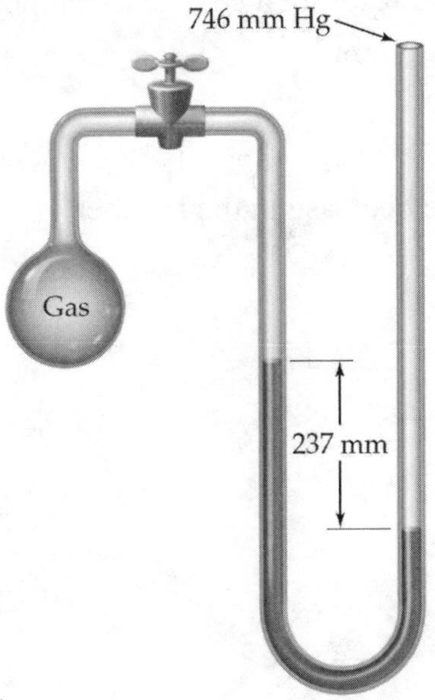

(b) (13.6 g/mL)/(0.822 g/mL) = 16.5

$P_{bulb} = 746 \text{ mm Hg} - \left(237 \text{ mm min oil} \times \dfrac{1 \text{ mm Hg}}{16.5 \text{ mm min oil}}\right) = 732 \text{ mm Hg}$

10.5 (a) Assume an initial volume of 1.00 L.

First consider the volume change resulting from a change in the number of moles with the pressure and temperature constant.

$$\frac{V_i}{n_i} = \frac{V_f}{n_f};\ V_f = \frac{V_i n_f}{n_i} = \frac{(1.00\ L)(0.225\ mol)}{0.3\ mol} = 0.75\ L$$

Now consider the volume change from 0.75 L as a result of a change in temperature with the number of moles and the pressure constant.

$$\frac{V_i}{T_i} = \frac{V_f}{T_f};\ V_f = \frac{V_i T_f}{T_i} = \frac{(0.75\ L)(400\ K)}{300\ K} = 1.0\ L$$

There is no net change in the volume as a result of the decrease in the number of moles of gas and a temperature increase.

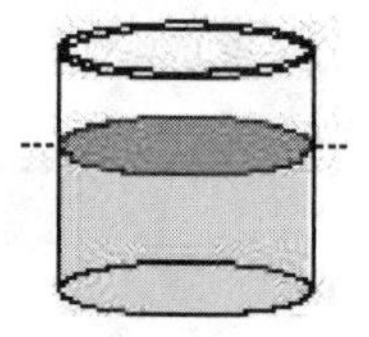

(b) Assume an initial volume of 1.00 L.

First consider the volume change resulting from a change in the number of moles with the pressure and temperature constant.

$$\frac{V_i}{n_i} = \frac{V_f}{n_f};\ V_f = \frac{V_i n_f}{n_i} = \frac{(1.00\ L)(0.225\ mol)}{0.3\ mol} = 0.75\ L$$

Now consider the volume change from 0.75 L as a result of a change in temperature with the number of moles and the pressure constant.

$$\frac{V_i}{T_i} = \frac{V_f}{T_f};\ V_f = \frac{V_i T_f}{T_i} = \frac{(0.75\ L)(200\ K)}{300\ K} = 0.5\ L$$

The volume would be cut in half as a result of the decrease in the number of moles of gas and a temperature decrease.

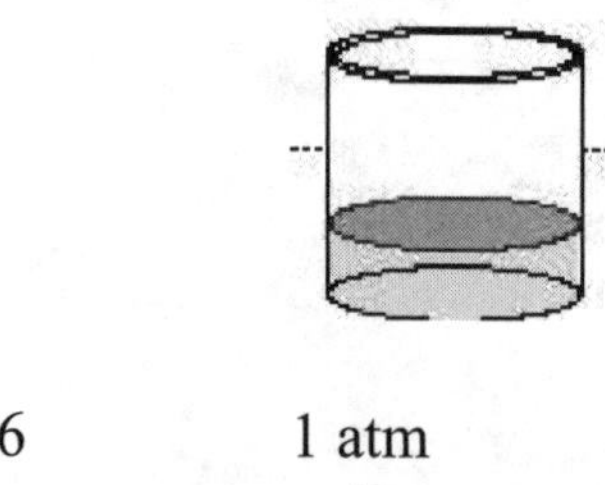

10.6 1 atm (a) (b)

10.7 $n = \frac{PV}{RT} = \frac{(1.000 \text{ atm})(1.000 \times 10^5 \text{ L})}{\left(0.082\ 06 \frac{\text{L} \cdot \text{atm}}{\text{K} \cdot \text{mol}}\right)(273.15 \text{ K})} = 4.461 \times 10^3 \text{ mol } CH_4$

CH_4, 16.04; mass $CH_4 = (4.461 \times 10^3 \text{ mol})\left(\frac{16.04 \text{ g}}{1 \text{ mol}}\right) = 7.155 \times 10^4 \text{ g } CH_4$

10.8 C_3H_8, 44.10; V = 350 mL = 0.350 L; T = 20 °C = 293 K

$n = 3.2 \text{ g} \times \frac{1 \text{ mol } C_3H_8}{44.10 \text{ g } C_3H_8} = 0.073 \text{ mol } C_3H_8$

$P = \frac{nRT}{V} = \frac{(0.073 \text{ mol})\left(0.082\ 06 \frac{\text{L} \cdot \text{atm}}{\text{K} \cdot \text{mol}}\right)(293 \text{ K})}{0.350 \text{ L}} = 5.0 \text{ atm}$

10.9 The volume and number of moles of gas remain constant.

$\frac{nR}{V} = \frac{P_i}{T_i} = \frac{P_f}{T_f}$; $T_f = \frac{P_f T_i}{P_i} = \frac{(2.37 \text{ atm})(273 \text{ K})}{2.15 \text{ atm}} = 301 \text{ K} = 28 \text{ °C}$

10.10 $\frac{P_i V_i}{T_i} = \frac{P_f V_f}{T_f}$

25.0 °C = 298 K; $V_f = \frac{P_i V_i T_f}{T_i P_f} = \frac{(745 \text{ mm Hg})(45.0 \text{ L})(225 \text{ K})}{(298 \text{ K})(178 \text{ mm Hg})} = 142 \text{ L}$

10.11 $CaCO_3(s) + 2\,HCl(aq) \rightarrow CaCl_2(aq) + CO_2(g) + H_2O(l)$

$CaCO_3$, 100.1; CO_2, 44.01

mole $CO_2 = 33.7 \text{ g } CaCO_3 \times \frac{1 \text{ mol } CaCO_3}{100.1 \text{ g } CaCO_3} \times \frac{1 \text{ mol } CO_2}{1 \text{ mol } CaCO_3} = 0.337 \text{ mol } CO_2$

mass $CO_2 = 0.337 \text{ mol } CO_2 \times \frac{44.01 \text{ g } CO_2}{1 \text{ mol } CO_2} = 14.8 \text{ g } CO_2$

$V = \frac{nRT}{P} = \frac{(0.337 \text{ mol})\left(0.082\ 06 \frac{\text{L} \cdot \text{atm}}{\text{K} \cdot \text{mol}}\right)(273 \text{ K})}{1.00 \text{ atm}} = 7.55 \text{ L}$

10.12 $3\,H_2(g) + N_2(g) \rightarrow 2\,NH_3(g)$

$V_{H_2} = 500.0 \text{ L } NH_3 \times \frac{3 \text{ mol } H_2}{2 \text{ mol } NH_3} = 750.0 \text{ L } H_2$

$V_{N_2} = 500.0 \text{ L } NH_3 \times \frac{1 \text{ mol } N_2}{2 \text{ mol } NH_3} = 250.0 \text{ L } N_2$

10.13 $n = \frac{PV}{RT} = \frac{(1.00 \text{ atm})(1.00 \text{ L})}{\left(0.082\ 06 \frac{\text{L} \cdot \text{atm}}{\text{K} \cdot \text{mol}}\right)(273 \text{ K})} = 0.0446 \text{ mol}$

molar mass = $\frac{1.52 \text{ g}}{0.0446 \text{ mol}} = 34.1$ g/mol; molecular mass = 34.1

$Na_2S(aq) + 2\ HCl(aq) \rightarrow H_2S(g) + 2\ NaCl(aq)$
The foul-smelling gas is H_2S, hydrogen sulfide.

10.14 (a) CO_2, 44.0

$d = \frac{PM}{RT} = \frac{(1 \text{ atm})(44.0 \text{ g/mol})}{\left(0.082\ 06 \frac{\text{L} \cdot \text{atm}}{\text{K} \cdot \text{mol}}\right)(298 \text{ K})} = 1.80 \text{ g/L}$

(b) Carbon dioxide has a higher molar mass (44.0 g/mol) than the major components of air N_2 (28.0 g/mol) and O_2 (32.0 g/mol).

10.15 $X_{O_2} = 0.36$; $X_{N_2} = 0.64$; 25.0 °C = 298 K

(a) $P_{O_2} = P_{tot} \cdot X_{O_2} = (50 \text{ atm})(0.36) = 18 \text{ atm}$

$P_{N_2} = P_{tot} \cdot X_{N_2} = (50 \text{ atm})(0.64) = 32 \text{ atm}$

(b) $n_{O_2} = \frac{PV}{RT} = \frac{(18 \text{ atm})(10.0 \text{ L})}{\left(0.082\ 06 \frac{\text{L} \cdot \text{atm}}{\text{K} \cdot \text{mol}}\right)(298 \text{ K})} = 7.36 \text{ mol } O_2$

$n_{N_2} = \frac{PV}{RT} = \frac{(32 \text{ atm})(10.0 \text{ L})}{\left(0.082\ 06 \frac{\text{L} \cdot \text{atm}}{\text{K} \cdot \text{mol}}\right)(298 \text{ K})} = 13.1 \text{mol } N_2$

10.16 $P_{O_2} = P_{tot} \cdot X_{O_2}$

$0.21 \text{ atm} = (8.38 \text{ atm}) \cdot X_{O_2}$

$X_{O_2} = \frac{0.21 \text{ atm}}{8.38 \text{ atm}} = 0.025$

10.17 (a) $\frac{\text{rate } O_2}{\text{rate Kr}} = \sqrt{\frac{M_{Kr}}{M_{O_2}}} = \sqrt{\frac{83.8}{32.0}};\ \frac{\text{rate } O_2}{\text{rate Kr}} = 1.62$

O_2 diffuses 1.62 times faster than Kr.

(b) $\frac{\text{rate } C_2H_2}{\text{rate } N_2} = \sqrt{\frac{M_{N_2}}{M_{C_2H_2}}} = \sqrt{\frac{28.0}{26.0}};\ \frac{\text{rate } C_2H_2}{\text{rate } N_2} = 1.04$

C_2H_2 diffuses 1.04 times faster than N_2.

10.18 SO_2, 64.06

$$\frac{rate_1}{rate_2} = 1.414 = \sqrt{\frac{M_{SO_2}}{M_x}}; \quad \sqrt{M_x} = \frac{\sqrt{M_x}}{1.414} = \frac{\sqrt{64.06}}{1.414} = 5.660$$

$M_x = (5.660)^2 = 32.00$; The unknown gas could be O_2.

10.19 $110 \text{ ppb} = X_{O_3} \text{ x } 10^9 \text{ ppb}; \quad X_{O_3} = \frac{110}{10^9} = 110 \text{ x } 10^{-9}$

volume % = X_{O_3} x 100 = 1.10 x 10^{-5}%

ppm = X_{O_3} x 10^6 = 0.110 ppm

O_3 molecules = $(X_{O_3})(3.3 \text{ x } 10^{-2} \text{ mol})(6.022 \text{ x } 10^{23} \text{ molecules/mol}) = 2.2 \text{ x } 10^{15}$ O_3 molecules

10.20 $95 \text{ ppb} = X_{O_3} \text{ x } 10^9 \text{ ppb}; \quad X_{O_3} = \frac{95}{10^9} = 95 \text{ x } 10^{-9}$

$P_{O_3} = P_{tot} \cdot X_{O_3} = (0.79 \text{ atm})(95 \text{ x } 10^{-9}) = 7.5 \text{ x } 10^{-8} \text{ atm}$

10.21 Nitrogen and oxygen are diatomic molecules that do not have a dipole moment. As the bond stretches in a vibration, the molecule still does not have a dipole moment. Since no change in dipole moment occurs, IR radiation will not be absorbed.

10.22 The water molecule has a bent geometry and thus all vibrations will result in a change in dipole moment and the absorption of IR radiation.

10.23 The symmetric stretch does not absorb IR radiation; the asymmetric stretch absorbs IR radiation.

10.24 1.82/0.48 = 3.8; The contribution from CO_2 is 3.8 times as large as the contribution from CH_4.

10.25 Although N_2O has a greater potential for warming on a per mass basis, the atmosphere has a much higher concentration of CO_2.

10.26 Water vapor is a potent greenhouse gas and when considering the greenhouse effect, increased levels would cause warming. However, increased levels of water vapor may also cause changes in some types of clouds that have a cooling effect.

Conceptual Problems

10.28 The gas pressure in the bulb in mm Hg is equal to the difference in the height of the Hg in the two arms of the manometer.

10.30 The picture on the right will be the same as that on the left, apart from random scrambling of the He and Ar atoms.

10.32 The sample remains a gas at 150 K. Drawing (c) represents the gas at this temperature. The gas molecules still fill the container.

10.34 n_{Total} = (1 black) + (3 blue) + (4 red) + (6 green) = 14
P_{Total} = 420 mm Hg

$$P_x = P_{Total} \times \frac{n_x}{n_{Total}}$$

$$P_{black} = (420 \text{ mm Hg}) \times \left(\frac{1}{14}\right) = 30.0 \text{ mm Hg}; \quad P_{blue} = (420 \text{ mm Hg}) \times \left(\frac{3}{14}\right) = 90.0 \text{ mm Hg}$$

$$P_{red} = (420 \text{ mm Hg}) \times \left(\frac{4}{14}\right) = 120 \text{ mm Hg}; \quad P_{green} = (420 \text{ mm Hg}) \times \left(\frac{6}{14}\right) = 180 \text{ mm Hg}$$

10.36 (a) The temperature has increased by about 10% (from 300 K to 325 K) while the amount and the pressure are unchanged. Thus, the volume should increase by about 10%.

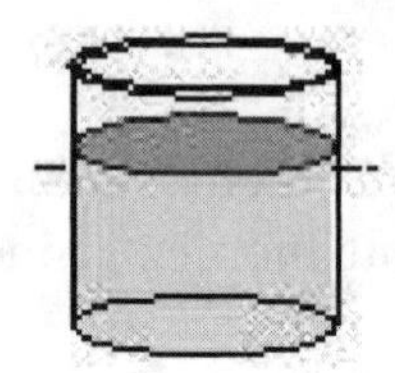

(b) The temperature has increased by a factor of 1.5 (from 300 K to 450 K) and the pressure has increased by a factor of 3 (from 0.9 atm to 2.7 atm) while the amount is unchanged. Thus, the volume should decrease by half (1.5/3 = 0.5).

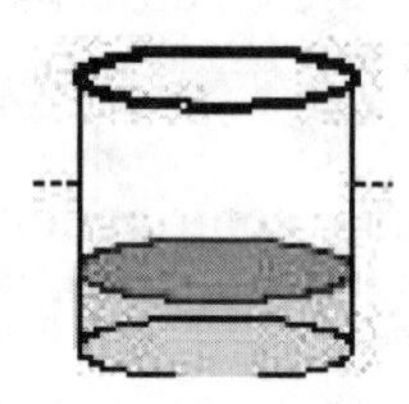

(c) Both the amount and the pressure have increased by a factor of 3 (from 0.075 mol to 0.22 mol and from 0.9 atm to 2.7 atm) while the temperature is unchanged. Thus, the volume is unchanged.

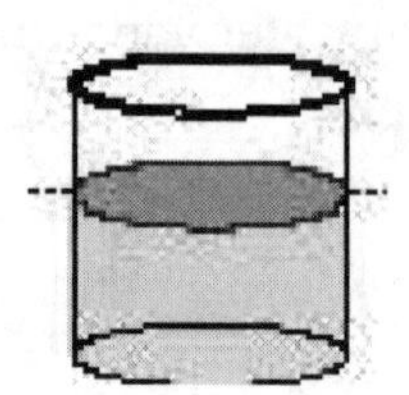

Section Problems

Gases and Gas Pressure (Section 10.1)

10.38 1.00 atm = 14.7 psi

$$1.00 \text{ mm Hg} \times \frac{1 \text{ atm}}{760 \text{ mm Hg}} \times \frac{14.7 \text{ psi}}{1 \text{ atm}} = 1.93 \times 10^{-2} \text{ psi}$$

10.40 Temperature is a measure of the average kinetic energy of gas particles.

10.42
$$P = 480 \text{ mm Hg} \times \frac{1.00 \text{ atm}}{760 \text{ mm Hg}} = 0.632 \text{ atm}$$

$$P = 480 \text{ mm Hg} \times \frac{101{,}325 \text{ Pa}}{760 \text{ mm Hg}} = 6.40 \times 10^4 \text{ Pa}$$

10.44 $P_{flask} > 754.3$ mm Hg; $P_{flask} = 754.3 \text{ mm Hg} + 176 \text{ mm Hg} = 930 \text{ mm Hg}$

10.46 $P_{flask} > 752.3$ mm Hg (see Figure 10.5); If the pressure in the flask can support a column of ethyl alcohol (d = 0.7893 g/mL) 55.1 cm high, then it can only support a column of Hg that is much shorter because of the higher density of Hg.

$$55.1 \text{ cm} \times \frac{0.7893 \text{ g/mL}}{13.546 \text{ g/mL}} = 3.21 \text{ cm Hg} = 32.1 \text{ mm Hg}$$

$$P_{flask} = 752.3 \text{ mm Hg} + 32.1 \text{ mm Hg} = 784.4 \text{ mm Hg}$$

$$P_{flask} = 784.4 \text{ mm Hg} \times \frac{101{,}325 \text{ Pa}}{760 \text{ mm Hg}} = 1.046 \times 10^5 \text{ Pa}$$

10.48

	% Volume
N_2	78.08
O_2	20.95
Ar	0.93
CO_2	0.037

The % volume for a particular gas is proportional to the number of molecules of that gas in a mixture of gases.

Average molecular mass of air

$= (0.7808)(\text{mol. mass } N_2) + (0.2095)(\text{mol. mass } O_2)$
$\quad + (0.0093)(\text{at. mass Ar}) + (0.000\ 37)(\text{mol. mass } CO_2)$

$= (0.7808)(28.01) + (0.2095)(32.00)$
$\quad + (0.0093)(39.95) + (0.000\ 37)(44.01) = 28.96$

The Gas Laws (Sections 10.2 and 10.3)

10.50 (a)
$$\frac{nR}{V} = \frac{P_i}{T_i} = \frac{P_f}{T_f}; \qquad \frac{P_i T_f}{T_i} = P_f$$

Let $P_i = 1$ atm, $T_i = 100$ K, $T_f = 300$ K

$$P_f = \frac{P_i T_f}{T_i} = \frac{(1\ \text{atm})(300\ \text{K})}{(100\ \text{K})} = 3\ \text{atm}$$

The pressure would triple.

(b) $\frac{RT}{V} = \frac{P_i}{n_i} = \frac{P_f}{n_f}$; $\quad \frac{P_i n_f}{n_i} = P_f$

Let $P_i = 1$ atm, $n_i = 3$ mol, $n_f = 1$ mol

$$P_f = \frac{P_i n_f}{n_i} = \frac{(1\ \text{atm})(1\ \text{mol})}{(3\ \text{mol})} = \frac{1}{3}\ \text{atm}$$

The pressure would be $\frac{1}{3}$ the initial pressure.

(c) $nRT = P_iV_i = P_fV_f$; $\quad \frac{P_iV_i}{V_f} = P_f$

Let $P_i = 1$ atm, $V_i = 1$ L, $V_f = 1 - 0.45\ \text{L} = 0.55$ L

$$P_f = \frac{P_iV_i}{V_f} = \frac{(1\ \text{atm})(1\ \text{L})}{(0.55\ \text{L})} = 1.8\ \text{atm}$$

The pressure would increase by 1.8 times.

(d) $nR = \frac{P_iV_i}{T_i} = \frac{P_fV_f}{T_f}$; $\quad \frac{P_iV_iT_f}{T_iV_f} = P_f$

Let $P_i = 1$ atm, $V_i = 1$ L, $T_i = 200$ K, $V_f = 3$ L, $T_i = 100$ K

$$P_f = \frac{P_iV_iT_f}{T_iV_f} = \frac{(1\ \text{atm})(1\ \text{L})(100\ \text{K})}{(200\ \text{K})(3\ \text{L})} = 0.17\ \text{atm}$$

The pressure would be 0.17 times the initial pressure.

10.52 They all contain the same number of gas molecules.

10.54 n and T are constant; therefore $nRT = P_iV_i = P_fV_f$

$$V_f = \frac{P_iV_i}{P_f} = \frac{(150\ \text{atm})(49.0\ \text{L})}{(1.02\ \text{atm})} = 7210\ \text{L}$$

n and P are constant; therefore $\frac{nR}{P} = \frac{V_i}{T_i} = \frac{V_f}{T_f}$

$$V_f = \frac{V_iT_f}{T_i} = \frac{(49.0\ \text{L})(308\ \text{K})}{(293\ \text{K})} = 51.5\ \text{L}$$

10.56 $15.0\text{ g }CO_2 \times \frac{1\text{ mol }CO_2}{44.0\text{ g }CO_2} = 0.341\text{ mol }CO_2$

$$P = \frac{nRT}{V} = \frac{(0.341\text{ mol})\left(0.082\ 06\ \frac{L\cdot atm}{K\cdot mol}\right)(300\text{ K})}{(0.30\text{ L})} = 27.98\text{ atm}$$

$$27.98\text{ atm} \times \frac{760\text{ mm Hg}}{1\text{ atm}} = 2.1 \times 10^4\text{ mm Hg}$$

10.58 $\frac{1\text{ H atom}}{cm^3} \times \frac{1\text{ mol H}}{6.02 \times 10^{23}\text{ atoms}} \times \frac{1000\text{ cm}^3}{1\text{ L}} = 1.7 \times 10^{-21}\text{ mol H/L}$

$$P = \frac{nRT}{V} = \frac{(1.7 \times 10^{-21}\text{ mol})\left(0.082\ 06\ \frac{L\cdot atm}{K\cdot mol}\right)(100\text{ K})}{(1\text{ L})} = 1.4 \times 10^{-20}\text{ atm}$$

$$P = 1.4 \times 10^{-20}\text{ atm} \times \frac{760\text{ mm Hg}}{1.0\text{ atm}} = 1 \times 10^{-17}\text{ mm Hg}$$

10.60 $n = \frac{PV}{RT} = \frac{\left(17{,}180\text{ kPa} \times \frac{1000\text{ Pa}}{1\text{ k Pa}} \times \frac{1\text{ atm}}{101{,}325\text{ Pa}}\right)(43.8\text{ L})}{\left(0.082\ 06\ \frac{L\cdot atm}{K\cdot mol}\right)(293K)} = 308.9\text{ mol}$

$$\text{mass Ar} = 308.9\text{ mol} \times \frac{39.948\text{ g}}{1\text{ mol}} = 12340\text{ g} = 1.23 \times 10^4\text{ g}$$

Gas Stoichiometry (Section 10.4)

10.62 For steam, T = 123.0 °C = 396 K

$$n = \frac{PV}{RT} = \frac{(0.93\text{ atm})(15.0\text{ L})}{\left(0.082\ 06\ \frac{L\cdot atm}{K\cdot mol}\right)(396\text{ K})} = 0.43\text{ mol steam}$$

For ice, H_2O, 18.02; $n = 10.5\text{ g} \times \frac{1\text{ mol}}{18.02\text{ g}} = 0.583\text{ mol ice}$

Because the number of moles of ice is larger than the number of moles of steam, the ice contains more H_2O molecules.

10.64 The containers are identical. Both containers contain the same number of gas molecules. Weigh the containers. Because the molecular mass for O_2 is greater than the molecular mass for H_2, the heavier container contains O_2.

10.66 room volume = $4.0 \text{ m} \times 5.0 \text{ m} \times 2.5 \text{ m} \times \dfrac{1 \text{ L}}{10^{-3} \text{ m}^3} = 5.0 \times 10^4 \text{ L}$

$$n_{total} = \frac{PV}{RT} = \frac{(1.0 \text{ atm})(5.0 \times 10^4 \text{ L})}{\left(0.082\ 06 \dfrac{\text{L}\cdot\text{atm}}{\text{K}\cdot\text{mol}}\right)(273 \text{ K})} = 2.23 \times 10^3 \text{ mol}$$

$n_{O_2} = (0.2095)n_{total} = (0.2095)(2.23 \times 10^3 \text{ mol}) = 467 \text{ mol } O_2$

mass $O_2 = 467 \text{ mol} \times \dfrac{32.0 \text{ g}}{1 \text{ mol}} = 1.5 \times 10^4 \text{ g } O_2$

10.68 (a) CH_4, 16.04; $d = \dfrac{16.04 \text{ g}}{22.4 \text{ L}} = 0.716 \text{ g/L}$

(b) CO_2, 44.01; $d = \dfrac{44.01 \text{ g}}{22.4 \text{ L}} = 1.96 \text{ g/L}$

(c) O_2, 32.00; $d = \dfrac{32.00 \text{ g}}{22.4 \text{ L}} = 1.43 \text{ g/L}$

10.70 $$n = \frac{PV}{RT} = \frac{\left(356 \text{ mm Hg} \times \dfrac{1.00 \text{ atm}}{760 \text{ mm Hg}}\right)(1.500 \text{ L})}{\left(0.082\ 06 \dfrac{\text{L}\cdot\text{atm}}{\text{K}\cdot\text{mol}}\right)(295.5 \text{ K})} = 0.0290 \text{ mol}$$

molar mass = $\dfrac{0.9847 \text{ g}}{0.0290 \text{ mol}} = 34.0 \text{ g/mol}$; molecular weight = 34.0

10.72 $2\ HgO(s) \rightarrow 2\ Hg(l) + O_2(g)$; HgO, 216.59

$10.57 \text{ g HgO} \times \dfrac{1 \text{ mol HgO}}{216.59 \text{ g HgO}} \times \dfrac{1 \text{ mol } O_2}{2 \text{ mol HgO}} = 0.024\ 40 \text{ mol } O_2$

$$V = \frac{nRT}{P} = \frac{(0.024\ 40 \text{ mol})\left(0.082\ 06 \dfrac{\text{L}\cdot\text{atm}}{\text{K}\cdot\text{mol}}\right)(273.15 \text{ K})}{1.000 \text{ atm}} = 0.5469 \text{ L}$$

10.74 $Zn(s) + 2\ HCl(aq) \rightarrow ZnCl_2(aq) + H_2(g)$

(a) $25.5 \text{ g Zn} \times \dfrac{1 \text{ mol Zn}}{65.39 \text{ g Zn}} \times \dfrac{1 \text{ mol } H_2}{1 \text{ mol Zn}} = 0.390 \text{ mol } H_2$

$$V = \frac{nRT}{P} = \frac{(0.390 \text{ mol})\left(0.082\ 06 \dfrac{\text{L}\cdot\text{atm}}{\text{K}\cdot\text{mol}}\right)(288 \text{ K})}{\left(742 \text{ mm Hg} \times \dfrac{1.00 \text{ atm}}{760 \text{ mm Hg}}\right)} = 9.44 \text{ L}$$

$$\text{(b) } n = \frac{PV}{RT} = \frac{\left(350 \text{ mm Hg x } \dfrac{1.00 \text{ atm}}{760 \text{ mm Hg}}\right)(5.00 \text{ L})}{\left(0.082\ 06 \dfrac{\text{L} \cdot \text{atm}}{\text{K} \cdot \text{mol}}\right)(303.15 \text{ K})} = 0.092\ 56 \text{ mol } H_2$$

$$0.092\ 56 \text{ mol } H_2 \text{ x } \frac{1 \text{ mol Zn}}{1 \text{ mol } H_2} \text{ x } \frac{65.39 \text{ g Zn}}{1 \text{ mol Zn}} = 6.05 \text{ g Zn}$$

10.76 (a) $V_{24h} = (4.50 \text{ L/min})(60 \text{ min/h})(24 \text{ h/day}) = 6480 \text{ L}$
$V_{CO_2} = (0.034)V_{24h} = (0.034)(6480 \text{ L}) = 220 \text{ L}$

$$n = \frac{PV}{RT} = \frac{\left(735 \text{ mm Hg x } \dfrac{1.00 \text{ atm}}{760 \text{ mm Hg}}\right)(220 \text{ L})}{\left(0.082\ 06 \dfrac{\text{L} \cdot \text{atm}}{\text{K} \cdot \text{mol}}\right)(298 \text{ K})} = 8.70 \text{ mol } CO_2$$

$$8.70 \text{ mol } CO_2 \text{ x } \frac{44.01 \text{ g } CO_2}{1 \text{ mol } CO_2} = 383 \text{ g} = 380 \text{ g } CO_2$$

(b) $2\ Na_2O_2(s) + 2\ CO_2(g) \rightarrow 2\ Na_2CO_3(s) + O_2(g)$; Na_2O_2, 77.98
3.65 kg = 3650 g

$$3650 \text{ g } Na_2O_2 \text{ x } \frac{1 \text{ mol } Na_2O_2}{77.98 \text{ g } Na_2O_2} \text{ x } \frac{2 \text{ mol } CO_2}{2 \text{ mol } Na_2O_2} \text{ x } \frac{1 \text{ day}}{8.70 \text{ mol } CO_2} = 5.4 \text{ days}$$

10.78 He, 4.00; 365 mL = 0.365 L; 25 °C = 298 K

$$n_{He} = \frac{PV}{RT} = \frac{(7.8 \text{ atm})(0.365 \text{ L})}{\left(0.082\ 06 \dfrac{\text{L} \cdot \text{atm}}{\text{K} \cdot \text{mol}}\right)(298 \text{ K})} = 0.116 \text{ mol He}$$

$$\text{mass He} = 0.116 \text{ mol He x } \frac{4.00 \text{ g He}}{1 \text{ mol He}} = 0.464 \text{ g He}$$

Dalton's Law and Mole Fraction (Section 10.5)

10.80 Because of Avogadro's Law ($V \propto n$), the % volumes are also % moles.

	% mole
N_2	78.08
O_2	20.95
Ar	0.93
CO_2	0.038

In decimal form, % mole = mole fraction.

$$P_{N_2} = X_{N_2} \cdot P_{total} = (0.7808)(1.000 \text{ atm}) = 0.7808 \text{ atm}$$
$$P_{O_2} = X_{O_2} \cdot P_{total} = (0.2095)(1.000 \text{ atm}) = 0.2095 \text{ atm}$$

$P_{Ar} = X_{Ar} \cdot P_{total} = (0.0093)(1.000 \text{ atm}) = 0.0093 \text{ atm}$

$P_{CO_2} = X_{CO_2} \cdot P_{total} = (0.000\ 38)(1.000 \text{ atm}) = 0.000\ 38 \text{ atm}$

Pressures of the rest are negligible.

10.82 Assume a 100.0 g sample. g CO_2 = 1.00 g and g O_2 = 99.0 g

$$\text{mol } CO_2 = 1.00 \text{ g } CO_2 \times \frac{1 \text{ mol } CO_2}{44.01 \text{ g } CO_2} = 0.0227 \text{ mol } CO_2$$

$$\text{mol } O_2 = 99.0 \text{ g } O_2 \times \frac{1 \text{ mol } O_2}{32.00 \text{ g } O_2} = 3.094 \text{ mol } O_2$$

$n_{total} = 3.094 \text{ mol} + 0.0227 \text{ mol} = 3.117 \text{ mol}$

$$X_{O_2} = \frac{3.094 \text{ mol}}{3.117 \text{ mol}} = 0.993; \quad X_{CO_2} = \frac{0.0227 \text{ mol}}{3.117 \text{ mol}} = 0.007\ 28$$

$P_{O_2} = X_{O_2} \cdot P_{total} = (0.993)(0.977 \text{ atm}) = 0.970 \text{ atm}$

$P_{CO_2} = X_{CO_2} \cdot P_{total} = (0.007\ 28)(0.977 \text{ atm}) = 0.007\ 11 \text{ atm}$

10.84 Assume a 100.0 g sample.

$$\text{g HCl} = (0.0500)(100.0 \text{ g}) = 5.00 \text{ g}; \quad 5.00 \text{ g HCl} \times \frac{1 \text{ mol HCl}}{36.5 \text{ g HCl}} = 0.137 \text{ mol HCl}$$

$$\text{g } H_2 = (0.0100)(100.0 \text{ g}) = 1.00 \text{ g}; \quad 1.00 \text{ g } H_2 \times \frac{1 \text{ mol } H_2}{2.016 \text{ g } H_2} = 0.496 \text{ mol } H_2$$

$$\text{g Ne} = (0.94)(100.0 \text{ g}) = 94 \text{ g}; \quad 94 \text{ g Ne} \times \frac{1 \text{ mol Ne}}{20.18 \text{ g Ne}} = 4.66 \text{ mol Ne}$$

$n_{total} = 0.137 + 0.496 + 4.66 = 5.3 \text{ mol}$

$$X_{HCl} = \frac{0.137 \text{ mol}}{5.3 \text{ mol}} = 0.026; \quad X_{H_2} = \frac{0.496 \text{ mol}}{5.3 \text{ mol}} = 0.094; \quad X_{Ne} = \frac{4.66 \text{ mol}}{5.3 \text{ mol}} = 0.88$$

10.86 (a) H_2, 2.016

$$P = \frac{nRT}{V} = \frac{\left(14.2 \text{ g} \times \frac{1 \text{ mol}}{2.016 \text{ g}}\right)\left(0.082\ 06 \frac{\text{L} \cdot \text{atm}}{\text{K} \cdot \text{mol}}\right)(290 \text{ K})}{(100.0 \text{ L})} = 1.68 \text{ atm}$$

(b) $$P = \frac{nRT}{V} = \frac{\left(36.7 \text{ g} \times \frac{1 \text{ mol}}{39.95 \text{ g}}\right)\left(0.082\ 06 \frac{\text{L} \cdot \text{atm}}{\text{K} \cdot \text{mol}}\right)(290 \text{ K})}{(100.0 \text{ L})} = 0.219 \text{ atm}$$

10.88 $P_{total} = P_{H_2} + P_{H_2O}$; $P_{H_2} = P_{total} - P_{H_2O}$ = 747 mm Hg – 23.8 mm Hg = 723 mm Hg

$$n = \frac{PV}{RT} = \frac{\left(723 \text{ mm Hg} \times \frac{1.00 \text{ atm}}{760 \text{ mm Hg}}\right)(3.557 \text{ L})}{\left(0.082\ 06 \frac{\text{L}\cdot\text{atm}}{\text{K}\cdot\text{mol}}\right)(298 \text{ K})} = 0.1384 \text{ mol } H_2$$

$$0.1384 \text{ mol } H_2 \times \frac{1 \text{ mol Mg}}{1 \text{ mol } H_2} \times \frac{24.3 \text{ g Mg}}{1 \text{ mol Mg}} = 3.36 \text{ g Mg}$$

Kinetic–Molecular Theory and Graham's Law (Sections 10.6 and 10.7)

10.90 The kinetic-molecular theory is based on the following assumptions:

1. A gas consists of tiny particles, either atoms or molecules, moving about at random.
2. The volume of the particles themselves is negligible compared with the total volume of the gas; most of the volume of a gas is empty space.
3. The gas particles act independently; there are no attractive or repulsive forces between particles.
4. Collisions of the gas particles, either with other particles or with the walls of the container, are elastic; that is, the total kinetic energy of the gas particles is constant at constant T.
5. The average kinetic energy of the gas particles is proportional to the Kelvin temperature of the sample.

10.92 Heat is the energy transferred from one object to another as the result of a temperature difference between them. Temperature is a measure of the kinetic energy of molecular motion.

10.94 $$u = \sqrt{\frac{3RT}{M}} = \sqrt{\frac{3 \times 8.314 \text{ kg m}^2/(\text{s}^2 \text{ K mol}) \times 220 \text{ K}}{28.0 \times 10^{-3} \text{ kg/mol}}} = 443 \text{ m/s}$$

10.96 For Br_2: $$u = \sqrt{\frac{3RT}{M}} = \sqrt{\frac{3 \times 8.314 \text{ kg m}^2/(\text{s}^2 \text{ K mol}) \times 293 \text{ K}}{159.8 \times 10^{-3} \text{ kg/mol}}} = 214 \text{ m/s}$$

For Xe: $$u = 214 \text{ m/s} = \sqrt{\frac{3 \times 8.314 \text{ kg m}^2/(\text{s}^2 \text{ K mol}) \times T}{131.3 \times 10^{-3} \text{ kg/mol}}}$$

Square both sides of the equation and solve for T.

$$45796 \text{ m}^2/\text{s}^2 = \frac{3 \times 8.314 \text{ kg m}^2/(\text{s}^2 \text{ K mol}) \times T}{131.3 \times 10^{-3} \text{ kg/mol}}$$

T = 241 K = –32 °C

10.98 For H_2, $u = \sqrt{\frac{3RT}{M}} = \sqrt{\frac{3 \text{ x } 8.314 \text{ kg m}^2/(\text{s}^2 \text{ K mol}) \text{ x } 150 \text{ K}}{2.02 \text{ x } 10^{-3} \text{ kg/mol}}} = 1360 \text{ m/s}$

For He, $u = \sqrt{\frac{3 \text{ x } 8.314 \text{ kg m}^2/(\text{s}^2 \text{ K mol}) \text{ x } 648 \text{ K}}{4.00 \text{ x } 10^{-3} \text{ kg/mol}}} = 2010 \text{ m/s}$

He at 375 °C has the higher average speed.

10.100 $\frac{\text{rate}_{H_2}}{\text{rate}_X} = \sqrt{\frac{M_X}{M_{H_2}}}$; $\frac{2.92}{1} = \frac{\sqrt{M_X}}{\sqrt{2.02}}$; $2.92\sqrt{2.02} = \sqrt{M_X}$

$M_X = (2.92\sqrt{2.02})^2 = 17.2$ g/mol; molecular weight = 17.2

10.102 HCl, 36.5; F_2, 38.0; Ar, 39.9

$\frac{\text{rate HCl}}{\text{rate Ar}} = \sqrt{\frac{M_{Ar}}{M_{HCl}}} = \sqrt{\frac{39.9}{36.5}} = 1.05$; $\frac{\text{rate } F_2}{\text{rate Ar}} = \sqrt{\frac{M_{Ar}}{M_{F_2}}} = \sqrt{\frac{39.9}{38.0}} = 1.02$

The relative rates of diffusion are HCl(1.05) > F_2(1.02) > Ar(1.00).

10.104 $u = 45 \text{ m/s} = \sqrt{\frac{3 \text{ x } 8.314 \text{ kg m}^2/(\text{s}^2 \text{ K mol}) \text{ x } T}{4.00 \text{ x } 10^{-3} \text{ kg/mol}}}$

Square both sides of the equation and solve for T.

$2025 \text{ m}^2/\text{s}^2 = \frac{3 \text{ x } 8.314 \text{ kg m}^2/(\text{s}^2 \text{ K mol}) \text{ x } T}{4.00 \text{ x } 10^{-3} \text{ kg/mol}}$

T = 0.325 K = –272.83 °C (near absolute zero)

Real Gases (Section 10.8)

10.106 (a) high (b) larger

10.108 $P = \frac{nRT}{V} = \frac{(0.500 \text{ mol})\left(0.082\ 06 \frac{\text{L}\cdot\text{atm}}{\text{K}\cdot\text{mol}}\right)(300 \text{ K})}{(0.600 \text{ L})} = 20.5 \text{ atm}$

$P = \frac{nRT}{V - nb} - \frac{an^2}{V^2}$

$$P = \frac{(0.500 \text{ mol})\left(0.082\ 06 \frac{\text{L}\cdot\text{atm}}{\text{K}\cdot\text{mol}}\right)(300 \text{ K})}{[(0.600 \text{ L}) - (0.500 \text{ mol})(0.0387 \text{ L/mol})]} - \frac{\left(1.35 \frac{\text{L}^2\cdot\text{atm}}{\text{mol}^2}\right)(0.500 \text{ mol})^2}{(0.600 \text{ L})^2} = 20.3 \text{ atm}$$

The Earth's Atmosphere and Pollution (Section 10.9)

10.110 Troposphere, stratosphere, mesosphere, and thermosphere. Temperature changes are used to distinguish between different regions of the atmosphere.

10.112 The force of gravity is strongest at the earth's surface and becomes weaker at higher altitude. Because of this, the troposphere contains about 75% of the mass of the entire atmosphere.

10.114 The more toxic pollutant is the one with the lower exposure limit, in this case $PM_{2.5}$.

10.116 $\% \ CO_2 = 0.040\% = X_{CO_2} \times 100; \quad \frac{0.040}{100} = X_{CO_2} = 4.0 \times 10^{-4}$

$CO_2 \text{ ppm} = X_{CO_2} \times 10^6 = (4.0 \times 10^{-4})(10^6) = 400 \text{ ppm}$

10.118 SO_2, 64.1

$$0.26 \times 10^{-6} \text{ g } SO_2 \times \frac{1 \text{ mol } SO_2}{64.1 \text{ g } SO_2} \times \frac{6.022 \times 10^{23} \ SO_2 \text{ molecules}}{1 \text{ mol } SO_2} = 2.44 \times 10^{15} \ SO_2 \text{ molecules/L}$$

$$\text{mol air/L} = n = \frac{PV}{RT} = \frac{(1.00 \text{ atm})(1.00 \text{ L})}{\left(0.082\ 06 \frac{\text{L} \cdot \text{atm}}{\text{K} \cdot \text{mol}}\right)(298 \text{ K})} = 0.0409 \text{ mol air/L}$$

$$0.0409 \text{ mol air/L} \times \frac{6.022 \times 10^{23} \text{ air molecules}}{1 \text{ mol air}} = 2.46 \times 10^{22} \text{ air molecules/L}$$

$$SO_2 \text{ ppb} = \frac{2.44 \times 10^{15} \ SO_2 \text{ molecules}}{2.46 \times 10^{22} \text{ air molecules}} \times 10^9 = 99 \text{ ppb}$$

This exceeds the 1-hour limit of 75 ppb.

10.120 S, 32.1

$$\text{mass S} = 1 \text{ kg} \times \frac{1000 \text{ g}}{1 \text{ kg}} \times 0.02 = 20 \text{ g S}$$

$$\text{mol S} = 20 \text{ g S} \times \frac{1 \text{ mol S}}{32.1 \text{ g S}} = 0.62 \text{ mol S} = 0.62 \text{ mol } SO_2$$

$$V_{SO_2} = \frac{nRT}{P} = \frac{(0.62 \text{ mol})\left(0.082\ 06 \frac{\text{L} \cdot \text{atm}}{\text{K} \cdot \text{mol}}\right)(298 \text{ K})}{1.0 \text{ atm}} = 15 \text{ L}$$

10.122 (a) CO and NO_2 (b) CO, NO_2, and SO_2

10.124 Secondary pollutants are formed by the chemical reaction of a primary pollutant and are not directly emitted from a source. Secondary pollutants are NO_2, O_3, and photochemical smog.

10.126 400 nm = 400 x 10^{-9} m

$$E = \frac{hc}{\lambda} = (6.626 \times 10^{-34}\ J\cdot s)\left(\frac{3.00 \times 10^{8}\ m/s}{400 \times 10^{-9}\ m}\right)\left(\frac{1\ kJ}{1000\ J}\right)(6.022 \times 10^{23}/mol)$$

$E = 299\ kJ/mol$

The N–O bond energy is 299 kJ/mol.

10.128 The drastic difference in air quality as measured by $PM_{2.5}$ and O_3 could be from a dramatic change in the weather. June 17 could have been a cool, cloudy, late Spring day with little or no photochemical smog whereas June 28 could have been a clear, bright, hot sunny Summer day that produced a significant amount of photochemical smog.

The Greenhouse Effect and Climate Change (Sections 10.10 and 10.11)

10.130 (a) visible and UV (b) UV (c) infrared (d) infrared

10.132 Because of nuclear fusion, the Sun emits all forms of electromagnetic radiation. The Earth absorbs visible radiation, warms up, and radiates infrared radiation back to space.

10.134 CO_2, N_2O, and CH_4

10.136 The Earth's average temperature has risen about 1 °C since 1900.

Chapter Problems

10.138 (a), (b), and (e) affect air quality; (c) and (d) affect climate

10.140 Average molecular mass of air = 28.96; CO_2, 44.01

$$P = 760\ mm\ Hg \times \frac{44.01\ g/mol}{28.96\ g/mol} = 1155\ mm\ Hg$$

10.142 $1\ atm = \left(\frac{14.70\ lb}{in.^2}\right)\left(\frac{1.00\ in.}{2.54\ cm}\right)^2\left(\frac{453.59\ g}{1\ lb}\right) = 1033.5\ g/cm^2$

column height = $(1033.5\ g/cm^2)(1\ cm^3/0.89 g)$ = 1161 cm = 1200 cm = 12 m

10.144 $15.0\ gal \times \frac{3.7854\ L}{1\ gal} = 56.8\ L$; 25 °C = 25 + 273 = 298 K

$$n = \frac{PV}{RT} = \frac{\left(743\ mm\ Hg \times \frac{1.00\ atm}{760\ mm\ Hg}\right)(56.8\ L)}{\left(0.082\ 06\ \frac{L\cdot atm}{K\cdot mol}\right)(298\ K)} = 2.27\ mol\ gasoline$$

$$2.27 \text{ mol gasoline} \times \frac{105 \text{ g gasoline}}{1 \text{ mol gasoline}} = 238 \text{ g gasoline}$$

$$238 \text{ g} \times \frac{1 \text{ mL}}{0.75 \text{ g}} \times \frac{1 \text{ L}}{1000 \text{ mL}} \times \frac{1 \text{ gal}}{3.7854 \text{ L}} \times \frac{20.0 \text{ mi}}{1 \text{ gal}} = 1.68 \text{ mi}$$

10.146 Both tanks contain the same mass of gas at the same temperature. There are more Ne atoms than N_2 molecules. Because Ne is lighter than N_2, the average Ne velocity is greater than the average N_2 velocity.
(a) $N_2 < Ne$ (b) $N_2 < Ne$ (c) $N_2 < Ne$ (d) Both are the same.

10.148 I_2, 253.8; Calculate the I_2 pressure before any I_2 dissociation.

$$P = \frac{nRT}{V} = \frac{\left(42.189 \text{ g} \times \frac{1 \text{ mol}}{253.8 \text{ g}}\right)\left(0.082\ 06 \frac{\text{L}\cdot\text{atm}}{\text{K}\cdot\text{mol}}\right)(1173 \text{ K})}{(10.00 \text{ L})} = 1.600 \text{ atm}$$

	$I_2(g)$	$\rightarrow$	$2\ I(g)$
before reaction (atm)	1.600		0
change (atm)	−x		+2x
after reaction (atm)	1.600 − x		2x

$1.733 = 1.600 - x + 2x$
$1.733 = 1.600 + x$
$x = 1.733 - 1.600 = 0.133$ atm
$P_{I_2} = 1.600 - x = 1.600 - 0.133 = 1.467$ atm
$P_I = 2x = 2(0.133) = 0.266$ atm

$$X_{I_2} = \frac{1.467 \text{ atm}}{1.733 \text{ atm}} = 0.8465 \text{ and } X_I = \frac{0.266 \text{ atm}}{1.733 \text{ atm}} = 0.153$$

10.150 This is initially a Boyle's Law problem, because only P and V are changing while n and T remain fixed. The initial volume for each gas is the volume of their individual bulbs. The final volume for each gas is the total volume of the three bulbs.
$nRT = P_iV_i = P_fV_f$; $V_f = 1.50 + 1.00 + 2.00 = 4.50$ L

For CO_2: $P_f = \frac{P_iV_i}{V_f} = \frac{(2.13 \text{ atm})(1.50 \text{ L})}{(4.50 \text{ L})} = 0.710 \text{ atm}$

For H_2: $P_f = \frac{P_iV_i}{V_f} = \frac{(0.861 \text{ atm})(1.00 \text{ L})}{(4.50 \text{ L})} = 0.191 \text{ atm}$

For Ar: $P_f = \frac{P_iV_i}{V_f} = \frac{(1.15 \text{ atm})(2.00 \text{ L})}{(4.50 \text{ L})} = 0.511 \text{ atm}$

From Dalton's Law, $P_{total} = P_{CO_2} + P_{H_2} + P_{Ar}$

$P_{total} = 0.710 \text{ atm} + 0.191 \text{ atm} + 0.511 \text{ atm} = 1.412 \text{ atm}$

10.152 $C_3H_5N_3O_9$, 227.1

(a) moles $C_3H_5N_3O_9 = 1.00 \text{ g} \times \dfrac{1 \text{ mol}}{227.1 \text{ g}} = 0.004\ 40 \text{ mol}$

$$n_{air} = \frac{PV}{RT} = \frac{(1.00 \text{ atm})(0.500 \text{ L})}{\left(0.082\ 06 \dfrac{\text{L} \cdot \text{atm}}{\text{K} \cdot \text{mol}}\right)(293 \text{ K})} = 0.0208 \text{ mol air}$$

(b) moles gas from $C_3H_5N_3O_9 = 0.004\ 40 \text{ mol} \times \dfrac{29 \text{ mol gas}}{4 \text{ mol nitro}}$

moles gas from $C_3H_5N_3O_9$ = 0.0319 mol gas from $C_3H_5N_3O_9$

n_{total} = 0.0319 mol + 0.0208 mol = 0.0527 mol

(c) $$P = \frac{nRT}{V} = \frac{(0.0527 \text{ mol})\left(0.082\ 06 \dfrac{\text{L} \cdot \text{atm}}{\text{K} \cdot \text{mol}}\right)(698 \text{ K})}{(0.500 \text{ L})} = 6.04 \text{ atm}$$

10.154 (a) $$n_{total} = \frac{PV}{RT} = \frac{\left(258 \text{ mm Hg} \times \dfrac{1.00 \text{ atm}}{760 \text{ mm Hg}}\right)(0.500 \text{ L})}{\left(0.082\ 06 \dfrac{\text{L} \cdot \text{atm}}{\text{K} \cdot \text{mol}}\right)(293 \text{ K})} = 0.007\ 06 \text{ mol}$$

(b) $$n_B = \frac{PV}{RT} = \frac{\left(344 \text{ mm Hg} \times \dfrac{1 \text{ atm}}{760 \text{ mm Hg}}\right)(0.250 \text{ L})}{\left(0.082\ 06 \dfrac{\text{L} \cdot \text{atm}}{\text{K} \cdot \text{mol}}\right)(293 \text{ K})} = 0.004\ 71 \text{ moles}$$

(c) $d = \dfrac{0.218 \text{ g}}{0.250 \text{ L}} = 0.872 \text{ g/L}$

(d) molar mass $= \dfrac{0.218 \text{ g}}{0.004\ 71 \text{ mol}} = 46.3$ g/mol, NO_2; molecular weight = 46.3

(e) $Hg_2CO_3(s) + 6\ HNO_3(aq) \rightarrow 2\ Hg(NO_3)_2(aq) + 3\ H_2O(l) + CO_2(g) + 2\ NO_2(g)$

10.156 (a) Let x = mol C_nH_{2n+2} in the reaction mixture.

Combustion of $C_nH_{2n+2} \rightarrow nCO_2 + (n+1)H_2O$ needs $n + \left(\dfrac{n+1}{2}\right) = \dfrac{3n+1}{2}$ mol O_2

Balanced equation is: $C_nH_{2n+2}(g) + \left(\dfrac{3n+1}{2}\right) O_2(g) \rightarrow nCO_2(g) + (n+1)H_2O(g)$

In going from reactants to products, the increase in the number of moles is

$[n + (n+1)] - \left[1 + \dfrac{3n+1}{2}\right] = \dfrac{n-1}{2}$ per mol of C_nH_{2n+2} reacted.

Before reaction: total mol $$= \frac{PV}{RT} = \frac{(2.000 \text{ atm})(0.4000 \text{ L})}{\left(0.082\ 06 \dfrac{\text{L} \cdot \text{atm}}{\text{K} \cdot \text{mol}}\right)(298.15 \text{ K})} = 0.032\ 70 \text{ mol}$$

After reaction: total mol $= \frac{PV}{RT} = \frac{(2.983\ \text{atm})(0.4000\ \text{L})}{\left(0.082\ 06\ \frac{\text{L}\cdot\text{atm}}{\text{K}\cdot\text{mol}}\right)(398.15\ \text{K})} = 0.036\ 52$ mol

Difference = 0.032 70 mol – 0.036 52 mol = 0.003 82 mol

Increase in number of mol $= \left(\frac{n-1}{2}\right)x = 0.003\ 82$ mol; $x = \frac{2(0.003\ 82)}{n-1}$

Also $x = \frac{\text{g } C_nH_{2n+2}}{\text{molar mass}} = \frac{0.148\text{ g}}{[12.01n + 1.008(2n+2)]\text{ g/mol}}$

So $\frac{2(0.003\ 82)}{n-1} = \frac{0.148}{14.026n + 2.016}$; $0.148\,n - 0.148 = 0.107\,n + 0.0154$

$0.041\,n = 0.163$; $n = \frac{0.163}{0.041} = 4.0$

C_nH_{2n+2} is C_4H_{10} (butane); molar mass = (4)(12.01) + (10)(1.008) = 58.12 g/mol

(b) $0.148\text{ g } C_4H_{10} \times \frac{1\text{ mol } C_4H_{10}}{58.12\text{ g } C_4H_{10}} = 0.002\ 55\text{ mol } C_4H_{10}$

mol O_2 initially = total mol – mol C_4H_{10} = 0.032 70 mol – 0.002 55 mol = 0.030 15 mol O_2

$$P_{C_4H_{10}} = \left(\frac{n_{C_4H_{10}}}{n_{total}}\right)P_{initial} = \left(\frac{0.002\ 55\text{ mol}}{0.032\ 70\text{ mol}}\right)(2.000\text{ atm}) = 0.156\text{ atm}$$

$$P_{O_2} = \left(\frac{n_{O_2}}{n_{total}}\right)P_{initial} = \left(\frac{0.030\ 15\text{ mol}}{0.032\ 70\text{ mol}}\right)(2.000\text{ atm}) = 1.844\text{ atm}$$

(c) $C_4H_{10}(g) + \frac{13}{2}O_2 \rightarrow 4\,CO_2(g) + 5\,H_2O(g)$

$0.002\ 55\text{ mol } C_4H_{10} \times \frac{4\text{ mol } CO_2}{1\text{ mol } C_4H_{10}} = 0.0102\text{ mol } CO_2$

$0.002\ 55\text{ mol } C_4H_{10} \times \frac{5\text{ mol } H_2O}{1\text{ mol } C_4H_{10}} = 0.012\ 75\text{ mol } H_2O$

mol O_2 unreacted = total mol after reaction – mol CO_2 – mol H_2O
= 0.03652 mol – 0.0102 mol – 0.01275 = 0.01357 mol O_2

$$P_{CO_2} = \left(\frac{n_{CO_2}}{n_{total}}\right)P_{final} = \left(\frac{0.0102\text{ mol}}{0.036\ 52\text{ mol}}\right)(2.983\text{ atm}) = 0.833\text{ atm}$$

$$P_{H_2O} = \left(\frac{n_{H_2O}}{n_{total}}\right)P_{final} = \left(\frac{0.012\ 75\text{ mol}}{0.036\ 52\text{ mol}}\right)(2.983\text{ atm}) = 1.041\text{ atm}$$

$$P_{O_2} = \left(\frac{n_{O_2}}{n_{total}}\right)P_{final} = \left(\frac{0.013\ 57\ mol}{0.036\ 52\ mol}\right)(2.983\ atm) = 1.108\ atm$$

10.158 $PV = nRT$

$$n_{total(initial)} = \frac{PV}{RT} = \frac{(3.00\ atm)(10.0\ L)}{\left(0.082\ 06\ \frac{L\cdot atm}{K\cdot mol}\right)(373.1\ K)} = 0.980\ mol$$

$$n_{total(final)} = \frac{PV}{RT} = \frac{(2.40\ atm)(10.0\ L)}{\left(0.082\ 06\ \frac{L\cdot atm}{K\cdot mol}\right)(373.1\ K)} = 0.784\ mol$$

	$CS_2(g)$ +	$3\ O_2(g)$ →	$CO_2(g)$ +	$2\ SO_2(g)$
before reaction (mol)	y	0.980 – y	0	0
change (mol)	–x	–3x	+x	+2x
after reaction (mol)	y – x = 0	0.980 – y – 3x	x	2x

$n_{total(final)} = (y - x) + (0.980 - y - 3x) + x + 2x = 0.784\ mol$

$0.980\ mol - 4x + 3x = 0.784\ mol$

$x = 0.980\ mol - 0.784\ mol = 0.196\ mol$

mol $CO_2 = x = 0.196\ mol$

$$P_{CO_2} = \frac{nRT}{V} = \frac{(0.196\ mol)\left(0.082\ 06\ \frac{L\cdot atm}{K\cdot mol}\right)(373.1\ K)}{(10.0\ L)} = 0.600\ atm$$

mol $SO_2 = 2x = 2(0.196\ mol) = 0.392\ mol$

$$P_{SO_2} = \frac{nRT}{V} = \frac{(0.392\ mol)\left(0.082\ 06\ \frac{L\cdot atm}{K\cdot mol}\right)(373.1\ K)}{(10.0\ L)} = 1.20\ atm$$

mol $O_2 = 0.980\ mol - y - 3x = 0.980\ mol - x - 3x = 0.980 - 4(0.196\ mol) = 0.196\ mol$

$P_{O_2} = P_{CO_2} = 0.600\ atm$

10.160 $Ca(ClO_3)_2$, 206.98; $Ca(ClO)_2$, 142.98

(a) $Ca(ClO_3)_2(s) \rightarrow CaCl_2(s) + 3\ O_2(g)$

$Ca(ClO)_2(s) \rightarrow CaCl_2(s) + O_2(g)$

(b) $T = 700\ °C = 700 + 273 = 973\ K$

$PV = nRT$

$$n_{O_2} = \frac{PV}{RT} = \frac{(1.00\ atm)(10.0\ L)}{\left(0.082\ 06\ \frac{L\cdot atm}{K\cdot mol}\right)(973\ K)} = 0.125\ mol\ O_2$$

Let X = mol $Ca(ClO_3)_2$ and let Y = mol $Ca(ClO)_2$

$X(206.98\ g/mol) + Y(142.98\ g/mol) = 10.0\ g$

$3X + Y = 0.125\ mol$, so $Y = 0.125\ mol - 3X$ (substitute for Y and solve for X)

$X(206.98\ g/mol) + (0.125\ mol - 3X)(142.98\ g/mol) = 10.0\ g$

$X(206.98\ g/mol) + 17.9\ g - X(428.94\ g/mol) = 10.0\ g$

$X(206.98 \text{ g/mol}) - X(428.94 \text{ g/mol}) = 10.0 \text{ g} - 17.9 \text{ g} = -7.9 \text{ g}$
$X(-221.96 \text{ g/mol}) = -7.9 \text{ g}$
$X = (-7.9 \text{ g})/(-221.96 \text{ g/mol}) = 0.0356 \text{ mol } Ca(ClO_3)_2$
$Y = 0.125 \text{ mol} - 3X;\ \ Y = 0.125 \text{ mol} - 3(0.0356 \text{ mol}) = 0.0182 \text{ mol } Ca(ClO)_2$

$$\text{mass } Ca(ClO_3)_2 = 0.0356 \text{ mol } Ca(ClO_3)_2 \times \frac{206.98 \text{ g } Ca(ClO_3)_2}{1 \text{ mol } Ca(ClO_3)_2} = 7.4 \text{ g } Ca(ClO_3)_2$$

$\text{mass } Ca(ClO)_2 = 10.0 \text{ g} - 7.4 \text{ g} = 2.6 \text{ g } Ca(ClO)_2$

10.162 (a) $T = 225\ °C = 225 + 273 = 498 \text{ K}$

$$PV = nRT \qquad P^o_{NOCl} = \frac{nRT}{V} = \frac{(2.00 \text{ mol})\left(0.082\ 06 \frac{L \cdot atm}{K \cdot mol}\right)(498 \text{ K})}{(400.0 \text{ L})} = 0.204 \text{ atm}$$

	$2\ NOCl(g)$	$\rightarrow$	$2\ NO(g)$	+	$Cl_2(g)$
before reaction (atm)	0.204		0		0
change (atm)	$-2x$		$+2x$		$+x$
after reaction (atm)	$0.204 - 2x$		$2x$		x

P_{total} (after rxn) $= (0.204 \text{ atm} - 2x) + 2x + x = 0.246 \text{ atm}$
$x = 0.246 \text{ atm} - 0.204 \text{ atm} = 0.042 \text{ atm}$
$P_{NO} = 2x = 2(0.042) = 0.084 \text{ atm};\ \ P_{Cl_2} = x = 0.042 \text{ atm}$
$P_{NOCl} = 0.204 - 2x = 0.204 - 2(0.042) = 0.120 \text{ atm}$

(b) $$\%\ NOCl \text{ decomposed} = \frac{2x}{P^o_{NOCl}} \times 100\% = \frac{0.084 \text{ atm}}{0.204 \text{ atm}} \times 100\% = 41\%$$

10.164 $CaCO_3$, 100.09; CaO, 56.08

$$\text{mol } CaO \text{ (or } CO_2) = 25.0 \text{ g } CaCO_3 \times \frac{1 \text{ mol } CaCO_3}{100.09 \text{ g } CaCO_3} \times \frac{1 \text{ mol CaO or } CO_2}{1 \text{ mol } CaCO_3} = 0.250 \text{ mol}$$

$$\text{mass CaO} = 0.250 \text{ mol CaO} \times \frac{56.08 \text{ g CaO}}{1 \text{ mol CaO}} = 14.02 \text{ g CaO}$$

(a) $500.0 \text{ mL} = 0.5000 \text{ L}$
$PV = nRT;\ \ n_{CO_2} = 0.250 \text{ mol}$

$$P_{CO_2} = \frac{nRT}{V} = \frac{(0.250 \text{ mol})\left(0.082\ 06 \frac{L \cdot atm}{K \cdot mol}\right)(1500 \text{ K})}{(0.5000 \text{ L})} = 61.5 \text{ atm}$$

(b) $V_{CaO} = (14.02 \text{ g})/(3.34 \text{ g/mL}) = 4.20 \text{ mL}$

$V = 500.0 \text{ mL} - 4.20 \text{ mL} = 495.8 \text{ mL} = 0.4958 \text{ L}$

$$P = \frac{nRT}{V - nb} - \frac{an^2}{V^2}$$

$$P = \frac{(0.250\ \text{mol})\left(0.082\ 06\ \frac{\text{L}\cdot\text{atm}}{\text{K}\cdot\text{mol}}\right)(1500\ \text{K})}{[(0.4958\ \text{L}) - (0.250\ \text{mol})(0.0427\ \text{L/mol})]} - \frac{\left(3.59\ \frac{\text{L}^2\cdot\text{atm}}{\text{mol}^2}\right)(0.250\ \text{mol})^2}{(0.4958\ \text{L})^2} = 62.5\ \text{atm}$$

10.166 Assume a container volume of 1.00 L with 3.309 g of gas.

25 °C = 25 + 273 = 298 K

$$n = \frac{PV}{RT} = \frac{(1.00\ \text{atm})(1.00\ \text{L})}{\left(0.082\ 06\ \frac{\text{L}\cdot\text{atm}}{\text{K}\cdot\text{mol}}\right)(298\ \text{K})} = 0.0409\ \text{mol}$$

$$\text{molar mass} = \frac{3.309\ \text{g}}{0.0409\ \text{mol}} = 80.9\ \text{g/mol}$$

The gas could be H_2Se, 81.0.

Multiconcept Problems

10.168 $X + 3\ O_2 \rightarrow 2\ CO_2 + 3\ H_2O$

(a) $X = C_2H_6O$

$C_2H_6O + 3\ O_2 \rightarrow 2\ CO_2 + 3\ H_2O$

(b) It is an empirical formula because it is the smallest whole number ratio of atoms. It is also a molecular formula because any higher multiple such as $C_4H_{12}O_2$ does not correspond to a stable electron-dot structure.

(c)

```
      H       H               H  H
      |  ..   |               |  |  ..
   H—C—O—C—H          H—C—C—O—H
      |  ..   |               |  |  ..
      H       H               H  H
```

(d) C_2H_6O, 46.07

$$\text{mol } C_2H_6O = 5.000\ \text{g } C_2H_6O \times \frac{1\ \text{mol } C_2H_6O}{46.07\ \text{g } C_2H_6O} = 0.1085\ \text{mol } C_2H_6O$$

$$\Delta H_{\text{combustion}} = \frac{-144.2\ \text{kJ}}{0.1085\ \text{mol}} = -1328.6\ \text{kJ/mol}$$

$\Delta H_{\text{combustion}} = [2\ \Delta H^\circ_f(CO_2) + 3\ \Delta H^\circ_f(H_2O)] - \Delta H^\circ_f(C_2H_6O)$

$\Delta H^\circ_f(C_2H_6O) = [2\ \Delta H^\circ_f(CO_2) + 3\ \Delta H^\circ_f(H_2O)] - \Delta H_{\text{combustion}}$

$= [(2\text{mol})(-393.5\ \text{kJ/mol}) + (3\ \text{mol})(-241.8\ \text{kJ/mol})] - (-1328.6\ \text{kJ})$

$= -183.8\ \text{kJ/mol}$

10.170 (a) Freezing point of H_2O on the Rankine scale is (9/5)(273.15) = 492 °R.

(b) $$R = \frac{PV}{nT} = \frac{(1.00\ \text{atm})(22.414\ \text{L})}{(1.00\ \text{mol})(492\ ^\circ\text{R})} = 0.0456\ \frac{\text{L}\cdot\text{atm}}{^\circ\text{R}\cdot\text{mol}}$$

(c) $$P = \frac{(2.50\ \text{mol})\left(0.0456\ \frac{\text{L}\cdot\text{atm}}{^\circ\text{R}\cdot\text{mol}}\right)(525\ ^\circ\text{R})}{[(0.4000\ \text{L}) - (2.50\ \text{mol})(0.04278\ \text{L/mol})]} - \frac{\left(2.253\ \frac{\text{L}^2\cdot\text{atm}}{\text{mol}^2}\right)(2.50\ \text{mol})^2}{(0.4000\ \text{L})^2}$$

$P = 204.2\ \text{atm} - 88.0\ \text{atm} = 116\ \text{atm}$

10.172 CO_2, 44.01; H_2O, 18.02

(a) $\text{mol C} = 0.3744\text{ g }CO_2 \times \dfrac{1\text{ mol }CO_2}{44.01\text{ g }CO_2} \times \dfrac{1\text{ mol C}}{1\text{ mol }CO_2} = 0.008\ 507\text{ mol C}$

$\text{mass C} = 0.008\ 507\text{ mol C} \times \dfrac{12.011\text{ g C}}{1\text{ mol C}} = 0.1022\text{ g C}$

$\text{mol H} = 0.1838\text{ g }H_2O \times \dfrac{1\text{ mol }H_2O}{18.02\text{ g }H_2O} \times \dfrac{2\text{ mol H}}{1\text{ mol }H_2O} = 0.020\ 400\text{ mol H}$

$\text{mass H} = 0.020\ 400\text{ mol H} \times \dfrac{1.008\text{ g H}}{1\text{ mol H}} = 0.02056\text{ g H}$

$\text{mass O} = 0.1500\text{ g} - 0.1022\text{ g} - 0.02056\text{ g} = 0.0272\text{ g O}$

$\text{mol O} = 0.0272\text{ g O} \times \dfrac{1\text{ mol O}}{16.00\text{ g O}} = 0.001\ 70\text{ mol O}$

$C_{0.008\ 507}H_{0.020\ 400}O_{0.001\ 70}$; divide each subscript by the smallest, 0.001 70.
$C_{0.008\ 507\ /\ 0.001\ 70}H_{0.020\ 400\ /\ 0.001\ 70}O_{0.001\ 70\ /\ 0.001\ 70}$
The empirical formula is $C_5H_{12}O$.
The empirical formula mass is 88 g/mol.

(b) 1 atm = 101,325 Pa; T = 54.8 °C = 54.8 + 273.15 = 327.9 K
PV = nRT

$$n = \frac{PV}{RT} = \frac{\left(100.0\text{ kPa} \times \dfrac{1.00\text{ atm}}{101.325\text{ kPa}}\right)(1.00\text{ L})}{\left(0.082\ 06\ \dfrac{\text{L}\cdot\text{atm}}{\text{K}\cdot\text{mol}}\right)(327.9\text{ K})} = 0.0367\text{ mol methyl } \textit{tert}\text{-butyl ether}$$

$\text{methyl } \textit{tert}\text{-butyl ether molar mass} = \dfrac{3.233\text{ g}}{0.0367\text{ mol}} = 88.1\text{ g/mol}$

The empirical formula weight and the molar mass are the same, so the molecular formula and empirical formula are the same. $C_5H_{12}O$ is the molecular formula and 88.15 is the molecular weight for methyl *tert*-butyl ether.

(c) $C_5H_{12}O(l) + 15/2\ O_2(g) \rightarrow 5\ CO_2(g) + 6\ H_2O(l)$
(d) $\Delta H^\circ_{\text{combustion}} = [5\ \Delta H^\circ_f(CO_2) + 6\ \Delta H^\circ_f(H_2O(l))] - \Delta H^\circ_f(C_5H_{12}O) = -3368.7\text{ kJ}$
$-3368.7\text{ kJ} = [(5\text{ mol})(-393.5\text{ kJ/mol}) + (6\text{ mol})(-285.8\text{ kJ/mol})] - (1\text{ mol})\Delta H^\circ_f(C_5H_{12}O)$
$(1\text{ mol})\Delta H^\circ_f(C_5H_{12}O) = [(5\text{ mol})(-393.5\text{ kJ/mol}) + (6\text{ mol})(-285.8\text{ kJ/mol})] + 3368.7\text{ kJ}$
$\Delta H^\circ_f(C_5H_{12}O) = -313.6\text{ kJ/mol}$

11 Liquids, Solids, and Phase Changes

11.1 boiling point = 78.4 °C = 351.6 K

$\Delta G = \Delta H_{vap} - T\Delta S_{vap}$; At the boiling point (phase change), $\Delta G = 0$

$\Delta H_{vap} = T\Delta S_{vap}$

$$\Delta S_{vap} = \frac{\Delta H_{vap}}{T} = \frac{38.56 \text{ kJ/mol}}{351.6 \text{ K}} = 0.1097 \text{ kJ/(K}\cdot\text{mol)} = 109.7 \text{ J/(K}\cdot\text{mol)}$$

11.2 $\Delta G = \Delta H - T\Delta S$; at the boiling point (phase change), $\Delta G = 0$.

$$\Delta H = T\Delta S; \quad T = \frac{\Delta H_{vap}}{\Delta S_{vap}} = \frac{29.2 \text{ kJ/mol}}{87.5 \times 10^{-3} \text{ kJ/(K}\cdot\text{mol)}} = 334 \text{ K}$$

11.3 C_6H_6, 78.1; $\quad 15.0 \text{ g } C_6H_6 \times \frac{1 \text{ mol } C_6H_6}{78.1 \text{ g } C_6H_6} = 0.192 \text{ mol } C_6H_6$

$q_1 = (0.192 \text{ mol})[136.0 \times 10^{-3} \text{ kJ/(K}\cdot\text{mol)}](80.1\text{ °C} - 50\text{ °C}) = 0.786 \text{ kJ}$

$q_2 = (0.192 \text{ mol})(30.72 \text{ kJ/mol}) = 5.90 \text{ kJ}$

$q_3 = (0.192 \text{ mol})(82.4 \times 10^{-3} \text{ kJ/(K}\cdot\text{mol)}](100\text{ °C} - 80.1\text{ °C}) = 0.315 \text{ kJ}$

$q_{total} = q_1 + q_2 + q_3 = 7.00 \text{ kJ}$; 7.00 kJ of heat is required.

11.4 H_2O, 18.0; $\quad 10.0 \text{ g } H_2O \times \frac{1 \text{ mol } H_2O}{18.0 \text{ g } H_2O} = 0.556 \text{ mol } H_2O$

$q_1 = (0.556 \text{ mol})[75.4 \times 10^{-3} \text{ kJ/(K}\cdot\text{mol)}](0\text{ °C} - 25\text{ °C}) = -1.05 \text{ kJ}$

$q_2 = (0.556 \text{ mol})(-6.01 \text{ kJ/mol}) = -3.34 \text{ kJ}$

$q_3 = (0.556 \text{ mol})(36.6 \times 10^{-3} \text{ kJ/(K}\cdot\text{mol)}](-10\text{ °C} - 0\text{ °C}) = -0.203 \text{ kJ}$

$q_{total} = q_1 + q_2 + q_3 = -4.59 \text{ kJ}$; 4.59 kJ of heat is removed.

11.5
$$\Delta H_{vap} = \frac{(\ln P_2 - \ln P_1)(R)}{\left(\frac{1}{T_1} - \frac{1}{T_2}\right)}$$

$P_1 = 400$ mm Hg; $\quad T_1 = 41.0\text{ °C} = 314.2 \text{ K}$

$P_2 = 760$ mm Hg; $\quad T_2 = 331.9 \text{ K}$

$$\Delta H_{vap} = \frac{[\ln(760) - \ln(400)]\left(8.3145 \frac{\text{J}}{\text{K}\cdot\text{mol}}\right)}{\left(\frac{1}{314.2 \text{ K}} - \frac{1}{331.9 \text{ K}}\right)} = 31{,}442 \text{ J/mol} = 31.4 \text{ kJ/mol}$$

11.6 $$\ln\left(\frac{P_2}{P_1}\right) = \ln P_2 - \ln P_1 = \frac{\Delta H_{vap}}{R}\left(\frac{1}{T_1} - \frac{1}{T_2}\right)$$

$$(\ln P_2 - \ln P_1)\left(\frac{R}{\Delta H_{vap}}\right) = \frac{1}{T_1} - \frac{1}{T_2}$$

$$\frac{1}{T_1} - (\ln P_2 - \ln P_1)\left(\frac{R}{\Delta H_{vap}}\right) = \frac{1}{T_2}$$

$P_1 = 760$ mm Hg; $T_1 = 100.0\ ^\circ C = 373.1$ K
$P_2 = 407$ mm Hg;
Solve for T_2, the boiling point for H_2O on top of Pikes Peak

$$\frac{1}{373.1\ K} - [\ln(407) - \ln(760)]\left(\frac{8.3145 \times 10^{-3} \frac{kJ}{K \cdot mol}}{40.7\ kJ/mol}\right) = \frac{1}{T_2}$$

$\frac{1}{T_2} = 0.002\ 808$; $T_2 = 356.1\ K = 83.0\ ^\circ C$

11.7 There are several possibilities. Here's one:

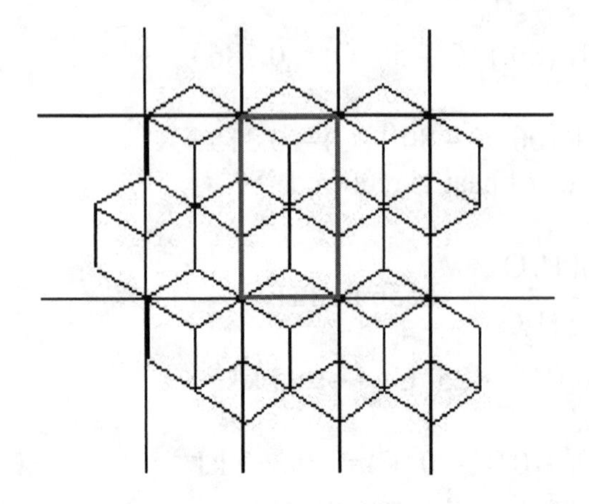

11.8 Here are two possibilities:

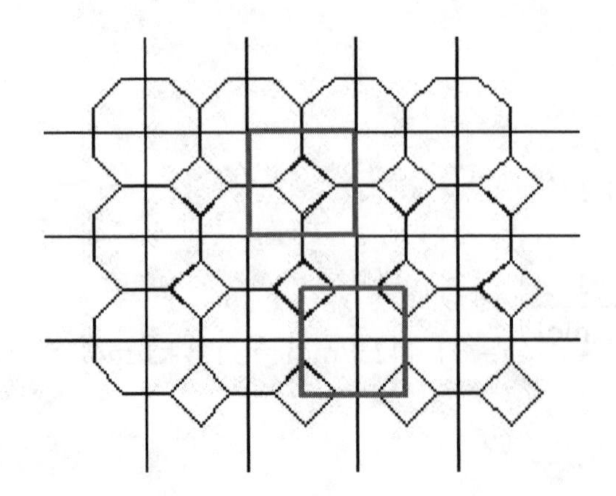

11.9 The face-centered cube is the unit cell for cubic closest packing.

$$r = \sqrt{\frac{d^2}{8}}; \quad d = r \cdot 2 \cdot \sqrt{2} = 197\text{ pm} \cdot 2 \cdot \sqrt{2} = 557\text{ pm}$$

11.10 For a simple cube, d = 2r; $r = \frac{d}{2} = \frac{334\text{ pm}}{2} = 167\text{ pm}$

11.11 For a simple cube, there is one atom per unit cell.

mass of one Po atom = $209\text{ g/mol} \times \frac{1\text{ mol}}{6.022 \times 10^{23}\text{ atoms}} = 3.4706 \times 10^{-22}\text{ g/atom}$

unit cell edge = d = 334 pm = 334×10^{-12} m = 3.34×10^{-8} cm

unit cell volume = $d^3 = (3.34 \times 10^{-8}\text{ cm})^3 = 3.7260 \times 10^{-23}\text{ cm}^3$

$$\text{density} = \frac{\text{mass}}{\text{volume}} = \frac{3.4706 \times 10^{-22}\text{ g}}{3.7260 \times 10^{-23}\text{ cm}^3} = 9.31\text{ g/cm}^3$$

11.12 The unit cell is a face-centered cube. There are four atoms in the unit cell.

$$\text{unit cell volume} = \left(383.3 \times 10^{-12}\text{ m} \times \frac{1\text{ cm}}{1 \times 10^{-2}\text{ m}}\right)^3 = 5.631 \times 10^{-23}\text{ cm}^3$$

unit cell mass = $(5.631 \times 10^{-23}\text{ cm}^3)(22.67\text{ g/cm}^3) = 1.277 \times 10^{-21}$ g

$$\text{atom mass} = \frac{1.227 \times 10^{-21}\text{ g}}{4\text{ atoms}} = 3.192 \times 10^{-22}\text{ g/atom}$$

atomic mass = $(3.192 \times 10^{-22}\text{ g/atom})(6.022 \times 10^{23}\text{ atoms/mol}) = 192.2$ g/mol

The metal is Ir.

11.13 (a) 1/8 S^{2-} at 8 corners = 1 S^{2-}; 1/2 S^{2-} at 6 faces = 3 S^{2-}; 4 Zn^{2+} inside
(b) The formula for zinc sulfide is ZnS.
(b) The oxidation state of Zn is 2+.
(c) The geometry around each zinc is tetrahedral.

11.14 (a) 1/8 Ca^{2+} at 8 corners = 1 Ca^{2+}; 1/2 O^{2-} at 6 faces = 3 O^{2-}; 1 Ti^{4+} inside
The formula for perovskite is $CaTiO_3$.
(b) The oxidation number of Ti is +4 to maintain charge neutrality in the unit cell.
(c) The geometry around titanium, oxygen, and calcium are all octahedral.

11.15 (a) solid → liquid
(b) liquid → gas
(c) gas → liquid → gas

11.16 (a)

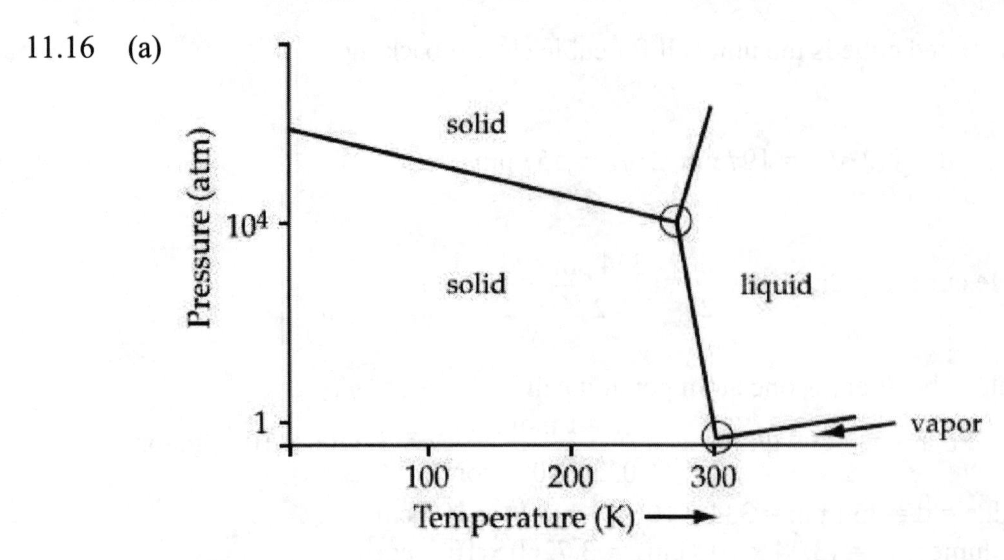

(b) Gallium has two triple points. The one below 1 atm is a solid, liquid, vapor triple point. The one at 10^4 atm is a solid(1), solid(2), liquid triple point.
(c) Increasing the pressure favors the liquid phase, giving the solid/liquid boundary a negative slope. At 1 atm pressure the liquid phase is more dense than the solid phase.

11.17 75 °F = 24 °C; At 70 atm and 24 °C, CO_2 is a liquid.

11.18 (a) $CO_2(s) \rightarrow CO_2(g)$
(b) $CO_2(l) \rightarrow CO_2(g)$
(c) $CO_2(g) \rightarrow CO_2(l) \rightarrow$ supercritical CO_2

11.19 (a) 5.11 atm (b) –56.4 °C (c) 31.1 °C

11.20 (a) For the phase transition, $CO_2(s) \rightarrow CO_2(g)$, the system is becoming more disordered, therefore $\Delta S > 0$.
(b) –78.5 °C
(c) $\Delta G = \Delta H - T\Delta S$; $\Delta G = 0$ at the sublimation temperature (–78.5 °C = 194.6 K). Set $\Delta G = 0$ and solve for ΔS.
$\Delta G = 0 = \Delta H - T\Delta S$

$$\Delta S = \frac{\Delta H}{T} = \frac{26.1 \text{ kJ/mol}}{194.6 \text{ K}} = 0.1341 \text{ kJ/(K·mol)} = 134.1 \text{ J/(K·mol)}$$

11.21 CO_2, 44.01

$$100.0 \text{ g x } \frac{1 \text{ mol } CO_2}{44.01 \text{ g } CO_2} = 2.272 \text{ mol } CO_2$$

$q_1 = (2.272 \text{ mol})(26.1 \text{ kJ/mol}) = 59.3 \text{ kJ}$
$q_2 = (2.272 \text{ mol})[35.0 \times 10^{-3} \text{ kJ/(mol · °C)}][33 \text{ °C} - (-78.5 \text{ °C})] = 8.86 \text{ kJ}$
$q_{total} = q_1 + q_2 = 68.2 \text{ kJ}$

Conceptual Problems

11.22 After the volume is decreased, equilibrium is reestablished and the equilibrium vapor pressure will still be 28.0 mm Hg. Vapor pressure depends only on temperature, not on volume.

11.24 (a) cubic closest-packed
(b) 1/8 S^{2-} at 8 corners and 1/2 S^{2-} at 6 faces = 4 S^{2-}; 4 Zn^{2+} inside

11.26 (a) normal boiling point ≈ 300 K; normal melting point ≈ 180 K
(b) (i) solid (ii) gas (iii) supercritical fluid

Section Problems
Properties of Liquids (Section 11.1)

11.28 It's water's surface tension that keeps the needle on top.

11.30 (a) CH_2Br_2 has the higher surface tension because it is polar while CCl_4 is nonpolar.
(b) Ethylene glycol has the higher surface tension because it can hydrogen bond from both ends of the molecule while ethanol can only hydrogen bond from one end of the molecule.

11.32 Oleic acid has a higher viscosity than H_2O because it is a much larger molecule than H_2O with larger dispersion forces. It can also hydrogen bond.

Vapor Pressure and Phase Changes (Sections 11.2 and 11.3)

11.34 ΔH_{vap} is usually larger than ΔH_{fusion} because ΔH_{vap} is the heat required to overcome all intermolecular forces.

11.36 (a) Hg(l) → Hg(g)
(b) no change of state, Hg remains a liquid
(c) Hg(g) → Hg(l) → Hg(s)

11.38 As the pressure over the liquid H_2O is lowered, H_2O vapor is removed by the pump. As H_2O vapor is removed, more of the liquid H_2O is converted to H_2O vapor. This conversion is an endothermic process and the temperature decreases. The combination of both a decrease in pressure and temperature takes the system across the liquid/solid boundary in the phase diagram so the H_2O that remains turns to ice.

11.40 H_2O, 18.02; $5.00 \text{ g } H_2O \times \dfrac{1 \text{ mol } H_2O}{18.02 \text{ g } H_2O} = 0.2775 \text{ mol } H_2O$

$q_1 = (0.2775 \text{ mol})[36.6 \times 10^{-3} \text{ kJ/(K} \cdot \text{mol)}](273 \text{ K} - 263 \text{ K}) = 0.1016 \text{ kJ}$
$q_2 = (0.2775 \text{ mol})(6.01 \text{ kJ/mol}) = 1.668 \text{ kJ}$

$q_3 = (0.2775 \text{ mol})(75.3 \times 10^{-3} \text{ kJ/(K} \cdot \text{mol)})(303 \text{ K} - 273 \text{ K}) = 0.6269 \text{ kJ}$
$q_{total} = q_1 + q_2 + q_3 = 2.40 \text{ kJ};$ 2.40 kJ of heat is required.

11.42 H_2O, 18.02; $7.55 \text{ g } H_2O \times \dfrac{1 \text{ mol } H_2O}{18.02 \text{ g } H_2O} = 0.4190 \text{ mol } H_2O$

$q_1 = (0.4190 \text{ mol})[75.3 \times 10^{-3} \text{ kJ/(K} \cdot \text{mol)}](273.15 \text{ K} - 306.65 \text{ K}) = -1.057 \text{ kJ}$
$q_2 = -(0.4190 \text{ mol})(6.01 \text{ kJ/mol}) = -2.518 \text{ kJ}$
$q_3 = (0.4190 \text{ mol})[36.6 \times 10^{-3} \text{ kJ/(K} \cdot \text{mol)}](263.15 \text{ K} - 273.15 \text{ K}) = -0.1534 \text{ kJ}$
$q_{total} = q_1 + q_2 + q_3 = -3.73 \text{ kJ}$
3.73 kJ of heat is released.

11.44

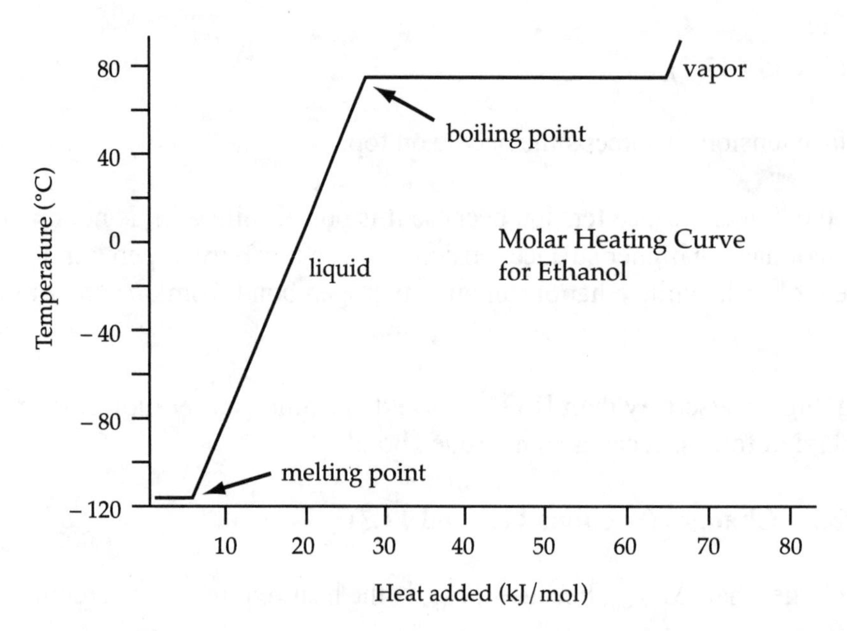

11.46 boiling point = 218 °C = 491 K
$\Delta G = \Delta H_{vap} - T\Delta S_{vap}$; At the boiling point (phase change), $\Delta G = 0$

$\Delta H_{vap} = T\Delta S_{vap}; \quad \Delta S_{vap} = \dfrac{\Delta H_{vap}}{T} = \dfrac{43.3 \text{ kJ/mol}}{491 \text{ K}} = 0.0882 \text{ kJ/(K} \cdot \text{mol)} = 88.2 \text{ J/(K} \cdot \text{mol)}$

11.48 $$\Delta H_{vap} = \frac{(\ln P_2 - \ln P_1)(R)}{\left(\dfrac{1}{T_1} - \dfrac{1}{T_2}\right)}$$

$T_1 = -5.1 \text{ °C} = 268.0 \text{ K};$ $P_1 = 100 \text{ mm Hg}$
$T_2 = 46.5 \text{ °C} = 319.6 \text{ K};$ $P_2 = 760 \text{ mm Hg}$

$$\Delta H_{vap} = \frac{[\ln(760) - \ln(100)][8.3145 \times 10^{-3} \text{ kJ/(K} \cdot \text{mol)}]}{\left(\dfrac{1}{268.0 \text{ K}} - \dfrac{1}{319.6 \text{ K}}\right)} = 28.0 \text{ kJ/mol}$$

11.50 $\ln P_2 = \ln P_1 + \frac{\Delta H_{vap}}{R}\left(\frac{1}{T_1} - \frac{1}{T_2}\right)$

$\Delta H_{vap} = 28.0$ kJ/mol
$P_1 = 100$ mm Hg; $T_1 = -5.1\ °C = 268.0$ K; $T_2 = 20.0\ °C = 293.2$ K
Solve for P_2.

$$\ln P_2 = \ln(100) + \frac{28.0\ \text{kJ/mol}}{[8.3145 \times 10^{-3}\ \text{kJ/(K}\cdot\text{mol)}]}\left(\frac{1}{268.0\ \text{K}} - \frac{1}{293.2\ \text{K}}\right)$$

$\ln P_2 = 5.6852$; $P_2 = e^{5.6852} = 294.5$ mm Hg $= 294$ mm Hg

11.52

T(K)	P_{vap}(mm Hg)	$\ln P_{vap}$	1/T
263	80.1	4.383	0.003 802
273	133.6	4.8949	0.003 663
283	213.3	5.3627	0.003 534
293	329.6	5.7979	0.003 413
303	495.4	6.2054	0.003 300
313	724.4	6.5853	0.003 195

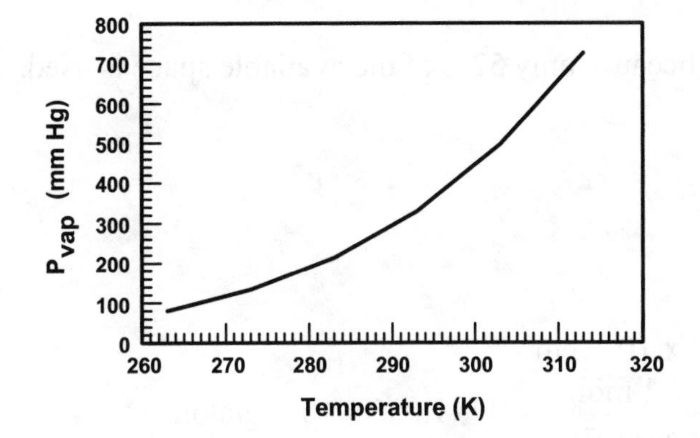

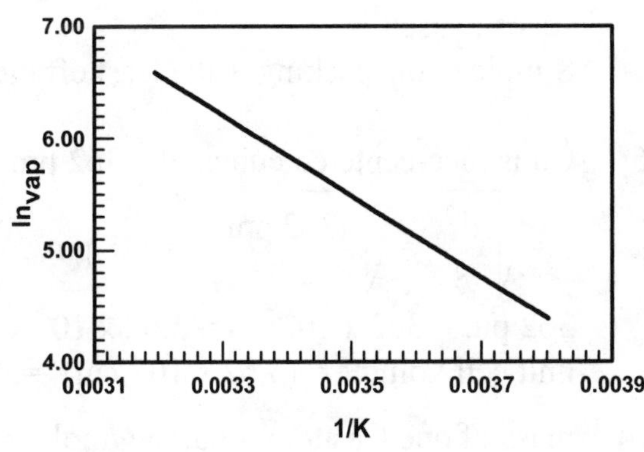

$$\ln P_{vap} = \left(-\frac{\Delta H_{vap}}{R}\right)\frac{1}{T} + C;\quad C = 18.2$$

$$\text{slope} = -3628\ \text{K} = -\frac{\Delta H_{vap}}{R}$$

$$\Delta H_{vap} = (3628\ \text{K})(R) = (3628\ \text{K})[8.3145 \times 10^{-3}\ \text{kJ/(K}\cdot\text{mol)}] = 30.1\ \text{kJ/mol}$$

11.54 $\Delta H_{vap} = 30.1$ kJ/mol

11.56 $\Delta H_{vap} = \frac{(\ln P_2 - \ln P_1)(R)}{\left(\frac{1}{T_1} - \frac{1}{T_2}\right)}$

$P_1 = 80.1$ mm Hg; $T_1 = 263$ K
$P_2 = 724.4$ mm Hg; $T_2 = 313$ K

$$\Delta H_{vap} = \frac{[\ln(724.4) - \ln(80.1)][8.3145 \times 10^{-3}\ kJ/(K \cdot mol)]}{\left(\frac{1}{263\ K} - \frac{1}{313\ K}\right)} = 30.1\ kJ/mol$$

The calculated ΔH_{vap} and that obtained from the plot in Problem 11.54 are the same.

Structures of Solids (Sections 11.4–11.8)

11.58 molecular solid, CO_2, I_2; metallic solid, any metallic element; covalent network solid, diamond; ionic solid, NaCl

11.60 (a) rubber (b) Na_3PO_4 (c) CBr_4 (d) quartz (e) Au

11.62 Silicon carbide is a covalent network solid.

11.64 $$d = \frac{n\lambda}{2\sin\theta} = \frac{(1)(154.2\ pm)}{2\sin(22.5^{\circ})} = 201\ pm$$

11.66 From Table 11.5.
Hexagonal and cubic closest packing are the most efficient because 74% of the available space is used.
Simple cubic packing is the least efficient because only 52% of the available space is used.

11.68 Cu is face-centered cubic. d = 362 pm

$$r = \sqrt{\frac{d^2}{8}} = \sqrt{\frac{(362\ pm)^2}{8}} = 128\ pm$$

362 pm = 362×10^{-12} m = 3.62×10^{-8} cm
unit cell volume = $(3.62 \times 10^{-8}\ cm)^3 = 4.74 \times 10^{-23}\ cm^3$

$$\text{mass of one Cu atom} = 63.55\ g/mol \times \frac{1\ mol}{6.022 \times 10^{23}\ atom} = 1.055 \times 10^{-22}\ g/atom$$

Cu is face-centered cubic; there are, therefore, four Cu atoms in the unit cell.
unit cell mass = (4 atoms)(1.055×10^{-22} g/atom) = 4.22×10^{-22} g

$$\text{density} = \frac{\text{mass}}{\text{volume}} = \frac{4.22 \times 10^{-22} g}{4.74 \times 10^{-23}\ cm^3} = 8.90\ g/cm^3$$

11.70 $$\text{mass of one Al atom} = 26.98\ g/mol \times \frac{1\ mol}{6.022 \times 10^{23}\ atom} = 4.480 \times 10^{-23}\ g/atom$$

Al is face-centered cubic; there are, therefore, four Al atoms in the unit cell.
unit cell mass = (4 atoms)(4.480×10^{-23} g/atom) = 1.792×10^{-22} g

$$\text{density} = \frac{\text{mass}}{\text{volume}}$$

$$\text{unit cell volume} = \frac{\text{unit cell mass}}{\text{density}} = \frac{1.792 \times 10^{-22}\ g}{2.699\ g/cm^3} = 6.640 \times 10^{-23}\ cm^3$$

unit cell edge = d = $\sqrt[3]{6.640 \times 10^{-23}\ cm^3}$ = 4.049 x 10^{-8} cm

d = 4.049 x 10^{-8} cm x $\frac{1 m}{100 cm}$ = 4.049 x 10^{-10} m = 404.9 x 10^{-12} m = 404.9 pm

11.72 unit cell body diagonal = 4r = 549 pm

For W, r = $\frac{549 pm}{4}$ = 137 pm

11.74 mass of one Ti atom = 47.88 g/mol x $\frac{1 mol}{6.022 \times 10^{23}\ atoms}$ = 7.951 x 10^{-23} g/atom

r = 144.8 pm = 144.8 x 10^{-12} m

r = 144.8 x 10^{-12} m x $\frac{100 cm}{1 m}$ = 1.448 x 10^{-8} cm

Calculate the volume and then the density for Ti assuming it is primitive cubic, body-centered cubic, and face-centered cubic. Compare the calculated density with the actual density to identify the unit cell.

For primitive cubic:

d = 2r; volume = d^3 = $[2(1.448 \times 10^{-8}\ cm)]^3$ = 2.429 x 10^{-23} cm^3

density = $\frac{\text{unit cell mass}}{\text{volume}}$ = $\frac{7.951 \times 10^{-23}\ g}{2.429 \times 10^{-23}\ cm^3}$ = 3.273 g/cm^3

For face-centered cubic:

d = $2\sqrt{2}$r; volume = d^3 = $[2\sqrt{2}(1.448 \times 10^{-8}\ cm)]^3$ = 6.870 x 10^{-23} cm^3

density = $\frac{4(7.951 \times 10^{-23}\ g)}{6.870 \times 10^{-23}\ cm^3}$ = 4.630 g/cm^3

For body-centered cubic:

d = $\frac{4r}{\sqrt{3}}$; volume = d^3 = $\left[\frac{4(1.448 \times 10^{-8}\ cm)}{\sqrt{3}}\right]^3$ = 3.739 x 10^{-23} cm^3

density = $\frac{2(7.951 \times 10^{-23}\ g)}{3.739 \times 10^{-23}\ cm^3}$ = 4.253 g/cm^3

The calculated density for a face-centered cube (4.630 g/cm^3) is closest to the actual density of 4.54 g/cm^3. Ti crystallizes in the face-centered cubic unit cell.

11.76 Six Na^+ ions touch each H^- ion and six H^- ions touch each Na^+ ion.

11.78 Na^+ H^- Na^+

← 488 pm → unit cell edge = d = 488 pm; Na–H bond = d/2 = 244 pm

11.80 mass of one Pb atom = $207.2 \text{ g/mol} \times \dfrac{1 \text{ mol}}{6.022 \times 10^{23} \text{ atom}} = 3.441 \times 10^{-22}$ g/atom

If the unit cell for Pb is the primitive cube it would contain one Pb atom and weigh 3.441×10^{-22} g.

If the unit cell for Pb is the face-centered cube it would contain four Pb atom and weigh $(4)(3.441 \times 10^{-22} \text{ g}) = 1.376 \times 10^{-21}$ g.

For a primitive cubic unit cell:
unit cell edge = d = 2r = 2(175 pm) = 3.50×10^{-8} cm
unit cell volume = $d^3 = (3.50 \times 10^{-8} \text{ cm})^3 = 4.29 \times 10^{-23} \text{ cm}^3$
unit cell mass = $(4.29 \times 10^{-23} \text{ cm}^3)(11.34 \text{ g/cm}^3) = 4.86 \times 10^{-22}$ g

For a face-centered cubic unit cell:
unit cell edge = d = $2\sqrt{2}\,r = 2\sqrt{2}\,(175 \text{ pm}) = 4.95 \times 10^{-8}$ cm
unit cell volume = $d^3 = (4.95 \times 10^{-8} \text{ cm})^3 = 1.21 \times 10^{-22} \text{ cm}^3$
unit cell mass = $(1.21 \times 10^{-22} \text{ cm}^3)(11.34 \text{ g/cm}^3) = 1.38 \times 10^{-21}$ g

The masses agree for the face-centered cube, therefore it is the unit cell for Pb.

Phase Diagrams (Section 11.9)

11.82 (a) gas (b) liquid (c) solid

11.84

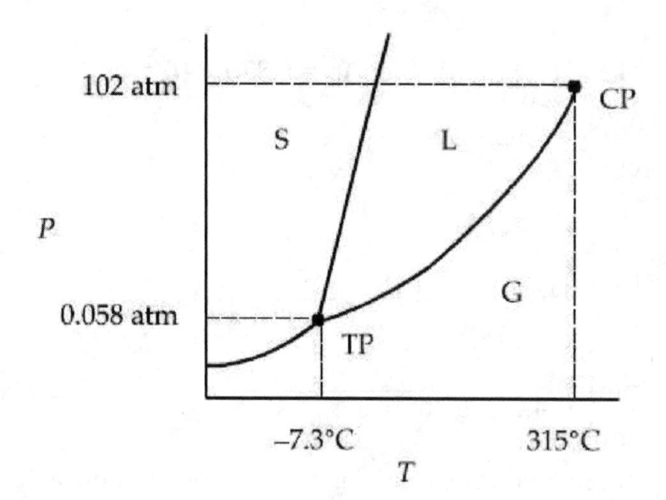

11.86 (a) $Br_2(s)$ (b) $Br_2(l)$

11.88 Solid O_2 does not melt when pressure is applied because the solid is denser than the liquid, and the solid/liquid boundary in the phase diagram slopes to the right.

11.90

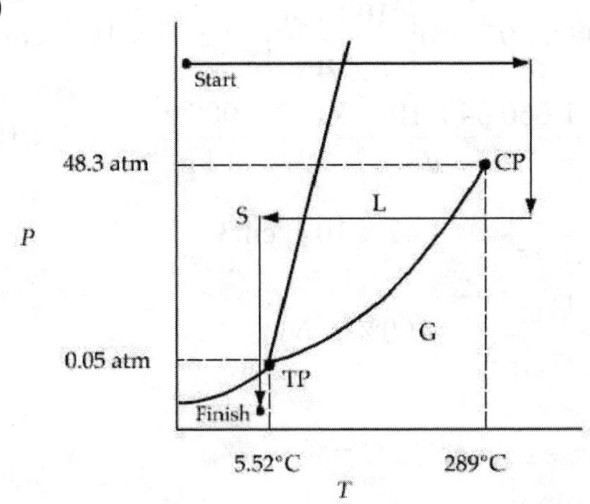

The starting phase is benzene as a solid, and the final phase is benzene as a gas.

11.92 solid → liquid → supercritical fluid → liquid → solid → gas

Chapter Problems

11.94 Because Ar crystallizes in a face-centered cubic unit cell, there are four Ar atoms in the unit cell.

$$\text{mass of one Ar atom} = 39.95 \text{ g/mol x } \frac{1 \text{ mol}}{6.022 \text{ x } 10^{23} \text{ atom}} = 6.634 \text{ x } 10^{-23} \text{ g/atom}$$

unit cell mass = 4 atoms x mass of one Ar atom

$$= 4 \text{ atoms x } 6.634 \text{ x } 10^{-23} \text{ g/atom} = 2.654 \text{ x } 10^{-22} \text{ g}$$

$$\text{density} = \frac{\text{mass}}{\text{volume}}$$

$$\text{unit cell volume} = \frac{\text{unit cell mass}}{\text{density}} = \frac{2.654 \text{ x } 10^{-22} \text{ g}}{1.623 \text{ g/cm}^3} = 1.635 \text{ x } 10^{-22} \text{ cm}^3$$

$$\text{unit cell edge} = d = \sqrt[3]{1.635 \text{ x } 10^{-22} \text{ cm}^3} = 5.468 \text{ x } 10^{-8} \text{ cm}$$

$$d = 5.468 \text{ x } 10^{-8} \text{ cm x } \frac{1 \text{ x } 10^{-2} \text{ m}}{1 \text{ cm}} = 5.468 \text{ x } 10^{-10} \text{ m} = 546.8 \text{ x } 10^{-12} \text{ m} = 546.8 \text{ pm}$$

$$r = \sqrt{\frac{d^2}{8}} = \sqrt{\frac{(546.8 \text{ pm})^2}{8}} = 193.3 \text{ pm}$$

11.96

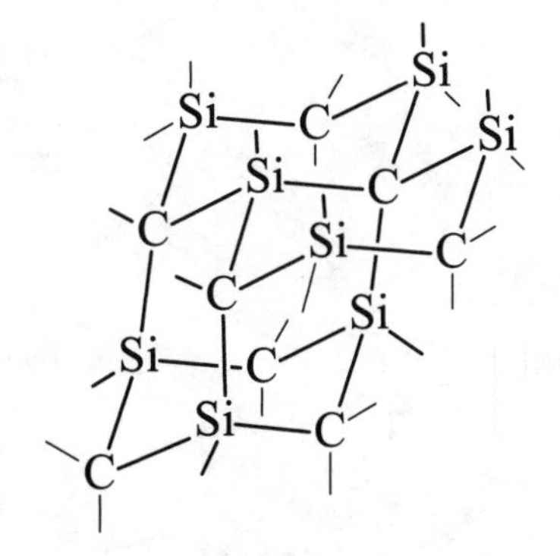

11.98 Unit cell volume $= (7900 \times 10^{-12}\ m)^2(3800 \times 10^{-12}\ m)\left(\frac{100\ cm}{1\ m}\right)^3 = 2.37 \times 10^{-19}\ cm^3$

total lysozyme mass $= 8 \times 1.44 \times 10^4\ u \times \frac{1.660\,54 \times 10^{-27}\ kg}{1\ u} \times \frac{1000\ g}{1\ kg} = 1.91 \times 10^{-19}\ g$

total lysozyme volume $= 1.91 \times 10^{-19}\ g \times \frac{1\ cm^3}{1.35\ g} = 1.42 \times 10^{-19}$ cm3

% occupied unit cell volume $= \frac{1.42 \times 10^{-19}\ cm^3}{2.37 \times 10^{-19}\ cm^3} \times 100\% = 60\%$

11.100 $\Delta G = \Delta H - T\Delta S$; at the melting point (phase change), $\Delta G = 0$.

$\Delta H = T\Delta S;\quad T = \frac{\Delta H_{fus}}{\Delta S_{fus}} = \frac{9.037\ kJ/mol}{9.79 \times 10^{-3}\ kJ/(K\cdot mol)} = 923\ K = 650\ ^\circ C$

11.102 $\Delta H_{vap} = \frac{(\ln P_2 - \ln P_1)(R)}{\left(\frac{1}{T_1} - \frac{1}{T_2}\right)}$

$P_1 = 40.0$ mm Hg; $T_1 = -81.6\ ^\circ C = 191.6\ K$
$P_2 = 400$ mm Hg; $T_2 = -43.9\ ^\circ C = 229.2\ K$

$$\Delta H_{vap} = \frac{[\ln(400) - \ln(40.0)]\left(8.3145 \times 10^{-3} \frac{kJ}{K\cdot mol}\right)}{\left(\frac{1}{191.6\ K} - \frac{1}{229.2\ K}\right)} = 22.36\ kJ/mol$$

Using $\Delta H_{vap} = 22.36$ kJ/mol

$$\ln P_2 = \ln P_1 + \frac{\Delta H_{vap}}{R}\left(\frac{1}{T_1} - \frac{1}{T_2}\right)$$

$$(\ln P_2 - \ln P_1)\left(\frac{R}{\Delta H_{vap}}\right) = \frac{1}{T_1} - \frac{1}{T_2}$$

$$\frac{1}{T_1} - (\ln P_2 - \ln P_1)\left(\frac{R}{\Delta H_{vap}}\right) = \frac{1}{T_2}$$

$P_1 = 40.0$ mm Hg; $T_1 = 191.6$ K
$P_2 = 760$ mm Hg
Solve for T_2, the normal boiling point.

$$\frac{1}{191.6\ K} - [\ln(760) - \ln(40.0)]\left(\frac{8.3145 \times 10^{-3} \frac{kJ}{K\cdot mol}}{22.36\ kJ/mol}\right) = \frac{1}{T_2}$$

$$\frac{1}{T_2} = 0.004\ 124\ 33;\ \ T_2 = 242.46\ K = -30.7\ °C$$

11.104 $$\Delta H_{vap} = \frac{(\ln P_2 - \ln P_1)(R)}{\left(\frac{1}{T_1} - \frac{1}{T_2}\right)}$$

$P_1 = 100$ mm Hg; $T_1 = -110.3\ °C = 162.85$ K
$P_2 = 760$ mm Hg; $T_2 = -88.5\ °C = 184.65$ K

$$\Delta H_{vap} = \frac{[\ln(760) - \ln(100)]\left(8.3145 \times 10^{-3} \frac{kJ}{K \cdot mol}\right)}{\left(\frac{1}{162.85\ K} - \frac{1}{184.65\ K}\right)} = 23.3\ kJ/mol$$

11.106

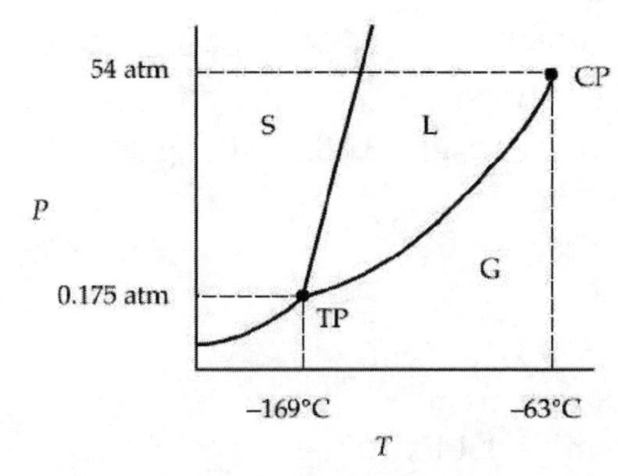

Kr cannot be liquefied at room temperature because room temperature is above T_c (–63 °C).

11.108 For a body-centered cube:

$4r = \sqrt{3}$ edge; edge $= \frac{4r}{\sqrt{3}}$

volume of sphere $= \frac{4}{3}\pi r^3$

volume of unit cell $= \left(\frac{4r}{\sqrt{3}}\right)^3 = \frac{64\,r^3}{3\sqrt{3}}$

volume of 2 spheres $= 2\left(\frac{4}{3}\pi r^3\right) = \frac{8}{3}\pi r^3$

$$\%\ \text{volume occupied} = \frac{\left(\frac{8}{3}\pi r^3\right)}{\left(\frac{64 r^3}{3\sqrt{3}}\right)} \times 100\% = 68\%$$

11.110 unit cell edge = d = 287 pm = 287×10^{-12} m = 2.87×10^{-8} cm
unit cell volume = $d^3 = (2.87 \times 10^{-8}\ \text{cm})^3 = 2.364 \times 10^{-23}\ \text{cm}^3$
unit cell mass = $(2.364 \times 10^{-23}\ \text{cm}^3)(7.86\ \text{g/cm}^3) = 1.858 \times 10^{-22}$ g
Fe is body-centered cubic; therefore there are two Fe atoms per unit cell.

$$\text{mass of one Fe atom} = \frac{1.858 \times 10^{-22}\ \text{g}}{2\ \text{Fe atoms}} = 9.290 \times 10^{-23}\ \text{g/atom}$$

$$\text{Avogadro's number} = 55.85\ \text{g/mol} \times \frac{1\ \text{atom}}{9.290 \times 10^{-23}\ \text{g}} = 6.01 \times 10^{23}\ \text{atoms/mol}$$

11.112 (a) unit cell edge = $2r_{Cl^-} + 2r_{Na^+} = 2(181\ \text{pm}) + 2(97\ \text{pm}) = 556$ pm
(b) unit cell edge = d = 556 pm = 556×10^{-12} m = 5.56×10^{-8} cm
unit cell volume = $(5.56 \times 10^{-8}\ \text{cm})^3 = 1.719 \times 10^{-22}\ \text{cm}^3$
The unit cell contains 4 Na^+ ions and 4 Cl^- ions.

$$\text{mass of one Na}^+\ \text{ion} = 22.99\ \text{g/mol} \times \frac{1\ \text{mol}}{6.022 \times 10^{23}\ \text{ions}} = 3.818 \times 10^{-23}\ \text{g/Na}^+$$

$$\text{mass of one Cl}^-\ \text{ion} = 35.45\ \text{g/mol} \times \frac{1\ \text{mol}}{6.022 \times 10^{23}\ \text{ions}} = 5.887 \times 10^{-23}\ \text{g/Cl}^-$$

unit cell mass = $4(3.818 \times 10^{-23}\ \text{g}) + 4(5.887 \times 10^{-23}\ \text{g}) = 3.882 \times 10^{-22}$ g

$$\text{density} = \frac{\text{unit cell mass}}{\text{unit cell volume}} = \frac{3.882 \times 10^{-22}\ \text{g}}{1.719 \times 10^{-22}\ \text{cm}^3} = 2.26\ \text{g/cm}^3$$

11.114 Ag_2Te, 343.33; 529 pm = 529×10^{-12} m = 529×10^{-10} cm
unit cell volume = $(529 \times 10^{-10}\ \text{cm})^3 = 1.48 \times 10^{-22}\ \text{cm}^3$
unit cell mass = $(1.48 \times 10^{-22}\ \text{cm}^3)(7.70\ \text{g/cm}^3) = 1.14 \times 10^{-21}$ g

$$\text{mass of one Ag}_2\text{Te} = \frac{343.33\ \text{g Ag}_2\text{Te/mol}}{6.022 \times 10^{23}\ \text{Ag}_2\text{Te formula units/mol}} = 5.70 \times 10^{-22}\ \text{g Ag}_2\text{Te/formula unit}$$

$$\text{Ag}_2\text{Te formula units/unit cell} = \frac{1.14 \times 10^{-21}\ \text{g/unit cell}}{5.70 \times 10^{-22}\ \text{g/Ag}_2\text{Te}} = 2\ \text{Ag}_2\text{Te/unit cell}$$

$$\text{Ag/unit cell} = \frac{2\ \text{Ag}_2\text{Te}}{\text{unit cell}} \times \frac{2\ \text{Ag}}{\text{Ag}_2\text{Te}} = 4\ \text{Ag/unit cell}$$

Multiconcept Problems

11.116 Al_2O_3, ionic (greater lattice energy than NaCl because of higher ion charges);
F_2, dispersion; H_2O, H–bonding, dipole-dipole; Br_2, dispersion (larger and more polarizable than F_2), ICl, dipole-dipole, NaCl, ionic
rank according to normal boiling points: $F_2 < Br_2 < ICl < H_2O < NaCl < Al_2O_3$

11.118 (a) Let the formula of magnetite be Fe_xO_y, then $Fe_xO_y + y\,CO \rightarrow x\,Fe + y\,CO_2$

$$n_{CO_2} = y = \frac{PV}{RT} = \frac{\left(751 \text{ mm Hg x } \dfrac{1.00 \text{ atm}}{760 \text{ mm Hg}}\right)(1.136 \text{ L})}{\left(0.082\ 06\ \dfrac{\text{L}\cdot\text{atm}}{\text{K}\cdot\text{mol}}\right)(298 \text{ K})} = 0.04590 \text{ mol } CO_2$$

0.04590 mol CO_2 = mol of O in Fe_xO_y

$$\text{mass of O in } Fe_xO_y = 0.04590 \text{ mol O x } \frac{16.0 \text{ g O}}{1 \text{ mol O}} = 0.7345 \text{ g O}$$

mass of Fe in Fe_xO_y = 2.660 g – 0.7345 g = 1.926 g Fe

(b) $$\text{mol Fe in magnetite} = 1.926 \text{ g Fe x } \frac{1 \text{ mol Fe}}{55.85 \text{ g Fe}} = 0.0345 \text{ mol Fe}$$

formula of magnetite: $Fe_{0.0345}O_{0.0459}$ (divide each subscript by the smaller)

$Fe_{0.0345/0.0345}O_{0.0459/0.0345}$

$FeO_{1.33}$ (multiply both subscripts by 3)

$Fe_{(1 \times 3)}O_{(1.33 \times 3)}$; Fe_3O_4

(c) unit cell edge = d = 839 pm = 839 x 10^{-12} m

$$d = 839 \times 10^{-12} \text{ m x } \frac{100 \text{ cm}}{1 \text{ m}} = 8.39 \times 10^{-8} \text{ cm}$$

unit cell volume = d^3 = $(8.39 \times 10^{-8} \text{ cm})^3 = 5.91 \times 10^{-22} \text{ cm}^3$

unit cell mass = $(5.91 \times 10^{-22} \text{ cm}^3)(5.20 \text{ g/cm}^3) = 3.07 \times 10^{-21}$ g

$$\text{mass of Fe in unit cell} = \left(\frac{1.926 \text{ g Fe}}{2.660 \text{ g}}\right)(3.07 \times 10^{-21} \text{ g}) = 2.22 \times 10^{-21} \text{ g Fe}$$

$$\text{mass of O in unit cell} = \left(\frac{0.7345 \text{ g O}}{2.660 \text{ g}}\right)(3.07 \times 10^{-21} \text{ g}) = 8.47 \times 10^{-22} \text{ g O}$$

$$\text{Fe atoms in unit cell} = 2.22 \times 10^{-21} \text{ g x } \frac{6.022 \times 10^{23} \text{ atoms/mol}}{55.847 \text{ g/mol}} = 24 \text{ Fe atoms}$$

$$\text{O atoms in unit cell} = 8.47 \times 10^{-22} \text{ g x } \frac{6.022 \times 10^{23} \text{ atoms/mol}}{16.00 \text{ g/mol}} = 32 \text{ O atoms}$$

11.120 (a) M = alkali metal; 500.0 mL = 0.5000 L; 802 °C = 1075 K

$$n_M = \frac{PV}{RT} = \frac{\left(12.5 \text{ mm Hg x } \dfrac{1.00 \text{ atm}}{760 \text{ mm Hg}}\right)(0.5000 \text{ L})}{\left(0.082\ 06\ \dfrac{\text{L}\cdot\text{atm}}{\text{K}\cdot\text{mol}}\right)(1075 \text{ K})} = 9.32 \times 10^{-5} \text{ mol M}$$

1.62 mm = 1.62 x 10^{-3} m; crystal volume = $(1.62 \times 10^{-3} \text{ m})^3 = 4.25 \times 10^{-9} \text{ m}^3$

M atoms in crystal = $(9.32 \times 10^{-5} \text{ mol})(6.022 \times 10^{23} \text{ atoms/mol}) = 5.61 \times 10^{19}$ M atoms

Because M is body-centered cubic, only 68% (Table 11.5) of the total volume is occupied by M atoms.

$$\text{volume of M atom} = \frac{(0.68)(4.25 \times 10^{-9} \text{ m})}{5.61 \times 10^{19} \text{ M atoms}} = 5.15 \times 10^{-29} \text{ m}^3\text{/M atom}$$

volume of a sphere = $\frac{4}{3}\pi r^3$

$r_M = \sqrt[3]{\frac{3(\text{volume})}{4\pi}} = \sqrt[3]{\frac{3(5.15 \times 10^{-29}\ \text{m}^3)}{4\pi}} = 2.31 \times 10^{-10}\ \text{m} = 231 \times 10^{-12}\ \text{m} = 231\ \text{pm}$

(b) The radius of 231 pm is closest to that of K.

(c) 1.62 mm = 0.162 cm

density of solid = $\frac{(9.32 \times 10^{-5}\ \text{mol})(39.1\ \text{g/mol})}{(0.162\ \text{cm})^3} = 0.857\ \text{g/cm}^3$

density of vapor = $\frac{(9.32 \times 10^{-5}\ \text{mol})(39.1\ \text{g/mol})}{500.0\ \text{cm}^3} = 7.29 \times 10^{-6}\ \text{g/cm}^3$

12 Solutions and Their Properties

12.1 KBr < 1,5 pentanediol < toluene

12.2 Vitamin E is more fat soluble.

12.3 NaCl, 58.44; 1.00 mol NaCl = 58.44 g
1.00 L H_2O = 1000 mL = 1000 g (assuming a density of 1.00 g/mL)

$$\text{mass \% NaCl} = \frac{58.44\text{ g}}{1000\text{ g} + 58.44\text{ g}} \times 100\% = 5.52\text{ mass \%}$$

12.4 $$7.50\% = \left(\frac{15.0\text{ g}}{15.0\text{ g} + x}\right)(100\%)$$

Let x equal the mass of water. Solve for x.
$(7.50\%)(15.0\text{ g} + x) = (15.0\text{ g})(100\%)$

$$x = \frac{(15.0\text{ g})(100\%)}{7.50\%} - 15.0\text{ g} = 185\text{ g H}_2\text{O}$$

12.5 $1.25\ \mu\text{g} = 1.25 \times 10^{-6}$ g; (50.0 mL)(1.00 g/mL) = 50.0 g

$$\frac{1.25 \times 10^{-6}\text{ g}}{50.0\text{ g}} \times 10^9 = 25.0\text{ ppb}$$

This exceeds the 10.0 ppb limit for As.

12.6 $$\text{ppm} = \frac{\text{mass of CO}_2}{\text{total mass of solution}} \times 10^6\text{ ppm}$$

total mass of solution = density x volume = (1.3 g/L)(1.0 L) = 1.3 g

$$35\text{ ppm} = \frac{\text{mass of CO}_2}{1.3\text{ g}} \times 10^6\text{ ppm}$$

$$\text{mass of CO}_2 = \frac{(35\text{ ppm})(1.3\text{ g})}{10^6\text{ ppm}} = 4.6 \times 10^{-5}\text{ g CO}_2$$

12.7 CH_3CO_2Na, 82.03

$$\text{kg H}_2\text{O} = (0.150\text{ mol CH}_3\text{CO}_2\text{Na})\left(\frac{1\text{ kg H}_2\text{O}}{0.500\text{ mol CH}_3\text{CO}_2\text{Na}}\right) = 0.300\text{ kg H}_2\text{O}$$

$$\text{mass CH}_3\text{CO}_2\text{Na} = 0.150\text{ mol CH}_3\text{CO}_2\text{Na} \times \frac{82.03\text{ g CH}_3\text{CO}_2\text{Na}}{1\text{ mol CH}_3\text{CO}_2\text{Na}} = 12.3\text{ g CH}_3\text{CO}_2\text{Na}$$

mass of solution needed = 300 g + 12.3 g = 312 g

12.8 $C_{27}H_{46}O$, 386.7; $CHCl_3$, 119.4; $40.0 \text{ g} \times \frac{1 \text{ kg}}{1000 \text{ g}} = 0.0400 \text{ kg}$

$$\text{molality} = \frac{\text{mol } C_{27}H_{46}O}{\text{kg } CHCl_3} = \frac{\left(0.385 \text{ g} \times \frac{1 \text{ mol}}{386.7 \text{ g}}\right)}{0.0400 \text{ kg}} = 0.0249 \text{ mol/kg} = 0.0249\ m$$

$$X_{C_{27}H_{46}O} = \frac{\text{mol } C_{27}H_{46}O}{\text{mol } C_{27}H_{46}O + \text{mol } CHCl_3}$$

$$X_{C_{27}H_{46}O} = \frac{\left(0.385 \text{ g} \times \frac{1 \text{ mol}}{386.7 \text{ g}}\right)}{\left[\left(0.385 \text{ g} \times \frac{1 \text{ mol}}{386.7 \text{ g}}\right) + \left(40.0 \text{ g} \times \frac{1 \text{ mol}}{119.4 \text{ g}}\right)\right]} = 2.96 \times 10^{-3}$$

12.9 H_2O, 18.02
Assume 1.00 L of solution.
mass of 1.00 L = (1.0042 g/mL)(1000 mL) = 1004.2 g of solution

$$0.500 \text{ mol } CH_3CO_2H \times \frac{60.05 \text{ g } CH_3CO_2H}{1 \text{ mol } CH_3CO_2H} = 30.02 \text{ g } CH_3CO_2H$$

1004.2 g – 30.02 g = 974.2 g = 0.9742 kg of H_2O

$$974.2 \text{ g } H_2O \times \frac{1 \text{ mol } H_2O}{18.02 \text{ g } H_2O} = 54.06 \text{ mol } H_2O$$

$$X_{CH_3CO_2H} = \frac{\text{mol } CH_3CO_2H}{\text{mol } CH_3CO_2H + \text{mol } H_2O} = \frac{0.500 \text{ mol}}{0.500 \text{ mol} + 54.06 \text{ mol}} = 0.00916$$

$$\text{mass \% } CH_3CO_2H = \frac{30.02 \text{ g}}{30.02 \text{ g} + 974.2 \text{ g}} \times 100\% = 2.99 \text{ mass \%}$$

$$\text{molality} = \frac{0.500 \text{ mol}}{0.9742 \text{ kg}} = 0.513\ m$$

12.10 Assume you have a solution with 1.000 kg (1000 g) of H_2O. If this solution is 0.258 *m*, then it must also contain 0.258 mol glucose.

$$\text{mass of glucose} = 0.258 \text{ mol} \times \frac{180.2 \text{ g}}{1 \text{ mol}} = 46.5 \text{ g glucose}$$

mass of solution = 1000 g + 46.5 g = 1046.5 g
density = 1.0173 g/mL

$$\text{volume of solution} = 1046.5 \text{ g} \times \frac{1 \text{ mL}}{1.0173 \text{ g}} = 1028.7 \text{ mL}$$

$$\text{volume} = 1028.7 \text{ mL} \times \frac{1 \text{ L}}{1000 \text{ mL}} = 1.029 \text{ L}; \qquad \text{molarity} = \frac{0.258 \text{ mol}}{1.029 \text{ L}} = 0.251 \text{ M}$$

12.11 $M = k \cdot P$; $k = \frac{M}{P} = \frac{3.2 \times 10^{-2} \text{ M}}{1.0 \text{ atm}} = 3.2 \times 10^{-2} \text{ mol/(L} \cdot \text{atm)}$

12.12 (a) $M = k \cdot P = [3.2 \times 10^{-2}\ mol/(L \cdot atm)](2.5\ atm) = 0.080\ M$

(b) $P_{CO_2} = (0.0004)(1\ atm) = 4.0 \times 10^{-4}\ atm$

$M = k \cdot P = [3.2 \times 10^{-2}\ mol/(L \cdot atm)](4.0 \times 10^{-4}\ atm) = 1.3 \times 10^{-5}\ M$

12.13 H_2O, 18.02; $CaCl_2$, 110.0

$$100.0\ g\ H_2O \times \frac{1\ mol\ H_2O}{18.02\ g\ H_2O} = 5.549\ mol\ H_2O$$

$$10.00 g\ CaCl_2 \times \frac{1\ mol\ CaCl_2}{110.0\ g\ CaCl_2} = 0.090\ 91\ mol\ CaCl_2$$

$$X_{H_2O} = \frac{5.549\ mol}{(2.7)(0.090\ 91\ mol) + 5.549\ mol} = 0.9567$$

$$P_{soln} = P_{H_2O} \times X_{H_2O} = (233.7\ mm\ Hg)(0.9576) = 223.8\ mm\ Hg$$

12.14 H_2O, 18.02; $MgCl_2$, 95.21

$$100.0\ g\ H_2O \times \frac{1\ mol\ H_2O}{18.02\ g\ H_2O} = 5.549\ mol\ H_2O$$

$$8.110 g\ MgCl_2 \times \frac{1\ mol\ MgCl_2}{95.21\ g\ MgCl_2} = 0.085\ 18\ mol\ MgCl_2$$

$$X_{H_2O} = \frac{P_{soln}}{P_{H_2O}} = \frac{224.7\ mm\ Hg}{233.7\ mm\ Hg} = 0.9615$$

$$X_{H_2O} = 0.9615 = \frac{5.549\ mol}{(i)(0.085\ 18\ mol) + 5.549\ mol};$$ solve for the van't Hoff factor, i.

$$(0.9615)[(i)(0.085\ 18\ mol) + 5.549\ mol)] = 5.549\ mol$$

$$i = \frac{\left(\frac{5.549\ mol}{0.9615} - 5.549\ mol\right)}{0.085\ 18} = 2.6$$

12.15 $P_{soln} = P_{solv} \cdot X_{solv}$; $X_{solv} = \frac{P_{soln}}{P_{solv}} = \frac{(55.3 - 1.30)\ mm\ Hg}{55.3\ mm\ Hg} = 0.976$

NaBr dissociates into two ions in aqueous solution.

$$X_{solv} = \frac{mol\ H_2O}{mol\ H_2O + mol\ Na^+ + mol\ Br^-}$$

$$X_{solv} = 0.976 = \frac{\left(250\ g \times \frac{1\ mol}{18.02\ g}\right)}{\left(250\ g \times \frac{1\ mol}{18.02\ g}\right) + x\ mol\ Na^+ + x\ mol\ Br^-}$$

$$0.976 = \frac{13.9\ mol}{13.9\ mol + 2x\ mol};$$ solve for x.

$0.976(13.9\text{ mol} + 2x\text{ mol}) = 13.9\text{ mol}$
$13.566\text{ mol} + 1.952\,x\text{ mol} = 13.9\text{ mol}$
$1.952\,x\text{ mol} = 13.9\text{ mol} - 13.566\text{ mol}$

$$x\text{ mol} = \frac{13.9\text{ mol} - 13.566\text{ mol}}{1.952} = 0.171\text{ mol}$$

$x = 0.171\text{ mol } Na^+ = 0.171\text{ mol } Br^- = 0.171\text{ mol NaBr}$

NaBr, 102.9; mass NaBr $= 0.171\text{ mol} \times \frac{102.9\text{ g}}{1\text{ mol}} = 17.6\text{ g NaBr}$

12.16 At any given temperature, the vapor pressure of a solution is lower than the vapor pressure of the pure solvent. The upper curve represents the vapor pressure of the pure solvent. The lower curve represents the vapor pressure of the solution.

12.17 $100\text{ g } C_2H_5OH \times \frac{1\text{ mol } C_2H_5OH}{46.07\text{ g } C_2H_5OH} = 2.171\text{ mol } C_2H_6O$

$$25.0\text{ g } H_2O \times \frac{1\text{ mol } H_2O}{18.02\text{ g } H_2O} = 1.387\text{ mol } H_2O$$

$$X_{C_2H_5OH} = \frac{2.171\text{ mol}}{2.171\text{ mol} + 1.387\text{ mol}} = 0.6102$$

$$X_{H_2O} = \frac{1.387\text{ mol}}{2.171\text{ mol} + 1.387\text{ mol}} = 0.3898$$

$$P_{soln} = X_{C_2H_5OH}P^o_{C_2H_5OH} + X_{H_2O}P^o_{H_2O}$$

$$P_{soln} = (0.6102)(61.2\text{ mm Hg}) + (0.3898)(23.8\text{ mm Hg}) = 46.6\text{ mm Hg}$$

12.18 The red and blue curves are the pure liquids and the green curve is the mixture.

12.19 $C_2H_6O_2$, 62.07; $\quad 500.0\text{ g} = 0.5000\text{ kg}$

$$616.9\text{ g } C_2H_6O_2 \times \frac{1\text{ mol } C_2H_6O_2}{62.07\text{ g } C_2H_6O_2} = 9.939\text{ mol } C_2H_6O_2$$

$$\Delta T = K_b \cdot m = \left(0.51\frac{^oC\cdot kg}{mol}\right)\left(\frac{9.939\text{ mol}}{0.5000\text{ kg}}\right) = 10.1\ ^oC$$

bp $= 100.0\ ^oC + 10.1\ ^oC = 110.1\ ^oC$

12.20 The red curve represents the vapor pressure of pure chloroform.
(a) The normal boiling point for a liquid is the temperature where the vapor pressure of the liquid equals 1 atm (760 mm Hg). The approximate boiling point of pure chloroform is 62 °C.
(b) The approximate boiling point of the solution is 69 °C.
$\Delta T_b = 69\ ^oC - 62\ ^oC = 7\ ^oC$
$\Delta T_b = K_b \cdot m$

$$m = \frac{\Delta T_b}{K_b} = \frac{7\ ^\circ C}{3.63 \frac{^\circ C \cdot kg}{mol}} = 2\ mol/kg = 2\ m$$

12.21 $C_6H_{12}O_6$, 180.2; 37.0 °C = 310.1 K

$$\Pi = \left(\frac{50.0\ g \times \frac{1\ mol}{180.2\ g}}{1.00\ L} \right) \left(0.082\ 06 \frac{L \cdot atm}{K \cdot mol} \right) (310.1\ K) = 7.06\ atm$$

12.22 (a) $\Pi = iMRT$; $M = \frac{\Pi}{iRT} = \frac{8.0\ atm}{(1.9)\left(0.082\ 06 \frac{L \cdot atm}{K \cdot mol} \right)(298\ K)} = 0.17\ M$

(b) There would be a net transfer of water from outside the cell to the inside of the cell and the cell would swell.

12.23 $\Pi = MRT$; $M = \frac{\Pi}{RT} = \frac{\left(423.1\ mm\ Hg \times \frac{1\ atm}{760\ mm\ Hg} \right)}{\left(0.082\ 06 \frac{L \cdot atm}{K \cdot mol} \right)(298\ K)} = 0.0228\ mol/L$

200.0 mL = 0.2000 L

mol = (0.0228 mol/L)(0.2000 L) = 0.004 553 mol

$$\text{molar mass} = \frac{0.8220\ g}{0.004\ 553\ mol} = 180.5\ g/mol$$

12.24 $C_{12}H_{22}O_{11}$, 342.3

$\Pi = MRT$; $M = \frac{\Pi}{RT} = \frac{\left(278\ mm\ Hg \times \frac{1\ atm}{760\ mm\ Hg} \right)}{\left(0.082\ 06 \frac{L \cdot atm}{K \cdot mol} \right)(298\ K)} = 0.014\ 96\ mol/L$

100.0 mL = 0.1000 L

mol = (0.014 96 mol/L)(0.1000 L) = 0.001 496 mol

$$\text{molar mass} = \frac{0.512\ g}{0.001\ 496\ mol} = 342.3\ g/mol$$

The white powder is sucrose.

12.25 Both solvent molecules and small solute particles can pass through a semipermeable dialysis membrane. Only large colloidal particles such as proteins can't pass through. Only solvent molecules can pass through a semipermeable membrane used for osmosis.

12.26 Solvent–solvent is hydrogen bonding, solvent–solute is hydrogen bonding, and solute–solute is hydrogen bonding.

12.27 (a) Total conc = (137 + 105 + 3.0 + 4.0 + 2.0 + 33 + 0.75 + 11.1) mmol/L = 295.85 mmol/L
295.85 mmol/L = 295.85 x 10^{-3} mol/L = 0.295 85 mol/L
25 °C = 298 K

$$\Pi = MRT = (0.295\ 85\ \text{mol/L})\left(0.082\ 06\ \frac{\text{L}\cdot\text{atm}}{\text{K}\cdot\text{mol}}\right)(298\text{K}) = 7.23\ \text{atm}$$

(b) Solvent moves from the dialysis solution to blood.

12.28 $$12.0\ \text{L} \times \frac{1\ \text{mL}}{1 \times 10^{-3}\ \text{L}} \times \frac{1\ \text{g}}{\text{mL}} = 12{,}000\ \text{g}$$

$$2\ \text{ppm} = \frac{\text{mass of F}^-}{12{,}000\ \text{g}} \times 10^6$$

$$\text{mass of F}^- = \frac{(2)(12{,}000\ \text{g})}{10^6} = 0.024\ \text{g} = 24\ \text{mg}$$

Conceptual Problems

12.30 (a) < (b) < (c)

12.32 The upper curve is pure ether.
(a) The normal boiling point for ether is the temperature where the upper curve intersects the 760 mm Hg line, ~ 37 °C.
(b) $\Delta T_b \approx 3$ °C

$$\Delta T_b = K_b \cdot m; \qquad m = \frac{\Delta T_b}{K_b} = \frac{3\ ^\circ\text{C}}{2.02\ \frac{^\circ\text{C}\cdot\text{kg}}{\text{mol}}} \approx 1.5\ \text{mol/kg} \approx 1.5\ m$$

12.34 (a) The red curve represents the solution of a volatile solute and the green curve represents the solution of a nonvolatile solute.
(b) & (d)

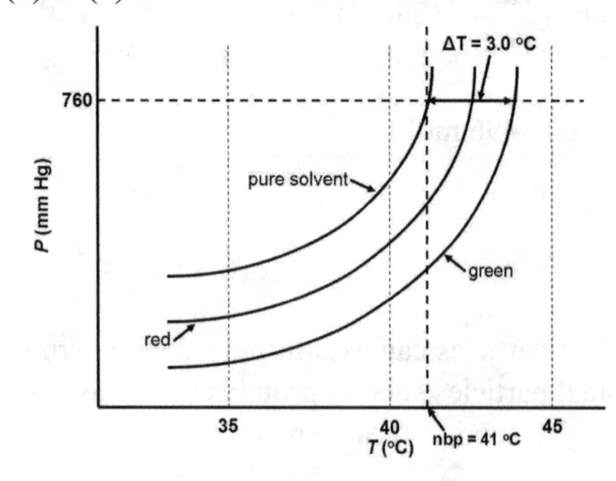

(c) $$\Delta T = K_f \cdot m; \quad m = \frac{\Delta T}{K_f} = \frac{3.0\ ^\circ\text{C}}{2.0\ ^\circ\text{C}/m} = 1.5\ m$$

12.36 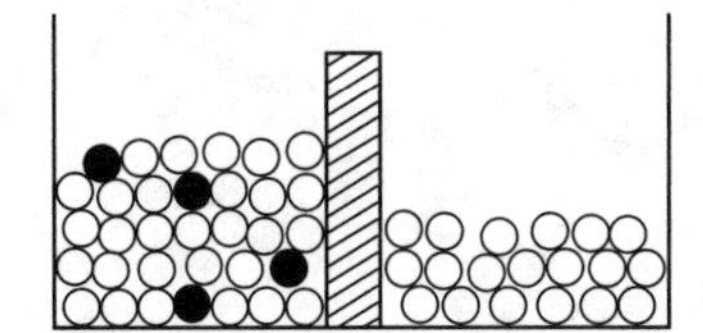

Assume that only the blue (open) spheres (solvent) can pass through the semipermeable membrane. There will be a net transfer of solvent from the right compartment (pure solvent) to the left compartment (solution) to achieve equilibrium.

12.38 (a) - (c)

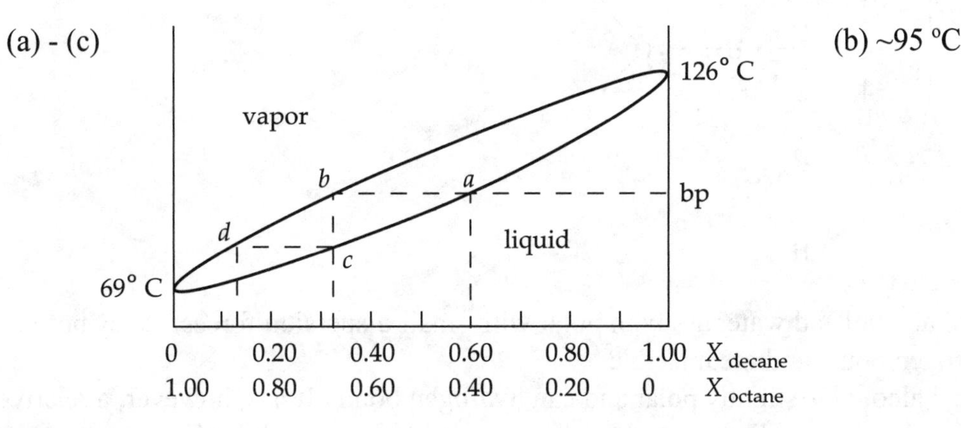

(b) ~95 °C

Section Problems

Solutions and Energy Changes (Sections 12.1 and 12.2)

12.40 The surface area of a solid plays an important role in determining how rapidly a solid dissolves. The larger the surface area, the more solid–solvent interactions, and the more rapidly the solid will dissolve. Powdered NaCl has a much larger surface area than a large block of NaCl, and it will dissolve more rapidly.

12.42 (a) Na^+ has the larger hydration energy because of its smaller size and higher charge density.
(b) Ba^{2+} has the larger hydration energy because of its higher charge.

12.44 Solvent-solvent is hydrogen bonding, solvent–solute is dispersion, and solute–solute is dispersion. I_2 is not soluble in water.

12.46 Both Br_2 and CCl_4 are nonpolar, and intermolecular forces for both are dispersion forces. H_2O is a polar molecule with dipole-dipole forces and hydrogen bonding. Therefore, Br_2 is more soluble in CCl_4.

12.48 CH_3COOH and water are both polar molecules and they both can hydrogen bond. C_6H_6 is nonpolar. The solubility of CH_3COOH is greater in water.

12.50 Toluene is nonpolar and is insoluble in water.
Br_2 is nonpolar but because of its size, is polarizable and is soluble in water.
KBr is an ionic compound and is very soluble in water.
toluene < Br_2 < KBr (solubility in H_2O)

12.52 There are three hydrogen bonds.

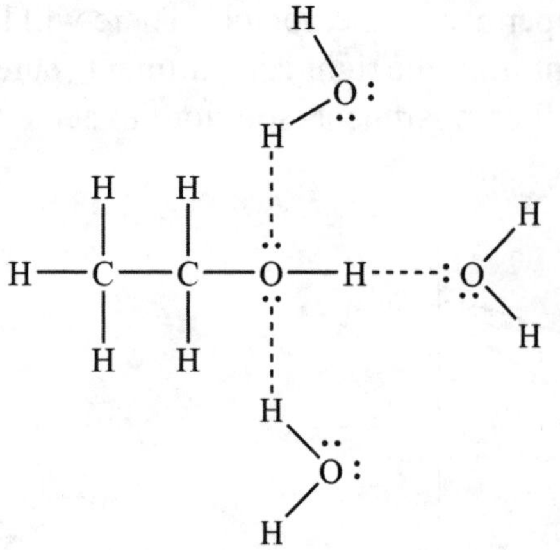

```
                 H
                  \
                   O:
                  /
                 H
       H    H    :
       |    |    :          H
                 ..        /
  H—C—C—O—H·····:O
       |    |    ..         \
       H    H    :          H
                 :
                 H
                  \
                   O:
                  /
                 H
```

12.54 Ethyl alcohol and water are both polar with small dispersion forces. They both can hydrogen bond, and are miscible.
Pentyl alcohol is slightly polar and can hydrogen bond. It has, however, a relatively large solute-solute dispersion force because of its size, which limits its water solubility.

Units of Concentration (Section 12.3)

12.56 $CaCl_2$, 110.98
For a 1.00 *m* solution:
heat released = 81,300 J
mass of solution = 1000 g H_2O + 110.98 g $CaCl_2$ = 1110.98 g

$$\Delta T = \frac{q}{\text{(specific heat)(mass of solution)}} = \frac{81{,}300\ \text{J}}{[4.18\ \text{J/(K}\cdot\text{g)}](1110.98\ \text{g})} = 17.5\ \text{K} = 17.5\ ^\circ\text{C}$$

Final temperature = 25.0 °C + 17.5 °C = 42.5 °C

12.58 Assume 1.00 L of seawater.
mass of 1.00 L = (1000 mL)(1.025 g/mL) = 1025 g

$$\frac{\text{mass NaCl}}{1025\ \text{g}} \times 100\% = 3.50\ \text{mass \%}; \qquad \text{mass NaCl} = \frac{1025\ \text{g} \times 3.50}{100} = 35.88\ \text{g}$$

There are 35.88 g NaCl per 1.00 L of solution.

$$M = \frac{\left(35.88\ \text{g NaCl} \times \frac{1\ \text{mol NaCl}}{58.44\ \text{g NaCl}}\right)}{1.00\ \text{L}} = 0.614\ \text{M}$$

12.60 $C_{16}H_{21}NO_2$, 259.3; 50 ng = 50×10^{-9} g
(a) mass of solution = (1.025 g/mL)(1000 mL) = 1025 g

$$\text{ppb} = \frac{50 \times 10^{-9}\ \text{g}}{1025\ \text{g}} \times 10^9 = 0.049\ \text{ppb}$$

(b) $50 \times 10^{-9}\ g \times \frac{1\ mol\ C_{16}H_{21}NO_2}{259.3\ g\ C_{16}H_{21}NO_2} = 1.9 \times 10^{-10}$ mol

$M = \frac{1.9 \times 10^{-10}\ mol}{1.0\ L} = 1.9 \times 10^{-10}$ mol/L

12.62 (a) Dissolve 0.150 mol of glucose in water; dilute to 1.00 L.
(b) Dissolve 1.135 mol of KBr in 1.00 kg of H_2O.
(c) Mix together 0.15 mol of CH_3OH with 0.85 mol of H_2O.

12.64 $C_7H_6O_2$, 122.12, 165 mL = 0.165 L
mol $C_7H_6O_2$ = (0.0268 mol/L)(0.165 L) = 0.004 42 mol
mass $C_7H_6O_2 = 0.004\ 42\ mol \times \frac{122.12\ g}{1\ mol} = 0.540$ g
Dissolve 4.42×10^{-3} mol (0.540 g) of $C_7H_6O_2$ in enough $CHCl_3$ to make 165 mL of solution.

12.66 (a) KCl, 74.6
A 0.500 M KCl solution contains 37.3 g of KCl per 1.00 L of solution.
A 0.500 mass % KCl solution contains 5.00 g of KCl per 995 g of water.
The 0.500 M KCl solution is more concentrated (that is, it contains more solute per amount of solvent).
(b) Both solutions contain the same amount of solute. The 1.75 M solution contains less solvent than the 1.75 m solution. The 1.75 M solution is more concentrated.

12.68 (a) $C_6H_8O_7$, 192.12
$0.655\ mol\ C_6H_8O_7 \times \frac{192.12\ g\ C_6H_8O_7}{1\ mol\ C_6H_8O_7} = 126\ g\ C_6H_8O_7$

mass % $C_6H_8O_7 = \frac{126\ g}{126\ g + 1000\ g} \times 100\% = 11.2$ mass %
(b) 0.135 mg = 0.135×10^{-3} g
(5.00 mL H_2O)(1.00 g/mL) = 5.00 g H_2O
mass % KBr = $\frac{0.135 \times 10^{-3}\ g}{(0.135 \times 10^{-3}\ g) + 5.00\ g} \times 100\% = 0.002\ 70$ mass % KBr
(c) mass % aspirin = $\frac{5.50\ g}{5.50\ g + 145\ g} \times 100\% = 3.65$ mass % aspirin

12.70 $P_{O_3} = P_{total} \cdot X_{O_3}$

$X_{O_3} = \frac{P_{O_3}}{P_{total}} = \frac{1.6 \times 10^{-9}\ atm}{1.3 \times 10^{-2}\ atm} = 1.2 \times 10^{-7}$
Assume one mole of air (29 g/mol)
mol $O_3 = n_{air} \cdot X_{O_3} = (1\ mol)(1.2 \times 10^{-7}) = 1.2 \times 10^{-7}$ mol O_3

O_3, 48.00; mass $O_3 = 1.2 \times 10^{-7}\ \text{mol} \times \dfrac{48.0\ \text{g}}{1\ \text{mol}} = 5.8 \times 10^{-6}\ \text{g}\ O_3$

$\text{ppm}\ O_3 = \dfrac{5.8 \times 10^{-6}\ \text{g}}{29\ \text{g}} \times 10^6 = 0.20\ \text{ppm}$

12.72 (a) H_2SO_4, 98.08; $\text{molality} = \dfrac{\left(25.0\ \text{g} \times \dfrac{1\ \text{mol}}{98.08\ \text{g}}\right)}{1.30\ \text{kg}} = 0.196\ \text{mol/kg} = 0.196\ m$

(b) $C_{10}H_{14}N_2$, 162.23; CH_2Cl_2, 84.93

$2.25\ \text{g}\ C_{10}H_{14}N_2 \times \dfrac{1\ \text{mol}\ C_{10}H_{14}N_2}{162.23\ \text{g}\ C_{10}H_{14}N_2} = 0.0139\ \text{mol}\ C_{10}H_{14}N_2$

$80.0\ \text{g}\ CH_2Cl_2 \times \dfrac{1\ \text{mol}\ CH_2Cl_2}{84.93\ \text{g}\ CH_2Cl_2} = 0.942\ \text{mol}\ CH_2Cl_2$

$X_{C_{10}H_{14}N_2} = \dfrac{0.0139\ \text{mol}}{0.942\ \text{mol} + 0.0139\ \text{mol}} = 0.0145;\quad X_{CH_2Cl_2} = \dfrac{0.942\ \text{mol}}{0.942\ \text{mol} + 0.0139\ \text{mol}} = 0.985$

12.74 $16.0\ \text{mass}\ \% = \dfrac{16.0\ \text{g}\ H_2SO_4}{16.0\ \text{g}\ H_2SO_4 + 84.0\ \text{g}\ H_2O}$

H_2SO_4, 98.08; density = 1.1094 g/mL

$\text{volume of solution} = 100.0\ \text{g} \times \dfrac{1\ \text{mL}}{1.1094\ \text{g}} = 90.14\ \text{mL} = 0.090\ 14\ \text{L}$

$\text{molarity} = \dfrac{\left(16.0\ \text{g} \times \dfrac{1\ \text{mol}}{98.08\ \text{g}}\right)}{0.090\ 14\ \text{L}} = 1.81\ \text{M}$

12.76 $\text{molality} = \dfrac{\left(40.0\ \text{g} \times \dfrac{1\ \text{mol}}{62.07\ \text{g}}\right)}{0.0600\ \text{kg}} = 10.7\ \text{mol/kg} = 10.7\ m$

12.78 $C_{19}H_{21}NO_3$, 311.38; $1.5\ \text{mg} = 1.5 \times 10^{-3}\ \text{g}$

$1.3 \times 10^{-3}\ \text{mol/kg} = \dfrac{\left(1.5 \times 10^{-3}\ \text{g} \times \dfrac{1\ \text{mol}}{311.38\ \text{g}}\right)}{\text{kg of solvent}}$; solve for kg of solvent.

$\text{kg of solvent} = \dfrac{\left(1.5 \times 10^{-3}\ \text{g} \times \dfrac{1\ \text{mol}}{311.38\ \text{g}}\right)}{1.3 \times 10^{-3}\ \text{mol/kg}} = 0.0037\ \text{kg}$

Because the solution is very dilute, kg of solvent ≈ kg of solution.

$\text{g of solution} = (0.0037\ \text{kg})\left(\dfrac{1000\ \text{g}}{1\ \text{kg}}\right) = 3.7\ \text{g}$

12.80 $C_6H_{12}O_6$, 180.16; H_2O, 18.02; Assume 1.00 L of solution.

mass of solution = (1000 mL)(1.0624 g/mL) = 1062.4 g

$$\text{mass of solute} = 0.944 \text{ mol} \times \frac{180.16 \text{ g}}{1 \text{ mol}} = 170.1 \text{ g } C_6H_{12}O_6$$

mass of H_2O = 1062.4 g – 170.1 g = 892.3 g H_2O

$$\text{mol } C_6H_{12}O_6 = 0.944 \text{ mol}; \quad \text{mol } H_2O = 892.3 \text{ g} \times \frac{1 \text{ mol}}{18.02 \text{ g}} = 49.5 \text{ mol}$$

$$\text{(a) } X_{C_6H_{12}O_6} = \frac{\text{mol } C_6H_{12}O_6}{\text{mol } C_6H_{12}O_6 + \text{mol } H_2O} = \frac{0.944 \text{ mol}}{0.944 \text{ mol} + 49.5 \text{ mol}} = 0.0187$$

$$\text{(b) mass \%} = \frac{\text{mass } C_6H_{12}O_6}{\text{total mass of solution}} \times 100\% = \frac{170.1 \text{ g}}{1062.4 \text{ g}} \times 100\% = 16.0\%$$

$$\text{(c) molality} = \frac{\text{mol } C_6H_{12}O_6}{\text{kg } H_2O} = \frac{0.944 \text{ mol}}{0.8923 \text{ kg}} = 1.06 \text{ mol/kg} = 1.06\ m$$

Solubility and Henry's Law (Section 12.4)

12.82 From Figure 12.6, first determine the solubility, in g/100 mL, for each compound.

(a) $CuSO_4$, 159.6, ~42 g/100 mL; NH_4Cl, 53.5, ~56 g/100 mL

$$M = \frac{\left(42 \text{ g} \times \frac{1 \text{ mol } CuSO_4}{159.6 \text{ g}}\right)}{0.100 \text{ L}} = 2.6 \text{ mol/L}$$

$$M = \frac{\left(56 \text{ g} \times \frac{1 \text{ mol } NH_4Cl}{53.5 \text{ g}}\right)}{0.100 \text{ L}} = 10.5 \text{ mol/L}$$

NH_4Cl has the higher molar solubility.

(b) CH_3CO_2Na, 82.0, ~48 g/100 mL; glucose ($C_6H_{12}O_6$), 180.2, ~90 g/100 mL

$$M = \frac{\left(48 \text{ g} \times \frac{1 \text{ mol } CH_3CO_2Na}{82.0 \text{ g}}\right)}{0.100 \text{ L}} = 5.9 \text{ mol/L}$$

$$M = \frac{\left(90 \text{ g} \times \frac{1 \text{ mol glucose}}{180.2 \text{ g}}\right)}{0.100 \text{ L}} = 5.0 \text{ mol/L}$$

CH_3CO_2Na has the higher molar solubility.

12.84 $M = k \cdot P = (0.091 \dfrac{\text{mol}}{\text{L} \cdot \text{atm}})(0.75 \text{ atm}) = 0.068 \text{ M}$

12.86 $M = k \cdot P$

$$k = \frac{M}{P} = \frac{2.21 \times 10^{-3} \text{ mol/L}}{1.00 \text{ atm}} = 2.21 \times 10^{-3} \frac{\text{mol}}{\text{L} \cdot \text{atm}}$$

Convert 4 mg/L to mol/L:
$4 \text{ mg} = 4 \times 10^{-3} \text{ g}$

$$O_2 \text{ molarity} = \frac{\left(4 \times 10^{-3} \text{ g} \times \frac{1 \text{ mol}}{32.00 \text{ g}}\right)}{1.00 \text{ L}} = 1.25 \times 10^{-4} \text{ M}$$

$$P_{O_2} = \frac{M}{k} = \frac{1.25 \times 10^{-4} \frac{\text{mol}}{\text{L}}}{2.21 \times 10^{-3} \frac{\text{mol}}{\text{L} \cdot \text{atm}}} = 0.06 \text{ atm}$$

12.88 $k = 2.4 \times 10^{-4} \text{ mol/(L} \cdot \text{atm)}$
$M = k \cdot P = [2.4 \times 10^{-4} \text{ mol/(L} \cdot \text{atm)}](2.00 \text{ atm}) = 4.8 \times 10^{-4} \text{ mol/L}$

Colligative Properties (Sections 12.5–12.8)

12.90 $FeCl_3$ and $CaCl_2$ are a strong electrolytes with a van't Hoff factors of i = 4 and 3, respectively. Glucose is a nonelectrolyte. The effective molality of $FeCl_3$ is (4)(0.10) = 0.40 *m*. The effective molality of $CaCl_2$ is (3)(0.15) = 0.45 *m*. Glucose = (1)(0.30) = 0.30 *m*.
Freezing point ranking: $CaCl_2 < FeCl_3 <$ glucose

12.92 $C_7H_6O_2$, 122.1; C_2H_6O, 46.07

$$X_{solv} = \frac{\text{mol } C_2H_6O}{\text{mol } C_2H_6O + \text{mol } C_7H_6O_2} = \frac{\left(100 \text{ g} \times \frac{1 \text{ mol}}{46.07 \text{ g}}\right)}{\left(100 \text{ g} \times \frac{1 \text{ mol}}{46.07 \text{ g}}\right) + \left(5.00 \text{ g} \times \frac{1 \text{ mol}}{122.1 \text{ g}}\right)} = 0.981$$

$P_{soln} = P_{solv} \cdot X_{solv} = (100.5 \text{ mm Hg})(0.981) = 98.6 \text{ mm Hg}$

12.94 $MgCl_2$, 95.21

$$110 \text{ g} \times \frac{1 \text{ kg}}{1000 \text{ g}} = 0.110 \text{ kg}$$

$$\Delta T_f = K_f \cdot m \cdot i = \left(1.86 \frac{°C \cdot \text{kg}}{\text{mol}}\right)\left(\frac{\left(7.40 \text{ g} \times \frac{1 \text{ mol}}{95.21 \text{ g}}\right)}{0.110 \text{ kg}}\right)(2.7) = 3.55 \text{ °C}$$

Solution freezing point = 0.00 °C – ΔT_f = 0.00 °C – 3.55 °C = –3.55 °C

12.96 HCl, 36.46; $\Delta T_f = K_f \cdot m \cdot i$

$$190 \text{ g} \times \frac{1 \text{ kg}}{1000 \text{ g}} = 0.190 \text{ kg}$$

Solution freezing point = – 4.65 °C = 0.00 °C – ΔT_f; ΔT_f = 4.65 °C

$$i = \frac{\Delta T_f}{K_f \cdot m} = \frac{4.65\ ^\circ C}{\left(1.86\ \frac{^\circ C \cdot kg}{mol}\right)\left(\frac{9.12\ g \times \frac{1\ mol}{36.46\ g}}{0.190\ kg}\right)} = 1.9$$

12.98 NaCl is a nonvolatile solute. Methyl alcohol is a volatile solute. When NaCl is added to water, the vapor pressure of the solution is decreased, which means that the boiling point of the solution will increase. When methyl alcohol is added to water, the vapor pressure of the solution is increased, which means that the boiling point of the solution will decrease.

12.100

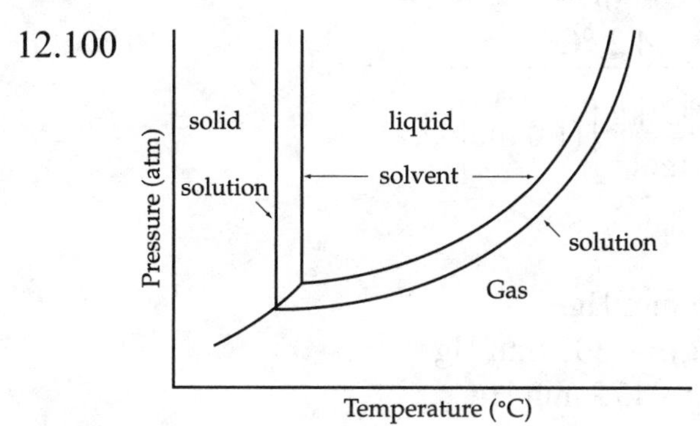

12.102 (a) CH_4N_2O, 60.06; H_2O, 18.02

$$10.0\ g\ CH_4N_2O \times \frac{1\ mol\ CH_4N_2O}{60.06\ g\ CH_4N_2O} = 0.167\ mol\ CH_4N_2O$$

$$150.0\ g\ H_2O \times \frac{1\ mol\ H_2O}{18.02\ g\ H_2O} = 8.32\ mol\ H_2O$$

$$X_{H_2O} = \frac{8.32\ mol}{8.32\ mol + 0.167\ mol} = 0.980$$

$$P_{soln} = P^o_{H_2O} \cdot X_{H_2O} = (71.93\ mm\ Hg)(0.980) = 70.5\ mm\ Hg$$

(b) LiCl, 42.39; $10.0\ g\ LiCl \times \frac{1\ mol\ LiCl}{42.39\ g\ LiCl} = 0.236\ mol\ LiCl$

LiCl dissociates into $Li^+(aq)$ and $Cl^-(aq)$ in H_2O.

mol Li^+ = mol Cl^- = mol LiCl = 0.236 mol

$$150.0\ g\ H_2O \times \frac{1\ mol\ H_2O}{18.02\ g\ H_2O} = 8.32\ mol\ H_2O$$

$$X_{H_2O} = \frac{8.32\ mol}{8.32\ mol + 0.236\ mol + 0.236\ mol} = 0.946$$

$$P_{soln} = P^o_{H_2O} \cdot X_{H_2O} = (71.93\ mm\ Hg)(0.946) = 68.0\ mm\ Hg$$

12.104 For H_2O, $K_b = 0.51 \frac{°C \cdot kg}{mol}$; 150.0 g = 0.1500 kg

(a) $\Delta T_b = K_b \cdot m = \left(0.51 \frac{°C \cdot kg}{mol}\right)\left(\frac{0.167\ mol}{0.1500\ kg}\right) = 0.57\ °C$

Solution boiling point = 100.00 °C + ΔT_b = 100.00 °C + 0.57 °C = 100.57 °C

(b) $\Delta T_b = K_b \cdot m = \left(0.51 \frac{°C \cdot kg}{mol}\right)\left(\frac{2(0.236\ mol)}{0.1500\ kg}\right) = 1.6\ °C$

Solution boiling point = 100.00 °C + ΔT_b = 100.00 °C + 1.6 °C = 101.6 °C

12.106 Solution freezing point = – 4.3 °C = 0.00 °C – ΔT_f; ΔT_f = 4.3 °C

$\Delta T_f = K_f \cdot m \cdot i$; $\quad i = \frac{\Delta T_f}{K_f \cdot m} = \frac{4.3\ °C}{\left(1.86 \frac{°C \cdot kg}{mol}\right)(1.0\ mol/kg)} = 2.3$

12.108 Let $X_{heptane} = x$ and $X_{octane} = 1 - x$

(428 mm Hg)x + (175 mm Hg)(1 – x) = 305 mm Hg

(428 mm Hg)x + 175 mm Hg – (175 mm Hg)x = 305 mm Hg

(428 mm Hg – 175 mm Hg)x = 305 mm Hg – 175 mm Hg

(253 mm Hg)x = 130 mm Hg

$x = X_{heptane} = \frac{130\ mm\ Hg}{253\ mm\ Hg} = 0.514$

12.110 Acetone, C_3H_6O, 58.08, $P^o_{C_3H_6O}$ = 285 mm Hg

Ethyl acetate, $C_4H_8O_2$, 88.11, $P^o_{C_4H_8O_2}$ = 118 mm Hg

$25.0\ g\ C_3H_6O \times \frac{1\ mol\ C_3H_6O}{58.08\ g\ C_3H_6O} = 0.430\ mol\ C_3H_6O$

$25.0\ g\ C_4H_8O_2 \times \frac{1\ mol\ C_4H_8O_2}{88.11\ g\ C_4H_8O_2} = 0.284\ mol\ C_4H_8O_2$

$X_{C_3H_6O} = \frac{0.430\ mol}{0.430\ mol + 0.284\ mol} = 0.602$; $X_{C_4H_8O_2} = \frac{0.284\ mol}{0.430\ mol + 0.284\ mol} = 0.398$

$P_{soln} = P^o_{C_3H_6O} \cdot X_{C_3H_6O} + P^o_{C_4H_8O_2} \cdot X_{C_4H_8O_2}$

P_{soln} = (285 mm Hg)(0.602) + (118 mm Hg)(0.398) = 219 mm Hg

12.112 In the liquid, $X_{acetone} = 0.602$ and $X_{ethyl\ acetate} = 0.398$

In the vapor, P_{Total} = 219 mm Hg

$P_{acetone} = P^o_{acetone} \cdot X_{acetone}$ = (285 mm Hg)(0.602) = 172 mm Hg

$P_{ethyl\ acetate} = P^o_{ethyl\ acetate} \cdot X_{ethyl\ acetate}$ = (118 mm Hg)(0.398) = 47 mm Hg

$$X_{acetone} = \frac{P_{acetone}}{P_{total}} = \frac{172 \text{ mm Hg}}{219 \text{ mm Hg}} = 0.785; \quad X_{ethyl\ acetate} = \frac{P_{ethyl\ acetate}}{P_{total}} = \frac{47 \text{ mm Hg}}{219 \text{ mm Hg}} = 0.215$$

12.114 $C_9H_8O_4$, 180.16; 215 g = 0.215 kg

$$\Delta T_b = K_b \cdot m = 0.47\ ^\circ\text{C}; \qquad K_b = \frac{\Delta T_b}{m} = \frac{0.47\ ^\circ\text{C}}{\left(\dfrac{5.00 \text{ g x } \dfrac{1 \text{ mol}}{180.16 \text{ g}}}{0.215 \text{ kg}}\right)} = 3.6\ \frac{^\circ\text{C} \cdot \text{kg}}{\text{mol}}$$

12.116 $\Delta T_b = K_b \cdot m = 1.76\ ^\circ\text{C}; \qquad m = \dfrac{\Delta T_b}{K_b} = \dfrac{1.76\ ^\circ\text{C}}{3.07\ \dfrac{^\circ\text{C} \cdot \text{kg}}{\text{mol}}} = 0.573\ m$

12.118 $\Pi = MRT$

(a) NaCl 58.44; 350.0 mL = 0.3500 L

There are 2 moles of ions/mole of NaCl

$$\Pi = (2)\left(\frac{5.00 \text{ g x } \dfrac{1 \text{ mol}}{58.44 \text{ g}}}{0.3500 \text{ L}}\right)\left(0.082\ 06\ \frac{\text{L} \cdot \text{atm}}{\text{K} \cdot \text{mol}}\right)(323 \text{ K}) = 13.0 \text{ atm}$$

(b) CH_3CO_2Na, 82.03; 55.0 mL = 0.0550 L

There are 2 moles of ions/mole of CH_3CO_2Na

$$\Pi = (2)\left(\frac{6.33 \text{ g x } \dfrac{1 \text{ mol}}{82.03 \text{ g}}}{0.0550 \text{ L}}\right)\left(0.082\ 06\ \frac{\text{L} \cdot \text{atm}}{\text{K} \cdot \text{mol}}\right)(283\text{K}) = 65.2 \text{ atm}$$

12.120 $\Pi = MRT; \qquad M = \dfrac{\Pi}{RT} = \dfrac{4.85 \text{ atm}}{\left(0.082\ 06\ \dfrac{\text{L} \cdot \text{atm}}{\text{K} \cdot \text{mol}}\right)(300 \text{ K})} = 0.197 \text{ M}$

12.122 K_f for snow (H_2O) is $1.86\ \dfrac{^\circ\text{C} \cdot \text{kg}}{\text{mol}}$. Reasonable amounts of salt are capable of lowering the freezing point (ΔT_f) of the snow below an air temperature of –2 °C. Reasonable amounts of salt, however, are not capable of causing a ΔT_f of more than 30 °C, which would be required if it is to melt snow when the air temperature is –30 °C.

12.124 $\Pi = 407.2 \text{ mm Hg x } \dfrac{1 \text{ atm}}{760 \text{ mm Hg}} = 0.5358 \text{ atm}$

$$\Pi = MRT; \quad M = \frac{\Pi}{RT} = \frac{0.5358 \text{ atm}}{\left(0.082\ 06\ \dfrac{\text{L} \cdot \text{atm}}{\text{K} \cdot \text{mol}}\right)(298.15 \text{ K})} = 0.021\ 90 \text{ M}$$

$200.0 \text{ mL x } \dfrac{1 \text{ L}}{1000 \text{ mL}} = 0.2000 \text{ L}$

mol cellobiose = (0.2000 L)(0.021 90 mol/L) = 4.380×10^{-3} mol

$\text{molar mass of cellobiose} = \dfrac{1.500 \text{ g cellobiose}}{4.380 \times 10^{-3} \text{ mol cellobiose}} = 342.5 \text{ g/mol}$

molecular weight = 342.5

12.126 HCl is a strong electrolyte in H_2O and completely dissociates into two solute particles per each HCl.
HF is a weak electrolyte in H_2O. Only a few percent of the HF molecules dissociates into ions.

12.128 First, determine the empirical formula:
Assume 100.0 g of β-carotene.

10.51% H $\quad 10.51 \text{ g H x } \dfrac{1 \text{ mol H}}{1.008 \text{ g H}} = 10.43 \text{ mol H}$

89.49% C $\quad 89.49 \text{ g C x } \dfrac{1 \text{ mol C}}{12.01 \text{ g C}} = 7.45 \text{ mol C}$

$C_{7.45}H_{10.43}$; divide each subscript by the smaller, 7.45.
$C_{7.45 / 7.45}H_{10.43 / 7.45}$
$CH_{1.4}$
Multiply each subscript by 5 to obtain integers.
Empirical formula is C_5H_7, 67.1.

Second, calculate the molecular weight:
$\Delta T_f = K_f \cdot m$

$$m = \frac{\Delta T_f}{K_f} = \frac{1.17\ ^\circ\text{C}}{37.7 \dfrac{^\circ\text{C} \cdot \text{kg}}{\text{mol}}} = 0.0310 \text{ mol/kg} = 0.0310\ m$$

$1.50 \text{ g x } \dfrac{1\text{kg}}{1000 \text{ g}} = 1.50 \times 10^{-3} \text{ kg}$

mol β-carotene = (1.50×10^{-3} kg)(0.0310 mol/kg) = 4.65×10^{-5} mol

$\text{molar mass of } \beta\text{-carotene} = \dfrac{0.0250 \text{ g } \beta\text{-carotene}}{4.65 \times 10^{-5} \text{ mol } \beta\text{-carotene}} = 538 \text{ g/mol}$

molecular weight = 538

Finally, determine the molecular formula:
Divide the molecular weight by the empirical formula weight.

$\dfrac{538}{67.1} = 8$

molecular formula is $C_{(8 \times 5)}H_{(8 \times 7)}$, or $C_{40}H_{56}$

Chapter Problems

12.130 $C_2H_6O_2$, 62.07; $\Delta T_f = 22.0\ ^\circ C$

$\Delta T_f = K_f \cdot m$

$$m = \frac{\Delta T_f}{K_f} = \frac{22.0\ ^\circ C}{1.86\ \frac{^\circ C \cdot kg}{mol}} = 11.8\ mol/kg = 11.8\ m$$

mol $C_2H_6O_2$ = (3.55 kg)(11.8 mol/kg) = 41.9 mol $C_2H_6O_2$

$$\text{mass } C_2H_6O_2 = 41.9\ mol\ C_2H_6O_2 \times \frac{62.07\ g\ C_2H_6O_2}{1\ mol\ C_2H_6O_2} = 2.60 \times 10^3\ g\ C_2H_6O_2$$

12.132 When solid $CaCl_2$ is added to liquid water, the temperature rises because ΔH_{soln} for $CaCl_2$ is exothermic.
When solid $CaCl_2$ is added to ice at 0 °C, some of the ice will melt (an endothermic process) and the temperature will fall because the $CaCl_2$ lowers the freezing point of an ice/water mixture.

12.134 Sucrose ($C_{12}H_{22}O_{11}$), 342.3; fructose ($C_6H_{12}O_6$), 180.2

$$\Pi = MRT; \quad M = \frac{\Pi}{RT} = \frac{0.1843\ atm}{\left(0.082\ 06\ \frac{L \cdot atm}{K \cdot mol}\right)(298.0\ K)} = 0.007\ 537\ M$$

n_{total} = (0.007 537 mol/L)(1.50 L) = 0.0113 mol

Let x = g of sucrose and 2.850 g – x = g of fructose

0.0113 mol = mol sucrose + mol fructose

$$0.0113\ mol = (x)\left(\frac{1\ mol\ sucrose}{342.3\ g}\right) + (2.850\ g - x)\left(\frac{1\ mol\ fructose}{180.2\ g}\right)$$

Solve for x.

$$0.0113\ mol = (x)\left(\frac{1\ mol}{342.3\ g}\right) + (2.850\ g - x)\left(\frac{1\ mol}{180.2\ g}\right)$$

$$0.0113\ mol = (x)\left(\frac{1\ mol}{342.3\ g}\right) + 0.0158\ mol - (x)\left(\frac{1\ mol}{180.2\ g}\right)$$

$$0.0113\ mol - 0.0158\ mol = (x)\left(\frac{1\ mol}{342.3\ g}\right) - (x)\left(\frac{1\ mol}{180.2\ g}\right)$$

$$-0.0045\ mol = (x)\left[\left(\frac{1\ mol}{342.3\ g}\right) - \left(\frac{1\ mol}{180.2\ g}\right)\right]$$

$$x = \frac{-0.0045\ mol}{\left[\left(\frac{1\ mol}{342.3\ g}\right) - \left(\frac{1\ mol}{180.2\ g}\right)\right]} = 1.71\ g\ sucrose$$

$$1.71\ g\ sucrose \times \frac{1\ mol\ sucrose}{342.3\ g\ sucrose} = 0.005\ 00\ mol\ sucrose$$

$$X_{sucrose} = \frac{n_{sucrose}}{n_{total}} = \frac{0.005\ 00\ mol}{0.0113\ mol} = 0.442$$

12.136 $C_{10}H_8$, 128.17; $\Delta T_f = 0.35\ ^\circ C$

$$\Delta T_f = K_f \cdot m; \qquad m = \frac{\Delta T_f}{K_f} = \frac{0.35\ ^\circ C}{5.12\ \frac{^\circ C \cdot kg}{mol}} = 0.0684\ mol/kg = 0.0684\ m$$

$$150.0\ g \times \frac{1 kg}{1000\ g} = 0.1500\ kg$$

mol $C_{10}H_8$ = (0.1500 kg)(0.0684 mol/kg) = 0.0103 mol $C_{10}H_8$

$$\text{mass } C_{10}H_8 = 0.0103\ mol\ C_{10}H_8 \times \frac{128.17\ g\ C_{10}H_8}{1\ mol\ C_{10}H_8} = 1.3\ g\ C_{10}H_8$$

12.138 NaCl, 58.44; there are 2 ions/NaCl
A 3.5 mass % aqueous solution of NaCl contains 3.5 g NaCl and 96.5 g H_2O.

$$\text{molality} = \frac{\left(3.5\ g \times \frac{1\ mol}{58.44\ g}\right)}{0.0965\ kg} = 0.62\ mol/kg = 0.62\ m$$

$$\Delta T_f = K_f \cdot 2 \cdot m = \left(1.86\ \frac{^\circ C \cdot kg}{mol}\right)(2)(0.62\ mol/kg) = 2.3\ ^\circ C$$

Solution freezing point = 0.0 °C – ΔT_f = 0.0 °C – 2.3 °C = –2.3 °C

$$\Delta T_b = K_b \cdot 2 \cdot m = \left(0.51\ \frac{^\circ C \cdot kg}{mol}\right)(2)(0.62\ mol/kg) = 0.63\ ^\circ C$$

Solution boiling point = 100.00 °C + ΔT_b = 100.00 °C + 0.63 °C = 100.63 °C

12.140 (a) $$90 \text{ mass \% isopropyl alcohol} = \frac{10.5\ g}{10.5\ g + \text{mass of } H_2O} \times 100\%$$

Solve for the mass of H_2O.

$$\text{mass of } H_2O = \left(10.5\ g \times \frac{100}{90}\right) - 10.5\ g = 1.2\ g$$

mass of solution = 10.5 g + 1.2 g = 11.7 g
11.7 g of rubbing alcohol contains 10.5 g of isopropyl alcohol.

(b) C_3H_8O, 60.10
mass C_3H_8O = (0.90)(50.0 g) = 45 g

$$45\ g\ C_3H_8O \times \frac{1\ mol\ C_3H_8O}{60.10\ g\ C_3H_8O} = 0.75\ mol\ C_3H_8O$$

12.142 $\Delta T_f = K_f \cdot i \cdot m$; $CaCl_2$, 111.0

$$m = \frac{\Delta T_f}{K_f \cdot i} = \frac{1.14\ ^\circ C}{\left(1.86\ \frac{^\circ C \cdot kg}{mol}\right)(2.71)} = 0.226\ mol/kg$$

$$0.226 \text{ mol } CaCl_2 \text{ x } \frac{111.0 \text{ g } CaCl_2}{1 \text{ mol } CaCl_2} = 25.1 \text{ g } CaCl_2$$

$$\text{mass \% } CaCl_2 = \frac{\text{mass } CaCl_2}{\text{mass } CaCl_2 + \text{mass } H_2O} \text{ x } 100\%$$

$$= \frac{25.1 \text{ g}}{25.1 \text{ g} + 1000.0 \text{ g}} \text{ x } 100\% = 2.45\%$$

12.144 H_2O, 18.02
A 0.62 *m* LiCl solution contains 0.62 mol of LiCl and 1.00 kg (= 1000 g) of H_2O.
(0.62 mol LiCl)(1.96) = 1.21 mol of solute particles

$$1000 \text{ g } H_2O \text{ x } \frac{1 \text{ mol } H_2O}{18.02 \text{ g } H_2O} = 55.5 \text{ mol } H_2O$$

$$P_{soln} = P^o_{H_2O} \cdot X_{H_2O} = 23.76 \text{ mm Hg x } \frac{55.5 \text{ mol } H_2O}{55.5 \text{ mol } H_2O + 1.21 \text{ mol solute}} = 23.25 \text{ mm Hg}$$

Vapor pressure depression = 23.76 mm Hg – 23.25 mm Hg = 0.51 mm Hg

12.146 First, determine the empirical formula.
3.47 mg = $3.47 \text{ x } 10^{-3}$ g sample
10.10 mg = $10.10 \text{ x } 10^{-3}$ g CO_2
2.76 mg = $2.76 \text{ x } 10^{-3}$ g H_2O

$$\text{mass C} = 10.10 \text{ x } 10^{-3} \text{ g } CO_2 \text{ x } \frac{12.01 \text{ g C}}{44.01 \text{ g } CO_2} = 2.76 \text{ x } 10^{-3} \text{ g C}$$

$$\text{mass H} = 2.76 \text{ x } 10^{-3} \text{ g } H_2O \text{ x } \frac{2 \text{ x } 1.008 \text{ g H}}{18.02 \text{ g } H_2O} = 3.09 \text{ x } 10^{-4} \text{ g H}$$

$$\text{mass O} = 3.47 \text{ x } 10^{-3} \text{ g} - 2.76 \text{ x } 10^{-3} \text{ g C} - 3.09 \text{ x } 10^{-4} \text{ g H} = 4.01 \text{ x } 10^{-4} \text{ g O}$$

$$2.76 \text{ x } 10^{-3} \text{ g C x } \frac{1 \text{ mol C}}{12.01 \text{ g C}} = 2.30 \text{ x } 10^{-4} \text{ mol C}$$

$$3.09 \text{ x } 10^{-4} \text{ g H x } \frac{1 \text{ mol H}}{1.008 \text{ g H}} = 3.07 \text{ x } 10^{-4} \text{ mol H}$$

$$4.01 \text{ x } 10^{-4} \text{ g O x } \frac{1 \text{ mol O}}{16.00 \text{ g O}} = 2.51 \text{ x } 10^{-5} \text{ mol O} = 0.251 \text{ x } 10^{-4} \text{ mol O}$$

To simplify the empirical formula, divide each mol quantity by 10^{-4}.
$C_{2.30}H_{3.07}O_{0.251}$; divide all subscripts by the smallest, 0.251.
$C_{2.30/0.251}H_{3.07/0.251}O_{0.251/0.251}$
$C_{9.16}H_{12.23}O$
Empirical formula is $C_9H_{12}O$, 136.

Second, determine the molecular weight.

$$7.55 \text{ mg} = 7.55 \text{ x } 10^{-3} \text{ g estradiol; } 0.500 \text{ g x } \frac{1 \text{ kg}}{1000 \text{ g}} = 5.00 \text{ x } 10^{-4} \text{ kg camphor}$$

$$\Delta T_f = K_f \cdot m; \ m = \frac{\Delta T_f}{K_f} = \frac{2.10\,^\circ\text{C}}{37.7\ \frac{^\circ\text{C}\cdot\text{kg}}{\text{mol}}} = 0.0557\ \text{mol/kg} = 0.0557\ m$$

$$m = \frac{\text{mol estradiol}}{\text{kg solvent}}$$

mol estradiol = m x (kg solvent) = (0.0557 mol/kg)(5.00 x 10^{-4} kg) = 2.79 x 10^{-5} mol

$$\text{molar mass} = \frac{7.55 \times 10^{-3}\ \text{g estradiol}}{2.79 \times 10^{-5}\ \text{mol estradiol}} = 271\ \text{g/mol; molecular weight} = 271$$

Finally, determine the molecular formula:
Divide the molecular weight by the empirical formula weight.

$$\frac{271}{136} = 2$$

molecular formula is $C_{(2 \times 9)}H_{(2 \times 12)}O_{(2 \times 1)}$, or $C_{18}H_{24}O_2$

12.148 (a) H_2SO_4, 98.08; 2.238 mol H_2SO_4 x $\frac{98.08\ \text{g}\ H_2SO_4}{1\ \text{mol}\ H_2SO_4}$ = 219.50 g H_2SO_4

mass of 2.238 m solution = 219.50 g H_2SO_4 + 1000 g H_2O = 1219.50 g

volume of 2.238 m solution = 1219.50 g x $\frac{1.0000\ \text{mL}}{1.1243\ \text{g}}$ = 1084.68 mL = 1.0847 L

molarity of 2.238 m solution = $\frac{2.238\ \text{mol}}{1.0847\ \text{L}}$ = 2.063 M

The molarity of the H_2SO_4 solution is less than the molarity of the $BaCl_2$ solution. Because equal volumes of the two solutions are mixed, H_2SO_4 is the limiting reactant and the number of moles of H_2SO_4 determines the number of moles of $BaSO_4$ produced as the white precipitate.

$$(0.05000\ \text{L}) \times (2.063\ \text{mol}\ H_2SO_4/\text{L}) \times \frac{1\ \text{mol}\ BaSO_4}{1\ \text{mol}\ H_2SO_4} \times \frac{233.39\ \text{g}\ BaSO_4}{1\ \text{mol}\ BaSO_4} = 24.07\ \text{g}\ BaSO_4$$

(b) More precipitate will form because of the excess $BaCl_2$ in the solution.

12.150 Let x = X_{H_2O} and y = X_{CH_3OH} and assume n_{total} = 1.00 mol

(14.5 mm Hg)x + (82.5 mm Hg)y = 39.4 mm Hg
(26.8 mm Hg)x + (140.3 mm Hg)y = 68.2 mm Hg

$$x = \frac{68.2 - 140.3y}{26.8}$$

$$\frac{14.5(68.2 - 140.3y)}{26.8} + 82.5y = 39.4$$

$$\frac{(988.9 - 2034.35y)}{26.8} + 82.5y = 39.4$$

$$36.90 - 75.91y + 82.5y = 39.4;\ 6.59y = 2.5;\ y = \frac{2.5}{6.59} = 0.3794$$

$$x = \frac{[68.2 - 140.3(0.3794)]}{26.8} = 0.5586$$

$X_{LiCl} = 1 - X_{H_2O} - X_{CH_3OH} = 1 - 0.5586 - 0.3794 = 0.0620$

The mole fraction equals the number of moles of each component because $n_{total} = 1.00$ mol.

$$\text{mass LiCl} = 0.0620 \text{ mol LiCl x } \frac{42.39 \text{ g LiCl}}{1 \text{ mol LiCl}} = 2.6 \text{ g LiCl}$$

$$\text{mass } H_2O = 0.5588 \text{ mol } H_2O \text{ x } \frac{18.02 \text{ g } H_2O}{1 \text{ mol } H_2O} = 10.1 \text{ g } H_2O$$

$$\text{mass } CH_3OH = 0.3794 \text{ mol } CH_3OH \text{ x } \frac{32.04 \text{ g } CH_3OH}{1 \text{ mol } CH_3OH} = 12.2 \text{ g } CH_3OH$$

total mass = 2.6 g + 10.1 g + 12.2 g = 24.9 g

$$\text{mass \% LiCl} = \frac{2.6 \text{ g}}{24.9 \text{ g}} \text{ x } 100\% = 10\%$$

$$\text{mass \% } H_2O = \frac{10.1 \text{ g}}{24.9 \text{ g}} \text{ x } 100\% = 41\%$$

$$\text{mass \% } CH_3OH = \frac{12.2 \text{ g}}{24.9 \text{ g}} \text{ x } 100\% = 49\%$$

12.152 Solution freezing point = $-1.03\ ^\circ C = 0.00\ ^\circ C - \Delta T_f$; $\Delta T_f = 1.03\ ^\circ C$

$$\Delta T_f = K_f \cdot m; \quad m = \frac{\Delta T_f}{K_f} = \frac{1.03\ ^\circ C}{1.86 \frac{^\circ C \cdot kg}{mol}} = 0.554 \text{ mol/kg} = 0.554\ m$$

$$\Pi = MRT; \quad M = \frac{\Pi}{RT} = \frac{(12.16 \text{ atm})}{\left(0.082\ 06 \frac{L \cdot atm}{K \cdot mol}\right)(298 \text{ K})} = 0.497 \frac{mol}{L}$$

Assume 1.000 L = 1000 mL of solution.

mass of solution = (1000 mL)(1.063 g/mL) = 1063 g

$$\text{mass of } H_2O \text{ in 1000 mL of solution} = \frac{1000 \text{ g } H_2O}{0.554 \text{ mol of solute}} \text{ x } 0.497 \text{ mol} = 897 \text{ g } H_2O$$

mass of solute = total mass – mass of H_2O = 1063 g – 897 g = 166 g solute

$$\text{molar mass} = \frac{166 \text{ g}}{0.497 \text{ mol}} = 334 \text{ g/mol}$$

12.154 (a) NaCl, 58.44; $CaCl_2$, 110.98; H_2O, 18.02

$$\text{mol NaCl} = 100.0 \text{ g NaCl x } \frac{1 \text{ mol NaCl}}{58.44 \text{ g NaCl}} = 1.711 \text{ mol NaCl}$$

$$\text{mol } CaCl_2 = 100.0 \text{ g } CaCl_2 \text{ x } \frac{1 \text{ mol } CaCl_2}{110.98 \text{ g } CaCl_2} = 0.9011 \text{ mol } CaCl_2$$

mass of solution = (1000 mL)(1.15 g/mL) = 1150 g

mass of H_2O in solution = mass of solution – mass NaCl – mass $CaCl_2$

$$= 1150\text{ g} - 100.0\text{ g} - 100.0\text{ g} = 950\text{ g}$$

$$= 950\text{ g x }\frac{1\text{ kg}}{1000\text{ g}} = 0.950\text{ kg}$$

$$\Delta T_b = K_b \cdot (m_{NaCl} \cdot i + m_{CaCl_2} \cdot i)$$

$$\Delta T_b = \left(0.51\frac{^\circ C \cdot kg}{mol}\right)\left(\frac{(1.711\text{ mol NaCl} \cdot 2) + (0.9011\text{ mol }CaCl_2 \cdot 3)}{0.950\text{ kg}}\right) = 3.3\ ^\circ C$$

solution boiling point = 100.0 °C + ΔT_b = 100.0 °C + 3.3 °C = 103.3 °C

(b) $$\text{mol }H_2O = 950\text{ g }H_2O\text{ x }\frac{1\text{ mol }H_2O}{18.02\text{ g }H_2O} = 52.7\text{ mol }H_2O$$

$$P_{Solution} = P^o \cdot X_{H_2O}$$

$$P_{Solution} = P^o \cdot \left(\frac{52.7\text{ mol }H_2O}{(52.7\text{ mol }H_2O) + (1.711\text{ mol NaCl} \cdot 2) + (0.9011\text{ mol }CaCl_2 \cdot 3)}\right)$$

$$P_{Solution} = (23.8\text{ mm Hg})(0.896) = 21.3\text{ mm Hg}$$

12.156 (a) KI, 166.00

Assume you have 1.000 L of 1.24 M solution.

mass of solution = (1000 mL)(1.15 g/mL) = 1150 g

$$\text{mass of KI in solution} = 1.24\text{ mol KI x }\frac{166.00\text{ g KI}}{1\text{ mol KI}} = 206\text{ g KI}$$

mass of H_2O in solution = mass of solution – mass KI = 1150 g – 206 g = 944 g

$$= 944\text{ g x }\frac{1\text{ kg}}{1000\text{ g}} = 0.944\text{ kg}$$

$$\text{molality} = \frac{1.24\text{ mol KI}}{0.944\text{ kg }H_2O} = 1.31\ m$$

(b) For KI, i = 2 assuming complete dissociation.

$$\Delta T_f = K_f \cdot m \cdot i = \left(1.86\frac{^\circ C \cdot kg}{mol}\right)(1.31\ m)(2) = 4.87\ ^\circ C$$

Solution freezing point = 0.00 °C – ΔT_f = 0.00 °C – 4.87 °C = – 4.87 °C

(c) $$i = \frac{\Delta T_f}{K_f \cdot m} = \frac{4.46\ ^\circ C}{\left(1.86\frac{^\circ C \cdot kg}{mol}\right)(1.31\text{ mol/kg})} = 1.83$$

Because the calculated i is only 1.83 and not 2, the percent dissociation for KI is 83%.

12.158 NaCl, 58.44; $C_{12}H_{22}O_{11}$, 342.3

Let X = mass NaCl and Y = mass $C_{12}H_{22}O_{11}$, then X + Y = 100.0 g.

$$500.0\text{ g x }\frac{1\text{ kg}}{1000\text{ g}} = 0.5000\text{ kg}$$

Solution freezing point = –2.25 °C = 0.00 °C – $\Delta T_{f;}$ = ΔT_f = 0.00 °C + 2.25 °C = 2.25 °C

$$\Delta T_f = K_f \cdot (m_{NaCl} \cdot i + m_{C_{12}H_{22}O_{11}})$$

$$\Delta T_b = \left(1.86 \frac{^\circ C \cdot kg}{mol}\right)\left(\frac{(mol\ NaCl \cdot 2) + (mol\ C_{12}H_{22}O_{11})}{0.5000\ kg}\right) = 2.25\ ^\circ C$$

$$mol\ NaCl = X\ g\ NaCl \times \frac{1\ mol\ NaCl}{58.44\ g\ NaCl} = X/58.44\ mol$$

$$mol\ C_{12}H_{22}O_{11} = Y\ g\ C_{12}H_{22}O_{11} \times \frac{1\ mol\ C_{12}H_{22}O_{11}}{342.3\ g\ C_{12}H_{22}O_{11}} = Y/342.3\ mol$$

$$\Delta T_b = \left(1.86 \frac{^\circ C \cdot kg}{mol}\right)\left(\frac{((X/58.44) \cdot 2\ mol) + ((Y/342.3)\ mol)}{0.5000\ kg}\right) = 2.25\ ^\circ C$$

$X = 100 - Y$

$$\left(1.86 \frac{^\circ C \cdot kg}{mol}\right)\left(\frac{\{[(100 - Y)/58.44] \cdot 2\ mol]\} + [(Y/342.3)\ mol]}{0.5000\ kg}\right) = 2.25\ ^\circ C$$

$$\left(\frac{[(200/58.44) - (2Y/58.44) + (Y/342.3)]\ mol}{0.5000\ kg}\right) = \frac{2.25\ ^\circ C}{\left(1.86 \frac{^\circ C \cdot kg}{mol}\right)} = 1.21\ mol/kg$$

$$\left(\frac{[(3.42) - (0.0313Y)]\ mol}{0.5000\ kg}\right) = 1.21\ mol/kg$$

$[(3.42) - (0.0313Y)] = (0.5000\ kg)(1.21) = 0.605$
$- 0.0313\ Y = 0.605 - 3.42 = - 2.81$
$Y = (- 2.81)/(- 0.0313) = 89.8$ g of $C_{12}H_{22}O_{11}$
$X = 100.0\ g - Y = 100.0\ g - 89.8\ g = 10.2$ g of NaCl

Multiconcept Problems

12.160 (a) $382.6\ mL = 0.3826\ L$; $20.0\ ^\circ C = 293.2\ K$

$PV = nRT$

$$n_{H_2} = \frac{PV}{RT} = \frac{\left(755\ mm\ Hg \times \frac{1.0\ atm}{760\ mm\ Hg}\right)(0.3826\ L)}{\left(0.082\ 06 \frac{L \cdot atm}{K \cdot mol}\right)(293.2\ K)} = 0.0158\ mol\ H_2$$

(b) $M + x\ HCl \rightarrow x/2\ H_2 + MCl_x$

$$moles\ HCl\ reacted = 0.0158\ mol\ H_2 \times \frac{x\ mol\ HCl}{x/2\ mol\ H_2} = 0.0316\ mol\ HCl$$

moles Cl reacted = moles HCl reacted = 0.0316 mol Cl

$$mass\ Cl = 0.0316\ mol\ Cl \times \frac{35.453\ g\ Cl}{1\ mol\ Cl} = 1.120\ g\ Cl$$

mass MCl_x = mass M + mass Cl = 1.385 g + 1.120 g = 2.505 g MCl_x

(c) $\Delta T_f = K_f \cdot m; \quad m = \dfrac{\Delta T_f}{K_f} = \dfrac{3.53\ ^\circ C}{1.86\ \dfrac{^\circ C \cdot kg}{mol}} = 1.90\ mol/kg = 1.90\ m$

(d) 25.0 g = 0.0250 kg

$1.90\ m = 1.90\ \dfrac{mol}{kg} = \dfrac{x\ mol\ ions}{0.0250\ kg}$

mol ions = (1.90 mol/kg)(0.0250 kg) = 0.0475 mol ions

(e) mol M = mol ions – mol Cl = 0.0475 mol – 0.0316 mol = 0.0159 mol M

$\dfrac{Cl}{M} = \dfrac{0.0316\ mol}{0.0159\ mol} = 2$, the formula is MCl_2.

molar mass = $\dfrac{2.505\ g}{0.0159\ mol}$ = 157.5 g/mol; molecular mass = 157.5

(f) atomic mass of M = 157.5 – 2(35.453) = 86.6; M = Sr

12.162 CO_2, 44.01; H_2O, 18.02

$$\text{mol C} = 106.43\ mg\ CO_2 \times \frac{1\ g}{1000\ mg} \times \frac{1\ mol\ CO_2}{44.01\ g\ CO_2} \times \frac{1\ mol\ C}{1\ mol\ CO_2} = 0.002\ 418\ mol\ C$$

$$\text{mass C} = 0.002\ 418\ mol\ C \times \frac{12.011\ g\ C}{1\ mol\ C} = 0.029\ 04\ g\ C$$

$$\text{mol H} = 32.100\ mg\ H_2O \times \frac{1\ g}{1000\ mg} \times \frac{1\ mol\ H_2O}{18.02\ g\ H_2O} \times \frac{2\ mol\ H}{1\ mol\ H_2O} = 0.003\ 563\ mol\ H$$

$$\text{mass H} = 0.003\ 563\ mol\ H \times \frac{1.008\ g\ H}{1\ mol\ H} = 0.003\ 592\ g\ H$$

$$\text{mass O} = \left(36.72\ mg \times \frac{1\ g}{1000\ mg}\right) - 0.029\ 04\ g\ C - 0.003\ 592\ g\ H = 0.004\ 088\ g\ O$$

$$\text{mol O} = 0.004\ 088\ g\ O \times \frac{1\ mol\ O}{16.00\ g\ O} = 0.000\ 255\ 5\ mol\ O$$

$C_{0.002\ 418}H_{0.003\ 563}O_{0.000\ 255\ 5}$; divide all subscripts by the smallest, 0.000 255 5.

$C_{0.002\ 418\ /\ 0.000\ 255\ 5}H_{0.003\ 563\ /\ 0.000\ 255\ 5}O_{0.000\ 255\ 5\ /\ 0.000\ 255\ 5}$

$C_{9.5}H_{14}O$; multiply all subscripts by 2.

$C_{(9.5 \times 2)}H_{(14 \times 2)}O_{(1 \times 2)}$

Empirical formula is $C_{19}H_{28}O_2$, 288

T = 25 °C = 25 + 273 = 298 K

$$\Pi = MRT; \quad M = \frac{\Pi}{RT} = \frac{\left(21.5\ mm\ Hg \times \dfrac{1\ atm}{760\ mm\ Hg}\right)}{\left(0.082\ 06\ \dfrac{L \cdot atm}{K \cdot mol}\right)(298\ K)} = 0.001\ 16\ mol/L$$

15.0 mL = 0.0150 L

mol solute = (0.001 16 mol/L)(0.0150 L) = 1.74×10^{-5} mol

$$\text{molar mass} = \frac{\left(5.00\text{ mg x }\dfrac{1\text{ g}}{1000\text{ mg}}\right)}{1.74\text{ x }10^{-5}\text{ mol}} = 287\text{ g/mol}$$

The molar mass and the empirical formula weight are essentially identical, so the molecular formula and the empirical formula are the same. The molecular formula is $C_{19}H_{28}O_2$.

12.164 CO_2, 44.01; H_2O, 18.02

$$\text{mol C} = 7.0950\text{ g }CO_2\text{ x }\frac{1\text{ mol }CO_2}{44.01\text{ g }CO_2}\text{ x }\frac{1\text{ mol C}}{1\text{ mol }CO_2} = 0.1612\text{ mol C}$$

$$\text{mass C} = 0.1612\text{ mol C x }\frac{12.011\text{ g C}}{1\text{ mol C}} = 1.9362\text{ g C}$$

$$\text{mol H} = 2.2668\text{ g }H_2O\text{ x }\frac{1\text{ mol }H_2O}{18.02\text{ g }H_2O}\text{ x }\frac{2\text{ mol H}}{1\text{ mol }H_2O} = 0.2516\text{ mol H}$$

$$\text{mass H} = 0.2516\text{ mol H x }\frac{1.008\text{ g H}}{1\text{ mol H}} = 0.2536\text{ g H}$$

$$\text{mass O} = 3.0078\text{ g} - 1.9362\text{ g C} - 0.2536\text{ g H} = 0.8180\text{ g O}$$

$$\text{mol O} = 0.8180\text{ g O x }\frac{1\text{ mol O}}{16.00\text{ g O}} = 0.05112\text{ mol O}$$

$C_{0.1612}H_{0.2516}O_{0.05112}$; divide all subscripts by the smallest, 0.05112.

$C_{0.1612\,/\,0.05112}H_{0.2516\,/\,0.05112}O_{0.05112\,/\,0.05112}$

$C_{3.159}H_{4.922}O$, mass of this unit is 58.90

$$\Pi = MRT$$

$$M = \frac{\Pi}{RT} = \frac{(0.026\,44\text{ atm})}{\left(0.082\,06\dfrac{\text{L}\cdot\text{atm}}{\text{K}\cdot\text{mol}}\right)(298\text{ K})} = 0.001\,081\text{ mol/L}$$

$$\text{mol solute} = (0.001\,081\text{ mol/L})(0.800\text{ L}) = 8.65\text{ x }10^{-4}\text{ mol}$$

$$\text{molar mass} = \frac{0.6617\text{ g}}{8.65\text{ x }10^{-4}\text{ mol}} = 765\text{ g/mol}$$

Divide the molar mass by the unit mass to find the appropriate multiplier.

$$\frac{765}{58.9} = 13$$

$C_{3.159}H_{4.922}O$, multiply each subscript by 13 to get the molecular formula.

$C_{(3.159\text{ x }13)}H_{(4.922\text{ x }13)}O_{(1\text{ x }13)}$

$C_{41}H_{64}O_{13}$ is the molecular formula.

13 Chemical Kinetics

13.1 $3\,I^-(aq) + H_3AsO_4(aq) + 2\,H^+(aq) \rightarrow I_3^-(aq) + H_3AsO_3(aq) + H_2O(l)$

(a) $-\dfrac{\Delta[I^-]}{\Delta t} = 4.8 \times 10^{-4}$ M/s

$$\frac{\Delta[I_3^-]}{\Delta t} = \frac{1}{3}\left(-\frac{\Delta[I^-]}{\Delta t}\right) = \left(\frac{1}{3}\right)(4.8 \times 10^{-4}\text{ M/s}) = 1.6 \times 10^{-4}\text{ M/s}$$

(b) $-\dfrac{\Delta[H^+]}{\Delta t} = 2\left(\dfrac{\Delta[I_3^-]}{\Delta t}\right) = (2)(1.6 \times 10^{-4}\text{ M/s}) = 3.2 \times 10^{-4}\text{ M/s}$

13.2 (a) $\dfrac{-\Delta[A]}{\Delta t} = \dfrac{0.04\text{ M} - 0.07\text{ M}}{500\text{ s}} = 6.0 \times 10^{-5}\text{ M/s}$

$$\frac{-\Delta[B]}{\Delta t} = \frac{0.01\text{ M} - 0.07\text{ M}}{500\text{ s}} = 1.2 \times 10^{-4}\text{ M/s}$$

$$\frac{\Delta[C]}{\Delta t} = \frac{0.03\text{ M} - 0.0\text{ M}}{500\text{ s}} = 6.0 \times 10^{-5}\text{ M/s}$$

(b) $A + 2B \rightarrow C$

13.3 Rate = $k[BrO_3^-][Br^-][H^+]^2$,
1st order in BrO_3^-, 1st order in Br^-, 2nd order in H^+, 4th order overall
Rate = $k[H_2][I_2]$, 1st order in H_2, 1st order in I_2, 2nd order overall

13.4 For NO, when the [NO] doubles, the rate doubles. This indicates that the reaction order for NO is 1. For Cl_2, when the $[Cl_2]$ is halved, the rate decreases by a factor of 4. This indicates that the reaction order for Cl_2 is 2. The reaction is 3rd order overall.

13.5 (a) Rate = $k[NO_2]^m[CO]^n$

$$m = \frac{\ln\left(\dfrac{\text{Rate}_2}{\text{Rate}_1}\right)}{\ln\left(\dfrac{[NO_2]_2}{[NO_2]_1}\right)} = \frac{\ln\left(\dfrac{1.13 \times 10^{-2}}{5.00 \times 10^{-3}}\right)}{\ln\left(\dfrac{0.150}{0.100}\right)} = 2$$

$$n = \frac{\ln\left(\frac{\text{Rate}_3 \cdot [NO_2]_2}{\text{Rate}_2 \cdot [NO_2]_3}\right)}{\ln\left(\frac{[CO]_3}{[CO]_2}\right)} = \frac{\ln\left(\frac{(2.00 \times 10^{-2})(0.150)^2}{(1.13 \times 10^{-2})(0.200)^2}\right)}{\ln\left(\frac{0.200}{0.100}\right)} = 0$$

Rate = $k[NO_2]^2$

(b) From Experiment 1:

$$k = \frac{\text{Rate}}{[NO_2]^2} = \frac{5.00 \times 10^{-3}\ \text{M/s}}{(0.100\ \text{M})^2} = 0.500\ 1/(\text{M} \cdot \text{s})$$

(c) Rate = $k[NO_2]^2 = [0.500\ 1/(\text{M} \cdot \text{s})](0.150\ \text{M})^2 = 1.13 \times 10^{-2}$ M/s

13.6 (a) Rate = $k[C_2H_4Br_2]^m[I^-]^n$

$$m = \frac{\ln\left(\frac{\text{Rate}_2}{\text{Rate}_1}\right)}{\ln\left(\frac{[C_2H_4Br_2]_2}{[C_2H_4Br_2]_1}\right)} = \frac{\ln\left(\frac{1.74 \times 10^{-4}}{6.45 \times 10^{-5}}\right)}{\ln\left(\frac{0.343}{0.127}\right)} = 1$$

$$n = \frac{\ln\left(\frac{\text{Rate}_3 \cdot [C_2H_4Br_2]_2}{\text{Rate}_2 \cdot [C_2H_4Br_2]_3}\right)}{\ln\left(\frac{[I^-]_3}{[I^-]_2}\right)} = \frac{\ln\left(\frac{(1.26 \times 10^{-4})(0.343)}{(1.74 \times 10^{-4})(0.203)}\right)}{\ln\left(\frac{0.125}{0.102}\right)} = 1$$

Rate = $k[C_2H_4Br_2][I^-]$

(b) From Experiment 1:

$$k = \frac{\text{Rate}}{[C_2H_4Br_2][I^-]} = \frac{6.45 \times 10^{-5}\ \text{M/s}}{(0.127\ \text{M})(0.102\ \text{M})} = 4.98 \times 10^{-3}/(\text{M} \cdot \text{s})$$

(c) $$\text{Rate} = \frac{\Delta[I_3^-]}{\Delta t} = k[C_2H_4Br_2][I^-] = [4.98 \times 10^{-3}/(\text{M} \cdot \text{s})](0.150\ \text{M})^2 = 1.12 \times 10^{-4}\ \text{M/s}$$

(d) $$-\frac{1}{3}\frac{\Delta[I^-]}{\Delta t} = \frac{1}{1}\frac{\Delta[I_3^-]}{\Delta t} = 1.12 \times 10^{-4}\ \text{M/s}$$

$$-3\frac{\Delta[I_3^-]}{\Delta t} = (-3)(1.12 \times 10^{-4}\ \text{M/s}) = -3.36 \times 10^{-4}\ \text{M/s}$$

13.7 (a) The reactions in vessels (a) and (b) have the same rate, the same number of B molecules, but different numbers of A molecules. Therefore, the rate does not depend on

A and its reaction order is zero. The same conclusion can be drawn from the reactions in vessels (c) and (d).
The rate for the reaction in vessel (c) is four times the rate for the reaction in vessel (a). Vessel (c) has twice as many B molecules than does vessel (a). Because the rate quadruples when the concentration of B doubles, the reaction order for B is two.
(b) rate = $k[B]^2$

13.8 The rate law is Rate = $k[A]^2[B]$. In vessel 1, the Rate = $k(2)^2(4) = 0.01$ M/s

$$k = \frac{0.01 \text{ M/s}}{(2)^2(4)} = 6.25 \times 10^{-4} \text{ M/s}$$

In vessel 2, Rate = $(6.25 \times 10^{-4}$ M/s$)(4)^2(2) = 0.02$ M/s
In vessel 3, Rate = $(6.25 \times 10^{-4}$ M/s$)(1)^2(8) = 0.005$ M/s

13.9 $k = 1.50 \times 10^{-6}$ M/s
$[A]_t = -kt + [A]_o$
$kt = [A]_o - [A]_t$

$$t = \frac{[A]_o - [A]_t}{k} = \frac{(5.00 \times 10^{-3} \text{ M}) - (1.00 \times 10^{-3} \text{ M})}{1.50 \times 10^{-6} \text{ M/s}} = 2667 \text{ s} = 2670 \text{ s}$$

$$t = 2667 \text{ s} \times \frac{1 \text{ min}}{60 \text{ s}} = 44.4 \text{ min}$$

13.10

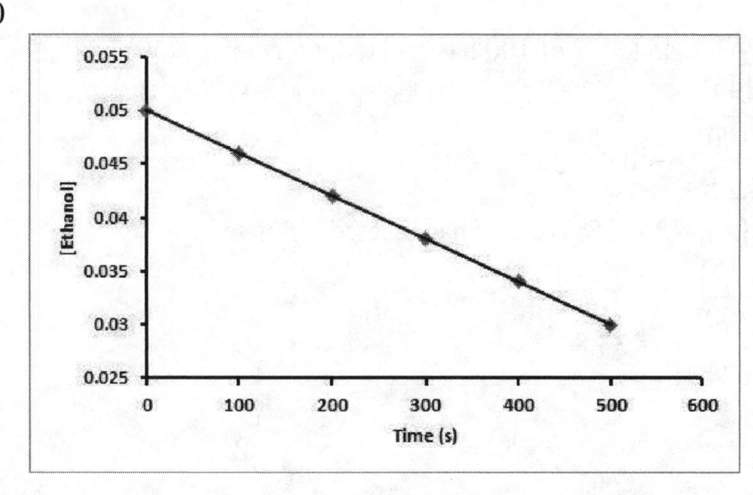

(a) A graph of $[C_2H_5OH]$ vs time is linear (slope = -4.0×10^{-5} M/s), which indicates that the reaction is zeroth order in $[C_2H_5OH]$.
(b) $k = -\text{slope} = 4.0 \times 10^{-5}$ M/s
(c) $t = 15.0 \text{ min s} \times \dfrac{60 \text{ s}}{1 \text{ min}} = 900 \text{ s}$

$[C_2H_5OH]_t = -kt + [C_2H_5OH]_o$
$[C_2H_5OH]_t = -(4.0 \times 10^{-5} \text{ M/s})(900 \text{ s}) + (5.0 \times 10^{-2} \text{ M}) = 0.014 \text{ M}$

13.11 (a) $\ln \dfrac{[Co(NH_3)_5Br^{2+}]_t}{[Co(NH_3)_5Br^{2+}]_o} = -kt$

$k = 6.3 \times 10^{-6}/s$; $t = 10.0\ h \times \dfrac{3600\ s}{1\ h} = 36{,}000\ s$

$\ln[Co(NH_3)_5Br^{2+}]_t = -kt + \ln[Co(NH_3)_5Br^{2+}]_o$

$\ln[Co(NH_3)_5Br^{2+}]_t = -(6.3 \times 10^{-6}/s)(36{,}000\ s) + \ln(0.100)$

$\ln[Co(NH_3)_5Br^{2+}]_t = -2.5294$; After 10.0 h, $[Co(NH_3)_5Br^{2+}] = e^{-2.5294} = 0.080\ M$

(b) $[Co(NH_3)_5Br^{2+}]_o = 0.100\ M$

If 75% of the $Co(NH_3)_5Br^{2+}$ reacts then 25% remains.

$[Co(NH_3)_5Br^{2+}]_t = (0.25)(0.100\ M) = 0.025\ M$

$$\ln \frac{[Co(NH_3)_5Br^{2+}]_t}{[Co(NH_3)_5Br^{2+}]_o} = -kt;\quad t = \frac{\ln \dfrac{[Co(NH_3)_5Br^{2+}]_t}{[Co(NH_3)_5Br^{2+}]_o}}{-k}$$

$$t = \frac{\ln\left(\dfrac{0.025}{0.100}\right)}{-(6.3 \times 10^{-6}/s)} = 2.2 \times 10^5\ s;\quad t = 2.2 \times 10^5\ s \times \frac{1\ h}{3600\ s} = 61\ h$$

13.12 $\ln \dfrac{[A]_t}{[A]_o} = -kt$

(a) Let $[A]_o = 100$ and $[A]_t = 45$ (55% of 100 has reacted, 45 is left)

$$k = -\frac{\ln \dfrac{[A]_t}{[A]_o}}{t} = -\frac{\ln \dfrac{[45]_t}{[100]_o}}{14.2\ h} = 0.0562\ h^{-1}$$

(b) Let $[A]_o = 100$ and $[A]_t = 15$ (85% of 100 has reacted, 15 is left)

$$t = -\frac{\ln \dfrac{[A]_t}{[A]_o}}{k} = -\frac{\ln \dfrac{[15]_t}{[100]_o}}{0.0562\ h^{-1}} = 33.8\ h$$

13.13

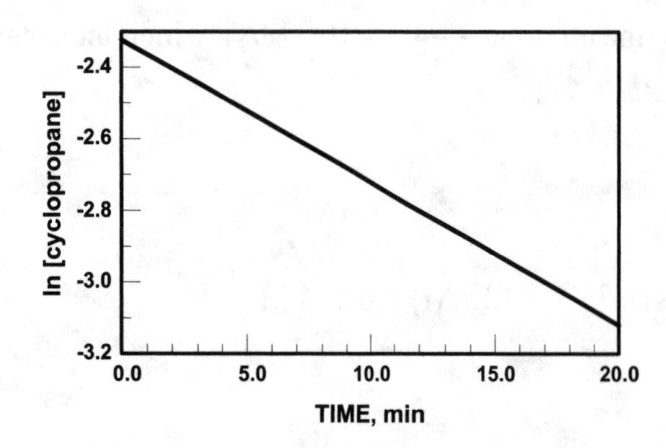

Slope = –0.03989/min = –6.6 x 10^{-4}/s and k = –slope
A plot of ln[cyclopropane] versus time is linear, indicating that the data fit the equation for a first-order reaction. k = 6.6 x 10^{-4}/s (0.040/min)

13.14 $\ln\frac{[N_2O_5]_t}{[N_2O_5]_0} = -kt$, k = – (slope) = – (– 9.8 x 10^{-4}/s) = 9.8 x 10^{-4}/s

(a) $t = 10 \text{ min} \times \frac{60 \text{ s}}{1 \text{ min}} = 600 \text{ s}$

$\ln[N_2O_5]_t = -kt + \ln[N_2O_5]_0 = -(9.8 \times 10^{-4}/\text{s})(600 \text{ s}) + \ln(0.100) = -2.891$
$[N_2O_5]_t = e^{-2.891} = 0.055$ M

	$2\ N_2O_5$	$\rightarrow$	$4\ NO_2$	+	O_2
initial (M)	0.100		0		0
change (M)	–2x		+4x		+x
2x = 0.100 – 0.055 = 0.045					
final (M)	0.055		0.090		0.023

After 10.0 min: $[N_2O_5]$ = 0.055 M, $[NO_2]$ = 0.090 M and $[O_2]$ = 0.023 M

13.15 (a) k = 1.8 x 10^{-5}/s

$t_{1/2} = \frac{0.693}{k} = \frac{0.693}{1.8 \times 10^{-5}/\text{s}} = 38{,}500 \text{ s}$; $\quad t_{1/2} = 38{,}500 \text{ s} \times \frac{1 \text{ h}}{3600 \text{ s}} = 11 \text{ h}$

(b) $0.30 \text{ M} \xrightarrow{t_{1/2}} 0.15 \text{ M} \xrightarrow{t_{1/2}} 0.075 \text{ M} \xrightarrow{t_{1/2}} 0.0375 \text{ M} \xrightarrow{t_{1/2}} 0.019 \text{ M}$
(c) Because 25% of the initial concentration corresponds to 1/4 or $(1/2)^2$ of the initial concentration, the time required is two half-lives: $t = 2t_{1/2} = 2(11 \text{ h}) = 22 \text{ h}$

13.16 After one half-life, there would be four A molecules remaining. After two half-lives, there would be two A molecules remaining. This is represented by the drawing at t = 10 min. 10 min is equal to two half-lives, therefore, $t_{1/2}$ = 5 min for this reaction. After 15 min (three half-lives) only one A molecule would remain.

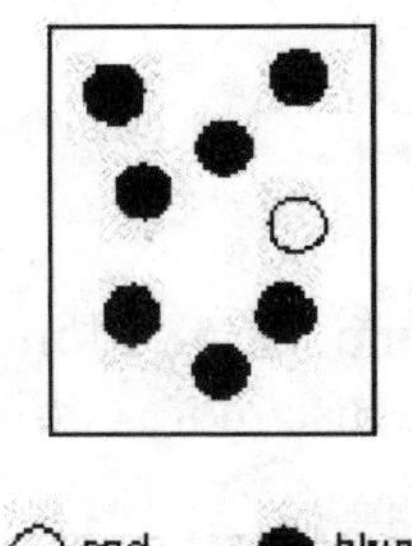

red blue

13.17

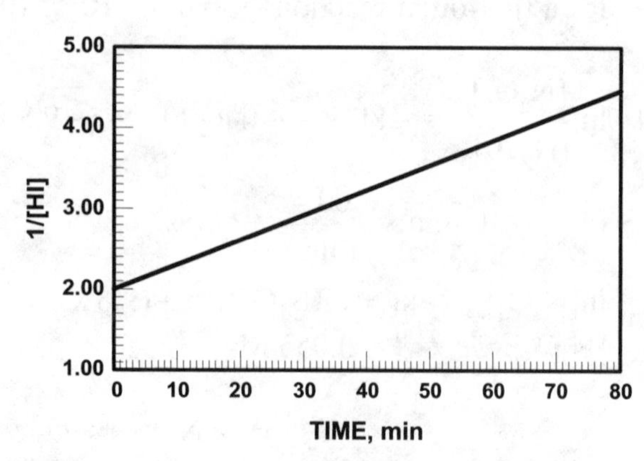

(a) A plot of $1/[HI]$ versus time is linear. The reaction is second-order.

(b) $k = \text{slope} = 0.0308/(M \cdot min)$

(c) $t = \frac{1}{k}\left[\frac{1}{[HI]_t} - \frac{1}{[HI]_o}\right] = \frac{1}{0.0308/(M \cdot min)}\left[\frac{1}{0.100\ M} - \frac{1}{0.500\ M}\right] = 260\ min$

(d) It requires one half-life ($t_{1/2}$) for the [HI] to drop from 0.400 M to 0.200 M.

$$t_{1/2} = \frac{1}{k[HI]_o} = \frac{1}{[0.0308/(M \cdot min)](0.400\ M)} = 81.2\ min$$

13.18

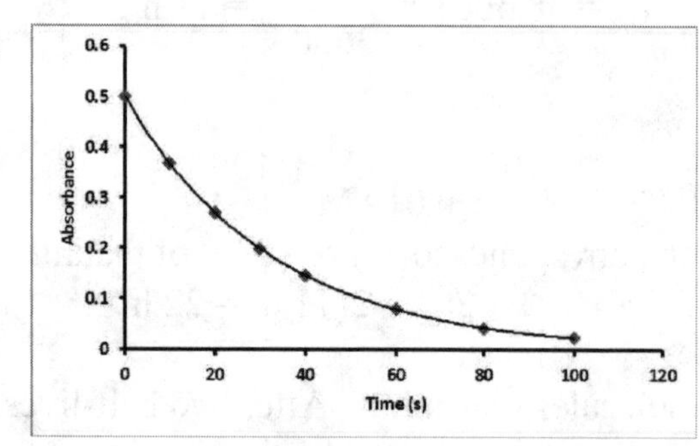

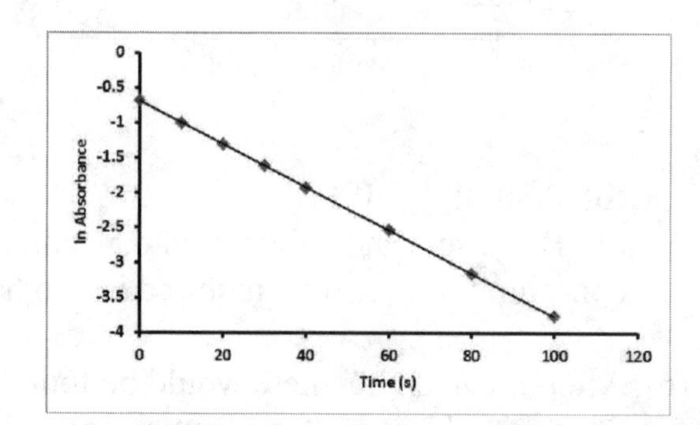

A graph of ln Abs vs time is linear (slope = –0.03071 /s), which indicates that the reaction is first-order in dye.

$k = -\text{slope} = 0.03071\ /s$

$$t_{1/2} = \frac{0.693}{k} = \frac{0.693}{0.03071/s} = 22.6\ s$$

13.19 (a) Because $\Delta E < 0$, the reaction is exothermic.

(b) E_a for the reverse reaction equals 132 kJ/mol + 226 kJ/mol = 358 kJ/mol

(c) The reaction rate increases as temperature increases because more collisions occur with an energy greater than the activation energy.

13.20 unsuccessful

$:\ddot{O}:$ — $N\cdots\cdots:C\equiv O:$ — $:\ddot{O}$ (N bonded to O and =O)

successful

$:\ddot{O}$—$\dot{N}$=$\ddot{O}:\cdots\cdots:C\equiv O:$

unsuccessful

$:\ddot{O}:$ — $N\cdots\cdots:O\equiv C:$ — $:\ddot{O}$ (N bonded to O and =O)

unsuccessful

$:\ddot{O}$—$\dot{N}$=$\ddot{O}:\cdots\cdots:O\equiv C:$

13.21 (a) $\ln\left(\frac{k_2}{k_1}\right) = \left(\frac{-E_a}{R}\right)\left(\frac{1}{T_2} - \frac{1}{T_1}\right)$

$k_1 = 3.7 \times 10^{-5}/s$, $T_1 = 25\ °C = 298\ K$
$k_2 = 1.7 \times 10^{-3}/s$, $T_2 = 55\ °C = 328\ K$

$$E_a = -\frac{[\ln k_2 - \ln k_1]R}{\left(\frac{1}{T_2} - \frac{1}{T_1}\right)}$$

$$E_a = -\frac{[\ln(1.7 \times 10^{-3}) - \ln(3.7 \times 10^{-5})][8.314 \times 10^{-3}\ kJ/(K\cdot mol)]}{\left(\frac{1}{328\ K} - \frac{1}{298\ K}\right)} = 104\ kJ/mol$$

(b) $k_1 = 3.7 \times 10^{-5}/s$, $T_1 = 25\ °C = 298\ K$
solve for k_2, $T_2 = 35\ °C = 308\ K$

$$\ln k_2 = \left(\frac{-E_a}{R}\right)\left(\frac{1}{T_2} - \frac{1}{T_1}\right) + \ln k_1$$

$$\ln k_2 = \left(\frac{-104\ kJ/mol}{8.314 \times 10^{-3}\ kJ/(K\cdot mol)}\right)\left(\frac{1}{308\ K} - \frac{1}{298\ K}\right) + \ln(3.7 \times 10^{-5})$$

$\ln k_2 = -8.84$; $k_2 = e^{-8.84} = 1.4 \times 10^{-4}/s$

13.22 $\ln\left(\frac{k_2}{k_1}\right) = \left(\frac{-E_a}{R}\right)\left(\frac{1}{T_2} - \frac{1}{T_1}\right)$

$k_1 = 1$, $T_1 = 25\ °C = 298\ K$
$k_2 = 3$, $T_2 = 36\ °C = 309\ K$

$$E_a = -\frac{[\ln k_2 - \ln k_1]R}{\left(\frac{1}{T_2} - \frac{1}{T_1}\right)}$$

$$E_a = -\frac{[\ln(3) - \ln(1)][8.314 \times 10^{-3}\ \text{kJ/(K}\cdot\text{mol)}]}{\left(\frac{1}{309\ \text{K}} - \frac{1}{298\ \text{K}}\right)} = 76.5\ \text{kJ/mol}$$

13.23 (a)

$NO_2(g) + F_2(g) \rightarrow NO_2F(g) + F(g)$
$F(g) + NO_2(g) \rightarrow NO_2F(g)$

Overall reaction $2\,NO_2(g) + F_2(g) \rightarrow 2\,NO_2F(g)$

Because F(g) is produced in the first reaction and consumed in the second, it is a reaction intermediate.

(b) In each reaction there are two reactants, so each elementary reaction is bimolecular.

13.24 (a)

$H_2O_2(aq) \rightarrow 2\,OH(aq)$
$H_2O_2(aq) + OH(aq) \rightarrow H_2O(l) + HO_2(aq)$
$HO_2(aq) + OH(aq) \rightarrow H_2O(l) + O_2(g)$

Overall reaction $2\,H_2O_2(aq) \rightarrow 2\,H_2O(l) + O_2(g)$

Because OH(aq) and HO_2(aq) are produced and then consumed, they are reaction intermediates.

(b) Step 1 is unimolecular because it has only one reactant. Steps 2 and 3 are bimolecular because in each reaction there are two reactants.

13.25 (a) Rate = $k[O_3][O]$ (b) Rate = $k[Br]^2[Ar]$ (c) Rate = $k[Co(CN)_5(H_2O)^{2-}]$

13.26 (a) ii) (b) iii) (c) i)

13.27

$NO(g) + Cl_2(g) \rightarrow NOCl(g) + Cl(g)$ (slow)
$NO(g) + Cl(g) \rightarrow NOCl(g)$ (fast)

Overall reaction $2\,NO(g) + Cl_2(g) \rightarrow 2\,NOCl(g)$

The predicted rate law for the overall reaction is the rate law for the first (slow) elementary reaction: Rate = $k[NO][Cl_2]$

The predicted rate law is in accord with the observed rate law.

13.28

$Co(CN)_5(H_2O)^{2-}(aq) \rightarrow Co(CN)_5^{2-}(aq) + H_2O(l)$ (slow)
$Co(CN)_5^{2-}(aq) + I^-(aq) \rightarrow Co(CN)_5I^{3-}(aq)$ (fast)

Overall reaction $Co(CN)_5(H_2O)^{2-}(aq) + I^-(aq) \rightarrow Co(CN)_5I^{3-}(aq) + H_2O(l)$

The predicted rate law for the overall reaction is the rate law for the first (slow) elementary reaction: Rate = $k[Co(CN)_5(H_2O)^{2-}]$

The predicted rate law is in accord with the observed rate law.

13.29

$$NO(g) + O_2(g) \underset{k_{-1}}{\overset{k_1}{\rightleftharpoons}} NO_3(g) \quad \text{fast}$$

$$NO_3(g) + NO(g) \xrightarrow{k_2} 2NO_2(g) \quad \text{slow}$$

(a) $2\,NO(g) + O_2(g) \rightarrow 2\,NO_2(g)$

(b) $Rate_{forward} = k_1[NO][O_2]$ and $Rate_{reverse} = k_{-1}[NO_3]$

Because of the equilibrium, $Rate_{forward} = Rate_{reverse}$, and $k_1[NO][O_2] = k_{-1}[NO_3]$.

$$[NO_3] = \frac{k_1}{k_{-1}}[NO][O_2]$$

The rate law for the rate determining step is $Rate = k_2[NO_3][NO]$. In this rate law substitute for $[NO_3]$.

$Rate = k_2 \dfrac{k_1}{k_{-1}}[NO]^2[O_2]$, which is consistent with the experimental rate law.

(c) $k = \dfrac{k_2 k_1}{k_{-1}}$

13.30

$$I_2(g) \underset{k_{-1}}{\overset{k_1}{\rightleftharpoons}} 2\,I(g) \quad \text{fast}$$

$$H_2(g) + I(g) \underset{k_{-2}}{\overset{k_2}{\rightleftharpoons}} H_2I(g) \quad \text{fast}$$

$$H_2I(g) + I(g) \xrightarrow{k_3} 2\,HI(g) \quad \text{slow}$$

(a) $H_2(g) + I_2(g) \rightarrow 2\,HI(g)$

(b) From step 1, $Rate_{forward} = k_1[I_2]$ and $Rate_{reverse} = k_{-1}[I]^2$

Because of the equilibrium, $Rate_{forward} = Rate_{reverse}$, and $k_1[I_2] = k_{-1}[I]^2$.

$$[I]^2 = \frac{k_1}{k_{-1}}[I_2]$$

From step 2, $Rate_{forward} = k_2[H_2][I]$ and $Rate_{reverse} = k_{-2}[H_2I]$

Because of the equilibrium, $Rate_{forward} = Rate_{reverse}$, and $k_2[H_2][I] = k_{-2}[H_2I]$.

$$[H_2I] = \frac{k_2}{k_{-2}}[H_2][I]$$

The rate law for the rate determining step (step 3) is $Rate = k_3[H_2I][I]$.

Substitute for $[H_2I]$. $Rate = k_3 \dfrac{k_2}{k_{-2}}[H_2][I][I] = k_3 \dfrac{k_2}{k_{-2}}[H_2][I]^2$

Substitute for $[I]^2$. $Rate = k_3 \dfrac{k_2}{k_{-2}}[H_2]\dfrac{k_1}{k_{-1}}[I_2] = k_3 \dfrac{k_1}{k_{-1}}\dfrac{k_2}{k_{-2}}[H_2][I_2]$, which is consistent with the experimental rate law.

(c) $k = \dfrac{k_1 k_2 k_3}{k_{-1} k_{-2}}$

13.31 Assume that concentration is proportional to the number of each molecule in a box.
(a) Comparing boxes (1) and (2), the concentration of A doubles, B and C_2 remain the same and the rate does not change. This means the reaction is zeroth-order in A.
Comparing boxes (1) and (3), the concentration of C_2 doubles, A and B remain the same and the rate doubles. This means the reaction is first-order in C_2.
Comparing boxes (1) and (4), the concentration of B triples, A and C_2 remain the same and the rate triples. This means the reaction is first-order in B.
(b) Rate = k $[B][C_2]$
(c) The mechanism agrees with the rate law. The rate law for the overall reaction is the rate law for the first (slow) elementary reaction: Rate = k$[B][C_2]$
(d) B doesn't appear in the overall reaction because it is consumed in the first step and regenerated in the third step. B is therefore a catalyst. C and CB are intermediates because they are formed in one step and then consumed in subsequent steps in the mechanism.

13.32 (a)

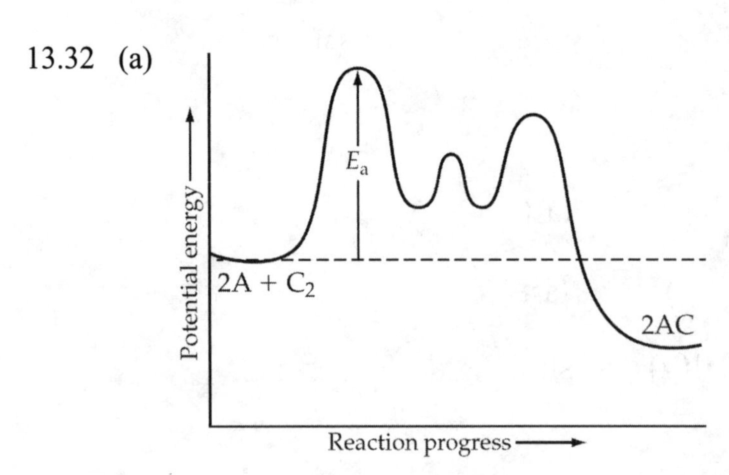

(b)

$$
\begin{array}{lll}
B + C_2 & \rightarrow & BC_2 \quad \text{(slow)} \\
A + BC_2 & \rightarrow & AC + BC \\
A + BC & \rightarrow & AC + B \\
\hline
2\,A + C_2 & \rightarrow & 2\,AC \quad \text{(overall)}
\end{array}
$$

B doesn't appear in the overall reaction because it is consumed in the first step and regenerated in the third step. B is therefore a catalyst. BC_2 and BC are intermediates because they are formed in one step and then consumed in a subsequent step in the reaction.

13.33 (a) Chlorine radicals are gas phase and homogeneous.
(b) Ice crystals in polar stratospheric clouds are heterogeneous.

13.34 $\ln\left(\frac{k_2}{k_1}\right) = \left(\frac{-E_a}{R}\right)\left(\frac{1}{T_2} - \frac{1}{T_1}\right)$

$k_1 = 8.57 \times 10^{-16}$ cm^3/molecules · s, $T_1 = 225$ K
$k_2 = 8.42 \times 10^{-15}$ cm^3/molecules · s, $T_2 = 300$ K

$$E_a = -\frac{[\ln k_2 - \ln k_1]R}{\left(\frac{1}{T_2} - \frac{1}{T_1}\right)}$$

$$E_a = -\frac{[\ln(8.42 \times 10^{-15}) - \ln(8.57 \times 10^{-16})][8.314 \times 10^{-3}\ \text{kJ/(K}\cdot\text{mol)}]}{\left(\frac{1}{300\ \text{K}} - \frac{1}{225\ \text{K}}\right)} = 17.1\ \text{kJ/mol}$$

$$k = A e^{-E_a/RT}$$

$$A = \frac{k}{e^{-E_a/RT}} = \frac{8.57 \times 10^{-16}\ \text{cm}^3\text{/molecules}\cdot\text{s}}{e^{-(17.1\ \text{kJ/mol})/[(8.314 \times 10^{-3}\ \text{kJ/mol}\cdot\text{K})(225\ \text{K})]}} = 8.0 \times 10^{-12}\ \text{cm}^3\text{/molecules}\cdot\text{s}$$

13.35 $O_3 + O \rightarrow 2\ O_2$

(a) Rate = $k[O_3][O]$

(b) $k = A e^{-E_a/RT} = (8.0 \times 10^{-12}\ \text{cm}^3\text{/molecules}\cdot\text{s})e^{-(17.1\ \text{kJ/mol})/[(8.314 \times 10^{-3}\ \text{kJ/mol}\cdot\text{K})(190\ \text{K})]}$

$k = 1.6 \times 10^{-16}\ \text{cm}^3\text{/molecules}\cdot\text{s}$

(c) Rate = $k[O_3][O]$

$= (1.6 \times 10^{-16}\ \text{cm}^3\text{/molecules}\cdot\text{s})(3.5 \times 10^{12}\ \text{molecules/cm}^3)(4.0 \times 10^5\ \text{molecules/cm}^3)$

$= 2.2 \times 10^2\ \text{molecules/cm}^3\cdot\text{s}$

13.36

	Reaction	
(1)	$2[Cl + O_3 \rightarrow O_2 + ClO]$	fast
(2)	$2\ ClO \rightarrow Cl_2O_2$	slow, Rate-determining step
(3)	$Cl_2O_2 + h\nu \rightarrow 2\ Cl + O_2$	fast
	$2\ O3 + h\nu \rightarrow 3\ O_2$	overall

(a) Rate = $k[ClO]^2$

(b) Rate = $(7.2 \times 10^{-13}\ \text{cm}^3\text{/molecules}\cdot\text{s})(2.4 \times 10^9\ \text{molecules/cm}^3)^2$

$= 4.1 \times 10^6\ \text{molecules/cm}^3\cdot\text{s}$

(c) $4.1 \times 10^6\ \text{molecules/cm}^3\cdot\text{s}$ because the rate of loss of ozone is determined by the rate limiting step and there are two O_3 molecules lost for every 2 ClO molecules.

13.37 $$\frac{\text{Rate}_{(\text{catalytic})}}{\text{Rate}_{(\text{natural})}} = \frac{4.1 \times 10^6\ \text{molecules/cm}^3\cdot\text{s}}{2.2 \times 10^2\ \text{molecules/cm}^3\cdot\text{s}} = 1.9 \times 10^4$$

The catalytic chlorine process has a rate that is approximately 19,000 times greater than the rate of natural loss with oxygen.

13.38

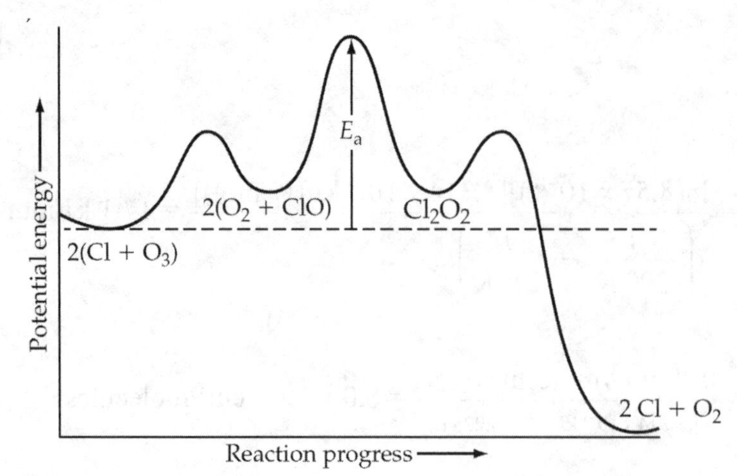

Conceptual Problems

13.40 (a) Because Rate = k[A][B], the rate is proportional to the product of the number of A molecules and the number of B molecules. The relative rates of the reaction in vessels (a) – (d) are 2 : 1 : 4 : 2.
(b) Because the same reaction takes place in each vessel, the k's are all the same.

13.42 (a) For the first-order reaction, half of the A molecules are converted to B molecules each minute.

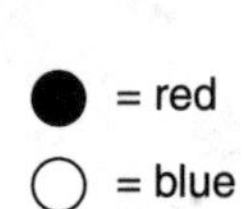

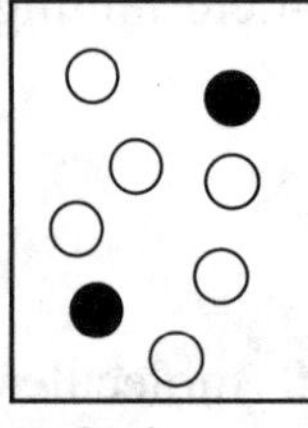

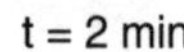

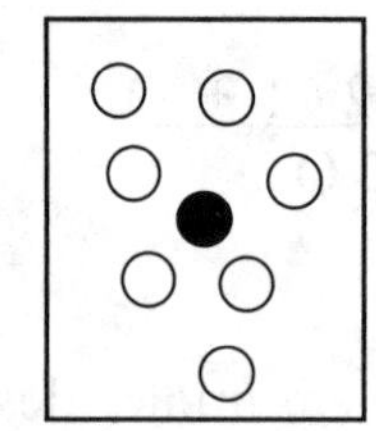

t = 3 min

(b) Because half of the A molecules are converted to B molecules in 1 min, the half-life is 1 minute.

13.44 (a) Because the half-life is inversely proportional to the concentration of A molecules, the reaction is second-order in A.
(b) Rate = $k[A]^2$
(c) The second box represents the passing of one half-life, and the third box represents the passing of a second half-life for a second-order reaction. A relative value of k can be calculated.

$$k = \frac{1}{t_{1/2}[A]} = \frac{1}{(1)(16)} = 0.0625$$

$t_{1/2}$ in going from box 3 to box 4 is: $t_{1/2} = \frac{1}{k[A]} = \frac{1}{(0.0625)(4)} = 4\text{ min}$

(For fourth box, t = 7 min)

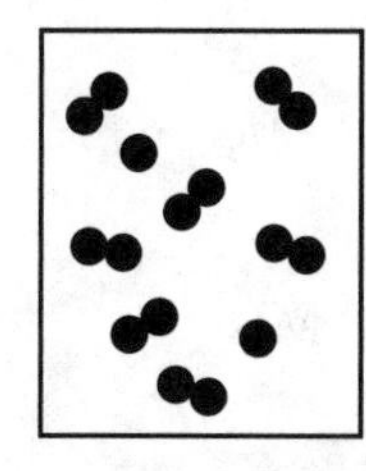

t = 3 min + 4 min = 7 min

13.46 Assume that concentration is proportional to the number of each molecule in a box.
(a) Comparing boxes (1) and (2), the concentration of B doubles, A remains the same and the rate does not change. This means the reaction is zeroth-order in B.
Comparing boxes (3) and (2), the concentration of A doubles, B remains the same and the rate quadruples. This means the reaction is second-order in A.
(b) Rate = k $[A]^2$.
(c)

$$\begin{array}{lcll} 2\,A & \rightarrow & A_2 & \text{(slow)} \\ \underline{A_2 + B} & \underline{\rightarrow} & \underline{AB + A} & \\ A + B & \rightarrow & AB & \text{(overall)} \end{array}$$

(d) A_2 is an intermediate because it is formed in one step and then consumed in a subsequent step in the reaction.

13.48 (a) BC + D → B + CD
(b) 1. B–C + D (reactants), A (catalyst); 2. B---C---A (transition state), D (reactant);
3. A–C (intermediate), B (product), D (reactant); 4. A---C---D (transition state),
B (product); 5. A (catalyst), C–D + B (products)
(c) The first step is rate determining because the first maximum in the potential energy curve is greater than the second (relative) maximum; Rate = k[A][BC]
(d) Endothermic

Section Problems
Reaction Rates (Section 13.1)

13.50 $2\,N_2O_5(g) \rightarrow 4\,NO_2(g) + O_2(g)$

time	$[N_2O_5]$	$[O_2]$
200 s	0.0142 M	0.0029 M
300 s	0.0120 M	0.0040 M

$$\text{Rate of decomposition of } N_2O_5 = -\frac{\Delta[N_2O_5]}{\Delta t} = -\frac{0.0120\text{ M} - 0.0142\text{ M}}{300\text{ s} - 200\text{ s}} = 2.2 \times 10^{-5}\text{ M/s}$$

$$\text{Rate of formation of } O_2 = \frac{\Delta[O_2]}{\Delta t} = \frac{0.0040\text{ M} - 0.0029\text{ M}}{300\text{ s} - 200\text{ s}} = 1.1 \times 10^{-5}\text{ M/s}$$

13.52 (a) $\text{Rate} = \dfrac{\Delta[NO_2]}{\Delta t} = \dfrac{(0.0278\text{ M}) - (0\text{ M})}{700\text{ s} - 0\text{ s}} = 4.0 \times 10^{-5}\text{ M/s}$

(b) $\text{Rate} = \dfrac{\Delta[NO_2]}{\Delta t} = \dfrac{(0.0256\text{ M}) - (0.0063\text{ M})}{600\text{ s} - 100\text{ s}} = 3.9 \times 10^{-5}\text{ M/s}$

(c) $\text{Rate} = \dfrac{\Delta[NO_2]}{\Delta t} = \dfrac{(0.0229\text{ M}) - (0.0115\text{ M})}{500\text{ s} - 200\text{ s}} = 3.8 \times 10^{-5}\text{ M/s}$

(d) $\text{Rate} = \dfrac{\Delta[NO_2]}{\Delta t} = \dfrac{(0.0197\text{ M}) - (0.0160\text{ M})}{400\text{ s} - 300\text{ s}} = 3.7 \times 10^{-5}\text{ M/s}$

Rate (d) is the best estimate of the instantaneous rate because it is determined from measurements taken over the smallest time interval.

13.54 (a) The instantaneous rate of decomposition of N_2O_5 at t = 200 s is determined from the slope of the curve at t = 200 s.

$\text{Rate} = -\dfrac{\Delta[N_2O_5]}{\Delta t} = -\text{slope} = -\dfrac{(1.20 \times 10^{-2}\text{ M}) - (1.69 \times 10^{-2}\text{ M})}{300\text{ s} - 100\text{ s}} = 2.4 \times 10^{-5}\text{ M/s}$

(b) The initial rate of decomposition of N_2O_5 is determined from the slope of the curve at t = 0 s. This is equivalent to the slope of the curve from 0 s to 100 s because in this time interval the curve is almost linear.

$\text{Initial rate} = -\dfrac{\Delta[N_2O_5]}{\Delta t} = -\text{slope} = -\dfrac{(1.69 \times 10^{-2}\text{ M}) - (2.00 \times 10^{-2}\text{ M})}{100\text{ s} - 0\text{ s}} = 3.1 \times 10^{-5}\text{ M/s}$

13.56 (a) $-\dfrac{\Delta[H_2]}{\Delta t} = -3\dfrac{\Delta[N_2]}{\Delta t}$; The rate of consumption of H_2 is 3 times faster.

(b) $\dfrac{\Delta[NH_3]}{\Delta t} = -2\dfrac{\Delta[N_2]}{\Delta t}$; The rate of formation of NH_3 is 2 times faster.

13.58 (a) $-\dfrac{1}{2}\dfrac{\Delta[Br_2]}{\Delta t} = \dfrac{\Delta[ClO_2^-]}{\Delta t} = -2.4 \times 10^{-6}\text{ M/s}$

(b) $\text{Rate} = -\dfrac{\Delta[Br^-]}{\Delta t} = -4\dfrac{\Delta[ClO_2^-]}{\Delta t} = -4(-2.4 \times 10^{-6}\text{ M/s}) = 9.6 \times 10^{-6}\text{ M/s}$

Rate Laws (Sections 13.2 and 13.3)

13.60 Rate = $k[H_2][ICl]$; units for k are $\dfrac{\text{L}}{\text{mol}\cdot\text{s}}$ or $1/(\text{M}\cdot\text{s})$

13.62 (a) Rate = $k[CH_3Br][OH^-]$

(b) Because the reaction is first-order in OH^-, if the $[OH^-]$ is decreased by a factor of 5, the rate will also decrease by a factor of 5.

(c) Because the reaction is first-order in each reactant, if both reactant concentrations are doubled, the rate will increase by a factor of 2 x 2 = 4.

13.64 (a) Rate = $k[Cu(C_{10}H_8N_2)_2^+]^2[O_2]$

(b) The overall reaction order is 2 + 1 = 3.

(c) Because the reaction is second-order in $Cu(C_{10}H_8N_2)_2^+$, if the $[Cu(C_{10}H_8N_2)_2^+]$ is decreased by a factor of four, the rate will decrease by a factor of 16 (1/4 x 1/4 = 1/16).

13.66 (a) Rate = $k[NH_4^+]^m[NO_2^-]^n$

$$m = \frac{\ln\left(\frac{\text{Rate}_2}{\text{Rate}_1}\right)}{\ln\left(\frac{[NH_4^+]_2}{[NH_4^+]_1}\right)} = \frac{\ln\left(\frac{3.6 \times 10^{-6}}{7.2 \times 10^{-6}}\right)}{\ln\left(\frac{0.12}{0.24}\right)} = 1; \quad n = \frac{\ln\left(\frac{\text{Rate}_3}{\text{Rate}_2}\right)}{\ln\left(\frac{[NO_2^-]_3}{[NO_2^-]_2}\right)} = \frac{\ln\left(\frac{5.4 \times 10^{-6}}{3.6 \times 10^{-6}}\right)}{\ln\left(\frac{0.15}{0.10}\right)} = 1$$

Rate = $k[NH_4^+][NO_2^-]$

(b) From Experiment 1: $k = \frac{\text{Rate}}{[NH_4^+][NO_2^-]} = \frac{7.2 \times 10^{-6}\ \text{M/s}}{(0.24\ \text{M})(0.10\ \text{M})} = 3.0 \times 10^{-4}/(\text{M} \cdot \text{s})$

(c) Rate = $k[NH_4^+][NO_2^-] = [3.0 \times 10^{-4}/(\text{M} \cdot \text{s})](0.39\ \text{M})(0.052\ \text{M}) = 6.1 \times 10^{-6}\ \text{M/s}$

Integrated Rate Law; Half-Life (Sections 13.4–13.6)

13.68 $\ln\frac{[C_3H_6]_t}{[C_3H_6]_0} = -kt,\ k = 6.7 \times 10^{-4}/\text{s}$

(a) $t = 30\ \text{min} \times \frac{60\ \text{s}}{1\ \text{min}} = 1800\ \text{s}$

$\ln[C_3H_6]_t = -kt + \ln[C_3H_6]_0 = -(6.7 \times 10^{-4}/\text{s})(1800\ \text{s}) + \ln(0.0500) = -4.202$

$[C_3H_6]_t = e^{-4.202} = 0.015\ \text{M}$

(b) $$t = \frac{\ln\frac{[C_3H_6]_t}{[C_3H_6]_0}}{-k} = \frac{\ln\left(\frac{0.0100}{0.0500}\right)}{-(6.7 \times 10^{-4}/\text{s})} = 2402\ \text{s}; \quad t = 2402\ \text{s} \times \frac{1\ \text{min}}{60\ \text{s}} = 40\ \text{min}$$

(c) $[C_3H_6]_0 = 0.0500$ M; If 25% of the C_3H_6 reacts then 75% remains.

$[C_3H_6]_t = (0.75)(0.0500\ \text{M}) = 0.0375\ \text{M}$

$$t = \frac{\ln\frac{[C_3H_6]_t}{[C_3H_6]_0}}{-k} = \frac{\ln\left(\frac{0.0375}{0.0500}\right)}{-(6.7 \times 10^{-4}/\text{s})} = 429\ \text{s}; \quad t = 429\ \text{s} \times \frac{1\ \text{min}}{60\ \text{s}} = 7.2\ \text{min}$$

13.70 $t_{1/2} = \frac{0.693}{k} = \frac{0.693}{6.7 \times 10^{-4}/\text{s}} = 1034\ \text{s} = 17\ \text{min}$

$$t = \frac{\ln\frac{[C_3H_6]_t}{[C_3H_6]_o}}{-k} = \frac{\ln\frac{(0.0625)(0.0500)}{(0.0500)}}{-6.7 \times 10^{-4}/\text{s}} = 4140\ \text{s}$$

$$t = 4140 \text{ s} \times \frac{1 \text{ min}}{60 \text{ s}} = 69 \text{ min}$$

This is also 4 half-lives. $100 \xrightarrow{t_{1/2}} 50 \xrightarrow{t_{1/2}} 25 \xrightarrow{t_{1/2}} 12.5 \xrightarrow{t_{1/2}} 6.25$

13.72 $$kt = \frac{1}{[C_4H_6]_t} - \frac{1}{[C_4H_6]_0}$$

$k = 4.0 \times 10^{-2}/(M \cdot s)$

(a) $$t = 1.00 \text{ h} \times \frac{60 \text{ min}}{1 \text{ hr}} \times \frac{60 \text{ s}}{1 \text{ min}} = 3600 \text{ s}$$

$$\frac{1}{[C_4H_6]_t} = kt + \frac{1}{[C_4H_6]_0} = (4.0 \times 10^{-2}/(M \cdot s))(3600 \text{ s}) + \frac{1}{0.0200 \text{ M}}$$

$$\frac{1}{[C_4H_6]_t} = 194/M \text{ and } [C_4H_6] = 5.2 \times 10^{-3} \text{ M}$$

(b) $$t = \frac{1}{k}\left[\frac{1}{[C_4H_6]_t} - \frac{1}{[C_4H_6]_0}\right]$$

$$t = \frac{1}{4.0 \times 10^{-2}/(M \cdot s)}\left[\frac{1}{(0.0020 \text{ M})} - \frac{1}{(0.0200 \text{ M})}\right] = 11{,}250 \text{ s}$$

$$t = 11{,}250 \text{ s} \times \frac{1 \text{ min}}{60 \text{ s}} \times \frac{1 \text{ hr}}{60 \text{ min}} = 3.1 \text{ h}$$

13.74 $$t_{1/2} = \frac{1}{k[C_4H_6]_o} = \frac{1}{[4.0 \times 10^{-2}/(M{\cdot}s)](0.0200 \text{ M})} = 1250 \text{ s} = 21 \text{ min}$$

$$t = t_{1/2} = \frac{1}{k[C_4H_6]_o} = \frac{1}{[4.0 \times 10^{-2}/(M{\cdot}s)](0.0100 \text{ M})} = 2500 \text{ s} = 42 \text{ min}$$

13.76

time (min)	$[N_2O]$	$\ln[N_2O]$	$1/[N_2O]$
0	0.250	−1.386	4.00
60	0.228	−1.478	4.39
90	0.216	−1.532	4.63
300	0.128	−2.056	7.81
600	0.0630	−2.765	15.9

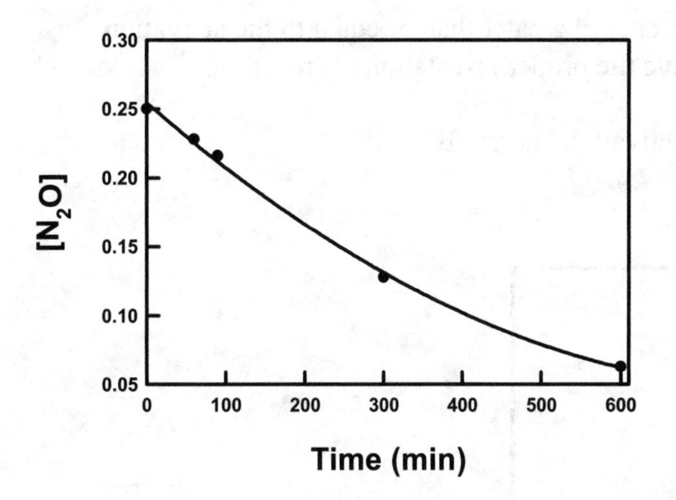

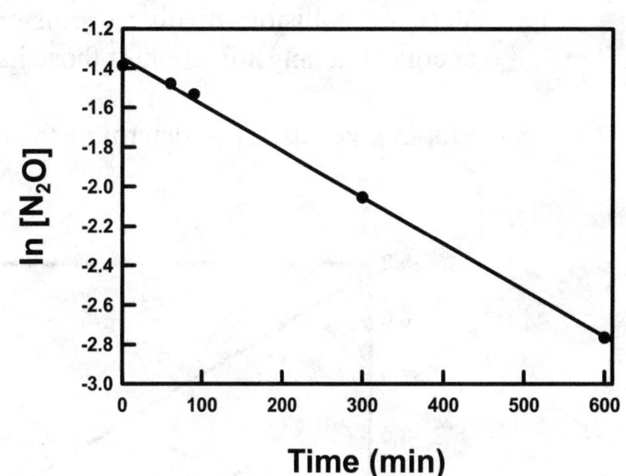

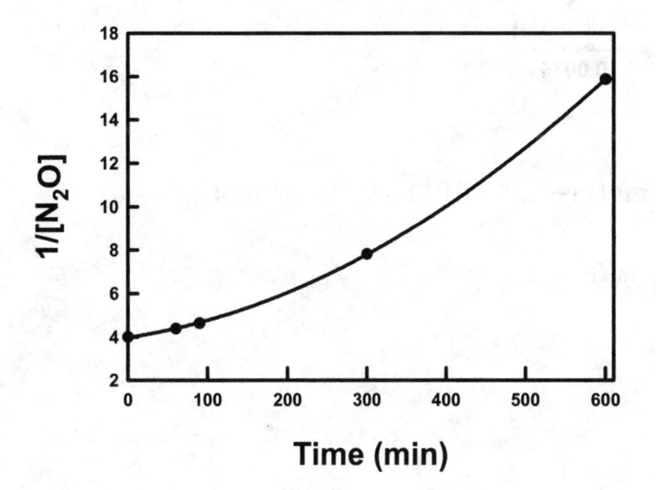

A plot of $\ln[N_2O]$ versus time is linear. The reaction is first-order in N_2O.

$k = -\text{slope} = -(-2.35 \times 10^{-3}/\text{min}) = 2.35 \times 10^{-3}/\text{min}$

$$k = 2.35 \times 10^{-3}/\text{min} \times \frac{1\ \text{min}}{60\ \text{s}} = 3.92 \times 10^{-5}/\text{s}$$

13.78 $k = \dfrac{0.693}{t_{1/2}} = \dfrac{0.693}{248\ \text{s}} = 2.79 \times 10^{-3}/\text{s}$

13.80 (a) The units for the rate constant, k, indicate the reaction is zeroth-order.

(b) For a zeroth-order reaction, $[A]_t - [A]_o = -kt$

$$t = 30\ \text{min} \times \frac{60\ \text{s}}{1\ \text{min}} = 1800\ \text{s}$$

$$[A]_t = -kt + [A]_o = -(3.6 \times 10^{-5}\ \text{M/s})(1800\ \text{s}) + 0.096\ \text{M} = 0.031\ \text{M}$$

(c) Let $[A]_t = [A]_o/2$, $t_{1/2} = \dfrac{[A]_o/2 - [A]_o}{-k} = \dfrac{0.096/2\ \text{M} - 0.096\ \text{M}}{-3.6 \times 10^{-5}\ \text{M/s}} = 1333\ \text{s}$

$$t_{1/2} = 1333\ \text{s} \times \frac{1\ \text{min}}{60\ \text{s}} = 22\ \text{min}$$

The Arrhenius Equation (Sections 13.7 and 13.8)

13.82 Very few collisions involve a collision energy greater than or equal to the activation energy, and only a fraction of those have the proper orientation for reaction.

13.84 Plot ln k versus 1/T to determine the activation energy, E_a.

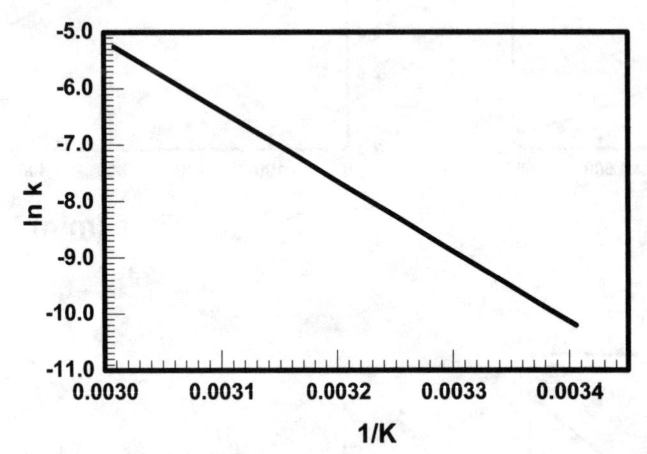

Slope = -1.25×10^4 K

$E_a = -R(\text{slope}) = -[8.314 \times 10^{-3}\ \text{kJ/(K} \cdot \text{mol)}](-1.25 \times 10^4\ \text{K}) = 104\ \text{kJ/mol}$

13.86 (a) $$\ln\left(\frac{k_2}{k_1}\right) = \left(\frac{-E_a}{R}\right)\left(\frac{1}{T_2} - \frac{1}{T_1}\right)$$

$k_1 = 1.3/(\text{M} \cdot \text{s})$, $T_1 = 700$ K

$k_2 = 23.0/(\text{M} \cdot \text{s})$, $T_2 = 800$ K

$$E_a = -\frac{[\ln k_2 - \ln k_1](R)}{\left(\frac{1}{T_2} - \frac{1}{T_1}\right)}$$

$$E_a = -\frac{[\ln(23.0) - \ln(1.3)][8.314 \times 10^{-3}\ \text{kJ/(K} \cdot \text{mol)}]}{\left(\frac{1}{800\ \text{K}} - \frac{1}{700\ \text{K}}\right)} = 134\ \text{kJ/mol}$$

(b) $k_1 = 1.3/(\text{M} \cdot \text{s})$, $T_1 = 700$ K
solve for k_2, $T_2 = 750$ K

$$\ln k_2 = \left(\frac{-E_a}{R}\right)\left(\frac{1}{T_2} - \frac{1}{T_1}\right) + \ln k_1$$

$$\ln k_2 = \left(\frac{-133.8 \text{ kJ/mol}}{8.314 \times 10^{-3} \text{ kJ/(K·mol)}}\right)\left(\frac{1}{750 \text{ K}} - \frac{1}{700 \text{ K}}\right) + \ln(1.3) = 1.795$$

$k_2 = e^{1.795} = 6.0/(M \cdot s)$

13.88 $$\ln\left(\frac{k_2}{k_1}\right) = \left(\frac{-E_a}{R}\right)\left(\frac{1}{T_2} - \frac{1}{T_1}\right)$$

assume $k_1 = 1.0/(M \cdot s)$ at $T_1 = 25\ ^\circ C = 298$ K
assume $k_2 = 15/(M \cdot s)$ at $T_2 = 50\ ^\circ C = 323$ K

$$E_a = -\frac{[\ln k_2 - \ln k_1](R)}{\left(\frac{1}{T_2} - \frac{1}{T_1}\right)}$$

$$E_a = -\frac{[\ln(15) - \ln(1.0)][8.314 \times 10^{-3} \text{ kJ/(K·mol)}]}{\left(\frac{1}{323 \text{ K}} - \frac{1}{298 \text{ K}}\right)} = 87 \text{ kJ/mol}$$

13.90

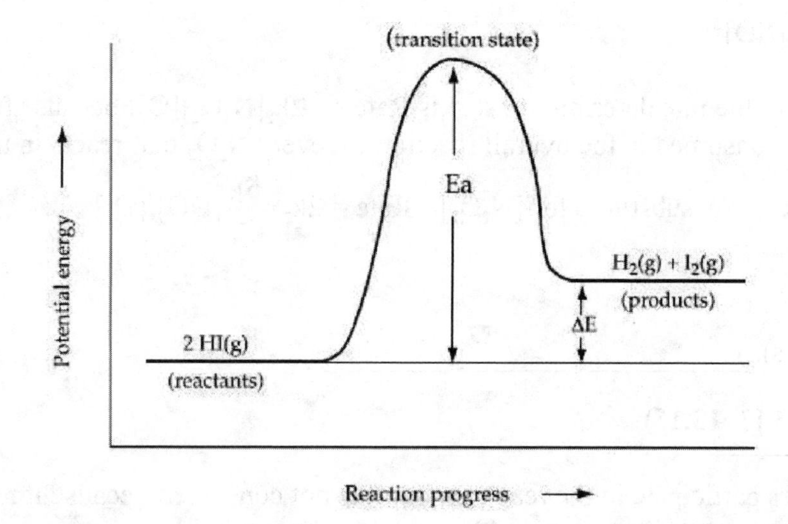

Reaction Mechanisms (Sections 13.9–13.11)

13.92 There is no relationship between the coefficients in a balanced chemical equation for an overall reaction and the exponents in the rate law unless the overall reaction occurs in a single elementary step, in which case the coefficients in the balanced equation are the exponents in the rate law.

13.94 (a)

$H_2(g) + ICl(g) \rightarrow HI(g) + HCl(g)$

$HI(g) + ICl(g) \rightarrow I_2(g) + HCl(g)$

Overall reaction $H_2(g) + 2\ ICl(g) \rightarrow I_2(g) + 2\ HCl(g)$

(b) Because HI(g) is produced in the first step and consumed in the second step, it is a reaction intermediate.
(c) In each reaction there are two reactant molecules, so each elementary reaction is bimolecular.

13.96 (a) bimolecular, Rate = $k[O_3][Cl]$ (b) unimolecular, Rate = $k[NO_2]$
(c) bimolecular, Rate = $k[ClO][O]$ (d) termolecular, Rate = $k[Cl]^2[N_2]$

13.98 (a)

$$NO_2Cl(g) \rightarrow NO_2(g) + Cl(g)$$
$$Cl(g) + NO_2Cl(g) \rightarrow NO_2(g) + Cl_2(g)$$

Overall reaction $2\ NO_2Cl(g) \rightarrow 2\ NO_2(g) + Cl_2(g)$

(b) 1. unimolecular; 2. bimolecular
(c) Rate = $k[NO_2Cl]$

13.100 $NO_2(g) + F_2(g) \rightarrow NO_2F(g) + F(g)$ (slow)
$F(g) + NO_2(g) \rightarrow NO_2F(g)$ (fast)

13.102 (a) $2\ NO(g) + O_2(g) \rightarrow 2\ NO_2(g)$
(b) $Rate_{forward} = k_1[NO]^2$ and $Rate_{reverse} = k_{-1}[N_2O_2]$
Because of the equilibrium, $Rate_{forward} = Rate_{reverse}$, and $k_1[NO]^2 = k_{-1}[N_2O_2]$.

$$[N_2O_2] = \frac{k_1}{k_{-1}}[NO]^2$$

The rate law for the rate determining step is Rate = $2k_2[N_2O_2][O_2]$ because two NO molecules are consumed in the overall reaction for every N_2O_2 that reacts in the second step. In this rate law substitute for $[N_2O_2]$. Rate = $2k_2 \frac{k_1}{k_{-1}}[NO]^2[O_2]$

(c) $k = \dfrac{2k_2k_1}{k_{-1}}$

Catalysis (Sections 13.13–13.15)

13.104 A catalyst does participate in the reaction, but it is not consumed because it reacts in one step of the reaction and is regenerated in a subsequent step.

13.106 (a) $O_3(g) + O(g) \rightarrow 2\ O_2(g)$
(b) Cl acts as a catalyst.
(c) ClO is a reaction intermediate.
(d) A catalyst reacts in one step and is regenerated in a subsequent step. A reaction intermediate is produced in one step and consumed in another.

13.108 (a)

$$NH_2NO_2(aq) + OH^-(aq) \rightarrow NHNO_2^-(aq) + H_2O(l)$$
$$NHNO_2^-(aq) \rightarrow N_2O(g) + OH^-(aq)$$

Overall reaction $NH_2NO_2(aq) \rightarrow N_2O(g) + H_2O(l)$

(b) OH^- acts as a catalyst because it is used in the first step and regenerated in the second. $NHNO_2^-$ is a reaction intermediate because it is produced in the first step and consumed in the second.
(c) The rate will decrease because added acid decreases the concentration of OH^-, which appears in the rate law since it is a catalyst.

13.110 Soluble enzymes are homogeneous catalysts and membrane-bound enzymes are heterogeneous.

13.112

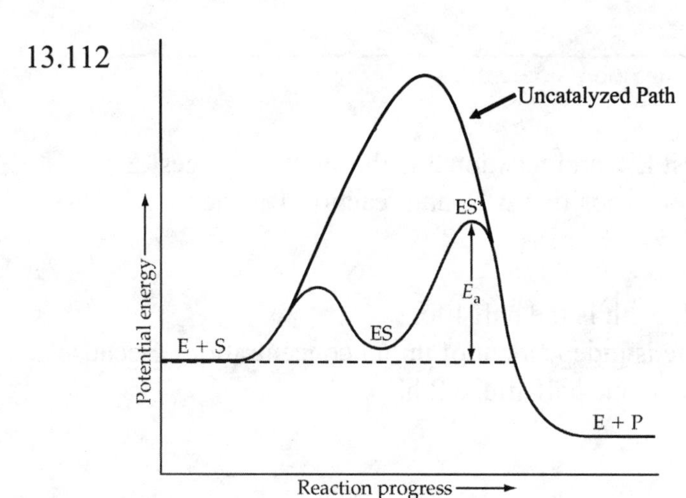

13.114 (a) Because the reaction is zeroth-order in substrate at high substrate concentration, there is no change in rate when the substrate concentration is changed from 2.8×10^{-3} M to 4.8×10^{-3} M.
(b) At high substrate concentrations, the enzyme becomes saturated (i.e., completely bound) with the substrate. At this point, the reaction rate reaches a maximum value and becomes independent of substrate concentration (zeroth order in substrate) because only substrate bound to the enzyme can react.

Chapter Problems

13.116 $A \rightarrow B + C$

(a) Measure the change in the concentration of A as a function of time at several different temperatures.

(b) Plot ln [A] versus time, for each temperature. Straight line graphs will result and k at each temperature equals –slope. Graph ln k versus 1/K, where K is the kelvin temperature. Determine the slope of the line. $E_a = -R(\text{slope})$ where $R = 8.314 \times 10^{-3}$ kJ/(K · mol).

13.118 (a)

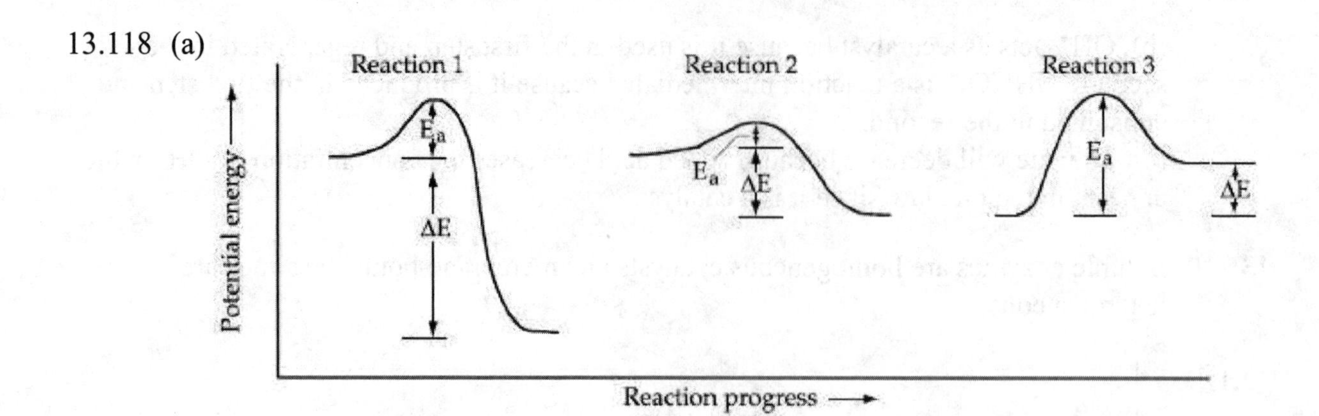

(b) Reaction 2 is the fastest (smallest E_a), and reaction 3 is the slowest (largest E_a).
(c) Reaction 3 is the most endothermic (positive ΔE), and reaction 1 is the most exothermic (largest negative ΔE).

13.120 Because 0.060 M is half of 0.120 M, 5.2 h is the half-life.
For a first-order reaction, the half-life is independent of initial concentration. Because 0.015 M is half of 0.030 M, it will take one half-life, 5.2 h.

$$k = \frac{0.693}{t_{1/2}} = \frac{0.693}{5.2\ h} = 0.133/h$$

$$\ln\frac{[N_2O_5]_t}{[N_2O_5]_0} = -kt$$

$$t = \frac{\ln\frac{[N_2O_5]_t}{[N_2O_5]_0}}{-k} = \frac{\ln\left(\frac{0.015}{0.480}\right)}{-(0.133/h)} = 26\ h \quad \text{(Note that t is five half-lives.)}$$

13.122 As the temperature of a gas is raised by 10 °C, even though the collision frequency increases by only ~2%, the reaction rate increases by 100% or more because there is an exponential increase in the fraction of the collisions that leads to products.

13.124 (a) Rate = $k[(CH_3)_3N]^m[ClO_2]^n$

$$m = \frac{\ln\left(\frac{Rate_3}{Rate_2}\right)}{\ln\left(\frac{[(CH_3)_3N]_3}{[(CH_3)_3N]_2}\right)} = \frac{\ln\left(\frac{1.79}{0.90}\right)}{\ln\left(\frac{1.3 \times 10^{-2}}{6.5 \times 10^{-3}}\right)} = 1$$

$$n = \frac{\ln\left(\dfrac{\text{Rate}_1 \cdot [(CH_3)_3N]_3}{\text{Rate}_3 \cdot [(CH_3)_3N]_1}\right)}{\ln\left(\dfrac{[ClO_2]_1}{[ClO_2]_3}\right)} = \frac{\ln\left(\dfrac{(0.90)(1.30 \text{ x } 10^{-2})}{(1.79)(3.25 \text{ x } 10^{-3})}\right)}{\ln\left(\dfrac{4.60 \text{ x } 10^{-3}}{2.30 \text{ x } 10^{-3}}\right)} = 1$$

Rate = $k[(CH_3)_3N][ClO_2]$

From Experiment 1:

$$k = \frac{\text{Rate}}{[(CH_3)_3N][ClO_2]} = \frac{0.90 \text{ M/s}}{(3.25 \text{ x } 10^{-3} \text{ M})(4.60 \text{ x } 10^{-3} \text{ M})} = 6.0 \text{ x } 10^4/(\text{M} \cdot \text{s})$$

(b) Rate = $k[(CH_3)_3N][ClO_2]$

Rate = $[6.0 \text{ x } 10^4/(\text{M} \cdot \text{s})](4.2 \text{ x } 10^{-2} \text{ M})(3.4 \text{ x } 10^{-2} \text{ M}) = 86 \text{ M/s}$

13.126 (a) From the data in the table for Experiment 1, we see that 0.20 mol of A reacts with 0.10 mol of B to produce 0.10 mol of D. The balanced equation for the reaction is:
2 A + B → D

(b) From the data in the table, initial Rates = $-\dfrac{\Delta A}{\Delta t}$ have been calculated.

For example, from Experiment 1:

$$\text{Initial rate} = -\frac{\Delta A}{\Delta t} = -\frac{(4.80 \text{ M} - 5.00 \text{ M})}{60 \text{ s}} = 3.33 \text{ x } 10^{-3} \text{ M/s}$$

Initial concentrations and initial rate data have been collected in the table below.

EXPT	$[A]_o$ (M)	$[B]_o$ (M)	$[C]_o$ (M)	Initial Rate (M/s)
1	5.00	2.00	1.00	$3.33 \text{ x } 10^{-3}$
2	10.00	2.00	1.00	$6.66 \text{ x } 10^{-3}$
3	5.00	4.00	1.00	$3.33 \text{ x } 10^{-3}$
4	5.00	2.00	2.00	$6.66 \text{ x } 10^{-3}$

Rate = $k[A]^m[B]^n[C]^p$
From Expts 1 and 2, [A] doubles and the initial rate doubles; therefore m = 1.
From Expts 1 and 3, [B] doubles but the initial rate does not change; therefore n = 0.
From Expts 1 and 4, [C] doubles and the initial rate doubles; therefore p = 1.
The reaction is: first-order in A; zeroth-order in B; first-order in C; second-order overall.
(c) Rate = k[A][C]
(d) C is a catalyst. C appears in the rate law, but it is not consumed in the reaction.
(e) A + C → AC (slow)
AC + B → AB + C (fast)
A + AB → D (fast)

(f) From data in Expt 1:

$$k = \frac{\text{Rate}}{[A][C]} = \frac{\Delta D/\Delta t}{[A][C]} = \frac{0.10\text{ M}/60\text{ s}}{(5.00\text{ M})(1.00\text{ M})} = 3.4 \times 10^{-4}/(\text{M}\cdot\text{s})$$

13.128 $\ln\left(\frac{k_2}{k_1}\right) = \left(\frac{-E_a}{R}\right)\left(\frac{1}{T_2} - \frac{1}{T_1}\right)$

$k_2 = 2.5k_1$
$k_1 = 1.0,\ \ T_1 = 20\ ^\circ\text{C} = 293\text{ K}$
$k_2 = 2.5,\ \ T_2 = 30\ ^\circ\text{C} = 303\text{ K}$

$$E_a = -\frac{[\ln k_2 - \ln k_1](R)}{\left(\frac{1}{T_2} - \frac{1}{T_1}\right)}$$

$$E_a = -\frac{[\ln(2.5) - \ln(1.0)][8.314 \times 10^{-3}\text{ kJ/(K}\cdot\text{mol)}]}{\left(\frac{1}{303\text{ K}} - \frac{1}{293\text{ K}}\right)} = 68\text{ kJ/mol}$$

$k_1 = 1.0,\ \ T_1 = 120\ ^\circ\text{C} = 393\text{ K}$
$k_2 = ?,\ \ T_2 = 130\ ^\circ\text{C} = 403\text{ K}$
Solve for k_2.

$$\ln k_2 = \frac{-E_a}{R}\left(\frac{1}{T_2} - \frac{1}{T_1}\right) + \ln k_1$$

$$\ln k_2 = \frac{-68\text{ kJ/mol}}{[8.314 \times 10^{-3}\text{ kJ/(K}\cdot\text{mol)}]}\left(\frac{1}{403\text{ K}} - \frac{1}{393\text{ K}}\right) + \ln(1.0) = 0.516$$

$k_2 = e^{0.516} = 1.7$; The rate increases by a factor of 1.7.

13.130 (a) $I^-(aq) + OCl^-(aq) \rightarrow Cl^-(aq) + OI^-(aq)$

(b) From the data in the table, initial rates $= -\frac{\Delta[I^-]}{\Delta t}$ have been calculated.

For example, from Experiment 1:

$$\text{Initial rate} = -\frac{\Delta[I^-]}{\Delta t} = -\frac{(2.17 \times 10^{-4}\text{ M} - 2.40 \times 10^{-4}\text{ M})}{10\text{ s}} = 2.30 \times 10^{-6}\text{ M/s}$$

Initial concentrations and initial rate data have been collected in the table below.

EXPT	$[I^-]_o$ (M)	$[OCl^-]_o$ (M)	$[OH^-]_o$ (M)	Initial Rate (M/s)
1	2.40×10^{-4}	1.60×10^{-4}	1.00	2.30×10^{-6}
2	1.20×10^{-4}	1.60×10^{-4}	1.00	1.20×10^{-6}
3	2.40×10^{-4}	4.00×10^{-5}	1.00	6.00×10^{-7}
4	1.20×10^{-4}	1.60×10^{-4}	2.00	6.00×10^{-7}

Rate = $k[I^-]^m[OCl^-]^n[OH^-]^p$

From Expts 1 and 2, $[I^-]$ is cut in half and the initial rate is cut in half; therefore m = 1.
From Expts 1 and 3, $[OCl^-]$ is reduced by a factor of four and the initial rate is reduced by a factor of four; therefore n = 1.
From Expts 2 and 4, $[OH^-]$ is doubled and the initial rate is cut in half; therefore p = –1.

$$\text{Rate} = k\frac{[I^-][OCl^-]}{[OH^-]}$$

From data in Expt 1:

$$k = \frac{\text{Rate}\,[OH^-]}{[I^-][OCl^-]} = \frac{(2.30 \times 10^{-6}\ \text{M/s})(1.00\ \text{M})}{(2.40 \times 10^{-4}\ \text{M})(1.60 \times 10^{-4}\ \text{M})} = 60/\text{s}$$

(c) The reaction does not occur by a single-step mechanism because OH^- appears in the rate law but not in the overall reaction.

(d)

	$OCl^-(aq) + H_2O(l) \rightleftarrows HOCl(aq) + OH^-(aq)$	(fast)
	$HOCl(aq) + I^-(aq) \rightarrow HOI(aq) + Cl^-(aq)$	(slow)
	$HOI(aq) + OH^-(aq) \rightarrow H_2O(l) + OI^-(aq)$	(fast)
Overall reaction	$I^-(aq) + OCl^-(aq) \rightarrow Cl^-(aq) + OI^-(aq)$	

Because the forward and reverse rates in step 1 are equal, $k_1[OCl^-][H_2O] = k_{-1}[HOCl][OH^-]$. Solving for [HOCl] and substituting into the rate law for the second step gives

$$\text{Rate} = k_2[HOCl][I^-] = \frac{k_1 k_2}{k_{-1}}\frac{[OCl^-][H_2O][I^-]}{[OH^-]}$$

$[H_2O]$ is constant and can be combined into k.
Because the rate law for the overall reaction is equal to the rate law for the rate-determining step, the rate law for the overall reaction is

$$\text{Rate} = k\frac{[OCl^-][I^-]}{[OH^-]} \quad \text{where } k = \frac{k_1 k_2 [H_2O]}{k_{-1}}$$

13.132 X → products is a first-order reaction

$$t = 60\ \text{min} \times \frac{60\ \text{s}}{1\ \text{min}} = 3600\ \text{s}$$

$$\ln\frac{[X]_t}{[X]_o} = -kt; \qquad k = \frac{\ln\frac{[X]_t}{[X]_o}}{-t}$$

At 25 °C, calculate k_1:

$$k_1 = \frac{\ln\left(\frac{0.600\ \text{M}}{1.000\ \text{M}}\right)}{-3600\ \text{s}} = 1.42 \times 10^{-4}\ \text{s}^{-1}$$

At 35 °C, calculate k_2: $$k_2 = \frac{\ln\left(\frac{0.200\ \text{M}}{0.600\ \text{M}}\right)}{-3600\ \text{s}} = 3.05 \times 10^{-4}\ \text{s}^{-1}$$

At an unknown temperature calculate k_3: $$k_3 = \frac{\ln\left(\frac{0.010\ \text{M}}{0.200\ \text{M}}\right)}{-3600\ \text{s}} = 8.32 \times 10^{-4}\ \text{s}^{-1}$$

$T_1 = 25\ °\text{C} = 25 + 273 = 298\ \text{K}$
$T_2 = 35\ °\text{C} = 35 + 273 = 308\ \text{K}$

Calculate E_a using k_1 and k_2.

$$\ln\left(\frac{k_2}{k_1}\right) = \left(\frac{-E_a}{R}\right)\left(\frac{1}{T_2} - \frac{1}{T_1}\right)$$

$$E_a = -\frac{[\ln k_2 - \ln k_1](R)}{\left(\frac{1}{T_2} - \frac{1}{T_1}\right)}$$

$$E_a = -\frac{[\ln(3.05 \times 10^{-4}) - \ln(1.42 \times 10^{-4})][8.314 \times 10^{-3}\ \text{kJ/(K}\cdot\text{mol)}]}{\left(\frac{1}{308\ \text{K}} - \frac{1}{298\ \text{K}}\right)} = 58.3\ \text{kJ/mol}$$

Use E_a, k_1, and k_3 to calculate T_3.

$$\frac{1}{T_3} = \frac{\ln\left(\frac{k_3}{k_1}\right)}{\left(\frac{-E_a}{R}\right)} + \frac{1}{T_1} = \frac{\ln\left(\frac{8.32 \times 10^{-4}}{1.42 \times 10^{-4}}\right)}{\left(\frac{-58.3\ \text{kJ/mol}}{8.314 \times 10^{-3}\ \text{kJ/(K}\cdot\text{mol)}}\right)} + \frac{1}{298\ \text{K}} = 0.003104/\text{K}$$

$$T_3 = \frac{1}{0.003104/\text{K}} = 322\ \text{K} = 322 - 273 = 49\ °\text{C}$$

At 3:00 p.m. raise the temperature to 49 °C to finish the reaction by 4:00 p.m.

13.134 (a) When equal volumes of two solutions are mixed, both concentrations are cut in half.
$[H_3O^+]_o = [OH^-]_o = 1.0\ \text{M}$
When 99.999% of the acid is neutralized, $[H_3O^+] = [OH^-] = 1.0\ \text{M} - (1.0\ \text{M} \times 0.99999)$
$= 1.0 \times 10^{-5}\ \text{M}$

Using the 2nd order integrated rate law:

$$kt = \frac{1}{[H_3O^+]_t} - \frac{1}{[H_3O^+]_o}; \qquad t = \frac{1}{k}\left[\frac{1}{[H_3O^+]_t} - \frac{1}{[H_3O^+]_o}\right]$$

$$t = \frac{1}{(1.3 \times 10^{11}\ \text{M}^{-1}\text{s}^{-1})}\left[\frac{1}{(1.0 \times 10^{-5}\ \text{M})} - \frac{1}{(1.0\ \text{M})}\right] = 7.7 \times 10^{-7}\ \text{s}$$

(b) The rate of an acid-base neutralization reaction would be limited by the speed of mixing, which is much slower than the intrinsic rate of the reaction itself.

13.136 Looking at the two experiments at 600 K, when the NO_2 concentration is doubled, the rate increased by a factor of 4. Therefore, the reaction is 2nd order.

Rate = $k\,[NO_2]^2$

Calculate k_1 at 600 K: $k_1 = \text{Rate}/[NO_2]^2 = 5.4 \times 10^{-7}\ M\ s^{-1}/(0.0010\ M)^2 = 0.54\ M^{-1}\ s^{-1}$

Calculate k_2 at 700 K: $k_2 = \text{Rate}/[NO_2]^2 = 5.2 \times 10^{-5}\ M\ s^{-1}/(0.0020\ M)^2 = 13\ M^{-1}\ s^{-1}$

Calculate E_a using k_1 and k_2.

$$\ln\left(\frac{k_2}{k_1}\right) = \left(\frac{-E_a}{R}\right)\left(\frac{1}{T_2} - \frac{1}{T_1}\right)$$

$$E_a = -\frac{[\ln k_2 - \ln k_1](R)}{\left(\frac{1}{T_2} - \frac{1}{T_1}\right)}$$

$$E_a = -\frac{[\ln(13) - \ln(0.54)][8.314 \times 10^{-3}\ \text{kJ/(K}\cdot\text{mol)}]}{\left(\frac{1}{700\ \text{K}} - \frac{1}{600\ \text{K}}\right)} = 111\ \text{kJ/mol}$$

Calculate k_3 at 650 K using E_a and k_1.

Solve for k_3.

$$\ln k_3 = \frac{-E_a}{R}\left(\frac{1}{T_3} - \frac{1}{T_1}\right) + \ln k_1$$

$$\ln k_3 = \frac{-111\ \text{kJ/mol}}{[8.314 \times 10^{-3}\ \text{kJ/(K}\cdot\text{mol)}]}\left(\frac{1}{650\ \text{K}} - \frac{1}{600\ \text{K}}\right) + \ln(0.54) = 1.0955$$

$$k_3 = e^{1.0955} = 3.0\ M^{-1}\ s^{-1}$$

$$k_3 t = \frac{1}{[NO_2]_t} - \frac{1}{[NO_2]_o}; \qquad t = \frac{1}{k_3}\left[\frac{1}{[NO_2]_t} - \frac{1}{[NO_2]_o}\right]$$

$$t = \frac{1}{(3.0\ M^{-1} s^{-1})}\left[\frac{1}{(0.0010\ M)} - \frac{1}{(0.0050\ M)}\right] = 2.7 \times 10^2\ s$$

13.138 $A \rightarrow C$ is a first-order reaction.

The reaction is complete at 200 s when the absorbance of C reaches 1.200.

Because there is a one to one stoichiometry between A and C, the concentration of A must be proportional to 1.200 – absorbance of C. Any two data points can be used to find k. Let $[A]_o \propto 1.200$ and at 100 s, $[A]_t \propto 1.200 - 1.188 = 0.012$

$$\ln\frac{[A]_t}{[A]_o} = -kt; \qquad k = \frac{\ln\frac{[A]_t}{[A]_o}}{-t}; \qquad k = \frac{\ln\left(\frac{0.012\ M}{1.200\ M}\right)}{-100\ s} = 0.0461\ s^{-1}$$

$$t_{1/2} = \frac{0.693}{k} = \frac{0.693}{0.0461\ s^{-1}} = 15\ s$$

13.140 $\ln\left(\frac{k_2}{k_1}\right) = \left(\frac{-E_a}{R}\right)\left(\frac{1}{T_2} - \frac{1}{T_1}\right)$

$k_1 = 0.267$ L/(mg · min), $T_1 = 5\ °C = 278$ K
$k_2 = 3.45$ L/(mg · min), $T_2 = 30\ °C = 303$ K

$$E_a = -\frac{[\ln k_2 - \ln k_1](R)}{\left(\frac{1}{T_2} - \frac{1}{T_1}\right)}$$

$$E_a = -\frac{[\ln(3.45) - \ln(0.267)][8.314 \times 10^{-3}\ kJ/(K \cdot mol)]}{\left(\frac{1}{303\ K} - \frac{1}{278\ K}\right)} = 71.7\ kJ/mol$$

13.142

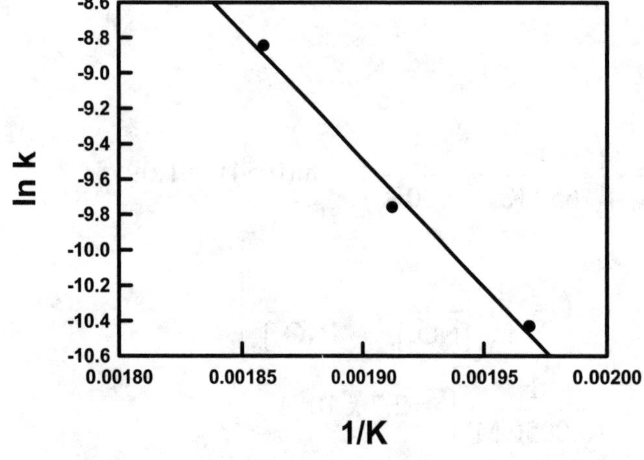

$$\text{Slope} = -\frac{E_a}{R} = -14418.2\ K$$

$E_a = [8.314 \times 10^{-3}\ kJ/(K \cdot mol)](14418.2\ K) = 120\ kJ/mol$

Multiconcept Problems

13.144 (a) $k = A e^{-\frac{E_a}{RT}} = (6.0 \times 10^8/(M \cdot s))\ e^{-\frac{6.3\ kJ/mol}{[8.314 \times 10^{-3}\ kJ/(K \cdot mol)](298\ K)}} = 4.7 \times 10^7/(M \cdot s)$

(b) :Ö=N̈—F̤̈: N has 3 electron clouds, is sp^2 hybridized, and the molecule is bent.

(c) O—N---F---F

(d) The reaction has such a low activation energy because the F–F bond is very weak and the N–F bond is relatively strong.

13.146 $2\ NO_2(g) \rightarrow 2\ NO(g) + O_2(g)$

$k = 4.7/(M \cdot s)$

(a) The units for k indicate a second-order reaction.

(b) 383 °C = 656 K

PV = nRT

$$[NO_2]_o = \frac{n}{V} = \frac{P}{RT} = \frac{\left(746 \text{ mm Hg x } \frac{1.000 \text{ atm}}{760 \text{ mm Hg}}\right)}{\left(0.082\ 06 \frac{L \cdot atm}{K \cdot mol}\right)(656\ K)} = 0.01823 \text{ mol/L}$$

$$\text{initial rate} = k[NO_2]_o^2 = [4.7/(M \cdot s)](0.01823 \text{ mol/L})^2 = 1.56 \times 10^{-3} \text{ mol/(L} \cdot \text{s)}$$

$$\text{initial rate for } O_2 = \frac{\text{initial rate for } NO_2}{2} = \frac{1.56 \times 10^{-3} \text{ mol/(L} \cdot \text{s)}}{2} = 7.80 \times 10^{-4} \text{ mol/(L} \cdot \text{s)}$$

$$\text{initial rate for } O_2 = [7.80 \times 10^{-4} \text{ mol/(L} \cdot \text{s)}](32.00 \text{ g/mol}) = 0.025 \text{ g/(L} \cdot \text{s)}$$

(c) $$\frac{1}{[NO_2]_t} = kt + \frac{1}{[NO_2]_0} = [4.7/(M \cdot s)](60\ s) + \frac{1}{0.01823\ M}$$

$$\frac{1}{[NO_2]_t} = 336.9/M \text{ and } [NO_2] = 0.00297\ M$$

	$2\ NO_2(g)$	$\rightarrow$	$2\ NO(g)$	+	$O_2(g)$
before reaction (M)	0.01823		0		0
change (M)	–2x		+2x		+x
after 1.00 min (M)	0.01823 – 2x		2x		x

after 1.00 min $[NO_2] = 0.00297\ M = 0.01823 - 2x$

$x = 0.00763\ M = [O_2]$

mass $O_2 = (0.00763 \text{ mol/L})(5.00\ L)(32.00 \text{ g/mol}) = 1.22 \text{ g } O_2$

13.148 $2\ N_2O(g) \rightarrow 2\ N_2(g) + O_2(g)$

P_{O_2}(in exit gas) = 1.0 mm Hg; P_{total} = 1.50 atm = 1140 mm Hg

From the reaction stoichiometry:

P_{N_2}(in exit gas) = 2 P_{O_2} = 2.0 mm Hg

P_{N_2O}(in exit gas) = $P_{total} - P_{N_2} - P_{O_2}$ = 1140 – 2.0 – 1.0 = 1137 mm Hg

Assume P_{N_2O}(initial) = P_{total} = 1140 mm Hg (In assuming a constant total pressure in the tube, we are neglecting the slight change in pressure due to the reaction.)

Volume of tube = $\pi r^2 l = \pi(1.25 \text{ cm})^2(20 \text{ cm}) = 98.2 \text{ cm}^3 = 0.0982$ L

$$\text{Time, t, gases are in the tube} = \frac{\text{volume of tube}}{\text{flow rate}} \text{ x } \frac{0.0982\ L}{0.75\ L/min} \text{ x } \frac{60\ s}{1\ min} = 7.86\ s$$

$$\text{At time t, } \frac{[N_2O]_t}{[N_2O]_0} = \frac{P_{N_2O}\text{ (in exit gas)}}{P_{N_2O}\text{(initial)}} = \frac{1137 \text{ mm Hg}}{1140 \text{ mm Hg}} = 0.997\ 37$$

Because $k = Ae^{-\frac{E_a}{RT}}$ and $A = 4.2 \times 10^9\ s^{-1}$, k has units of s^{-1}. Therefore, this is a first-order reaction and the appropriate integrated rate law is $\ln\frac{[N_2O]_t}{[N_2O]_0} = -kt$.

$$k = \frac{-\ln\frac{[N_2O]_t}{[N_2O]_0}}{t} = \frac{-\ln(0.99737)}{7.86\ s} = 3.35 \times 10^{-4}\ s^{-1}$$

From the Arrhenius equation, $\ln k = \ln A - \frac{E_a}{RT}$

$$T = \frac{E_a}{(R)[\ln A - \ln k]} = \frac{222\ kJ/mol}{(8.314 \times 10^{-3}\ kJ/(K \cdot mol))[(22.16) - (-8.00)]} = 885\ K$$

13.150 (a)

	$CH_3CHO(g)$	$\rightarrow$	$CH_4(g)$	+	$CO(g)$
before (atm)	0.500		0		0
change (atm)	–x		+x		+x
after (atm)	0.500 – x		x		x

At 605 s, $P_{total} = P_{CH_3CHO} + P_{CH_4} + P_{CO} = (0.500\ atm - x) + x + x = 0.808\ atm$

$x = 0.808\ atm - 0.500\ atm = 0.308\ atm$

The integrated rate law for a second-order reaction in terms of molar concentrations is $\frac{1}{[A]_t} = kt + \frac{1}{[A]_o}$. The ideal gas law, PV = nRT, can be rearranged to show how P is proportional to the molar concentration of a gas.

$P = \frac{n}{V}RT$ (R and T are constant), so $P \propto \frac{n}{V}$ = molar concentration

Because of this relationship, the second-order integrated rate law can be rewritten in terms of partial pressures.

$$\frac{1}{P_t} = kt + \frac{1}{P_o}; \quad \frac{1}{P_t} - \frac{1}{P_o} = kt; \quad \frac{\left(\frac{1}{P_t} - \frac{1}{P_o}\right)}{t} = k$$

P is the partial pressure of CH_3CHO.

At t = 0, $P_o = 0.500$ and at t = 605 s, $P_t = 0.500\ atm - 0.308\ atm = 0.192\ atm$

$$k = \frac{\left(\frac{1}{0.192\ atm} - \frac{1}{0.500\ atm}\right)}{605\ s} = 5.30 \times 10^{-3}\ atm^{-1}\ s^{-1}$$

(b) Use the ideal gas law to convert atm^{-1} to M^{-1}.

$$P = \frac{n}{V}RT; \quad \frac{P}{RT} = \frac{n}{V}; \quad \frac{1}{P}RT = \frac{V}{n} = M^{-1}$$

So, multiply k by RT to convert $atm^{-1}\ s^{-1}$ to $M^{-1}\ s^{-1}$.

$k = (5.30 \times 10^{-3}\ \text{atm}^{-1}\ \text{s}^{-1})RT$

$k = (5.30 \times 10^{-3}\ \text{atm}^{-1}\ \text{s}^{-1})\left(0.082\ 06\ \frac{\text{L} \cdot \text{atm}}{\text{K} \cdot \text{mol}}\right)(791\ \text{K}) = 0.344\ \frac{\text{L}}{\text{mol} \cdot \text{s}} = 0.344\ \text{M}^{-1}\ \text{s}^{-1}$

(c) $CH_3CHO(g) \rightarrow CH_4(g) + CO(g)$

$\Delta H^\circ_{rxn} = [\Delta H^\circ_f(CH_4) + \Delta H^\circ_f(CO)] - \Delta H^\circ_f(CH_3CHO)]$

$\Delta H^\circ_{rxn} = [(1\ \text{mol})(-74.8\ \text{kJ/mol}) + (1\ \text{mol})(-110.5\ \text{kJ/mol})] - (1\ \text{mol})(-166.2\ \text{kJ/mol})$

$\Delta H^\circ_{rxn} = -19.1$ kJ per mole of CH_3CHO that decomposes

$PV = nRT$

$\text{mol } CH_3CHO \text{ reacted} = \frac{PV}{RT} = \frac{(0.308\ \text{atm})(1.00\ \text{L})}{\left(0.082\ 06\ \frac{\text{L} \cdot \text{atm}}{\text{K} \cdot \text{mol}}\right)(791\ \text{K})} = 0.004\ 74\ \text{mol}$

$q = (0.004\ 74\ \text{mol})(19.1\ \text{kJ/mol})(1000\ \text{J/kJ}) = 90.6$ J liberated after a reaction time of 605 s.

14 Chemical Equilibrium

14.1 (a) $K_c = \dfrac{[SO_3]^2}{[SO_2]^2[O_2]}$ (b) $K_c = \dfrac{[H^+][CH_3COO^-]}{[CH_3COOH]}$

(c) $K_c' = \dfrac{[SO_2]^4[O_2]^2}{[SO_3]^4}$; $K_c' = \left(\dfrac{1}{K_c}\right)^2$

14.2 (a) $K_c(\text{overall}) = \dfrac{[NO_2]^2}{[N_2][O_2]^2}$

(b) $K_c(\text{overall}) = K_{c1} \times K_{c2} = (4.3 \times 10^{-25})(6.4 \times 10^9) = 2.8 \times 10^{-15}$

14.3 (a) $K_c = \dfrac{[SO_3]^2}{[SO_2]^2[O_2]} = \dfrac{(5.0 \times 10^{-2})^2}{(3.0 \times 10^{-3})^2(3.5 \times 10^{-3})} = 7.9 \times 10^4$

(b) $K_c' = \dfrac{1}{K_c} = \dfrac{[SO_2]^2[O_2]}{[SO_3]^2} = \dfrac{(3.0 \times 10^{-3})^2(3.5 \times 10^{-3})}{(5.0 \times 10^{-2})^2} = 1.3 \times 10^{-5}$

14.4 (a) $K_c = \dfrac{[H^+][C_3H_5O_3^-]}{[C_3H_6O_3]} = \dfrac{(3.65 \times 10^{-3})(3.65 \times 10^{-3})}{(9.64 \times 10^{-2})} = 1.38 \times 10^{-4}$

(b) $[C3H6O3] = \dfrac{[H^+][C_3H_5O_3^-]}{K_c} = \dfrac{(1.17 \times 10^{-2})(1.17 \times 10^{-2})}{(1.38 \times 10^{-4})} = 0.992\ M$

14.5 From (1), $K_c = \dfrac{[AB][B]}{[A][B_2]} = \dfrac{(1)(2)}{(1)(2)} = 1$

For a mixture to be at equilibrium, $\dfrac{[AB][B]}{[A][B_2]}$ must be equal to 1.

For (2), $\dfrac{[AB][B]}{[A][B_2]} = \dfrac{(2)(1)}{(2)(1)} = 1$. This mixture is at equilibrium.

For (3), $\dfrac{[AB][B]}{[A][B_2]} = \dfrac{(1)(1)}{(4)(2)} = 0.125$. This mixture is not at equilibrium.

For (4), $\dfrac{[AB][B]}{[A][B_2]} = \dfrac{(2)(2)}{(4)(1)} = 1$. This mixture is at equilibrium.

14.6 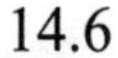

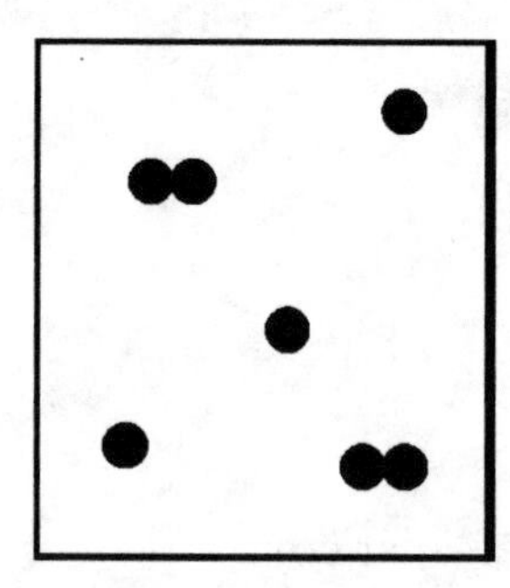

14.7 $K_p = \dfrac{(P_{CO_2})(P_{H_2})}{(P_{CO})(P_{H_2O})} = \dfrac{(6.12)(20.3)}{(1.31)(10.0)} = 9.48$

14.8 $K_p = 25 = \dfrac{(P_{HI})^2}{(P_{H_2})(P_{I_2})}$

$P_{HI} = \sqrt{(25)(P_{H_2})(P_{I_2})} = \sqrt{(25)(0.286)(0.286)} = 1.43$ atm

14.9 $2\ NO(g) + O_2(g) \rightleftarrows 2\ NO_2(g)$; $\Delta n = 2 - 3 = -1$

$K_p = K_c(RT)^{\Delta n}$, $K_c = K_p(1/RT)^{\Delta n}$

at 500 K: $K_p = (6.9 \times 10^5)[(0.082\ 06)(500)]^{-1} = 1.7 \times 10^4$

at 1000 K: $K_c = (1.3 \times 10^{-2})\left(\dfrac{1}{(0.082\ 06)(1000)}\right)^{-1} = 1.1$

14.10 $K_c' = \dfrac{1}{K_c} = \dfrac{1}{245}$

$2\ SO_3(g) \rightleftarrows 2\ SO_2(g) + O_2(g)$; $\Delta n = 3 - 2 = 1$

$K_p = K_c'(RT)^{\Delta n}$

$K_p = (1/245)[(0.082\ 06)(1{,}000)] = 0.335$

14.11 (a) $K_c = \dfrac{[H_2]^3}{[H_2O]^3}$, $K_p = \dfrac{(P_{H_2})^3}{(P_{H_2O})^3}$, $\Delta n = (3) - (3) = 0$ and $K_p = K_c$

(b) $K_c = [H_2]^2[O_2]$, $K_p = (P_{H_2})^2(P_{O_2})$, $\Delta n = (3) - (0) = 3$ and $K_p = K_c(RT)^3$

(c) $K_c = \dfrac{[HCl]^4}{[SiCl_4][H_2]^2}$, $K_p = \dfrac{(P_{HCl})^4}{(P_{SiCl_4})(P_{H_2})^2}$, $\Delta n = (4) - (3) = 1$ and $K_p = K_c(RT)$

(d) $K_c = \dfrac{1}{[Hg_2^{2+}][Cl^-]^2}$

14.12 $Mg(OH)_2(s) \rightleftarrows Mg^{2+}(aq) + 2\ OH^-(aq)$

$Kc = [Mg^{2+}][OH^-]^2 = (1.65 \times 10^{-4})(3.30 \times 10^{-4})^2 = 1.80 \times 10^{-11}$

14.13 $K_c = 1.2 \times 10^{-42}$. Because K_c is very small, the equilibrium mixture contains mostly H_2 molecules. H is in periodic group 1A. A very small value of K_c is consistent with strong bonding between 2 H atoms, each with one valence electron.

14.14 $K_c = 1.2 \times 10^{82}$ is very large. When equilibrium is reached, very little if any ethanol will remain because the reaction goes to completion.

14.15 The container volume of 5.0 L must be included to calculate molar concentrations.

(a) $$Q_c = \frac{[NO_2]_t^2}{[NO]_t^2[O_2]_t} = \frac{(0.80 \text{ mol}/5.0 \text{ L})^2}{(0.060 \text{ mol}/5.0 \text{ L})^2(1.0 \text{ mol}/5.0 \text{ L})} = 890$$

Because $Q_c < K_c$, the reaction is not at equilibrium. The reaction will proceed to the right to reach equilibrium.

(b) $$Q_c = \frac{[NO_2]_t^2}{[NO]_t^2[O_2]_t} = \frac{(4.0 \text{ mol}/5.0 \text{ L})^2}{(5.0 \times 10^{-3} \text{ mol}/5.0 \text{ L})^2(0.20 \text{ mol}/5.0 \text{ L})} = 1.6 \times 10^7$$

Because $Q_c > K_c$, the reaction is not at equilibrium. The reaction will proceed to the left to reach equilibrium.

14.16 $K_c = \dfrac{[AB]^2}{[A_2][B_2]} = 4$; For a mixture to be at equilibrium, $\dfrac{[AB]^2}{[A_2][B_2]}$ must be equal to 4.

For (1), $Q_c = \dfrac{[AB]^2}{[A_2][B_2]} = \dfrac{(6)^2}{(1)(1)} = 36$, $Q_c > K_c$

For (2), $Q_c = \dfrac{[AB]^2}{[A_2][B_2]} = \dfrac{(4)^2}{(2)(2)} = 4$, $Q_c = K_c$

For (3), $Q_c = \dfrac{[AB]^2}{[A_2][B_2]} = \dfrac{(2)^2}{(3)(3)} = 0.44$, $Q_c < K_c$

(a) (2) (b) (1), reverse; (3), forward

14.17

	$CO(g)$ +	$H_2O(g)$ ⇄	$CO_2(g)$ +	$H_2(g)$
initial (M)	0.150	0.150	0	0
change (M)	–x	–x	+x	+x
equil (M)	0.150 – x	0.150 – x	x	x

$$K_c = 4.24 = \frac{[CO_2][H_2]}{[CO][H_2O]} = \frac{x^2}{(0.150 - x)^2}$$

Take the square root of both sides and solve for x.

$$\sqrt{4.24} = \sqrt{\frac{x^2}{(0.150 - x)^2}}; \quad 2.06 = \frac{x}{0.150 - x}; \quad x = 0.101$$

At equilibrium, $[CO_2] = [H_2] = x = 0.101$ M

$[CO] = [H_2O] = 0.150 - x = 0.150 - 0.101 = 0.049$ M

14.18

	$CO(g)$ +	$H_2O(g)$ ⇄	$CO_2(g)$ +	$H_2(g)$
initial (M)	0.100	0.100	0.100	0.100
change (M)	–x	–x	+x	+x
equil (M)	0.100 – x	0.100 – x	0.100 + x	0.100 + x

$$K_c = 4.24 = \frac{[CO_2][H_2]}{[CO][H_2O]} = \frac{(0.100 + x)^2}{(0.100 - x)^2}$$

Take the square root of both sides and solve for x.

$$\sqrt{4.24} = \sqrt{\frac{(0.100 + x)^2}{(0.100 - x)^2}}; \quad 2.06 = \frac{0.100 + x}{0.100 - x}; \quad x = 0.035$$

At equilibrium, $[CO_2] = [H_2] = 0.100 + x = 0.100 + 0.035 = 0.135$ M

$[CO] = [H_2O] = 0.100 - x = 0.100 - 0.035 = 0.065$ M

14.19

	$H_2(g)$ +	$I_2(g)$ ⇄	$2\ HI(g)$
initial (M)	0.100	0.300	0
change (M)	–x	–x	+2x
equil (M)	0.100 – x	0.300 – x	2x

$$K_c = \frac{[HI]^2}{[H_2][I_2]} = \frac{(2x)^2}{(0.100 - x)(0.300 - x)} = 57.0$$

$53x^2 - 22.8x + 1.71 = 0$

Use the quadratic formula to solve for x.

$$x = \frac{-(-22.8) \pm \sqrt{(-22.8)^2 - 4(53)(1.71)}}{2(53)} = \frac{22.8 \pm 12.54}{106}$$

x = 0.333 and 0.0968

Discard 0.333 because it leads to negative concentrations and that is impossible.

$[H_2] = 0.100 - x = 0.100 - 0.0968 = 0.0032$ M

$[I_2] = 0.300 - x = 0.300 - 0.0968 = 0.2032$

$[HI] = 2x = (2)(0.0968) = 0.1936$ M

14.20

	$N_2O_4(g)$ ⇄	$2\ NO_2(g)$
initial (M)	0.500	0
change (M)	–x	+2x
equil (M)	0.500 – x	2x

$$K_c = \frac{[NO_2]^2}{[N_2O_4]} = \frac{(2x)^2}{(0.500 - x)} = 4.64 \times 10^{-3}$$

$4x^2 + (4.64 \times 10^{-3})x - (2.32 \times 10^{-3}) = 0$

Use the quadratic formula to solve for x.

$$x = \frac{-(4.64 \times 10^{-3}) \pm \sqrt{(4.64 \times 10^{-3})^2 - 4(4)(-2.32 \times 10^{-3})}}{2(4)} = \frac{-0.004\ 64 \pm 0.1927}{8}$$

x = –0.0247 and 0.0235

Discard the negative solution (–0.0247) because it leads to a negative concentration of NO_2 and that is impossible.

$[N_2O_4] = 0.500 - x = 0.500 - 0.0235 = 0.476$ M
$[NO_2] = 2x = 2(0.0235) = 0.0470$ M

14.21

	C(s) +	$H_2O(g)$ ⇄	CO(g) +	$H_2(g)$
initial (atm)		0	1.00	1.40
change (atm)		+x	−x	−x
equil (atm)		x	1.00 − x	1.40 − x

$$K_p = \frac{(P_{CO})(P_{H_2})}{(P_{H_2O})} = 2.44 = \frac{(1.00 - x)(1.40 - x)}{x}$$

$x^2 - 4.84x + 1.40 = 0$

Use the quadratic formula to solve for x.

$$x = \frac{-(-4.84) \pm \sqrt{(-4.84)^2 - 4(1)(1.40)}}{2(1)} = \frac{4.84 \pm 4.22}{2}$$

x = 4.53 and 0.310

Discard 4.53 because it leads to negative partial pressures and that is impossible.

$P_{H_2O} = x = 0.310$ atm

$P_{CO} = 1.00 - x = 1.00 - 0.310 = 0.69$ atm

$P_{H_2} = 1.40 - x = 1.40 - 0.310 = 1.09$ atm

14.22 $K_p = \frac{(P_{CO})(P_{H_2})}{(P_{H_2O})} = 2.44$, $Q_p = \frac{(1.00)(1.40)}{(1.20)} = 1.17$, $Q_p < K_p$ and the reaction goes to the right to reach equilibrium.

	C(s) +	$H_2O(g)$ ⇄	CO(g) +	$H_2(g)$
initial (atm)		1.20	1.00	1.40
change (atm)		−x	+x	+x
equil (atm)		1.20 − x	1.00 + x	1.40 + x

$$K_p = \frac{(P_{CO})(P_{H_2})}{(P_{H_2O})} = 2.44 = \frac{(1.00 + x)(1.40 + x)}{(1.20 - x)}$$

$x^2 + 4.84x - 1.53 = 0$

Use the quadratic formula to solve for x.

$$x = \frac{-(4.84) \pm \sqrt{(4.84)^2 - 4(1)(-1.53)}}{2(1)} = \frac{-4.84 \pm 5.44}{2}$$

x = −5.14 and 0.300

Discard the negative solution (−5.14) because it leads to negative partial pressures and that is impossible.

$P_{H_2O} = 1.20 - x = 1.20 - 0.300 = 0.90$ atm

$P_{CO} = 1.00 + x = 1.00 + 0.300 = 1.30$ atm

$P_{H_2} = 1.40 + x = 1.40 + 0.300 = 1.70$ atm

14.23 (a) CO (reactant) added, H_2 concentration increases.
(b) CO_2 (product) added, H_2 concentration decreases.
(c) H_2O (reactant) removed, H_2 concentration decreases.
(d) CO_2 (product) removed, H_2 concentration increases.
At equilibrium, $Q_c = K_c = \dfrac{[CO_2][H_2]}{[CO][H_2O]}$. If some CO_2 is removed from the equilibrium mixture, the numerator in Q_c is decreased, which means that $Q_c < K_c$ and the reaction will shift to the right, increasing the H_2 concentration.

14.24 (a) The concentration of both Ca^{2+} and $C_2O_4^{2-}$ will increase as fluid is lost and the reaction will proceed toward products.
(b) Increasing the concentration of Ca^{2+} will cause the reaction to proceed toward products.
(c) Decreasing the concentration of $C_2O_4^{2-}$ will cause the reaction to proceed toward reactants and the kidney stone will dissolve.
(d) The concentration of both Ca^{2+} and $C_2O_4^{2-}$ will decrease causing the reaction proceed toward reactants and the kidney stone will dissolve.

14.25 (a) Because there are 2 mol of gas on both sides of the balanced equation, the composition of the equilibrium mixture is unaffected by a change in pressure. The number of moles of reaction products remains the same.
(b) Because there are 2 mol of gas on the left side and 1 mol of gas on the right side of the balanced equation, the stress of an increase in pressure is relieved by a shift in the reaction to the side with fewer moles of gas (in this case, to products). The number of moles of reaction products increases.
(c) Because there is 1 mol of gas on the left side and 2 mol of gas on the right side of the balanced equation, the stress of an increase in pressure is relieved by a shift in the reaction to the side with fewer moles of gas (in this case, to reactants). The number of moles of reaction product decreases.

14.26 (a)

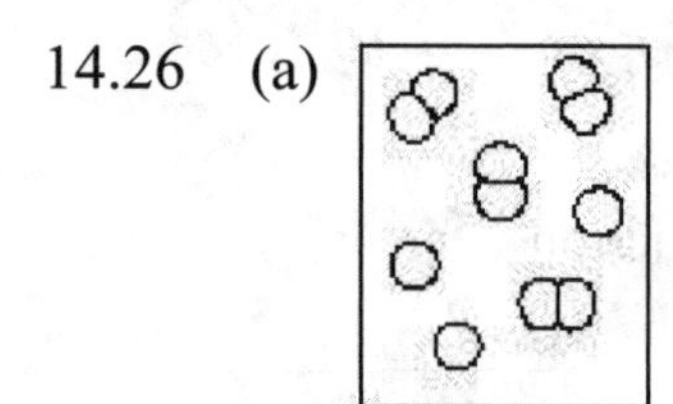

(b) If there is no change in temperature, K_c remains the same.

14.27 Le Châtelier's principle predicts that a stress of added heat will be relieved by net reaction in the direction that absorbs the heat. Since the reaction is endothermic, the equilibrium will shift from left to right (K_c will increase) with an increase in temperature. Therefore, the equilibrium mixture will contain more of the offending NO, the higher the temperature.

14.28 The reaction is exothermic. As the temperature is increased the reaction shifts from right to left. The amount of ethyl acetate decreases.

$$K_c = \frac{[CH_3CO_2C_2H_5][H_2O]}{[CH_3CO_2H][C_2H_5OH]}$$

As the temperature is decreased, the reaction shifts from left to right. The product concentrations increase, and the reactant concentrations decrease. This corresponds to an increase in K_c.

14.29 There are more AB(g) molecules at the higher temperature. The equilibrium shifted to the right at the higher temperature, which means the reaction is endothermic.

14.30 Heat + $BaCO_3(s) \rightleftarrows BaO(s) + CO_2(g)$

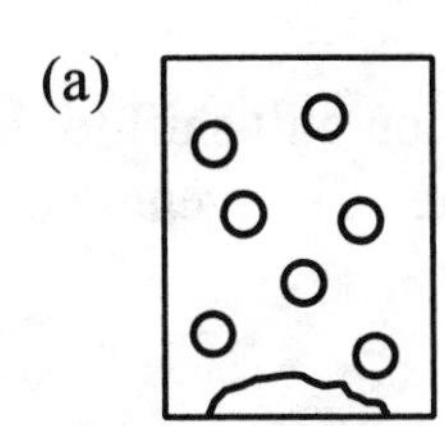

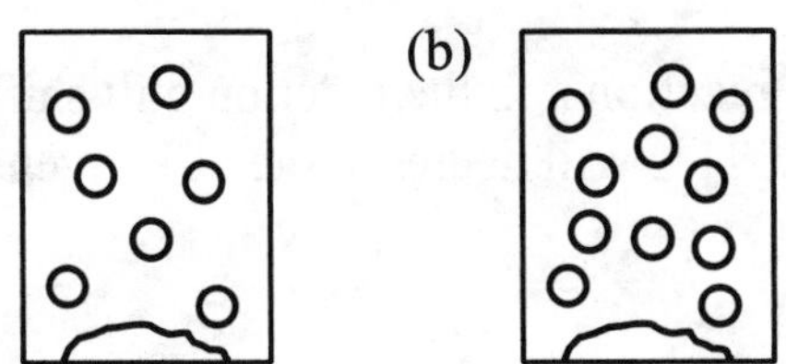

14.31 (a) Because K_c is so large, k_f is larger than k_r.

(b) $K_c = \frac{k_f}{k_r}$; $\quad k_r = \frac{k_f}{K_c} = \frac{8.5 \times 10^6\ M^{-1}\,s^{-1}}{3.4 \times 10^{34}} = 2.5 \times 10^{-28}\ M^{-1}\ s^{-1}$

(c) Because the reaction is exothermic, E_a (forward) is less than E_a (reverse). Consequently, as the temperature decreases, k_r decreases more than k_f decreases, and therefore $K_c = \frac{k_f}{k_r}$ increases.

14.32 (a) If A is similar for forward and reverse reactions, then k_r is larger because the activation energy for the reverse reaction is lower.

(b) Because $K_c = \frac{k_f}{k_r}$ and $k_r > k_f$, then $K_c < 1$.

(c) The rates of forward and reverse reactions both decrease. However, the rate of the forward reaction has a larger decrease than the rate of reverse reaction. K_c decreases when temperature is lowered in a endothermic process.

14.33 (a) 75% (b) 20%

(c) An efficient unloading of oxygen occurs in working muscle tissue when the partial pressure of oxygen drops.

14.34 An increase in temperature will shift the position of the equilibrium toward reactants, releasing oxygen from hemoglobin and making it available for use by the muscles.

14.35 (a) Equilibrium constants for each sequential step in binding oxygen become larger.

(b) Each step becomes more product favored, meaning that when oxygen binds it increases the affinity of the remaining heme groups for other oxygen molecules.

(c) The oxygen carrying capacity is greatly increased.

14.36 (a) $K_c = K_{c1} \times K_{c2} \times K_{c3} \times K_{c4}$

$K_c = (1.5 \times 10^4)(3.5 \times 10^4)(5.9 \times 10^4)(1.7 \times 10^6) = 5.3 \times 10^{19}$

(b) $Hb + O_2 \rightleftarrows Hb(O_2)_4$

$[O_2] = (1.61\ \mu M/mm\ Hg)(95\ mm\ Hg) = 153\ \mu M = 153 \times 10^{-6}\ M$

$$K_c = \frac{[Hb(O_2)_4]}{[Hb][O_2]^4} = 5.3 \times 10^{19}$$

$$\frac{[Hb(O_2)_4]}{[Hb]} = K_c[O_2]^4 = (5.3 \times 10^{19})(153 \times 10^{-6})^4 = 2.9 \times 10^4$$

14.37 $Hb + O_2 \rightleftarrows Hb(O_2)$

If CO binds to Hb, Hb is removed from the reaction and the reaction will shift to the left, resulting in O_2 being released from $Hb(O_2)$. This will decrease the effectiveness of Hb for carrying O_2.

14.38 $Hb(O_2)(aq) + CO(g) \rightleftarrows Hb(CO)(aq) + O_2(g)$ $K = 207$

(a) $P_{O_2} = 0.20$ atm and $P_{CO} = 0.0015$ atm

$$K = \frac{[Hb(CO)](P_{O_2})}{[Hb(O_2)](P_{CO})} = 207$$

$$\frac{[Hb(CO)]}{[Hb(O_2)]} = \frac{(P_{CO})(207)}{(P_{O_2})} = \frac{(0.0015)(207)}{(0.20)} = 1.6$$

(b) An increase in the concentration of oxygen in the blood will shift the equilibrium toward reactants increasing the concentration of oxygen bound to hemoglobin.

Conceptual Problems

14.40 (a) $A_2 + C_2 \rightleftarrows 2\ AC$ (most product molecules)

(b) $A_2 + B_2 \rightleftarrows 2\ AB$ (fewest product molecules)

14.42 (a) $A_2 + 2\ B \rightleftarrows 2\ AB$

(b) The number of AB molecules will increase, because as the volume is decreased at constant temperature, the pressure will increase and the reaction will shift to the side of fewer molecules to reduce the pressure.

14.44 As the temperature is raised, the reaction proceeds in the reverse direction. This is consistent with an exothermic reaction where "heat" can be considered as a product.

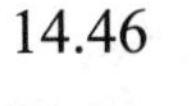

14.46 (a) 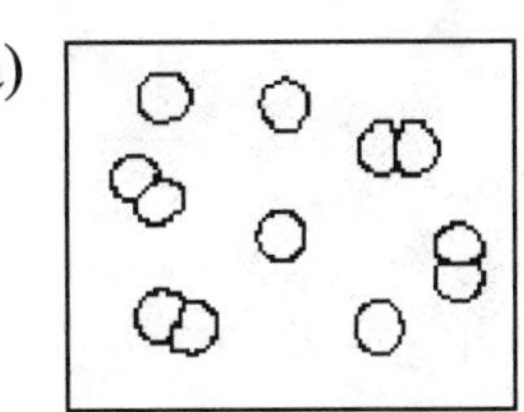 (b) 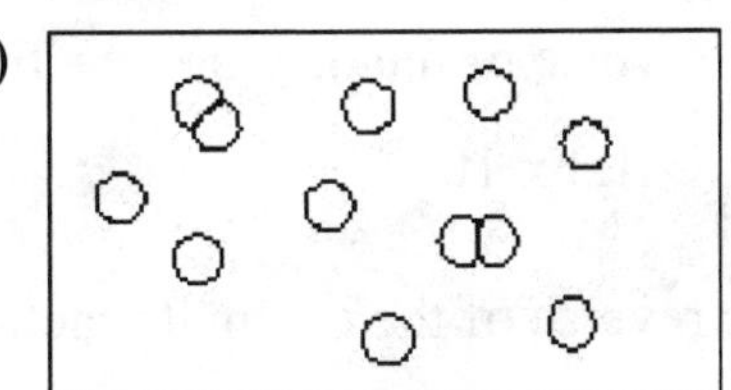 (c) 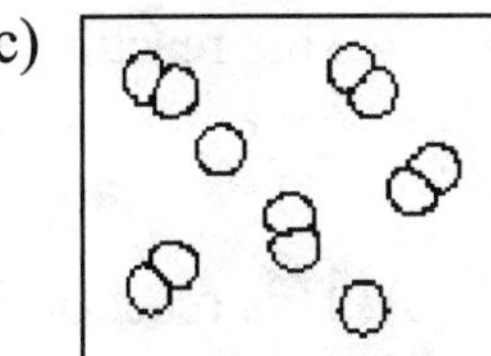

14.48 (a) A $\rightarrow$ 2 B

(b) (1) The reaction is exothermic. As the temperature is increased, the reaction shifts to the left. A increases, B decreases, and K_c decreases.

(2) When the volume decreases, the reaction shifts to the side with fewer gas molecules, which is towards the reactant. The amount of A increases.

(3) If there is no volume change, there is no change in the equilibrium composition, and the amount of A remains the same.

(4) A catalyst does not change the equilibrium composition, and the amount of A remains the same.

Section Problems

Equilibrium Constant Expressions and Equilibrium Constants (Sections 14.1–14.4)

14.50 (d) The rate of the forward reaction is equal to the rate of the reverse reaction.

14.52 $K_c = \dfrac{[C_2H_5OC_2H_5][H_2O]}{[C_2H_5OH]^2}$

14.54 (a) $K_c = \dfrac{[CO][H_2]^3}{[CH_4][H_2O]}$ (b) $K_c = \dfrac{[ClF_3]^2}{[F_2]^3[Cl_2]}$ (c) $K_c = \dfrac{[HF]^2}{[H_2][F_2]}$

14.56 (a) $K_p = \dfrac{(P_{CO})(P_{H_2})^3}{(P_{CH_4})(P_{H_2O})}$, $\Delta n = 2$ and $K_p = K_c(RT)^2$

(b) $K_p = \dfrac{(P_{ClF_3})^2}{(P_{F_2})^3(P_{Cl_2})}$, $\Delta n = -2$ and $K_p = K_c(RT)^{-2}$

(c) $K_p = \dfrac{(P_{HF})^2}{(P_{H_2})(P_{F_2})}$, $\Delta n = 0$ and $K_p = K_c$

14.58 (a) $K_c = \dfrac{[CO_2]^3}{[CO]^3}$, $K_p = \dfrac{(P_{CO_2})^3}{(P_{CO})^3}$ (b) $K_c = \dfrac{1}{[O_2]^3}$, $K_p = \dfrac{1}{(P_{O_2})^3}$

(c) $K_c = [SO_3]$, $K_p = P_{SO_3}$ (d) $K_c = [Ba^{2+}][SO_4^{2-}]$

14.60 (a) This reaction is the reverse of the original reaction.

$$K_c' = \frac{1}{K_c} = \frac{1}{7.5 \times 10^{-9}} = 1.3 \times 10^{8}$$

(b) This reaction is the reverse of the original reaction multiplied by ½.

$$K_c' = \sqrt{\frac{1}{K_c}} = \sqrt{\frac{1}{7.5 \times 10^{-9}}} = 1.2 \times 10^{4}$$

(c) This reaction is the original reaction multiplied by 2.

$$K_c' = (K_c)^2 = (7.5 \times 10^{-9})^2 = 5.6 \times 10^{-17}$$

14.62 $K_p(1) = 7.2 \times 10^{7}$

(a) $K_p(2) = \dfrac{1}{K_p(1)} = \dfrac{1}{7.2 \times 10^{7}} = 1.4 \times 10^{-8}$

(b) $K_p(3) = (K_p(1))^2 = (7.2 \times 10^{7})^2 = 5.2 \times 10^{15}$

(c) $K_p(4) = \dfrac{1}{(K_p(1))^3} = \dfrac{1}{(7.2 \times 10^{7})^3} = 2.7 \times 10^{-24}$

14.64

$2\,Na(l) + O_2(g) \rightleftarrows Na_2O_2(s)$ $\quad K_c' = 1/K_c = 1/(5 \times 10^{-29})$

$Na_2O(s) \rightleftarrows 2\,Na(l) + \tfrac{1}{2}\,O_2(g)$ $\quad K_c = 2 \times 10^{-25}$

Overall $Na_2O(s) + \tfrac{1}{2}\,O_2(g) \rightleftarrows Na_2O_2(s)$

$$K_c(\text{overall}) = K_c' \times K_c = \frac{2 \times 10^{-25}}{5 \times 10^{-29}} = 4 \times 10^{3}$$

14.66 $K_c = \dfrac{[PCl_3][Cl_2]}{[PCl_5]} = \dfrac{(1.5 \times 10^{-2})(3.2 \times 10^{-2})}{(8.3 \times 10^{-3})} = 0.058$

14.68 $\Delta n = (1) - (1 + 1) = -1$

$$K_p = K_c(RT)^{\Delta n} = (2.2 \times 10^{5})[(0.08206)(298)]^{-1} = 9.0 \times 10^{3}$$

14.70 $K_p = P_{H_2O} = 0.0313$ atm; $\Delta n = 1$

$$K_c = K_p\left(\frac{1}{RT}\right)^{\Delta n} = (0.0313)\left(\frac{1}{(0.082\,06)(298)}\right) = 1.28 \times 10^{-3}$$

14.72 $2\,ClO(g) \rightleftarrows Cl_2O_2(g)$ $\quad K_c = 4.96 \times 10^{11}$

$$K_c = \frac{[Cl_2O_2]}{[ClO]^2} = \frac{(6.00 \times 10^{-6})}{[ClO]^2} = 4.96 \times 10^{11}$$

$$[ClO]^2 = \frac{(6.00 \times 10^{-6})}{(4.96 \times 10^{11})}; \qquad [ClO] = \sqrt{\frac{(6.00 \times 10^{-6})}{(4.96 \times 10^{11})}} = 3.48 \times 10^{-9}\ M$$

Using the Equilibrium Constant (Section 14.5)

14.74 (a) Because K_c is very large, the equilibrium mixture contains mostly product.
(b) Because K_c is very small, the equilibrium mixture contains mostly reactants.

14.76 K_p is large. The equilibrium lies far to the right and much SO_2 will be present at equilibrium.

14.78 The container volume of 10.0 L must be included to calculate molar concentrations.

$$Q_c = \frac{[CS_2]_t[H_2]_t^4}{[CH_4]_t[H_2S]_t^2} = \frac{(3.0\ \text{mol}/10.0\ \text{L})(3.0\ \text{mol}/10.0\ \text{L})^4}{(2.0\ \text{mol}/10.0\ \text{L})(4.0\ \text{mol}/10.0\ \text{L})^2} = 7.6 \times 10^{-2}; \quad K_c = 2.5 \times 10^{-3}$$

The reaction is not at equilibrium because $Q_c > K_c$. The reaction will proceed in the reverse direction to attain equilibrium.

14.80 (a) $$Q_p = \frac{(P_{P_2})(P_{H_2})^3}{(P_{PH_3})^2} = \frac{(0.871)(0.517)^3}{(0.0260)^2} = 178; \qquad K_p = 398$$

Because $Q_p < K_p$, the reaction will proceed in the forward direction to attain equilibrium.

(b) $$K_p = \frac{(P_{P_2})(P_{H_2})^3}{(P_{PH_3})^2} = 398$$

$$(P_{PH_3})^2 = \frac{(P_{P_2})(P_{H_2})^3}{398}$$

$$P_{PH_3} = \sqrt{\frac{(P_{P_2})(P_{H_2})^3}{398}} = \sqrt{\frac{(0.412)(0.822)^3}{398}} = 0.0240\ \text{atm}$$

14.82

	$N_2O_4(g)$ ⇄	$2\ NO_2(g)$
initial (M)	0.0500	0
change (M)	–x	+2x
equil (M)	0.0500 – x	2x

$$K_c = 4.64 \times 10^{-3} = \frac{[NO_2]^2}{[N_2O_4]} = \frac{(2x)^2}{(0.0500 - x)}$$

$4x^2 + (4.64 \times 10^{-3})x - (2.32 \times 10^{-4}) = 0$

Use the quadratic formula to solve for x.

$$x = \frac{-(4.64 \times 10^{-3}) \pm \sqrt{(4.64 \times 10^{-3})^2 - 4(4)(-2.32 \times 10^{-4})}}{2(4)} = \frac{-0.00464 \pm 0.06110}{8}$$

x = –0.008 22 and 0.007 06

Discard the negative solution (–0.008 22) because it leads to a negative concentration of NO_2 and that is impossible.

$[N_2O_4] = 0.0500 - x = 0.0500 - 0.007\ 06 = 0.0429$ M
$[NO_2] = 2x = 2(0.007\ 06) = 0.0141$ M

14.84 The container volume of 2.00 L must be included to calculate molar concentrations.
Initial [HI] = 9.30×10^{-3} mol/2.00 L = 4.65×10^{-3} M = 0.004 65 M

	$H_2(g)$ +	$I_2(g)$ ⇄	2 HI(g)
initial (M)	0	0	0.004 65
change (M)	+x	+x	−2x
equil (M)	x	x	0.004 65 − 2x

$x = [H_2] = [I_2] = 6.29 \times 10^{-4}$ M = 0.000 629 M
[HI] = 0.004 65 − 2x = 0.004 65 − 2(0.000 629) = 0.003 39 M

$$K_c = \frac{[HI]^2}{[H_2][I_2]} = \frac{(0.003\ 39)^2}{(0.000\ 629)^2} = 29.0$$

14.86 $CH_3CO_2C_2H_5(soln) + H_2O(soln) \rightleftharpoons CH_3CO_2H(soln) + C_2H_5OH(soln)$

$$K_c(\text{hydrolysis}) = \frac{1}{K_c(\text{forward})} = \frac{1}{3.4} = 0.29$$

14.88 (a) $$Q_c = \frac{[Br_2][Cl_2]}{[BrCl]^2} = \frac{(0.035)(0.035)}{(0.050)^2} = 0.49; \quad K_c = 0.145$$

Because $Q_c > K_c$, the reaction will proceed in the reverse direction to attain equilibrium.

(b)	2 BrCl(*soln*) ⇄	Br_2(*soln*) +	Cl_2(*soln*)
initial (M)	0.050	0.035	0.035
change (M)	+2x	−x	−x
equil (M)	0.050 + 2x	0.035 − x	0.035 − x

$$K_c = \frac{[Br_2][Cl_2]}{[BrCl]^2} = 0.145 = \frac{(0.035 - x)(0.035 - x)}{(0.050 + 2x)^2}$$

Take the square root of both sides and solve for x.

$$\sqrt{0.145} = \sqrt{\frac{(0.035 - x)^2}{(0.050 + 2x)^2}}$$

$$0.381 = \frac{(0.035 - x)}{(0.050 + 2x)}; \quad x = 0.009$$

[BrCl] = 0.050 + 2x = 0.050 + 2(0.009) = 0.068 M
$[Br_2] = [Cl_2] = 0.035 - x = 0.035 - 0.009 = 0.026$ M

14.90 $$K_c = 2.7 \times 10^2 = \frac{[SO_3]^2}{[SO_2]^2[O_2]}$$

Because $[SO_3] = [SO_2]$, then $2.7 \times 10^2 = \dfrac{1}{[O_2]}$

$[O_2] = 3.7 \times 10^{-3}$ M

14.92

	$N_2(g)$ +	$O_2(g)$ ⇄	$2\ NO(g)$
initial (M)	2.24	0.56	0
change (M)	–x	–x	+2x
equil (M)	2.24 – x	0.56 – x	2x

$$K_c = \frac{[NO]^2}{[N_2][O_2]} = 1.7 \times 10^{-3} = \frac{(2x)^2}{(2.24 - x)(0.56 - x)}$$

$4x^2 + (4.8 \times 10^{-3})x - (2.1 \times 10^{-3}) = 0$

Use the quadratic formula to solve for x.

$$x = \frac{-(4.8 \times 10^{-3}) \pm \sqrt{(4.8 \times 10^{-3})^2 - 4(4)(-2.1 \times 10^{-3})}}{2(4)} = \frac{-0.0048 \pm 0.1834}{8}$$

x = –0.0235 and 0.0223

Discard the negative solution (–0.0235) because it gives a negative NO concentration and that is impossible.

$[N_2] = 2.24 - x = 2.24 - 0.0223 = 2.22$ M; $[O_2] = 0.56 - x = 0.56 - 0.0223 = 0.54$ M

$[NO] = 2x = 2(0.0223) = 0.045$ M

14.94 (a) $K_c = \frac{[CH_3CO_2C_2H_5][H_2O]}{[CH_3CO_2H][C_2H_5OH]} = 3.4 = \frac{(x)(12.0)}{(4.0)(6.0)}$; x = 6.8 moles $CH_3CO_2C_2H_5$

Note that the volume cancels because the same number of molecules appear on both sides of the chemical equation.

(b)

	$CH_3CO_2H(soln)$ +	$C_2H_5OH(soln)$ ⇄	$CH_3CO_2C_2H_5(soln)$ +	$H_2O(soln)$
initial (mol)	1.00	10.00	0	0
change (mol)	–x	–x	+x	+x
equil (mol)	1.00 – x	10.00 – x	x	x

$$K_c = 3.4 = \frac{x^2}{(1.00 - x)(10.00 - x)}$$

$2.4x^2 - 37.4x + 34 = 0$

Use the quadratic formula to solve for x.

$$x = \frac{-(-37.4) \pm \sqrt{(-37.4)^2 - 4(2.4)(34)}}{2(2.4)} = \frac{37.4 \pm 32.75}{4.8}$$

x = 0.969 and 14.6

Discard the larger solution (14.6) because it leads to negative concentrations and that is impossible.

mol $CH_3CO_2H = 1.00 - x = 1.00 - 0.969 = 0.03$ mol

mol $C_2H_5OH = 10.00 - x = 10.00 - 0.969 = 9.03$ mol

mol $CH_3CO_2C_2H_5$ = mol H_2O = x = 0.97 mol

14.96

	$ClF_3(g)$ ⇄	$ClF(g)$ +	$F_2(g)$
initial (atm)	1.47	0	0
change (atm)	–x	+x	+x
equil (atm)	1.47 – x	x	x

$$K_p = \frac{(P_{ClF})(P_{F_2})}{(P_{ClF_3})} = 0.140 = \frac{(x)(x)}{1.47 - x}; \text{ solve for x.}$$

$x^2 + 0.140x - 0.2058 = 0$

Use the quadratic formula to solve for x.

$$x = \frac{-(0.140) \pm \sqrt{(0.140)^2 - (4)(1)(-0.2058)}}{2(1)}$$

$$x = \frac{-0.140 \pm 0.918}{2}$$

x = 0.389 and –0.529

Discard the negative solution (–0.529) because it gives negative partial pressures and that is impossible.

$P_{ClF} = P_{F_2} = x = 0.389$ atm

$P_{ClF_3} = 1.47 - x = 1.47 - 0.389 = 1.08$ atm

Le Châtelier's Principle (Sections 14.6–14.9)

14.98 (a) Cl^- (reactant) added, AgCl(s) increases
(b) Ag^+ (reactant) added, AgCl(s) increases
(c) Ag^+ (reactant) removed, AgCl(s) decreases
(d) Cl^- (reactant) removed, AgCl(s) decreases

Disturbing the equilibrium by decreasing $[Cl^-]$ increases Q_c $\left(Q_c = \frac{1}{[Ag^+]_t[Cl^-]_t}\right)$ to a value greater than K_c. To reach a new state of equilibrium, Q_c must decrease, which means that the denominator must increase; that is, the reaction must go from right to left, thus decreasing the amount of solid AgCl.

14.100 (a) Because there are 2 mol of gas on the left side and 3 mol of gas on the right side of the balanced equation, the stress of an increase in pressure is relieved by a shift in the reaction to the side with fewer moles of gas (in this case, to reactants). The number of moles of reaction products decreases.
(b) Because there are 2 mol of gas on both sides of the balanced equation, the composition of the equilibrium mixture is unaffected by a change in pressure. The number of moles of reaction product remains the same.
(c) Because there are 2 mol of gas on the left side and 1 mol of gas on the right side of the balanced equation, the stress of an increase in pressure is relieved by a shift in the reaction to the side with fewer moles of gas (in this case, to products). The number of moles of reaction products increases.

14.102 $CO(g) + H_2O(g) \rightleftharpoons CO_2(g) + H_2(g)$ $\Delta H° = -41.2$ kJ

The reaction is exothermic. $[H_2]$ decreases when the temperature is increased.
As the temperature is decreased, the reaction shifts to the right. $[CO_2]$ and $[H_2]$ increase, $[CO]$ and $[H_2O]$ decrease, and K_c increases.

14.104 (a) HCl is a source of Cl^- (product), the reaction shifts left, the equilibrium $[CoCl_4^{2-}]$ increases.
(b) $Co(NO_3)_2$ is a source of $Co(H_2O)_6^{2+}$ (product), the reaction shifts left, the equilibrium $[CoCl_4^{2-}]$ increases.
(c) All concentrations will initially decrease and the reaction will shift to the right; the equilibrium $[CoCl_4^{2-}]$ decreases.
(d) For an exothermic reaction, the reaction shifts to the left when the temperature is increased; the equilibrium $[CoCl_4^{2-}]$ increases.

14.106 (a) The reaction is exothermic. The amount of CH_3OH (product) decreases as the temperature increases.
(b) When the volume decreases, the reaction shifts to the side with fewer gas molecules. The amount of CH_3OH increases.
(c) Addition of an inert gas (He) does not affect the equilibrium composition. There is no change.
(d) Addition of CO (reactant) shifts the reaction toward product. The amount of CH_3OH increases.
(e) Addition or removal of a catalyst does not affect the equilibrium composition. There is no change.

14.108 (a) add Au (a solid); no shift
(b) OH^- (product) increased; shift toward reactants.
(c) O_2 (reactant) partial pressure increased; shift toward products.
(d) Fe^{3+} decreases CN^- (reactant) by forming $Fe(CN)_6^{3-}$; shift toward reactants.

Chemical Equilibrium and Chemical Kinetics (Sections 14.10 and 14.11)

14.110 (a) A catalyst does not affect the equilibrium composition. The amount of CO remains the same.
(b) The reaction is exothermic. An increase in temperature shifts the reaction toward reactants. The amount of CO increases.
(c) Because there are 3 mol of gas on the left side and 2 mol of gas on the right side of the balanced equation, the stress of an increase in pressure is relieved by a shift in the reaction to the side with fewer moles of gas (in this case, to products). The amount of CO decreases.
(d) An increase in pressure as a result of the addition of an inert gas (with no volume change) does not affect the equilibrium composition. The amount of CO remains the same.
(e) Adding O_2 increases the O_2 concentration and shifts the reaction toward products. The amount of CO decreases.

14.112 Because the rate of the forward reaction is greater than the rate of the reverse reaction, the reaction will proceed in the forward direction to attain equilibrium.

14.114 $A + B \rightleftarrows C$

$\text{rate}_f = k_f[A][B]$ and $\text{rate}_r = k_r[C]$; at equilibrium, $\text{rate}_f = \text{rate}_r$

$$k_f[A][B] = k_r[C]; \quad \frac{k_f}{k_r} = \frac{[C]}{[A][B]} = K_c$$

14.116 $K_c = \frac{k_f}{k_r} = \frac{0.13}{6.2 \times 10^{-4}} = 210$

14.118 k_r increases more than k_f, this means that E_a (reverse) is greater than E_a (forward). The reaction is exothermic when E_a (reverse) > E_a (forward).

Chapter Problems

14.120 $2\ HI(g) \rightleftarrows H_2(g) + I_2(g)$

Calculate K_c. $K_c = \frac{[H_2][I_2]}{[HI]^2} = \frac{(0.13)(0.70)}{(2.1)^2} = 0.0206$

$[HI] = \frac{0.20\ mol}{0.5000\ L} = 0.40\ M$

	$2\ HI(g)$	$\rightleftarrows$	$H_2(g)$	+	$I_2(g)$
initial (M)	0.40		0		0
change (M)	–2x		+x		+x
equil (M)	0.40 – 2x		x		x

$K_c = 0.0206 = \frac{[H_2][I_2]}{[HI]^2} = \frac{x^2}{(0.40 - 2x)^2}$

Take the square root of both sides, and solve for x.

$\sqrt{0.0206} = \sqrt{\frac{x^2}{(0.40 - 2x)^2}}$; $0.144 = \frac{x}{0.40 - 2x}$; $x = 0.045$

At equilibrium, $[H_2] = [I_2] = x = 0.045\ M$; $[HI] = 0.40 - 2x = 0.40 - 2(0.045) = 0.31\ M$

14.122 $[H_2O] = \frac{6.00\ mol}{5.00\ L} = 1.20\ M$

	$C(s)$	+	$H_2O(g)$	$\rightleftarrows$	$CO(g)$	+	$H_2(g)$
initial (M)			1.20		0		0
change (M)			–x		+x		+x
equil (M)			1.20 – x		x		x

$K_c = \frac{[CO][H_2]}{[H_2O]} = 3.0 \times 10^{-2} = \frac{x^2}{1.20 - x}$

$x^2 + (3.0 \times 10^{-2})x - 0.036 = 0$

Use the quadratic formula to solve for x.

$x = \frac{-(0.030) \pm \sqrt{(0.030)^2 - 4(-0.036)}}{2(1)} = \frac{-0.030 \pm 0.381}{2}$

x = 0.176 and –0.206

Discard the negative solution (–0.206) because it leads to negative concentrations and that is impossible.

$[CO] = [H_2] = x = 0.18\ M$

$[H_2O] = 1.20 - x = 1.20 - 0.18 = 1.02\ M$

14.124 (a) $[PCl_5] = 1.000\ mol/5.000\ L = 0.2000\ M$

	$PCl_5(g)$	$\rightleftarrows$	$PCl_3(g)$	+	$Cl_2(g)$
initial (M)	0.2000		0		0
change (M)	−(0.2000)(0.7850)		+(0.2000)(0.7850)		+(0.2000)(0.7850)
equil (M)	0.0430		0.1570		0.1570

$$K_c = \frac{[PCl_3][Cl_2]}{[PCl_5]} = \frac{(0.1570)(0.1570)}{(0.0430)} = 0.573$$

$\Delta n = 1$ and $K_p = K_c(RT) = (0.573)(0.082\ 06)(500) = 23.5$

(b) $$Q_c = \frac{[PCl_3][Cl_2]}{[PCl_5]} = \frac{(0.150)(0.600)}{(0.500)} = 0.18$$

Because $Q_c < K_c$, the reaction proceeds to the right to reach equilibrium.

	$PCl_5(g)$	$\rightleftarrows$	$PCl_3(g)$	+	$Cl_2(g)$
initial (M)	0.500		0.150		0.600
change (M)	− x		+x		+x
equil (M)	0.500 − x		0.150 + x		0.600 + x

$$K_c = \frac{[PCl_3][Cl_2]}{[PCl_5]} = 0.573 = \frac{(0.150 + x)(0.600 + x)}{(0.500 - x)};\ \text{solve for x.}$$

$x^2 + 1.323x - 0.1965 = 0$

$$x = \frac{-(1.323) \pm \sqrt{(1.323)^2 - (4)(1)(-0.1965)}}{2(1)} = \frac{-1.323 \pm 1.593}{2}$$

$x = -1.458$ and 0.135

Discard the negative solution (−1.458) because it will lead to negative concentrations and that is impossible.

$[PCl_5] = 0.500 - x = 0.500 - 0.135 = 0.365\ M$

$[PCl_3] = 0.150 + x = 0.150 + 0.135 = 0.285\ M$

$[Cl_2] = 0.600 + x = 0.600 + 0.135 = 0.735\ M$

14.126 (a) $$K_c = \frac{[C_2H_6][C_2H_4]}{[C_4H_{10}]} \qquad K_p = \frac{(P_{C_2H_6})(P_{C_2H_4})}{P_{C_4H_{10}}}$$

(b) $K_p = 12$; $\Delta n = 1$; $$K_c = K_p\left(\frac{1}{RT}\right) = (12)\left(\frac{1}{(0.082\ 06)(773)}\right) = 0.19$$

(c)

	$C_4H_{10}(g)$	$\rightleftarrows$	$C_2H_6(g)$	+	$C_2H_4(g)$
initial (atm)	50		0		0
change (atm)	−x		+x		+x
equil (atm)	50 − x		x		x

$$K_p = 12 = \frac{x^2}{50 - x}; \qquad x^2 + 12x - 600 = 0$$

Use the quadratic formula to solve for x.

$$x = \frac{(-12) \pm \sqrt{(12)^2 - 4(1)(-600)}}{2(1)} = \frac{-12 \pm 50.44}{2}$$

x = –31.22 and 19.22

Discard the negative solution (–31.22) because it leads to negative concentrations and that is impossible.

$$\% \ C_4H_{10} \text{ converted} = \frac{19.22}{50} \times 100\% = 38\%$$

$$P_{total} = P_{C_4H_{10}} + P_{C_2H_6} + P_{C_2H_4} = (50 - x) + x + x = (50 - 19) + 19 + 19 = 69 \text{ atm}$$

(d) A decrease in volume would decrease the % conversion of C_4H_{10}.

14.128 (a) $K_p = 3.45$; $\Delta n = 1$; $K_c = K_p\left(\frac{1}{RT}\right) = (3.45)\left(\frac{1}{(0.082\ 06)(500)}\right) = 0.0840$

(b) $[(CH_3)_3CCl] = 1.00 \text{ mol}/5.00 \text{ L} = 0.200 \text{ M}$

	$(CH_3)_3CCl(g)$	⇄	$(CH_3)_2C{=}CH_2(g)$	+	$HCl(g)$
initial (M)	0.200		0		0
change (M)	–x		+x		+x
equil (M)	0.200 – x		x		x

$$K_c = 0.0840 = \frac{x^2}{0.200 - x}; \quad x^2 + 0.0840x - 0.0168 = 0$$

Use the quadratic formula to solve for x.

$$x = \frac{(-0.0840) \pm \sqrt{(0.0840)^2 - 4(1)(-0.0168)}}{2(1)} = \frac{-0.0840 \pm 0.272}{2}$$

x = –0.178 and 0.094

Discard the negative solution (–0.178) because it leads to negative concentrations and that is impossible.

$[(CH_3)_2C{=}CCH_2] = [HCl] = x = 0.094 \text{ M}$

$[(CH_3)_3CCl] = 0.200 - x = 0.200 - 0.094 = 0.106 \text{ M}$

(c) $K_p = 3.45$

	$(CH_3)_3CCl(g)$	⇄	$(CH_3)_2C{=}CH_2(g)$	+	$HCl(g)$
initial (atm)	0		0.400		0.600
change (atm)	+x		–x		–x
equil (atm)	x		0.400 – x		0.600 – x

$$K_p = 3.45 = \frac{(0.400 - x)(0.600 - x)}{x}$$

$x^2 - 4.45x + 0.240 = 0$

Use the quadratic formula to solve for x.

$$x = \frac{-(-4.45) \pm \sqrt{(-4.45)^2 - 4(1)(0.240)}}{2(1)} = \frac{4.45 \pm 4.34}{2}$$

x = 0.055 and 4.40

Discard the larger solution (4.40) because it leads to negative partial pressures and that is impossible.

$P_{t\text{–butyl chloride}} = x = 0.055 \text{ atm}$; $\quad P_{isobutylene} = 0.400 - x = 0.400 - 0.055 = 0.345 \text{ atm}$

$P_{HCl} = 0.600 - x = 0.600 - 0.055 = 0.545 \text{ atm}$

14.130 The activation energy (E_a) is positive, and for an exothermic reaction, $E_{a,r} > E_{a,f}$.

$k_f = A_f\, e^{-E_{a,f}/RT}$, $k_r = A_r\, e^{-E_{a,r}/RT}$

$$K_c = \frac{k_f}{k_r} = \frac{A_f\, e^{-E_{a,f}/RT}}{A_r\, e^{-E_{a,r}/RT}} = \frac{A_f}{A_r}\, e^{(E_{a,r} - E_{a,f})/RT}$$

($E_{a,r} - E_{a,f}$) is positive, so the exponent is always positive. As the temperature increases, the exponent, $(E_{a,r} - E_{a,f})/RT$, decreases and the value for K_c decreases as well.

14.132 (a) $PV = nRT$, $n_{total} = \frac{PV}{RT} = \frac{(0.588\ \text{atm})(1.00\ \text{L})}{\left(0.082\ 06\ \frac{\text{L}\cdot\text{atm}}{\text{K}\cdot\text{mol}}\right)(300\ \text{K})} = 0.0239\ \text{mol}$

	$2\ NOBr(g)$	⇄	$2\ NO(g)$	+	$Br_2(g)$
initial (mol)	0.0200		0		0
change (mol)	–2x		+2x		+x
equil (mol)	0.0200 – 2x		2x		x

$n_{total} = 0.0239\ \text{mol} = (0.0200 - 2x) + 2x + x = 0.0200 + x$
$x = 0.0239 - 0.0200 = 0.0039\ \text{mol}$
Because the volume is 1.00 L, the molarity equals the number of moles.
$[NOBr] = 0.0200 - 2x = 0.0200 - 2(0.0039) = 0.0122\ \text{M}$
$[NO] = 2x = 2(0.0039) = 0.0078\ \text{M}$
$[Br_2] = x = 0.0039\ \text{M}$

$$K_c = \frac{[NO]^2[Br_2]}{[NOBr]^2} = \frac{(0.0078)^2(0.0039)}{(0.0122)^2} = 1.6 \times 10^{-3}$$

(b) $\Delta n = (3) - (2) = 1$, $K_p = K_c(RT) = (1.6 \times 10^{-3})(0.082\ 06)(300) = 0.039$

14.134 (a) $W(s) + 4\ Br(g) \rightleftarrows WBr_4(g)$

$$K_p = \frac{P_{WBr_4}}{(P_{Br})^4} = 100,\quad P_{WBr_4} = (P_{Br})^4(100) = (0.010\ \text{atm})^4(100) = 1.0 \times 10^{-6}\ \text{atm}$$

(b) Because K_p is smaller at the higher temperature, the reaction has shifted toward reactants at the higher temperature, which means the reaction is exothermic.

(c) At 2800 K, $Q_p = \frac{(1.0 \times 10^{-6})}{(0.010)^4} = 100$, $Q_p > K_p$ so the reaction will go from products to reactants, depositing tungsten back onto the filament.

14.136 $2\ NO_2(g) \rightleftarrows N_2O_4(g)$

$\Delta n = (1) - (2) = -1$ and $K_p = K_c(RT)^{-1} = (216)[(0.082\ 06)(298)]^{-1} = 8.83$

$$K_p = \frac{P_{N_2O_4}}{(P_{NO_2})^2} = 8.83$$

Let $X = P_{N_2O_4}$ and $Y = P_{NO_2}$

$P_{total} = 1.50 \text{ atm} = X + Y$ and $\dfrac{X}{Y^2} = 8.83$. Use these two equations to solve for X and Y.

$X = 1.50 - Y$

$$\frac{1.50 - Y}{Y^2} = 8.83$$

$8.83Y^2 + Y - 1.50 = 0$

Use the quadratic formula to solve for Y.

$$Y = \frac{-(1) \pm \sqrt{(1)^2 - 4(8.83)(-1.50)}}{2(8.83)} = \frac{-1 \pm 7.35}{17.7}$$

$Y = -0.472$ and 0.359

Discard the negative solution (–0.472) because it leads to a negative partial pressure of NO_2 and that is impossible.

$Y = P_{NO_2} = 0.359$ atm; $X = P_{N_2O_4} = 1.50 \text{ atm} - Y = 1.50 \text{ atm} - 0.359 \text{ atm} = 1.14$ atm

14.138 (a) $$P_{NO} = X_{NO} \cdot P_{total} = \frac{2.0 \text{ mol NO}}{(2.0 \text{ mol NO} + 3.0 \text{ mol } NO_2)} \times (1.65 \text{ atm}) = 0.66 \text{ atm}$$

$$P_{NO_2} = X_{NO_2} \cdot P_{total} = \frac{3.0 \text{ mol } NO_2}{(2.0 \text{ mol NO} + 3.0 \text{ mol } NO_2)} \times (1.65 \text{ atm}) = 0.99 \text{ atm}$$

	$NO(g)$ +	$NO_2(g)$ ⇄	$N_2O_3(g)$
initial (atm)	0.66	0.99	0
change (atm)	–x	–x	+x
equil (atm)	0.66 – x	0.99 – x	x

$K_c = 13$; $\Delta n = (1) - (1 + 1) = -1$; $K_p = K_c(RT)^{-1} = (13)[(0.082\,06)(298)]^{-1} = 0.53$

$$K_p = \frac{P_{N_2O_3}}{(P_{NO})(P_{NO_2})} = 0.53 = \frac{x}{(0.66 - x)(0.99 - x)}$$

$0.53x^2 - 1.87x + 0.35 = 0$

Use the quadratic formula to solve for x.

$$x = \frac{-(-1.87) \pm \sqrt{(-1.87)^2 - 4(0.53)(0.35)}}{2(0.53)} = \frac{1.87 \pm 1.66}{1.06}$$

$x = 3.33$ and 0.20

Discard the larger solution (3.33) because it leads to a negative concentrations of NO and NO_2 and that is impossible.

$P_{NO} = 0.66 - x = 0.66 - 0.20 = 0.46$ atm

$P_{NO_2} = 0.99 - x = 0.99 - 0.20 = 0.79$ atm

$P_{N_2O_3} = x = 0.20$ atm

(b) $PV = nRT$

$$V = \frac{nRT}{P} = \frac{(2.0 \text{ mol} + 3.0 \text{ mol})\left(0.082\,06 \dfrac{\text{L} \cdot \text{atm}}{\text{K} \cdot \text{mol}}\right)(298 \text{ K})}{1.65 \text{ atm}} = 74 \text{ L}$$

14.140

	$N_2(g)$ +	$3\ H_2(g)$ ⇄	$2\ NH_3(g)$
initial (mol)	0	0	X
change (mol)	+y	+3y	−2y
equil (mol)	y	3y	X − 2y

y = 0.200 mol

Because the volume is 1.00 L, the molarity equals the number of moles.

$[N_2] = y = 0.200\ M$; $[H_2] = 3y = 3(0.200) = 0.600\ M$

$$K_c = \frac{[NH_3]^2}{[N_2][H_2]^3} = \frac{[NH_3]^2}{(0.200)(0.600)^3} = 4.20\text{, solve for } [NH_3]_{eq}$$

$$[NH_3]_{eq}^2 = [N_2][H_2]^3(4.20) = (0.200)(0.600)^3(4.20)$$

$$[NH_3]_{eq} = \sqrt{[N_2][H_2]^3(4.20)} = \sqrt{(0.200)(0.600)^3(4.20)} = 0.426\ M$$

$[NH_3]_{eq} = 0.426\ M = X - 2(0.200) = [NH_3]_o - 2(0.200)$

$[NH_3]_o = 0.426 + 2(0.200) = 0.826\ M$

0.826 mol of NH_3 were placed in the 1.00 L reaction vessel.

14.142 ClF_3, 92.45

(a) $\text{mol } ClF_3 = 9.25\ g \times \frac{1\ \text{mol } ClF_3}{92.45\ g} = 0.100\ \text{mol } ClF_3$

$$[ClF_3] = \frac{0.100\ \text{mol } ClF_3}{2.00\ L} = 0.0500\ M$$

	$ClF_3(g)$ ⇄	$ClF(g)$ +	$F_2(g)$	
initial (M)	0.0500	0	0	
change (M)	−x	+x	+x	where x = 0.0500 x 0.198 = 0.009 90
equil (M)	0.0500 − x	x	x	
	0.0401	0.009 90	0.009 90	

$$K_c = \frac{[ClF][F_2]}{[ClF_3]} = \frac{(0.009\ 90)^2}{0.0401} = 0.002\ 44$$

(b) $K_p = K_c(RT)^{\Delta n}$; $\Delta n = 2 - 1 = 1$; $K_p = K_c(RT) = (0.002\ 44)(0.082\ 06)(700) = 0.140$

(c) $\text{mol } ClF_3 = 39.4\ g \times \frac{1\ \text{mol } ClF_3}{92.45\ g} = 0.426\ \text{mol } ClF_3$

$$[ClF_3] = \frac{0.426\ \text{mol } ClF_3}{2.00\ L} = 0.213\ M$$

	$ClF_3(g)$ ⇄	$ClF(g)$ +	$F_2(g)$
initial (M)	0.213	0	0
change (M)	−x	+x	+x
equil (M)	0.213 − x	x	x

$$K_c = \frac{[ClF][F_2]}{[ClF_3]} = 0.00\ 244 = \frac{x^2}{0.213 - x}$$

$(0.002\ 44)(0.213 - x) = x^2$
$5.20 \times 10^{-4} - 0.002\ 44x = x^2$
$x^2 + 0.002\ 44x - (5.20 \times 10^{-4}) = 0$
Use the quadratic formula to solve for x.

$$x = \frac{-(0.002\ 44) \pm \sqrt{(0.002\ 44)^2 - 4(1)(-5.20 \times 10^{-4})}}{2(1)} = \frac{-0.002\ 44 \pm 0.0457}{2}$$

$x = -0.0241$ and 0.0216
Discard the negative solution (–0.0241) because it leads to a negative partial pressure and that is impossible.
$[ClF_3] = 0.213 - x = 0.213 - 0.0216 = 0.191$ M; $[ClF] = [F_2] = x = 0.0216$ M

14.144

	Fumarate	⇄	L-Malate
initial (M)	0.001 56		0.002 27
change (M)	–x		+x
equil (M)	0.001 56 – x		0.002 27 + x

$$K_c = \frac{[\text{L-Malate}]}{[\text{Fumarate}]} = 3.3 = \frac{0.002\ 27 + x}{0.001\ 56 - x}$$

$(3.3)(0.001\ 56 - x) = 0.002\ 27 + x$
$0.005\ 15 - 3.3x = 0.002\ 27 + x$
$0.002\ 88 = 4.3x$

$$x = \frac{0.002\ 88}{4.3} = 6.7 \times 10^{-4} = 0.000\ 67$$

$[\text{Fumarate}] = 0.001\ 56 - x = 0.001\ 56 - 0.000\ 67 = 8.9 \times 10^{-4}$ M
$[\text{L-Malate}] = 0.002\ 27 + x = 0.002\ 27 + 0.000\ 67 = 2.94 \times 10^{-3}$ M

14.146 (a) Addition of a solid does not affect the equilibrium composition. There is no change in the number of moles of CO_2.
(b) Adding a product causes the reaction to shift toward reactants. The number of moles of CO_2 decreases.
(c) Decreasing the volume causes the reaction to shift toward the side with fewer mol of gas (reactant side). The number of moles of CO_2 decreases.
(d) The reaction is endothermic. An increase in temperature shifts the reaction toward products. The number of moles of CO_2 increases.

Multiconcept Problems

14.148 (a) $[N_2O_4] = \dfrac{0.500 \text{ mol}}{4.00 \text{ L}} = 0.125$ M

	$N_2O_4(g)$	⇄	$2\ NO_2(g)$
initial (M)	0.125		0
change (M)	–(0.793)(0.125)		+(2)(0.793)(0.125)
equil (M)	0.125 – (0.793)(0.125)		(2)(0.793)(0.125)

At equilibrium, $[N_2O_4] = 0.125 - (0.793)(0.125) = 0.0259$ M

$[NO_2] = (2)(0.793)(0.125) = 0.198$ M

$$K_c = \frac{[NO_2]^2}{[N_2O_4]} = \frac{(0.198)^2}{(0.0259)} = 1.51$$

$\Delta n = 2 - 1 = 1$ and $K_p = K_c(RT)^{\Delta n}$; $K_p = K_c(RT) = (1.51)(0.082\ 06)(400) = 49.6$

(b) [Lewis structures of NO_2 and N_2O_4]

14.150 2 monomer ⇄ dimer

(a) In benzene, $K_c = 1.51 \times 10^2$

	2 monomer	⇄	dimer
initial (M)	0.100		0
change (M)	–2x		+x
equil (M)	0.100 – 2x		x

$$K_c = \frac{[\text{dimer}]}{[\text{monomer}]^2} = 1.51 \times 10^2 = \frac{x}{(0.100 - 2x)^2}$$

$604x^2 - 61.4x + 1.51 = 0$

Use the quadratic formula to solve for x.

$$x = \frac{-(-61.4) \pm \sqrt{(-61.4)^2 - (4)(604)(1.51)}}{2(604)} = \frac{61.4 \pm 11.04}{1208}$$

x = 0.0600 and 0.0417

Discard the larger solution (0.0600) because it gives a negative concentration of the monomer and that is impossible.

[monomer] = 0.100 – 2x = 0.100 – 2(0.0417) = 0.017 M; [dimer] = x = 0.0417 M

$$\frac{[\text{dimer}]}{[\text{monomer}]} = \frac{0.0417\text{ M}}{0.017\text{ M}} = 2.5$$

(b) In H_2O, $K_c = 3.7 \times 10^{-2}$

	2 monomer	⇄	dimer
initial (M)	0.100		0
change (M)	–2x		+x
equil (M)	0.100 – 2x		x

$$K_c = \frac{[\text{dimer}]}{[\text{monomer}]^2} = 3.7 \times 10^{-2} = \frac{x}{(0.100 - 2x)^2}$$

$0.148x^2 - 1.0148x + 0.000\ 37 = 0$

Use the quadratic formula to solve for x.

$$x = \frac{-(-1.0148) \pm \sqrt{(-1.0148)^2 - (4)(0.148)(0.00037)}}{2(0.148)} = \frac{1.0148 \pm 1.0147}{0.296}$$

x = 6.86 and 3.4×10^{-4}

Discard the larger solution (6.86) because it gives a negative concentration of the monomer and that is impossible.

$[\text{monomer}] = 0.100 - 2x = 0.100 - 2(3.4 \times 10^{-4}) = 0.099\ \text{M}$; $[\text{dimer}] = x = 3.4 \times 10^{-4}\ \text{M}$

$$\frac{[\text{dimer}]}{[\text{monomer}]} = \frac{3.4 \times 10^{-4}\ \text{M}}{0.099\ \text{M}} = 0.0034$$

(c) K_c for the water solution is so much smaller than K_c for the benzene solution because H_2O can hydrogen bond with acetic acid, thus preventing acetic acid dimer formation. Benzene cannot hydrogen bond with acetic acid.

14.152 (a) CO_2, 44.01; CO, 28.01

$$79.2\ \text{g}\ CO_2 \times \frac{1\ \text{mol}\ CO_2}{44.01\ \text{g}\ CO_2} = 1.80\ \text{mol}\ CO_2$$

	$CO_2(g)$ + $C(s)$	⇄	$2\ CO(g)$
initial (mol)	1.80		0
change (mol)	$-x$		$+2x$
equil (mol)	$1.80 - x$		$2x$

total mass of gas in flask = (16.3 g/L)(5.00 L) = 81.5 g

$81.5 = (1.80 - x)(44.01) + (2x)(28.01)$

$81.5 = 79.22 - 44.01x + 56.02x$; $2.28 = 12.01x$; $x = 2.28/12.01 = 0.19$

$n_{CO_2} = 1.80 - x = 1.80 - 0.19 = 1.61\ \text{mol}\ CO_2$; $n_{CO} = 2x = 2(0.19) = 0.38\ \text{mol CO}$

$$P_{CO_2} = \frac{nRT}{V} = \frac{(1.61\ \text{mol})\left(0.082\ 06\ \frac{\text{L}\cdot\text{atm}}{\text{K}\cdot\text{mol}}\right)(1000\ \text{K})}{5.0\ \text{L}} = 26.4\ \text{atm}$$

$$P_{CO} = \frac{nRT}{V} = \frac{(0.38\ \text{mol})\left(0.082\ 06\ \frac{\text{L}\cdot\text{atm}}{\text{K}\cdot\text{mol}}\right)(1000\ \text{K})}{5.0\ \text{L}} = 6.24\ \text{atm}$$

$$K_p = \frac{(P_{CO})^2}{(P_{CO_2})} = \frac{(6.24)^2}{(26.4)} = 1.47$$

(b) At 1100K, the total mass of gas in flask = (16.9 g/L)(5.00 L) = 84.5 g

$84.5 = (1.80 - x)(44.01) + (2x)(28.01)$

$84.5 = 79.22 - 44.01x + 56.02x$; $5.28 = 12.01x$; $x = 5.28/12.01 = 0.44$

$n_{CO_2} = 1.80 - x = 1.80 - 0.44 = 1.36\ \text{mol}\ CO_2$; $n_{CO} = 2x = 2(0.44) = 0.88\ \text{mol CO}$

$$P_{CO_2} = \frac{nRT}{V} = \frac{(1.36\ \text{mol})\left(0.082\ 06\ \frac{\text{L}\cdot\text{atm}}{\text{K}\cdot\text{mol}}\right)(1100\ \text{K})}{5.0\ \text{L}} = 24.6\ \text{atm}$$

$$P_{CO} = \frac{nRT}{V} = \frac{(0.88\ \text{mol})\left(0.082\ 06\ \frac{\text{L}\cdot\text{atm}}{\text{K}\cdot\text{mol}}\right)(1100\ \text{K})}{5.0\ \text{L}} = 15.9\ \text{atm}$$

$$K_p = \frac{(P_{CO})^2}{(P_{CO_2})} = \frac{(15.9)^2}{(24.6)} = 10.3$$

(c) In agreement with Le Châtelier's principle, the reaction is endothermic because K_p increases with increasing temperature.

14.154 (a) N_2O_4, 92.01

$$14.58 \text{ g } N_2O_4 \times \frac{1 \text{ mol } N_2O_4}{92.01 \text{ g } N_2O_4} = 0.1585 \text{ mol } N_2O_4$$

$$PV = nRT \quad P_{N_2O_4} = \frac{nRT}{V} = \frac{(0.1585 \text{ mol})\left(0.082\,06 \frac{L \cdot atm}{K \cdot mol}\right)(400 \text{ K})}{1.000 \text{ L}} = 5.20 \text{ atm}$$

	$N_2O_4(g)$	⇄	$2\,NO_2(g)$
initial (atm)	5.20		0
change (atm)	–x		+2x
equil (atm)	5.20 – x		2x

$P_{total} = P_{N_2O_4} + P_{NO_2} = (5.20 - x) + (2x) = 9.15$ atm

$5.20 + x = 9.15$ atm

$x = 3.95$ atm

$P_{N_2O_4} = 5.20 - x = 5.20 - 3.95 = 1.25$ atm

$P_{NO_2} = 2x = 2(3.95) = 7.90$ atm

$$K_p = \frac{(P_{NO_2})^2}{(P_{N_2O_4})} = \frac{(7.90)^2}{(1.25)} = 49.9$$

$$\Delta n = 1 \text{ and } K_c = K_p\left(\frac{1}{RT}\right) = \frac{(49.9)}{(0.082\,06)(400)} = 1.52$$

(b) $\Delta H^\circ_{rxn} = [2\,\Delta H^\circ_f(NO_2)] - \Delta H^\circ_f(N_2O_4)$

$\Delta H^\circ_{rxn} = [(2 \text{ mol})(33.2 \text{ kJ/mol})] - [(1\text{mol})(11.1 \text{ kJ/mol})] = 55.3$ kJ

$$\text{moles } N_2O_4 \text{ reacted} = n = \frac{PV}{RT} = \frac{(3.95 \text{ atm})(1.000 \text{ L})}{\left(0.082\,06 \frac{L \cdot atm}{K \cdot mol}\right)(400 \text{ K})} = 0.1203 \text{ mol } N_2O_4$$

$q = (55.3 \text{ kJ/mol } N_2O_4)(0.1203 \text{ mol } N_2O_4) = 6.65$ kJ

14.156 The atmosphere is 21% (0.21) O_2; $P_{O_2} = (0.21)\left(720 \text{ mm Hg} \times \frac{1 \text{ atm}}{760 \text{ mm Hg}}\right) = 0.199$ atm

$2\,O_3(g) \rightleftarrows 3\,O_2(g)$

$$K_p = \frac{(P_{O_2})^3}{(P_{O_3})^2}$$

$$P_{O_3} = \sqrt{\frac{(P_{O_2})^3}{K_p}} = \sqrt{\frac{(0.199)^3}{1.3 \times 10^{57}}} = 2.46 \times 10^{-30} \text{ atm}$$

$$\text{vol} = 10 \times 10^6 \text{ m}^3 \times \left(\frac{100 \text{ cm}}{1 \text{ m}}\right)^3 \times \frac{1 \text{ L}}{1000 \text{ cm}^3} = 1.0 \times 10^{10} \text{ L}$$

$$n_{O_3} = \frac{PV}{RT} = \frac{(2.46 \times 10^{-30} \text{ atm})(1.0 \times 10^{10} \text{ L})}{\left(0.082\ 06 \frac{\text{L} \cdot \text{atm}}{\text{K} \cdot \text{mol}}\right)(298 \text{ K})} = 1.0 \times 10^{-21} \text{ mol } O_3$$

$$O_3 \text{ molecules} = 1.0 \times 10^{-21} \text{ mol } O_3 \times \frac{6.022 \times 10^{23} \text{ } O_3 \text{ molecules}}{1 \text{ mol } O_3} = 6.0 \times 10^2 \text{ } O_3 \text{ molecules}$$

14.158 $PCl_5(g) \rightleftarrows PCl_3(g) + Cl_2(g)$

$\Delta n = (2) - (1) = 1$ and at 700 K, $K_p = K_c(RT) = (46.9)(0.082\ 06)(700) = 2694$

(a) Because K_p is larger at the higher temperature, the reaction has shifted toward products at the higher temperature, which means the reaction is endothermic. Because the reaction involves breaking two P–Cl bonds and forming just one Cl–Cl bond, it should be endothermic.

(b) PCl_5, 208.24

$$\text{mol } PCl_5 = 1.25 \text{ g } PCl_5 \times \frac{1 \text{ mol } PCl_5}{208.24 \text{ g } PCl_5} = 6.00 \times 10^{-3} \text{ mol}$$

$$PV = nRT, \quad P_{PCl_5} = \frac{nRT}{V} = \frac{(6.00 \times 10^{-3} \text{ mol})\left(0.082\ 06 \frac{\text{L} \cdot \text{atm}}{\text{K} \cdot \text{mol}}\right)(700 \text{ K})}{0.500 \text{ L}} = 0.689 \text{ atm}$$

Because K_p is so large, first assume the reaction goes to completion and then allow for a small back reaction.

	$PCl_5(g)$	$\rightleftarrows$	$PCl_3(g)$	+	$Cl_2(g)$
before rxn (atm)	0.689		0		0
100% rxn (atm)	–0.689		+0.689		+0.689
after rxn (atm)	0		0.689		0.689
back rxn (atm)	+x		–x		–x
equil (atm)	x		0.689 – x		0.689 – x

$$K_p = \frac{(P_{PCl_3})(P_{Cl_2})}{P_{PCl_5}} = 2694 = \frac{(0.689 - x)^2}{x} \approx \frac{(0.689)^2}{x}$$

$$x = P_{PCl_5} = \frac{(0.689)^2}{2694} = 1.76 \times 10^{-4} \text{ atm}$$

$$P_{total} = P_{PCl_5} + P_{PCl_3} + P_{Cl_2}$$

$$P_{total} = x + (0.689 - x) + (0.689 - x) = 0.689 + 0.689 - 1.76 \times 10^{-4} = 1.38 \text{ atm}$$

$$\% \text{ dissociation} = \frac{(P_{PCl_5})_o - (P_{PCl_5})}{(P_{PCl_5})_o} \times 100\% = \frac{0.689 - (1.76 \times 10^{-4})}{0.689} \times 100\% = 99.97\%$$

(c) [Lewis structure of PCl_5: P bonded to five Cl atoms, each with three lone pairs]

The molecular geometry is trigonal bipyramidal. There is no dipole moment because of a symmetrical distribution of Cl's around the central P.

[Lewis structure of PCl_3: P with one lone pair bonded to three Cl atoms, each with three lone pairs]

The molecular geometry is trigonal pyramidal. There is a dipole moment because of the lone pair of electrons on the P and an unsymmetrical distribution of Cl's around the central P.

15 Aqueous Equilibria: Acids and Bases

15.1 (a) $H_2SO_4(aq) + H_2O(l) \rightleftharpoons H_3O^+(aq) + HSO_4^-(aq)$
conjugate base

(b) $HSO_4^-(aq) + H_2O(l) \rightleftharpoons H_3O^+(aq) + SO_4^{2-}(aq)$
conjugate base

(c) $H_3O^+(aq) + H_2O(l) \rightleftharpoons H_3O^+(aq) + H_2O(l)$
conjugate base

(d) $NH_4^+(aq) + H_2O(l) \rightleftharpoons H_3O^+(aq) + NH_3(aq)$
conjugate base

15.2 (a) $HCO_3^-(aq) + H_2O(l) \rightleftharpoons H_2CO_3(aq) + OH^-(aq)$
conjugate acid

(b) $CO_3^{2-}(aq) + H_2O(l) \rightleftharpoons HCO_3^-(aq) + OH^-(aq)$
conjugate acid

(c) $OH^-(aq) + H_2O(l) \rightleftharpoons H_2O(l) + OH^-(aq)$
conjugate acid

(d) $H_2PO_4^-(aq) + H_2O(l) \rightleftharpoons H_3PO_4(aq) + OH^-(aq)$
conjugate acid

15.3 $HCl(aq) + NH_3(aq) \rightleftharpoons NH_4^+(aq) + Cl^-(aq)$
acid base acid base
conjugate acid-base pairs

15.4 (a) $CH_3NH_2(aq) + H_2O(l) \rightleftharpoons OH^-(aq) + CH_3NH_3^+(aq)$
base acid base acid
conjugate acid-base pairs

(b) $HNO_3(aq) + H_2O(l) \rightleftharpoons H_3O^+(aq) + NO_3^-(aq)$
acid base acid base
conjugate acid-base pairs

15.5 (a) $HF(aq) + NO_3^-(aq) \rightleftharpoons HNO_3(aq) + F^-(aq)$
HNO_3 is a stronger acid than HF, and F^- is a stronger base than NO_3^- (see Table 15.1). Because proton transfer occurs from the stronger acid to the stronger base, the reaction proceeds from right to left.

(b) $NH_4^+(aq) + CO_3^{2-}(aq) \rightleftarrows HCO_3^-(aq) + NH_3(aq)$
NH_4^+ is a stronger acid than HCO_3^-, and CO_3^{2-} is a stronger base than NH_3 (see Table 15.1). Because proton transfer occurs from the stronger acid to the stronger base, the reaction proceeds from left to right.

15.6 (a) Both HX and HY have the same initial concentration. HY is more dissociated than HX. Therefore, HY is the stronger acid.
(b) The conjugate base (X^-) of the weaker acid (HX) is the stronger base.
(c) $HX + Y^- \rightleftarrows HY + X^-$; Proton transfer occurs from the stronger acid to the stronger base. The reaction proceeds to the left.

15.7 (a) H_2Se is a stronger acid than H_2S because Se is below S in the 6A group and the H–Se bond is weaker than the H–S bond.
(b) HI is a stronger acid than H_2Te because I is to the right of Te in the same row of the periodic table, I is more electronegative than Te, and the H–I bond is more polar.
(c) HNO_3 is a stronger acid than HNO_2 because acid strength increases with increasing oxidation number of N. The oxidation number for N is +5 in HNO_3 and +3 in HNO_2.
(d) H_2SO_3 is a stronger acid than H_2SeO_3 because acid strength increases with increasing electronegativity of the central atom. S is more electronegative than Se.

15.8 The weaker acid has the stronger conjugate base.
(a) H_2Se is a stronger acid than H_2S; $HS^- > HSe^-$
(b) HI is a stronger acid than H_2Te; $HTe^- > I^-$
(c) HNO_3 is a stronger acid than HNO_2; $NO_2^- > NO_3^-$
(d) H_2SO_3 is a stronger acid than H_2SeO_3; $HSeO_3^- > HSO_3^-$

15.9 $$[OH^-] = \frac{K_w}{[H_3O^+]} = \frac{1.0 \times 10^{-14}}{1.4 \times 10^{-4}} = 7.1 \times 10^{-11}\ M$$
Because $[H_3O^+] > [OH^-]$, the solution is acidic.

15.10 $K_w = [H_3O^+][OH^-]$; In a neutral solution, $[H_3O^+] = [OH^-]$
At 50 °C, $[H_3O^+] = [OH^-] = \sqrt{K_w} = \sqrt{5.5 \times 10^{-14}} = 2.3 \times 10^{-7}\ M$

15.11 (a) $$[H_3O^+] = \frac{K_w}{[OH^-]} = \frac{1.0 \times 10^{-14}}{1.58 \times 10^{-6}} = 6.3 \times 10^{-9}\ M$$
$pH = -\log[H_3O^+] = -\log(6.3 \times 10^{-9}) = 8.20$
(b) $pH = -\log[H_3O^+] = -\log(6.0 \times 10^{-5}) = 4.22$

15.12 $pH = -\log[H_3O^+] = -\log(6.3) = -0.80$

15.13 (a) $[H_3O^+] = 10^{-pH} = 10^{-7.40} = 4.0 \times 10^{-8}\ M$
$$[OH^-] = \frac{K_w}{[H_3O^+]} = \frac{1.0 \times 10^{-14}}{4.0 \times 10^{-8}} = 2.5 \times 10^{-7}\ M$$

(b) $[H_3O^+] = 10^{-pH} = 10^{-2.8} = 2 \times 10^{-3}$ M

$$[OH^-] = \frac{K_w}{[H_3O^+]} = \frac{1.0 \times 10^{-14}}{2 \times 10^{-3}} = 5 \times 10^{-12} \text{ M}$$

15.14 milk: $[H_3O^+] = 10^{-pH} = 10^{-6.6} = 2.5 \times 10^{-7}$ M

black coffee: $[H_3O^+] = 10^{-pH} = 10^{-5.0} = 1.0 \times 10^{-5}$ M

$$\frac{[H_3O^+]_{\text{black coffee}}}{[H_3O^+]_{\text{milk}}} = \frac{1.0 \times 10^{-5}}{2.5 \times 10^{-7}} = 40$$

The $[H_3O^+]$ in coffee is 40 times greater.

15.15 (a) Because $HClO_4$ is a strong acid, $[H_3O^+] = 0.050$ M.

$pH = -\log[H_3O^+] = -\log(0.050) = 1.30$

(b) Because HCl is a strong acid, $[H_3O^+] = 6.0$ M.

$pH = -\log[H_3O^+] = -\log(6.0) = -0.78$

(c) Because KOH is a strong base, $[OH^-] = 4.0$ M.

$$[H_3O^+] = \frac{K_w}{[OH^-]} = \frac{1.0 \times 10^{-14}}{4.0} = 2.5 \times 10^{-15} \text{ M}$$

$pH = -\log[H_3O^+] = -\log(2.5 \times 10^{-15}) = 14.60$

(d) Because $Ba(OH)_2$ is a strong base, $[OH^-] = 2(0.010 \text{ M}) = 0.020$ M.

$$[H_3O^+] = \frac{K_w}{[OH^-]} = \frac{1.0 \times 10^{-14}}{0.020} = 5.0 \times 10^{-13} \text{ M}$$

$pH = -\log[H_3O^+] = -\log(5.0 \times 10^{-13}) = 12.30$

15.16 $CaO(s) + H_2O(l) \rightarrow Ca(OH)_2(aq)$; CaO, 56.08

$$0.25 \text{ g CaO} \times \frac{1 \text{ mol CaO}}{56.08 \text{ g CaO}} \times \frac{1 \text{ mol Ca(OH)}_2}{1 \text{ mol CaO}} \times \frac{2 \text{ mol OH}^-}{1 \text{ mol Ca(OH)}_2} = 8.92 \times 10^{-3} \text{ mol OH}^-$$

$$[OH^-] = \frac{8.92 \times 10^{-3} \text{ mol OH}^-}{1.50 \text{ L}} = 5.94 \times 10^{-3} \text{ M}$$

$$[H_3O^+] = \frac{K_w}{[OH^-]} = \frac{1.0 \times 10^{-14}}{5.94 \times 10^{-3}} = 1.68 \times 10^{-12} \text{ M}$$

$pH = -\log[H_3O^+] = -\log(1.68 \times 10^{-12}) = 11.77$

15.17

	$HOCl(aq) + H_2O(l)$	$\rightleftarrows$	$H_3O^+(aq)$	$+ OCl^-(aq)$
initial (M)	0.10		~0	0
change (M)	–x		+x	+x
equil (M)	0.10 – x		x	x

$x = [H_3O^+] = 10^{-pH} = 10^{-4.23} = 5.9 \times 10^{-5}$ M

$[OCl^-] = x = 5.9 \times 10^{-5}$ M; $[HOCl] = 0.10 - x = (0.10 - 5.9 \times 10^{-5})$ M

$$K_a = \frac{[H_3O^+][OCl^-]}{[HOCl]} = \frac{(5.9 \times 10^{-5})(5.9 \times 10^{-5})}{(0.10 - 5.9 \times 10^{-5})} = 3.5 \times 10^{-8}$$

This value of K_a agrees with the value in Table 15.2.

15.18 (a) HZ is completely dissociated. HX and HY are at the same concentration and HX is more dissociated than HY. The strongest acid is HZ, the weakest is HY.
K_a (HY) < K_a (HX) < K_a (HZ)
(b) HZ
(c) HY has the highest pH; HX has the lowest pH (highest $[H_3O^+]$).

15.19 (a)

	$CH_3CO_2H(aq) + H_2O(l)$	$\rightleftarrows$	$H_3O^+(aq)$	$+ CH_3CO_2^-(aq)$
initial (M)	1.00		~0	0
change (M)	–x		+x	+x
equil (M)	1.00 – x		x	x

$$K_a = \frac{[H_3O^+][CH_3CO_2^-]}{[CH_3CO_2H]} = 1.8 \times 10^{-5} = \frac{x^2}{1.00 - x} \approx \frac{x^2}{1.00}$$

Solve for x. $x = [H_3O^+] = 4.2 \times 10^{-3}$ M
$pH = -\log[H_3O^+] = -\log(4.2 \times 10^{-3}) = 2.38$
$[CH_3CO_2^-] = x = 4.2 \times 10^{-3}$ M; $[CH_3CO_2H] = 1.00 - x = 1.00$ M

$$[OH^-] = \frac{K_w}{[H_3O^+]} = \frac{1.0 \times 10^{-14}}{4.2 \times 10^{-3}} = 2.4 \times 10^{-12} \text{ M}$$

(b)

	$CH_3CO_2H(aq) + H_2O(l)$	$\rightleftarrows$	$H_3O^+(aq)$	$+ CH_3CO_2^-(aq)$
initial (M)	0.0100		~0	0
change (M)	–x		+x	+x
equil (M)	0.0100 – x		x	x

$$K_a = \frac{[H_3O^+][CH_3CO_2^-]}{[CH_3CO_2H]} = 1.8 \times 10^{-5} = \frac{x^2}{0.0100 - x}$$

$x^2 + (1.8 \times 10^{-5})x - (1.8 \times 10^{-7}) = 0$
Use the quadratic formula to solve for x.

$$x = \frac{-(1.8 \times 10^{-5}) \pm \sqrt{(1.8 \times 10^{-5})^2 - 4(-1.8 \times 10^{-7})}}{2(1)} = \frac{(-1.8 \times 10^{-5}) \pm (8.5 \times 10^{-4})}{2}$$

$x = 4.2 \times 10^{-4}$ and -4.3×10^{-4}
Of the two solutions for x, only the positive value of x has physical meaning because x is the $[H_3O^+]$.
$x = [H_3O^+] = 4.2 \times 10^{-4}$ M
$pH = -\log[H_3O^+] = -\log(4.2 \times 10^{-4}) = 3.38$
$[CH_3CO_2^-] = x = 4.2 \times 10^{-4}$ M
$[CH_3CO_2H] = 0.0100 - x = 0.0100 - (4.2 \times 10^{-4}) = 0.0096$ M

$$[OH^-] = \frac{K_w}{[H_3O^+]} = \frac{1.0 \times 10^{-14}}{4.2 \times 10^{-4}} = 2.4 \times 10^{-11} \text{ M}$$

15.20 $[H_3O^+] = 10^{-pH} = 10^{-2.00} = 0.010$ M

	$HCO_2H(aq) + H_2O(l)$	$\rightleftarrows$	$H_3O^+(aq)$	$+ HCO_2^-(aq)$	$Ka = 1.8 \times 10^{-4}$
(M)	x – 0.010		0.010	0.010	

$$K_a = \frac{[H_3O^+][HCO_2^-]}{[HCO_2H]} = 1.8 \text{ x } 10^{-4} = \frac{(0.010)^2}{(x - 0.010)}$$

$1.8 \text{ x } 10^{-4} x - 1.8 \text{ x } 10^{-6} = 1.0 \text{ x } 10^{-4}$

$x = (1.0 \text{ x } 10^{-4} + 1.8 \text{ x } 10^{-6})/(1.8 \text{ x } 10^{-4}) = 0.57$ M

15.21

	$H_2SO_3(aq)$ + $H_2O(l)$	⇄	$H_3O^+(aq)$	+ $HSO_3^-(aq)$
initial (M)	0.10		~0	0
change (M)	–x		+x	+x
equil (M)	0.10 – x		x	x

$$K_{a1} = \frac{[H_3O^+][HSO_3^-]}{[H_2SO_3]} = 1.5 \text{ x } 10^{-2} = \frac{x^2}{0.10 - x}$$

$x^2 + 0.015x - 0.0015 = 0$

Use the quadratic formula to solve for x.

$$x = \frac{-(0.015) \pm \sqrt{(0.015)^2 - (4)(-0.0015)}}{2(1)} = \frac{-0.015 \pm 0.079}{2}$$

x = 0.032 and –0.047

Of the two solutions for x, only the positive value of x has physical meaning since x is the $[H_3O^+]$.

$x = [H_3O^+] = [HSO_3^-] = 0.032$ M; $[H_2SO_3] = 0.10 - x = 0.10 - 0.032 = 0.07$ M

The second dissociation of H_2SO_3 produces a negligible amount of H_3O^+ compared with that from the first dissociation.

$HSO_3^-(aq) + H_2O(l) \rightleftarrows H_3O^+(aq) + SO_3^{2-}(aq)$

$$K_{a2} = \frac{[H_3O^+][SO_3^{2-}]}{[HSO_3^-]} = 6.3 \text{ x } 10^{-8} = \frac{(0.032)[SO_3^{2-}]}{(0.032)}$$

$[SO_3^{2-}] = K_{a2} = 6.3 \text{ x } 10^{-8}$ M

$$[OH^-] = \frac{K_w}{[H_3O^+]} = \frac{1.0 \text{ x } 10^{-14}}{0.032} = 3.1 \text{ x } 10^{-13} \text{ M}$$

$pH = -\log[H_3O^+] = -\log(0.032) = 1.49$

15.22 Solubility $= k \cdot P = [3.2 \text{ x } 10^{-2} \text{ mol/(L} \cdot \text{atm)}](4.5 \text{ atm}) = 0.14$ M

	$H_2CO_3(aq)$ + $H_2O(l)$	⇄	$H_3O^+(aq)$	+ $HCO_3^-(aq)$
initial (M)	0.14		~0	0
change (M)	–x		+x	+x
equil (M)	0.14 – x		x	x

$$K_{a1} = \frac{[H_3O^+][HCO_3^-]}{[H_2CO_3]} = 4.3 \text{ x } 10^{-7} = \frac{x^2}{0.14 - x} \approx \frac{x^2}{0.14}$$

Solve for x. $x = 2.5 \text{ x } 10^{-4}$

$[H_3O^+] = [HCO_3^-] = x = 2.5 \text{ x } 10^{-4}$ M; $[H_2CO_3] = 0.14 - x = 0.14$ M

The second dissociation of H_2CO_3 produces a negligible amount of H_3O^+ compared with that from the first dissociation.

$HCO_3^-(aq) + H_2O(l) \rightleftarrows H_3O^+(aq) + CO_3^{2-}(aq)$

$$K_{a2} = \frac{[H_3O^+][CO_3^{2-}]}{[HCO_3^-]} = 5.6 \times 10^{-11} = \frac{(2.5 \times 10^{-4})[CO_3^{2-}]}{(2.5 \times 10^{-4})}$$

$[CO_3^{2-}] = K_{a2} = 5.6 \times 10^{-11}$ M

$$[OH^-] = \frac{K_w}{[H_3O^+]} = \frac{1.0 \times 10^{-14}}{2.5 \times 10^{-4}} = 4.0 \times 10^{-11} \text{ M}$$

$pH = -\log[H_3O^+] = -\log(2.5 \times 10^{-4}) = 3.60$

15.23

	$NH_3(aq)$ + $H_2O(l)$	$\rightleftarrows$	$NH_4^+(aq)$	+ $OH^-(aq)$
initial (M)	0.40		0	~0
change (M)	–x		+x	+x
equil (M)	0.40 – x		x	x

$$K_b = \frac{[NH_4^+][OH^-]}{[NH_3]} = 1.8 \times 10^{-5} = \frac{x^2}{0.40 - x} \approx \frac{x^2}{0.40}$$

Solve for x. $x = [OH^-] = 2.7 \times 10^{-3}$ M

$[NH_4^+] = x = 2.7 \times 10^{-3}$ M; $[NH_3] = 0.40 - x = 0.40$ M

$$[H_3O^+] = \frac{K_w}{[OH^-]} = \frac{1.0 \times 10^{-14}}{2.7 \times 10^{-3}} = 3.7 \times 10^{-12} \text{ M}$$

$pH = -\log[H_3O^+] = -\log(3.7 \times 10^{-12}) = 11.43$

15.24 $[H_3O^+] = 10^{-pH} = 10^{-8.15} = 7.1 \times 10^{-9}$ M

$$[OH^-] = \frac{K_w}{[H_3O^+]} = \frac{1.0 \times 10^{-14}}{7.1 \times 10^{-9}} = 1.4 \times 10^{-6} \text{ M}$$

	$C_3H_5O_3^-(aq)$ + $H_2O(l)$	$\rightleftarrows$	$C_3H_5O_3H(aq)$	+ $OH^-(aq)$
(M)	$0.028 - 1.4 \times 10^{-6}$		1.4×10^{-6}	1.4×10^{-6}

$$K_b = \frac{[C_3H_5O_3H][OH^-]}{[C_3H_5O_3^-]} = \frac{(1.4 \times 10^{-6})^2}{(0.028 - 1.4 \times 10^{-6})} \approx \frac{(1.4 \times 10^{-6})^2}{0.028} = 7.0 \times 10^{-11}$$

15.25 (a) $$K_a = \frac{K_w}{K_b \text{ for } C_5H_{11}N} = \frac{1.0 \times 10^{-14}}{1.3 \times 10^{-3}} = 7.7 \times 10^{-12}$$

(b) $$K_b = \frac{K_w}{K_a \text{ for HOCl}} = \frac{1.0 \times 10^{-14}}{3.5 \times 10^{-8}} = 2.9 \times 10^{-7}$$

(c) pK_b for $HCO_2^- = 14.00 - pK_a = 14.00 - 3.74 = 10.26$

15.26 (a) The strongest acid has the largest K_a. HX is the strongest acid because it is the most dissociated. Therefore, HX has the largest K_a.

(b) The weakest acid has the strongest conjugate base. HY is the weakest acid because it is the least dissociated. Therefore, Y^- has the largest K_b.
(c) The strongest acid has the weakest conjugate base. HX is the strongest acid because it is the most dissociated. Therefore, X^- is the weakest base.

15.27 (a) 0.25 M NH_4Br
NH_4^+ is an acidic cation. Br^- is a neutral anion. The salt solution is acidic.

For NH_4^+, $K_a = \dfrac{K_w}{K_b \text{ for } NH_3} = \dfrac{1.0 \times 10^{-14}}{1.8 \times 10^{-5}} = 5.6 \times 10^{-10}$

	$NH_4^+(aq)$ + $H_2O(l)$	⇄ $H_3O^+(aq)$	+ $NH_3(aq)$
initial (M)	0.25	~0	0
change (M)	–x	+x	+x
equil (M)	0.25 – x	x	x

$$K_a = \frac{[H_3O^+][NH_3]}{[NH_4^+]} = 5.6 \times 10^{-10} = \frac{x^2}{0.25 - x} \approx \frac{x^2}{0.25}$$

Solve for x. $x = [H_3O^+] = 1.2 \times 10^{-5}$ M
$pH = -\log[H_3O^+] = -\log(1.2 \times 10^{-5}) = 4.92$

(b) 0.20 M $NaNO_2$

For NO_2^-, $K_b = \dfrac{K_w}{K_a \text{ for } HNO_2} = \dfrac{1.0 \times 10^{-14}}{4.6 \times 10^{-4}} = 2.2 \times 10^{-11}$

	$NO_2^-(aq)$ + $H_2O(l)$	⇄ $HNO_2(aq)$	+ $OH^-(aq)$
initial (M)	0.20	0	~0
change (M)	–x	+x	+x
equil (M)	0.20 – x	x	x

$$K_b = \frac{[HNO_2][OH^-]}{[NO_2^-]} = 2.2 \times 10^{-11} = \frac{x^2}{0.20 - x} \approx \frac{x^2}{0.20}$$

Solve for x. $x = [OH^-] = 2.1 \times 10^{-6}$ M

$$[H_3O^+] = \frac{K_w}{[OH^-]} = \frac{1.0 \times 10^{-14}}{2.1 \times 10^{-6}} = 4.8 \times 10^{-9} \text{ M}$$

$pH = -\log[H_3O^+] = -\log(4.8 \times 10^{-9}) = 8.32$

15.28 (a) Zn^{2+} is an acidic cation. Cl^- is a neutral anion. The salt solution is acidic.

	$Zn(H_2O)_6^{2+}(aq)$ + $H_2O(l)$	⇄ $H_3O^+(aq)$	+ $Zn(H_2O)_5(OH)^+(aq)$
initial (M)	0.40	~0	0
change (M)	–x	+x	+x
equil(M)	0.40 – x	x	x

$$K_a = \frac{[H_3O^+][Zn(H_2O)_5(OH)^+]}{[Zn(H_2O)_6^{2+}]} = 2.5 \times 10^{-10} = \frac{x^2}{0.40 - x} \approx \frac{x^2}{0.40}$$

Solve for x. $x = [H_3O^+] = 1.0 \times 10^{-5}$ M
$pH = -\log[H_3O^+] = -\log(1.0 \times 10^{-5}) = 5.00$

$$\% \text{ dissociation} = \frac{[Zn(H_2O)_6{}^{2+}]_{diss}}{[Zn(H_2O)_6{}^{2+}]_{initial}} \times 100\% = \frac{1.0 \times 10^{-5}\ M}{0.40\ M} \times 100\% = 2.5 \times 10^{-3}\ \%$$

(b) $Fe(H_2O)_6{}^{3+}$ is a stronger acid than $Zn(H_2O)_6{}^{2+}$ because of the higher 3+ charge. The stronger acid, $Fe(H_2O)_6{}^{3+}$, has the higher percent dissociation.

15.29 (a) Lewis acid, $AlCl_3$; Lewis base, Cl^- (b) Lewis acid, Ag^+; Lewis base, NH_3
(c) Lewis acid, SO_2; Lewis base, OH^- (d) Lewis acid, Cr^{3+}; Lewis base, H_2O

15.30

:C̈l—Be—C̈l: + 2 :C̈l:⁻ ⟶ [Cl—Be(—Cl)(—Cl)—Cl]2−

15.31 (a) $\Delta pH = 5.6 - 3.6 = 2.0$
A decrease of 2 pH units represents an increase in $[H_3O^+]$ by a factor of 100.
(b) $[H_3O^+] = 10^{-pH} = 10^{-5.10} = 7.9 \times 10^{-6}$ M

$$[OH^-] = \frac{K_w}{[H_3O^+]} = \frac{1.0 \times 10^{-14}}{7.9 \times 10^{-6}} = 1.3 \times 10^{-9}\ M$$

15.32 (a) $P_{CO_2} = \left(\frac{750}{10^6} \cdot 1.0\ atm\right) = 7.5 \times 10^{-4}$ atm

Solubility $= k \cdot P = [3.2 \times 10^{-2}\ mol/(L \cdot atm)](7.5 \times 10^{-4}\ atm) = 2.4 \times 10^{-5}$ M

(b)	$H_2CO_3(aq)$ + $H_2O(l)$	⇄	$H_3O^+(aq)$	+ $HCO_3^-(aq)$
initial (M)	2.4×10^{-5}		~0	0
change (M)	−x		+x	+x
equil (M)	$2.4 \times 10^{-5} - x$		x	x

$$K_{a1} = \frac{[H_3O^+][HCO_3^-]}{[H_2CO_3]} = 4.3 \times 10^{-7} = \frac{x^2}{2.4 \times 10^{-5} - x}$$

$x^2 + 4.3 \times 10^{-7}x - 1.0 \times 10^{-11} = 0$

Use the quadratic formula to solve for x.

$$x = \frac{-(4.3 \times 10^{-7}) \pm \sqrt{(4.3 \times 10^{-7})^2 - (4)(1)(-1.0 \times 10^{-11})}}{2(1)} = \frac{(-4.3 \times 10^{-7}) \pm (6.3 \times 10^{-6})}{2}$$

$x = -3.4 \times 10^{-6}$ and 3.0×10^{-6}

Of the two solutions for x, only the positive value of x has physical meaning because x is the $[H_3O^+]$.

$x = [H_3O^+] = 3.0 \times 10^{-6}$ M

The second dissociation of H_2CO_3 produces a negligible amount of H_3O^+ compared with that from the first dissociation.

$pH = -\log[H_3O^+] = -\log(3.0 \times 10^{-6}) = 5.52$

(c) The acidity of rain will increase but only slightly.

15.33 NO_2, 46.0

$$5.50\text{ mg} \times \frac{1 \times 10^{-3}\text{ g}}{1\text{ mg}} \times \frac{1\text{ mol } NO_2}{46.0\text{ g } NO_2} = 1.20 \times 10^{-4}\text{ mol } NO_2$$

$4\ NO_2(g) + 2\ H_2O(l) + O_2(g) \rightarrow 4\ HNO_3(aq)$

mol HNO_3 = mol NO_2 = 1.20×10^{-4} mol
$[HNO_3] = 1.20 \times 10^{-4}$ mol/1.00 L = 1.20×10^{-4} M
Because HNO_3 is a strong acid, $[H_3O^+] = [HNO_3] = 1.20 \times 10^{-4}$ M
pH = $-\log[H_3O^+] = -\log(1.20 \times 10^{-4}) = 3.921$

15.34 Lewis acids include not only H^+ but also other cations and neutral molecules having vacant valence orbitals that can accept a share in a pair of electrons donated by a Lewis base. The O^{2-} from CaO is the Lewis base and SO_2 is the Lewis acid.

:Ö—S̤—Ö: + :Ö:$^{2-}$ ⟶ :$SO_3^{\,2-}$

15.35 For NH_4^+, $K_a = \dfrac{K_w}{K_b \text{ for } NH_3} = \dfrac{1.0 \times 10^{-14}}{1.8 \times 10^{-5}} = 5.6 \times 10^{-10}$

For SO_4^{2-}, $K_b = \dfrac{K_w}{K_a \text{ for } HSO_4^-} = \dfrac{1.0 \times 10^{-14}}{1.2 \times 10^{-2}} = 8.3 \times 10^{-13}$

NH_4^+ is acidic, NO_3^- is neutral. NH_4NO_3 is an acidic salt.
NH_4^+ is acidic, SO_4^{2-} is basic. $K_a > K_b$, $(NH_4)_2SO_4$ is an acidic salt.

Conceptual Problems

15.36 (a) acids, HCO_3^- and H_3O^+; bases, H_2O and CO_3^{2-}
(b) acids, HF and H_2CO_3; bases HCO_3^- and F^-

15.38 (c) represents a solution of a weak diprotic acid, H_2A. Because K_{a2} is always less than K_{a1}, (a) and (d) represent impossible situations. (b) contains no H_2A.

15.40 (a) $Y^- < Z^- < X^-$
(b) The weakest base, Y^-, has the strongest conjugate acid.
(c) X^- is the strongest conjugate base and has the smallest pK_b.
(d) The numbers of HA molecules and OH^- ions are equal because the reaction of A^- with water has a 1:1 stoichiometry: $A^- + H_2O \rightleftharpoons HA + OH^-$

15.42

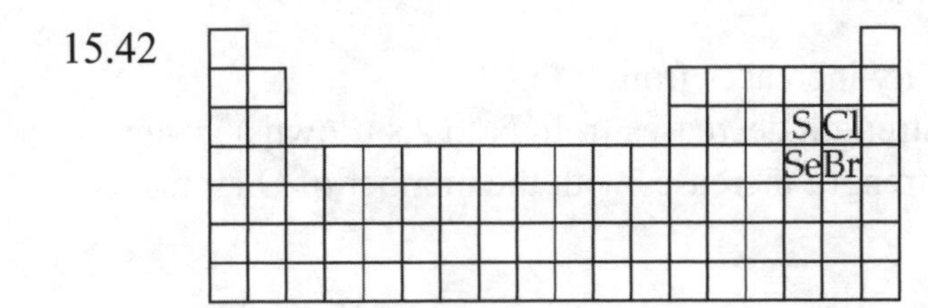

(a) H_2S, weakest; HBr, strongest. Acid strength for H_nX increases with increasing polarity of the H–X bond and with increasing size of X.
(b) H_2SeO_3, weakest; $HClO_3$, strongest. Acid strength for H_nYO_3 increases with increasing electronegativity of Y.

15.44 (a) $H_3BO_3(aq) + H_2O(l) \rightleftharpoons H_3O^+(aq) + H_2BO_3^-(aq)$
(b) $H_3BO_3(aq) + 2\,H_2O(l) \rightleftharpoons H_3O^+(aq) + B(OH)_4^-(aq)$

Section Problems
Acid–Base Concepts (Sections 15.1 and 15.2)

15.46 A Brønsted-Lowry base is a proton acceptor. An Arrhenius base dissociates producing OH^-.
(a) NH_3 and (c) HCO_3^-

15.48 (a) SO_4^{2-} (b) HSO_3^- (c) HPO_4^{2-} (d) NH_3 (e) OH^- (f) NH_2^-

15.50 (a) $CH_3CO_2H(aq) + NH_3(aq) \rightleftharpoons NH_4^+(aq) + CH_3CO_2^-(aq)$
acid base —— acid base

(b) $CO_3^{2-}(aq) + H_3O^+(aq) \rightleftharpoons H_2O(l) + HCO_3^-(aq)$
base acid —— base acid

(c) $HSO_3^-(aq) + H_2O(l) \rightleftharpoons H_3O^+(aq) + SO_3^{2-}(aq)$
acid base —— acid base

(d) $HSO_3^-(aq) + H_2O(l) \rightleftharpoons H_2SO_3(aq) + OH^-(aq)$
base acid acid base

15.52 From data in Table 15.1: Strong acids: HNO_3 and H_2SO_4; Strong bases: H^- and O^{2-}

15.54 The direction of the reaction for $K_c > 1$, is proton transfer from the stronger acid to the stronger base to give the weaker base and the weaker acid:
$HSO_4^- + NO_2^- \rightarrow SO_4^{2-} + HNO_2$

Factors That Affect Acid Strength (Section 15.3)

15.56 (a) $PH_3 < H_2S < HCl$; electronegativity increases from P to Cl
(b) $NH_3 < PH_3 < AsH_3$; X–H bond strength decreases from N to As (down a group)
(c) $HBrO < HBrO_2 < HBrO_3$; acid strength increases with the number of O atoms

15.58 (a) HCl; The strength of a binary acid H_nA increases as A moves from left to right and from top to bottom in the periodic table.
(b) $HClO_3$; The strength of an oxoacid increases with increasing electronegativity and increasing oxidation state of the central atom.
(c) HBr; The strength of a binary acid H_nA increases as A moves from left to right and from top to bottom in the periodic table.

15.60 (a) H_2Te, weaker X–H bond
(b) H_3PO_4, P has higher electronegativity
(c) $H_2PO_4^-$, lower negative charge
(d) NH_4^+, higher positive charge and N is more electronegative than C

15.62 (b)

```
        O
        ‖
H — C — O — H
```

The HCO group is more electronegative than the CH_3 group, (b) is more acidic.

Dissociation of Water; pH (Sections 15.4–15.6)

15.64 $[H_3O^+] = \dfrac{K_w}{[OH^-]} = \dfrac{1.0 \times 10^{-14}}{2.0 \times 10^{-6}} = 5.0 \times 10^{-9}$ M

Because $[OH^-] > [H_3O^+]$, the solution is basic.

15.66 If $[H_3O^+] > 1.0 \times 10^{-7}$ M, solution is acidic.
If $[H_3O^+] < 1.0 \times 10^{-7}$ M, solution is basic.
If $[H_3O^+] = [OH^-] = 1.0 \times 10^{-7}$ M, solution is neutral.
If $[OH^-] > 1.0 \times 10^{-7}$ M, solution is basic
If $[OH^-] < 1.0 \times 10^{-7}$ M, solution is acidic.

(a) $[OH^-] = \dfrac{K_w}{[H_3O^+]} = \dfrac{1.0 \times 10^{-14}}{3.4 \times 10^{-9}} = 2.9 \times 10^{-6}$ M, basic

(b) $[H_3O^+] = \dfrac{K_w}{[OH^-]} = \dfrac{1.0 \times 10^{-14}}{0.010} = 1.0 \times 10^{-12}$ M, basic

(c) $[H_3O^+] = \dfrac{K_w}{[OH^-]} = \dfrac{1.0 \times 10^{-14}}{1.0 \times 10^{-10}} = 1.0 \times 10^{-4}$ M, acidic

(d) $[OH^-] = \dfrac{K_w}{[H_3O^+]} = \dfrac{1.0 \times 10^{-14}}{1.0 \times 10^{-7}} = 1.0 \times 10^{-7}$ M, neutral

(e) $[OH^-] = \dfrac{K_w}{[H_3O^+]} = \dfrac{1.0 \times 10^{-14}}{8.6 \times 10^{-5}} = 1.2 \times 10^{-10}$ M, acidic

15.68 At 200 °C and 750 atm, $K_w = [H_3O^+][OH^-] = 1.5 \times 10^{-11}$ and in pure water $[H_3O^+] = [OH^-]$.
$[H_3O^+]^2 = K_w$

$[H_3O^+] = [OH^-] = \sqrt{K_w} = \sqrt{1.5 \times 10^{-11}} = 3.9 \times 10^{-6}$ M
Because $[H_3O^+] = [OH^-]$, the solution is neutral.

15.70 (a) $pH = -\log[H_3O^+] = -\log(2.0 \times 10^{-5}) = 4.70$

(b) $[H_3O^+] = \dfrac{K_w}{[OH^-]} = \dfrac{1.0 \times 10^{-14}}{4 \times 10^{-3}} = 2.5 \times 10^{-12}$ M

$pH = -\log[H_3O^+] = -\log(2.5 \times 10^{-12}) = 11.6$

(c) $pH = -\log[H_3O^+] = -\log(3.56 \times 10^{-9}) = 8.449$

(d) $pH = -\log[H_3O^+] = -\log(10^{-3}) = 3$

(e) $[H_3O^+] = \dfrac{K_w}{[OH^-]} = \dfrac{1.0 \times 10^{-14}}{12} = 8.3 \times 10^{-16}$ M

$pH = -\log[H_3O^+] = -\log(8.3 \times 10^{-16}) = 15.08$

15.72 $[H_3O^+] = 10^{-pH}$; (a) 8×10^{-5} M (b) 1.5×10^{-11} M (c) 1.0 M
(d) 5.6×10^{-15} M (e) 10 M (f) 5.78×10^{-6} M

15.74 (a) chlorphenol red (b) thymol blue (c) methyl orange

Strong Acids and Strong Bases (Section 15.7)

15.76 (a) Because $Sr(OH)_2$ is a strong base, $[OH^-] = 2(1.0 \times 10^{-3}\text{ M}) = 2.0 \times 10^{-3}$ M.

$[H_3O^+] = \dfrac{K_w}{[OH^-]} = \dfrac{1.0 \times 10^{-14}}{2.0 \times 10^{-3}} = 5.0 \times 10^{-12}$ M

$pH = -\log[H_3O^+] = -\log(5.0 \times 10^{-12}) = 11.30$

(b) Because HNO_3 is a strong acid, $[H_3O^+] = 0.015$ M.

$pH = -\log[H_3O^+] = -\log(0.015) = 1.82$

(c) Because NaOH is a strong base, $[OH^-] = 0.035$ M.

$[H_3O^+] = \dfrac{K_w}{[OH^-]} = \dfrac{1.0 \times 10^{-14}}{0.035} = 2.9 \times 10^{-13}$ M

$pH = -\log[H_3O^+] = -\log(2.9 \times 10^{-13}) = 12.54$

15.78 (a) LiOH, 23.95; 250 mL = 0.250 L

$$\text{molarity of LiOH(aq)} = \frac{\left(4.8\text{ g} \times \dfrac{1\text{ mol}}{23.95\text{ g}}\right)}{0.250\text{ L}} = 0.80\text{ M}$$

LiOH is a strong base; therefore $[OH^-] = 0.80$ M

$[H_3O^+] = \dfrac{K_w}{[OH^-]} = \dfrac{1.0 \times 10^{-14}}{0.80} = 1.25 \times 10^{-14}$ M

$pH = -\log[H_3O^+] = -\log(1.25 \times 10^{-14}) = 13.90$

(b) HCl, 36.46

$$\text{molarity of HCl(aq)} = \frac{\left(0.93\text{ g x } \dfrac{1\text{ mol}}{36.46\text{ g}}\right)}{0.40\text{ L}} = 0.064\text{ M}$$

HCl is a strong acid; therefore $[H_3O^+] = 0.064$ M

$pH = -\log[H_3O^+] = -\log(0.064) = 1.19$

(c) $M_f \cdot V_f = M_i \cdot V_i$

$$M_f = \frac{M_i \cdot V_i}{V_f} = \frac{(0.10\text{ M})(50\text{ mL})}{(1000\text{ mL})} = 5.0 \times 10^{-3}\text{ M}$$

$pH = -\log[H_3O^+] = -\log(5.0 \times 10^{-3}) = 2.30$

$$\text{(d) For HCl, } M_f = \frac{M_i \cdot V_i}{V_f} = \frac{(2.0 \times 10^{-3}\text{ M})(100\text{ mL})}{(500\text{ mL})} = 4.0 \times 10^{-4}\text{ M}$$

$$\text{For HClO}_4\text{, } M_f = \frac{M_i \cdot V_i}{V_f} = \frac{(1.0 \times 10^{-3}\text{ M})(400\text{ mL})}{(500\text{ mL})} = 8.0 \times 10^{-4}\text{ M}$$

$[H_3O^+] = (4.0 \times 10^{-4}\text{ M}) + (8.0 \times 10^{-4}\text{ M}) = 1.2 \times 10^{-3}$ M

$pH = -\log[H_3O^+] = -\log(1.2 \times 10^{-3}) = 2.92$

15.80 CaO, 56.08

$[H_3O^+] = 10^{-pH} = 10^{-(10.50)} = 3.16 \times 10^{-11}$ M

$$[OH^-] = \frac{K_w}{[H_3O^+]} = \frac{1.0 \times 10^{-14}}{3.16 \times 10^{-11}} = 3.16 \times 10^{-4}\text{ M} = 3.16 \times 10^{-4}\text{ mol/L}$$

$CaO(s) + H_2O(l) \rightarrow Ca^{2+}(aq) + 2\,OH^-(aq)$

$$3.16 \times 10^{-4}\text{ mol OH}^- \text{ x } \frac{1\text{ mol CaO}}{2\text{ mol OH}^-} \text{ x } \frac{56.08\text{ g CaO}}{1\text{ mol CaO}} = 0.0089\text{ g CaO}$$

Weak Acids (Sections 15.8–15.10)

15.82 (a) The larger the K_a, the stronger the acid.

$C_6H_5OH < HOCl < CH_3CO_2H < HNO_3$

(b) The larger the K_a, the larger the percent dissociation for the same concentration.

$HNO_3 > CH_3CO_2H > HOCl > C_6H_5OH$

1 M HNO_3, $[H_3O^+] = 1$ M

1 M CH_3CO_2H, $[H_3O^+] = \sqrt{[HA] \times K_a} = \sqrt{(1\text{ M})(1.8 \times 10^{-5})} = 4 \times 10^{-3}$ M

1 M HOCl, $[H_3O^+] = \sqrt{[HA] \times K_a} = \sqrt{(1\text{ M})(3.5 \times 10^{-8})} = 2 \times 10^{-4}$ M

1 M C_6H_5OH, $[H_3O^+] = \sqrt{[HA] \times K_a} = \sqrt{(1\text{ M})(1.3 \times 10^{-10})} = 1 \times 10^{-5}$ M

15.84

	$HOBr(aq)$	+	$H_2O(l)$	$\rightleftarrows$	$H_3O^+(aq)$	+	$OBr^-(aq)$
initial (M)	0.040				~0		0
change (M)	–x				+x		+x
equil (M)	0.040 – x				x		x

$x = [H_3O^+] = 10^{-pH} = 10^{-5.05} = 8.9 \times 10^{-6}$ M

$$K_a = \frac{[H_3O^+][OBr^-]}{[HOBr]} = \frac{x^2}{0.040 - x} = \frac{(8.9 \times 10^{-6})^2}{0.040 - (8.9 \times 10^{-6})} = 2.0 \times 10^{-9}$$

15.86 $C_6H_8O_6$, 176.13; 250 mg = 0.250 g; 250 mL = 0.250 L

$$[C_6H_8O_6] = \frac{\left(0.250 \text{ g} \times \frac{1 \text{ mol}}{176.13 \text{ g}}\right)}{0.250 \text{ L}} = 5.68 \times 10^{-3} \text{ M}$$

	$C_6H_8O_6(aq)$ + $H_2O(l)$	⇄ $H_3O^+(aq)$	+ $C_6H_7O_6^-(aq)$
initial (M)	5.68×10^{-3}	~0	0
change (M)	−x	+x	+x
equil (M)	$(5.68 \times 10^{-3}) - x$	x	x

$$K_a = \frac{[H_3O^+][C_6H_7O_6^-]}{[C_6H_8O_6]} = 8.0 \times 10^{-5} = \frac{x^2}{(5.68 \times 10^{-3}) - x}$$

$x^2 + (8.0 \times 10^{-5})x - (4.54 \times 10^{-7}) = 0$

Use the quadratic formula to solve for x.

$$x = \frac{-(8.0 \times 10^{-5}) \pm \sqrt{(8.0 \times 10^{-5})^2 - (4)(-4.54 \times 10^{-7})}}{2(1)} = \frac{(-8.0 \times 10^{-5}) \pm 0.001\ 35}{2}$$

$x = 6.35 \times 10^{-4}$ and -7.15×10^{-4}

Of the two solutions for x, only the positive value of x has physical meaning because x is the $[H_3O^+]$.

$x = [H_3O^+] = 6.35 \times 10^{-4}$ M

$pH = -\log[H_3O^+] = -\log(6.35 \times 10^{-4}) = 3.20$

15.88 $K_a = 10^{-pK_a} = 10^{-4.25} = 5.6 \times 10^{-5}$

	$HC_3H_3O_2(aq)$ + $H_2O(l)$	⇄ $H_3O^+(aq)$	+ $C_3H_3O_2^-(aq)$
initial (M)	0.150	~0	0
change (M)	−x	+x	+x
equil (M)	0.150 − x	x	x

$$K_a = \frac{[H_3O^+][C_3H_3O_2^-]}{[HC_3H_3O_2]} = 5.6 \times 10^{-5} = \frac{x^2}{0.150 - x} \approx \frac{x^2}{0.150}$$

Solve for x. $x = 0.0029 \text{ M} = [H_3O^+] = [C_3H_3O_2^-]$

$[HC_3H_3O_2] = 0.150 - x = 0.150 - 0.0029 = 0.147$ M

$pH = -\log[H_3O^+] = -\log(0.0029) = 2.54$

$$[OH^-] = \frac{K_w}{[H_3O^+]} = \frac{1.0 \times 10^{-14}}{0.0029} = 3.4 \times 10^{-12} \text{ M}$$

(b)	$HC_3H_3O_2(aq)$ + $H_2O(l)$	⇄ $H_3O^+(aq)$	+ $C_3H_3O_2^-(aq)$
initial (M)	0.0500	~0	0
change (M)	−x	+x	+x
equil (M)	0.0500 − x	x	x

$$K_a = \frac{[H_3O^+][C_3H_5O_2^-]}{[HC_3H_5O_2]} = 5.6 \times 10^{-5} = \frac{x^2}{0.0500 - x} \approx \frac{x^2}{0.0500}$$

Solve for x. $x = 0.001\,67\ M = [H_3O^+] = [HC_3H_5O_2]_{diss}$

$$\%\ \text{dissociation} = \frac{[HC_3H_5O_2]_{diss}}{[HC_3H_5O_2]_{initial}} \times 100\% = \frac{0.001\,67\ M}{0.0500\ M} \times 100\% = 3.3\%$$

15.90

	$HNO_2(aq)$ + $H_2O(l)$	⇄ $H_3O^+(aq)$	+ $NO_2^-(aq)$
initial (M)	1.5	~0	0
change (M)	–x	+x	+x
equil (M)	1.5 – x	x	x

$$K_a = \frac{[H_3O^+][NO_2^-]}{[HNO_2]} = 4.5 \times 10^{-4} = \frac{x^2}{1.5 - x} \approx \frac{x^2}{1.5}$$

Solve for x. $x = 0.026\ M = [H_3O^+]$; $pH = -\log[H_3O^+] = -\log(0.026) = 1.59$

$$\%\ \text{dissociation} = \frac{[HNO_2]_{diss}}{[HNO_2]_{initial}} \times 100\% = \frac{0.026\ M}{1.5\ M} \times 100\% = 1.7\%$$

15.92 (a) From Example 15.10 in the text:

$[H_3O^+] = [HF]_{diss} = 4.0 \times 10^{-3}\ M$

$$\%\ \text{dissociation} = \frac{[HF]_{diss}}{[HF]_{initial}} \times 100\% = \frac{4.0 \times 10^{-3}\ M}{0.050\ M} \times 100\% = 8.0\%\ \text{dissociation}$$

(b)

	$HF(aq)$ + $H_2O(l)$	⇄ $H_3O^+(aq)$	+ $F^-(aq)$
initial (M)	0.50	~0	0
change (M)	–x	+x	+x
equil (M)	0.50 – x	x	x

$$K_a = \frac{[H_3O^+][F^-]}{[HF]} = 3.5 \times 10^{-4} = \frac{x^2}{0.50 - x}$$

$x^2 + (3.5 \times 10^{-4})x - (1.75 \times 10^{-4}) = 0$

Use the quadratic formula to solve for x.

$$x = \frac{-(3.5 \times 10^{-4}) \pm \sqrt{(3.5 \times 10^{-4})^2 - 4(1)(-1.75 \times 10^{-4})}}{2(1)} = \frac{(-3.5 \times 10^{-4}) \pm 0.0265}{2}$$

x = 0.0131 and –0.0134

Of the two solutions for x, only the positive value of x has physical meaning, because x is the $[H_3O^+]$. $[H_3O^+] = [HF]_{diss} = 0.013\ M$

$$\%\ \text{dissociation} = \frac{[HF]_{diss}}{[HF]_{initial}} \times 100\% = \frac{0.013\ M}{0.50\ M} \times 100\% = 2.6\%\ \text{dissociation}$$

Polyprotic Acids (Section 15.11)

15.94 $H_2SeO_4(aq) + H_2O(l) \rightleftharpoons H_3O^+(aq) + HSeO_4^-(aq)$; $K_{a1} = \frac{[H_3O^+][HSeO_4^-]}{[H_2SeO_4]}$

$HSeO_4^-(aq) + H_2O(l) \rightleftarrows H_3O^+(aq) + SeO_4^{2-}(aq)$; $K_{a2} = \dfrac{[H_3O^+][SeO_4^{2-}]}{[HSeO_4^-]}$

15.96

	$H_2CO_3(aq) + H_2O(l)$	$\rightleftarrows$	$H_3O^+(aq)$	+	$HCO_3^-(aq)$
initial (M)	0.010		~0		0
change (M)	–x		+x		+x
equil (M)	0.010 – x		x		x

$$K_{a1} = \frac{[H_3O^+][HCO_3^-]}{[H_2CO_3]} = 4.3 \times 10^{-7} = \frac{x^2}{0.010 - x} \approx \frac{x^2}{0.010}$$

Solve for x. $x = 6.6 \times 10^{-5}$

$[H_3O^+] = [HCO_3^-] = x = 6.6 \times 10^{-5}$ M; $[H_2CO_3] = 0.010 - x = 0.010$ M

The second dissociation of H_2CO_3 produces a negligible amount of H_3O^+ compared with that from the first dissociation.

$HCO_3^-(aq) + H_2O(l) \rightleftarrows H_3O^+(aq) + CO_3^{2-}(aq)$

$$K_{a2} = \frac{[H_3O^+][CO_3^{2-}]}{[HCO_3^-]} = 5.6 \times 10^{-11} = \frac{(6.6 \times 10^{-5})[CO_3^{2-}]}{(6.6 \times 10^{-5})}$$

$[CO_3^{2-}] = K_{a2} = 5.6 \times 10^{-11}$ M

$$[OH^-] = \frac{K_w}{[H_3O^+]} = \frac{1.0 \times 10^{-14}}{6.6 \times 10^{-5}} = 1.5 \times 10^{-10} \text{ M}$$

$pH = -\log[H_3O^+] = -\log(6.6 \times 10^{-5}) = 4.18$

15.98 For the dissociation of the first proton, the following equilibrium must be considered:

	$H_2C_2O_4(aq) + H_2O(l)$	$\rightleftarrows$	$H_3O^+(aq)$	+	$HC_2O_4^-(aq)$
initial (M)	0.20		~0		0
change (M)	–x		+x		+x
equil (M)	0.20 – x		x		x

$$K_{a1} = \frac{[H_3O^+][HC_2O_4^-]}{[H_2C_2O_4]} = 5.9 \times 10^{-2} = \frac{x^2}{0.20 - x}$$

$x^2 + 0.059x - 0.0118 = 0$

Use the quadratic formula to solve for x.

$$x = \frac{-(0.059) \pm \sqrt{(0.059)^2 - 4(1)(-0.0118)}}{2(1)} = \frac{-0.059 \pm 0.225}{2}$$

$x = 0.083$ and -0.142

Of the two solutions for x, only the positive value of x has physical meaning, because x is the $[H_3O^+]$.

$[H_3O^+] = [HC_2O_4^-] = 0.083$ M

For the dissociation of the second proton, the following equilibrium must be considered:

$$HC_2O_4^-(aq) + H_2O(l) \rightleftarrows H_3O^+(aq) + C_2O_4^{2-}(aq)$$

	$HC_2O_4^-(aq)$	$H_3O^+(aq)$	$C_2O_4^{2-}(aq)$
initial (M)	0.083	0.083	0
change (M)	−x	+x	+x
equil (M)	0.083 − x	0.083 + x	x

$$K_{a2} = \frac{[H_3O^+][C_2O_4^{2-}]}{[HC_2O_4^-]} = 6.4 \times 10^{-5} = \frac{(0.083 + x)(x)}{0.083 - x} \approx \frac{(0.083)(x)}{0.083} = x$$

$[H_3O^+] = 0.083 + x = 0.083$ M
pH = $-\log[H_3O^+] = -\log(0.083) = 1.08$
$[C_2O_4^{2-}] = x = 6.4 \times 10^{-5}$ M

15.100 From the complete dissociation of the first proton, $[H_3O^+] = [HSeO_4^-] = 0.50$ M.
For the dissociation of the second proton, the following equilibrium must be considered:

$$HSeO_4^-(aq) + H_2O(l) \rightleftarrows H_3O^+(aq) + SeO_4^{2-}(aq)$$

	$HSeO_4^-(aq)$	$H_3O^+(aq)$	$SeO_4^{2-}(aq)$
initial (M)	0.50	0.50	0
change (M)	−x	+x	+x
equil (M)	0.50 − x	0.50 + x	x

$$K_{a2} = \frac{[H_3O^+][SeO_4^{2-}]}{[HSeO_4^-]} = 1.2 \times 10^{-2} = \frac{(0.50 + x)(x)}{0.50 - x}$$

$x^2 + 0.512x - 0.0060 = 0$
Use the quadratic formula to solve for x.

$$x = \frac{-(0.512) \pm \sqrt{(0.512)^2 - 4(-0.0060)}}{2(1)} = \frac{-0.512 \pm 0.535}{2}$$

x = 0.011 and −0.524
Of the two solutions for x, only the positive value of x has physical meaning, since x is the $[SeO_4^{2-}]$.
$[H_2SeO_4] = 0$ M; $[HSeO_4^-] = 0.50 - x = 0.49$ M; $[SeO_4^{2-}] = x = 0.011$ M
$[H_3O^+] = 0.50 + x = 0.51$ M
pH = $-\log[H_3O^+] = -\log(0.51) = 0.29$

$$[OH^-] = \frac{K_w}{[H_3O^+]} = \frac{1.0 \times 10^{-14}}{0.51} = 2.0 \times 10^{-14} \text{ M}$$

Weak Bases; Relation Between K_a and K_b (Sections 15.12 and 15.13)

15.102 (a) $(CH_3)_2NH(aq) + H_2O(l) \rightleftarrows (CH_3)_2NH_2^+(aq) + OH^-(aq)$; $K_b = \frac{[(CH_3)_2NH_2^+][OH^-]}{[(CH_3)_2NH]}$

(b) $C_6H_5NH_2(aq) + H_2O(l) \rightleftarrows C_6H_5NH_3^+(aq) + OH^-(aq)$; $K_b = \frac{[C_6H_5NH_3^+][OH^-]}{[C_6H_5NH_2]}$

(c) $CN^-(aq) + H_2O(l) \rightleftarrows HCN(aq) + OH^-(aq)$; $K_b = \frac{[HCN][OH^-]}{[CN^-]}$

15.104 $C_{21}H_{22}N_2O_2$, 334.42; 16 mg = 0.016 g

$$\text{molarity} = \frac{\left(0.016 \text{ g x } \frac{1 \text{ mol}}{334.42 \text{ g}}\right)}{0.100 \text{ L}} = 4.8 \text{ x } 10^{-4} \text{ M}$$

	$C_{21}H_{22}N_2O_2(aq)$ + $H_2O(l)$	⇄ $C_{21}H_{23}N_2O_2^+(aq)$	+ $OH^-(aq)$
initial (M)	$4.8 \text{ x } 10^{-4}$	0	~0
change (M)	–x	+x	+x
equil (M)	$(4.8 \text{ x } 10^{-4}) - x$	x	x

$$K_b = \frac{[C_{21}H_{23}N_2O_2^+][OH^-]}{[C_{21}H_{22}N_2O_2]} = 1.8 \text{ x } 10^{-6} = \frac{x^2}{(4.8 \text{ x } 10^{-4}) - x}$$

$x^2 + (1.8 \text{ x } 10^{-6})x - (8.6 \text{ x } 10^{-10}) = 0$

Use the quadratic formula to solve for x.

$$x = \frac{-(1.8 \text{ x } 10^{-6}) \pm \sqrt{(1.8 \text{ x } 10^{-6})^2 - (4)(-8.6 \text{ x } 10^{-10})}}{2(1)} = \frac{(-1.8 \text{ x } 10^{-6}) \pm (5.87 \text{ x } 10^{-5})}{2}$$

$x = 2.84 \text{ x } 10^{-5}$ and $-3.02 \text{ x } 10^{-5}$

Of the two solutions for x, only the positive value of x has physical meaning, because x is the $[OH^-]$.

$[OH^-] = 2.84 \text{ x } 10^{-5}$ M

$$[H_3O^+] = \frac{K_w}{[OH^-]} = \frac{1.0 \text{ x } 10^{-14}}{2.84 \text{ x } 10^{-5}} = 3.52 \text{ x } 10^{-10} \text{ M}$$

$pH = -\log[H_3O^+] = -\log(3.52 \text{ x } 10^{-10}) = 9.45$

15.106 $[H_3O^+] = 10^{-pH} = 10^{-9.5} = 3.16 \text{ x } 10^{-10}$ M

$$[OH^-] = \frac{K_w}{[H_3O^+]} = \frac{1.0 \text{ x } 10^{-14}}{3.16 \text{ x } 10^{-10}} = 3.16 \text{ x } 10^{-5} \text{ M}$$

	$C_{17}H_{19}NO_3(aq)$ + $H_2O(l)$	⇄ $C_{17}H_{20}NO_3^+(aq)$	+ $OH^-(aq)$
initial (M)	$7.0 \text{ x } 10^{-4}$	0	~0
change (M)	–x	+x	+x
equil (M)	$(7.0 \text{ x } 10^{-4}) - x$	x	x

$x = [OH^-] = 3.16 \text{ x } 10^{-5}$ M

$$K_b = \frac{[C_{17}H_{20}NO_3^+][OH^-]}{[C_{17}H_{19}NO_3]} = \frac{x^2}{(7.0 \text{ x } 10^{-4}) - x} = \frac{(3.16 \text{ x } 10^{-5})^2}{(7.0 \text{ x } 10^{-4}) - (3.16 \text{ x } 10^{-5})} = 1.49 \text{ x } 10^{-6}$$

$K_b = 1 \text{ x } 10^{-6}$

$pK_b = -\log K_b = -\log(1.49 \text{ x } 10^{-6}) = 5.827 = 5.8$

15.108 $K_b = 10^{-pK_b} = 10^{-5.47} = 3.4 \text{ x } 10^{-6}$

	$C_{18}H_{21}NO_4(aq)$ + $H_2O(l)$	⇄ $HC_{18}H_{21}NO_4^+(aq)$	+ $OH^-(aq)$
initial (M)	$2.50 \text{ x } 10^{-3}$	0	~0
change (M)	–x	+x	+x
equil (M)	$(2.50 \text{ x } 10^{-3}) - x$	x	x

$$K_b = \frac{[HC_{18}H_{21}NO_4^+][OH^-]}{[C_{18}H_{21}NO_4]} = 3.4 \times 10^{-6} = \frac{x^2}{(2.50 \times 10^{-3}) - x}$$

$x^2 + (3.4 \times 10^{-6})x - (8.5 \times 10^{-9}) = 0$

Use the quadratic formula to solve for x.

$$x = \frac{-(3.4 \times 10^{-6}) \pm \sqrt{(3.4 \times 10^{-6})^2 - 4(1)(-8.5 \times 10^{-9})}}{2(1)}$$

$$x = = \frac{-(3.4 \times 10^{-6}) \pm (1.84 \times 10^{-4})}{2}$$

$x = 9.0 \times 10^{-5}$ and -1.9×10^{-4}

Of the two solutions for x, only the positive value of x has physical meaning, because x is the $[OH^-]$.

$x = 9.0 \times 10^{-5}\ M = [OH^-] = [HC_{18}H_{21}NO_4^+]$

$[C_{18}H_{21}NO_4] = (2.50 \times 10^{-3}) - x = (2.50 \times 10^{-3}) - (9.0 \times 10^{-5}) = 0.0024\ M$

$$[H_3O^+] = \frac{K_w}{[OH^-]} = \frac{1.0 \times 10^{-14}}{9.0 \times 10^{-5}} = 1.1 \times 10^{-10}\ M$$

$pH = -\log[H_3O^+] = -\log(1.1 \times 10^{-10}) = 9.96$

15.110 (a) $K_a = \dfrac{K_w}{K_b \text{ for } C_3H_7NH_2} = \dfrac{1.0 \times 10^{-14}}{5.1 \times 10^{-4}} = 2.0 \times 10^{-11}$

(b) $K_a = \dfrac{K_w}{K_b \text{ for } NH_2OH} = \dfrac{1.0 \times 10^{-14}}{9.1 \times 10^{-9}} = 1.1 \times 10^{-6}$

(c) $K_a = \dfrac{K_w}{K_b \text{ for } C_6H_5NH_2} = \dfrac{1.0 \times 10^{-14}}{4.3 \times 10^{-10}} = 2.3 \times 10^{-5}$

(d) $K_a = \dfrac{K_w}{K_b \text{ for } C_5H_5N} = \dfrac{1.0 \times 10^{-14}}{1.8 \times 10^{-9}} = 5.6 \times 10^{-6}$

Acid–Base Properties of Salts (Section 15.14)

15.112 (a) $CH_3NH_3^+(aq) + H_2O(l) \rightleftarrows H_3O^+(aq) + CH_3NH_2(aq)$

acid base —— acid base (conjugate pairs: $CH_3NH_3^+$/CH_3NH_2; H_2O/H_3O^+)

(b) $Cr(H_2O)_6^{3+}(aq) + H_2O(l) \rightleftarrows H_3O^+(aq) + Cr(H_2O)_5(OH)^{2+}(aq)$

acid base —— acid base (conjugate pairs: $Cr(H_2O)_6^{3+}$/$Cr(H_2O)_5(OH)^{2+}$; H_2O/H_3O^+)

(c) $CH_3CO_2^-(aq) + H_2O(l) \rightleftarrows CH_3CO_2H(aq) + OH^-(aq)$

base acid acid base (conjugate pairs: $CH_3CO_2^-$/CH_3CO_2H; H_2O/OH^-)

(d) $PO_4^{3-}(aq) + H_2O(l) \rightleftarrows HPO_4^{2-}(aq) + OH^-(aq)$

base acid acid base

15.114 (a) F^- (conjugate base of a weak acid), basic solution
(b) Br^- (anion of a strong acid), neutral solution
(c) NH_4^+ (conjugate acid of a weak base), acidic solution
(d) $K(H_2O)_6^+$ (neutral cation), neutral solution
(e) SO_3^{2-} (conjugate base of a weak acid), basic solution
(f) $Cr(H_2O)_6^{3+}$ (acidic cation), acidic solution

15.116 (a) $(C_2H_5NH_3)NO_3$: $C_2H_5NH_3^+$, acidic cation; NO_3^-, neutral anion
$C_2H_5NH_2$, $K_b = 6.4 \times 10^{-4}$

$$C_2H_5NH_3^+,\ K_a = \frac{K_w}{K_b \text{ for } C_2H_5NH_2} = \frac{1.0 \times 10^{-14}}{6.4 \times 10^{-4}} = 1.56 \times 10^{-11}$$

	$C_2H_5NH_3^+(aq)$ + $H_2O(l)$	$\rightleftarrows$ $H_3O^+(aq)$	+ $C_2H_5NH_2(aq)$
initial (M)	0.10	~0	0
change (M)	–x	+x	+x
equil (M)	0.10 – x	x	x

$$K_a = \frac{[H_3O^+][C_2H_5NH_2]}{[C_2H_5NH_3^+]} = 1.56 \times 10^{-11} = \frac{x^2}{0.10 - x} \approx \frac{x^2}{0.10}$$

Solve for x. $x = 1.25 \times 10^{-6}\ M = 1.2 \times 10^{-6}\ M = [H_3O^+] = [C_2H_5NH_2]$
$pH = -\log[H_3O^+] = -\log(1.25 \times 10^{-6}) = 5.90$
$[C_2H_5NH_3^+] = 0.10 - x = 0.10\ M$; $[NO_3^-] = 0.10\ M$

$$[OH^-] = \frac{K_w}{[H_3O^+]} = \frac{1.0 \times 10^{-14}}{1.25 \times 10^{-6}\ M} = 8.0 \times 10^{-9}$$

(b) $Na(CH_3CO_2)$: Na^+, neutral cation; $CH_3CO_2^-$, basic anion
CH_3CO_2H, $K_a = 1.8 \times 10^{-5}$

$$CH_3CO_2^-,\ K_b = \frac{K_w}{K_a \text{ for } CH_3CO_2H} = \frac{1.0 \times 10^{-14}}{1.8 \times 10^{-5}} = 5.6 \times 10^{-10}$$

	$CH_3CO_2^-(aq)$ + $H_2O(aq)$	$\rightleftarrows$ $CH_3CO_2H(aq)$	+ $OH^-(aq)$
initial (M)	0.10	0	~0
change (M)	–x	+x	+x
equil (M)	0.10 – x	x	x

$$K_b = \frac{[CH_3CO_2H][OH^-]}{[CH_3CO_2^-]} = 5.6 \times 10^{-10} = \frac{x^2}{0.10 - x} \approx \frac{x^2}{0.10}$$

Solve for x. $x = 7.5 \times 10^{-6}\ M = [CH_3CO_2H] = [OH^-]$
$[CH_3CO_2^-] = 0.10 - x = 0.10\ M$; $[Na^+] = 0.10\ M$

$$[H_3O^+] = \frac{K_w}{[OH^-]} = \frac{1.0 \times 10^{-14}}{7.5 \times 10^{-6}} = 1.3 \times 10^{-9}\ M$$

$pH = -\log[H_3O^+] = -\log(1.3 \times 10^{-9}) = 8.89$
(c) $NaNO_3$: Na^+, neutral cation; NO_3^-, neutral anion
$[Na^+] = [NO_3^-] = 0.10$ M
$[H_3O^+] = [OH^-] = 1.0 \times 10^{-7}$ M; pH = 7.00

15.118 For NH_4^+, $K_a = \dfrac{K_w}{K_b \text{ for } NH_3} = \dfrac{1.0 \times 10^{-14}}{1.8 \times 10^{-5}} = 5.6 \times 10^{-10}$

For CN^-, $K_b = \dfrac{K_w}{K_a \text{ for HCN}} = \dfrac{1.0 \times 10^{-14}}{4.9 \times 10^{-10}} = 2.0 \times 10^{-5}$

Because $K_b > K_a$, the solution is basic.

Lewis Acids and Bases (Section 15.16)

15.120 (a) Lewis acid, SiF_4; Lewis base, F^-
(b) Lewis acid, Zn^{2+}; Lewis base, NH_3
(c) Lewis acid, $HgCl_2$; Lewis base, Cl^-
(d) Lewis acid, CO_2; Lewis base, H_2O

15.122 (a) $2\ :\ddot{F}:^- + SiF_4 \longrightarrow SiF_6^{2-}$
(b) $4\ \ddot{N}H_3 + Zn^{2+} \longrightarrow Zn(NH_3)_4^{2+}$
(c) $2\ :\ddot{Cl}:^- + HgCl_2 \longrightarrow HgCl_4^{2-}$
(d) $H_2\ddot{O}: + CO_2 \longrightarrow H_2CO_3$

15.124 (a) CN^-, Lewis base
(b) H^+, Lewis acid
(c) H_2O, Lewis base
(d) Fe^{3+}, Lewis acid
(e) OH^-, Lewis base
(f) CO_2, Lewis acid
(g) $P(CH_3)_3$, Lewis base
(h) $B(CH_3)_3$, Lewis acid

Chapter Problems

15.126 In aqueous solution:
H_2S acts as an acid only.
HS^- can act as both an acid and a base.
S^{2-} can act as a base only.
H_2O can act as both an acid and a base.
H_3O^+ acts as an acid only.
OH^- acts as a base only.

15.128 $HCO_3^-(aq) + Al(H_2O)_6^{3+}(aq) \rightarrow H_2O(l) + CO_2(g) + Al(H_2O)_5(OH)^{2+}(aq)$

15.130 H_2O, 18.02

$$\text{at } 0\ ^\circ\text{C}, [H_2O] = \frac{\left(0.9998 \text{ g} \times \dfrac{1 \text{ mol}}{18.02 \text{ g}}\right)}{0.001 \text{ L}} = 55.48 \text{ M}$$

$K_w = [H_3O^+][OH^-]$, for a neutral solution $[H_3O^+] = [OH^-]$

$[H_3O^+] = \sqrt{K_w} = \sqrt{1.14 \times 10^{-15}} = 3.376 \times 10^{-8}$ M

$pH = -\log[H_3O^+] = -\log(3.376 \times 10^{-8}) = 7.472$

$$\text{fraction dissociated} = \frac{[H_2O]_{diss}}{[H_2O]_{initial}} = \frac{3.376 \times 10^{-8}\text{ M}}{55.48\text{ M}} = 6.09 \times 10^{-10}$$

$$\%\text{ dissociation} = \frac{[H_2O]_{diss}}{[H_2O]_{initial}} \times 100\% = \frac{3.376 \times 10^{-8}\text{ M}}{55.48\text{ M}} \times 100\% = 6.09 \times 10^{-8}\%$$

15.132

	$HA(aq)$ + $H_2O(l)$ ⇄	$H_3O^+(aq)$ +	$A^-(aq)$
initial (M)	0.050	~0	0
change (M)	–x	+x	+x
equil (M)	0.050 – x	x	x

$x = [H_3O^+] = 10^{-pH} = 10^{-2.86} = 1.38 \times 10^{-3}$ M

$$K_a = \frac{[H_3O^+][A^-]}{[HA]} = \frac{x^2}{0.050 - x} = \frac{(1.38 \times 10^{-3})^2}{0.050 - (1.38 \times 10^{-3})} = 3.92 \times 10^{-5} = 3.9 \times 10^{-5}$$

$pK_a = -\log K_a = -\log(3.92 \times 10^{-5}) = 4.41$

15.134 For $C_{10}H_{14}N_2H^+$, $$K_{a1} = \frac{K_w}{K_{b1}\text{ for }C_{10}H_{14}N_2} = \frac{1.0 \times 10^{-14}}{1.0 \times 10^{-6}} = 1.0 \times 10^{-8}$$

For $C_{10}H_{14}N_2H_2^{2+}$, $$K_{a2} = \frac{K_w}{K_{b2}\text{ for }C_{10}H_{14}N_2H^+} = \frac{1.0 \times 10^{-14}}{1.3 \times 10^{-11}} = 7.7 \times 10^{-4}$$

15.136 (a) $A^-(aq) + H_2O(l) \rightleftarrows HA(aq) + OH^-(aq)$; basic

(b) $M(H_2O)_6^{3+}(aq) + H_2O(l) \rightleftarrows H_3O^+(aq) + M(H_2O)_5(OH)^{2+}(aq)$; acidic

(c) $2\,H_2O(l) \rightleftarrows H_3O^+(aq) + OH^-(aq)$; neutral

(d) $M(H_2O)_6^{3+}(aq) + A^-(aq) \rightleftarrows HA(aq) + M(H_2O)_5(OH)^{2+}(aq)$;
acidic because K_a for $M(H_2O)_6^{3+}$ (10^{-4}) is greater than K_b for A^- (10^{-9})

15.138

	$HIO_3(aq)$ + $H_2O(l)$ ⇄	$H_3O^+(aq)$ +	$IO_3^-(aq)$
initial (M)	0.0500	~0	0
change (M)	–x	+x	+x
equil (M)	0.0500 – x	x	x

$$K_a = \frac{[H_3O^+][IO_3^-]}{[HIO_3]} = 1.7 \times 10^{-1} = \frac{x^2}{0.0500 - x}$$

$x^2 + 0.17x - 0.0085 = 0$

Use the quadratic formula to solve for x.

$$x = \frac{-(0.17) \pm \sqrt{(0.17)^2 - (4)(1)(-0.0085)}}{2(1)} = \frac{(-0.17) \pm 0.251}{2}$$

x = –0.210 and 0.0405

Of the two solutions for x, only the positive value of x has physical meaning because x is the $[H_3O^+]$.

$x = [H_3O^+] = 0.0405\ M = 0.040\ M$

$pH = -\log[H_3O^+] = -\log(0.040) = 1.39$

$[HIO_3] = 0.0500 - x = 0.0500 - 0.040 = 0.010\ M$

$[IO_3^-] = x = 0.040\ M$

$$[OH^-] = \frac{K_w}{[H_3O^+]} = \frac{1.0 \times 10^{-14}}{0.040} = 2.5 \times 10^{-13}\ M$$

15.140 $K_{a1} = 10^{-pK_{a1}} = 10^{-2.89} = 1.3 \times 10^{-3}$; $K_{a2} = 10^{-pK_{a2}} = 10^{-5.51} = 3.1 \times 10^{-6}$

	$H_2C_8H_4O_4(aq)$ + $H_2O(l)$	⇄	$H_3O^+(aq)$	+ $HC_8H_4O_4^-(aq)$
initial (M)	0.0250		~0	0
change (M)	–x		+x	+x
equil (M)	0.0250 – x		x	x

$$K_{a1} = \frac{[H_3O^+][HC_8H_4O_4^-]}{[H_2C_8H_4O_4]} = 1.3 \times 10^{-3} = \frac{x^2}{0.0250 - x}$$

$x^2 + (1.3 \times 10^{-3})x - (3.25 \times 10^{-5}) = 0$

Use the quadratic formula to solve for x.

$$x = \frac{-(1.3 \times 10^{-3}) \pm \sqrt{(1.3 \times 10^{-3})^2 - (4)(1)(-3.25 \times 10^{-5})}}{2(1)} = \frac{-(1.3 \times 10^{-3}) \pm 0.0115}{2}$$

$x = 0.0051$ and -0.0064

Of the two solutions for x, only the positive value of x has physical meaning because x is the $[H_3O^+]$.

$x = 0.0051\ M = [H_3O^+] = [HC_8H_4O_4^-]$

$[H_2C_8H_4O_4] = 0.0250 - x = 0.0250 - 0.0051 = 0.020\ M$

The second dissociation of $H_2C_7H_3NO_4$ produces a negligible amount of H_3O^+ compared with that from the first dissociation.

$HC_8H_4O_4^-(aq) + H_2O(l) \rightleftharpoons H_3O^+(aq) + C_8H_4O_4^{2-}(aq)$

$$K_{a2} = \frac{[H_3O^+][C_8H_4O_4^{2-}]}{[HC_8H_4O_4^-]} = 3.1 \times 10^{-6} = \frac{(0.0051)[C_8H_4O_4^{2-}]}{(0.0051)}$$

$[C_8H_4O_4^{2-}] = K_{a2} = 3.1 \times 10^{-6}\ M$

$$[OH^-] = \frac{K_w}{[H_3O^+]} = \frac{1.0 \times 10^{-14}}{0.0051} = 2.0 \times 10^{-12}\ M$$

$pH = -\log[H_3O^+] = -\log(0.0051) = 2.29$

15.142 (a) NH_4F; For NH_4^+, $K_a = 5.6 \times 10^{-10}$ and for F^-, $K_b = 2.9 \times 10^{-11}$

Because $K_a > K_b$, the salt solution is acidic.

(b) $(NH_4)_2SO_3$; For NH_4^+, $K_a = 5.6 \times 10^{-10}$ and for SO_3^{2-}, $K_b = 1.6 \times 10^{-7}$

Because $K_b > K_a$, the salt solution is basic.

15.144 Fraction dissociated = $\dfrac{[HA]_{diss}}{[HA]_{initial}}$

For a weak acid, $[HA]_{diss} = [H_3O^+] = [A^-]$

$$K_a = \frac{[H_3O^+][A^-]}{[HA]} = \frac{[H_3O^+]^2}{[HA]}; \quad [H_3O^+] = \sqrt{K_a[HA]}$$

$$\text{Fraction dissociated} = \frac{[HA]_{diss}}{[HA]} = \frac{[H_3O^+]}{[HA]} = \frac{\sqrt{K_a[HA]}}{[HA]} = \sqrt{\frac{K_a}{[HA]}}$$

When the concentration of HA that dissociates is negligible compared with its initial concentration, the equilibrium concentration, [HA], equals the initial concentration, $[HA]_{initial}$.

$$\%\text{ dissociation} = \sqrt{\frac{K_a}{[HA]_{initial}}} \times 100\%$$

15.146 Both reactions occur together.

Let x = $[H_3O^+]$ from CH_3CO_2H and y = $[H_3O^+]$ from $C_6H_5CO_2H$

The following two equilibria must be considered:

	$CH_3CO_2H(aq)$ + $H_2O(l)$	⇄	$H_3O^+(aq)$ +	$CH_3CO_2^-(aq)$
initial (M)	0.10		y	0
change (M)	–x		+x	+x
equil (M)	0.10 – x		x + y	x

	$C_6H_5CO_2H(aq)$ + $H_2O(l)$	⇄	$H_3O^+(aq)$ +	$C_6H_5CO_2^-(aq)$
initial (M)	0.10		x	0
change (M)	–y		+y	+y
equil (M)	0.10 – y		x + y	y

$$K_a(\text{for } CH_3CO_2H) = \frac{[H_3O^+][CH_3CO_2^-]}{[CH_3CO_2H]} = 1.8 \times 10^{-5} = \frac{(x+y)(x)}{0.10 - x} \approx \frac{(x+y)(x)}{0.10}$$

$1.8 \times 10^{-6} = (x + y)(x)$

$$K_a(\text{for } C_6H_5CO_2H) = \frac{[H_3O^+][C_6H_5CO_2^-]}{[C_6H_5CO_2H]} = 6.5 \times 10^{-5} = \frac{(x+y)(y)}{0.10 - y} \approx \frac{(x+y)(y)}{0.10}$$

$6.5 \times 10^{-6} = (x + y)(y)$

$1.8 \times 10^{-6} = (x + y)(x)$

$6.5 \times 10^{-6} = (x + y)(y)$

These two equations must be solved simultaneously for x and y. Divide the first equation by the second.

$$\frac{x}{y} = \frac{1.8 \times 10^{-6}}{6.5 \times 10^{-6}}; \quad x = 0.277y$$

$6.5 \times 10^{-6} = (x + y)(y)$; substitute x = 0.277y into this equation and solve for y.

$6.5 \times 10^{-6} = (0.277y + y)(y) = 1.277y^2$
$y = 0.002\,256$
$x = 0.277y = (0.277)(0.002\,256) = 0.000\,624\,9$
$[H_3O^+] = (x + y) = (0.000\,624\,9 + 0.002\,256) = 0.002\,881$ M
$pH = -\log[H_3O^+] = -\log(0.002\,881) = 2.54$

15.148 $2\ NO_2(g) + H_2O(l) \rightarrow HNO_3(aq) + HNO_2(aq)$

$$\text{mol } HNO_3 = 0.0500 \text{ mol } NO_2 \times \frac{1 \text{ mol } HNO_3}{2 \text{ mol } NO_2} = 0.0250 \text{ mol } HNO_3$$

$$\text{mol } HNO_2 = 0.0500 \text{ mol } NO_2 \times \frac{1 \text{ mol } HNO_2}{2 \text{ mol } NO_2} = 0.0250 \text{ mol } HNO_2$$

Because the volume is 1.00 L, mol and molarity are the same.
HNO_3 is a strong acid and completely dissociated. From HNO_3, $[NO_3^-] = [H_3O^+] = 0.0250$ M.

	$HNO_2(aq)$ + $H_2O(l)$	$\rightleftarrows$ $H_3O^+(aq)$	+ $NO_2^-(aq)$
initial (M)	0.0250	0.0250	0
change (M)	–x	+x	+x
equil (M)	0.0250 – x	0.0250 + x	x

$$K_a = \frac{[H_3O^+][NO_2^-]}{[HNO_2]} = 4.5 \times 10^{-4} = \frac{(0.0250 + x)x}{0.0250 - x}$$

$x^2 + 0.02545x - 1.125 \times 10^{-5} = 0$
Solve for x using the quadratic formula.

$$x = \frac{-(0.025\,45) \pm \sqrt{(0.025\,45)^2 - (4)(-1.125 \times 10^{-5})}}{2(1)} = \frac{(-0.025\,45) \pm (0.026\,32)}{2}$$

$x = -0.0259$ and 4.35×10^{-4}
Of the two solutions for x, only the positive value of x has physical meaning, because x is the $[NO_2^-]$.
$[H_3O^+] = (0.0250) + x = (0.0250) + (4.35 \times 10^{-4}) = 0.0254$ M
$pH = -\log[H_3O^+] = -\log(0.02543) = 1.59$

$$[OH^-] = \frac{K_w}{[H_3O^+]} = \frac{1.0 \times 10^{-14}}{0.0254} = 3.9 \times 10^{-13} \text{ M}$$

$[NO_3^-] = 0.0250$ M
$[HNO_2] = 0.0250 - x = 0.0250 - 4.35 \times 10^{-4} = 0.0246$ M; $[NO_2^-] = x = 4.3 \times 10^{-4}$ M

Multiconcept Problems

15.150 H_3PO_4, 98.00
Assume 1.000 L of solution.

$$\text{Mass of solution} = 1.000 \text{ L} \times \frac{1000 \text{ mL}}{1 \text{ L}} \times \frac{1.0353 \text{ g}}{1 \text{ mL}} = 1035.3 \text{ g}$$

Mass $H_3PO_4 = (0.070)(1035.3 \text{ g}) = 72.47$ g H_3PO_4

$\text{mol } H_3PO_4 = 72.47 \text{ g } H_3PO_4 \times \dfrac{1 \text{ mol } H_3PO_4}{98.00 \text{ g } H_3PO_4} = 0.740 \text{ mol } H_3PO_4$

$[H_3PO_4] = \dfrac{0.740 \text{ mol } H_3PO_4}{1.000 \text{ L}} = 0.740 \text{ M}$

For the dissociation of the first proton, the following equilibrium must be considered:

	$H_3PO_4(aq)$ + $H_2O(l)$ ⇄	$H_3O^+(aq)$ +	$H_2PO_4^-(aq)$
initial (M)	0.740	~0	0
change (M)	–x	+x	+x
equil (M)	0.740 – x	x	x

$K_{a1} = \dfrac{[H_3O^+][H_2PO_4^-]}{[H_3PO_4]} = 7.5 \times 10^{-3} = \dfrac{x^2}{0.740 - x}$

$x^2 + (7.5 \times 10^{-3})x - (5.55 \times 10^{-3}) = 0$

Solve for x using the quadratic formula.

$x = \dfrac{-(7.5 \times 10^{-3}) \pm \sqrt{(7.5 \times 10^{-3})^2 - (4)(-5.55 \times 10^{-3})}}{2(1)} = \dfrac{(-7.5 \times 10^{-3}) \pm 0.149}{2}$

x = 0.0708 and –0.0783

Of the two solutions for x, only the positive value of x has physical meaning, because x is the $[H_3O^+]$.

$x = 0.0708 \text{ M} = [H_2PO_4^-] = [H_3O^+]$

For the dissociation of the second proton, the following equilibrium must be considered:

	$H_2PO_4^-(aq)$ + $H_2O(l)$ ⇄	$H_3O^+(aq)$ +	$HPO_4^{2-}(aq)$
initial (M)	0.0708	0.0708	0
change (M)	–y	+y	+y
equil (M)	0.0708 – y	0.0708 + y	y

$K_{a2} = \dfrac{[H_3O^+][HPO_4^{2-}]}{[H_2PO_4^-]} = 6.2 \times 10^{-8} = \dfrac{(0.0708 + y)(y)}{0.0708 - y} \approx \dfrac{(0.0708)(y)}{0.0708} = y$

$y = 6.2 \times 10^{-8} \text{ M} = [HPO_4^{2-}]$

For the dissociation of the third proton, the following equilibrium must be considered:

	$HPO_4^{2-}(aq)$ + $H_2O(l)$ ⇄	$H_3O^+(aq)$ +	$PO_4^{3-}(aq)$
initial (M)	6.2×10^{-8}	0.0708	0
change (M)	–z	+z	+z
equil (M)	$(6.2 \times 10^{-8}) - z$	0.0708 + z	z

$K_{a3} = \dfrac{[H_3O^+][PO_4^{3-}]}{[HPO_4^{2-}]} = 4.8 \times 10^{-13} = \dfrac{(0.0708 + z)(z)}{(6.2 \times 10^{-8}) - z} \approx \dfrac{(0.0708)(z)}{6.2 \times 10^{-8}}$

$z = 4.2 \times 10^{-19} \text{ M} = [PO_4^{3-}]$

$[H_3PO_4] = 0.740 - x = 0.740 - 0.0708 = 0.67 \text{ M}$

$[H_2PO_4^-] = [H_3O^+] = 0.0708 \text{ M} = 0.071 \text{ M}$

$[HPO_4^{2-}] = 6.2 \times 10^{-8} \text{ M}$

$[PO_4^{3-}] = 4.2 \times 10^{-19} \text{ M}$

$$[OH^-] = \frac{K_w}{[H_3O^+]} = \frac{1.0 \times 10^{-14}}{0.0708} = 1.4 \times 10^{-13}\ M$$

$pH = -\log[H_3O^+] = -\log(0.0708) = 1.15$

15.152 $[H_3O^+] = 10^{-pH} = 10^{-9.07} = 8.51 \times 10^{-10}\ M$

$[H_3O^+][OH^-] = K_w$

$$[OH^-] = \frac{K_w}{[H_3O^+]} = \frac{1.0 \times 10^{-14}}{8.51 \times 10^{-10}} = 1.18 \times 10^{-5}\ M = x \text{ below.}$$

$K_a = 1.8 \times 10^{-5}$ for CH_3CO_2H and $K_b = \frac{K_w}{K_a} = \frac{1.0 \times 10^{-14}}{1.8 \times 10^{-5}} = 5.56 \times 10^{-10}$

Use the equilibrium associated with a weak base to solve for $[CH_3CO_2^-] = y$ below.

	$CH_3CO_2^-(aq)$ + $H_2O(l)$	⇄	$CH_3CO_2H(aq)$	+ $OH^-(aq)$
initial (M)	y		~0	0
change (M)	−x		+x	+x
equil (M)	y − x		x	x

$$K_b = \frac{[CH_3CO_2H][OH^-]}{[CH_3CO_2^-]} = 5.56 \times 10^{-10} = \frac{(1.18 \times 10^{-5})^2}{[y - (1.18 \times 10^{-5})]}$$

Solve for y.

$(5.56 \times 10^{-10})[y - (1.18 \times 10^{-5})] = (1.18 \times 10^{-5})^2$

$(5.56 \times 10^{-10})y - 6.56 \times 10^{-15} = 1.39 \times 10^{-10}$

$(5.56 \times 10^{-10})y = 1.39 \times 10^{-10}$

$y = (1.39 \times 10^{-10})/(5.56 \times 10^{-10}) = [CH_3CO_2^-] = 0.25\ M$

In 1.00 L of solution, the mass of CH_3CO_2Na solute $= (0.25\ mol/L)\left(\frac{82.035\ g\ CH_3CO_2Na}{1\ mol\ CH_3CO_2Na}\right) = 20.5\ g$

mass of solution $= (1000\ mL)\left(\frac{1.0085\ g}{1\ mL}\right) = 1008.5\ g$

mass of solvent = 1008.5 g − 20.5 g = 988 g = 0.988 kg

$$m = \frac{0.25\ mol\ CH_3CO_2Na}{0.988\ kg} = 0.25\ m$$

Because CH_3CO_2Na is a strong electrolyte, the ionic compound is completely dissociated and $[CH_3CO_2^-] = [Na^+]$. The contribution of CH_3CO_2H and OH^- to the total molality of the solution is negligible.

$\Delta T_f = K_f \cdot (2 \cdot m) = (1.86\ °C/m)(2)(0.25\ m) = 0.93\ °C$

Solution freezing point $= 0.00\ °C - \Delta T_f = 0.00\ °C - 0.93\ °C = -0.93\ °C$

15.154 Na_3PO_4, 163.94

$$3.28\ g\ Na_3PO_4 \times \frac{1\ mol\ Na_3PO_4}{163.94\ g\ Na_3PO_4} = 0.0200\ mol = 20.0\ mmol\ Na_3PO_4$$

300.0 mL x 0.180 mmol/mL = 54.0 mmol HCl

	$H_3O^+(aq)$ +	$PO_4^{3-}(aq)$ ⇄	$HPO_4^{2-}(aq)$ + $H_2O(l)$
before (mmol)	54.0	20.0	0
change (mmol)	–20.0	–20.0	+20.0
after (mmol)	34.0	0	20.0

	$H_3O^+(aq)$ +	$HPO_4^{2-}(aq)$ ⇄	$H_2PO_4^-(aq)$ + $H_2O(l)$
before (mmol)	34.0	20.0	0
change (mmol)	–20.0	–20.0	+20.0
after (mmol)	14.0	0	20.0

	$H_3O^+(aq)$ +	$H_2PO_4^-(aq)$ ⇄	$H_3PO_4(aq)$ + $H_2O(l)$
before (mmol)	14.0	20.0	0
change (mmol)	–14.0	–14.0	+14.0
after (mmol)	0	6.0	14.0

$$[H_3PO_4] = \frac{14.0 \text{ mmol}}{300.0 \text{ mL}} = 0.047 \text{ M}; \quad [H_2PO_4^-] = \frac{6.0 \text{ mmol}}{300.0 \text{ mL}} = 0.020 \text{ M}$$

	$H_3PO_4(aq)$ + $H_2O(l)$ ⇄	$H_3O^+(aq)$ +	$H_2PO_4^-(aq)$
initial (M)	0.047	~0	0.020
change (M)	–x	+x	+x
equil (M)	0.047 – x	x	0.020 + x

$$K_a = \frac{[H_3O^+][H_2PO_4^{2-}]}{[H_3PO_4]} = 7.5 \times 10^{-3} = \frac{x(0.020 + x)}{(0.047 - x)}$$

$x^2 + 0.0275x - (3.525 \times 10^{-4}) = 0$

Solve for x using the quadratic formula.

$$x = \frac{-(0.0275) \pm \sqrt{(0.0275)^2 - (4)(-3.525 \times 10^{-4})}}{2(1)} = \frac{-0.0275 \pm 0.0465}{2}$$

$x = 0.009\ 52$ and -0.0370

Of the two solutions for x, only the positive value of x has physical meaning, because x is the $[H_3O^+]$.

$pH = -\log[H_3O^+] = -\log(0.009\ 52) = 2.02$

15.156 (a) $PV = nRT$; $n = \frac{PV}{RT} = \frac{(0.601 \text{ atm})(1.000 \text{ L})}{\left(0.082\ 06 \frac{L \cdot atm}{K \cdot mol}\right)(293.1 \text{ K})} = 0.0250 \text{ mol HF}$

$$50.0 \text{ mL} \times \frac{1.00 \text{ L}}{1000 \text{ mL}} = 0.0500 \text{ L}$$

$$[HF] = \frac{0.0250 \text{ mol HF}}{0.0500 \text{ L}} = 0.500 \text{ M}$$

	$HF(aq)$ + $H_2O(l)$ ⇄	$H_3O^+(aq)$ +	$F^-(aq)$
initial (M)	0.500	~0	0
change (M)	–x	+x	+x
equil (M)	0.500 – x	x	x

$$K_a = \frac{[H_3O^+][F^-]}{[HF]} = 3.5 \text{ x } 10^{-4} = \frac{x^2}{0.500 - x}$$

$x^2 + (3.5 \text{ x } 10^{-4})x - (1.75 \text{ x } 10^{-4}) = 0$

Solve for x using the quadratic formula.

$$x = \frac{-(3.5 \text{ x } 10^{-4}) \pm \sqrt{(3.5 \text{ x } 10^{-4})^2 - (4)(-1.75 \text{ x } 10^{-4})}}{2(1)} = \frac{(-3.5 \text{ x } 10^{-4}) \pm 0.0265}{2}$$

x = –0.0134 and 0.0131

Of the two solutions for x, only the positive value of x has physical meaning, because x is the $[H_3O^+]$.

pH = $-\log[H_3O^+] = -\log(0.0131) = 1.883 = 1.88$

(b) $\% \text{ dissociation} = \frac{0.0131 \text{ M}}{0.500} \text{ x } 100\% = 2.62\% = 2.6\%$

New % dissociation = (3)(2.62%) = 7.86%

Let X equal the concentration of HF dissociated and Y the new volume (in liters) that would triple the % dissociation.

$$K_a = \frac{X^2}{(0.0250/Y) - X} = 3.5 \text{ x } 10^{-4}$$

$$\% \text{ dissociation} = \frac{X}{(0.0250/Y)} \text{ x } 100\% = 7.86\% \text{ and } \frac{X}{(0.0250/Y)} = 0.0786$$

$X = 1.965 \text{ x } 10^{-3}/Y$

Substitute X into the K_a equation.

$$\frac{(1.965 \text{ x } 10^{-3}/Y)^2}{(0.0250/Y) - (1.965 \text{ x } 10^{-3}/Y)} = 3.5 \text{ x } 10^{-4}$$

$$\frac{3.861 \text{ x } 10^{-6}/Y^2}{0.0230/Y} = 3.5 \text{ x } 10^{-4}$$

$$\frac{3.861 \text{ x } 10^{-6}/Y}{0.0230} = 3.5 \text{ x } 10^{-4}$$

$$\frac{3.861 \text{ x } 10^{-6}}{8.05 \text{ x } 10^{-6}} = Y = 0.48 \text{ L}$$

The result in Problem 15.144 can't be used here because the concentration of HF that dissociates can't be neglected compared with the initial HF concentration.

15.158 (a) Rate = $k[OCl^-]^x[NH_3]^y[OH^-]^z$

From experiments 1 & 2, the $[OCl^-]$ doubles and the rate doubles, therefore x = 1.

From experiments 2 & 3, the $[NH_3]$ triples and the rate triples, therefore y = 1.

From experiments 3 & 4, the $[OH^-]$ goes up by a factor of 10 and the rate goes down by a factor of 10, therefore z = –1.

$$\text{Rate} = k\frac{[OCl^-][NH_3]}{[OH^-]}$$

$[H_3O^+] = 10^{-pH} = 10^{-12} = 1 \times 10^{-12}$ M

$$[OH^-] = \frac{K_w}{[H_3O^+]} = \frac{1.0 \times 10^{-14}}{1 \times 10^{-12}} = 0.01 \text{ M}$$

$$\text{Using experiment 1: } k = \frac{(\text{Rate})[OH^-]}{[OCl^-][NH_3]} = \frac{(0.017 \text{ M/s})(0.01 \text{ M})}{(0.001 \text{ M})(0.01 \text{ M})} = 17 \text{ s}^{-1}$$

$$\text{(b) } K_1 = \frac{[HOCl][OH^-]}{[OCl^-]} = K_b(OCl^-) = \frac{K_w}{K_a(HOCl)} = \frac{1.0 \times 10^{-14}}{3.5 \times 10^{-8}} = 2.9 \times 10^{-7}$$

For second step, Rate = k_2 [HOCl][NH_3]

$$\text{Multiply the Rate by } \frac{K_1}{K_1} = \frac{K_1}{\left(\frac{[HOCl][OH^-]}{[OCl^-]}\right)}$$

$$\text{Rate} = K_1 k_2 \frac{[HOCl][NH_3]}{\left(\frac{[HOCl][OH^-]}{[OCl^-]}\right)} = K_1 k_2 \frac{[HOCl][NH_3][OCl^-]}{[HOCl][OH^-]} = K_1 k_2 \frac{[OCl^-][NH_3]}{[OH^-]}$$

$K_1 k_2 = k = 17 \text{ s}^{-1}$

$$k_2 = \frac{17 \text{ s}^{-1}}{2.9 \times 10^{-7} \text{ M}} = 5.9 \times 10^7 \text{ M}^{-1}\text{s}^{-1}$$

16 Applications of Aqueous Equilibria

16.1 (a) $HNO_2(aq) + OH^-(aq) \rightleftarrows NO_2^-(aq) + H_2O(l)$; NO_2^- (basic anion), pH > 7.00
(b) $H_3O^+(aq) + NH_3(aq) \rightleftarrows NH_4^+(aq) + H_2O(l)$; NH_4^+ (acidic cation), pH < 7.00
(c) $OH^-(aq) + H_3O^+(aq) \rightleftarrows 2\ H_2O(l)$; pH = 7.00

16.2 (a) $HF(aq) + OH^-(aq) \rightleftarrows H_2O(l) + F^-(aq)$

$$K_n = \frac{K_a}{K_w} = \frac{3.5 \times 10^{-4}}{1.0 \times 10^{-14}} = 3.5 \times 10^{10}$$

(b) $H_3O^+(aq) + OH^-(aq) \rightleftarrows 2\ H_2O(l)$

$$K_n = \frac{1}{K_w} = \frac{1}{1.0 \times 10^{-14}} = 1.0 \times 10^{14}$$

(c) $HF(aq) + NH_3(aq) \rightleftarrows NH_4^+(aq) + F^-(aq)$

$$K_n = \frac{K_aK_b}{K_w} = \frac{(3.5 \times 10^{-4})(1.8 \times 10^{-5})}{1.0 \times 10^{-14}} = 6.3 \times 10^5$$

The tendency to proceed to completion is determined by the magnitude of K_n. The larger the value of K_n, the further does the reaction proceed to completion.
The tendency to proceed to completion is: reaction (c) < reaction (a) < reaction (b)

16.3

	$HCN(aq)$ + $H_2O(l)$	$\rightleftarrows$ $H_3O^+(aq)$	+ $CN^-(aq)$
initial (M)	0.025	~0	0.010
change (M)	–x	+x	+x
equil (M)	0.025 – x	x	0.010 + x

$$K_a = \frac{[H_3O^+][CN^-]}{[HCN]} = 4.9 \times 10^{-10} = \frac{x(0.010 + x)}{0.025 - x} \approx \frac{x(0.010)}{0.025}$$

Solve for x. $x = 1.23 \times 10^{-9}$ M = 1.2×10^{-9} M = $[H_3O^+]$
$pH = -\log[H_3O^+] = -\log(1.23 \times 10^{-9}) = 8.91$

$$[OH^-] = \frac{K_w}{[H_3O^+]} = \frac{1.0 \times 10^{-14}}{1.23 \times 10^{-9}} = 8.1 \times 10^{-6}\ M$$

$[Na^+] = [CN^-] = 0.010$ M; $[HCN] = 0.025$ M

$$\%\ \text{dissociation} = \frac{[HCN]_{diss}}{[HCN]_{initial}} \times 100\% = \frac{1.23 \times 10^{-9}\ M}{0.025\ M} \times 100\% = 4.9 \times 10^{-6}\ \%$$

16.4 On mixing equal volumes of two solutions, both concentrations are cut in half.
$[CH_3NH_2] = 0.10$ M; $[CH_3NH_3Cl] = 0.30$ M

	$CH_3NH_2(aq)$ + $H_2O(l)$	⇄	$CH_3NH_3^+(aq)$	+ $OH^-(aq)$
initial (M)	0.10		0.30	~0
change (M)	–x		+x	+x
equil (M)	0.10 – x		0.30 + x	x

$$K_b = \frac{[CH_3NH_3^+][OH^-]}{[CH_3NH_2]} = 3.7 \times 10^{-4} = \frac{(0.30 + x)x}{0.10 - x} \approx \frac{(0.30)x}{0.10}$$

Solve for x. $x = [OH^-] = 1.2 \times 10^{-4}$ M

$$[H_3O^+] = \frac{K_w}{[OH^-]} = \frac{1.0 \times 10^{-14}}{1.2 \times 10^{-4}} = 8.3 \times 10^{-11} \text{ M}$$

$pH = -\log[H_3O^+] = -\log(8.3 \times 10^{-11}) = 10.08$

16.5 Both solutions contain the same number of HF molecules but solution 2 also contains five F^- ions. The dissociation equilibrium lies farther to the left for solution 2, and therefore solution 2 has the lower $[H_3O^+]$ and the higher pH. For solution 1, no common ion is present to suppress the dissociation of HF, and therefore solution 1 has the larger percent dissociation.

16.6 Each solution contains the same number of B molecules. The presence of BH^+ from BHCl lowers the percent dissociation of B. Solution (2) contains no BH^+, therefore it has the largest percent dissociation. BH^+ is the conjugate acid of B. Solution (1) has the largest amount of BH^+, and it would be the most acidic solution and have the lowest pH.

16.7

	$HF(aq)$ + $H_2O(l)$	⇄	$H_3O^+(aq)$	+ $F^-(aq)$
initial (M)	0.25		~0	0.50
change (M)	–x		+x	+x
equil (M)	0.25 – x		x	0.50 + x

$$K_a = \frac{[H_3O^+][F^-]}{[HF]} = 3.5 \times 10^{-4} = \frac{x(0.50 + x)}{0.25 - x} \approx \frac{x(0.50)}{0.25}$$

Solve for x. $x = 1.75 \times 10^{-4}$ M $= [H_3O^+]$
For the buffer, $pH = -\log[H_3O^+] = -\log(1.75 \times 10^{-4}) = 3.76$

(a) mol HF = 0.025 mol; mol F^- = 0.050 mol; vol = 0.100 L

	$F^-(aq)$	+ $H_3O^+(aq)$	$\xrightarrow{100\%}$	$HF(aq)$	+ $H_2O(l)$
before (mol)	0.050	0.002		0.025	
change (mol)	–0.002	–0.002		+0.002	
after (mol)	0.048	0		0.027	

$$[H_3O^+] = K_a \frac{[HF]}{[F^-]} = (3.5 \times 10^{-4})\left(\frac{0.27}{0.48}\right) = 1.97 \times 10^{-4} \text{ M}$$

$pH = -\log[H_3O^+] = -\log(1.97 \times 10^{-4}) = 3.71$

(b) mol HF = 0.025 mol; mol F^- = 0.050 mol; vol = 0.100 L

	$HF(aq)$	+ $OH^-(aq)$	$\xrightarrow{100\%}$ $F^-(aq)$	+ $H_2O(l)$
before (mol)	0.025	0.004	0.050	
change (mol)	–0.004	–0.004	+0.004	
after (mol)	0.021	0	0.054	

$$[H_3O^+] = K_a \frac{[HF]}{[F^-]} = (3.5 \times 10^{-4})\left(\frac{0.21}{0.54}\right) = 1.36 \times 10^{-4}\text{ M}$$

$pH = -\log[H_3O^+] = -\log(1.36 \times 10^{-4}) = 3.87$

16.8 (a)

	$HF(aq)$ + $H_2O(l)$	$\rightleftarrows$ $H_3O^+(aq)$	+ $F^-(aq)$
initial (M)	0.050	~0	0.100
change (M)	–x	+x	+x
equil (M)	0.050 – x	x	0.100 + x

$$K_a = \frac{[H_3O^+][F^-]}{[HF]} = 3.5 \times 10^{-4} = \frac{x(0.100 + x)}{0.050 - x} \approx \frac{x(0.100)}{0.050}$$

Solve for x. $x = [H_3O^+] = 1.75 \times 10^{-4}$ M

$pH = -\log[H_3O^+] = -\log(1.75 \times 10^{-4}) = 3.76$

mol HF = 0.050 mol/L x 0.100 L = 0.0050 mol HF

mol F^- = 0.100 mol/L x 0.100 L = 0.0100 mol F^-

mol HNO_3 = mol H_3O^+ = 0.002 mol

Neutralization reaction:	$F^-(aq)$	+ $H_3O^+(aq)$	$\xrightarrow{100\%}$ $HF(aq)$	+ $H_2O(l)$
before reaction (mol)	0.0100	0.002	0.0050	
change (mol)	–0.002	–0.002	+0.002	
after reaction (mol)	0.008	0	0.007	

$$[HF] = \frac{0.007\text{ mol}}{0.100\text{ L}} = 0.07\text{ M}; \qquad [F^-] = \frac{0.008\text{ mol}}{0.100\text{ L}} = 0.08\text{ M}$$

$$[H_3O^+] = K_a \frac{[HF]}{[F^-]} = (3.5 \times 10^{-4})\frac{(0.07)}{(0.08)} = 3 \times 10^{-4}\text{ M}$$

$pH = -\log[H_3O^+] = -\log(3 \times 10^{-4}) = 3.5$

(b) When the solution is diluted, the acid to conjugate base ratio remains the same and therefore the pH does not change.

16.9 (a) (1) and (3). Both pictures show equal concentrations of HA and A^-.
(b) (3). It contains a higher concentration of HA and A^-.

16.10 When equal volumes of two solutions are mixed together, the concentration of each solution is cut in half.

$$pH = pK_a + \log\frac{[\text{base}]}{[\text{acid}]} = pK_a + \log\frac{[CO_3^{2-}]}{[HCO_3^-]}$$

For HCO_3^-, $K_a = 5.6 \times 10^{-11}$, $pK_a = -\log K_a = -\log(5.6 \times 10^{-11}) = 10.25$

$$pH = 10.25 + \log\left(\frac{0.050}{0.10}\right) = 10.25 - 0.30 = 9.95$$

16.11 (a) $pH = pK_a + \log\frac{[\text{base}]}{[\text{acid}]} = 9.15 + \log\left(\frac{1}{50}\right) = 7.45$

(b) $\frac{[\text{base}]}{[\text{acid}]} = \frac{1}{50}$; % dissociation $= \frac{1}{(50+1)} \times 100\% = 1.96\%$

16.12 $pH = pK_a + \log\frac{[\text{base}]}{[\text{acid}]} = pK_a + \log\frac{[CO_3^{2-}]}{[HCO_3^-]}$

For HCO_3^-, $K_a = 5.6 \times 10^{-11}$, $pK_a = -\log K_a = -\log(5.6 \times 10^{-11}) = 10.25$

$10.40 = 10.25 + \log\frac{[CO_3^{2-}]}{[HCO_3^-]}$; $\log\frac{[CO_3^{2-}]}{[HCO_3^-]} = 10.40 - 10.25 = 0.15$

$$\frac{[CO_3^{2-}]}{[HCO_3^-]} = 10^{0.15} = 1.4$$

To obtain a buffer solution with pH 10.40, make the Na_2CO_3 concentration 1.4 times the concentration of $NaHCO_3$.

16.13 (a) HOCl, $K_a = 3.5 \times 10^{-8}$, $pK_a = 7.46$; HOBr, $K_a = 2.0 \times 10^{-9}$, $pK_a = 8.70$
Choose an acid with a pK_a within 1 pH unit of the desired pH. The desired pH is 7.00, so HOCl–NaOCl is the buffer of choice.

(b) $pH = 7.00 = pK_a + \log\frac{[\text{base}]}{[\text{acid}]} = 7.46 + \log\frac{[OCl^-]}{[HOCl]}$

$7.00 = 7.46 + \log\frac{[OCl^-]}{[HOCl]}$; $\log\frac{[OCl^-]}{[HOCl]} = 7.00 - 7.46 = -0.46$

$$\frac{[OCl^-]}{[HOCl]} = 10^{-0.46} = 0.35$$

16.14 (a) mol HCl = mol H_3O^+ = 0.100 mol/L x 0.0400 L = 0.004 00 mol
mol NaOH = mol OH^- = 0.100 mol/L x 0.0350 L = 0.003 50 mol

Neutralization reaction:	$H_3O^+(aq)$	+ $OH^-(aq)$	→ $2\,H_2O(l)$
before reaction (mol)	0.004 00	0.003 50	
change (mol)	−0.003 50	−0.003 50	
after reaction (mol)	0.000 50	0	

$$[H_3O^+] = \frac{0.000\ 50\ \text{mol}}{(0.0400\ \text{L} + 0.0350\ \text{L})} = 6.7 \times 10^{-3}\ \text{M}$$

$pH = -\log[H_3O^+] = -\log(6.7 \times 10^{-3}) = 2.17$

(b) mol HCl = mol H_3O^+ = 0.100 mol/L x 0.0400 L = 0.004 00 mol
mol NaOH = mol OH^- = 0.100 mol/L x 0.0450 L = 0.004 50 mol

Neutralization reaction:	$H_3O^+(aq)$ +	$OH^-(aq)$	$\rightarrow$ $2\ H_2O(l)$
before reaction (mol)	0.004 00	0.004 50	
change (mol)	–0.004 00	–0.004 00	
after reaction (mol)	0	0.000 50	

$$[OH^-] = \frac{0.000\ 50\ \text{mol}}{(0.0400\ \text{L} + 0.0450\ \text{L})} = 5.9 \times 10^{-3}\ \text{M}$$

$$[H_3O^+] = \frac{K_w}{[OH^-]} = \frac{1.0 \times 10^{-14}}{5.9 \times 10^{-3}} = 1.7 \times 10^{-12}\ \text{M}$$

$pH = -\log[H_3O^+] = -\log(1.7 \times 10^{-12}) = 11.77$

The results obtained here are consistent with the pH data in Table 16.1.

16.15 (a) mol NaOH = mol OH^- = 0.100 mol/L x 0.0400 L = 0.004 00 mol

mol HCl = mol H_3O^+ = 0.0500 mol/L x 0.0600 L = 0.003 00 mol

Neutralization reaction:	$H_3O^+(aq)$ +	$OH^-(aq)$	$\rightarrow$ $2\ H_2O(l)$
before reaction (mol)	0.003 00	0.004 00	
change (mol)	–0.003 00	–0.003 00	
after reaction (mol)	0	0.001 00	

$$[OH^-] = \frac{0.001\ 00\ \text{mol}}{(0.0400\ \text{L} + 0.0600\ \text{L})} = 1.0 \times 10^{-2}\ \text{M}$$

$$[H_3O^+] = \frac{K_w}{[OH^-]} = \frac{1.0 \times 10^{-14}}{1.0 \times 10^{-2}} = 1.0 \times 10^{-12}\ \text{M}$$

$pH = -\log[H_3O^+] = -\log(1.0 \times 10^{-12}) = 12.00$

(b) mol NaOH = mol OH^- = 0.100 mol/L x 0.0400 L = 0.004 00 mol

mol HCl = mol H_3O^+ = 0.0500 mol/L x 0.0802 L = 0.004 01 mol

Neutralization reaction:	$H_3O^+(aq)$ +	$OH^-(aq)$	$\rightarrow$ $2\ H_2O(l)$
before reaction (mol)	0.004 01	0.004 00	
change (mol)	–0.004 00	–0.004 00	
after reaction (mol)	0.000 01	0	

$$[H_3O^+] = \frac{0.000\ 01\ \text{mol}}{(0.0400\ \text{L} + 0.0802\ \text{L})} = 8.3 \times 10^{-5}\ \text{M}$$

$pH = -\log[H_3O^+] = -\log(8.3 \times 10^{-5}) = 4.08$

(c) mol NaOH = mol OH^- = 0.100 mol/L x 0.0400 L = 0.004 00 mol

mol HCl = mol H_3O^+ = 0.0500 mol/L x 0.1000 L = 0.005 00 mol

Neutralization reaction:	$H_3O^+(aq)$ +	$OH^-(aq)$	$\rightarrow$ $2\ H_2O(l)$
before reaction (mol)	0.005 00	0.004 00	
change (mol)	–0.004 00	–0.004 00	
after reaction (mol)	0.001 00	0	

$$[H_3O^+] = \frac{0.001\ 00\ \text{mol}}{(0.0400\ \text{L} + 0.1000\ \text{L})} = 7.1 \times 10^{-3}\ \text{M}$$

$pH = -\log[H_3O^+] = -\log(7.1 \times 10^{-3}) = 2.15$

16.16 mol NaOH required = $\left(\dfrac{0.016 \text{ mol HOCl}}{\text{L}}\right)(0.100 \text{ L})\left(\dfrac{1 \text{ mol NaOH}}{1 \text{ mol HOCl}}\right) = 0.0016 \text{ mol}$

vol NaOH required = $(0.0016 \text{ mol})\left(\dfrac{1 \text{ L}}{0.0400 \text{ mol}}\right) = 0.040 \text{ L} = 40 \text{ mL}$

40 mL of 0.0400 M NaOH are required to reach the equivalence point.

(a) mmol HOCl = 0.016 mmol/mL x 100.0 mL = 1.6 mmol

mmol NaOH = mmol OH^- = 0.0400 mmol/mL x 10.0 mL = 0.400 mmol

Neutralization reaction:	$HOCl(aq)$	+ $OH^-(aq)$	→ $OCl^-(aq)$ + $H_2O(l)$
before reaction (mmol)	1.6	0.400	0
change (mmol)	–0.400	–0.400	+0.400
after reaction (mmol)	1.2	0	0.400

$[HOCl] = \dfrac{1.2 \text{ mmol}}{(100.0 \text{ mL} + 10.0 \text{ mL})} = 1.09 \times 10^{-2} \text{ M}$

$[OCl^-] = \dfrac{0.400 \text{ mmol}}{(100.0 \text{ mL} + 10.0 \text{ mL})} = 3.64 \times 10^{-3} \text{ M}$

	$HOCl(aq)$ + $H_2O(l)$	⇄ $H_3O^+(aq)$	+ $OCl^-(aq)$
initial (M)	0.0109	~0	0.003 64
change (M)	–x	+x	+x
equil (M)	0.0109 – x	x	0.003 64 + x

$K_a = \dfrac{[H_3O^+][OCl^-]}{[HOCl]} = 3.5 \times 10^{-8} = \dfrac{x(0.003\ 64 + x)}{0.0109 - x} \approx \dfrac{x(0.003\ 64)}{0.0109}$

Solve for x. x = $[H_3O^+]$ = 1.05×10^{-7} M

pH = $-\log[H_3O^+] = -\log(1.05 \times 10^{-7}) = 6.98$

(b) Halfway to the equivalence point, $[OCl^-] = [HOCl]$

pH = $pK_a = -\log K_a = -\log(3.5 \times 10^{-8}) = 7.46$

(c) At the equivalence point the solution contains the salt, NaOCl.

mol NaOCl = initial mol HOCl = 0.0016 mol = 1.6 mmol

$[OCl^-] = \dfrac{1.6 \text{ mmol}}{(100.0 \text{ mL} + 40.0 \text{ mL})} = 1.1 \times 10^{-2} \text{ M}$

For OCl^-, $K_b = \dfrac{K_w}{K_a \text{ for HOCl}} = \dfrac{1.0 \times 10^{-14}}{3.5 \times 10^{-8}} = 2.9 \times 10^{-7}$

	$OCl^-(aq)$ + $H_2O(l)$	⇄ $HOCl(aq)$	+ $OH^-(aq)$
initial (M)	0.011	0	~0
change (M)	–x	+x	+x
equil (M)	0.011 – x	x	x

$K_b = \dfrac{[HOCl][OH^-]}{[OCl^-]} = 2.9 \times 10^{-7} = \dfrac{x^2}{0.011 - x} \approx \dfrac{x^2}{0.011}$

Solve for x. x = $[OH^-]$ = 5.65×10^{-5} M

$[H_3O^+] = \dfrac{K_w}{[OH^-]} = \dfrac{1.0 \times 10^{-14}}{5.65 \times 10^{-5}} = 1.77 \times 10^{-10} = 1.8 \times 10^{-10} \text{ M}$

pH = $-\log[H_3O^+] = -\log(1.77 \times 10^{-10}) = 9.75$

(d) pH = 9.75 at the equivalence point.
Use thymolphthalein (pH 9.4–10.6). Bromthymol blue (6.0–7.6) is unacceptable because it changes color halfway to the equivalence point. Alizarin yellow (10.1–12.0) could be used, but thymolphthalein is better.

16.17 (a) (3), only HA present (b) (1), HA and A^- present
(c) (4), only A^- present (d) (2), A^- and OH^- present

16.18 (a) mol NaOH required to reach first equivalence point

$$= \left(\frac{0.0800 \text{ mol } H_2SO_3}{\text{L}}\right)(0.0400 \text{ L})\left(\frac{1 \text{ mol NaOH}}{1 \text{ mol } H_2SO_3}\right) = 0.003\ 20 \text{ mol}$$

vol NaOH required to reach first equivalence point

$$= (0.003\ 20 \text{ mol})\left(\frac{1 \text{ L}}{0.160 \text{ mol}}\right) = 0.020 \text{ L} = 20.0 \text{ mL}$$

20.0 mL is enough NaOH solution to reach the first equivalence point for the titration of the diprotic acid, H_2SO_3.
For H_2SO_3,

$K_{a1} = 1.5 \times 10^{-2}$, $pK_{a1} = -\log K_{a1} = -\log(1.5 \times 10^{-2}) = 1.82$

$K_{a2} = 6.3 \times 10^{-8}$, $pK_{a2} = -\log K_{a2} = -\log(6.3 \times 10^{-8}) = 7.20$

At the first equivalence point, $pH = \frac{pK_{a1} + pK_{a2}}{2} = \frac{1.82 + 7.20}{2} = 4.51$

(b) mol NaOH required to reach second equivalence point

$$= \left(\frac{0.0800 \text{ mol } H_2SO_3}{\text{L}}\right)(0.0400 \text{ L})\left(\frac{2 \text{ mol NaOH}}{1 \text{ mol } H_2SO_3}\right) = 0.006\ 40 \text{ mol}$$

vol NaOH required to reach second equivalence point

$$= (0.006\ 40 \text{ mol})\left(\frac{1 \text{ L}}{0.160 \text{ mol}}\right) = 0.040 \text{ L} = 40.0 \text{ mL}$$

30.0 mL is enough NaOH solution to reach halfway to the second equivalent point.
Halfway to the second equivalence point

$pH = pK_{a2} = -\log K_{a2} = -\log(6.3 \times 10^{-8}) = 7.20$

(c) mmol HSO_3^- = 0.0800 mmol/mL x 40.0 mL = 3.20 mmol
volume NaOH added after first equivalence point = 35.0 mL – 20.0 mL = 15.0 mL
mmol NaOH = mmol OH^- = 0.160 mmol/L x 15.0 mL = 2.40 mmol

Neutralization reaction:	$HSO_3^-(aq)$	+ $OH^-(aq)$	⇄ $SO_3^{2-}(aq)$	+ $H_2O(l)$
before reaction (mmol)	3.20	2.40	0	
change (mmol)	–2.40	–2.40	+2.40	
after reaction (mmol)	0.80	0	2.40	

$$[HSO_3^-] = \frac{0.80 \text{ mmol}}{(40.0 \text{ mL} + 35.0 \text{ mL})} = 0.0107 \text{ M}$$

$$[SO_3^{2-}] = \frac{2.40 \text{ mmol}}{(40.0 \text{ mL} + 35.0 \text{ mL})} = 0.0320 \text{ M}$$

	$HSO_3^-(aq)$ + $H_2O(l)$	⇌ $H_3O^+(aq)$	+ $SO_3^{2-}(aq)$
initial (M)	0.0107	~0	0.0320
change (M)	−x	+x	+x
equil (M)	0.0107 − x	x	0.0320 + x

$$K_a = \frac{[H_3O^+][SO_3^{2-}]}{[HSO_3^-]} = 6.3 \times 10^{-8} = \frac{x(0.0320 + x)}{0.0107 - x} \approx \frac{x(0.0320)}{0.0107}$$

Solve for x. $x = [H_3O^+] = 2.1 \times 10^{-8}$ M

$pH = -\log[H_3O^+] = -\log(2.1 \times 10^{-8}) = 7.68$

16.19 Let H_2A^+ = valine cation

(a) mol NaOH required to reach first equivalence point

$$= \left(\frac{0.0250 \text{ mol } H_2A^+}{L}\right)(0.0400 \text{ L})\left(\frac{1 \text{ mol NaOH}}{1 \text{ mol } H_2A^+}\right) = 0.001\ 00 \text{ mol}$$

vol NaOH required to reach first equivalence point

$$= (0.001\ 00 \text{ mol})\left(\frac{1 \text{ L}}{0.100 \text{ mol}}\right) = 0.0100 \text{ L} = 10.0 \text{ mL}$$

10.0 mL is enough NaOH solution to reach the first equivalence point for the titration of the diprotic acid, H_2A^+.

For H_2A^+,

$K_{a1} = 4.8 \times 10^{-3}$, $pK_{a1} = -\log K_{a1} = -\log(4.8 \times 10^{-3}) = 2.32$

$K_{a2} = 2.4 \times 10^{-10}$, $pK_{a2} = -\log K_{a2} = -\log(2.4 \times 10^{-10}) = 9.62$

At the first equivalence point, $pH = \frac{pK_{a1} + pK_{a2}}{2} = \frac{2.32 + 9.62}{2} = 5.97$

(b) mol NaOH required to reach second equivalence point

$$= \left(\frac{0.0250 \text{ mol } H_2A^+}{L}\right)(0.0400 \text{ L})\left(\frac{2 \text{ mol NaOH}}{1 \text{ mol } H_2A^+}\right) = 0.002\ 00 \text{ mol}$$

vol NaOH required to reach second equivalence point

$$= (0.002\ 00 \text{ mol})\left(\frac{1 \text{ L}}{0.100 \text{ mol}}\right) = 0.0200 \text{ L} = 20.0 \text{ mL}$$

15.0 mL is enough NaOH solution to reach halfway to the second equivalent point.

Halfway to the second equivalence point

$pH = pK_{a2} = -\log K_{a2} = -\log(2.4 \times 10^{-10}) = 9.62$

(c) 20.0 mL is enough NaOH to reach the second equivalence point.

At the second equivalence point

mmol A^- = (0.0250 mmol/mL)(40.0 mL) = 1.00 mmol A^-

solution volume = 40.0 mL + 20.0 mL = 60.0 mL

$$[A^-] = \frac{1.00 \text{ mmol}}{60.0 \text{ mL}} = 0.0167 \text{ M}$$

	$A^-(aq)$ + $H_2O(l)$	⇄ $HA(aq)$	+ $OH^-(aq)$
initial (M)	0.0167	0	~0
change (M)	–x	+x	+x
equil (M)	0.0167 – x	x	x

$$K_b = \frac{K_w}{K_a \text{ for HA}} = \frac{K_w}{K_{a2}} = \frac{1.0 \times 10^{-14}}{2.4 \times 10^{-10}} = 4.17 \times 10^{-5}$$

$$K_b = \frac{[HA][OH^-]}{[A^-]} = 4.17 \times 10^{-5} = \frac{x^2}{0.0167 - x}$$

$x^2 + (4.17 \times 10^{-5})x - (6.964 \times 10^{-7}) = 0$

Use the quadratic formula to solve for x.

$$x = \frac{-(4.17 \times 10^{-5}) \pm \sqrt{(4.17 \times 10^{-5})^2 - (4)(1)(-6.964 \times 10^{-7})}}{2(1)} = \frac{(-4.17 \times 10^{-5}) \pm (1.67 \times 10^{-3})}{2}$$

$x = 8.14 \times 10^{-4}$ and -8.56×10^{-4}

Of the two solutions for x, only the positive value has physical meaning because x is the $[OH^-]$.

$x = [OH^-] = 8.14 \times 10^{-4}$ M

$$[H_3O^+] = \frac{K_w}{[OH^-]} = \frac{1.0 \times 10^{-14}}{8.14 \times 10^{-4}} = 1.23 \times 10^{-11} \text{ M}$$

$pH = -\log[H_3O^+] = -\log(1.23 \times 10^{-11}) = 10.91$

16.20 (a) $K_{sp} = [Ag^+][Cl^-]$ (b) $K_{sp} = [Pb^{2+}][I^-]^2$
(c) $K_{sp} = [Ca^{2+}]^3[PO_4^{3-}]^2$ (d) $K_{sp} = [Cr^{3+}][OH^-]^3$

16.21 Let the number of ions be proportional to its concentration.
For AgX, $K_{sp} = [Ag^+][X^-] \propto (4)(4) = 16$
For AgY, $K_{sp} = [Ag^+][Y^-] \propto (1)(9) = 9$
For AgZ, $K_{sp} = [Ag^+][Z^-] \propto (3)(6) = 18$
(a) AgZ (b) AgY

16.22 $K_{sp} = [Ca^{2+}]^3[PO_4^{3-}]^2 = (2.01 \times 10^{-8})^3(1.6 \times 10^{-5})^2 = 2.1 \times 10^{-33}$

16.23 CaC_2O_4, $K_{sp} = 2.3 \times 10^{-9}$; $[Ca^{2+}] = 3.0 \times 10^{-8}$ M
$K_{sp} = [Ca^{2+}][C_2O_4^{2-}] = (3.0 \times 10^{-8})[C_2O_4^{2-}] = 2.3 \times 10^{-9}$

$$[C_2O_4^{2-}] = \frac{2.3 \times 10^{-9}}{3.0 \times 10^{-8}} = 0.077 \text{ M}$$

minimum $[Na_2C_2O_4] = 0.077$ M

16.24 (a) $AgCl(s) \rightleftharpoons Ag^+(aq) + Cl^-(aq)$
equil (M) x x
$K_{sp} = [Ag^+][Cl^-] = 1.8 \times 10^{-10} = (x)(x)$
molar solubility $= x = \sqrt{K_{sp}} = 1.3 \times 10^{-5}$ mol/L

AgCl, 143.32, $\text{solubility} = \frac{\left(1.3 \times 10^{-5}\ \text{mol} \times \frac{143.32\ \text{g}}{1\ \text{mol}}\right)}{1\ \text{L}} = 0.0019\ \text{g/L}$

(b) $Ag_2CrO_4(s) \rightleftarrows 2\ Ag^+(aq) + CrO_4^{2-}(aq)$

equil (M) 2x x

$K_{sp} = [Ag^+]^2[CrO_4^{2-}] = 1.1 \times 10^{-12} = (2x)^2(x) = 4x^3$

$\text{molar solubility} = x = \sqrt[3]{\frac{1.1 \times 10^{-12}}{4}} = 6.5 \times 10^{-5}\ \text{mol/L}$

Ag_2CrO_4, 331.73, $\text{solubility} = \frac{\left(6.5 \times 10^{-5}\ \text{mol} \times \frac{331.73\ \text{g}}{1\ \text{mol}}\right)}{1\ \text{L}} = 0.022\ \text{g/L}$

Ag_2CrO_4 has both the higher molar and gram solubility, despite its smaller value of K_{sp}.

16.25 $[Ba^{2+}] = [SO_4^{2-}] = 1.05 \times 10^{-5}$ M; $K_{sp} = [Ba^{2+}][SO_4^{2-}] = (1.05 \times 10^{-5})^2 = 1.10 \times 10^{-10}$

16.26 $[Mg^{2+}]_0$ is from 0.10 M $MgCl_2$.

	$MgF_2(s)$	$\rightleftarrows$	$Mg^{2+}(aq)$	+	$2\ F^-(aq)$
initial (M)			0.10		0
change (M)			+x		+2x
equil (M)			0.10 + x		2x

$K_{sp} = 7.4 \times 10^{-11} = [Mg^{2+}][F^-]^2 = (0.10 + x)(2x)^2 \approx (0.10)(4x^2)$

$x = 1.4 \times 10^{-5}$, molar solubility = x = 1.4×10^{-5} M

16.27 pH = 11; $[H_3O^+] = 10^{-pH} = 10^{-11} = 1.0 \times 10^{-11}$ M

$[OH^-] = \frac{K_w}{[H_3O^+]} = \frac{1.0 \times 10^{-14}}{1.0 \times 10^{-11}} = 0.0010\ \text{M}$

	Zn(OH)2(s)	$\rightleftarrows$	$Zn^{2+}(aq)$	+	$2\ OH^-(aq)$
equil (M)			x		0.0010

$K_{sp} = 4.1 \times 10^{-17} = [Zn^{2+}][OH^-]^2 = (x)(0.0010)^2$

$x = 4.1 \times 10^{-11}$, molar solubility = x = 4.1×10^{-11} M

16.28 Compounds that contain basic anions are more soluble in acidic solution than in pure water. AgCN, $Al(OH)_3$, and ZnS all contain basic anions.

16.29 $[Cu^{2+}] = (5.0 \times 10^{-3}\ \text{mol})/(0.500\ \text{L}) = 0.010$ M

	$Cu^{2+}(aq)$	+	$4\ NH_3(aq)$	$\rightleftarrows$	$Cu(NH_3)_4^{2+}(aq)$
before reaction (M)	0.010		0.40		0
assume 100% reaction (M)	–0.010		– 4(0.010)		+0.010
after reaction (M)	0		0.36		0.010
assume small back reaction (M)	+x		+4x		–x
equil (M)	x		0.36 + 4x		0.010 – x

$$K_f = \frac{[Cu(NH_3)_4^{2+}]}{[Cu^{2+}][NH_3]^4} = 5.6 \times 10^{11} = \frac{(0.010 - x)}{(x)(0.36 + 4x)^4} \approx \frac{0.010}{x(0.36)^4}$$

Solve for x. $x = [Cu^{2+}] = 1.1 \times 10^{-12}$ M

16.30 Total solution volume = 25.0 mL + 35.0 mL = 60.0 mL

$$[Au^{3+}] = \frac{(25.0\ \text{mL})(3.0 \times 10^{-2}\ \text{M})}{(60.0\ \text{mL})} = 0.0125\ \text{M}$$

$$[CN^-] = \frac{(35.0\ \text{mL})(1.0\ \text{M})}{(60.0\ \text{mL})} = 0.583\ \text{M}$$

	$Au^{3+}(aq)$ +	$2\ CN^-(aq)$ ⇄	$Au(CN)_2^-(aq)$
before reaction (M)	0.0125	0.583	0
assume 100% reaction (M)	–0.0125	–2(0.0125)	+0.0125
after reaction (M)	0	0.558	0.0125
assume small back reaction (M)	+x	+2x	–x
equil (M)	x	0.558 + 2x	0.0125 – x

$$K_f = \frac{[Au(CN)_2^-]}{[Au^{3+}][CN^-]^2} = 2 \times 10^{38} = \frac{(0.0125 - x)}{(x)(0.558 + 2x)^2} \approx \frac{0.0125}{x(0.558)^2}$$

Solve for x. $x = [Au^{3+}] = 2 \times 10^{-40}$ M

16.31

$AgBr(s) \rightleftarrows Ag^+(aq) + Br^-(aq)$ $\quad K_{sp} = 5.4 \times 10^{-13}$

$Ag^+(aq) + 2\ S_2O_3^{2-} \rightarrow Ag(S_2O_3)_2^{3-}(aq)$ $\quad K_f = 4.7 \times 10^{13}$

dissolution reaction $\quad AgBr(s) + 2\ S_2O_3^{2-}(aq) \rightleftarrows Ag(S_2O_3)_2^{3-}(aq) + Br^-(aq)$

$K = (K_{sp})(K_f) = (5.4 \times 10^{-13})(4.7 \times 10^{13}) = 25.4$

	$AgBr(s) + 2\ S_2O_3^{2-}(aq)$ ⇄	$Ag(S_2O_3)_2^{3-}(aq)$ +	$Br^-(aq)$
initial (M)	0.10	0	0
change (M)	–2x	+x	+x
equil (M)	0.10 – 2x	x	x

$$K = \frac{[Ag(S_2O_3)_2^{3-}][Br^-]}{[S_2O_3^{2-}]^2} = 25.4 = \frac{x^2}{(0.10 - 2x)^2}$$

Take the square root of both sides and solve for x.

$$\sqrt{25.4} = \sqrt{\frac{x^2}{(0.10 - 2x)^2}}; \quad 5.04 = \frac{x}{0.10 - 2x}; \quad x = \text{molar solubility} = 0.045\ \text{mol/L}$$

16.32 Step 1: The precipitate is $Cu(OH)_2(s)$. NH_3 is a base and OH^- ions are present in aqueous solution to react with the Cu^{2+} ions.

$$Cu^{2+}(aq) + 2\ OH^-(aq) \rightleftarrows Cu(OH)_2(s)$$

Step 2: In the presence of additional NH_3, the complex ion $Cu(NH_3)_4^{2+}$ forms.

$$Cu(OH)_2(s) + 4\ NH_3(aq) \rightleftarrows Cu(NH_3)_4^{2+}(aq) + 2\ OH^-(aq)$$

16.33 On mixing equal volumes of two solutions, the concentrations of both solutions are cut in half.

For $BaCO_3$, $K_{sp} = 2.6 \times 10^{-9}$

(a) $IP = [Ba^{2+}][CO_3^{2-}] = (1.5 \times 10^{-3})(1.0 \times 10^{-3}) = 1.5 \times 10^{-6}$

$IP > K_{sp}$; a precipitate of $BaCO_3$ will form.

(b) $IP = [Ba^{2+}][CO_3^{2-}] = (5.0 \times 10^{-6})(2.0 \times 10^{-5}) = 1.0 \times 10^{-10}$

$IP < K_{sp}$; no precipitate will form.

16.34 $$pH = pK_a + \log\frac{[\text{base}]}{[\text{acid}]} = pK_a + \log\frac{[NH_3]}{[NH_4^+]}$$

For NH_4^+, $K_a = 5.6 \times 10^{-10}$, $pK_a = -\log K_a = -\log(5.6 \times 10^{-10}) = 9.25$

$$pH = 9.25 + \log\frac{(0.20)}{(0.20)} = 9.25; \quad [H_3O^+] = 10^{-pH} = 10^{-9.25} = 5.6 \times 10^{-10}\ M$$

$$[OH^-] = \frac{K_w}{[H_3O^+]} = \frac{1.0 \times 10^{-14}}{5.6 \times 10^{-10}} = 1.8 \times 10^{-5}\ M$$

$$[Fe^{2+}] = [Mn^{2+}] = \frac{(25\ mL)(1.0 \times 10^{-3}\ M)}{250\ mL} = 1.0 \times 10^{-4}\ M$$

For $Mn(OH)_2$, $K_{sp} = 2.1 \times 10^{-13}$

$IP = [Mn^{2+}][OH^-]^2 = (1.0 \times 10^{-4})(1.8 \times 10^{-5})^2 = 3.2 \times 10^{-14}$

$IP < K_{sp}$; no precipitate will form.

For $Fe(OH)_2$, $K_{sp} = 4.9 \times 10^{-17}$

$IP = [Fe^{2+}][OH^-]^2 = (1.0 \times 10^{-4})(1.8 \times 10^{-5})^2 = 3.2 \times 10^{-14}$

$IP > K_{sp}$; a precipitate of $Fe(OH)_2$ will form.

16.35 $MS(s) + 2\ H_3O^+(aq) \rightleftarrows M^{2+}(aq) + H_2S(aq) + 2\ H_2O(l)$

$$K_{spa} = \frac{[M^{2+}][H_2S]}{[H_3O^+]^2}$$

For ZnS, $K_{spa} = 3 \times 10^{-2}$; for CdS, $K_{spa} = 8 \times 10^{-7}$

$[Cd^{2+}] = [Zn^{2+}] = 0.005\ M$

Because the two cation concentrations are equal, Q_c is the same for both.

$$Q_c = \frac{[M^{2+}]_t[H_2S]_t}{[H_3O^+]_t^2} = \frac{(0.005)(0.10)}{(0.3)^2} = 6 \times 10^{-3}$$

$Q_c > K_{spa}$ for CdS; CdS will precipitate. $Q_c < K_{spa}$ for ZnS; Zn^{2+} will remain in solution.

16.36 According to Le Châtelier's Principle removing a product will shift the equilibrium toward the products. In the three equilibrium reactions shown, CO_3^{2-} is a product in the last reaction. As organisms use CO_3^{2-} to make their shells, all three equilibrium reactions shift toward the products.

16.37 (a) pH = 8.2; $[H_3O^+] = 10^{-pH} = 10^{-8.2} = 6.3 \times 10^{-9}$ M

$$150 = \frac{[H_3O^+]_{final} - [H_3O^+]_{initial}}{[H_3O^+]_{initial}} \times 100 = \frac{[H_3O^+]_{final} - 6.3 \times 10^{-9}\text{ M}}{6.3 \times 10^{-9}\text{ M}} \times 100$$

$$[H_3O^+]_{final} = 1.6 \times 10^{-8}\text{ M}$$

$pH = -\log[H_3O^+] = -\log(1.6 \times 10^{-8}) = 7.8$

(b) HCO_3^-, $K_a = 5.6 \times 10^{-11}$, $pK_a = 10.25$

$$pH = pK_a + \log\frac{[base]}{[acid]} = 10.25 - \log\frac{[HCO_3^-]}{[CO_3^{2-}]}$$

$$pH = 8.2 = pK_a + \log\frac{[base]}{[acid]} = 8.2 = 10.25 - \log\frac{[HCO_3^-]}{[CO_3^{2-}]}$$

$$\log\frac{[HCO_3^-]}{[CO_3^{2-}]} = 10.25 - 8.2 = 2.05; \quad \frac{[HCO_3^-]}{[CO_3^{2-}]} = 10^{2.05} = 112$$

$$pH = 7.8 = pK_a + \log\frac{[base]}{[acid]} = 7.8 = 10.25 - \log\frac{[HCO_3^-]}{[CO_3^{2-}]}$$

$$\log\frac{[HCO_3^-]}{[CO_3^{2-}]} = 10.25 - 7.8 = 2.45; \quad \frac{[HCO_3^-]}{[CO_3^{2-}]} = 10^{2.45} = 282$$

16.38

	$CaCO_3(s) \rightleftarrows Ca^{2+}(aq) + CO_3^{2-}(aq)$	
equil (M)	x	x

$K_{sp} = [Ca^{2+}][CO_3^{2-}] = 5.0 \times 10^{-9} = (x)(x)$

molar solubility = x = $\sqrt{K_{sp}} = 7.1 \times 10^{-5}$ mol/L

16.39 (a)

Reaction	K
$CaCO_3(s) \rightleftarrows Ca^{2+}(aq) + CO_3^{2-}(aq)$	$K_{sp} = 5.0 \times 10^{-9}$
$H_2CO_3(aq) + H_2O(l) \rightleftarrows H_3O^+(aq) + HCO_3^-(aq)$	$K_{a1} = 4.3 \times 10^{-7}$
$CO_3^{2-}(aq) + H_3O^+(aq) \rightleftarrows HCO_3^-(aq) + H_2O(l)$	$1/K_{a2} = 1/5.6 \times 10^{-11}$
$CaCO_3(s) + H_2CO_3(aq) \rightleftarrows Ca^{2+}(aq) + 2\ HCO_3^-(aq)$	$K = K_{sp} \cdot K_{a1} \cdot (1/K_{a2})$

$K = K_{sp} \cdot K_{a1} \cdot (1/K_{a2}) = (5.0 \times 10^{-9})(4.3 \times 10^{-7})(1/5.6 \times 10^{-11}) = 3.8 \times 10^{-5}$

(b) As atmospheric CO_2 levels rise, the concentration of H_2CO_3 will increase, shifting the equilibrium of the overall reaction to the products. The molar solubility of $CaCO_3$ will increase.

Conceptual Problems

16.40 (4); only A^- and water should be present

16.42 A buffer solution contains a conjugate acid-base pair in about equal concentrations.
(a) (1), (3), and (4)
(b) (4) because it has the highest buffer concentration.

16.44 (a) (i) (1), only B present (ii) (4), equal amounts of B and BH^+ present
(iii) (3), only BH^+ present (iv) (2), BH^+ and H_3O^+ present
(b) The pH is less than 7 because BH^+ is an acidic cation.

16.46 (a) The lower curve represents the titration of a strong acid; the upper curve represents the titration of a weak acid.
(b) pH = 7 for titration of the strong acid; pH = 10 for titration of the weak acid.
(c) Halfway to the equivalence point, the $pH = pK_a \sim 6.3$.

16.48 (2) is supersaturated; (3) is unsaturated; (4) is unsaturated

Section Problems

Neutralization Reactions (Section 16.1)

16.50 (a) $HNO_3(aq) + KOH(aq) \rightarrow H_2O(l) + KNO_3(aq)$
net ionic equation: $H_3O^+(aq) + OH^-(aq) \rightarrow 2\ H_2O(l)$
The solution at neutralization contains a neutral salt (KNO_3); pH = 7.00.
(b) $2\ HOI(aq) + Ba(OH)_2(aq) \rightarrow 2\ H_2O(l) + Ba(OI)_2(aq)$
net ionic equation: $HOI(aq) + OH^-(aq) \rightarrow H_2O(l) + OI^-(aq)$
The solution at neutralization contains a basic anion (OI^-); pH > 7.00
(c) $HBr(aq) + C_6H_5NH_2(aq) \rightarrow C_6H_5NH_3Br(aq)$
net ionic equation: $H_3O^+(aq) + C_6H_5NH_2(aq) \rightarrow H_2O(l) + C_6H_5NH_3^+(aq)$
The solution at neutralization contains an acidic cation ($C_6H_5NH_3^+$); pH < 7.00.
(d) $HNO_2(aq) + KOH(aq) \rightarrow H_2O(l) + KNO_2(aq)$
net ionic equation: $HNO_2(aq) + OH^-(aq) \rightarrow H_2O(l) + NO_2^-(aq)$
The solution at neutralization contains a basic anion (NO_2^-); pH > 7.00.

16.52 (a) After mixing, the solution contains the basic salt, NaCN; pH > 7.00
(b) After mixing, the solution contains the neutral salt, $NaClO_4$; pH = 7.00
Solution (a) has the higher pH.

16.54 Weak acid - weak base reaction $K_n = \dfrac{K_a K_b}{K_w} = \dfrac{(1.3 \times 10^{-10})(1.8 \times 10^{-9})}{1.0 \times 10^{-14}} = 2.3 \times 10^{-5}$

K_n is small so the neutralization reaction does not proceed very far to completion.

16.56 $K_n = \dfrac{K_a K_b}{K_w} = 2.1 \times 10^{-4}$; $K_b = \dfrac{K_n K_w}{K_a} = \dfrac{(2.1 \times 10^{-4})(1.0 \times 10^{-14})}{(1.4 \times 10^{-4})} = 1.5 \times 10^{-14}$

The Common-Ion Effect (Section 16.2)

16.58 (a) $HF(aq) + H_2O(l) \rightleftarrows H_3O^+(aq) + F^-(aq)$
LiF is a source of F^- (reaction product). The equilibrium shifts toward reactants, and the $[H_3O^+]$ decreases. The pH increases.
(b) Because HI is a strong acid, addition of KI, a neutral salt, does not change the pH.

(c) $NH_3(aq) + H_2O(l) \rightleftharpoons NH_4^+(aq) + OH^-(aq)$
NH_4Cl is a source of NH_4^+ (reaction product). The equilibrium shifts toward reactants, and the $[OH^-]$ decreases. The pH decreases.

16.60 For 0.25 M HF and 0.10 M NaF

	$HF(aq) + H_2O(l)$	$\rightleftharpoons$ $H_3O^+(aq)$	$+ F^-(aq)$
initial (M)	0.25	~0	0.10
change (M)	–x	+x	+x
equil (M)	0.25 – x	x	0.10 + x

$$K_a = \frac{[H_3O^+][F^-]}{[HF]} = 3.5 \times 10^{-4} = \frac{x(0.10 + x)}{0.25 - x} \approx \frac{x(0.10)}{0.25}$$

Solve for x. $x = [H_3O^+] = 8.8 \times 10^{-4}$ M
$pH = -\log[H_3O^+] = -\log(8.8 \times 10^{-4}) = 3.06$

16.62 $pH = 4.86$; $[H_3O^+] = 10^{-pH} = 10^{-4.86} = 1.38 \times 10^{-5}$ M
$HN_3(aq) + H_2O(l) \rightleftharpoons H_3O^+(aq) + N_3^-(aq)$

$$K_a = \frac{[H_3O^+][N_3^-]}{[HN_3]} = 1.9 \times 10^{-5} = \frac{(1.38 \times 10^{-5})[N_3^-]}{(0.016)}$$

$[N_3^-] = 0.022$ M

16.64 For 0.10 M HN_3:

	$HN_3(aq) + H_2O(l)$	$\rightleftharpoons$ $H_3O^+(aq)$	$+ N_3^-(aq)$
initial (M)	0.10	~0	0
change (M)	–x	+x	+x
equil (M)	0.10 – x	x	x

$$K_a = \frac{[H_3O^+][N_3^-]}{[HN_3]} = 1.9 \times 10^{-5} = \frac{x^2}{0.10 - x} \approx \frac{x^2}{0.10}$$

Solve for x. $x = 1.4 \times 10^{-3}$ M

$$\% \text{ dissociation} = \frac{[HN_3]_{diss}}{[HN_3]_{initial}} \times 100\% = \frac{1.4 \times 10^{-3}\text{ M}}{0.10\text{ M}} \times 100\% = 1.4\%$$

For 0.10 M HN_3 in 0.10 M HCl:

	$HN_3(aq) + H_2O(l)$	$\rightleftharpoons$ $H_3O^+(aq)$	$+ N_3^-(aq)$
initial (M)	0.10	0.10	0
change (M)	–x	+x	+x
equil (M)	0.10 – x	0.10 + x	x

$$K_a = \frac{[H_3O^+][N_3^-]}{[HN_3]} = 1.9 \times 10^{-5} = \frac{(0.10 + x)(x)}{0.10 - x} \approx \frac{(0.10)(x)}{0.10} = x$$

Solve for x. $x = 1.9 \times 10^{-5}$ M

$$\% \text{ dissociation} = \frac{[HN_3]_{diss}}{[HN_3]_{initial}} \times 100\% = \frac{1.9 \times 10^{-5}\text{ M}}{0.10\text{ M}} \times 100\% = 0.019\%$$

The % dissociation is less because of the common ion (H_3O^+) effect.

Buffer Solutions (Sections 16.3 and 16.4)

16.66 Solutions (a), (c), and (d) are buffer solutions. Neutralization reactions for (c) and (d) result in solutions with equal concentrations of HF and F^-.

16.68 Both solutions buffer at the same pH because in both cases the $[NO_2^-]/[HNO_2] = 1$. Solution (a), however, has a higher concentration of both HNO_2 and NO_2^-, and therefore it has the greater buffer capacity.

16.70 $\text{pH} = \text{p}K_a + \log\frac{[\text{base}]}{[\text{acid}]} = \text{p}K_a + \log\frac{[CN^-]}{[HCN]}$

For HCN, $K_a = 4.9 \times 10^{-10}$, $\text{p}K_a = -\log K_a = -\log(4.9 \times 10^{-10}) = 9.31$

$$\text{pH} = 9.31 + \log\left(\frac{0.12}{0.20}\right) = 9.09$$

The pH of a buffer solution will not change on dilution because the acid and base concentrations will change by the same amount and their ratio will remain the same.

16.72

	$HCO_2H(aq)$	+ $H_2O(l)$ ⇄	$H_3O^+(aq)$	+ $HCO_2^-(aq)$
initial (M)	0.36		~0	0.30
change (M)	–x		+x	+x
equil (M)	0.36 – x		x	0.30 + x

$$K_a = \frac{[H_3O^+][HCO_2^-]}{[HCO_2H]} = 1.8 \times 10^{-4} = \frac{x(0.30 + x)}{0.36 - x} \approx \frac{x(0.30)}{0.36}$$

Solve for x. $x = 2.16 \times 10^{-4}\text{ M} = [H_3O^+]$

For the buffer, $\text{pH} = -\log[H_3O^+] = -\log(2.16 \times 10^{-4}) = 3.67$

mol HCO_2H = (0.36 mol/L)(0.250 L) = 0.090 mol HCO_2H

mol HCO_2^- = (0.30 mol/L)(0.250 L) = 0.075 mol HCO_2^-

(a)

	$HCO_2H(aq)$	+ $OH^-(aq)$ $\xrightarrow{100\%}$	$HCO_2^-(aq)$	+ $H_2O(l)$
before (mol)	0.090	0.0050	0.075	
change (mol)	–0.0050	–0.0050	+0.0050	
after (mol)	0.085	0	0.080	

$$[H_3O^+] = K_a\frac{[HCO_2H]}{[HCO_2^-]} = (1.8 \times 10^{-4})\left(\frac{0.085}{0.080}\right) = 1.91 \times 10^{-4}\text{ M}$$

$\text{pH} = -\log[H_3O^+] = -\log(1.91 \times 10^{-4}) = 3.72$

(b)

	$HCO_2^-(aq)$	+	$H_3O^+(aq)$	$\xrightarrow{100\%}$	$HCO_2H(aq)$	+	$H_2O(l)$
before (mol)	0.075		0.0050		0.090		
change (mol)	–0.0050		–0.0050		+0.0050		
after (mol)	0.070		0		0.095		

$$[H_3O^+] = K_a \frac{[HCO_2H]}{[HCO_2^-]} = (1.8 \times 10^{-4})\left(\frac{0.095}{0.070}\right) = 2.44 \times 10^{-4}\ M$$

$pH = -\log[H_3O^+] = -\log(2.44 \times 10^{-4}) = 3.61$

16.74 $$pH = pK_a + \log\frac{[base]}{[acid]} = pK_a + \log\frac{[HCO_2^-]}{[HCO_2H]}$$

For HCO_2H, $K_a = 1.8 \times 10^{-4}$; $pK_a = -\log K_a = -\log(1.8 \times 10^{-4}) = 3.74$

$$pH = 3.74 + \log\frac{(0.50)}{(0.25)} = 4.04$$

16.76 $$pH = pK_a + \log\frac{[base]}{[acid]} = pK_a + \log\frac{[HCO_3^-]}{[H_2CO_3]}$$

For H_2CO_3, at 37 °C, $K_a = 7.9 \times 10^{-7}$; $pK_a = -\log K_a = -\log(7.9 \times 10^{-7}) = 6.10$

$$7.40 = 6.10 + \log\frac{[HCO_3^-]}{[H_2CO_3]};\quad 1.30 = \log\frac{[HCO_3^-]}{[H_2CO_3]}$$

$$\frac{[HCO_3^-]}{[H_2CO_3]} = 10^{1.30} = 20.0$$

16.78 $$pH = pK_a + \log\frac{[base]}{[acid]} = pK_a + \log\frac{[NH_3]}{[NH_4^+]}$$

For NH_4^+, $K_a = 5.6 \times 10^{-10}$; $pK_a = -\log K_a = -\log(5.6 \times 10^{-10}) = 9.25$

$$9.80 = 9.25 + \log\frac{[NH_3]}{[NH_4^+]};\quad 0.550 = \log\frac{[NH_3]}{[NH_4^+]};\quad \frac{[NH_3]}{[NH_4^+]} = 10^{0.55} = 3.5$$

The volume of the 1.0 M NH_3 solution should be 3.5 times the volume of the 1.0 M NH_4Cl solution so that the mixture will buffer at pH 9.80.

16.80 H_3PO_4, $K_{a1} = 7.5 \times 10^{-3}$; $pK_{a1} = -\log K_{a1} = 2.12$

$H_2PO_4^-$, $K_{a2} = 6.2 \times 10^{-8}$; $pK_{a2} = -\log K_{a2} = 7.21$

HPO_4^{2-}, $K_{a3} = 4.8 \times 10^{-13}$; $pK_{a3} = -\log K_{a3} = 12.32$

The buffer system of choice for pH 7.00 is (b) $H_2PO_4^- - HPO_4^{2-}$ because the pK_a for $H_2PO_4^-$ (7.21) is closest to 7.00.

16.82 $pH = pK_a + \log \dfrac{[base]}{[acid]}$

$$9.46 = pK_a + \log\left(\frac{34.5}{100 - 34.5}\right)$$

$$pK_a = 9.46 - \log\left(\frac{34.5}{100 - 34.5}\right) = 9.74$$

$$K_a = 10^{-pK_a} = 10^{-9.74} = 1.8 \times 10^{-10}$$

pH Titration Curves (Sections 16.5–16.9)

16.84 (a) (0.060 L)(0.150 mol/L)(1000 mmol/mol) = 9.00 mmol HNO_3

(b) vol NaOH = $(9.00 \text{ mmol } HNO_3)\left(\dfrac{1 \text{ mmol NaOH}}{1 \text{ mmol } HNO_3}\right)\left(\dfrac{1 \text{ mL NaOH}}{0.450 \text{ mmol NaOH}}\right) = 20.0 \text{ mL NaOH}$

(c) At the equivalence point the solution contains the neutral salt $NaNO_3$. The pH is 7.00.

(d)

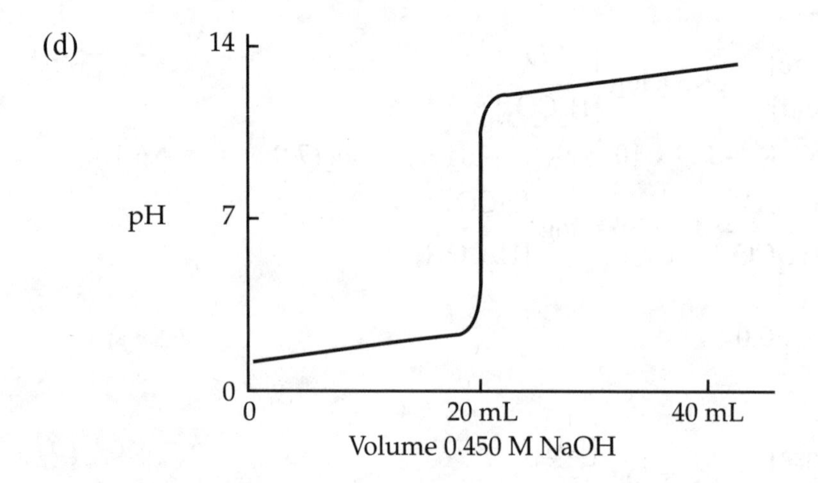

16.86 (0.0250 L)(0.125 mol/L)(1000 mmol/mol) = 3.125 mmol HCl

(a) (0.0030 L)(0.100 mol/L)(1000 mmol/mol) = 0.300 mmol NaOH

Neutralization reaction:	$H_3O^+(aq)$	+ $OH^-(aq)$	→ $H_2O(l)$
before reaction (mmol)	3.125	0.300	
change (mmol)	–0.300	–0.300	
after reaction (mmol)	2.825	0	

$$[H_3O^+] = \frac{2.825 \text{ mmol}}{(25.0 \text{ mL} + 3.0 \text{ mL})} = 0.1009 \text{ M}$$

$pH = -\log[H_3O^+] = -\log(0.1009) = 0.996$

(b) (0.020 L)(0.100 mol/L)(1000 mmol/mol) = 2.0 mmol NaOH

Neutralization reaction:	$H_3O^+(aq)$	+ $OH^-(aq)$	→ $H_2O(l)$
before reaction (mmol)	3.125	2.0	
change (mmol)	–2.0	–2.0	
after reaction (mmol)	1.125	0	

$$[H_3O^+] = \frac{1.125 \text{ mmol}}{(25.0 \text{ mL} + 20 \text{ mL})} = 0.025 \text{ M}$$

$pH = -\log[H_3O^+] = -\log(0.025) = 1.60$

(c) (0.065 L)(0.100 mol/L)(1000 mmol/mol) = 6.5 mmol NaOH

Neutralization reaction:	$H_3O^+(aq)$	+ $OH^-(aq)$	→ $H_2O(l)$
before reaction (mmol)	3.125	6.5	
change (mmol)	−3.125	−3.125	
after reaction (mmol)	0	3.375	

$$[OH^-] = \frac{3.375 \text{ mmol}}{(25.0 \text{ mL} + 65 \text{ mL})} = 0.0375 \text{ M}$$

$$[H_3O^+] = \frac{K_w}{[OH^-]} = \frac{1.0 \times 10^{-14}}{0.0375} = 2.7 \times 10^{-13} \text{ M}$$

$pH = -\log[H_3O^+] = -\log(2.7 \times 10^{-13}) = 12.57$

16.88 mmol HF = (40.0 mL)(0.250 mmol/mL) = 10.0 mmol

mmol NaOH required = mmol HF = 10.0 mmol

$$\text{mL NaOH required} = (10.0 \text{ mmol})\left(\frac{1.00 \text{ mL}}{0.200 \text{ mmol}}\right) = 50.0 \text{ mL}$$

50.0 mL of 0.200 M NaOH is required to reach the equivalence point.

For HF, $K_a = 3.5 \times 10^{-4}$; $pK_a = -\log K_a = -\log(3.5 \times 10^{-4}) = 3.46$

(a) mmol HF = 10.0 mmol

mmol NaOH = (0.200 mmol/mL)(10.0 mL) = 2.00 mmol

Neutralization reaction:	$HF(aq)$	+ $OH^-(aq)$	→ $F^-(aq)$	+ $H_2O(l)$
before reaction (mmol)	10.0	2.00	0	
change (mmol)	−2.00	−2.00	+2.00	
after reaction (mmol)	8.0	0	2.00	

$$[HF] = \frac{8.0 \text{ mmol}}{(40.0 \text{ mL} + 10.0 \text{ mL})} = 0.16 \text{ M}; \quad [F^-] = \frac{2.00 \text{ mmol}}{(40.0 \text{ mL} + 10.0 \text{ mL})} = 0.0400 \text{ M}$$

	$HF(aq)$	+ $H_2O(l)$	⇌ $H_3O^+(aq)$	+ $F^-(aq)$
initial (M)	0.16		~0	0.0400
change (M)	−x		+x	+x
equil (M)	0.16 − x		x	0.0400 + x

$$K_a = \frac{[H_3O^+][F^-]}{[HF]} = 3.5 \times 10^{-4} = \frac{x(0.0400 + x)}{0.16 - x} \approx \frac{x(0.0400)}{0.16}$$

Solve for x. $x = [H_3O^+] = 1.4 \times 10^{-3}$ M

$pH = -\log[H_3O^+] = -\log(1.4 \times 10^{-3}) = 2.85$

(b) Halfway to the equivalence point,

$pH = pK_a = -\log K_a = -\log(3.5 \times 10^{-4}) = 3.46$

(c) At the equivalence point only the salt NaF is in solution.

$$[F^-] = \frac{10.0 \text{ mmol}}{(40.0 \text{ mL} + 50.0 \text{ mL})} = 0.111 \text{ M}$$

	$F^-(aq)$ + $H_2O(l)$ ⇄	$HF(aq)$ +	$OH^-(aq)$
initial (M)	0.111	0	~0
change (M)	–x	+x	+x
equil (M)	0.111 – x	x	x

For F^-, $K_b = \dfrac{K_w}{K_a \text{ for HF}} = \dfrac{1.0 \times 10^{-14}}{3.5 \times 10^{-4}} = 2.9 \times 10^{-11}$

$$K_b = \frac{[HF][OH^-]}{[F^-]} = 2.9 \times 10^{-11} = \frac{x^2}{0.111 - x} \approx \frac{x^2}{0.111}$$

Solve for x. $x = [OH^-] = 1.8 \times 10^{-6}$ M

$$[H_3O^+] = \frac{K_w}{[OH^-]} = \frac{1.0 \times 10^{-14}}{1.8 \times 10^{-6}} = 5.6 \times 10^{-9} \text{ M}$$

$pH = -\log[H_3O^+] = -\log(5.6 \times 10^{-9}) = 8.25$

(d) mmol HF = 10.0 mmol

mmol NaOH = (0.200 mmol/mL)(80.0 mL) = 16.0 mmol

Neutralization reaction:	$HF(aq)$ +	$OH^-(aq)$ →	$F^-(aq)$ + $H_2O(l)$
before reaction (mmol)	10.0	16.0	0
change (mmol)	–10.0	–10.0	+10.0
after reaction (mmol)	0	6.0	10.0

After the equivalence point, the pH of the solution is determined by the $[OH^-]$.

$$[OH^-] = \frac{6.0 \text{ mmol}}{(40.0 \text{ mL} + 80.0 \text{ mL})} = 5.0 \times 10^{-2} \text{ M}$$

$$[H_3O^+] = \frac{K_w}{[OH^-]} = \frac{1.0 \times 10^{-14}}{5.0 \times 10^{-2}} = 2.0 \times 10^{-13} \text{ M}$$

$pH = -\log[H_3O^+] = -\log(2.0 \times 10^{-13}) = 12.70$

16.90 mmol CH_3NH_2 = (100.0 mL)(0.100 mmol/mL) = 10.0 mmol

mmol HNO_3 required = mmol CH_3NH_2 = 10.0 mmol

$$\text{vol } HNO_3 \text{ required} = (10.0 \text{ mmol})\left(\frac{1.00 \text{ mL}}{0.250 \text{ mmol}}\right) = 40.0 \text{ mL}$$

40.0 mL of 0.250 M HNO_3 are required to reach the equivalence point.

(a)	$CH_3NH_2(aq)$ + $H_2O(l)$ ⇄	$CH_3NH_3^+(aq)$ +	$OH^-(aq)$
initial (M)	0.100	0	~0
change (M)	–x	+x	+x
equil (M)	0.100 – x	x	x

$$K_b = \frac{[CH_3NH_3^+][OH^-]}{[CH_3NH_2]} = 3.7 \times 10^{-4} = \frac{x^2}{0.100 - x}$$

$x^2 + (3.7 \times 10^{-4})x - (3.7 \times 10^{-5}) = 0$

Use the quadratic formula to solve for x.

$$x = \frac{-(3.7 \times 10^{-4}) \pm \sqrt{(3.7 \times 10^{-4})^2 - (4)(-3.7 \times 10^{-5})}}{2(1)} = \frac{-3.7 \times 10^{-4} \pm 0.0122}{2}$$

x = 0.0059 and –0.0063

Of the two solutions for x, only the positive value of x has physical meaning because x is the $[OH^-]$.

$[OH^-] = x = 0.0059$ M

$$[H_3O^+] = \frac{K_w}{[OH^-]} = \frac{1.0 \times 10^{-14}}{5.9 \times 10^{-3}} = 1.7 \times 10^{-12} \text{ M}$$

$pH = -\log[H_3O^+] = -\log(1.7 \times 10^{-12}) = 11.77$

(b) 20.0 mL of HNO_3 is halfway to the equivalence point.

$$\text{For } CH_3NH_3^+,\ K_a = \frac{K_w}{K_b \text{ for } CH_3NH_2} = \frac{1.0 \times 10^{-14}}{3.7 \times 10^{-4}} = 2.7 \times 10^{-11}$$

$pH = pK_a = -\log(2.7 \times 10^{-11}) = 10.57$

(c) At the equivalence point only the salt $CH_3NH_3NO_3$ is in solution.

mmol $CH_3NH_3NO_3$ = (0.100 mmol/mL)(100.0 mL) = 10.0 mmol

$$[CH_3NH_3^+] = \frac{10.0 \text{ mmol}}{(100.0 \text{ mL} + 40.0 \text{ mL})} = 0.0714 \text{ M}$$

	$CH_3NH_3^+(aq)$ + $H_2O(l)$	⇄	$H_3O^+(aq)$	+	$CH_3NH_2(aq)$
initial (M)	0.0714		~0		0
change (M)	–x		+x		+x
equil (M)	0.0714 – x		x		x

$$K_a = \frac{[H_3O^+][CH_3NH_2]}{[CH_3NH_3^+]} = 2.7 \times 10^{-11} = \frac{x^2}{0.0714 - x} \approx \frac{x^2}{0.0714}$$

Solve for x. $x = [H_3O^+] = 1.4 \times 10^{-6}$ M; $pH = -\log[H_3O^+] = -\log(1.4 \times 10^{-6}) = 5.85$

(d) mmol CH_3NH_2 = (0.100 mmol/mL)(100.0 mL) = 10.0 mmol

mmol HNO_3 = (0.250 mmol/mL)(60.0 mL) = 15.0 mmol

Neutralization reaction:	$CH_3NH_2(aq)$	+ $H_3O^+(aq)$	→	$CH_3NH_3^+(aq)$	+ $H_2O(l)$
before reaction (mmol)	10.0	15.0		0	
change (mmol)	–10.0	–10.0		+10.0	
after reaction (mmol)	0	5.0		10.0	

After the equivalence point the pH of the solution is determined by the $[H_3O^+]$.

$$[H_3O^+] = \frac{5.0 \text{ mmol}}{(100.0 \text{ mL} + 60.0 \text{ mL})} = 3.1 \times 10^{-2} \text{ M}$$

$pH = -\log[H_3O^+] = -\log(3.1 \times 10^{-2}) = 1.51$

16.92 For H_2A^+, $K_{a1} = 4.6 \times 10^{-3}$ and $K_{a2} = 2.0 \times 10^{-10}$

(a) (10.0 mL)(0.100 mmol/mL) = 1.00 mmol NaOH added = 1.00 mmol HA produced.

(50.0 mL)(0.100 mmol/mL) = 5.00 mmol H_2A^+

5.00 mmol H_2A^+ – 1.00 mmol NaOH = 4.00 mmol H_2A^+ after neutralization

$$[H_2A^+] = \frac{4.00 \text{ mmol}}{(50.0 \text{ mL} + 10.0 \text{ mL})} = 6.67 \times 10^{-2} \text{ M}$$

$$[HA] = \frac{1.00 \text{ mmol}}{(50.0 \text{ mL} + 10.0 \text{ mL})} = 1.67 \times 10^{-2} \text{ M}$$

$$pH = pK_{a1} + \log\frac{[HA]}{[H_2A^+]} = -\log(4.6 \times 10^{-3}) + \log\left(\frac{1.67 \times 10^{-2}}{6.67 \times 10^{-2}}\right) = 1.74$$

(b) Halfway to the first equivalence point, $pH = pK_{a1} = 2.34$

(c) At the first equivalence point, $pH = \dfrac{pK_{a1} + pK_{a2}}{2} = 6.02$

(d) Halfway between the first and second equivalence points, $pH = pK_{a2} = 9.70$

(e) At the second equivalence point only the basic salt, NaA, is in solution.

$$K_b = \frac{K_w}{K_a \text{ for HA}} = \frac{K_w}{K_{a2}} = \frac{1.0 \times 10^{-14}}{2.0 \times 10^{-10}} = 5.0 \times 10^{-5}$$

mmol A^- = (50.0 mL)(0.100 mmol/mL) = 5.00 mmol

$$[A^-] = \frac{5.0 \text{ mmol}}{(50.0 \text{ mL} + 100.0 \text{ mL})} = 3.3 \times 10^{-2} \text{ M}$$

	$A^-(aq)$ + $H_2O(l)$	⇄ $HA(aq)$	+ $OH^-(aq)$
initial (M)	0.033	0	~0
change (M)	–x	+x	+x
equil (M)	0.033 – x	x	x

$$K_b = \frac{[HA][OH^-]}{[A^-]} = 5.0 \times 10^{-5} = \frac{(x)(x)}{0.033 - x} \approx \frac{x^2}{0.033}$$

Solve for x.

$x = [OH^-] = \sqrt{(5.0 \times 10^{-5})(0.033)} = 1.3 \times 10^{-3}$ M

$$[H_3O^+] = \frac{K_w}{[OH^-]} = \frac{1.0 \times 10^{-14}}{1.3 \times 10^{-3}} = 7.7 \times 10^{-12} \text{ M}$$

$pH = -\log[H_3O^+] = -\log(7.7 \times 10^{-12}) = 11.11$

16.94 (a) The strongest acid has the lowest pH at the equivalence point, C is the strongest acid.
(b) The weakest acid has the highest pH at the equivalence point, A is the weakest acid.

16.96 When equal volumes of acid and base react, all concentrations are cut in half.
(a) At the equivalence point, only the salt $NaNO_2$ is in solution.
$[NO_2^-] = 0.050$ M

$$\text{For } NO_2^-, K_b = \frac{K_w}{K_a \text{ for } HNO_2} = \frac{1.0 \times 10^{-14}}{4.5 \times 10^{-4}} = 2.2 \times 10^{-11}$$

	$NO_2^-(aq)$ + $H_2O(l)$	⇄ $HNO_2(aq)$	+ $OH^-(aq)$
initial (M)	0.050	0	~0
change (M)	–x	+x	+x
equil (M)	0.050 – x	x	x

$$K_b = \frac{[HNO_2][OH^-]}{[NO_2^-]} = 2.2 \times 10^{-11} = \frac{(x)(x)}{0.050 - x} \approx \frac{x^2}{0.050}$$

Solve for x. $x = [OH^-] = 1.1 \times 10^{-6}$ M

$$[H_3O^+] = \frac{K_w}{[OH^-]} = \frac{1.0 \times 10^{-14}}{1.1 \times 10^{-6}} = 9.1 \times 10^{-9}\ M$$

$pH = -\log[H_3O^+] = -\log(9.1 \times 10^{-9}) = 8.04$

Phenol red would be a suitable indicator. (see Figure 15.5)

(b) The pH is 7.00 at the equivalence point for the titration of a strong acid (HI) with a strong base (NaOH).

Bromthymol blue or phenol red would be suitable indicators. (Any indicator that changes color in the pH range 4 – 10 is satisfactory for a strong acid – strong base titration.)

(c) At the equivalence point only the salt CH_3NH_3Cl is in solution.

$[CH_3NH_3^+] = 0.050\ M$

$$\text{For } CH_3NH_3^+,\ K_a = \frac{K_w}{K_b \text{ for } CH_3NH_2} = \frac{1.0 \times 10^{-14}}{3.7 \times 10^{-4}} = 2.7 \times 10^{-11}$$

	$CH_3NH_3^+(aq)$ + $H_2O(l)$	⇄	$H_3O^+(aq)$	+	$CH_3NH_2(aq)$
initial (M)	0.050		~0		0
change (M)	–x		+x		+x
equil (M)	0.050 – x		x		x

$$K_a = \frac{[H_3O^+][CH_3NH_2]}{[CH_3NH_3^+]} = 2.7 \times 10^{-11} = \frac{(x)(x)}{0.050 - x} \approx \frac{x^2}{0.050}$$

Solve for x. $x = [H_3O^+] = 1.2 \times 10^{-6}\ M$

$pH = -\log[H_3O^+] = -\log(1.2 \times 10^{-6}) = 5.92$

Chlorphenol red would be a suitable indicator.

Solubility Equilibria (Sections 16.10 and 16.11)

16.98 (a) $Ag_2CO_3(s) \rightleftarrows 2\ Ag^+(aq) + CO_3^{2-}(aq)$ $K_{sp} = [Ag^+]^2[CO_3^{2-}]$

(b) $PbCrO_4(s) \rightleftarrows Pb^{2+}(aq) + CrO_4^{2-}(aq)$ $K_{sp} = [Pb^{2+}][CrO_4^{2-}]$

(c) $Al(OH)_3(s) \rightleftarrows Al^{3+}(aq) + 3\ OH^-(aq)$ $K_{sp} = [Al^{3+}][OH^-]^3$

(d) $Hg_2Cl_2(s) \rightleftarrows Hg_2^{2+}(aq) + 2\ Cl^-(aq)$ $K_{sp} = [Hg_2^{2+}][Cl^-]^2$

16.100 (a) $K_{sp} = [Pb^{2+}][I^-]^2 = (5.0 \times 10^{-3})(1.3 \times 10^{-3})^2 = 8.5 \times 10^{-9}$

(b) $$[I^-] = \sqrt{\frac{K_{sp}}{[Pb^{2+}]}} = \sqrt{\frac{(8.5 \times 10^{-9}}{(2.5 \times 10^{-4})}} = 5.8 \times 10^{-3}\ M$$

(c) $$[Pb^{2+}] = \frac{K_{sp}}{[I^-]^2} = \frac{(8.5 \times 10^{-9})}{(2.5 \times 10^{-4})^2} = 0.14\ M$$

16.102 $Ag_2CO_3(s) \rightleftarrows 2\ Ag^+(aq) + CO_3^{2-}(aq)$

equil (M) 2x x

$[Ag^+] = 2x = 2.56 \times 10^{-4}\ M$; $[CO_3^{2-}] = x = (2.56 \times 10^{-4}\ M)/2 = 1.28 \times 10^{-4}\ M$

$K_{sp} = [Ag^+]^2[CO_3^{2-}] = (2.56 \times 10^{-4})^2(1.28 \times 10^{-4}) = 8.39 \times 10^{-12}$

16.104 (a) $SrF_2(s) \rightleftarrows Sr^{2+}(aq) + 2\ F^-(aq)$

equil (M) x 2x

$[Sr^{2+}] = x = 1.03 \times 10^{-3}$ M; $[F^-] = 2x = 2(1.03 \times 10^{-3}\ M) = 2.06 \times 10^{-3}$ M

$K_{sp} = [Sr^{2+}][F^-]^2 = (1.03 \times 10^{-3})(2.06 \times 10^{-3})^2 = 4.37 \times 10^{-9}$

(b) $CuI(s) \rightleftarrows Cu^+(aq) + I^-(aq)$

equil (M) x x

$[Cu^+] = [I^-] = x = 1.05 \times 10^{-6}$ M

$K_{sp} = [Cu^+][I^-] = (1.05 \times 10^{-6})^2 = 1.10 \times 10^{-12}$

(c) MgC_2O_4, 112.32

$$[Mg^{2+}] = \text{molarity of } MgC_2O_4 = \frac{\left(0.094\ g \times \frac{1\ mol}{112.32\ g}\right)}{1\ L} = 8.37 \times 10^{-4}\ M$$

$MgC_2O_4(s) \rightleftarrows Mg^{2+}(aq) + C_2O_4^-(aq)$

equil (M) x x

$[Mg^{2+}] = [C_2O_4^-] = x = 8.37 \times 10^{-4}$ M

$K_{sp} = [Mg^{2+}][C_2O_4^-] = (8.37 \times 10^{-4})^2 = 7.0 \times 10^{-7}$

(d) $Zn(CN)_2$, 117.41

$$[Zn^{2+}] = \text{molarity of } Zn(CN)_2 = \frac{\left(4.95 \times 10^{-4}\ g \times \frac{1\ mol}{117.41\ g}\right)}{1\ L} = 4.22 \times 10^{-6}\ M$$

$Zn(CN)_2(s) \rightleftarrows Zn^{2+}(aq) + 2\ CN^-(aq)$

equil (M) x 2x

$[Zn^{2+}] = x = 4.22 \times 10^{-6}$ M; $[CN^-] = 2x = 2(4.22 \times 10^{-6}\ M) = 8.44 \times 10^{-6}$ M

$K_{sp} = [Zn^{2+}][CN^-]^2 = (4.22 \times 10^{-6})(8.44 \times 10^{-6})^2 = 3.01 \times 10^{-16}$

16.106 (a) $BaCrO_4(s) \rightleftarrows Ba^{2+}(aq) + CrO_4^{2-}(aq)$

equil (M) x x

$K_{sp} = [Ba^{2+}][CrO_4^{2-}] = 1.2 \times 10^{-10} = (x)(x)$

molar solubility = $x = \sqrt{1.2 \times 10^{-10}} = 1.1 \times 10^{-5}$ M

(b) $Mg(OH)_2(s) \rightleftarrows Mg^{2+}(aq) + 2\ OH^-(aq)$

equil (M) x 2x

$K_{sp} = [Mg^{2+}][OH^-]^2 = 5.6 \times 10^{-12} = x(2x)^2 = 4x^3$

molar solubility = $x = \sqrt[3]{\frac{5.6 \times 10^{-12}}{4}} = 1.1 \times 10^{-4}$ M

(c) $Ag_2SO_3(s) \rightleftarrows 2\ Ag^+(aq) + SO_3^{2-}(aq)$

equil (M) 2x x

$K_{sp} = [Ag^+]^2[SO_3^{2-}] = 1.5 \times 10^{-14} = (2x)^2x = 4x^3$

molar solubility = x = $\sqrt[3]{\dfrac{1.5 \times 10^{-14}}{4}}$ = 1.6×10^{-5} M

Factors That Affect Solubility (Section 16.12)

16.108 $Ag_2CO_3(s) \rightleftarrows 2\ Ag^+(aq) + CO_3^{2-}(aq)$
(a) $AgNO_3$, source of Ag^+; equilibrium shifts left
(b) HNO_3, source of H_3O^+, removes CO_3^{2-}; equilibrium shifts right
(c) Na_2CO_3, source of CO_3^{2-}; equilibrium shifts left
(d) NH_3, forms $Ag(NH_3)_2^+$, removes Ag^+; equilibrium shifts right

16.110 (a)

	$PbCrO_4(s) \rightleftarrows$	$Pb^{2+}(aq)$	+ $CrO_4^{2-}(aq)$
equil (M)		x	x

$K_{sp} = [Pb^{2+}][CrO_4^{2-}] = 2.8 \times 10^{-13} = (x)(x)$
molar solubility = x = $\sqrt{2.8 \times 10^{-13}}$ = 5.3×10^{-7} M

(b)

	$PbCrO_4(s) \rightleftarrows$	$Pb^{2+}(aq)$	+ $CrO_4^{2-}(aq)$
initial(M)		0	1.0×10^{-3}
equil (M)		x	$1.0 \times 10^{-3} + x$

$K_{sp} = [Pb^{2+}][CrO_4^{2-}] = 2.8 \times 10^{-13} = (x)(1.0 \times 10^{-3} + x) \approx (x)(1.0 \times 10^{-3})$

molar solubility = x = $\dfrac{2.8 \times 10^{-13}}{1 \times 10^{-3}}$ = 2.8×10^{-10} M

16.112 (b), (c), and (d) are more soluble in acidic solution.
(a) $AgBr(s) \rightleftarrows Ag^+(aq) + Br^-(aq)$
(b) $CaCO_3(s) + H_3O^+(aq) \rightleftarrows Ca^{2+}(aq) + HCO_3^-(aq) + H_2O(l)$
(c) $Ni(OH)_2(s) + 2\ H_3O^+(aq) \rightleftarrows Ni^{2+}(aq) + 4\ H_2O(l)$
(d) $Ca_3(PO_4)_2(s) + 2\ H_3O^+(aq) \rightleftarrows 3\ Ca^{2+}(aq) + 2\ HPO_4^{2-}(aq) + 2\ H_2O(l)$

16.114 (a) Because $Zn(OH)_2$ contains a basic anion, it becomes more soluble as the acidity of the solution increases. $Zn(OH)_2(s) + 2\ H_3O^+(aq) \rightleftarrows Zn^{2+}(aq) + 4\ H_2O(l)$

(b) Because $Zn(OH)_2$ forms the complex anion, $Zn(OH)_4^{2-}$, $Zn(OH)_2$ becomes more soluble in basic solution. $Zn(OH)_2(s) + 2\ OH^-(aq) \rightleftarrows Zn(OH)_4^{2-}(aq)$

(c) Because $Zn(OH)_2$ forms the complex anion, $Zn(CN)_4^{2-}$, $Zn(OH)_2$ becomes more soluble in the presence of CN^-.
$Zn(OH)_2(s) + 4\ CN^-(aq) \rightleftarrows Zn(CN)_4^{2-}(aq) + 2\ OH^-(aq)$

16.116 On mixing equal volumes of two solutions, the concentrations of both solutions are cut in half.

	$Ag^+(aq)$	+	$2\ CN^-(aq)$	⇄	$Ag(CN)_2^-(aq)$
before reaction (M)	0.0010		0.10		0
assume 100% reaction	–0.0010		–2(0.0010)		0.0010
after reaction (M)	0		0.098		0.0010
assume small back rxn	+x		+2x		–x
equil (M)	x		0.098 + 2x		0.0010 – x

$$K_f = 3.0 \times 10^{20} = \frac{[Ag(CN)_2^-]}{[Ag^+][CN^-]^2} = \frac{(0.0010 - x)}{x(0.098 + 2x)^2} \approx \frac{0.0010}{x(0.098)^2}$$

Solve for x. $x = [Ag^+] = 3.5 \times 10^{-22}$ M

16.118 (a)

$AgI(s) \rightleftarrows Ag^+(aq) + I^-(aq)$ $\qquad K_{sp} = 8.5 \times 10^{-17}$

$Ag^+(aq) + 2\ CN^-(aq) \rightarrow Ag(CN)_2^-(aq)$ $\qquad K_f = 3.0 \times 10^{20}$

dissolution rxn $AgI(s) + 2\ CN^-(aq) \rightleftarrows Ag(CN)_2^-(aq) + I^-(aq)$

$K = (K_{sp})(K_f) = (8.5 \times 10^{-17})(3.0 \times 10^{20}) = 2.6 \times 10^4$

(b)

$Al(OH)_3(s) \rightleftarrows Al^{3+}(aq) + 3\ OH^-(aq)$ $\qquad K_{sp} = 1.9 \times 10^{-33}$

$Al^{3+}(aq) + 4\ OH^-(aq) \rightarrow Al(OH)_4^-(aq)$ $\qquad K_f = 3 \times 10^{33}$

dissolution rxn $Al(OH)_3(s) + OH^-(aq) \rightleftarrows Al(OH)_4^-(aq)$

$K = (K_{sp})(K_f) = (1.9 \times 10^{-33})(3 \times 10^{33}) = 6$

(c)

$Zn(OH)_2(s) \rightleftarrows Zn^{2+}(aq) + 2\ OH^-(aq)$ $\qquad K_{sp} = 4.1 \times 10^{-17}$

$Zn^{2+}(aq) + 4\ NH_3(aq) \rightarrow Zn(NH_3)_4^{2+}(aq)$ $\qquad K_f = 7.8 \times 10^8$

dissolution rxn $Zn(OH)_2(s) + 4\ NH_3(aq) \rightleftarrows Zn(NH_3)_4^{2+} + 2\ OH^-(aq)$

$K = (K_{sp})(K_f) = (4.1 \times 10^{-17})(7.8 \times 10^8) = 3.2 \times 10^{-8}$

16.120 (a)

	$AgI(s)$	⇄	$Ag^+(aq)$	+	$I^-(aq)$
equil (M)			x		x

$K_{sp} = [Ag^+][I^-] = 8.5 \times 10^{-17} = (x)(x)$

molar solubility $= x = \sqrt{8.5 \times 10^{-17}} = 9.2 \times 10^{-9}$ M

(b)

	$AgI(s)$	+	$2\ CN^-(aq)$	⇄	$Ag(CN)_2^-(aq)$	+	$I^-(aq)$
initial (M)			0.10		0		0
change (M)			–2x		+x		+x
equil (M)			0.10 – 2x		x		x

$K = (K_{sp})(K_f) = (8.5 \times 10^{-17})(3.0 \times 10^{20}) = 2.6 \times 10^4$

$$K = 2.6 \times 10^4 = \frac{[Ag(CN)_2^-][I^-]}{[CN^-]^2} = \frac{x^2}{(0.10 - 2x)^2}$$

Take the square root of both sides and solve for x.

molar solubility = x = 0.050 M

Precipitation; Qualitative Analysis (Sections 16.13–16.15)

16.122 $BaSO_4$, $K_{sp} = 1.1 \times 10^{-10}$; $Fe(OH)_3$, $K_{sp} = 2.6 \times 10^{-39}$

Total volume = 80 mL + 20 mL = 100 mL

$$[Ba^{2+}] = \frac{(1.0 \times 10^{-5}\ M)(80\ mL)}{(100\ mL)} = 8.0 \times 10^{-6}\ M$$

$[OH^-] = 2[Ba^{2+}] = 2(8.0 \times 10^{-6}) = 1.6 \times 10^{-5}\ M$

$$[Fe^{3+}] = \frac{2(1.0 \times 10^{-5}\ M)(20\ mL)}{(100\ mL)} = 4.0 \times 10^{-6}\ M$$

$$[SO_4^{2-}] = \frac{3(1.0 \times 10^{-5}\ M)(20\ mL)}{(100\ mL)} = 6.0 \times 10^{-6}\ M$$

For $BaSO_4$, IP = $[Ba^{2+}]_t[SO_4^{2-}]_t = (8.0 \times 10^{-6})(6.0 \times 10^{-6}) = 4.8 \times 10^{-11}$

IP < K_{sp}; $BaSO_4$ will not precipitate.

For $Fe(OH)_3$, IP = $[Fe^{3+}]_t[OH^-]_t^3 = (4.0 \times 10^{-6})(1.6 \times 10^{-5})^3 = 1.6 \times 10^{-20}$

IP > K_{sp}; $Fe(OH)_3(s)$ will precipitate.

16.124 pH = 10.80; $[H_3O^+] = 10^{-pH} = 10^{-10.80} = 1.6 \times 10^{-11}\ M$

$$[OH^-] = \frac{K_w}{[H_3O^+]} = \frac{1.0 \times 10^{-14}}{1.6 \times 10^{-11}} = 6.3 \times 10^{-4}\ M$$

For $Mg(OH)_2$, $K_{sp} = 5.6 \times 10^{-12}$

IP = $[Mg^{2+}]_t[OH^-]_t^2 = (2.5 \times 10^{-4})(6.3 \times 10^{-4})^2 = 9.9 \times 10^{-11}$

IP > K_{sp}; $Mg(OH)_2(s)$ will precipitate

16.126 $K_{spa} = \dfrac{[M^{2+}][H_2S]}{[H_3O^+]^2}$; FeS, $K_{spa} = 6 \times 10^2$; SnS, $K_{spa} = 1 \times 10^{-5}$

Fe^{2+} and Sn^{2+} can be separated by bubbling H_2S through an acidic solution containing the two cations because their K_{spa} values are so different.

For FeS and SnS, $Q_c = \dfrac{(0.01)(0.10)}{(0.3)^2} = 1.1 \times 10^{-2}$

For FeS, $Q_c < K_{spa}$, and FeS will not precipitate.

For SnS, $Q_c > K_{spa}$, and SnS will precipitate.

16.128 FeS, $K_{spa} = \dfrac{[Fe^{2+}][H_2S]}{[H_3O^+]^2} = 6 \times 10^2$

(i) In 0.4 M HCl, $[H_3O^+] = 0.4\ M$

$Q_c = \dfrac{[Fe^{2+}]_t[H_2S]_t}{[H_3O^+]_t^2} = \dfrac{(0.10)(0.10)}{(0.4)^2} = 0.0625$; $Q_c < K_{spa}$; FeS will not precipitate

(ii) pH = 8; $[H_3O^+] = 10^{-pH} = 10^{-8} = 1 \times 10^{-8}\ M$

$Q_c = \dfrac{[Fe^{2+}]_t[H_2S]_t}{[H_3O^+]_t^2} = \dfrac{(0.10)(0.10)}{(1 \times 10^{-8})^2} = 1 \times 10^{14}$; $Q_c > K_{spa}$; FeS(s) will precipitate

16.130 (a) add Cl^- to precipitate AgCl
(b) add CO_3^{2-} to precipitate $CaCO_3$
(c) add H_2S to precipitate MnS
(d) add NH_3 and NH_4Cl to precipitate $Cr(OH)_3$
(Need buffer to control $[OH^-]$; excess OH^- produces the soluble $Cr(OH)_4^-$.)

Chapter Problems

16.132

$$Ca_5(PO_4)_3(OH)(s) \rightleftarrows 5\,Ca^{2+}(aq) + 3\,PO_4^{3-}(aq) + OH^-(aq)$$

equil (M) 5x 3x x

$Ksp = [Ca^{2+}]^5[PO_4^{3-}]^3[OH^-] = 2.3 \times 10^{-59} = (5x)^5(3x)^3(x) = 84{,}375\,x^9$

Solve for x, x = molar solubility = 8.6×10^{-8} M

$$Ca_5(PO_4)_3(F)(s) \rightleftarrows 5\,Ca^{2+}(aq) + 3\,PO_4^{3-}(aq) + F^-(aq)$$

equil (M) 5x 3x x

$Ksp = [Ca^{2+}]^5[PO_4^{3-}]^3[F^-] = 3.2 \times 10^{-60} = (5x)^5(3x)^3(x) = 84{,}375\,x^9$

Solve for x, x = molar solubility = 7.0×10^{-8} M

16.134 Prepare aqueous solutions of the three salts. Add a solution of $(NH_4)_2HPO_4$. If a white precipitate forms, the solution contains Mg^{2+}. Perform flame test on the other two solutions. A yellow flame test indicates Na^+. A violet flame test indicates K^+.

16.136 (a)

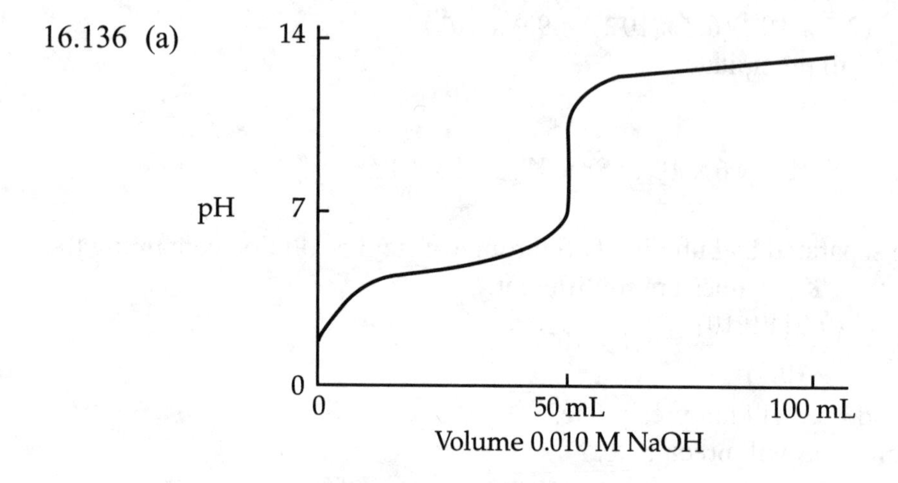

(b) $$\text{mol NaOH required} = \left(\frac{0.010 \text{ mol HA}}{\text{L}}\right)(0.0500 \text{ L})\left(\frac{1 \text{ mol NaOH}}{1 \text{ mol HA}}\right) = 0.000\ 50 \text{ mol}$$

$$\text{vol NaOH required} = (0.000\ 50 \text{ mol})\left(\frac{1 \text{ L}}{0.010 \text{ mol}}\right) = 0.050 \text{ L} = 50 \text{ mL}$$

(c) A basic salt is present at the equivalence point; pH > 7.00
(d) Halfway to the equivalence point, the pH = pK_a = 4.00

16.138 pH = 10.35; $[H_3O^+] = 10^{-pH} = 10^{-10.35} = 4.5 \times 10^{-11}$ M

$$[OH^-] = \frac{K_w}{[H_3O^+]} = \frac{1.0 \times 10^{-14}}{4.5 \times 10^{-11}} = 2.2 \times 10^{-4}\text{ M}$$

$$[Mg^{2+}] = \frac{[OH^-]}{2} = \frac{2.2 \times 10^{-4}}{2} = 1.1 \times 10^{-4}\text{ M}$$

$$K_{sp} = [Mg^{2+}][OH^-]^2 = (1.1 \times 10^{-4})(2.2 \times 10^{-4})^2 = 5.3 \times 10^{-12}$$

16.140 NaOH, 40.0; $20\text{ g} \times \frac{1\text{ mol}}{40.0\text{ g}} = 0.50$ mol NaOH

(0.500 L)(1.5 mol/L) = 0.75 mol NH_4Cl

	$NH_4^+(aq)$ +	$OH^-(aq)$ ⇄	$NH_3(aq)$ + $H_2O(l)$
before reaction (mol)	0.75	0.50	0
change (mol)	–0.50	–0.50	+0.50
after reaction (mol)	0.25	0	0.50

This reaction produces a buffer solution.

$[NH_4^+] = 0.25$ mol/0.500 L = 0.50 M; $[NH_3] = 0.50$ mol/0.500 L = 1.0 M

$$pH = pK_a + \log\frac{[\text{base}]}{[\text{acid}]} = pK_a + \log\frac{[NH_3]}{[NH_4^+]}$$

$$\text{For } NH_4^+,\ K_a = \frac{K_w}{K_b \text{ for } NH_3} = \frac{1.0 \times 10^{-14}}{1.8 \times 10^{-5}} = 5.6 \times 10^{-10};\ pK_a = -\log K_a = 9.25$$

$$pH = 9.25 + \log\left(\frac{1.0}{0.5}\right) = 9.55$$

16.142 $\text{For } NH_4^+,\ K_a = \frac{K_w}{K_b \text{ for } NH_3} = \frac{1.0 \times 10^{-14}}{1.8 \times 10^{-5}} = 5.6 \times 10^{-10};\ pK_a = -\log K_a = 9.25$

$$pH = pK_a + \log\frac{[NH_3]}{[NH_4^+]} = 9.25 + \log\frac{(0.50)}{(0.30)} = 9.47$$

$$[H_3O^+] = 10^{-pH} = 10^{-9.47} = 3.4 \times 10^{-10}\text{ M}$$

$$\text{For MnS, } K_{spa} = \frac{[Mn^{2+}][H_2S]}{[H_3O^+]^2} = 3 \times 10^7$$

$$\text{molar solubility} = [Mn^{2+}] = \frac{K_{spa}[H_3O^+]^2}{[H_2S]} = \frac{(3 \times 10^7)(3.4 \times 10^{-10})^2}{(0.10)} = 3.5 \times 10^{-11}\text{ M}$$

MnS, 87.00; solubility = $(3.5 \times 10^{-11}$ mol/L)(87.00 g/mol) = 3×10^{-9} g/L

16.144 60.0 mL = 0.0600 L

$$\text{mol } H_3PO_4 = 0.0600\text{ L} \times \frac{1.00\text{ mol } H_3PO_4}{1.00\text{ L}} = 0.0600\text{ mol } H_3PO_4$$

$$\text{mol LiOH} = 1.00 \text{ L} \times \frac{0.100 \text{ mol LiOH}}{1.00 \text{ L}} = 0.100 \text{ mol LiOH}$$

	$H_3PO_4(aq)$ +	$OH^-(aq)$ →	$H_2PO_4^-(aq)$ + $H_2O(l)$
before reaction (mol)	0.0600	0.100	0
change (mol)	–0.0600	–0.0600	+0.0600
after reaction (mol)	0	0.040	0.0600

	$H_2PO_4^-(aq)$ +	$OH^-(aq)$ →	$HPO_4^{2-}(aq)$ + $H_2O(l)$
before reaction (mol)	0.0600	0.040	0
change (mol)	–0.040	–0.040	+0.040
after reaction (mol)	0.020	0	0.040

The resulting solution is a buffer because it contains the conjugate acid-base pair, $H_2PO_4^-$ and HPO_4^{2-}, at acceptable buffer concentrations.
For $H_2PO_4^-$, $K_{a2} = 6.2 \times 10^{-8}$ and $pK_{a2} = -\log K_{a2} = -\log(6.2 \times 10^{-8}) = 7.21$

$$pH = pK_{a2} + \log\frac{[HPO_4^{2-}]}{[H_2PO_4^-]} = 7.21 + \log\frac{(0.040 \text{ mol}/1.06 \text{ L})}{(0.020 \text{ mol}/1.06 \text{ L})}$$

$$pH = 7.21 + \log\frac{(0.040)}{(0.020)} = 7.21 + 0.30 = 7.51$$

16.146 For CH_3CO_2H, $K_a = 1.8 \times 10^{-5}$ and $pK_a = -\log K_a = -\log(1.8 \times 10^{-5}) = 4.74$
The mixture will be a buffer solution containing the conjugate acid-base pair, CH_3CO_2H and $CH_3CO_2^-$, having a pH near the pK_a of CH_3CO_2H.

$$pH = pK_a + \log\frac{[CH_3CO_2^-]}{[CH_3CO_2H]}$$

$$4.85 = 4.74 + \log\frac{[CH_3CO_2^-]}{[CH_3CO_2H]}; \quad 4.85 - 4.74 = \log\frac{[CH_3CO_2^-]}{[CH_3CO_2H]}$$

$$0.11 = \log\frac{[CH_3CO_2^-]}{[CH_3CO_2H]}; \quad \frac{[CH_3CO_2^-]}{[CH_3CO_2H]} = 10^{0.11} = 1.3$$

In the Henderson-Hasselbalch equation, moles can be used in place of concentrations because both components are in the same volume so the volume terms cancel.
20.0 mL = 0.0200 L

Let X equal the volume of 0.10 M CH_3CO_2H and Y equal the volume of 0.15 M $CH_3CO_2^-$. Therefore, X + Y = 0.0200 L and

$$\frac{Y \times [CH_3CO_2^-]}{X \times [CH_3CO_2H]} = \frac{Y(0.15 \text{ mol/L})}{X(0.10 \text{ mol/L})} = 1.3$$

$$X = 0.0200 - Y$$

$$\frac{Y(0.15\ \text{mol/L})}{(0.020 - Y)(0.10\ \text{mol/L})} = 1.3$$

$$\frac{0.15Y}{0.0020 - 0.10Y} = 1.3$$

$0.15Y = 1.3(0.0020 - 0.10Y)$
$0.15Y = 0.0026 - 0.13Y$
$0.15Y + 0.13Y = 0.0026$
$0.28Y = 0.0026$
$Y = 0.0026/0.28 = 0.0093\ \text{L}$
$X = 0.0200 - Y = 0.0200 - 0.0093 = 0.0107\ \text{L}$
$X = 0.0107\ \text{L} = 10.7\ \text{mL}$ and $Y = 0.0093\ \text{L} = 9.3\ \text{mL}$
You need to mix together 10.7 mL of 0.10 M CH_3CO_2H and 9.3 mL of 0.15 M $NaCH_3CO_2$ to prepare 20.0 mL of a solution with a pH of 4.85.

16.148 (a) HCl is a strong acid. HCN is a weak acid with $K_a = 4.9 \times 10^{-10}$. Before the titration, the $[H_3O^+] = 0.100$ M. The HCN contributes an insignificant amount of additional H_3O^+, so the $pH = -\log[H_3O^+] = -\log(0.100) = 1.00$

(b) 100.0 mL = 0.1000 L

$$\text{mol } H_3O^+ = 0.1000\ \text{L} \times \frac{0.100\ \text{mol HCl}}{1.00\ \text{L}} = 0.0100\ \text{mol } H_3O^+$$

add 75.0 mL of 0.100 M NaOH; 75.0 mL = 0.0750 L

$$\text{mol } OH^- = 0.0750\ \text{L} \times \frac{0.100\ \text{mol NaOH}}{1.00\ \text{L}} = 0.00750\ \text{mol } OH^-$$

	$H_3O^+(aq)$ +	$OH^-(aq)$ →	$2\ H_2O(l)$
before reaction (mol)	0.0100	0.0075	
change (mol)	–0.0075	–0.0075	
after reaction (mol)	0.0025	0	

$$[H_3O^+] = \frac{0.0025\ \text{mol } H_3O^+}{0.1000\ \text{L} + 0.0750\ \text{L}} = 0.0143\ \text{M}$$

$pH = -\log[H_3O^+] = -\log(0.0143) = 1.84$

(c) 100.0 mL of 0.100 M NaOH will completely neutralize all of the H_3O^+ from 100.0 mL of 0.100 M HCl. Only NaCl and HCN remain in the solution. NaCl is a neutral salt and does not affect the pH of the solution. [HCN] changes because of dilution. Because the solution volume is doubled, [HCN] is cut in half.

[HCN] = 0.100 M/2 = 0.0500 M

	HCN(aq) +	$H_2O(l)$ ⇄	$H_3O^+(aq)$ +	$CN^-(aq)$
initial (M)	0.0500		~0	0
change (M)	–x		+x	+x
equil (M)	0.0500 – x		x	x

$$K_a = \frac{[H_3O^+][CN^-]}{HCN} = 4.9 \times 10^{-10} = \frac{x^2}{0.0500 - x} \approx \frac{x^2}{0.0500}$$

$[H_3O^+] = x = \sqrt{(0.0500)(4.9 \times 10^{-10})} = 4.95 \times 10^{-6}$ M
pH = $-\log[H_3O^+] = -\log(4.95 \times 10^{-6}) = 5.31$
(d) Add an additional 25.0 mL of 0.100 M NaOH.
25.0 mL = 0.0250 L

$$\text{additional mol } OH^- = 0.0250 \text{ L} \times \frac{0.100 \text{ mol NaOH}}{1.00 \text{ L}} = 0.00250 \text{ mol } OH^-$$

$$\text{mol HCN} = 0.200 \text{ L} \times \frac{0.0500 \text{ mol HCN}}{1.00 \text{ L}} = 0.0100 \text{ mol HCN}$$

	$HCN(aq)$	+	$OH^-(aq)$	→	$CN^-(aq)$	+	$H_2O(l)$
before reaction (mol)	0.0100		0.00250		0		
change (mol)	–0.00250		–0.00250		+0.00250		
after reaction (mol)	0.0075		0		0.00250		

The resulting solution is a buffer because it contains the conjugate acid-base pair, HCN and CN^-, at acceptable buffer concentrations.

For HCN, $K_a = 4.9 \times 10^{-10}$ and $pK_a = -\log K_a = -\log(4.9 \times 10^{-10}) = 9.31$

$$pH = pK_a + \log\frac{[CN^-]}{[HCN]} = 9.31 + \log\frac{(0.00250 \text{ mol}/0.2250 \text{ L})}{(0.0075 \text{ mol}/0.2250 \text{ L})}$$

$$pH = 9.31 + \log\frac{(0.00250)}{(0.0075)} = 9.31 - 0.48 = 8.83$$

16.150 (a)

	$Zn(OH)_2(s)$	⇄	$Zn^{2+}(aq)$	+	$2\,OH^-(aq)$
initial (M)			0		~0
equil (M)			x		2x

$K_{sp} = [Zn^{2+}][OH^-]^2 = 4.1 \times 10^{-17} = (x)(2x)^2 = 4x^3$

$$\text{molar solubility} = x = \sqrt[3]{\frac{4.1 \times 10^{-17}}{4}} = 2.2 \times 10^{-6} \text{ M}$$

(b) $[OH^-] = 2x = 2(2.2 \times 10^{-6} \text{ M}) = 4.4 \times 10^{-6}$ M

$$[H_3O^+] = \frac{1.0 \times 10^{-14}}{4.4 \times 10^{-6}} = 2.3 \times 10^{-9} \text{ M}; \qquad pH = -\log[H_3O^+] = -\log(2.3 \times 10^{-9}) = 8.64$$

(c)

$Zn(OH)_2(s) \rightleftarrows Zn^{2+}(aq) + 2\,OH^-(aq)$ $\qquad K_{sp} = 4.1 \times 10^{-17}$
$\underline{Zn^{2+}(aq) + 4\,OH^-(aq) \rightleftarrows Zn(OH)_4^{2-}(aq)}$ $\qquad K_f = 3 \times 10^{15}$
$Zn(OH)_2(s) + 2\,OH^-(aq) \rightleftarrows Zn(OH)_4^{2-}(aq)$ $\qquad K = K_{sp} \cdot K_f = 0.123$

	$2\,OH^-(aq)$	$Zn(OH)_4^{2-}(aq)$
initial (M)	0.10	0
change (M)	–2x	+x
equil (M)	0.10 – 2x	x

$$K = \frac{[Zn(OH)_4^{2-}]}{[OH^-]^2} = 0.123 = \frac{x}{(0.10 - 2x)^2}$$

$0.492x^2 - 1.0492x + 0.00123 = 0$
Use the quadratic formula to solve for x.

$$x = \frac{-(-1.0492) \pm \sqrt{(-1.0492)^2 - (4)(0.492)(0.00123)}}{2(0.492)} = \frac{1.0492 \pm 1.0480}{0.984}$$

x = 2.1 and 1.2×10^{-3}

Of the two solutions for x, only 1.2×10^{-3} has physical meaning because the other solution leads to a negative $[OH^-]$.

molar solubility of $Zn(OH)_4^{2-}$ in 0.10 M NaOH = x = 1.2×10^{-3} M

Multiconcept Problems

16.152 (a) pH = 5.5; $[H_3O^+] = 10^{-pH} = 10^{-5.5} = 3.2 \times 10^{-6}$ M

	$H_2C_2O_4(aq)$	+ $H_2O(l)$ ⇄	$H_3O^+(aq)$	+ $HC_2O_4^-(aq)$
equil (M)	$1.1\text{x}10^{-4} - 3.2\text{x}10^{-6}$		$3.2\text{x}10^{-6}$	$3.2\text{x}10^{-6}$

The second dissociation of $H_2C_2O_3$ produces a negligible amount of H_3O^+ compared with that from the first dissociation.

$HC_2O_4^-(aq)$	+ $H_2O(l)$ ⇄	$H_3O^+(aq)$	+ $C_2O_4^{2-}(aq)$
$3.2\text{x}10^{-6}$		$3.2\text{x}10^{-6}$	x

$$K_{a2} = \frac{[H_3O^+][C_2O_4^{2-}]}{[HC_2O_4^-]} = 6.4 \times 10^{-5} = \frac{(3.2 \times 10^{-6})[C_2O_4^{2-}]}{(3.2 \times 10^{-6})}$$

$x = [C_2O_4^{2-}] = K_{a2} = 6.4 \times 10^{-5}$

$Ksp = [Ca^{2+}][C_2O_4^{2-}] = 2.3 \times 10^{-9}$

For CaC_2O_4, IP = $[Ca^{2+}][C_2O_4^{2-}] = (2.5 \times 10^{-3})(6.4 \times 10^{-5}) = 1.6 \times 10^{-7}$

IP > K_{sp}; CaC_2O_4 will precipitate.

(b) An ionic compound that contains a basic anion becomes less soluble as the acidity of the solution decreases (i.e., pH increases). Kidney stones are made of CaC_2O_4. CaC_2O_4 contains a basic anion. Kidney stones would be more likely to form in urine with a higher pH.

16.154 (a) (i)

	en(aq)	+ $H_2O(l)$ ⇄	$enH^+(aq)$	+ $OH^-(aq)$
initial (M)	0.100		0	~0
change (M)	–x		+x	+x
equil (M)	0.100 – x		x	x

$$K_b = \frac{[enH^+][OH^-]}{[en]} = 5.2 \times 10^{-4} = \frac{(x)(x)}{0.100 - x}$$

$x^2 + (5.2 \times 10^{-4})x - (5.2 \times 10^{-5}) = 0$

Use the quadratic formula to solve for x.

$$x = \frac{-(5.2 \times 10^{-4}) \pm \sqrt{(5.2 \times 10^{-4})^2 - 4(1)(-5.2 \times 10^{-5})}}{2(1)} = \frac{-5.2 \times 10^{-4} \pm 0.01443}{2}$$

x = –0.0075 and 0.0070

Of the two solutions for x, only the positive value of x has physical meaning because x is the $[OH^-]$.

$[OH^-] = x = 0.0070$ M; $\quad [H_3O^+] = \dfrac{K_w}{[OH^-]} = \dfrac{1.0 \times 10^{-14}}{0.0070} = 1.43 \times 10^{-12}$ M

$pH = -\log[H_3O^+] = -\log(1.43 \times 10^{-12}) = 11.84$

(ii) (30.0 mL)(0.100 mmol/mL) = 3.00 mmol en

(15.0 mL)(0.100 mmol/mL) = 1.50 mmol HCl

Halfway to the first equivalence point, $[OH^-] = K_{b1}$

$$[H_3O^+] = \frac{K_w}{[OH^-]} = \frac{1.0 \times 10^{-14}}{5.2 \times 10^{-4}} = 1.92 \times 10^{-11} \text{ M}$$

$pH = -\log[H_3O^+] = -\log(1.92 \times 10^{-11}) = 10.72$

(iii) At the first equivalence point $pH = \dfrac{pK_{a1} + pK_{a2}}{2} = 9.14$

(iv) Halfway between the first and second equivalence points, $[OH^-] = K_{b2} = 3.7 \times 10^{-7}$ M

$$[H_3O^+] = \frac{K_w}{[OH^-]} = \frac{1.0 \times 10^{-14}}{3.7 \times 10^{-7}} = 2.70 \times 10^{-8} \text{ M}$$

$pH = -\log[H_3O^+] = -\log(2.70 \times 10^{-8}) = 7.57$

(v) At the second equivalence point only the acidic enH_2Cl_2 is in solution.

For enH_2^{2+}, $K_a = \dfrac{K_w}{K_b \text{ for } enH^+} = \dfrac{K_w}{K_{b2}} = \dfrac{1.0 \times 10^{-14}}{3.7 \times 10^{-7}} = 2.70 \times 10^{-8}$

$$[enH_2^{2+}] = \frac{3.00 \text{ mmol}}{(30.0 \text{ mL} + 60.0 \text{ mL})} = 0.0333 \text{ M}$$

	$enH_2^{2+}(aq)$ + $H_2O(l)$	⇄	$H_3O^+(aq)$	+	$enH^+(aq)$
initial (M)	0.0333		~0		0
change (M)	–x		+x		+x
equil (M)	0.0333 – x		x		x

$$K_a = \frac{[H_3O^+][enH^+]}{[enH_2^{2+}]} = 2.70 \times 10^{-8} = \frac{(x)(x)}{0.0333 - x} \approx \frac{x^2}{0.0333}$$

Solve for x. $x = [H_3O^+] = \sqrt{(2.70 \times 10^{-8})(0.0333)} = 3.00 \times 10^{-5}$ M

$pH = -\log[H_3O^+] = -\log(3.00 \times 10^{-5}) = 4.52$

(vi) excess HCl

(75.0 mL – 60.0 mL)(0.100 mmol/mL) = 1.50 mmol HCl = 1.50 mmol H_3O^+

$$[H_3O^+] = \frac{1.50 \text{ mmol}}{(30.0 \text{ mL} + 75.0 \text{ mL})} = 0.0143 \text{ M}$$

$pH = -\log[H_3O^+] = -\log(0.0143) = 1.84$

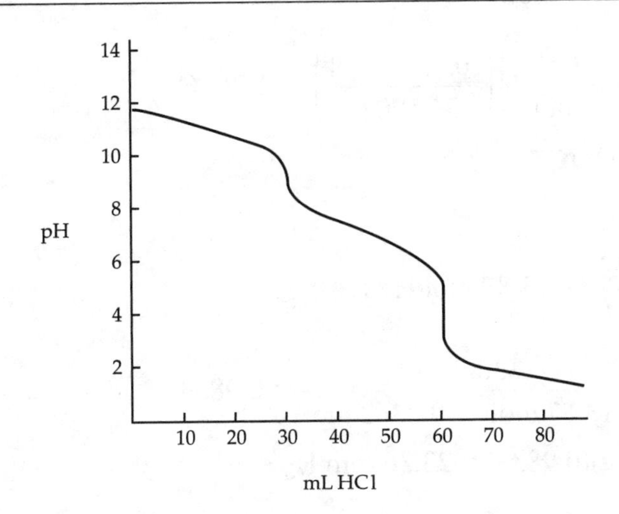

(b)

```
        H   H
  ..    |   |   ..
H—N—C—C—N—H
    |   |   |   |
    H   H   H   H
```

Each of the two nitrogens in ethylenediamine can accept a proton.

(c) Each nitrogen is sp^3 hybridized.

16.156 (a) PV = nRT; 25 °C = 298 K

$$n_{HCl} = \frac{PV}{RT} = \frac{\left(732 \text{ mm Hg x } \dfrac{1.00 \text{ atm}}{760 \text{ mm Hg}}\right)(1.000 \text{ L})}{\left(0.082\ 06 \dfrac{\text{L}\cdot\text{atm}}{\text{K}\cdot\text{mol}}\right)(298 \text{ K})} = 0.0394 \text{ mol HCl}$$

Na_2CO_3, 105.99

$$\text{mol } Na_2CO_3 = 6.954 \text{ g } Na_2CO_3 \text{ x } \frac{1 \text{ mol } Na_2CO_3}{105.99 \text{ g } Na_2CO_3} = 0.0656 \text{ mol } Na_2CO_3$$

	$CO_3^{2-}(aq)$ +	$H_3O^+(aq)$ →	$HCO_3^-(aq)$ +	$H_2O(l)$
before reaction (mol)	0.0656	0.0394	0	
change (mol)	–0.0394	–0.0394	+0.0394	
after reaction (mol)	0.0656 – 0.0394	0	0.0394	

mol CO_3^{2-} = 0.0656 – 0.0394 = 0.0262 mol and mol HCO_3^- = 0.0394 mol
Therefore, we have an HCO_3^-/CO_3^{2-} buffer solution.

$$pH = pK_{a2} + \log \frac{[CO_3^{2-}]}{[HCO_3^-]} = -\log(5.6 \text{ x } 10^{-11}) + \log \frac{0.0262 \text{ mol/V}}{0.0394 \text{ mol/V}}$$

pH = 10.25 – 0.177 = 10.08

(b) mol Na^+ = 2(0.0656 mol) = 0.1312 mol
mol CO_3^{2-} = 0.0262 mol
mol HCO_3^- = 0.0394 mol
mol Cl^- = 0.0394 mol
total ion moles = 0.2362 mol

$\Delta T_f = K_f \cdot m$; $\quad \Delta T_f = \left(1.86\ \frac{°C \cdot kg}{mol}\right)\left(\frac{0.2362\ mol}{0.2500\ kg}\right) = 1.76\ °C$

Solution freezing point = 0 °C – ΔT_f = –1.76 °C

(c) H_2O, 18.02

$$\text{mol } H_2O = 250.0 \text{ g x } \frac{1 \text{ mol } H_2O}{18.02 \text{ g } H_2O} = 13.87 \text{ mol } H_2O$$

$$X_{solv} = \frac{\text{mol } H_2O}{\text{mol } H_2O + \text{mol ions}} = \frac{13.87 \text{ mol}}{13.87 \text{ mol} + 0.2362 \text{ mol}} = 0.9833$$

$P_{soln} = P_{solv} \cdot X_{solv}$ = (23.76 mm Hg)(0.9833) = 23.36 mm Hg

16.158 (a) $HCO_3^-(aq) + OH^-(aq) \rightarrow CO_3^{2-}(aq) + H_2O(l)$

(b) mol HCO_3^- = (0.560 mol/L)(0.0500 L) = 0.0280 mol HCO_3^-

mol OH^- = (0.400 mol/L)(0.0500 L) = 0.0200 mol OH^-

	$HCO_3^-(aq)$ +	$OH^-(aq)$ →	$CO_3^{2-}(aq)$ +	$H_2O(l)$
before reaction (mol)	0.0280	0.0200	0	
change (mol)	–0.0200	–0.0200	+0.0200	
after reaction (mol)	0.0280 – 0.0200	0	0.0200	

mol HCO_3^- = 0.0280 – 0.0200 = 0.0080 mol

$[HCO_3^-] = \frac{0.0080 \text{ mol}}{0.1000 \text{ L}} = 0.080$ M; $\quad [CO_3^{2-}] = \frac{0.0200 \text{ mol}}{0.1000 \text{ L}} = 0.200$ M

	$HCO_3^-(aq)$ + $H_2O(l)$ ⇄	$H_3O^+(aq)$ +	$CO_3^{2-}(aq)$
initial (M)	0.080	~0	0.200
change (M)	–x	+x	+x
equil (M)	0.080 – x	x	0.200 + x

$$K_a = \frac{[H_3O^+][CO_3^{2-}]}{[HCO_3^-]} = 5.6 \times 10^{-11} = \frac{x(0.200 + x)}{0.080 - x} \approx \frac{x(0.200)}{0.080}$$

Solve for x. x = $[H_3O^+] = 2.24 \times 10^{-11}$ M

pH = $-\log[H_3O^+] = -\log(2.24 \times 10^{-11}) = 10.65$

Because this solution contains both a weak acid (HCO_3^-) and its conjugate base, the solution is a buffer.

(c) $HCO_3^-(aq) + OH^-(aq) \rightarrow CO_3^{2-}(aq) + H_2O(l)$

$\Delta H°_{rxn} = [\Delta H°_f(CO_3^{2-}) + \Delta H°_f(H_2O)] - [\Delta H°_f(HCO_3^-) + \Delta H°_f(OH^-)]$

$\Delta H°_{rxn}$ = [(1 mol)(–677.1 kJ/mol) + (1 mol)(–285.8 kJ/mol)]

– [(1 mol)(–692.0 kJ/mol) + (1 mol)(–230 kJ/mol)]

$\Delta H°_{rxn}$ = – 40.9 kJ

0.0200 moles each of HCO_3^- and OH^- reacted.

heat produced = q = (0.0200 mol)(40.9 kJ/mol) = 0.818 kJ = 818 J

(d) q = m x specific heat x ΔT

$$\Delta T = \frac{q}{m \text{ x specific heat}} = \frac{818 \text{ J}}{(100.0 \text{ g})[4.18 \text{ J/(g}\cdot{}^\circ\text{C)}]} = 2.0\ {}^\circ\text{C}$$

Final temperature = 25 °C + 2.0 °C = 27 °C

16.160 (a) H_2SO_4, 98.08

Assume 1.00 L = 1000 mL of solution.

mass of solution = (1000 mL)(1.836 g/mL) = 1836 g

mass H_2SO_4 = (0.980)(1836 g) = 1799 g H_2SO_4

$$\text{mol } H_2SO_4 = 1799 \text{ g } H_2SO_4 \text{ x } \frac{1 \text{ mol } H_2SO_4}{98.08 \text{ g } H_2SO_4} = 18.3 \text{ mol } H_2SO_4$$

$[H_2SO_4]$ = 18.3 mol/ 1.00 L = 18.3 M

(b) Na_2CO_3, 105.99; 1 kg = 1000 g = 2.2046 lb

$H_2SO_4(aq) + Na_2CO_3(s) \rightarrow Na_2SO_4(aq) + H_2O(l) + CO_2(g)$

$$\text{mass } H_2SO_4 = (0.980)(36 \text{ tons}) \text{ x } \frac{2000 \text{ lb}}{1 \text{ ton}} \text{ x } \frac{1000 \text{ g}}{2.2046 \text{ lb}} = 3.20 \text{ x } 10^7 \text{ g } H_2SO_4$$

$$\text{mol } H_2SO_4 = 3.20 \text{ x } 10^7 \text{ g } H_2SO_4 \text{ x } \frac{1 \text{ mol } H_2SO_4}{98.08 \text{ g } H_2SO_4} = 3.26 \text{ x } 10^5 \text{ mol } H_2SO_4$$

$$\text{mass } Na_2CO_3 = 3.26 \text{ x } 10^5 \text{ mol } H_2SO_4 \text{ x } \frac{1 \text{ mol } Na_2CO_3}{1 \text{ mol } H_2SO_4} \text{ x } \frac{105.99 \text{ g } Na_2CO_3}{1 \text{ mol } Na_2CO_3} \text{ x } \frac{1 \text{ kg}}{1000 \text{ g}} = 3.5 \text{ x } 10^4 \text{ kg } Na_2CO_3$$

$$\text{(c) mol } CO_2 = 3.26 \text{ x } 10^5 \text{ mol } H_2SO_4 \text{ x } \frac{1 \text{ mol } CO_2}{1 \text{ mol } H_2SO_4} = 3.26 \text{ x } 10^5 \text{ mol } CO_2$$

18 °C = 18 + 273 = 291 K

PV = nRT

$$V = \frac{nRT}{P} = \frac{(3.26 \text{ x } 10^5 \text{ mol})\left(0.082\ 06 \frac{\text{L}\cdot\text{atm}}{\text{K}\cdot\text{mol}}\right)(291 \text{ K})}{\left(745 \text{ mm Hg x } \frac{1.00 \text{ atm}}{760 \text{ mm Hg}}\right)} = 7.9 \text{ x } 10^6 \text{ L}$$

17 Thermodynamics: Entropy, Free Energy, and Equilibrium

17.1 (a) spontaneous; (b), (c), and (d) $Q_p > K_p$, are nonspontaneous

17.2 (a) $H_2O(g) \rightarrow H_2O(l)$
A liquid has less randomness than a gas. Therefore, ΔS is negative.
(b) $I_2(g) \rightarrow 2\ I(g)$
ΔS is positive because the reaction increases the number of gaseous particles from 1 mol to 2 mol.
(c) $CaCO_3(s) \rightarrow CaO(s) + CO_2(g)$
ΔS is positive because the reaction increases the number of gaseous molecules.
(d) $Ag^+(aq) + Br^-(aq) \rightarrow AgBr(s)$
A solid has less randomness than +1 and –1 charged ions in an aqueous solution. Therefore, ΔS is negative.
(e) Deposition of frost on a cold morning, $H_2O(g) \rightarrow H_2O(s)$
Deposition is the formation of a solid from a gas. A solid has less randomness than a gas. Therefore, ΔS is negative.

17.3 (a) $A_2(g) + 2\ B(g) \rightarrow 2\ AB(g)$.
(b) Because the reaction decreases the number of gaseous particles from 3 mol to 2 mol, the entropy change is negative.

17.4 $S = k \ln W$, $k = 1.38 \times 10^{-23}$ J/K
(a) $S = (1.38 \times 10^{-23} \text{ J/K}) \ln (3^{100}) = 1.52 \times 10^{-21}$ J/K
(b) $S = (1.38 \times 10^{-23} \text{ J/K}) \ln (3^{6.02 \times 10^{23}}) = (1.38 \times 10^{-23} \text{ J/K})(6.022 \times 10^{23}) \ln 3 = 9.13$ J/K

17.5 $S = k \ln W$, $k = 1.38 \times 10^{-23}$ J/K
(a) $W = 1$; $S = (1.38 \times 10^{-23} \text{ J/K}) \ln (1) = 0$
(b) $W = 3^{10}$; $S = (1.38 \times 10^{-23} \text{ J/K}) \ln (3^{10}) = 1.52 \times 10^{-22}$ J/K

17.6 (a) 1 mole N_2 at STP (larger volume, more randomness)
(b) $\Delta S = R \ln \left(\frac{V_{\text{State B}}}{V_{\text{State A}}}\right) = R \ln \left(\frac{11.2 \text{ L}}{22.4 \text{ L}}\right) = (8.314 \text{ J/K}) \ln(1/2) = -5.76$ J/K

17.7 $CaCO_3(s) \rightarrow CaO(s) + CO_2(g)$
$\Delta S° = [S°(CaO) + S°(CO_2)] - S°(CaCO_3)$
$\Delta S° = [(1 \text{ mol})(38.1 \text{ J/(K}\cdot\text{mol)}) + (1 \text{ mol})(213.6 \text{ J/(K}\cdot\text{mol)})]$
$- (1 \text{ mol})(91.7 \text{ J/(K}\cdot\text{mol)}) = +160.0$ J/K

17.8 (a) $C_3H_8(g) + 5\ O_2(g) \rightarrow 3\ CO_2(g) + 4\ H_2O(l)$

$\Delta S°$ is negative because 6 mol of gas in the reactants are converted to 3 mol of gas in the products.

(b) $\Delta S° = -376.6\ J/K = [3\ S°(CO2) + 4\ S°(H_2O(l))] - [S°(C_3H_8) + 5\ S°(O_2)]$

$S°(C_3H_8) = [3\ S°(CO2) + 4\ S°(H_2O(l))] - [5\ S°(O_2)] + 376.6\ J/K$

$S°(C_3H_8) = [(3\ mol)(213.6\ J/(K \cdot mol)) + (4\ mol)(69.9\ J/(K \cdot mol))]$
$\qquad - [(5\ mol)(205.0\ J/(K \cdot mol))] + 376.6\ J/K$

$S°(C_3H_8) = 272.0\ J/(K \cdot mol)$

17.9 From Problem 17.7, $\Delta S_{sys} = \Delta S° = 160.0\ J/K$

$CaCO_3(s) \rightarrow CaO(s) + CO_2(g)$

$\Delta H° = [\Delta H°_f(CaO) + \Delta H°_f(CO_2)] - \Delta H°_f(CaCO_3)$

$\Delta H° = [(1\ mol)(-634.9\ kJ/mol) + (1\ mol)(-393.5\ kJ/mol)]$
$\qquad - (1\ mol)(-1207.6\ kJ/mol) = +179.2\ kJ$

$$\Delta S_{surr} = \frac{-\Delta H°}{T} = \frac{-179{,}200\ J}{298\ K} = -601\ J/K$$

$\Delta S_{total} = \Delta S_{sys} + \Delta S_{surr} = 160.0\ J/K + (-601\ J/K) = -441\ J/K$

Because ΔS_{total} is negative, the reaction is not spontaneous under standard-state conditions at 25 °C.

17.10 $Br_2(l) \rightarrow Br_2(g)$

$\Delta H° = \Delta H°_f(Br_2(g)) = (1\ mol)(30.9\ kJ/mol) = 30.9\ kJ$

$\Delta S° = S°(Br_2(g)) - S°(Br_2(l))$

$\Delta S° = (1\ mol)(245.4\ J/(K \cdot mol)) - (1\ mol)(152.2\ J/(K \cdot mol)) = 93.2\ J/K = 93.2 \times 10^{-3}\ kJ/K$

$\Delta G = \Delta H - T\Delta S$

The boiling point (phase change) is associated with an equilibrium. Set $\Delta G = 0$ and solve for T, the boiling point.

$$0 = \Delta H - T\Delta S; \quad T_{bp} = \frac{\Delta H}{\Delta S} = \frac{30.9\ kJ}{93.2 \times 10^{-3}\ kJ/K} = 331.5\ K = 58.4\ °C$$

17.11 (a) $\Delta G = \Delta H - T\Delta S = 55.3\ kJ - (298\ K)(0.1757\ kJ/K) = +2.9\ kJ$

Because $\Delta G > 0$, the reaction is nonspontaneous at 25 °C (298 K)

(b) Set $\Delta G = 0$ and solve for T.

$$0 = \Delta H - T\Delta S; \quad T = \frac{\Delta H}{\Delta S} = \frac{55.3\ kJ}{0.1757\ kJ/K} = 315\ K = 42\ °C$$

17.12 $\Delta H < 0$ (reaction involves bond making - exothermic)

$\Delta S < 0$ (the reaction has less randomness in going from reactants (2 atoms) to products (1 molecule)

$\Delta G < 0$ (the reaction is spontaneous)

17.13 From Problems 17.7 and 17.9: $\Delta H° = 179.2\ kJ$ and $\Delta S° = 160.0\ J/K = 0.1600\ kJ/K$

(a) $\Delta G° = \Delta H° - T\Delta S° = 179.2\ kJ - (298\ K)(0.1600\ kJ/K) = +131.5\ kJ$

(b) Because $\Delta G > 0$, the reaction is nonspontaneous at 25 °C (298 K).

(c) Set $\Delta G = 0$ and solve for T, the temperature above which the reaction becomes spontaneous.

$0 = \Delta H - T\Delta S$; $\quad T = \dfrac{\Delta H}{\Delta S} = \dfrac{179.2 \text{ kJ}}{0.1600 \text{ kJ/K}} = 1120 \text{ K} = 847\ ^\circ\text{C}$

17.14 $2\ AB_2 \rightarrow A_2 + 2\ B_2$

(a) ΔS° is positive because the reaction increases the number of molecules.
(b) ΔH° is positive because the reaction is endothermic.
$\Delta G^\circ = \Delta H^\circ - T\Delta S^\circ$
For the reaction to be spontaneous, ΔG° must be negative. This will only occur at high temperature where $T\Delta S^\circ$ is greater than ΔH°.

17.15 (a) $CaC_2(s) + 2\ H_2O(l) \rightarrow C_2H_2(g) + Ca(OH)_2(s)$

$\Delta G^\circ = [\Delta G^\circ_f(C_2H_2) + \Delta G^\circ_f(Ca(OH)_2)] - [\Delta G^\circ_f(CaC_2) + 2\ \Delta G^\circ_f(H_2O)]$
$\Delta G^\circ = [(1 \text{ mol})(209.9 \text{ kJ/mol}) + (1 \text{ mol})(-897.5 \text{ kJ/mol})]$
$\quad - [(1 \text{ mol})(-64.8 \text{ kJ/mol}) + (2 \text{ mol})(-237.2 \text{ kJ/mol})] = -148.4 \text{ kJ}$
This reaction can be used for the synthesis of C_2H_2 because $\Delta G < 0$.
(b) It is not possible to synthesize acetylene from solid graphite and gaseous H_2 at 25 °C and 1 atm because $\Delta G^\circ_f(C_2H_2) > 0$.

17.16 $C_{diamond}(s) \rightarrow C_{graphite}(s)$

(a) $\Delta G^\circ = \Delta G^\circ_f(C_{graphite}) - \Delta G^\circ_f(C_{diamond}) = 0 - 2.9 \text{ kJ} = -2.9 \text{ kJ}$
(b) The reaction is spontaneous because $\Delta G^\circ < 0$.
(c) The reaction is spontaneous, but the reaction rate is extremely low.

17.17 $C(s) + 2\ H_2(g) \rightarrow C_2H_4(g)$

$$Q_p = \frac{P_{C_2H_4}}{(P_{H_2})^2} = \frac{(0.10)}{(100)^2} = 1.0 \times 10^{-5}$$

$\Delta G = \Delta G^\circ + RT \ln Q_p$
$\Delta G = 68.1 \text{ kJ/mol} + [8.314 \times 10^{-3} \text{ kJ/(K} \cdot \text{mol)}](298 \text{ K})\ln(1.0 \times 10^{-5}) = +39.6 \text{ kJ/mol}$
Because $\Delta G > 0$, the reaction is spontaneous in the reverse direction.

17.18 $\Delta G = \Delta G^\circ + RT \ln Q$ and $\Delta G^\circ = 15$ kJ

For $A_2(g) + B_2(g) \rightleftarrows 2\ AB(g)$, $Q_p = \dfrac{(P_{AB})^2}{(P_{A_2})(P_{B_2})}$

Let the number of molecules be proportional to the partial pressure.
(1) $Q_p = 1.0$ (2) $Q_p = 0.0667$ (3) $Q_p = 18$
(a) Reaction (3) has the largest ΔG because Q_p is the largest. Reaction (2) has the smallest ΔG because Q_p is the smallest.
(b) $\Delta G = \Delta G^\circ = 15$ kJ because $Q_p = 1$ and $\ln(1) = 0$.

17.19 From Problem 17.13, $\Delta G^\circ = +131.5$ kJ

$\Delta G^\circ = -RT \ln K_p$

$$\ln K_p = \frac{-\Delta G^o}{RT} = \frac{-131.5 \text{ kJ/mol}}{[8.314 \times 10^{-3} \text{ kJ/(K} \cdot \text{mol)}](298 \text{ K})} = -53.1$$

$K_p = e^{-53.1} = 9 \times 10^{-24}$

17.20 $\Delta G^o = -RT \ln K = -[8.314 \times 10^{-3} \text{ kJ/(K} \cdot \text{mol)}](298 \text{ K}) \ln (1.0 \times 10^{-14}) = 80 \text{ kJ/mol}$

17.21 $H_2O(l) \rightleftarrows H_2O(g)$

$K_p = P_{H_2O}$; K_p is equal to the vapor pressure for H_2O.

$\Delta G^o = \Delta G^o_f(H_2O(g)) - \Delta G^o_f(H_2O(l))$

$\Delta G^o = (1 \text{ mol})(-228.6 \text{ kJ/mol}) - (1 \text{ mol})(-237.2 \text{ kJ/mol}) = +8.6 \text{ kJ}$

$\Delta G^o = -RT \ln K_p$

$$\ln K_p = \frac{-\Delta G^o}{RT} = \frac{-8.6 \text{ kJ/mol}}{[8.314 \times 10^{-3} \text{ kJ/(K} \cdot \text{mol)}](298 \text{ K})} = -3.5$$

$K_p = P_{H_2O} = e^{-3.5} = 0.03 \text{ atm}$

7.22 $P_{EtOH} = 60.6 \text{ mm Hg} \times \frac{1.00 \text{ atm}}{760 \text{ mm Hg}} = 0.0797 \text{ atm}$

$K_p = P_{EtOH} = 0.0797 \text{ atm}$

$\Delta G^o = -RT \ln K_p = -[8.314 \text{ J/K}](298 \text{ K}) \ln (0.0797) = 6267 \text{ J} = 6.27 \text{ kJ}$

17.23 The growth of a human adult from a single cell does not violate the second law of thermodynamics. The energy an human obtains from glucose is used to build and organize complex molecules, resulting in a decrease in entropy for the human. At the same time, however, the entropy of the surroundings increases as the human releases small, simple waste products such as CO_2 and H_2O. Furthermore, heat is released by the human, further increasing the entropy of the surroundings. Thus, an organism pays for its decrease in entropy by increasing the entropy of the rest of the universe.

17.24 (a) $C_6H_{12}O_6(s) + 6\ O_2(g) \rightleftarrows 6\ CO_2(g) + 6\ H_2O(l)$ $\quad \Delta G^{o'} = -2870 \text{ kJ}$

$ADP^{3-}(aq) + H_2PO_4^-(aq) \rightleftarrows ATP^{4-}(aq) + H_2O(l)$ $\quad \Delta G^{o'} = +30.5 \text{ kJ}$

$C_6H_{12}O_6(s) + 6\ O_2(g) \rightleftarrows 6\ CO_2(g) + 6\ H_2O(l)$

$\underline{32\ [ADP^{3-}(aq) + H_2PO_4^-(aq) \rightleftarrows ATP^{4-}(aq) + H_2O(l)]}$

$C_6H_{12}O_6(s) + 6\ O_2(g) + 32\ ADP^{3-}(aq) + 32\ H_2PO_4^-(aq) \rightleftarrows 6\ CO_2(g) + 32\ ATP^{4-}(aq) + 38\ H_2O(l)$

$\Delta G^{o'} = (-2870 \text{ kJ}) + 32(30.5 \text{ kJ}) = -1894 \text{ kJ}$

(b) Because $\Delta G^{o'} < 0$, the reaction is spontaneous.

17.25 (a) Because $\Delta G^{o'} > 0$, the reaction is not spontaneous.

(b) $\Delta G^{o'} = (13.8 \text{ kJ}) + (-30.5 \text{ kJ}) = -16.7 \text{ kJ}$

Because the overall $\Delta G^{o'} < 0$, the reaction is spontaneous.

(c) $\Delta G^\circ = -RT \ln K$

$$\ln K = \frac{-\Delta G^\circ}{RT} = \frac{-(-16.7 \text{ kJ})}{[8.314 \times 10^{-3} \text{ kJ/K}](310 \text{ K})} = 6.48; \quad K = e^{6.48} = 652$$

17.26 (a) $ATP^{4-}(aq) + H_2O(l) \rightleftarrows ADP^{3-}(aq) + H_2PO_4^-(aq)$ $\quad \Delta G^{\circ'} = -30.5$ kJ

$$Q = \frac{[ADP^{3-}][H_2PO_4^-]}{[ATP^{4-}]} = \frac{(8.0 \times 10^{-3})(0.9 \times 10^{-3})}{(8.0 \times 10^{-3})}$$

$$\Delta G = \Delta G^{\circ'} + RT \ln Q$$

$$\Delta G = -30.5 \text{ kJ} + (8.314 \times 10^{-3} \text{ kJ/K})(310 \text{ K}) \ln\left(\frac{(8.0 \times 10^{-3})(0.9 \times 10^{-3})}{(8.0 \times 10^{-3})}\right) = -48.6 \text{ kJ}$$

(b) The amount of free-energy released increases at the concentrations present in muscle cells.

17.27 creatine phosphate(aq) + $H_2O(l) \rightleftarrows$ creatine(aq) + $H_2PO_4^-(aq)$ $\quad \Delta G^{\circ'} = -43.1$ kJ
$\underline{ADP^{3-}(aq) + H_2PO_4^-(aq) \rightleftarrows ATP^{4-}(aq) + H_2O(l)}$ $\quad \Delta G^{\circ'} = +30.5$ kJ
creatine phosphate(aq) + $ADP^{3-}(aq) \rightleftarrows$ creatine(aq) + $ATP^{4-}(aq)$
$\Delta G^{\circ'} = (-43.1 \text{ kJ}) + (30.5 \text{ kJ}) = -12.6$ kJ

Conceptual Problems

17.28 (a) $A_2 + AB_3 \rightarrow 3\,AB$
(b) ΔS is positive because the reaction increases the number of gaseous molecules.

17.30 $\Delta H > 0$ (heat is absorbed during sublimation)
$\Delta S > 0$ (gas has more randomness than solid)
$\Delta G < 0$ (the reaction is spontaneous)

17.32 $\Delta H = 0$ (system is an ideal gas at constant temperature)
$\Delta S < 0$ (there is less randomness in the smaller volume)
$\Delta G > 0$ (compression of a gas is not spontaneous)

17.34 (a) For initial state 1, $Q_p < K_p$
(more reactant (A_2) than product (A) compared to the equilibrium state)
For initial state 2, $Q_p > K_p$
(more product (A) than reactant (A_2) compared to the equilibrium state)
(b) $\Delta H > 0$ (reaction involves bond breaking - endothermic)
$\Delta S > 0$ (equilibrium state has more randomness than initial state 1)
$\Delta G < 0$ (reaction spontaneously proceeds toward equilibrium)
(c) $\Delta H < 0$ (reaction involves bond making - exothermic)
$\Delta S < 0$ (equilibrium state has less randomness than initial state 2)
$\Delta G < 0$ (reaction spontaneously proceeds toward equilibrium)
(d) State 1 lies to the left of the minimum in Figure 17.11. State 2 lies to the right of the minimum.

17.36 (a) Because the free energy decreases as pure reactants form products and also decreases as pure products form reactants, the free energy curve must go through a minimum somewhere between pure reactants and pure products. At the minimum point, $\Delta G = 0$ and the system is at equilibrium.
(b) The minimum in the plot is on the left side of the graph because $\Delta G^\circ > 0$ and the equilibrium composition is rich in reactants.

17.38 The equilibrium mixture is richer in reactant A at the higher temperature. This means the reaction is exothermic ($\Delta H < 0$). At 25 °C, $\Delta G^\circ < 0$ because $K > 1$ and at 45 °C, $\Delta G^\circ > 0$ because $K < 1$. Using the relationship. $\Delta G^\circ = \Delta H^\circ - T\Delta S^\circ$, with $\Delta H^\circ < 0$, ΔG° will become positive at the higher temperature only if ΔS° is negative.

Section Problems

Spontaneous Processes (Section 17.1)

17.40 (a) and (d) nonspontaneous; (b) and (c) spontaneous

17.42 (b) and (d) spontaneous (because of the large positive K_p's)

Entropy (Sections 17.2–17.4)

17.44 Molecular randomness is called entropy. For the following reaction, the entropy increases: $H_2O(s) \rightarrow H_2O(l)$ at 25 °C.

17.46 (a) + (solid → gas)
(b) – (liquid → solid)
(c) – (aqueous ions → solid)
(d) + ($CO_2(aq) \rightarrow CO_2(g)$)

17.48 (a) – (liquid → solid)
(b) – (decrease in number of O_2 molecules)
(c) + (gas has more randomness in larger volume)
(d) – (aqueous ions → solid)

17.50 $S = k \ln W$, $k = 1.38 \times 10^{-23}$ J/K
(a) $S = (1.38 \times 10^{-23}\ \text{J/K}) \ln (4^{12}) = 2.30 \times 10^{-22}$ J/K
(b) $S = (1.38 \times 10^{-23}\ \text{J/K}) \ln (4^{120}) = 2.30 \times 10^{-21}$ J/K
(c) $S = (1.38 \times 10^{-23}\ \text{J/K}) \ln (4^{6.02 \times 10^{23}}) = (1.38 \times 10^{-23}\ \text{J/K})(6.022 \times 10^{23}) \ln 4 = 11.5$ J/K
If all C–D bonds point in the same direction, $S = 0$.

17.52 $S = k \ln W$, $k = 1.38 \times 10^{-23}$ J/K
$W = 1000^{100}$; $S = (1.38 \times 10^{-23}\ \text{J/K}) \ln (1000^{100}) = 9.53 \times 10^{-21}$ J/K

17.54 $S = k \ln W$
$S_i = k \ln W_i = k \ln (1.00 \times 10^6)^{1000}$ and $S_f = k \ln W_f = k \ln (1.00 \times 10^7)^{1000}$

$$\frac{S_f}{S_i} = \frac{k \ln (1.00 \times 10^7)^{1000}}{k \ln (1.00 \times 10^6)^{1000}} = \frac{k (1000) \ln (1.00 \times 10^7)}{k (1000) \ln (1.00 \times 10^6)} = \frac{\ln(1.00 \times 10^7)}{\ln(1.00 \times 10^6)} = 1.17$$

and

$S_i = k \ln W_i = k \ln (1.00 \times 10^{16})^{1000}$ and $S_f = k \ln W_f = k \ln (1.00 \times 10^{17})^{1000}$

$$\frac{S_f}{S_i} = \frac{k \ln (1.00 \times 10^{17})^{1000}}{k \ln (1.00 \times 10^{16})^{1000}} = \frac{k (1000) \ln (1.00 \times 10^{17})}{k (1000) \ln (1.00 \times 10^{16})} = \frac{\ln(1.00 \times 10^{17})}{\ln(1.00 \times 10^{16})} = 1.06$$

17.56 (a) disordered N_2O (more randomness)
(b) quartz glass (amorphous solid, more randomness)

17.58 (a) H_2 at 25 °C in 50 L (larger volume)
(b) O_2 at 25 °C, 1 atm (larger volume)
(c) H_2 at 100 °C, 1 atm (larger volume and higher T)
(d) CO_2 at 100 °C, 0.1 atm (larger volume and higher T)

17.60 $\Delta S = nR \ln\left(\frac{V_f}{V_i}\right) = (0.050 \text{ mol})(8.314 \text{ J/K} \cdot \text{mol})\ln\left(\frac{3.5 \text{ L}}{2.5 \text{ L}}\right) = 0.14 \text{ J/K}$

Standard Molar Entropies and Standard Entropies of Reaction (Section 17.5)

17.62 (a) $C_2H_6(g)$; more atoms/molecule
(b) $CO_2(g)$; more atoms/molecule
(c) $I_2(g)$; gas has more randomness than the solid
(d) $CH_3OH(g)$; gas has more randomness than the liquid

17.64 $2\ CO(g) + O_2(g) \rightarrow 2\ CO_2(g)$
$\Delta S° = [2\ S°(CO_2)] - [2\ S°(CO) + S°(O_2)]$
$\Delta S° = [(2 \text{ mol})(213.6 \text{ J/(K} \cdot \text{mol}))]$
$- [(2 \text{ mol})(197.6 \text{ J/(K} \cdot \text{mol})) + (1 \text{ mol})(205.0 \text{ J/(K} \cdot \text{mol}))] = -173.0 \text{ J/K}$

17.66 (a) $2\ H_2O_2(l) \rightarrow 2\ H_2O(l) + O_2(g)$
$\Delta S° = [2\ S°(H_2O(l)) + S°(O_2)] - 2\ S°(H_2O_2)$
$\Delta S° = [(2 \text{ mol})(69.9 \text{ J/(K} \cdot \text{mol})) + (1 \text{ mol})(205.0 \text{ J/(K} \cdot \text{mol}))]$
$- (2 \text{ mol})(110 \text{ J/(K} \cdot \text{mol})) = +125 \text{ J/K}$ (+, because moles of gas increase)
(b) $2\ Na(s) + Cl_2(g) \rightarrow 2\ NaCl(s)$
$\Delta S° = 2\ S°(NaCl) - [2\ S°(Na) + S°(Cl_2)]$
$\Delta S° = (2 \text{ mol})(72.1 \text{ J/(K} \cdot \text{mol})) - [(2 \text{ mol})(51.2 \text{ J/(K} \cdot \text{mol})) + (1 \text{ mol})(223.0 \text{ J/(K} \cdot \text{mol}))]$
$\Delta S° = -181.2 \text{ J/K}$ (–, because moles of gas decrease)
(c) $2\ O_3(g) \rightarrow 3\ O_2(g)$
$\Delta S° = 3\ S°(O_2) - 2\ S°(O_3)$
$\Delta S° = (3 \text{ mol})(205.0 \text{ J/(K} \cdot \text{mol})) - (2 \text{ mol})(238.8 \text{ J/(K} \cdot \text{mol}))$
$\Delta S° = +137.4 \text{ J/K}$ (+, because moles of gas increase)

(d) $4\ Al(s) + 3\ O_2(g) \rightarrow 2\ Al_2O_3(s)$
$\Delta S° = 2\ S°(Al_2O_3) - [4\ S°(Al) + 3\ S°(O_2)]$
$\Delta S° = (2\ mol)(50.9\ J/(K \cdot mol)) - [(4\ mol)(28.3\ J/(K \cdot mol)) + (3\ mol)(205.0\ J/(K \cdot mol))]$
$\Delta S° = -626.4\ J/K$ (–, because moles of gas decrease)

Entropy and the Second Law of Thermodynamics (Section 17.6)

17.68 In any spontaneous process, the total entropy of a system and its surroundings always increases.

17.70 $\Delta S_{surr} = \frac{-\Delta H}{T}$; the temperature (T) is always positive.

(a) For an exothermic reaction, ΔH is negative and ΔS_{surr} is positive.
(b) For an endothermic reaction, ΔH is positive and ΔS_{surr} is negative.

17.72 $HgO(s) + Zn(s) \rightarrow ZnO(s) + Hg(l)$

(a) $\Delta S_{surr} = \frac{-\Delta H°}{T} = \frac{-(-259.7 \times 10^3\ J)}{298\ K} = +871.5\ J/K$

$\Delta S_{total} = \Delta S° + \Delta S_{surr} = +7.8\ J/K + 871.5\ J/K = +879.3\ J/K$
The reaction is spontaneous because ΔS_{total} is > 0.
(b) Because $\Delta S° > 0$ and $\Delta H° < 0$, there is no temperature at which the reaction is not spontaneous.

17.74 $3\ O_2(g) \rightarrow 2\ O_3(g)$
$\Delta H° = 2\ \Delta H°_f(O_3) = (2\ mol)(143\ kJ/mol) = 286\ kJ = 286 \times 10^3\ J$
$\Delta S° = 2\ S°(O_3) - 3\ S°(O_2)]$
$\Delta S° = (2\ mol)(238.8\ J/(K \cdot mol)) - (3\ mol)(205.0\ J/(K \cdot mol))]$
$\Delta S° = -137.4\ J/K$

$$\Delta S_{total} = \Delta S° + \Delta S_{surr} = \Delta S° + \frac{-\Delta H°}{T} = -137.4\ J/K + \frac{-(286 \times 10^3\ J)}{298\ K} = -1097\ J/K$$

Because $\Delta S_{total} < 0$, the reaction is not spontaneous under standard-state conditions at 25 °C.

17.76 $2\ HgO(s) \rightarrow 2\ Hg(l) + O_2(g)$
$\Delta H° = 0 - 2\ \Delta H°_f(HgO)$
$\Delta H° = -(2\ mol)(-90.8\ kJ/mol) = +181.6\ kJ = +181.6 \times 10^3\ J$
$\Delta S° = [2\ S°(Hg) + S°(O_2)] - 2\ S°(HgO)$
$\Delta S° = [(2\ mol)(76.0\ J/(K \cdot mol)) + (1\ mol)(205.0\ J/(K \cdot mol))] - (2\ mol)(70.3\ J/(K \cdot mol))$
$\Delta S° = \Delta S_{sys} = +216.4\ J/K$

$$\Delta S_{surr} = \frac{-\Delta H°}{T} = \frac{-(181.6 \times 10^3\ J)}{298\ K} = -609.4\ J/K$$

$\Delta S_{total} = \Delta S° + \Delta S_{surr} = 216.4\ J/K + (-609.4\ J/K) = -393.0\ J/K$
Because $\Delta S_{total} < 0$, the reaction is not spontaneous under standard-state conditions at 25 °C.

(b) $\Delta S_{total} = \Delta S° + \Delta S_{surr} = \Delta S° + \frac{-\Delta H°}{T}$

To find the temperature where the reaction becomes spontaneous, set $\Delta S_{total} = 0$ and solve for T.

$$0 = \Delta S^o + \frac{-\Delta H^o}{T} = +216.4\ J/K + \frac{-(181.6 \times 10^3\ J)}{T}$$

$$T = \frac{-(181.6 \times 10^3\ J)}{-216.4\ J/K} = 839.2\ K$$

17.78 (a) $\Delta S_{surr} = \frac{-\Delta H_{vap}}{T} = \frac{-30{,}700\ J/mol}{343\ K} = -89.5\ J/(K \cdot mol)$

$\Delta S_{total} = \Delta S_{vap} + \Delta S_{surr} = 87.0\ J/(K \cdot mol) + (-89.5\ J/(K \cdot mol)) = -2.5\ J/(K \cdot mol)$

(b) $\Delta S_{surr} = \frac{-\Delta H_{vap}}{T} = \frac{-30{,}700\ J/mol}{353\ K} = -87.0\ J/(K \cdot mol)$

$\Delta S_{total} = \Delta S_{vap} + \Delta S_{surr} = 87.0\ J/(K \cdot mol) + (-87.0\ J/(K \cdot mol)) = 0$

(c) $\Delta S_{surr} = \frac{-\Delta H_{vap}}{T} = \frac{-30{,}700\ J/mol}{363\ K} = -84.6\ J/(K \cdot mol)$

$\Delta S_{total} = \Delta S_{vap} + \Delta S_{surr} = 87.0\ J/(K \cdot mol) + (-84.6\ J/(K \cdot mol)) = +2.4\ J/(K \cdot mol)$
Benzene does not boil at 70 °C (343 K) because ΔS_{total} is negative.
The normal boiling point for benzene is 80 °C (353 K), where $\Delta S_{total} = 0$.

Free Energy (Section 17.7)

17.80

ΔH	ΔS	$\Delta G = \Delta H - T\Delta S$	Reaction Spontaneity
–	+	–	Spontaneous at all temperatures
–	–	– or +	Spontaneous at low temperatures where $\lvert\Delta H\rvert > \lvert T\Delta S\rvert$ Nonspontaneous at high temperatures where $\lvert\Delta H\rvert < \lvert T\Delta S\rvert$
+	–	+	Nonspontaneous at all temperatures
+	+	– or +	Spontaneous at high temperatures where $T\Delta S > \Delta H$ Nonspontaneous at low temperature where $T\Delta S < \Delta H$

17.82 (e) a negative free-energy change.

17.84 (a) 0 °C (temperature is below mp); $\Delta H > 0,\ \Delta S > 0,\ \Delta G > 0$
(b) 15 °C (temperature is above mp); $\Delta H > 0,\ \Delta S > 0,\ \Delta G < 0$

17.86 $\Delta H_{vap} = 30.7\ kJ/mol$
$\Delta S_{vap} = 87.0\ J/(K \cdot mol) = 87.0 \times 10^{-3}\ kJ/(K \cdot mol)$
$\Delta G_{vap} = \Delta H_{vap} - T\Delta S_{vap}$
(a) $\Delta G_{vap} = 30.7\ kJ/mol - (343\ K)(87.0 \times 10^{-3}\ kJ/(K \cdot mol)) = +0.9\ kJ/mol$
At 70 °C (343 K), benzene does not boil because ΔG_{vap} is positive.

(b) $\Delta G_{vap} = 30.7\ kJ/mol - (353\ K)(87.0 \times 10^{-3}\ kJ/(K \cdot mol)) = 0$
80 °C (353 K) is the boiling point for benzene because $\Delta G_{vap} = 0$
(c) $\Delta G_{vap} = 30.7\ kJ/mol - (363\ K)(87.0 \times 10^{-3}\ kJ/(K \cdot mol)) = -0.9\ kJ/mol$
At 90 °C (363 K), benzene boils because ΔG_{vap} is negative.

17.88 At the melting point (phase change), $\Delta G_{fusion} = 0$.
$\Delta G_{fusion} = \Delta H_{fusion} - T\Delta S_{fusion}$

$$0 = \Delta H_{fusion} - T\Delta S_{fusion}; \quad T = \frac{\Delta H_{fusion}}{\Delta S_{fusion}} = \frac{18.02\ kJ/mol}{45.56 \times 10^{-3}\ kJ/(K \cdot mol)} = 395.5\ K = 122.4\ °C$$

Standard Free-Energy Changes and Standard Free Energies of Formation (Sections 17.8 and 17.9)

17.90 (a) $\Delta G°$ is the change in free energy that occurs when reactants in their standard states are converted to products in their standard states.
(b) $\Delta G°_f$ is the free-energy change for formation of one mole of a substance in its standard state from the most stable form of the constituent elements in their standard states.

17.92 (a) $N_2(g) + 2\ O_2(g) \rightarrow 2\ NO_2(g)$
$\Delta H° = 2\ \Delta H°_f(NO_2) = (2\ mol)(33.2\ kJ/mol) = 66.4\ kJ$
$\Delta S° = 2\ S°(NO_2) - [S°(N_2) + 2\ S°(O_2)]$
$\Delta S° = (2\ mol)(240.0\ J/(K \cdot mol)) - [(1\ mol)(191.5\ J/(K \cdot mol)) + (2\ mol)(205.0\ J/(K \cdot mol))]$
$\Delta S° = -121.5\ J/K = -121.5 \times 10^{-3}\ kJ/K$
$\Delta G° = \Delta H° - T\Delta S° = 66.4\ kJ - (298\ K)(-121.5 \times 10^{-3}\ kJ/K) = +102.6\ kJ$
Because $\Delta G°$ is positive, the reaction is nonspontaneous under standard-state conditions at 25 °C.
(b) $2\ KClO_3(s) \rightarrow 2\ KCl(s) + 3\ O_2(g)$
$\Delta H° = 2\ \Delta H°_f(KCl) - 2\ \Delta H°_f(KClO_3)$
$\Delta H° = (2\ mol)(-436.5\ kJ/mol) - (2\ mol)(-397.7\ kJ/mol) = -77.6\ kJ$
$\Delta S° = [2\ S°(KCl) + 3\ S°(O_2)] - 2\ S°(KClO_3)$
$\Delta S° = [(2\ mol)(82.6\ J/(K \cdot mol)) + (3\ mol)(205.0\ J/(K \cdot mol))] - (2\ mol)(143.1\ J/(K \cdot mol))$
$\Delta S° = 494.0\ J/K = 494.0 \times 10^{-3}\ kJ/K$
$\Delta G° = \Delta H° - T\Delta S° = -77.6\ kJ - (298\ K)(494.0 \times 10^{-3}\ kJ/K) = -224.8\ kJ$
Because $\Delta G°$ is negative, the reaction is spontaneous under standard-state conditions at 25 °C.
(c) $CH_3CH_2OH(l) + O_2(g) \rightarrow CH_3CO_2H(l) + H_2O(l)$
$\Delta H° = [\Delta H°_f(CH_3CO_2H) + \Delta H°_f(H_2O)] - \Delta H°_f(CH_3CH_2OH)$
$\Delta H° = [(1\ mol)(-484.5\ kJ/mol) + (1\ mol)(-285.8\ kJ/mol)] - (1\ mol)(-277.7\ kJ/mol) = -492.6\ kJ$
$\Delta S° = [S°(CH_3CO_2H) + S°(H_2O)] - [S°(CH_3CH_2OH) + S°(O_2)]$
$\Delta S° = [(1\ mol)(160\ J/(K \cdot mol)) + (1\ mol)(69.9\ J/(K \cdot mol))]$
$- [(1\ mol)(161\ J/(K \cdot mol)) + (1\ mol)(205.0\ J/(K \cdot mol))]$
$\Delta S° = -136.1\ J/K = -136.1 \times 10^{-3}\ kJ/K$
$\Delta G° = \Delta H° - T\Delta S° = -492.6\ kJ - (298\ K)(-136.1 \times 10^{-3}\ kJ/K) = -452.0\ kJ$
Because $\Delta G°$ is negative, the reaction is spontaneous under standard-state conditions at 25 °C.

17.94 (a) $N_2(g) + 2\ O_2(g) \rightarrow 2\ NO_2(g)$
$\Delta G° = 2\ \Delta G°_f(NO_2) = (2\ mol)(51.3\ kJ/mol) = +102.6\ kJ$

(b) $2\ KClO_3(s) \rightarrow 2\ KCl(s) + 3\ O_2(g)$
$\Delta G° = 2\ \Delta G°_f(KCl) - 2\ \Delta G°_f(KClO_3)$
$\Delta G° = (2\ mol)(-408.5\ kJ/mol) - (2\ mol)(-296.3\ kJ/mol) = -224.4\ kJ$
(c) $CH_3CH_2OH(l) + O_2(g) \rightarrow CH_3CO_2H(l) + H_2O(l)$
$\Delta G° = [\Delta G°_f(CH_3CO_2H) + \Delta G°_f(H_2O)] - \Delta G°_f(CH_3CH_2OH)$
$\Delta G° = [(1\ mol)(-390\ kJ/mol) + (1\ mol)(-237.2\ kJ/mol)] - (1\ mol)(-174.9\ kJ/mol) = -452\ kJ$

17.96 A compound is thermodynamically stable with respect to its constituent elements at 25 °C if $\Delta G°_f$ is negative.

	$\Delta G°_f$ (kJ/mol)	Stable
(a) $BaCO_3(s)$	−1134.4	yes
(b) $HBr(g)$	−53.4	yes
(c) $N_2O(g)$	+104.2	no
(d) $C_2H_4(g)$	+68.1	no

17.98 $C_2H_4(g) + Cl_2(g) \rightarrow CH_2ClCH_2Cl(l)$
$\Delta G° = \Delta G°_f(CH_2ClCH_2Cl) - \Delta G°_f(C_2H_4)$
$\Delta G° = (1\ mol)(-79.6\ kJ/mol) - (1\ mol)(68.1\ kJ/mol) = -147.7\ kJ$
Because $\Delta G° < 0$, dichloroethane can be synthesized from gaseous C_2H_4 and Cl_2, each at 25 °C and 1 atm pressure.

17.100 $CH_2{=}CH_2(g) + H_2O(l) \rightarrow CH_3CH_2OH(l)$
$\Delta H° = \Delta H°_f(CH_3CH_2OH) - [\Delta H°_f(CH_2{=}CH_2) + \Delta H°_f(H_2O)]$
$\Delta H° = (1\ mol)(-277.7\ kJ/mol) - [(1\ mol)(52.3\ kJ/mol) + (1\ mol)(-285.8\ kJ/mol)]$
$\Delta H° = -44.2\ kJ$
$\Delta S° = S°(CH_3CH_2OH) - [S°(CH_2{=}CH_2) + S°(H_2O)]$
$\Delta S° = (1\ mol)(161\ J/(K \cdot mol)) - [(1\ mol)(219.5\ J/(K \cdot mol)) + (1\ mol)(69.9\ J/(K \cdot mol))]$
$\Delta S° = -128\ J/(K \cdot mol) = -128 \times 10^{-3}\ kJ/(K \cdot mol)$
$\Delta G° = \Delta H° - T\Delta S° = -44.2\ kJ - (298\ K)(-128 \times 10^{-3}\ kJ/K) = -6.1\ kJ$
Because $\Delta G°$ is negative, the reaction is spontaneous under standard-state conditions at 25 °C.
The reaction becomes nonspontaneous at high temperatures because $\Delta S°$ is negative.
To find the crossover temperature, set $\Delta G = 0$ and solve for T.

$$T = \frac{\Delta H°}{\Delta S°} = \frac{-44{,}200\ J}{-128\ J/K} = 345\ K = 72\ °C$$

The reaction becomes nonspontaneous at 72 °C.

17.102 $3\ C_2H_2(g) \rightarrow C_6H_6(l)$
$\Delta G° = \Delta G°_f(C_6H_6) - 3\ \Delta G°_f(C_2H_2)$
$\Delta G° = (1\ mol)(124.5\ kJ/mol) - (3\ mol)(209.9\ kJ/mol) = -505.2\ kJ$
Because $\Delta G°$ is negative, the reaction is possible. Look for a catalyst.
Because $\Delta G°_f$ for benzene is positive (+124.5 kJ/mol), the synthesis of benzene from graphite and gaseous H_2 at 25 °C and 1 atm pressure is not possible.

Free Energy, Composition, and Chemical Equilibrium (Sections 17.10 and 17.11)

17.104 $\Delta G = \Delta G^\circ + RT \ln Q$

17.106 $2\ NO(g) + Cl_2(g) \rightarrow 2\ NOCl(g)$

$\Delta G^\circ = 2\ \Delta G^\circ_f(NOCl) - 2\ \Delta G^\circ_f(NO)$

$\Delta G^\circ = (2\text{ mol})(66.1\text{ kJ/mol}) - (2\text{ mol})(87.6\text{ kJ/mol}) = -43.0\text{ kJ}$

$$\Delta G = \Delta G^\circ + RT \ln\left[\frac{(P_{NOCl})^2}{(P_{NO})^2(P_{Cl_2})}\right]$$

$$\Delta G = (-43.0\text{ kJ/mol}) + [8.314 \times 10^{-3}\text{ kJ/(K}\cdot\text{mol)}](298\text{ K})\ln\left[\frac{(2.00)^2}{(1.00 \times 10^{-3})^2(1.00 \times 10^{-3})}\right]$$

$\Delta G = +11.8\text{ kJ/mol}$

The reaction is spontaneous in the reverse direction.

17.108 $$\Delta G = \Delta G^\circ + RT \ln\left[\frac{(P_{SO_3})^2}{(P_{SO_2})^2(P_{O_2})}\right]$$

(a) $\Delta G = (-141.8\text{ kJ/mol}) + [8.314 \times 10^{-3}\text{ kJ/(K}\cdot\text{mol)}](298\text{ K})\ln\left[\frac{(1.0)^2}{(100)^2(100)}\right] = -176.0\text{ kJ/mol}$

(b) $\Delta G = (-141.8\text{ kJ/mol}) + [8.314 \times 10^{-3}\text{ kJ/(K}\cdot\text{mol)}](298\text{ K})\ln\left[\frac{(10)^2}{(2.0)^2(1.0)}\right] = -133.8\text{ kJ/mol}$

(c) $Q = 1$, $\ln Q = 0$, $\Delta G = \Delta G^\circ = -141.8\text{ kJ/mol}$

17.110 $\Delta G^\circ = -RT \ln K$

(a) If $K > 1$, ΔG° is negative. (b) If $K = 1$, $\Delta G^\circ = 0$. (c) If $K < 1$, ΔG° is positive.

17.112 $\Delta G^\circ = -RT \ln K_p = -141.8\text{ kJ}$

$$\ln K_p = \frac{-\Delta G^\circ}{RT} = \frac{-(-141.8\text{ kJ/mol})}{[8.314 \times 10^{-3}\text{ kJ/(K}\cdot\text{mol)}](298\text{ K})} = 57.23$$

$K_p = e^{57.23} = 7.2 \times 10^{24}$

17.114 $C_2H_5OH(l) \rightleftarrows C_2H_5OH(g)$

$\Delta G^\circ = \Delta G^\circ_f(C_2H_5OH(g)) - \Delta G^\circ_f(C_2H_5OH(l))$

$\Delta G^\circ = (1\text{ mol})(-167.9\text{ kJ/mol}) - (1\text{ mol})(-174.9\text{ kJ/mol}) = +7.0\text{ kJ}$

$\Delta G^\circ = -RT \ln K$

$$\ln K = \frac{-\Delta G^\circ}{RT} = \frac{-(7.0\text{ kJ/mol})}{[8.314 \times 10^{-3}\text{ kJ/(K}\cdot\text{mol)}](298\text{ K})} = -2.83$$

$K = e^{-2.83} = 0.059$; $K = K_p = P_{C_2H_5OH} = 0.059\text{ atm}$

17.116 $Br_2(l) \rightleftharpoons Br_2(g)$

$\Delta G° = \Delta G°_f(Br_2(g)) = 3.14$ kJ/mol

$\Delta G° = -RT \ln K$

$$\ln K = \frac{-\Delta G°}{RT} = \frac{-(3.14 \text{ kJ/mol})}{[8.314 \times 10^{-3} \text{ kJ/(K} \cdot \text{mol)}](298 \text{ K})} = -1.267$$

$K = e^{-1.267} = 0.282$; $K = K_p = P_{Br_2} = 0.282 \text{ atm} = 0.28 \text{ atm}$

17.118 $2\ CH_2{=}CH_2(g) + O_2(g) \rightarrow 2\ C_2H_4O(g)$

$\Delta G° = 2\ \Delta G°_f(C_2H_4O) - 2\ \Delta G°_f(CH_2{=}CH_2)$

$\Delta G° = (2 \text{ mol})(-13.1 \text{ kJ/mol}) - (2 \text{ mol})(68.1 \text{ kJ/mol}) = -162.4 \text{ kJ}$

$\Delta G° = -RT \ln K$

$$\ln K = \frac{-\Delta G°}{RT} = \frac{-(-162.4 \text{ kJ/mol})}{[8.314 \times 10^{-3} \text{ kJ/(K} \cdot \text{mol)}](298 \text{ K})} = 65.55$$

$K = K_p = e^{65.55} = 2.9 \times 10^{28}$

Chapter Problems

17.120 C_3H_8, 44.10; 20 °C = 293 K

$$\text{mol } C_3H_8 = 1.32 \text{ g} \times \frac{1 \text{ mol } C_3H_8}{44.10 \text{ g}} = 0.0300 \text{ mol } C_3H_8$$

$$V = \frac{nRT}{P} = \frac{(0.0300 \text{ mol})\left(0.082\ 06 \frac{\text{L} \cdot \text{atm}}{\text{K} \cdot \text{mol}}\right)(293 \text{ K})}{0.100 \text{ atm}} = 7.21 \text{ L}$$

Compress 7.21 L by a factor of 5 (7.21/5) to 1.44 L

$$\Delta S = nR \ln\left(\frac{V_f}{V_i}\right) = (0.0300 \text{ mol})(8.314 \text{ J/K} \cdot \text{mol}) \ln\left(\frac{1.44 \text{ L}}{7.21 \text{ L}}\right) = -0.402 \text{ J/K}$$

17.122 (a) Spontaneous does not mean fast, just possible.

(b) For a spontaneous reaction $\Delta S_{total} > 0$. ΔS_{sys} can be positive or negative.

(c) An endothermic reaction can be spontaneous if $\Delta S_{sys} > 0$.

(d) True, because the sign of ΔG changes when the direction of a reaction is reversed.

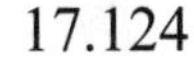

17.124

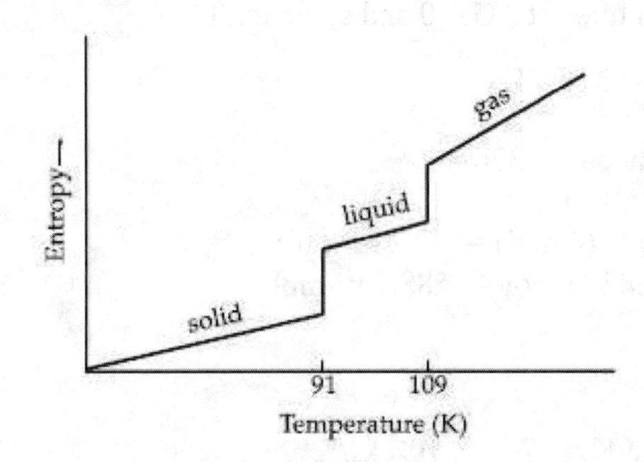

17.126 At the normal boiling point, $\Delta G = 0$.

$$\Delta G_{vap} = \Delta H_{vap} - T\Delta S_{vap}; \qquad T = \frac{\Delta H_{vap}}{\Delta S_{vap}} = \frac{38{,}600 \text{ J}}{110 \text{ J/K}} = 351 \text{ K} = 78\ ^{\circ}\text{C}$$

17.128 $\Delta G = \Delta H - T\Delta S$

(a) ΔH must be positive (endothermic) and greater than $T\Delta S$ in order for ΔG to be positive (nonspontaneous reaction).

(b) Set $\Delta G = 0$ and solve for ΔH.

$\Delta G = 0 = \Delta H - T\Delta S = \Delta H - (323 \text{ K})(104 \text{ J/K}) = \Delta H - (33592 \text{ J}) = \Delta H - (33.6 \text{ kJ})$

$\Delta H = 33.6$ kJ

ΔH must be greater than 33.6 kJ.

17.130 For $PbCrO_4$, $K_{sp} = 2.8 \times 10^{-13}$

$\Delta G^{\circ} = -RT \ln K_{sp}$

$\Delta G^{\circ} = -[8.314 \times 10^{-3} \text{ kJ/K}](298 \text{ K}) \ln(2.8 \times 10^{-13}) = +71.6 \text{ kJ}$

17.132 (a)

	$\Delta H_{vap}/T_{bp}$
ammonia	98 J/K
benzene	87 J/K
carbon tetrachloride	85 J/K
chloroform	87 J/K
mercury	94 J/K

(b) All processes are the conversion of a liquid to a gas at the boiling point. They should all have similar ΔS values. $\Delta H_{vap}/T_{bp}$ is equal to ΔS_{vap}.

(c) NH_3 deviates from Trouton's rule because of hydrogen bonding. $NH_3(l)$ has less randomness and ΔS_{vap} is larger. Hg has metallic bonding which also leads to less randomness of the liquid.

17.134 $Ni(s) + 4\ CO(g) \rightarrow Ni(CO)_4(l)$

(a) $\Delta H^{\circ} = \Delta H^{\circ}_f(Ni(CO)_4) - 4\ \Delta H^{\circ}_f(CO)$

$\Delta H^{\circ} = (1 \text{ mol})(-633.0 \text{ kJ/mol}) - (4 \text{ mol})(-110.5 \text{ kJ/mol}) = -191.0 \text{ kJ}$

$\Delta S^{\circ} = S^{\circ}(Ni(CO)_4) - [S^{\circ}(Ni) + 4\ S^{\circ}(CO)]$

$\Delta S^{\circ} = (1 \text{ mol})(313.4 \text{ J/(K} \cdot \text{mol)}) - [(1 \text{ mol})(29.9 \text{ J/(K} \cdot \text{mol)}) + (4 \text{ mol})(197.6 \text{ J/(K} \cdot \text{mol)})]$

$\Delta S^{\circ} = -506.9 \text{ J/K} = -506.9 \times 10^{-3} \text{ kJ/K}$

$\Delta G^{\circ} = \Delta H^{\circ} - T\Delta S^{\circ} = -191.0 \text{ kJ} - (298 \text{ K})(-506.9 \times 10^{-3} \text{ kJ/K}) = -39.9 \text{ kJ}$

(b) To find the crossover temperature set $\Delta G = 0$ and solve for T.

$$T = \frac{\Delta H^{\circ}}{\Delta S^{\circ}} = \frac{-191.0 \text{ kJ}}{-506.9 \times 10^{-3} \text{ kJ/K}} = 376.8 \text{ K}$$

The reaction becomes nonspontaneous at 376.8 K

(c) $\Delta G^{\circ} = \Delta G^{\circ}_f(Ni(CO)_4) - 4\ \Delta G^{\circ}_f(CO)$

$-39.9 \text{ kJ} = (1 \text{ mol})\Delta G^{\circ}_f(Ni(CO)_4) - (4 \text{ mol})(-137.2 \text{ kJ/mol})$

$\Delta G^{\circ}_f(Ni(CO)_4) = (-39.9 \text{ kJ} - 548.8 \text{ kJ})/\text{mol} = -588.7 \text{ kJ/mol}$

17.136 $MgCO_3(s) \rightarrow MgO(s) + CO_2(g)$

From Problem 17.131(b)

$\Delta H^{\circ} = +101$ kJ; $\qquad \Delta S^{\circ} = 174.8 \text{ J/K} = 174.8 \times 10^{-3} \text{ kJ/K}$

The equilibrium pressure of CO_2 is equal to $K_p = P_{CO_2}$. K_p is not affected by the quantities of $MgCO_3$ and MgO present. K_p can be calculated from $\Delta G°$.

$\Delta G° = \Delta H° - T\Delta S°$

$\Delta G° = -RT \ln K_p$

(a) $\Delta G° = 101 \text{ kJ} - (298 \text{ K})(174.8 \times 10^{-3} \text{ kJ/K}) = +49 \text{ kJ}$

$$\ln K_p = \frac{-\Delta G°}{RT} = \frac{-49 \text{ kJ/mol}}{[8.314 \times 10^{-3} \text{ kJ/(K·mol)}](298 \text{ K})} = -19.8$$

$K_p = P_{CO_2} = e^{-19.8} = 3 \times 10^{-9}$ atm

(b) $\Delta G° = 101 \text{ kJ} - (553 \text{ K})(174.8 \times 10^{-3} \text{ kJ/K}) = 4.3 \text{ kJ}$

$$\ln K_p = \frac{-\Delta G°}{RT} = \frac{-4.3 \text{ kJ/mol}}{[8.314 \times 10^{-3} \text{ kJ/(K·mol)}](553 \text{ K})} = -0.94$$

$K_p = P_{CO_2} = e^{-0.94} = 0.39$ atm

(c) $P_{CO_2} = 0.39$ atm because the temperature is the same as in (b).

17.138 (a) $\Delta H° = 2\,\Delta H°_f(NH_3) = (2 \text{ mol})(-46.1 \text{ kJ/mol}) = -92.2 \text{ kJ}$

$\Delta G° = 2\,\Delta G°_f(NH_3) = (2 \text{ mol})(-16.5 \text{ kJ/mol}) = -33.0 \text{ kJ}$

$\Delta G° = \Delta H° - T\Delta S°$

$\Delta H° - \Delta G° = T\Delta S°$

$$\Delta S° = \frac{\Delta H° - \Delta G°}{T} = \frac{-92.2 \text{ kJ} - (-33.0 \text{ kJ})}{298 \text{ K}} = -0.199 \text{ kJ/K} = -199 \text{ J/K}$$

(b) $\Delta S°$ is negative because the number of mol of gas molecules decreases from 4 mol to 2 mol on going from reactants to products.

(c) The reaction is spontaneous because $\Delta G°$ is negative.

(d) $\Delta G° = \Delta H° - T\Delta S° = -92.2 \text{ kJ} - (350 \text{ K})(-0.199 \text{ kJ/K}) = -22.55 \text{ kJ}$

$\Delta G° = -RT \ln K_p$

$$\ln K_p = \frac{-\Delta G°}{RT} = \frac{-(-22.55 \text{ kJ/mol})}{[8.314 \times 10^{-3} \text{ kJ/(K·mol)}](350 \text{ K})} = 7.749$$

$K_p = e^{7.749} = 2.3 \times 10^3$

$\Delta n = 2 - (1 + 3) = -2$

$$K_c = K_p\left(\frac{1}{RT}\right)^{\Delta n} = (2.3 \times 10^3)\left(\frac{1}{RT}\right)^{-2} = (2.3 \times 10^3)(RT)^2$$

$K_c = (2.3 \times 10^3)[(0.082\,06)(350)]^2 = 1.9 \times 10^6$

17.140 (a) $\Delta G° = \Delta H° - T\Delta S°$ and $\Delta G° = -RT \ln K$

Set the two equations equal to each other.

$-RT \ln K = \Delta H° - T\Delta S°$

$$\ln K = \frac{\Delta H° - T\Delta S°}{-RT}$$

$$\ln K = \frac{-\Delta H°}{RT} + \frac{T\Delta S°}{RT}$$

$$\ln K = \frac{-\Delta H^\circ}{RT} + \frac{\Delta S^\circ}{R}$$

$$\ln K = \frac{-\Delta H^\circ}{R}\left(\frac{1}{T}\right) + \frac{\Delta S^\circ}{R}$$ This is the equation for a straight line (y = mx + b).

$y = \ln K$; $m = -\frac{\Delta H^\circ}{R}$ = slope; $x = \frac{1}{T}$; $b = \frac{\Delta S^\circ}{R}$ = intercept

(b) Plot ln K versus 1/T

$\Delta H^\circ = -R(\text{slope})$ $\Delta S^\circ = R(\text{intercept})$

(c) For a reaction where K increases with increasing temperature, the following plot would be obtained:

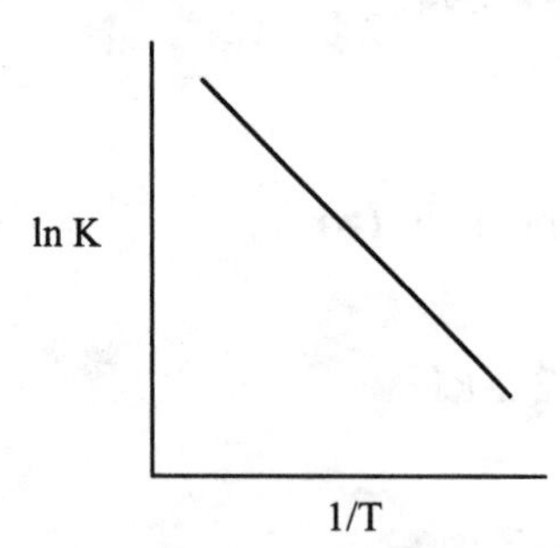

The slope is negative.
Because $\Delta H^\circ = -R(\text{slope})$, ΔH° is positive, and the reaction is endothermic.

This prediction is in accord with Le Châtelier's principle because when you add heat (raise the temperature) for an endothermic reaction, the reaction in the forward direction takes place, the product concentrations increase and the reactant concentrations decrease. This results in an increase in K.

17.142 For PbI_2, $K_{sp} = [Pb^{2+}][I^-]^2$

	$PbI_2(s)$ ⇄	$Pb^{2+}(aq)$	+ $2\ I^-(aq)$
initial (M)		0	0
equil (M)		x	2x

$K_{sp} = x(2x)^2 = 4x^3$, where x = molar solubility

At 20 °C = 20 + 273 = 293 K, $K_{sp} = 4(1.45 \times 10^{-3})^3 = 1.22 \times 10^{-8}$

At 80 °C = 80 + 273 = 353 K, $K_{sp} = 4(6.85 \times 10^{-3})^3 = 1.29 \times 10^{-6}$

From problem 17.140, $\ln K = \frac{-\Delta H^\circ}{RT} + \frac{\Delta S^\circ}{R}$

$$\ln K_1 - \ln K_2 = \frac{-\Delta H^\circ}{RT_1} + \frac{\Delta S^\circ}{R} - \left(\frac{-\Delta H^\circ}{RT_2} + \frac{\Delta S^\circ}{R}\right)$$

$$\ln\frac{K_1}{K_2} = \frac{-\Delta H^\circ}{RT_1} - \frac{-\Delta H^\circ}{RT_2} = \frac{-\Delta H^\circ}{R}\left(\frac{1}{T_1} - \frac{1}{T_2}\right) = \frac{\Delta H^\circ}{R}\left(\frac{1}{T_2} - \frac{1}{T_1}\right)$$

$$\Delta H^\circ = \frac{[\ln K_1 - \ln K_2]R}{\left(\frac{1}{T_2} - \frac{1}{T_1}\right)}$$

$$\Delta H^\circ = \frac{[\ln(1.22 \times 10^{-8}) - \ln(1.29 \times 10^{-6})][8.314 \times 10^{-3}\ \text{kJ/(K} \cdot \text{mol)}]}{\left(\dfrac{1}{353\ \text{K}} - \dfrac{1}{293\ \text{K}}\right)} = 66.8\ \text{kJ/mol}$$

$\Delta G^\circ = -RT \ln K_{sp} = -[8.314 \times 10^{-3}\ \text{kJ/(K} \cdot \text{mol)}](293\ \text{K}) \ln(1.22 \times 10^{-8}) = 44.4\ \text{kJ/mol}$

$\Delta G^\circ = \Delta H^\circ - T\Delta S^\circ$; $\quad \Delta H^\circ - \Delta G^\circ = T\Delta S^\circ$; $\quad \Delta S^\circ = \dfrac{\Delta H^\circ - \Delta G^\circ}{T}$

$$\Delta S^\circ = \frac{66.8\ \text{kJ/mol} - 44.4\ \text{kJ/mol}}{293\ \text{K}} = 0.0765\ \text{kJ/(K} \cdot \text{mol)} = 76.5\ \text{J/(K} \cdot \text{mol)}$$

17.144 $CS_2(l) \rightleftarrows CS_2(g)$

$\Delta H^\circ = \Delta H^\circ_f(CS_2(g)) - \Delta H^\circ_f(CS_2(l))$

$\Delta H^\circ = [(1\ \text{mol})(116.7\ \text{kJ/mol})] - [(1\ \text{mol})(89.0\ \text{kJ/mol})] = 27.7\ \text{kJ}$

$\Delta S^\circ = S^\circ(CS_2(g)) - S^\circ(CS_2(l))$

$\Delta S^\circ = [(1\ \text{mol})(237.7\ \text{J/(K} \cdot \text{mol)})] - [(1\ \text{mol})(151.3\ \text{J/(K} \cdot \text{mol)})] = 86.4\ \text{J/K}$

$\Delta G = \Delta H^\circ - T\Delta S^\circ$

At the boiling point, $\Delta G = 0$.

$0 = \Delta H^\circ - T_{bp}\Delta S^\circ$

$$T_{bp} = \frac{\Delta H^\circ}{\Delta S^\circ} = \frac{27.7\ \text{kJ}}{86.4 \times 10^{-3}\ \text{kJ/K}} = 321\ \text{K}$$

$T_{bp} = 321\ \text{K} = 321 - 273 = 48\ ^\circ\text{C}$

17.146 $2\ KClO_3(s) \rightarrow 2\ KCl(s) + 3\ O_2(g)$

$\Delta H^\circ = 2\ \Delta H^\circ_f(KCl) - 2\ \Delta H^\circ_f(KClO_3)$

$\Delta H^\circ = (2\ \text{mol})(-436.5\ \text{kJ}) - (2\ \text{mol})(-397.7\ \text{kJ}) = -77.6\ \text{kJ}$

$25\ ^\circ\text{C} = 25 + 273 = 298\ \text{K}$

$\Delta G^\circ = \Delta H^\circ - T\Delta S^\circ$

$\Delta H^\circ - \Delta G^\circ = T\Delta S^\circ$

$$\Delta S^\circ = \frac{\Delta H^\circ - \Delta G^\circ}{T} = \frac{-77.6\ \text{kJ} - (-224.4\ \text{kJ})}{298\ \text{K}} = 0.493\ \text{kJ/K} = 493\ \text{J/K}$$

$\Delta S^\circ = [2\ S^\circ(KCl) + 3\ S^\circ(O_2)] - 2\ S^\circ(KClO_3)$

$493\ \text{J/K} = [(2\ \text{mol})(82.6\ \text{J/(K} \cdot \text{mol)}) + (3\ \text{mol})S^\circ(O_2)] - (2\ \text{mol})(143.1\ \text{J/(K} \cdot \text{mol)})$

$(3\ \text{mol})S^\circ(O_2) = 493\ \text{J/K} - (2\ \text{mol})(82.6\ \text{J/(K} \cdot \text{mol)}) + (2\ \text{mol})(143.1\ \text{J/(K} \cdot \text{mol)})$

$(3\ \text{mol})S^\circ(O_2) = 614\ \text{J/K}$

$S^\circ(O_2) = (614\ \text{J/K})/(3\ \text{mol}) = 204.7\ \text{J/(K} \cdot \text{mol)} = 205\ \text{J/(K} \cdot \text{mol)}$

Multiconcept Problems

17.148 The kinetic parameters [(a), (b), and (h)] are affected by a catalyst. The thermodynamic and equilibrium parameters [(c), (d), (e), (f), and (g)] are not affected by a catalyst.

17.150 (a) $2\ SO_2(g) + O_2(g) \rightleftarrows 2\ SO_3(g)$

$\Delta H^\circ = 2\ \Delta H^\circ_f(SO_3) - 2\ \Delta H^\circ_f(SO_2)$

$\Delta H^\circ = (2\ \text{mol})(-395.7\ \text{kJ/mol}) - (2\ \text{mol})(-296.8\ \text{kJ/mol}) = -197.8\ \text{kJ}$

$\Delta S° = 2\ S°(SO_3) - [2\ S°(SO_2) + S°(O_2)]$
$\Delta S° = (2\ mol)(256.6\ J/(K \cdot mol)) - [(2\ mol)(248.1\ J/(K \cdot mol)) + (1\ mol)(205.0\ J/(K \cdot mol))]$
$\Delta S° = -188.0\ J/K = -188.0 \times 10^{-3}\ kJ/K$
$\Delta G° = \Delta H° - T\Delta S° = -197.8\ kJ - (800\ K)(-188.0 \times 10^{-3}\ kJ/K) = -47.4\ kJ$
$\Delta G° = -RT \ln K_p$

$$\ln K_p = \frac{-\Delta G°}{RT} = \frac{-(-47.4\ kJ/mol)}{[8.314 \times 10^{-3}\ kJ/(K \cdot mol)](800\ K)} = 7.13$$

$K_p = e^{7.13} = 1249$
SO_2, 64.06; O_2, 32.00

At 800 K:

$$P_{SO_2} = \frac{nRT}{V} = \frac{\left(192\ g \times \frac{1\ mol}{64.06\ g}\right)\left(0.082\ 06\ \frac{L \cdot atm}{K \cdot mol}\right)(800\ K)}{15.0\ L} = 13.1\ atm$$

$$P_{O_2} = \frac{nRT}{V} = \frac{\left(48.0\ g \times \frac{1\ mol}{32.00\ g}\right)\left(0.082\ 06\ \frac{L \cdot atm}{K \cdot mol}\right)(800\ K)}{15.0\ L} = 6.56\ atm$$

	$2\ SO_2(g)$	+ $O_2(g)$	⇄ $2\ SO_3(g)$
initial (atm)	13.1	6.56	0
assume complete rxn (atm)	0	0	13.1
assume a small back rxn	+2x	+x	−2x
equil (atm)	2x	x	13.1 − 2x

$$K_p = 1249 = \frac{[SO_3]^2}{[SO_2]^2[O_2]} = \frac{(13.1 - 2x)^2}{(2x)^2(x)} \approx \frac{(13.1)^2}{(2x)^2(x)}$$

Solve for x. $x^3 = 0.0343$; $x = 0.325$
Use successive approximations to solve for x because 2x is not negligible compared with 13.1.

Second approximation:

$$1249 = \frac{[13.1 - (2)(0.325)]^2}{(2x)^2(x)};$$ Solve for x. $x^3 = 0.0310$; $x = 0.314$

Third approximation:

$$1249 = \frac{[13.1 - (2)(0.314)]^2}{(2x)^2(x)};$$ Solve for x. $x^3 = 0.0311$; $x = 0.315$ (x has converged)

$P_{SO_2} = 2x = 2(0.315) = 0.63\ atm$

$P_{O_2} = x = 0.32\ atm$

$P_{SO_3} = 13.1 - 2x = 13.1 - 2(0.315) = 12.5\ atm$

(b) The % yield of SO_3 decreases with increasing temperature because $\Delta S°$ is negative. $\Delta G°$ becomes less negative and K_p gets smaller as the temperature increases.

(c) At 1000 K:

$\Delta G^\circ = \Delta H^\circ - T\Delta S^\circ = -197.8 \text{ kJ} - (1000 \text{ K})(-188.0 \times 10^{-3} \text{ kJ/K}) = -9.8 \text{ kJ}$

$\Delta G^\circ = -RT \ln K_p$

$$\ln K_p = \frac{-\Delta G^\circ}{RT} = \frac{-(-9.8 \text{ kJ/mol})}{[8.314 \times 10^{-3} \text{ kJ/(K}\cdot\text{mol)}](1000 \text{ K})} = 1.179$$

$K_p = e^{1.179} = 3.25$

$$P_{SO_2} = \frac{nRT}{V} = \frac{\left(192 \text{ g} \times \frac{1 \text{ mol}}{64.06 \text{ g}}\right)\left(0.082\,06 \frac{\text{L}\cdot\text{atm}}{\text{K}\cdot\text{mol}}\right)(1000 \text{ K})}{15.0 \text{ L}} = 16.4 \text{ atm}$$

$$P_{O_2} = \frac{nRT}{V} = \frac{\left(48.0 \text{ g} \times \frac{1 \text{ mol}}{32.00 \text{ g}}\right)\left(0.082\,06 \frac{\text{L}\cdot\text{atm}}{\text{K}\cdot\text{mol}}\right)(1000 \text{ K})}{15.0 \text{ L}} = 8.2 \text{ atm}$$

	$2\ SO_2(g)$	+ $O_2(g)$	⇄ $2\ SO_3(g)$
initial (atm)	16.4	8.2	0
assume complete rxn (atm)	0	0	16.4
assume a small back rxn	+2x	+x	−2x
equil (atm)	2x	x	16.4 − 2x

$$K_p = 3.25 = \frac{[SO_3]^2}{[SO_2]^2[O_2]} = \frac{(16.4 - 2x)^2}{(2x)^2(x)} \approx \frac{(16.4)^2}{(2x)^2(x)}$$

Solve for x. $x^3 = 20.7$; $x = 2.7$

Use successive approximations to solve for x because 2x is not negligible compared with 16.4.

Second approximation:

$$3.25 = \frac{[16.4 - (2)(2.7)]^2}{(2x)^2(x)}$$; Solve for x. $x^3 = 9.31$; $x = 2.1$

Third approximation:

$$3.25 = \frac{[16.4 - (2)(2.1)]^2}{(2x)^2(x)}$$; Solve for x. $x^3 = 11.4$; $x = 2.3$

Fourth approximation:

$$3.25 = \frac{[16.4 - (2)(2.3)]^2}{(2x)^2(x)}$$; Solve for x. $x^3 = 10.7$; $x = 2.2$ (x has converged)

$P_{SO_2} = 2x = 2(2.2) = 4.4$ atm

$P_{O_2} = x = 2.2$ atm

$P_{SO_3} = 16.4 - 2x = 16.4 - 2(2.2) = 12.0$ atm

$P_{total} = P_{SO_2} + P_{O_2} + P_{SO_3} = 4.4 + 2.2 + 12.0 = 18.6$ atm

On going from 800 K to 1000 K, P_{total} increases to 18.6 atm (because K_p decreases, but P increases with temperature at constant volume).

17.152 $CaCO_3(s) \rightleftarrows Ca^{2+}(aq) + CO_3^{2-}(aq)$

$\Delta H^\circ = [\Delta H^\circ_f(Ca^{2+}) + \Delta H^\circ_f(CO_3^{2-})] - \Delta H^\circ_f(CaCO_3)$

$\Delta H^\circ = [(1\ mol)(-542.8\ kJ/mol) + (1\ mol)(-677.1\ kJ/mol)] - (1\ mol)(-1207.6\ kJ/mol)$

$\Delta H^\circ = -12.3\ kJ$

$\Delta S^\circ = [S^\circ(Ca^{2+}) + S^\circ(CO_3^{2-})] - S^\circ(CaCO_3)$

$\Delta S^\circ = [(1\ mol)(-53.1\ J/(K \cdot mol)) + (1\ mol)(-56.9\ J/(K \cdot mol))] - (1\ mol)(91.7\ J/(K \cdot mol))$

$\Delta S^\circ = -201.7\ J/K = -201.7 \times 10^{-3}\ kJ/K$

50 °C = 50 + 273 = 323 K

$\Delta G = \Delta H^\circ - T\Delta S^\circ = -12.3\ kJ - (323\ K)(-201.7 \times 10^{-3}\ kJ/K) = +52.85\ kJ$

$\Delta G = -RT \ln K_{sp}$

$$\ln K_{sp} = \frac{-\Delta G}{RT} = \frac{-52.85\ J/mol}{[8.314 \times 10^{-3}\ kJ/(K \cdot mol)](323\ K)} = -19.68$$

$K_{sp} = e^{-19.68} = 2.8 \times 10^{-9}$

20 °C = 20 +273 = 293 K

$$n_{CO_2} = \frac{PV}{RT} = \frac{\left(731\ mm\ Hg \times \frac{1.00\ atm}{760\ mm\ Hg}\right)(1.000\ L)}{\left(0.082\ 06\ \frac{L \cdot atm}{K \cdot mol}\right)(293\ K)} = 0.0400\ mol\ CO_2$$

$Ca(OH)_2$, 74.09

$$mol\ Ca(OH)_2 = 3.335\ g\ Ca(OH)_2 \times \frac{1\ mol\ Ca(OH)_2}{74.09\ g\ Ca(OH)_2} = 0.0450\ mol\ Ca(OH)_2$$

$CO_2(g) + H_2O(l) \rightarrow H_2CO_3(aq)$

	$Ca(OH)_2(aq)$ +	$H_2CO_3(aq)$ →	$CaCO_3(s)$ +	$2\ H_2O(l)$
before (mol)	0.0450	0.0400	0	
change (mol)	−0.0400	−0.0400	+0.0400	
after (mol)	0.0050	0	0.0400	

500.0 mL = 0.5000 L

$[Ca(OH)_2] = [Ca^{2+}] = 0.0050\ mol/0.5000\ L = 0.010\ M$

	$CaCO_3(s) \rightleftarrows$	$Ca^{2+}(aq)$ +	$CO_3^{2-}(aq)$
initial (M)		0.010	0
change (M)		+x	+x
equil (M)		0.010 + x	x

$K_{sp} = [Ca^{2+}][CO_3^{2-}] = 2.8 \times 10^{-9} = (0.010 + x)x \approx 0.010x$

x = molar solubility $= 2.8 \times 10^{-9}/0.010 = 2.8 \times 10^{-7}\ M$

Because ΔH° is negative (exothermic), the solubility of $CaCO_3$ is lower at 50 °C.

17.154 (a) $I_2(s) \rightarrow 2\,I^-(aq)$

$[I_2(s) + 2\,e^- \rightarrow 2\,I^-(aq)] \times 5$ reduction half reaction

$I_2(s) \rightarrow 2\,IO_3^-(aq)$

$I_2(s) + 6\,H_2O(l) \rightarrow 2\,IO_3^-(aq)$

$I_2(s) + 6\,H_2O(l) \rightarrow 2\,IO_3^-(aq) + 12\,H^+(aq)$

$I_2(s) + 6\,H_2O(l) \rightarrow 2\,IO_3^-(aq) + 12\,H^+(aq) + 10\,e^-$ oxidation half reaction

Combine the two half reactions.

$6\,I_2(s) + 6\,H_2O(l) \rightarrow 10\,I^-(aq) + 2\,IO_3^-(aq) + 12\,H^+(aq)$

Divide all coefficients by 2.

$3\,I_2(s) + 3\,H_2O(l) \rightarrow 5\,I^-(aq) + IO_3^-(aq) + 6\,H^+(aq)$

$3\,I_2(s) + 3\,H_2O(l) + 6\,OH^-(aq) \rightarrow 5\,I^-(aq) + IO_3^-(aq) + 6\,H^+(aq) + 6\,OH^-(aq)$

$3\,I_2(s) + 3\,H_2O(l) + 6\,OH^-(aq) \rightarrow 5\,I^-(aq) + IO_3^-(aq) + 6\,H_2O(l)$

$3\,I_2(s) + 6\,OH^-(aq) \rightarrow 5\,I^-(aq) + IO_3^-(aq) + 3\,H_2O(l)$

(b) $\Delta G^\circ = [5\,\Delta G^\circ_f(I^-) + \Delta G^\circ_f(IO_3^-) + 3\,\Delta G^\circ_f(H_2O(l))] - 6\,\Delta G^\circ_f(OH^-)$

$\Delta G^\circ = [(5\text{ mol})(-51.6\text{ kJ/mol}) + (1\text{ mol})(-128.0\text{ kJ/mol}) + (3\text{ mol})(-237.2\text{ kJ/mol})]$
$- (6\text{ mol})(-157.3\text{ kJ/mol}) = -153.8\text{ kJ}$

(c) The reaction is spontaneous because ΔG° is negative.

(d) 25 °C = 25 + 273 = 298 K

$\Delta G^\circ = -RT \ln K_c$

$$\ln K_c = \frac{-\Delta G^\circ}{RT} = \frac{-(-153.8\text{ kJ/mol})}{[8.314 \times 10^{-3}\text{ kJ/(K}\cdot\text{mol)}](298\text{ K})} = 62.077$$

$K_c = e^{62.077} = 9.1 \times 10^{26}$

$$K_c = \frac{[I^-]^5[IO_3^-]}{[OH^-]^6} = 9.1 \times 10^{26} = \frac{(0.10)^5(0.50)}{[OH^-]^6}$$

$$[OH^-] = \sqrt[6]{\frac{(0.10)^5(0.50)}{9.1 \times 10^{26}}} = 4.2 \times 10^{-6}\text{ M};\quad [H_3O^+] = \frac{1.0 \times 10^{-14}}{4.2 \times 10^{-6}} = 2.38 \times 10^{-9}\text{ M}$$

$pH = -\log[H_3O^+] = -\log(2.38 \times 10^{-9}) = 8.62$

18 Electrochemistry

18.1 (a) $MnO_4^-(aq) \rightarrow MnO_2(s)$ (reduction)
$IO_3^-(aq) \rightarrow IO_4^-(aq)$ (oxidation)
(b) $NO_3^-(aq) \rightarrow NO_2(g)$ (reduction)
$SO_2(aq) \rightarrow SO_4^{2-}(aq)$ (oxidation)

18.2 $ClO^-(aq) \rightarrow Cl^-(aq)$ Cl goes from +1 to –1 (reduction)
$I^-(aq) \rightarrow I_2(aq)$ I goes from –1 to 0 (oxidation)

$ClO^-(aq) + I^-(aq) \rightarrow Cl^-(aq) + I_2(aq)$ overall reaction (unbalanced)

18.3 $Fe(OH)_2(s) + O_2(g) \rightarrow Fe(OH)_3(s)$
$[Fe(OH)_2(s) + OH^-(aq) \rightarrow Fe(OH)_3(s) + e^-] \times 4$ (oxidation half reaction)

$O_2(g) \rightarrow 2\,H_2O(l)$
$4\,H^+(aq) + O_2(g) \rightarrow 2\,H_2O(l)$
$4\,e^- + 4\,H^+(aq) + O_2(g) \rightarrow 2\,H_2O(l)$
$4\,e^- + 4\,H^+(aq) + 4\,OH^-(aq) + O_2(g) \rightarrow 2\,H_2O(l) + 4\,OH^-(aq)$
$4\,e^- + 4\,H_2O(l) + O_2(g) \rightarrow 2\,H_2O(l) + 4\,OH^-(aq)$
$4\,e^- + 2\,H_2O(l) + O_2(g) \rightarrow 4\,OH^-(aq)$ (reduction half reaction)

Combine the two half reactions.
$4\,Fe(OH)_2(s) + 4\,OH^-(aq) + 2\,H_2O(l) + O_2(g) \rightarrow 4\,Fe(OH)_3(s) + 4\,OH^-(aq)$
$4\,Fe(OH)_2(s) + 2\,H_2O(l) + O_2(g) \rightarrow 4\,Fe(OH)_3(s)$

18.4 $NO_3^-(aq) + Cu(s) \rightarrow NO(g) + Cu^{2+}(aq)$
$[Cu(s) \rightarrow Cu^{2+}(aq) + 2\,e^-] \times 3$ (oxidation half reaction)

$NO_3^-(aq) \rightarrow NO(g)$
$NO_3^-(aq) \rightarrow NO(g) + 2\,H_2O(l)$
$4\,H^+(aq) + NO_3^-(aq) \rightarrow NO(g) + 2\,H_2O(l)$
$[3\,e^- + 4\,H^+(aq) + NO_3^-(aq) \rightarrow NO(g) + 2\,H_2O(l)] \times 2$ (reduction half reaction)

Combine the two half reactions.
$2\,NO_3^-(aq) + 8\,H^+(aq) + 3\,Cu(s) \rightarrow 3\,Cu^{2+}(aq) + 2\,NO(g) + 4\,H_2O(l)$

18.5 $2\,Ag^+(aq) + Ni(s) \rightarrow 2\,Ag(s) + Ni^{2+}(aq)$
There is a Ni anode in an aqueous solution of Ni^{2+}, and a Ag cathode in an aqueous solution of Ag^+. A salt bridge connects the anode and cathode compartment. The electrodes are connected through an external circuit.

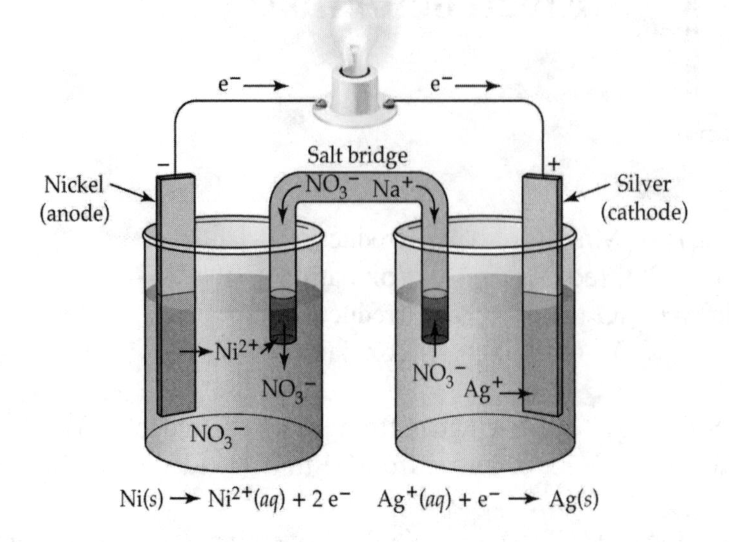

18.6

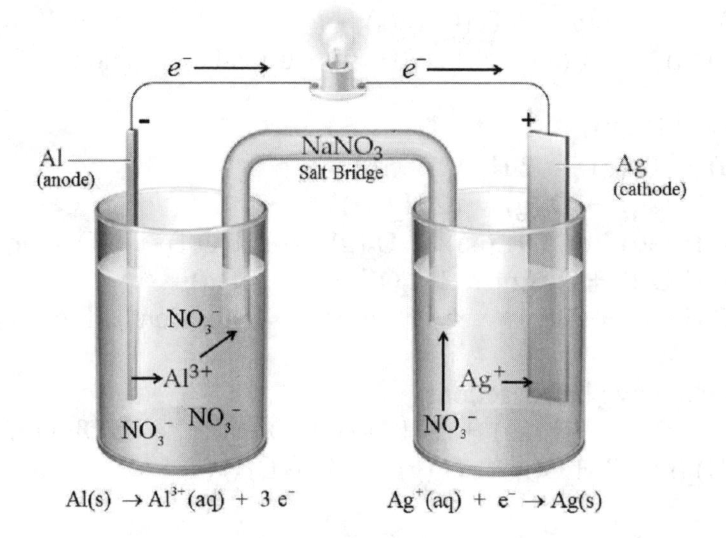

anode reaction	$Al(s) \rightarrow Al^{3+}(aq) + 3\,e^-$
cathode reaction	$\underline{3\,Ag^+(aq) + 3\,e^- \rightarrow 3\,Ag(s)}$
overall reaction	$Al(s) + 3\,Ag^+(aq) \rightarrow Al^{3+}(aq) + 3\,Ag(s)$

18.7 $Pb(s) + Br_2(l) \rightarrow Pb^{2+}(aq) + 2\,Br^-(aq)$
There is a Pb anode in an aqueous solution of Pb^{2+}. The cathode is a Pt wire that dips into a pool of liquid Br_2 and an aqueous solution that is saturated with Br_2. A salt bridge connects the anode and cathode compartment. The electrodes are connected through an external circuit.

18.8 (a) and (b)

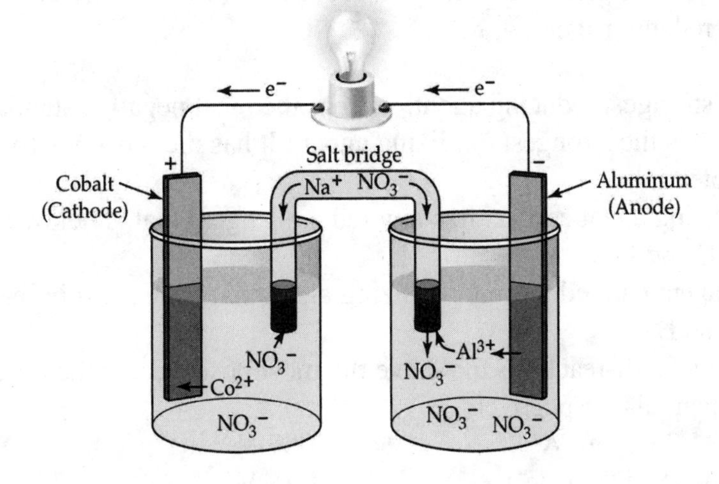

(c) $2\ Al(s) + 3\ Co^{2+}(aq) \rightarrow 2\ Al^{3+}(aq) + 3\ Co(s)$

(d) $Al(s)|Al^{3+}(aq)||Co^{2+}(aq)|Co(s)$

18.9 $Cr_2O_7^{2-}(aq) + 3\ Sn^{2+}(aq) + 14\ H^+(aq) \rightarrow 2\ Cr^{3+}(aq) + 3\ Sn^{4+}(aq) + 7\ H_2O(l)$

$n = 6$ mol e^-

$$\Delta G^\circ = -nFE^\circ = -(6\text{ mol e}^-)\left(\frac{96{,}500\text{ C}}{1\text{ mol e}^-}\right)(1.21\text{ V})\left(\frac{1\text{ J}}{1\text{ C}\cdot\text{V}}\right) = -700{,}590\text{ J} = -701\text{ kJ}$$

18.10 $Hg(l) + I_2(s) \rightarrow Hg^{2+}(aq) + 2\ I^-(aq)$

$n = 2$ mol e^- and 1 J/C $= 1$V

$\Delta G^\circ = 59.8$ kJ $= 59{,}800$ J $= -nFE^\circ$

$$E^\circ = \frac{-\Delta G^\circ}{nF} = \frac{-(59{,}800\text{ J})}{(2\text{ mol e}^-)\left(\frac{96{,}500\text{ C}}{1\text{ mol e}^-}\right)} = -0.310\text{ J/C} = -0.310\text{ V}$$

Because $E^\circ < 0$, the reaction is nonspontaneous.

18.11

oxidation:	$Al(s) \rightarrow Al^{3+}(aq) + 3\ e^-$	$E^\circ = 1.66$ V
reduction:	$Cr^{3+}(aq) + 3\ e^- \rightarrow Cr(s)$	$E^\circ = ?$
overall	$Al(s) + Cr^{3+}(aq) \rightarrow Al^{3+}(aq) + Cr(s)$	$E^\circ = 0.92$ V

The standard reduction potential for the Cr^{3+}/Cr half cell is:

$E^\circ = 0.92 - 1.66 = -0.74$ V

18.12 (a) $Cl_2(g) + 2\ e^- \rightarrow 2\ Cl^-(aq)$ $E^\circ = 1.36$ V

$Ag^+(aq) + e^- \rightarrow Ag(s)$ $E^\circ = 0.80$ V

Cl_2 has the greater tendency to be reduced (larger E°). The species that has the greater tendency to be reduced is the stronger oxidizing agent. Cl_2 is the stronger oxidizing agent.

(b) $Fe^{2+}(aq) + 2\ e^- \rightarrow Fe(s)$ $E^\circ = -0.45$ V

$Mg^{2+}(aq) + 2\ e^- \rightarrow Mg(s)$ $E^\circ = -2.37$ V

The second half-reaction has the lesser tendency to occur in the forward direction (more negative E°) and the greater tendency to occur in the reverse direction. Therefore, Mg is the stronger reducing agent.

18.13 (a) D is the strongest reducing agent. D^+ has the most negative standard reduction potential. A^{3+} is the strongest oxidizing agent. It has the most positive standard reduction potential.
(b) An oxidizing agent can oxidize any reducing agent that is below it in the table. B^{2+} can oxidize C and D.
A reducing agent can reduce any oxidizing agent that is above it in the table. C can reduce A^{3+} and B^{2+}.
(c) Use the two half-reactions that have the most positive and the most negative standard reduction potentials, respectively.

$$A^{3+} + 2e^- \rightarrow A^+ \qquad 1.47\ V$$
$$\underline{2 \times (D \rightarrow D^+ + e^-)} \qquad \underline{1.38\ V}$$
$$A^{3+} + 2D \rightarrow A^+ + 2D^+ \qquad 2.85\ V$$

18.14 (a) $2\,Fe^{3+}(aq) + 2\,I^-(aq) \rightarrow 2\,Fe^{2+}(aq) + I_2(s)$

reduction: $Fe^{3+}(aq) + e^- \rightarrow Fe^{2+}(aq)$ $E° = 0.77\ V$
oxidation: $2\,I^-(aq) \rightarrow I_2(s) + 2e^-$ $\underline{E° = -0.54\ V}$
overall $E° = 0.23\ V$

Because E° for the overall reaction is positive, this reaction can occur under standard-state conditions.

(b) $3\,Ni(s) + 2\,Al^{3+}(aq) \rightarrow 3\,Ni^{2+}(aq) + 2\,Al(s)$

oxidation: $Ni(s) \rightarrow Ni^{2+}(aq) + 2e^-$ $E° = 0.26\ V$
reduction: $Al^{3+}(aq) + 3e^- \rightarrow Al(s)$ $\underline{E° = -1.66\ V}$
overall $E° = -1.40\ V$

Because E° for the overall reaction is negative, this reaction cannot occur under standard-state conditions. This reaction can occur in the reverse direction.

18.15 (a) $Ni(s) + 2\,Ag^+(aq) \rightarrow Ni^{2+}(aq) + 2\,Ag(s)$

oxidation: $Ni(s) \rightarrow Ni^{2+}(aq) + 2e^-$ $E° = 0.26\ V$
reduction: $Ag^+(aq) + e^- \rightarrow Ag(s)$ $\underline{E° = 0.80\ V}$
overall $E° = 1.06\ V$

(b) $Ni(s)|Ni^{2+}(aq)(1.0\ M)\|Ag^+(aq)(1.0\ M)|Ag(s)$
(c) Ni(s) is the anode and Ag(s) is the cathode.

18.16 $Cu(s) + 2\,Fe^{3+}(aq) \rightarrow Cu^{2+}(aq) + 2\,Fe^{2+}(aq)$

$E° = E°_{Cu \rightarrow Cu^{2+}} + E°_{Fe^{3+} \rightarrow Fe^{2+}} = -0.34\ V + 0.77\ V = 0.43\ V;$ $n = 2$ mol e^-

$$E = E° - \frac{0.0592\ V}{n}\log\frac{[Cu^{2+}][Fe^{2+}]^2}{[Fe^{3+}]^2} = 0.43\ V - \frac{(0.0592\ V)}{2}\log\frac{(0.25)(0.20)^2}{(1.0 \times 10^{-4})^2} = 0.25\ V$$

18.17 (a) anode: $4[Al(s) \rightarrow Al^{3+}(aq) + 3e^-]$ $E° = 1.66\ V$
cathode: $\underline{3[O_2(g) + 4H^+(aq) + 4e^- \rightarrow 2H_2O(l)]}$ $\underline{E° = 1.23\ V}$
overall: $4\,Al(s) + 3\,O_2(g) + 12\,H^+(aq) \rightarrow 4\,Al^{3+}(aq) + 6\,H_2O(l)$ $E° = 2.89\ V$

(b) & (c) $E = E^\circ - \frac{2.303\,RT}{nF}\log\frac{[Al^{3+}]^4}{(P_{O_2})^3[H^+]^{12}}$

$$E = 2.89\text{ V} - \frac{(2.303)\left(8.314\frac{J}{K\cdot mol}\right)(310\text{ K})}{(12\text{ mol e}^-)(96{,}500\text{ C/mol e}^-)}\log\left(\frac{(1.0\times10^{-9})^4}{(0.20)^3(1.0\times10^{-7})^{12}}\right)$$

$E = 2.89\text{ V} - 0.257\text{ V} = 2.63\text{ V}$

18.18 $5\,[Cu(s) \rightarrow Cu^{2+}(aq) + 2\,e^-]$ (oxidation half reaction)
$2\,[5\,e^- + 8\,H^+(aq) + MnO_4^-(aq) \rightarrow Mn^{2+}(aq) + 4\,H_2O(l)]$ (reduction half reaction)

$5\,Cu(s) + 16\,H^+(aq) + 2\,MnO_4^-(aq) \rightarrow 5\,Cu^{2+}(aq) + 2\,Mn^{2+}(aq) + 8\,H_2O(l)$

$$\Delta E = -\frac{0.0592\text{ V}}{n}\log\frac{[Cu^{2+}]^5[Mn^{2+}]^2}{[MnO_4^-]^2[H^+]^{16}}$$

(a) The anode compartment contains Cu^{2+}.

$$\Delta E = -\frac{0.0592\text{ V}}{10}\log\frac{(0.01)^5(1)^2}{(1)^2(1)^{16}} = +0.059\text{ V}$$

(b) The cathode compartment contains Mn^{2+}, MnO_4^-, and H^+.

$$\Delta E = -\frac{0.0592\text{ V}}{10}\log\frac{(1)^5(0.01)^2}{(0.01)^2(0.01)^{16}} = -0.19\text{ V}$$

18.19 $Zn(s) + Cu^{2+}(aq) \rightarrow Zn^{2+}(aq) + Cu(s)$

oxidation: $Zn(s) \rightarrow Zn^{2+}(aq) + 2\,e^-$ $E^\circ = 0.76\text{ V}$
reduction: $Cu^{2+}(aq) + 2\,e^- \rightarrow Cu(s)$ $E^\circ = 0.34\text{ V}$
overall $E^\circ = 1.10\text{ V}$

$$E = 1.16\text{ V} = E^\circ - \frac{0.0592\text{ V}}{n}\log\frac{[Zn^{2+}]}{[Cu^{2+}]} = 1.10\text{ V} - \frac{(0.0592\text{ V})}{2}\log\frac{[Zn^{2+}]}{[Cu^{2+}]}$$

$$\frac{(1.16\text{ V} - 1.10\text{ V})}{(-0.0592\text{ V}/2)} = \log\frac{[Zn^{2+}]}{[Cu^{2+}]}$$

$$\log\frac{[Zn^{2+}]}{[Cu^{2+}]} = -2.03 \quad\text{and}\quad \frac{[Zn^{2+}]}{[Cu^{2+}]} = 10^{-2.03} = 9.3\times10^{-3}$$

18.20 $H_2(g) + Pb^{2+}(aq) \rightarrow 2\,H^+(aq) + Pb(s)$

$E^\circ = E^\circ_{H_2\rightarrow H^+} + E^\circ_{Pb^{2+}\rightarrow Pb} = 0\text{ V} + (-0.13\text{ V}) = -0.13\text{ V}$; $n = 2\text{ mol e}^-$

$$E = E^\circ - \frac{0.0592\text{ V}}{n}\log\frac{[H_3O^+]^2}{[Pb^{2+}](P_{H_2})}$$

$$0.28\text{ V} = -0.13\text{ V} - \frac{(0.0592\text{ V})}{2}\log\frac{[H_3O^+]^2}{(1)(1)} = -0.13\text{ V} - (0.0592\text{ V})\log[H_3O^+]$$

$pH = -\log[H_3O^+]$ therefore $0.28\ V = -0.13\ V + (0.0592\ V)\ pH$

$$pH = \frac{(0.28\ V\ +\ 0.13\ V)}{0.0592\ V} = 6.9$$

18.21 $H_2(g)\ +\ Hg_2Cl_2(s)\ \rightarrow\ 2\ H^+(aq)\ +\ 2\ Hg(l)\ +\ 2Cl^-(aq)$

$E^o = E^o_{H_2 \rightarrow H^+} + E^o_{Hg_2Cl_2 \rightarrow Hg} = 0\ V + 0.28\ V = 0.28\ V;\qquad n = 2\ mol\ e^-$

$$E = E^o - \frac{0.0592\,V}{n}\log\frac{[H_3O^+]^2[Cl^-]^2}{(P_{H_2})}$$

$$E = 0.28\ V - \frac{0.0592\,V}{2}\log\frac{(1.0 \times 10^{-7})^2(1.0)^2}{(1.0)} = 0.69\ V$$

18.22 $4\ Fe^{2+}(aq)\ +\ O_2(g)\ +\ 4\ H^+(aq)\ \rightarrow\ 4\ Fe^{3+}(aq)\ +\ 2\ H_2O(l)$

$E^o = E^o_{Fe^{2+} \rightarrow Fe^{3+}} + E^o_{O_2 \rightarrow H_2O} = -0.77\ V + 1.23\ V = 0.46\ V;\qquad n = 4\ mol\ e^-$

$$E^o = \frac{0.0592\ V}{n}\log K;\quad \log K = \frac{nE^o}{0.0592\ V} = \frac{(4)(0.46\ V)}{0.0592\ V} = 31;\quad K = 10^{31}\ \text{at}\ 25\ ^oC$$

18.23 $$E^o = \frac{0.0592\,V}{n}\log K = \frac{0.0592\,V}{2}\log(1.8 \times 10^{-5}) = -0.140\ V$$

18.24 (a) $Zn(s) + 2\ MnO_2(s) + 2\ NH_4^+(aq)\ \rightarrow\ Zn^{2+}(aq) + Mn_2O_3(s) + 2\ NH_3(aq) + H_2O(l)$

(b) $Zn(s)\ +\ 2\ MnO_2(s)\ \rightarrow\ ZnO(s)\ +\ Mn_2O_3(s)$

(c) $Cd(s) + 2\ NiO(OH)(s) + 2\ H_2O(l)\ \rightarrow\ Cd(OH)_2(s) + 2\ Ni(OH)_2(s)$

(d) $x\ Li(s)\ +\ MnO_2(s)\ \rightarrow\ Li_xMnO_2(s)$

(e) $Li_xC_6(s)\ +\ Li_{1-x}CoO_2(s)\ \rightarrow\ 6\ C(s)\ +\ LiCoO_2(s)$

18.25 (a) $[Mg(s)\ \rightarrow\ Mg^{2+}(aq)\ +\ 2\ e^-]\ x\ 2$

$\underline{O_2(g)\ +\ 4\ H^+(aq)\ +4\ e^-\ \rightarrow\ 2\ H_2O(l)}$

$2\ Mg(s)\ +\ O_2(g)\ +\ 4\ H^+(aq)\ \rightarrow\ 2\ Mg^{2+}(aq)\ +\ 2\ H_2O(l)$

(b) $[Fe(s)\ \rightarrow\ Fe^{2+}(aq)\ +\ 2\ e^-]\ x\ 4$

$[O_2(g)\ +\ 4\ H^+(aq)\ +4\ e^-\ \rightarrow\ 2\ H_2O(l)]\ x\ 2$

$4\ Fe^{2+}(aq)\ +\ O_2(g)\ +\ 4\ H^+(aq)\ \rightarrow\ 4\ Fe^{3+}(aq)\ +\ 2\ H_2O(l)$

$\underline{[2\ Fe^{3+}(aq)\ +\ 4\ H_2O(l)\ \rightarrow\ Fe_2O_3 \cdot H_2O(s)\ +\ 6\ H^+(aq)]\ x\ 2}$

$4\ Fe(s)\ +\ 3\ O_2(g)\ +\ 2\ H_2O(l)\ \rightarrow\ 2\ Fe_2O_3 \cdot H_2O(s)$

18.26 (a)

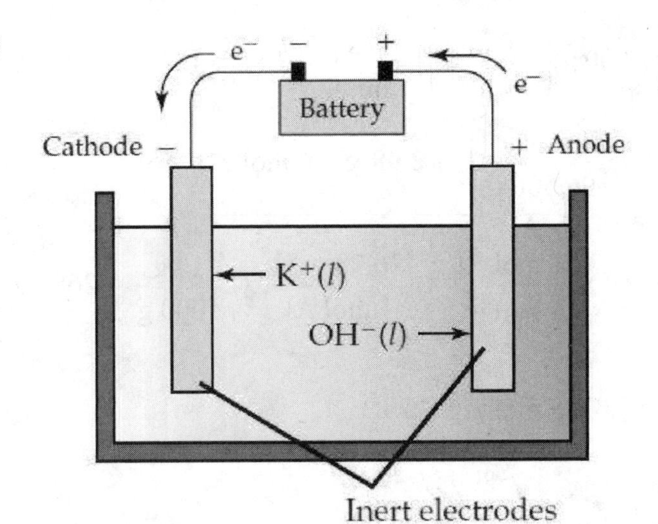

(b) anode reaction $4\ OH^-(l) \rightarrow O_2(g) + 2\ H_2O(l) + 4\ e^-$
cathode reaction $4\ K^+(l) + 4\ e^- \rightarrow 4\ K(l)$
overall reaction $4\ K^+(l) + 4\ OH^-(l) \rightarrow 4\ K(l) + O_2(g) + 2\ H_2O(l)$

18.27 (a) anode reaction $2\ Cl^-(aq) \rightarrow Cl_2(g) + 2\ e^-$
cathode reaction $2\ H_2O(l) + 2\ e^- \rightarrow H_2(g) + 2\ OH^-(aq)$
overall reaction $2\ Cl^-(aq) + 2\ H_2O(l) \rightarrow Cl_2(g) + H_2(g) + 2\ OH^-(aq)$

(b) anode reaction $2\ H_2O(l) \rightarrow O_2(g) + 4\ H^+(aq) + 4\ e^-$
cathode reaction $2\ Cu^{2+}(aq) + 4\ e^- \rightarrow 2\ Cu(s)$
overall reaction $2\ Cu^{2+}(aq) + 2\ H_2O(l) \rightarrow 2\ Cu(s) + O_2(g) + 4\ H^+(aq)$

(c) anode reaction $2\ H_2O(l) \rightarrow O_2(g) + 4\ H^+(aq) + 4\ e^-$
cathode reaction $4\ H_2O(l) + 4\ e^- \rightarrow 2\ H_2(g) + 4\ OH^-(aq)$
overall reaction $2\ H_2O(l) \rightarrow 2\ H_2(g) + O_2(g)$

18.28

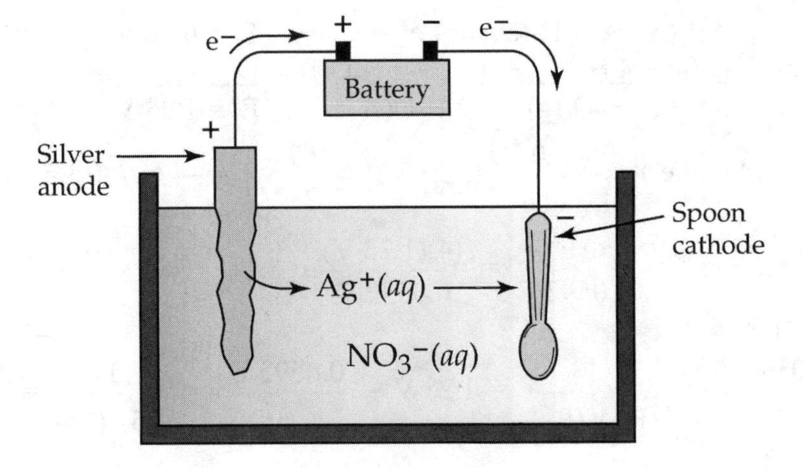

anode reaction $Ag(s) \rightarrow Ag^+(aq) + e^-$
cathode reaction $Ag^+(aq) + e^- \rightarrow Ag(s)$
The overall reaction is transfer of silver metal from the silver anode to the spoon.

18.29 Charge = $\left(1.00 \text{ x } 10^5 \frac{C}{s}\right)(8.00 \text{ h})\left(\frac{60 \text{ min}}{h}\right)\left(\frac{60 \text{ s}}{\text{min}}\right) = 2.88 \text{ x } 10^9 \text{ C}$

Moles of e^- = $(2.88 \text{ x } 10^9 \text{ C})\left(\frac{1 \text{ mol } e^-}{96{,}500 \text{ C}}\right) = 2.98 \text{ x } 10^4 \text{ mol } e^-$

cathode reaction: $Al^{3+} + 3 e^- \rightarrow Al$

mass Al = $(2.98 \text{ x } 10^4 \text{ mol } e^-) \text{ x } \frac{1 \text{ mol Al}}{3 \text{ mol } e^-} \text{ x } \frac{26.98 \text{ g Al}}{1 \text{ mol Al}} \text{ x } \frac{1 \text{ kg}}{1000 \text{ g}} = 268 \text{ kg Al}$

18.30 $3.00 \text{ g Ag x } \frac{1 \text{ mol Ag}}{107.9 \text{ g Ag}} = 0.0278 \text{ mol Ag}$

cathode reaction: $Ag^+(aq) + e^- \rightarrow Ag(s)$

Charge = $(0.0278 \text{ mol Ag})\left(\frac{1 \text{ mol } e^-}{1 \text{ mol Ag}}\right)\left(\frac{96{,}500 \text{ C}}{1 \text{ mol } e^-}\right) = 2682.7 \text{ C}$

Time = $\frac{C}{A} = \left(\frac{2682.7 \text{ C}}{0.100 \text{ C/s}} \text{ x } \frac{1 \text{ h}}{3600 \text{ s}}\right) = 7.45 \text{ h}$

18.31 A fuel cell and a battery are both galvanic cells that convert chemical energy into electrical energy utilizing a spontaneous redox reaction. A fuel cell differs from an ordinary battery in that the reactants are not contained within the cell but instead are continuously supplied from an external reservoir.

18.32 (a)

anode reaction	$2 H_2(g) \rightarrow 4 H^+(aq) + 4 e^-$	$E^\circ = 0.00 \text{ V}$
cathode reaction	$\underline{O_2(g) + 4 H^+(aq) + 4 e^- \rightarrow 2 H_2O(l)}$	$\underline{E^\circ = 1.23 \text{ V}}$
overall reaction	$2 H_2(g) + O_2(g) \rightarrow 2 H_2O(l)$	$E^\circ = 1.23 \text{ V}$

(b) $E = E^\circ - \frac{0.0592 \text{ V}}{n} \log \frac{1}{(P_{H_2})^2(P_{O_2})} = 1.23 \text{ V} - \frac{0.0592 \text{ V}}{4} \log \frac{1}{(6)^2(0.2)} = 1.24 \text{ V}$

18.33 (a)

anode reaction	$2 H_2(g) \rightarrow 4 H^+(aq) + 4 e^-$	$E^\circ = 0.00 \text{ V}$
cathode reaction	$\underline{O_2(g) + 4 H^+(aq) + 4 e^- \rightarrow 2 H_2O(l)}$	$\underline{E^\circ = 1.23 \text{ V}}$
overall reaction	$2 H_2(g) + O_2(g) \rightarrow 2 H_2O(l)$	$E^\circ = 1.23 \text{ V}$

(b) $\Delta G^\circ = -nFE^\circ = -(4 \text{ mol } e^-)\left(\frac{96{,}500 \text{ C}}{1 \text{ mol } e^-}\right)(1.23 \text{ V})\left(\frac{1 \text{ J}}{1 \text{ C} \cdot \text{V}}\right) = -474{,}780 \text{ J} = -475 \text{ kJ}$

$E^\circ = \frac{0.0592 \text{ V}}{n} \log K; \ \log K = \frac{nE^\circ}{0.0592 \text{ V}} = \frac{(4)(1.23 \text{ V})}{0.0592 \text{ V}} = 83.1$

$K = 10^{83.1} = 1.28 \text{ x } 10^{83}$

(c) $E = E^\circ - \frac{0.0592 \text{ V}}{n} \log \frac{1}{(P_{H_2})^2(P_{O_2})} = 1.23 \text{ V} - \frac{0.0592 \text{ V}}{4} \log \frac{1}{(25)^2(25)} = 1.29 \text{ V}$

18.34 $2 CH_3OH(l) + 3 O_2(g) \rightarrow 2 CO_2(g) + 4 H_2O(l)$

$\Delta G^\circ = [2 \Delta G^\circ_f(CO_2) + 4 \Delta G^\circ_f(H_2O)] - [2 \Delta G^\circ_f(CH_3OH)]$

$\Delta G^\circ = [(2 \text{ mol})(-394.4 \text{ kJ/mol}) + (4 \text{ mol})(-237.2 \text{ kJ/mol})] - (2 \text{ mol})(-166.6 \text{ kJ/mol})$

$\Delta G^\circ = -1404$ kJ

anode: $2\ CH_3OH(l) + 2\ H_2O(l) \rightarrow 2\ CO_2(g) + 12\ H^+(aq) + 12\ e^-$

cathode: $3\ O_2(g) + 12\ H^+(aq) + 12\ e^- \rightarrow 6\ H_2O(l)$

n = 12 mol e^- and 1 J = 1 C x 1 V

$\Delta G^\circ = -nFE^\circ$

$$E^\circ = \frac{-\Delta G^\circ}{nF} = \frac{-(-1{,}404{,}000\ J)}{(12\ mol\ e^-)\left(\frac{96{,}500\ C}{1\ mol\ e^-}\right)} = +1.21\ J/C = +1.21\ V$$

$$E^\circ = \frac{0.0592\ V}{n}\log K$$

$$\log K = \frac{nE^\circ}{0.0592\ V} = \frac{(12)(1.21\ V)}{0.0592\ V} = 245;\ K = 10^{245} = 1 \times 10^{245}$$

18.35 (a) H is reduced and C is oxidized.

(b) H is reduced and C is oxidized. H_2O is the oxidizing agent and CO is the reducing agent.

(c) The drawback of the process is the high reaction temperatures, which require a lot of energy and the greenhouse gas CO_2 is also produced.

18.36 $6\ H_2O(l) \rightarrow 2\ H_2(g) + O_2(g) + 4\ H^+(aq) + 4\ OH^-(aq)$

1 A = 1 C/s

(a) $$\text{mol } e^- = 250.0\ \frac{C}{s} \times 30\ min \times \frac{60\ s}{min} \times \frac{1\ mol\ e^-}{96{,}500\ C} = 4.66\ mol\ e^-$$

$$\text{mass } H_2 = 4.66\ mol\ e^- \times \frac{2\ mol\ H_2}{4\ mol\ e^-} \times \frac{2.02\ g\ H_2}{1\ mol\ H_2} = 4.71\ g\ H_2$$

(b) $$\text{charge} = 25\ mol\ O_2 \times \frac{4\ mol\ e^-}{1\ mol\ O_2} \times \frac{96{,}500\ C}{1\ mol\ e^-} = 9.65 \times 10^6\ C$$

$$\text{time} = \frac{9.65 \times 10^6\ C}{500.0\ C/s} \times \frac{1\ min}{60\ s} \times \frac{1\ h}{60\ min} = = 5.36\ h$$

Conceptual Problems

18.38 (a) anode is Ni; cathode is Pt

(b) anode reaction $3\ Ni(s) \rightarrow 3\ Ni^{2+}(aq) + 6\ e^-$

cathode reaction $Cr_2O_7^{2-}(aq) + 14\ H^+(aq) + 6\ e^- \rightarrow 2\ Cr^{3+}(aq) + 7\ H_2O(l)$

overall reaction $Cr_2O_7^{2-}(aq) + 3\ Ni(s) + 14\ H^+(aq) \rightarrow 2\ Cr^{3+}(aq) + 3\ Ni^{2+}(aq) + 7\ H_2O(l)$

(c) $Ni(s)|Ni^{2+}(aq)\|Cr_2O_7^{2-}(aq), Cr^{3+}|Pt(s)$

18.40 (a) - (b)

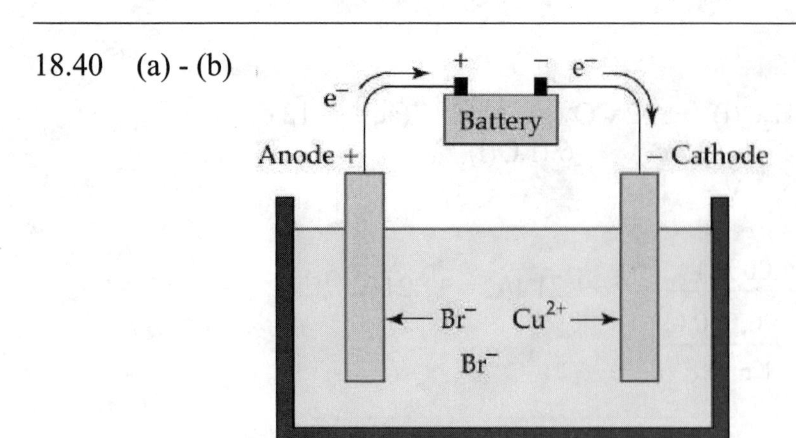

(c) anode reaction $2\ Br^-(aq) \rightarrow Br_2(aq) + 2\ e^-$

cathode reaction $Cu^{2+}(aq) + 2\ e^- \rightarrow Cu(s)$

overall reaction $Cu^{2+}(aq) + 2\ Br^-(aq) \rightarrow Cu(s) + Br_2(aq)$

18.42 (a) & (b)

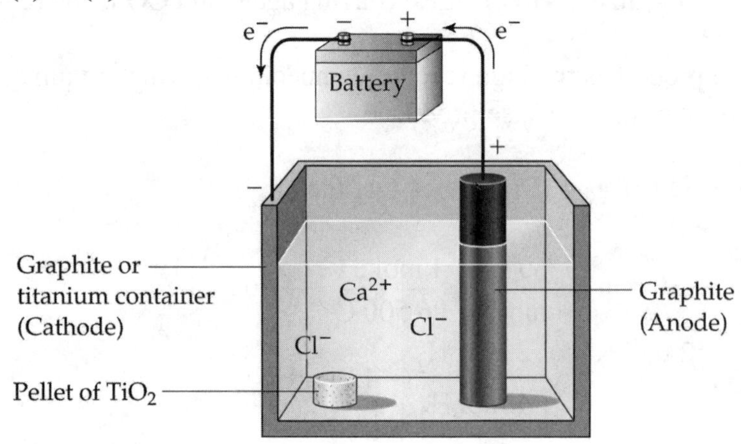

(c) anode reaction $2\ O^{2-} \rightarrow O_2(g) + 4\ e^-$

cathode reaction $TiO_2(s) + 4\ e^- \rightarrow Ti(s) + 2\ O^{2-}$

overall reaction $TiO_2(s) \rightarrow Ti(s) + O_2(g)$

18.44 $Cu(s) + 2\ Ag^+(aq) \rightarrow Cu^{2+}(aq) + 2\ Ag(s)$; $E = E^\circ - \dfrac{0.0592\ V}{2} \log \dfrac{[Cu^{2+}]}{[Ag^+]^2}$

(a) E decreases because addition of NaCl precipitates AgCl, which decreases $[Ag^+]$ and increases $\log \dfrac{[Cu^{2+}]}{[Ag^+]^2}$.

(b) E increases because addition of NaCl increases the volume, which decreases $[Cu^{2+}]$ and decreases $\log \dfrac{[Cu^{2+}]}{[Ag^+]^2}$.

(c) E decreases because addition of NH_3 complexes Ag^+, yielding $Ag(NH_3)_2{}^+$, which decreases $[Ag^+]$ and increases $\log \dfrac{[Cu^{2+}]}{[Ag^+]^2}$.

(d) E increases because addition of NH_3 complexes Cu^{2+}, yielding $Cu(NH_3)_4^{2+}$, which decreases $[Cu^{2+}]$ and decreases $\log \frac{[Cu^{2+}]}{[Ag^+]^2}$.

Section Problems

Balancing Redox Reactions (Section 18.1)

18.46 (a) N oxidation number decreases from +5 to +2; reduction.
(b) Zn oxidation number increases from 0 to +2; oxidation.
(c) Ti oxidation number increases from +3 to +4; oxidation.
(d) Sn oxidation number decreases from +4 to +2; reduction.

18.48 (a) $NO_3^-(aq) \rightarrow NO(g)$
$NO_3^-(aq) \rightarrow NO(g) + 2\ H_2O(l)$
$4\ H^+(aq) + NO_3^-(aq) \rightarrow NO(g) + 2\ H_2O(l)$
$3\ e^- + 4\ H^+(aq) + NO_3^-(aq) \rightarrow NO(g) + 2\ H_2O(l)$

(b) $Zn(s) \rightarrow Zn^{2+}(aq) + 2\ e^-$

(c) $Ti^{3+}(aq) \rightarrow TiO_2(s)$
$Ti^{3+}(aq) + 2\ H_2O(l) \rightarrow TiO_2(s)$
$Ti^{3+}(aq) + 2\ H_2O(l) \rightarrow TiO_2(s) + 4\ H^+(aq)$
$Ti^{3+}(aq) + 2\ H_2O(l) \rightarrow TiO_2(s) + 4\ H^+(aq) + e^-$

(d) $Sn^{4+}(aq) + 2\ e^- \rightarrow Sn^{2+}(aq)$

18.50 (a) $Te(s) + NO_3^-(aq) \rightarrow TeO_2(s) + NO(g)$
oxidation: $Te(s) \rightarrow TeO_2(s)$
reduction: $NO_3^-(aq) \rightarrow NO(g)$

(b) $H_2O_2(aq) + Fe^{2+}(aq) \rightarrow Fe^{3+}(aq) + H_2O(l)$
oxidation: $Fe^{2+}(aq) \rightarrow Fe^{3+}(aq)$
reduction: $H_2O_2(aq) \rightarrow H_2O(l)$

18.52 (a) $Cr_2O_7^{2-}(aq) \rightarrow Cr^{3+}(aq)$
$Cr_2O_7^{2-}(aq) \rightarrow 2\ Cr^{3+}(aq)$
$Cr_2O_7^{2-}(aq) \rightarrow 2\ Cr^{3+}(aq) + 7\ H_2O(l)$
$14\ H^+(aq) + Cr_2O_7^{2-}(aq) \rightarrow 2\ Cr^{3+}(aq) + 7\ H_2O(l)$
$14\ H^+(aq) + Cr_2O_7^{2-}(aq) + 6\ e^- \rightarrow 2\ Cr^{3+}(aq) + 7\ H_2O(l)$

(b) $CrO_4^{2-}(aq) \rightarrow Cr(OH)_4^-(aq)$
$4\ H^+(aq) + CrO_4^{2-}(aq) \rightarrow Cr(OH)_4^-(aq)$
$4\ H^+(aq) + 4\ OH^-(aq) + CrO_4^{2-}(aq) \rightarrow Cr(OH)_4^-(aq) + 4\ OH^-(aq)$
$4\ H_2O(l) + CrO_4^{2-}(aq) \rightarrow Cr(OH)_4^-(aq) + 4\ OH^-(aq)$
$4\ H_2O(l) + CrO_4^{2-}(aq) + 3\ e^- \rightarrow Cr(OH)_4^-(aq) + 4\ OH^-(aq)$

(c) $Bi^{3+}(aq) \rightarrow BiO_3^-(aq)$
$Bi^{3+}(aq) + 3\ H_2O(l) \rightarrow BiO_3^-(aq)$
$Bi^{3+}(aq) + 3\ H_2O(l) \rightarrow BiO_3^-(aq) + 6\ H^+(aq)$
$Bi^{3+}(aq) + 3\ H_2O(l) + 6\ OH^-(aq) \rightarrow BiO_3^-(aq) + 6\ H^+(aq) + 6\ OH^-(aq)$
$Bi^{3+}(aq) + 3\ H_2O(l) + 6\ OH^-(aq) \rightarrow BiO_3^-(aq) + 6\ H_2O(l)$
$Bi^{3+}(aq) + 6\ OH^-(aq) \rightarrow BiO_3^-(aq) + 3\ H_2O(l)$
$Bi^{3+}(aq) + 6\ OH^-(aq) \rightarrow BiO_3^-(aq) + 3\ H_2O(l) + 2\ e^-$

(d) $ClO^-(aq) \rightarrow Cl^-(aq)$
$ClO^-(aq) \rightarrow Cl^-(aq) + H_2O(l)$
$2\ H^+(aq) + ClO^-(aq) \rightarrow Cl^-(aq) + H_2O(l)$
$2\ H^+(aq) + 2\ OH^-(aq) + ClO^-(aq) \rightarrow Cl^-(aq) + H_2O(l) + 2\ OH^-(aq)$
$2\ H_2O(l) + ClO^-(aq) \rightarrow Cl^-(aq) + H_2O(l) + 2\ OH^-(aq)$
$H_2O(l) + ClO^-(aq) \rightarrow Cl^-(aq) + 2\ OH^-(aq)$
$H_2O(l) + ClO^-(aq) + 2\ e^- \rightarrow Cl^-(aq) + 2\ OH^-(aq)$

18.54 (a) $MnO_4^-(aq) \rightarrow MnO_2(s)$
$MnO_4^-(aq) \rightarrow MnO_2(s) + 2\ H_2O(l)$
$4\ H^+(aq) + MnO_4^-(aq) \rightarrow MnO_2(s) + 2\ H_2O(l)$
$[4\ H^+(aq) + MnO_4^-(aq) + 3\ e^- \rightarrow MnO_2(s) + 2\ H_2O(l)] \times 2$ (reduction half reaction)

$IO_3^-(aq) \rightarrow IO_4^-(aq)$
$H_2O(l) + IO_3^-(aq) \rightarrow IO_4^-(aq)$
$H_2O(l) + IO_3^-(aq) \rightarrow IO_4^-(aq) + 2\ H^+(aq)$
$[H_2O(l) + IO_3^-(aq) \rightarrow IO_4^-(aq) + 2\ H^+(aq) + 2\ e^-] \times 3$ (oxidation half reaction)

Combine the two half reactions.
$8\ H^+(aq) + 3\ H_2O(l) + 2\ MnO_4^-(aq) + 3\ IO_3^-(aq) \rightarrow$
$6\ H^+(aq) + 4\ H_2O(l) + 2\ MnO_2(s) + 3\ IO_4^-(aq)$
$2\ H^+(aq) + 2\ MnO_4^-(aq) + 3\ IO_3^-(aq) \rightarrow 2\ MnO_2(s) + 3\ IO_4^-(aq) + H_2O(l)$
$2\ H^+(aq) + 2\ OH^-(aq) + 2\ MnO_4^-(aq) + 3\ IO_3^-(aq) \rightarrow$
$2\ MnO_2(s) + 3\ IO_4^-(aq) + H_2O(l) + 2\ OH^-(aq)$
$2\ H_2O(l) + 2\ MnO_4^-(aq) + 3\ IO_3^-(aq) \rightarrow$
$2\ MnO_2(s) + 3\ IO_4^-(aq) + H_2O(l) + 2\ OH^-(aq)$
$H_2O(l) + 2\ MnO_4^-(aq) + 3\ IO_3^-(aq) \rightarrow 2\ MnO_2(s) + 3\ IO_4^-(aq) + 2\ OH^-(aq)$

(b) $Cu(OH)_2(s) \rightarrow Cu(s)$
$Cu(OH)_2(s) \rightarrow Cu(s) + 2\ H_2O(l)$
$2\ H^+(aq) + Cu(OH)_2(s) \rightarrow Cu(s) + 2\ H_2O(l)$
$[2\ H^+(aq) + Cu(OH)_2(s) + 2\ e^- \rightarrow Cu(s) + 2\ H_2O(l)] \times 2$ (reduction half reaction)

$N_2H_4(aq) \rightarrow N_2(g)$
$N_2H_4(aq) \rightarrow N_2(g) + 4\ H^+(aq)$
$N_2H_4(aq) \rightarrow N_2(g) + 4\ H^+(aq) + 4\ e^-$ (oxidation half reaction)

Combine the two half reactions.
$4\ H^+(aq) + 2\ Cu(OH)_2(s) + N_2H_4(aq) \rightarrow 2\ Cu(s) + 4\ H_2O(l) + N_2(g) + 4\ H^+(aq)$
$2\ Cu(OH)_2(s) + N_2H_4(aq) \rightarrow 2\ Cu(s) + 4\ H_2O(l) + N_2(g)$

(c) $Fe(OH)_2(s) \rightarrow Fe(OH)_3(s)$
$Fe(OH)_2(s) + H_2O(l) \rightarrow Fe(OH)_3(s)$
$Fe(OH)_2(s) + H_2O(l) \rightarrow Fe(OH)_3(s) + H^+(aq)$
$[Fe(OH)_2(s) + H_2O(l) \rightarrow Fe(OH)_3(s) + H^+(aq) + e^-] \times 3$ (oxidation half reaction)

$CrO_4^{2-}(aq) \rightarrow Cr(OH)_4^-(aq)$
$4\ H^+(aq) + CrO_4^{2-}(aq) \rightarrow Cr(OH)_4^-(aq)$
$4\ H^+(aq) + CrO_4^{2-}(aq) + 3\ e^- \rightarrow Cr(OH)_4^-(aq)$ (reduction half reaction)

Combine the two half reactions.
$3\ Fe(OH)_2(s) + 3\ H_2O(l) + 4\ H^+(aq) + CrO_4^{2-}(aq) \rightarrow$
$3\ Fe(OH)_3(s) + 3\ H^+(aq) + Cr(OH)_4^-(aq)$
$3\ Fe(OH)_2(s) + 3\ H_2O(l) + H^+(aq) + CrO_4^{2-}(aq) \rightarrow 3\ Fe(OH)_3(s) + Cr(OH)_4^-(aq)$
$3\ Fe(OH)_2(s) + 3\ H_2O(l) + H^+(aq) + OH^-(aq) + CrO_4^{2-}(aq) \rightarrow$
$3\ Fe(OH)_3(s) + Cr(OH)_4^-(aq) + OH^-(aq)$
$3\ Fe(OH)_2(s) + 4\ H_2O(l) + CrO_4^{2-}(aq) \rightarrow 3\ Fe(OH)_3(s) + Cr(OH)_4^-(aq) + OH^-(aq)$

(d) $ClO_4^-(aq) \rightarrow ClO_2^-(aq)$
$ClO_4^-(aq) \rightarrow ClO_2^-(aq) + 2\ H_2O(l)$
$4\ H^+(aq) + ClO_4^-(aq) \rightarrow ClO_2^-(aq) + 2\ H_2O(l)$
$4\ H^+(aq) + ClO_4^-(aq) + 4\ e^- \rightarrow ClO_2^-(aq) + 2\ H_2O(l)$ (reduction half reaction)

$H_2O_2(aq) \rightarrow O_2(g)$
$H_2O_2(aq) \rightarrow O_2(g) + 2\ H^+(aq)$
$[H_2O_2(aq) \rightarrow O_2(g) + 2\ H^+(aq) + 2\ e^-] \times 2$ (oxidation half reaction)

Combine the two half reactions.
$4\ H^+(aq) + ClO_4^-(aq) + 2\ H_2O_2(aq) \rightarrow ClO_2^-(aq) + 2\ H_2O(l) + 2\ O_2(g) + 4\ H^+(aq)$
$ClO_4^-(aq) + 2\ H_2O_2(aq) \rightarrow ClO_2^-(aq) + 2\ H_2O(l) + 2\ O_2(g)$

18.56 (a) $Zn(s) \rightarrow Zn^{2+}(aq)$
$Zn(s) \rightarrow Zn^{2+}(aq) + 2\ e^-$ (oxidation half reaction)

$VO^{2+}(aq) \rightarrow V^{3+}(aq)$
$VO^{2+}(aq) \rightarrow V^{3+}(aq) + H_2O(l)$
$2\ H^+(aq) + VO^{2+}(aq) \rightarrow V^{3+}(aq) + H_2O(l)$
$[2\ H^+(aq) + VO^{2+}(aq) + e^- \rightarrow V^{3+}(aq) + H_2O(l)] \times 2$ (reduction half reaction)

Combine the two half reactions.
$Zn(s) + 2\ VO^{2+}(aq) + 4\ H^+(aq) \rightarrow Zn^{2+}(aq) + 2\ V^{3+}(aq) + 2\ H_2O(l)$

(b) $Ag(s) \rightarrow Ag^+(aq)$
$Ag(s) \rightarrow Ag^+(aq) + e^-$ (oxidation half reaction)

$NO_3^-(aq) \rightarrow NO_2(g)$
$NO_3^-(aq) \rightarrow NO_2(g) + H_2O(l)$
$2\ H^+(aq) + NO_3^-(aq) \rightarrow NO_2(g) + H_2O(l)$
$2\ H^+(aq) + NO_3^-(aq) + e^- \rightarrow NO_2(g) + H_2O(l)$ (reduction half reaction)

Combine the two half reactions.
$2\ H^+(aq) + Ag(s) + NO_3^-(aq) \rightarrow Ag^+(aq) + NO_2(g) + H_2O(l)$

(c) $Mg(s) \rightarrow Mg^{2+}(aq)$
$[Mg(s) \rightarrow Mg^{2+}(aq) + 2\ e^-] \times 3$ (oxidation half reaction)

$VO_4^{3-}(aq) \rightarrow V^{2+}(aq)$
$VO_4^{3-}(aq) \rightarrow V^{2+}(aq) + 4\ H_2O(l)$
$8\ H^+(aq) + VO_4^{3-}(aq) \rightarrow V^{2+}(aq) + 4\ H_2O(l)$
$[8\ H^+(aq) + VO_4^{3-}(aq) + 3\ e^- \rightarrow V^{2+}(aq) + 4\ H_2O(l)] \times 2$ (reduction half reaction)

Combine the two half reactions.
$3\ Mg(s) + 16\ H^+(aq) + 2\ VO_4^{3-}(aq) \rightarrow 3\ Mg^{2+}(aq) + 2\ V^{2+}(aq) + 8\ H_2O(l)$

(d) $I^-(aq) \rightarrow I_3^-(aq)$
$3\ I^-(aq) \rightarrow I_3^-(aq)$
$[3\ I^-(aq) \rightarrow I_3^-(aq) + 2\ e^-] \times 8$ (oxidation half reaction)

$IO_3^-(aq) \rightarrow I_3^-(aq)$
$3\ IO_3^-(aq) \rightarrow I_3^-(aq)$
$3\ IO_3^-(aq) \rightarrow I_3^-(aq) + 9\ H_2O(l)$
$18\ H^+(aq) + 3\ IO_3^-(aq) \rightarrow I_3^-(aq) + 9\ H_2O(l)$
$18\ H^+(aq) + 3\ IO_3^-(aq) + 16\ e^- \rightarrow I_3^-(aq) + 9\ H_2O(l)$ (reduction half reaction)

Combine the two half reactions.
$18\ H^+(aq) + 3\ IO_3^-(aq) + 24\ I^-(aq) \rightarrow 9\ I_3^-(aq) + 9\ H_2O(l)$
Divide each coefficient by 3.
$6\ H^+(aq) + IO_3^-(aq) + 8\ I^-(aq) \rightarrow 3\ I_3^-(aq) + 3\ H_2O(l)$

Galvanic Cells (Sections 18.2 and 18.3)

18.58 The cathode of a galvanic cell is considered to be the positive electrode because electrons flow through the external circuit toward the positive electrode (the cathode).

18.60 (a) $Cd(s) + Sn^{2+}(aq) \rightarrow Cd^{2+}(aq) + Sn(s)$

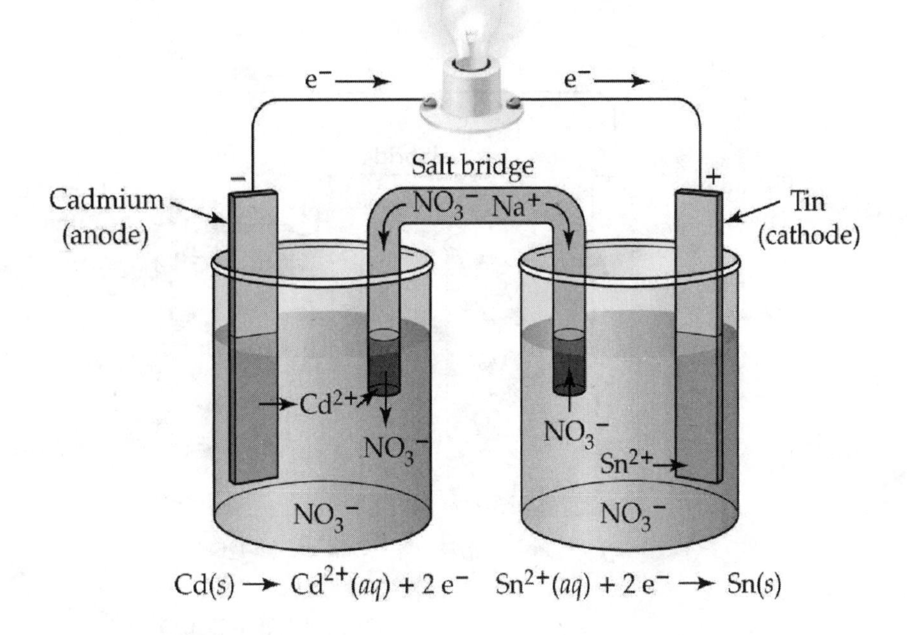

(b) $2\ Al(s) + 3\ Cd^{2+}(aq) \rightarrow 2\ Al^{3+}(aq) + 3\ Cd(s)$

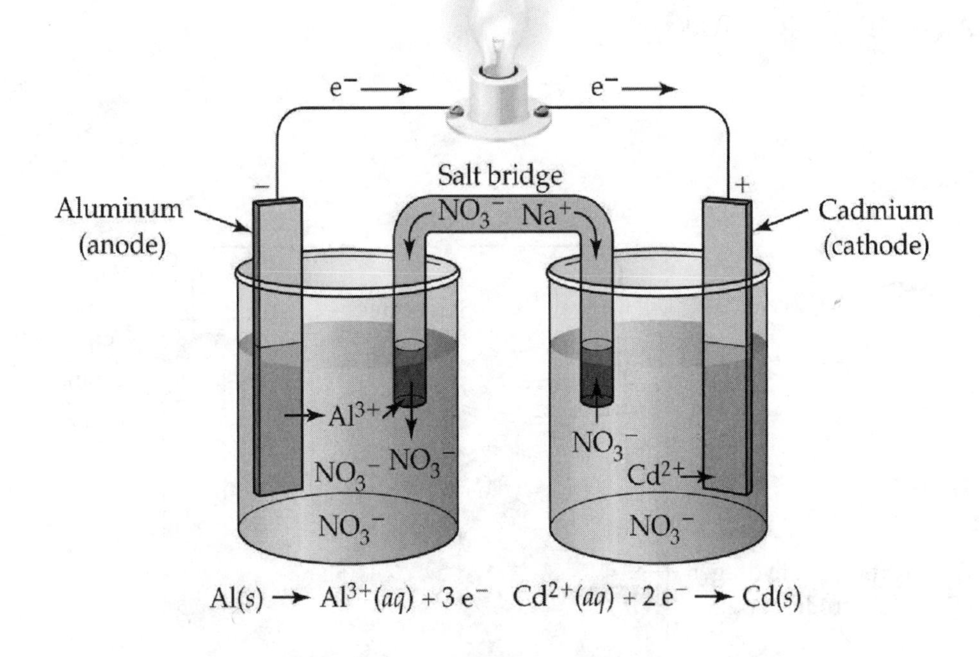

(c) $6\ Fe^{2+}(aq) + Cr_2O_7^{2-}(aq) + 14\ H^+(aq) \rightarrow 6\ Fe^{3+}(aq) + 2\ Cr^{3+}(aq) + 7\ H_2O(l)$

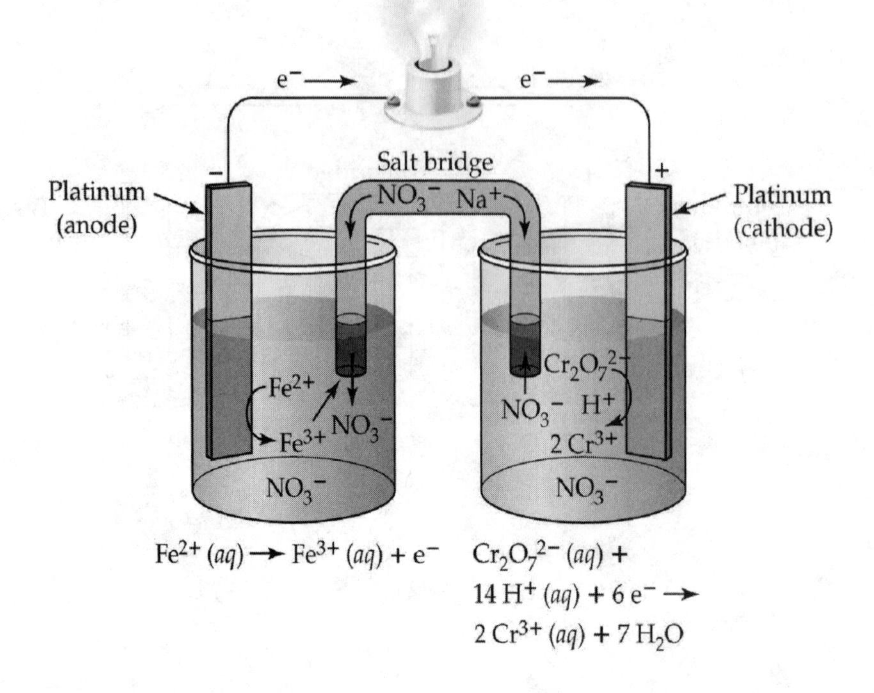

18.62 $2\ Br^-(aq) + Cl_2(g) \rightarrow Br_2(l) + 2\ Cl^-(aq)$
Inert electrodes are required because none of the reactants or products is an electrical conductor.

18.64 $Al(s)|Al^{3+}(aq)||Cd^{2+}|Cd(s)$

18.66 (a)

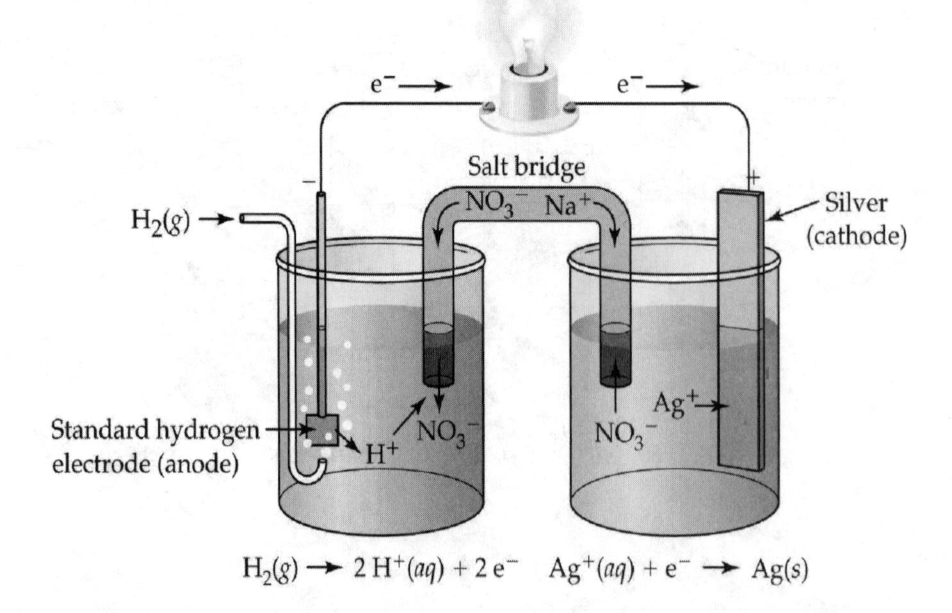

(b) anode reaction $H_2(g) \rightarrow 2\,H^+(aq) + 2\,e^-$
cathode reaction $2\,Ag^+(aq) + 2\,e^- \rightarrow 2\,Ag(s)$
overall reaction $H_2(g) + 2\,Ag^+(aq) \rightarrow 2\,H^+(aq) + 2\,Ag(s)$

(c) $Pt(s)|H_2(g)|H^+(aq)\|Ag^+(aq)|Ag(s)$

18.68 (a) anode reaction $Co(s) \rightarrow Co^{2+}(aq) + 2\,e^-$
cathode reaction $Cu^{2+}(aq) + 2\,e^- \rightarrow Cu(s)$
overall reaction $Co(s) + Cu^{2+}(aq) \rightarrow Co^{2+}(aq) + Cu(s)$

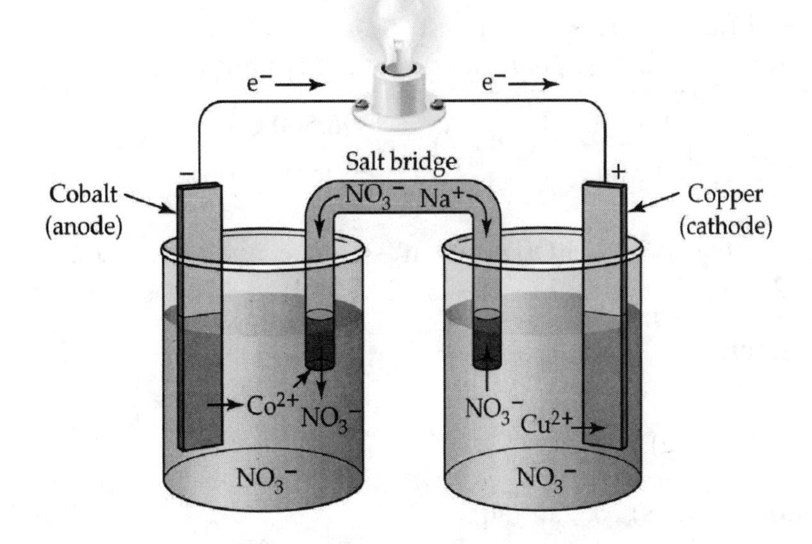

(b) anode reaction $2\,Fe(s) \rightarrow 2\,Fe^{2+}(aq) + 4\,e^-$
cathode reaction $O_2(g) + 4\,H^+(aq) + 4\,e^- \rightarrow 2\,H_2O(l)$
overall reaction $2\,Fe(s) + O_2(g) + 4\,H^+(aq) \rightarrow 2\,Fe^{2+}(aq) + 2\,H_2O(l)$

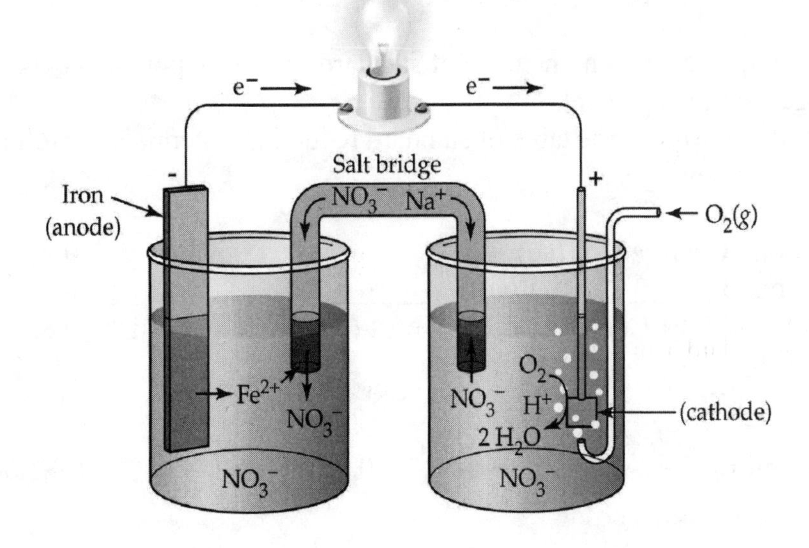

Cell Potentials and Free-Energy Changes; Standard Reduction Potentials (Sections 18.4–18.6)

18.70 E is the standard cell potential (E°) when all reactants and products are in their standard states–solutes at 1 M concentrations, gases at a partial pressure of 1 atm, solids and liquids in pure form, all at 25 °C.

18.72 $Zn(s) + Ag_2O(s) \rightarrow ZnO(s) + 2\ Ag(s)$; $n = 2\text{ mol } e^-$

$$\Delta G = -nFE = -(2\text{ mol } e^-)\left(\frac{96{,}500\text{ C}}{1\text{ mol } e^-}\right)(1.60\text{ V})\left(\frac{1\text{ J}}{1\text{ C}\cdot\text{V}}\right) = -308{,}800\text{ J} = -309\text{ kJ}$$

18.74 $\Delta G^o = -nFE^o$; $1\text{ J} = \text{C}\cdot\text{V}$

$$n = \frac{-\Delta G^o}{FE^o} = \frac{-(-414{,}000\text{ J})}{\left(\frac{96{,}500\text{ C}}{1\text{ mol } e^-}\right)(1.43\text{ V})} = \frac{-(-414{,}000\text{ C}\cdot\text{V})}{\left(\frac{96{,}500\text{ C}}{1\text{ mol } e^-}\right)(1.43\text{ V})} = 3\text{ mol } e^-$$

18.76 $2\ H_2(g) + O_2(g) \rightarrow 2\ H_2O(l)$; $n = 4\text{ mol } e^-$ and $1\text{ V} = 1\text{ J/C}$

$\Delta G^o = 2\ \Delta G^o_f(H_2O(l)) = (2\text{ mol})(-237.2\text{ kJ/mol}) = -474.4\text{ kJ}$

$$\Delta G^o = -nFE^o \qquad E^o = \frac{-\Delta G^o}{nF} = \frac{-(-474{,}400\text{ J})}{(4\text{ mol } e^-)\left(\frac{96{,}500\text{ C}}{1\text{ mol } e^-}\right)} = +1.23\text{ J/C} = +1.23\text{ V}$$

18.78

oxidation:	$Zn(s) \rightarrow Zn^{2+}(aq) + 2\ e^-$	$E^o = 0.76\text{ V}$
reduction:	$Eu^{3+}(aq) + e^- \rightarrow Eu^{2+}(aq)$	$E^o = ?$
overall	$Zn(s) + 2\ Eu^{3+}(aq) \rightarrow Zn^{2+}(aq) + 2\ Eu^{2+}(aq)$	$E^o = 0.40\text{ V}$

The standard reduction potential for the Eu^{3+}/Eu^{2+} half cell is:
$E^o = 0.40 - 0.76 = -0.36\text{ V}$

18.80 $Sn^{4+}(aq) < Br_2(aq) < MnO_4^-$

18.82 $Cr_2O_7^{2-}(aq)$ is highest in the table of standard reduction potentials, therefore it is the strongest oxidizing agent.
$Fe^{2+}(aq)$ is lowest in the table of standard reduction potentials, therefore it is the weakest oxidizing agent.

18.84

oxidation:	$Co(s) \rightarrow Co^{2+}(aq) + 2\ e^-$	$E^o = 0.28\text{ V}$
reduction:	$I_2(s) + 2\ e^- \rightarrow 2\ I^-(aq)$	$E^o = 0.54\text{ V}$
overall	$I_2(s) + Co(s) \rightarrow Co^{2+}(aq) + 2\ I^-(aq)$	$E^o = 0.82\text{ V}$

$n = 2\text{ mol } e^-$

$$\Delta G^o = -nFE^o = -(2\text{ mol } e^-)\left(\frac{96{,}500\text{ C}}{1\text{ mol } e^-}\right)(0.82\text{ V})\left(\frac{1\text{ J}}{1\text{ C}\cdot\text{V}}\right) = -158{,}260\text{ J} = -1.6 \times 10^2\text{ kJ}$$

18.86 oxidation: $2\ Al(s) \rightarrow 2\ Al^{3+}(aq) + 6\ e^-$ $E^\circ = 1.66\ V$
reduction: $3\ Cd^{2+}(aq) + 6\ e^- \rightarrow 3\ Cd(s)$ $E^\circ = -0.40\ V$
overall $2\ Al(s) + 3\ Cd^{2+}(aq) \rightarrow Al^{3+}(aq) + 3\ Cd(s)$ $E^\circ = 1.26\ V$
$n = 6\ mol\ e^-$

$$\Delta G^\circ = -nFE^\circ = -(6\ mol\ e^-)\left(\frac{96{,}500\ C}{1\ mol\ e^-}\right)(1.26\ V)\left(\frac{1\ J}{1\ C \cdot V}\right) = -729{,}540\ J = -730\ kJ$$

18.88 (a) $2\ Fe^{2+}(aq) + Pb^{2+}(aq) \rightarrow 2\ Fe^{3+}(aq) + Pb(s)$
oxidation: $2\ Fe^{2+}(aq) \rightarrow 2\ Fe^{3+}(aq) + 2\ e^-$ $E^\circ = -0.77\ V$
reduction: $Pb^{2+}(aq) + 2\ e^- \rightarrow Pb(s)$ $E^\circ = -0.13\ V$
overall $E^\circ = -0.90\ V$
Because the overall E° is negative, this reaction is nonspontaneous.

(b) $Mg(s) + Ni^{2+}(aq) \rightarrow Mg^{2+}(aq) + Ni(s)$
oxidation: $Mg(s) \rightarrow Mg^{2+}(aq) + 2\ e^-$ $E^\circ = 2.37\ V$
reduction: $Ni^{2+}(aq) + 2\ e^- \rightarrow Ni(s)$ $E^\circ = -0.26\ V$
overall $E^\circ = 2.11\ V$
Because the overall E° is positive, this reaction is spontaneous.

18.90 (a) oxidation: $Sn^{2+}(aq) \rightarrow Sn^{4+}(aq) + 2\ e^-$ $E^\circ = -0.15\ V$
reduction: $Br_2(aq) + 2\ e^- \rightarrow 2\ Br^-(aq)$ $E^\circ = 1.09\ V$
overall $E^\circ = +0.94\ V$
Because the overall E° is positive, $Sn^{2+}(aq)$ can be oxidized by $Br_2(aq)$.

(b) oxidation: $Sn^{2+}(aq) \rightarrow Sn^{4+}(aq) + 2\ e^-$ $E^\circ = -0.15\ V$
reduction: $Ni^{2+}(aq) + 2\ e^- \rightarrow Ni(s)$ $E^\circ = -0.26\ V$
overall $E^\circ = -0.41\ V$
Because the overall E° is negative, $Ni^{2+}(aq)$ cannot be reduced by $Sn^{2+}(aq)$.

(c) oxidation: $2\ Ag(s) \rightarrow 2\ Ag^+(aq) + 2\ e^-$ $E^\circ = -0.80\ V$
reduction: $Pb^{2+}(aq) + 2\ e^- \rightarrow Pb(s)$ $E^\circ = -0.13\ V$
overall $E^\circ = -0.93\ V$
Because the overall E° is negative, Ag(s) cannot be oxidized by $Pb^{2+}(aq)$.

(d) oxidation: $H_2SO_3(aq) + H_2O(l) \rightarrow SO_4^{2-}(aq) + 4\ H^+(aq) + 2\ e^-$ $E^\circ = -0.17\ V$
reduction: $I_2(s) + 2\ e^- \rightarrow 2\ I^-(aq)$ $E^\circ = 0.54\ V$
overall $E^\circ = +0.37\ V$
Because the overall E° is positive, $I_2(s)$ can be reduced by H_2SO_3.

18.92 (a) oxidation: $2\ Cr^{3+}(aq) + 7\ H_2O(l) \rightarrow Cr_2O_7^{2-}(aq) + 14\ H^+(aq) + 6\ e^-$ $E^\circ = -1.36\ V$
reduction: $O_2(g) + 4\ H^+(aq) + 4\ e^- \rightarrow 2\ H_2O(l)$ $E^\circ = 1.23\ V$
overall $E^\circ = -0.13\ V$

There is no reaction because the overall E° is negative.

(b) oxidation: $Pb(s) \rightarrow Pb^{2+}(aq) + 2\,e^-$ $E^\circ = 0.13\text{ V}$
reduction: $2\,Ag^+(aq) + 2\,e^- \rightarrow 2\,Ag(s)$ $E^\circ = 0.80\text{ V}$
overall $E^\circ = +0.93\text{ V}$

$Pb(s) + 2\,Ag^+(aq) \rightarrow Pb^{2+}(aq) + 2\,Ag(s)$
The reaction is spontaneous because the overall E° is positive.

(c) oxidation: $H_2C_2O_4(aq) \rightarrow 2\,CO_2(g) + 2\,H^+(aq) + 2\,e^-$ $E^\circ = 0.49\text{ V}$
reduction: $Cl_2(g) + 2\,e^- \rightarrow 2\,Cl^-(aq)$ $E^\circ = 1.36\text{ V}$
overall $E^\circ = +1.85\text{ V}$

$Cl_2(g) + H_2C_2O_4(aq) \rightarrow 2\,Cl^-(aq) + 2\,CO_2(g) + 2\,H^+(aq)$
The reaction is spontaneous because the overall E° is positive.

(d) oxidation: $Ni(s) \rightarrow Ni^{2+}(aq) + 2\,e^-$ $E^\circ = 0.26\text{ V}$
reduction: $2\,HClO(aq) + 2\,H^+(aq) + 2\,e^- \rightarrow Cl_2(g) + H_2O(l)$ $E^\circ = 1.61\text{ V}$
overall $E^\circ = +1.87\text{ V}$

$Ni(s) + 2\,HClO(aq) + 2\,H^+(aq) \rightarrow Ni^{2+}(aq) + Cl_2(g) + H_2O(l)$
The reaction is spontaneous because the overall E° is positive.

The Nernst Equation (Sections 18.7 and 18.8)

18.94 $2\,Ag^+(aq) + Sn(s) \rightarrow 2\,Ag(s) + Sn^{2+}(aq)$
oxidation: $Sn(s) \rightarrow Sn^{2+}(aq) + 2\,e^-$ $E^\circ = 0.14\text{ V}$
reduction: $2\,Ag^+(aq) + 2\,e^- \rightarrow 2\,Ag(s)$ $E^\circ = 0.80\text{ V}$
overall $E^\circ = 0.94\text{ V}$

$$E = E^\circ - \frac{0.0592\text{ V}}{n}\log\frac{[Sn^{2+}]}{[Ag^+]^2} = 0.94\text{ V} - \frac{(0.0592\text{ V})}{2}\log\frac{(0.020)}{(0.010)^2} = 0.87\text{ V}$$

18.96 $Pb(s) + Cu^{2+}(aq) \rightarrow Pb^{2+}(aq) + Cu(s)$
oxidation: $Pb(s) \rightarrow Pb^{2+}(aq) + 2\,e^-$ $E^\circ = 0.13\text{ V}$
reduction: $Cu^{2+}(aq) + 2\,e^- \rightarrow Cu(s)$ $E^\circ = 0.34\text{ V}$
overall $E^\circ = 0.47\text{ V}$

$$E = E^\circ - \frac{0.0592\text{ V}}{n}\log\frac{[Pb^{2+}]}{[Cu^{2+}]} = 0.47\text{ V} - \frac{(0.0592\text{ V})}{2}\log\frac{1.0}{(1.0 \text{ x } 10^{-4})} = 0.35\text{ V}$$

When E = 0, $0 = E^\circ - \frac{0.0592\text{ V}}{n}\log\frac{[Pb^{2+}]}{[Cu^{2+}]} = 0.47\text{ V} - \frac{(0.0592\text{ V})}{2}\log\frac{1.0}{[Cu^{2+}]}$

$$0 = 0.47\text{ V} + \frac{(0.0592\text{ V})}{2}\log[Cu^{2+}]$$

$$\log[Cu^{2+}] = (-0.47\text{ V})\left(\frac{2}{0.0592\text{ V}}\right) = -15.88$$

$[Cu^{2+}] = 10^{-15.88} = 1 \text{ x } 10^{-16}\text{ M}$

18.98 $Zn(s) + Cu^{2+}(aq) \rightarrow Zn^{2+}(aq) + Cu(s)$

oxidation: $Zn(s) \rightarrow Zn^{2+}(aq) + 2\,e^-$ $E° = 0.76$ V

reduction: $Cu^{2+}(aq) + 2\,e^- \rightarrow Cu(s)$ $E° = 0.34$ V

overall $E° = 1.10$ V

$$E = 1.07\text{ V} = E° - \frac{0.0592\text{ V}}{n}\log\frac{[Zn^{2+}]}{[Cu^{2+}]} = 1.10\text{ V} - \frac{(0.0592\text{ V})}{2}\log\left(\frac{[Zn^{2+}]}{[Cu^{2+}]}\right)$$

$$1.07\text{ V} - 1.10\text{ V} = -\frac{(0.0592\text{ V})}{2}\log\left(\frac{[Zn^{2+}]}{[Cu^{2+}]}\right)$$

$$\frac{0.03\text{ V}}{\frac{(0.0592\text{ V})}{2}} = \log\left(\frac{[Zn^{2+}]}{[Cu^{2+}]}\right) = 1; \qquad \frac{[Zn^{2+}]}{[Cu^{2+}]} = 10^1 = 10$$

18.100 (a) $E = E° - \frac{0.0592\text{ V}}{n}\log[I^-]^2 = 0.54\text{ V} - \frac{(0.0592\text{ V})}{2}\log(0.020)^2 = 0.64\text{ V}$

(b) $E = E° - \frac{0.0592\text{ V}}{n}\log\frac{[Fe^{2+}]}{[Fe^{3+}]} = 0.77\text{ V} - \frac{(0.0592\text{ V})}{1}\log\left(\frac{0.10}{0.10}\right) = 0.77\text{ V}$

(c) $E = E° - \frac{0.0592\text{ V}}{n}\log\frac{[Sn^{4+}]}{[Sn^{2+}]} = -0.15\text{ V} - \frac{(0.0592\text{ V})}{2}\log\left(\frac{0.40}{0.0010}\right) = -0.23\text{ V}$

(d) $E = E° - \frac{0.0592\text{ V}}{n}\log\frac{[Cr_2O_7^{2-}][H^+]^{14}}{[Cr^{3+}]^2} = -1.36\text{ V} - \frac{(0.0592\text{ V})}{6}\log\left(\frac{(1.0)(0.010)^{14}}{1.0}\right)$

$$E = -1.36\text{ V} - \frac{(0.0592\text{ V})}{6}(14)\log(0.010) = -1.08\text{ V}$$

18.102 $H_2(g) + Ni^{2+}(aq) \rightarrow 2\,H^+(aq) + Ni(s)$

$$E° = E°_{H_2 \rightarrow H^+} + E°_{Ni^{2+} \rightarrow Ni} = 0\text{ V} + (-0.26\text{ V}) = -0.26\text{ V}$$

$$E = E° - \frac{0.0592\text{ V}}{n}\log\frac{[H_3O^+]^2}{[Ni^{2+}](P_{H_2})}$$

$$0.27\text{ V} = -0.26\text{ V} - \frac{(0.0592\text{ V})}{2}\log\frac{[H_3O^+]^2}{(1)(1)}$$

$$0.27\text{ V} = -0.26\text{ V} - (0.0592\text{ V})\log[H_3O^+]$$

$pH = -\log[H_3O^+]$ therefore $0.27\text{ V} = -0.26\text{ V} + (0.0592\text{ V})\,pH$

$$pH = \frac{(0.27\text{ V} + 0.26\text{ V})}{0.0592\text{ V}} = 9.0$$

Standard Cell Potentials and Equilibrium Constants (Section 18.9)

18.104 $\Delta G^\circ = -nFE^\circ$

Because n and F are always positive, ΔG° is negative when E° is positive because of the negative sign in the equation.

$$E^\circ = \frac{0.0592\ V}{n}\log K; \quad \log K = \frac{nE^\circ}{0.0592\ V}; \qquad K = 10^{\frac{nE^\circ}{0.0592}}$$

If E° is positive, the exponent is positive (because n is positive), and K is greater than 1.

18.106 $Ni(s) + 2\ Ag^{+}(aq) \rightarrow Ni^{2+}(aq) + 2\ Ag(s)$

oxidation: $Ni(s) \rightarrow Ni^{2+}(aq) + 2\ e^{-}$ $\quad E^\circ = 0.26\ V$
reduction: $2\ Ag^{+}(aq) + 2\ e^{-} \rightarrow 2\ Ag(s)$ $\quad \underline{E^\circ = 0.80\ V}$
overall $E^\circ = 1.06\ V$

$$E^\circ = \frac{0.0592\ V}{n}\log K; \ \log K = \frac{nE^\circ}{0.0592\ V} = \frac{(2)(1.06\ V)}{0.0592\ V} = 35.8; \ K = 10^{35.8} = 6 \times 10^{35}$$

18.108 $Cd(s) + Sn^{2+}(aq) \rightarrow Cd^{2+}(aq) + Sn(s)$

oxidation $\quad Cd(s) \rightarrow Cd^{2+}(aq) + 2\ e^{-}$ $\quad E^\circ = 0.40\ V$
reduction: $\quad Sn^{2+}(aq) + 2\ e^{-} \rightarrow Sn(s)$ $\quad \underline{E^\circ = -0.14\ V}$
overall $E^\circ = 0.26\ V$

$$E^\circ = \frac{0.0592\ V}{n}\log K; \ \log K = \frac{nE^\circ}{0.0592\ V} = \frac{(2)(0.26\ V)}{0.0592\ V} = 8.8; \ K = 10^{8.8} = 6.3 \times 10^{8}$$

18.110 $Hg_2^{2+}(aq) \rightarrow Hg(l) + Hg^{2+}(aq)$

oxidation: $\quad \frac{1}{2}[Hg_2^{2+}(aq) \rightarrow 2\ Hg^{2+}(aq) + 2\ e^{-}]$ $\quad E^\circ = -0.92\ V$
reduction: $\quad \frac{1}{2}[Hg_2^{2+}(aq) + 2\ e^{-} \rightarrow 2\ Hg(l)]$ $\quad \underline{E^\circ = 0.80\ V}$
overall $E^\circ = -0.12\ V$

$$E^\circ = \frac{0.0592\ V}{n}\log K$$

$$\log K = \frac{nE^\circ}{0.0592\ V} = \frac{(1)(-0.12\ V)}{0.0592\ V} = -2.027; \qquad K = 10^{-2.027} = 9 \times 10^{-3}$$

Batteries; Corrosion (Sections 18.10 and 18.11)

18.112 (a)

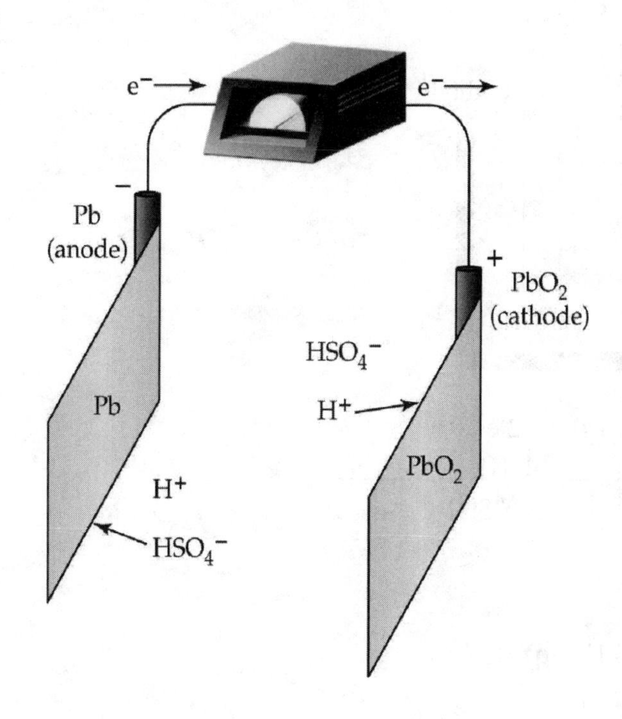

(b) anode: $Pb(s) + HSO_4^-(aq) \rightarrow PbSO_4(s) + H^+(aq) + 2\ e^-$ $E° = 0.296\ V$

cathode: $PbO_2(s) + 3\ H^+(aq) + HSO_4^-(aq) + 2\ e^- \rightarrow PbSO_4(s) + 2\ H_2O(l)$ $E° = 1.628\ V$

overall $Pb(s) + PbO_2(s) + 2\ H^+(aq) + 2\ HSO_4^-(aq) \rightarrow 2\ PbSO_4(s) + 2\ H_2O(l)$ $E° = 1.924\ V$

(c) $E° = \frac{0.0592\ V}{n} \log K$; $\log K = \frac{n E°}{0.0592\ V} = \frac{(2)(1.924\ V)}{0.0592\ V} = 65.0$; $K = 1 \times 10^{65}$

(d) When the cell reaction reaches equilibrium the cell voltage = 0.

18.114 Rust is a hydrated form of iron(III) oxide ($Fe_2O_3 \cdot H_2O$). Rust forms from the oxidation of Fe in the presence of O_2 and H_2O. Rust can be prevented by coating Fe with Zn (galvanizing).

18.116 Cr forms a protective oxide coating similar to Al.

18.118 (d) A strip of magnesium is attached to steel because the magnesium is more easily oxidized than iron.

18.120 Mn and Al

Electrolysis (Sections 18.12–18.14)

18.122 (a)

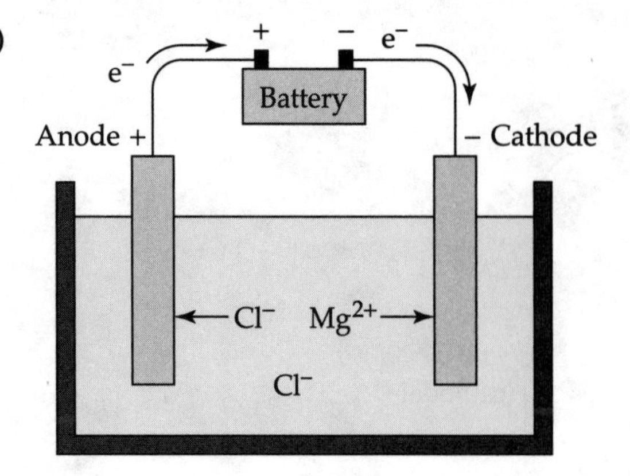

(b)

anode:	$2\,Cl^-(l) \rightarrow Cl_2(g) + 2\,e^-$
cathode:	$Mg^{2+}(l) + 2\,e^- \rightarrow Mg(l)$
overall:	$Mg^{2+}(l) + 2\,Cl^-(l) \rightarrow Mg(l) + Cl_2(g)$

18.124 Possible anode reactions:

$2\,Cl^-(aq) \rightarrow Cl_2(g) + 2\,e^-$

$2\,H_2O(l) \rightarrow O_2(g) + 4\,H^+(aq) + 4\,e^-$

Possible cathode reactions:

$2\,H_2O(l) + 2\,e^- \rightarrow H_2(g) + 2\,OH^-(aq)$

$Mg^{2+}(aq) + 2\,e^- \rightarrow Mg(s)$

Actual reactions:

anode:	$2\,Cl^-(aq) \rightarrow Cl_2(g) + 2\,e^-$
cathode:	$2\,H_2O(l) + 2\,e^- \rightarrow H_2(g) + 2\,OH^-(aq)$

This anode reaction takes place instead of $2\,H_2O(l) \rightarrow O_2(g) + 4\,H^+(aq) + 4\,e^-$ because of a high overvoltage for formation of gaseous O_2.
This cathode reaction takes place instead of $Mg^{2+}(aq) + 2\,e^- \rightarrow Mg(s)$ because H_2O is easier to reduce than Mg^{2+}.

18.126 (a) NaBr

anode:	$2\,Br^-(aq) \rightarrow Br_2(l) + 2\,e^-$
cathode:	$2\,H_2O(l) + 2\,e^- \rightarrow H_2(g) + 2\,OH^-(aq)$
overall:	$2\,H_2O(l) + 2\,Br^-(aq) \rightarrow Br_2(l) + H_2(g) + 2\,OH^-(aq)$

(b) $CuCl_2$

anode:	$2\,Cl^-(aq) \rightarrow Cl_2(g) + 2\,e^-$
cathode:	$Cu^{2+}(aq) + 2\,e^- \rightarrow Cu(s)$
overall:	$Cu^{2+}(aq) + 2\,Cl^-(aq) \rightarrow Cu(s) + Cl_2(g)$

(c) LiOH

anode:	$4\,OH^-(aq) \rightarrow O_2(g) + 2\,H_2O(l) + 4\,e^-$
cathode:	$4\,H_2O(l) + 4\,e^- \rightarrow 2\,H_2(g) + 4\,OH^-(aq)$
overall:	$2\,H_2O(l) \rightarrow O_2(g) + 2\,H_2(g)$

18.128 $Ag^+(aq) + e^- \rightarrow Ag(s)$; 1 A = 1 C/s

$$\text{mass Ag} = 2.40\ \frac{C}{s} \times 20.0\ \text{min} \times \frac{60\ s}{1\ \text{min}} \times \frac{1\ \text{mol } e^-}{96{,}500\ C} \times \frac{1\ \text{mol Ag}}{1\ \text{mol } e^-} \times \frac{107.87\ \text{g Ag}}{1\ \text{mol Ag}} = 3.22\ g$$

18.130 $2\ Na^+(l) + 2\ Cl^-(l) \rightarrow 2\ Na(l) + Cl_2(g)$

$Na^+(l) + e^- \rightarrow Na(l)$; 1 A = 1 C/s; 1.00×10^3 kg = 1.00×10^6 g

$$\text{Charge} = 1.00 \times 10^6\ \text{g Na} \times \frac{1\ \text{mol Na}}{22.99\ \text{g Na}} \times \frac{1\ \text{mol } e^-}{1\ \text{mol Na}} \times \frac{96{,}500\ C}{1\ \text{mol } e^-} = 4.20 \times 10^9\ C$$

$$\text{Time} = \frac{4.20 \times 10^9\ C}{30{,}000\ C/s} \times \frac{1\ h}{3600\ s} = 38.9\ h$$

$$1.00 \times 10^6\ \text{g Na} \times \frac{1\ \text{mol Na}}{22.99\ \text{g Na}} \times \frac{1\ \text{mol } Cl_2}{2\ \text{mol Na}} = 21{,}748.6\ \text{mol } Cl_2$$

PV = nRT

$$V = \frac{nRT}{P} = \frac{(21{,}748.6\ \text{mol})\left(0.082\ 06\ \frac{L \cdot atm}{K \cdot mol}\right)(273.15\ K)}{1.00\ atm} = 4.87 \times 10^5\ L\ Cl_2$$

18.132 $M^{2+} + 2\ e^- \rightarrow M$; 20.0 A = 20.0 C/s

$$\text{mol } e^- = 20.0\ \frac{C}{s} \times 325\ \text{min} \times \frac{60\ s}{\text{min}} \times \frac{1\ \text{mol } e^-}{96{,}500\ C} = 4.04\ \text{mol } e^-$$

$$4.04\ \text{mol } e^- \times \frac{1\ \text{mol M}}{2\ \text{mol } e^-} = 2.02\ \text{mol M}$$

$$\text{molar mass} = \frac{111\ \text{g M}}{2.02\ \text{mol M}} = 54.9\ \text{g/mol},\ M^{2+} = Mn^{2+}$$

Chapter Problems

18.134

$$\text{volume} = \left(0.0100\ \text{mm} \times \frac{1\ cm}{10\ mm}\right)(10.0\ cm)^2 = 0.100\ cm^3$$

$$\text{mol } Al_2O_3 = (0.100\ cm^3)(3.97\ g/cm^3)\frac{1\ \text{mol } Al_2O_3}{102.0\ \text{g } Al_2O_3} = 3.892 \times 10^{-3}\ \text{mol } Al_2O_3$$

$$\text{mole } e^- = 3.892 \times 10^{-3}\ \text{mol } Al_2O_3 \times \frac{6\ \text{mol } e^-}{1\ \text{mol } Al_2O_3} = 0.02335\ \text{mol } e^-$$

$$\text{coulombs} = 0.02335\ \text{mol } e^- \times \frac{96{,}500\ C}{1\ \text{mol } e^-} = 2253\ C$$

$$\text{time} = \frac{C}{A} = \frac{2253\ C}{0.600\ C/s} \times \frac{1\ \text{min}}{60\ s} = 62.6\ \text{min}$$

18.136

$PbSO_4(s) + 2\ H_2O(l) \rightarrow PbO_2(s) + 4\ H^+(aq) + SO_4^-(aq) + 2\ e^-$

$PbSO_4(s) + 2\ e^- \rightarrow Pb(s) + SO_4^-(aq)$

$2\ PbSO_4(s) + 2\ H_2O(l) \rightarrow Pb(s) + PbO_2(s) + 2\ H^+(aq) + 2\ HSO_4^-(aq)$

(a) The reaction represents an electrolytic cell.
(b) During the recharging (electrolysis) 250.0 g of $PbSO_4$ are oxidized to PbO_2 and 250.0 g of $PbSO_4$ are reduced to Pb.

$$\text{mol } PbSO_4 = 250.0 \text{ g } PbSO_4 \times \frac{1 \text{ mol } PbSO_4}{303.3 \text{ g } PbSO_4} = 0.824 \text{ mol } PbSO_4$$

$$\text{mole } e^- = 0.824 \text{ mol } PbSO_4 \times \frac{2 \text{ mol } e^-}{1 \text{ mol } PbSO_4} = 1.648 \text{ mol } e^-$$

$$\text{coulombs} = 1.648 \text{ mol } e^- \times \frac{96{,}500 \text{ C}}{1 \text{ mol } e^-} = 1.591 \times 10^5 \text{ C}$$

(c) $$\text{time} = \frac{C}{A} = \frac{1.591 \times 10^5 \text{ C}}{500 \text{ C/s}} \times \frac{1 \text{ min}}{60 \text{ s}} = 5.3 \text{ min}$$

18.138 $2\,Mn^{3+}(aq) + 2\,H_2O(l) \rightarrow Mn^{2+}(aq) + MnO_2(s) + 4\,H^+(aq)$
$E^\circ = 1.54 \text{ V} + (-0.95 \text{ V}) = +0.59 \text{ V}$
Because E° is positive, the disproportionation is spontaneous under standard-state conditions.

18.140 For Pb^{2+}, $E = -0.13 - \frac{0.0592 \text{ V}}{2} \log \frac{1}{[Pb^{2+}]}$

For Cd^{2+}, $E = -0.40 - \frac{0.0592 \text{ V}}{2} \log \frac{1}{[Cd^{2+}]}$

Set these two equations for E equal to each other and solve for $[Cd^{2+}]/[Pb^{2+}]$.

$$-0.13 - \frac{0.0592 \text{ V}}{2} \log \frac{1}{[Pb^{2+}]} = -0.40 - \frac{0.0592 \text{ V}}{2} \log \frac{1}{[Cd^{2+}]}$$

$$0.27 = \frac{0.0592 \text{ V}}{2}(\log[Cd^{2+}] - \log[Pb^{2+}]) = \frac{0.0592 \text{ V}}{2} \log \frac{[Cd^{2+}]}{[Pb^{2+}]}$$

$$\log \frac{[Cd^{2+}]}{[Pb^{2+}]} = \frac{(0.27)(2)}{0.0592} = 9.1; \qquad \frac{[Cd^{2+}]}{[Pb^{2+}]} = 10^{9.1} = 1 \times 10^9$$

18.142 $Ni(OH)_2$, 92.71

(a) $$3.35 \times 10^{-2} \text{ g } Ni(OH)_2 \times \frac{1 \text{ mol } Ni(OH)_2}{92.71 \text{ g } Ni(OH)_2} \times \frac{1 \text{ mol Zn}}{2 \text{ mol } Ni(OH)_2} \times \frac{65.38 \text{ g Zn}}{1 \text{ mol Zn}} = 0.0118 \text{ g Zn}$$

(b) 0.100 A = 0.100 C/s

$$6.17 \times 10^{-2} \text{ g Zn} \times \frac{1 \text{ mol Zn}}{65.38 \text{ g Zn}} \times \frac{2 \text{ mol } e^-}{1 \text{ mol Zn}} = 1.89 \times 10^{-3} \text{ mol } e^-$$

$$1.89 \times 10^{-3} \text{ mol } e^- \times \frac{96{,}500 \text{ C}}{1 \text{ mol } e^-} \times \frac{1 \text{ s}}{0.100 \text{ C}} \times \frac{1 \text{ min}}{60 \text{ s}} = 30.4 \text{ min}$$

18.144 (a) From: $B + A^+ \rightarrow B^+ + A$, A^+ is reduced more easily than B^+
From: $C + A^+ \rightarrow C^+ + A$, A^+ is reduced more easily than C^+
From: $B + C^+ \rightarrow B^+ + C$, C^+ is reduced more easily than B^+

$A^+ + e^- \rightarrow A$
$C^+ + e^- \rightarrow C$
$B^+ + e^- \rightarrow B$

(b) A^+ is the strongest oxidizing agent; B is the strongest reducing agent

(c) $A^+ + B \rightarrow B^+ + A$

18.146 (a) $2\ Na(soln) + S(soln) \rightarrow 2\ Na^+(soln) + S^{2-}(soln)$

(b) $1\ W = \dfrac{1\ C \cdot 1\ V}{s}; \quad \dfrac{1\ W}{1\ V} = 1\ C/s; \quad \dfrac{25\ kW}{2.0\ V} = \dfrac{25{,}000\ W}{2.0\ V} = 12{,}500\ C/s$

$$12{,}500\ C/s \times 32\ min \times \frac{60\ s}{1\ min} \times \frac{1\ mol\ e^-}{96{,}500\ C} \times \frac{1\ mol\ Na^+}{1\ mol\ e^-} \times \frac{22.99\ g\ Na}{1\ mol\ Na} = 5.7 \times 10^3\ g = 5.7\ kg$$

18.148 (a) $\Delta G^\circ = -nFE^\circ$

$\Delta G^\circ_3 = \Delta G^\circ_1 + \Delta G^\circ_2$ therefore $-n_3FE^\circ_3 = -n_1FE^\circ_1 + (-n_2FE^\circ_2)$

$n_3E^\circ_3 = n_1E^\circ_1 + n_2E^\circ_2$

$$E^\circ_3 = \frac{n_1E^\circ_1 + n_2E^\circ_2}{n_3}$$

(b) $E^\circ_3 = \dfrac{(3)(-0.04\ V) + (2)(0.45\ V)}{1} = 0.78\ V$

(c) E° values would be additive ($E^\circ_3 = E^\circ_1 + E^\circ_2$) if reaction (3) is an overall cell reaction because the electrons in the two half reactions, (1) and (2), cancel. That is, $n_1 = n_2 = n_3$ in the equation for E°_3.

18.150 (a)

anode:	$Cu(s) \rightarrow Cu^{2+}(aq) + 2\ e^-$	$E^\circ = -0.34\ V$
cathode:	$2\ Ag^+(aq) + 2\ e^- \rightarrow 2\ Ag(s)$	$E^\circ = 0.80\ V$
overall:	$2\ Ag^+(aq) + Cu(s) \rightarrow Cu^{2+}(aq) + 2\ Ag(s)$	$E^\circ = 0.46\ V$

$$E = E^\circ - \frac{0.0592\ V}{n}\log\frac{[Cu^{2+}]}{[Ag^+]^2} = 0.46\ V - \frac{(0.0592\ V)}{2}\log\left(\frac{1.0}{(0.050)^2}\right) = 0.38\ V$$

(b) $[Ag^+] = \dfrac{K_{sp}}{[Br^-]} = \dfrac{5.4 \times 10^{-13}}{1.0\ M} = 5.4 \times 10^{-13}\ M$

$$E = E^\circ - \frac{0.0592\ V}{n}\log\frac{[Cu^{2+}]}{[Ag^+]^2} = 0.46\ V - \frac{(0.0592\ V)}{2}\log\left(\frac{1.0}{(5.4 \times 10^{-13})^2}\right) = -0.27\ V$$

The cell potential for the spontaneous reaction is E = 0.27 V.

The spontaneous reaction is: $Cu^{2+}(aq) + 2\ Ag(s) + 2\ Br^-(aq) \rightarrow 2\ AgBr(s) + Cu(s)$

(c)

$Cu^{2+}(aq) + 2\ e^- \rightarrow Cu(s)$	$E^\circ = 0.34\ V$
$2\ Ag(s) + 2\ Br^-(aq) \rightarrow 2\ AgBr(s) + 2\ e^-$	$E^\circ = ?$
$Cu^{2+}(aq) + 2\ Ag(s) + 2\ Br^-(aq) \rightarrow 2\ AgBr(s) + Cu(s)$	$E^\circ = 0.27\ V$

$E^\circ = ? = 0.27\ V - 0.34\ V = -0.07\ V$

For: $AgBr(s) + e^- \rightarrow Ag(s) + Br^-(aq)$

the standard reduction potential is $E^\circ = 0.07\ V$

18.152 $H_2MoO_4(aq) + As(s) \rightarrow Mo^{3+}(aq) + H_3AsO_4(aq)$

$H_2MoO_4(aq) \rightarrow Mo^{3+}(aq)$
$H_2MoO_4(aq) \rightarrow Mo^{3+}(aq) + 4\ H_2O(l)$
$6\ H^+(aq) + H_2MoO_4(aq) \rightarrow Mo^{3+}(aq) + 4\ H_2O(l)$
$[3\ e^- + 6\ H^+(aq) + H_2MoO_4(aq) \rightarrow Mo^{3+}(aq) + 4\ H_2O(l)] \times 5$
(reduction half reaction)
$As(s) \rightarrow H_3AsO_4(aq)$
$As(s) + 4\ H_2O(l) \rightarrow H_3AsO_4(aq)$
$As(s) + 4\ H_2O(l) \rightarrow H_3AsO_4(aq) + 5\ H^+(aq)$
$[As(s) + 4\ H_2O(l) \rightarrow H_3AsO_4(aq) + 5\ H^+(aq) + 5\ e^-] \times 3$ (oxidation half reaction)

Combine the two half reactions.
$30\ H^+(aq) + 5\ H_2MoO_4(aq) + 3\ As(s) + 12\ H_2O(l) \rightarrow$
$5\ Mo^{3+}(aq) + 3\ H_3AsO_4(aq) + 15\ H^+(aq) + 20\ H_2O(l)$
$15\ H^+(aq) + 5\ H_2MoO_4(aq) + 3\ As(s) \rightarrow 5\ Mo^{3+}(aq) + 3\ H_3AsO_4(aq) + 8\ H_2O(l)$

$5 \times [H_2MoO_4(aq) + 2\ H^+(aq) + 2\ e^- \rightarrow MoO_2(s) + 2\ H_2O(l)]$ $E° = +0.646\ V$
$5 \times [MoO_2(s) + 4\ H^+(aq) + e^- \rightarrow Mo^{3+}(aq) + 2\ H_2O(l)]$ $E° = -0.008\ V$
$3 \times [As(s) + 3\ H_2O(l) \rightarrow H_3AsO_3(aq) + 3\ H^+(aq) + 3\ e^-]$ $E° = -0.240\ V$
$3 \times [H_3AsO_3(aq) + H_2O(l) \rightarrow H_3AsO_4(aq) + 2\ H^+(aq) + 2\ e^-]$ $E° = -0.560\ V$

$15\ H^+(aq) + 5\ H_2MoO_4(aq) + 3\ As(s) \rightarrow 5\ Mo^{3+}(aq) + 3\ H_3AsO_4(aq) + 8\ H_2O(l)$

$$\Delta G° = -nFE° = -(10\ mol\ e^-)\left(\frac{96{,}500\ C}{1\ mol\ e^-}\right)(0.646\ V)\left(\frac{1\ J}{1\ C \cdot V}\right) = -623{,}390\ J = -623.4\ kJ$$

$$\Delta G° = -nFE° = -(5\ mol\ e^-)\left(\frac{96{,}500\ C}{1\ mol\ e^-}\right)(-0.008\ V)\left(\frac{1\ J}{1\ C \cdot V}\right) = 3{,}860\ J = +3.9\ kJ$$

$$\Delta G° = -nFE° = -(9\ mol\ e^-)\left(\frac{96{,}500\ C}{1\ mol\ e^-}\right)(-0.240\ V)\left(\frac{1\ J}{1\ C \cdot V}\right) = 208{,}440\ J = +208.4\ kJ$$

$$\Delta G° = -nFE° = -(6\ mol\ e^-)\left(\frac{96{,}500\ C}{1\ mol\ e^-}\right)(-0.560\ V)\left(\frac{1\ J}{1\ C \cdot V}\right) = 324{,}240\ J = +324.2\ kJ$$

$\Delta G°(total) = -623.4\ kJ + 3.9\ kJ + 208.4\ kJ + 324.2\ kJ = -86.9\ kJ = -86{,}900\ J$

$1\ V = 1\ J/C$

$$\Delta G° = -nFE°;\ E° = \frac{-\Delta G°}{nF} = \frac{-(-86{,}900\ J)}{(15\ mol\ e^-)\left(\frac{96{,}500\ C}{1\ mol\ e^-}\right)} = +0.060\ J/C = +0.060\ V$$

18.154

oxidation:	$Cu^+(aq) \rightarrow Cu^{2+}(aq) + e^-$	$E° = -0.15\ V$
reduction:	$Cu^{2+}(aq) + 2\ CN^-(aq) + e^- \rightarrow Cu(CN)_2^-(aq)$	$E° = 1.103\ V$
overall:	$Cu^+(aq) + 2\ CN^-(aq) \rightarrow Cu(CN)_2^-(aq)$	$E° = 0.953\ V$

$E^\circ = \frac{0.0592\ V}{n} \log K; \quad \log K = \frac{nE^\circ}{0.0592\ V} = \frac{(1)(0.953\ V)}{0.0592\ V} = 16.1$

$K = K_f = 10^{16.1} = 1 \times 10^{16}$

Multiconcept Problems

18.156 (a) $4\ CH_2{=}CHCN + 2\ H_2O \rightarrow 2\ NC(CH_2)_4CN + O_2$

(b) $\text{mol } e^- = 3000\ C/s \times 10.0\ h \times \frac{3600\ s}{1\ h} \times \frac{1\ mol\ e^-}{96{,}500\ C} = 1119.2\ mol\ e^-$

mass adiponitrile =

$1119.2\ mol\ e^- \times \frac{1\ mol\ adiponitrile}{2\ mol\ e^-} \times \frac{108.14\ g\ adiponitrile}{1\ mol\ adiponitrile} \times \frac{1.0\ kg}{1000\ g} = 60.5\ kg$

(c) $1119.2\ mol\ e^- \times \frac{1\ mol\ O_2}{4\ mol\ e^-} = 279.8\ mol\ O_2$

$$PV = nRT; \quad V = \frac{nRT}{P} = \frac{(279.8\ mol)\left(0.082\,06\ \frac{L\cdot atm}{K\cdot mol}\right)(298\ K)}{\left(740\ mm\ Hg \times \frac{1\ atm}{760\ mm\ Hg}\right)} = 7030\ L\ O_2$$

18.158

$Ba(s) + Cl_2(g) \rightarrow BaCl_2(s)$	$\Delta G^\circ_f = -806.7\ kJ/mol$
$BaCl_2(s) \rightarrow Ba^{2+}(aq) + 2\ Cl^-(aq)$	$\Delta G^\circ_1 = -16.7\ kJ/mol$
$Ba(s) + Cl_2(g) \rightarrow Ba^{2+}(aq) + 2\ Cl^-(aq)$	$\Delta G^\circ_2 = \Delta G^\circ_f + \Delta G^\circ_1 = -823.4\ kJ/mol$

$$E^\circ = \frac{-\Delta G^\circ}{nF} = \frac{-(-823{,}400\ J)}{(2\ mol\ e^-)\left(\frac{96{,}500\ C}{1\ mol\ e^-}\right)} = +4.266\ J/C = +4.266\ V$$

$Ba(s) + Cl_2(g) \rightarrow Ba^{2+}(aq) + 2\ Cl^-(aq)$	$E^\circ = +4.266\ V$
$2\ Cl^-(aq) \rightarrow Cl_2(g) + 2\ e^-$	$E^\circ = -1.36\ V$
$Ba(s) \rightarrow Ba^{2+}(aq) + 2\ e^-$	$E^\circ = 2.91\ V$
$Ba^{2+}(aq) + 2\ e^- \rightarrow Ba(s)$	$E^\circ = -2.91\ V$

18.160 (a) $Cr_2O_7^{2-}(aq) + 6\ Fe^{2+}(aq) + 14\ H^+(aq) \rightarrow 2\ Cr^{3+}(aq) + 6\ Fe^{3+}(aq) + 7\ H_2O(l)$

(b) The two half reactions are:

oxidation: $Fe^{2+}(aq) \rightarrow Fe^{3+}(aq) + e^-$ $\quad E^\circ = -0.77\ V$

reduction: $Cr_2O_7^{2-}(aq) + 14\ H^+(aq) + 6\ e^- \rightarrow 2\ Cr^{3+}(aq) + 7\ H_2O(l)$ $\quad E^\circ = 1.36\ V$

At the equivalence point the potential is given by either of the following expressions:

(1) $E = 1.36\ V - \frac{0.0592\ V}{6} \log \frac{[Cr^{3+}]^2}{[Cr_2O_7^{2-}][H^+]^{14}}$

(2) $E = 0.77\ V - \dfrac{0.0592\ V}{1}\log\dfrac{[Fe^{2+}]}{[Fe^{3+}]}$

where E is the same in both because equilibrium is reached and the solution can have only one potential. Multiplying (1) by 6, adding it to (2), and using some stoichiometric relationships at the equivalence point will simplify the log term.

$$7E = [(6 \times 1.36\ V) + 0.77\ V] - (0.0592\ V)\log\frac{[Fe^{2+}][Cr^{3+}]^2}{[Fe^{3+}][Cr_2O_7^{2-}][H^+]^{14}}$$

At the equivalence point, $[Fe^{2+}] = 6[Cr_2O_7^{2-}]$ and $[Fe^{3+}] = 3[Cr^{3+}]$. Substitute these equalities into the previous equation.

$$7E = [(6 \times 1.36\ V) + 0.77\ V] - (0.0592\ V)\log\frac{6[Cr_2O_7^{2-}][Cr^{3+}]^2}{3[Cr^{3+}][Cr_2O_7^{2-}][H^+]^{14}}$$

Cancel identical terms.

$$7E = [(6 \times 1.36\ V) + 0.77\ V] - (0.0592\ V)\log\frac{6[Cr^{3+}]}{3[H^+]^{14}}$$

mol Fe^{2+} = (0.120 L)(0.100 mol/L) = 0.0120 mol Fe^{2+}

mol $Cr_2O_7^{2-}$ = 0.0120 mol Fe^{2+} x $\dfrac{1\ mol\ Cr_2O_7^{2-}}{6\ mol\ Fe^{2+}}$ = 0.002 00 mol $Cr_2O_7^{2-}$

volume $Cr_2O_7^{2-}$ = 0.002 00 mol x $\dfrac{1\ L}{0.120\ mol}$ = 0.0167 L

At the equivalence point assume mol Fe^{3+} = initial mol Fe^{2+} = 0.0120 mol

Total volume at the equivalence point is 0.120 L + 0.0167 L = 0.1367 L

$[Fe^{3+}] = \dfrac{0.0120\ mol}{0.1367\ L}$ = 0.0878 M; $[Cr^{3+}] = [Fe^{3+}]/3$ = (0.0878 M)/3 = 0.0293 M

$[H^+] = 10^{-pH} = 10^{-2.00}$ = 0.010 M

$$7E = [(6 \times 1.36\ V) + 0.77\ V] - (0.0592\ V)\log\frac{6(0.0293)}{3(0.010)^{14}} = 8.93 - 1.585 = 7.345\ V$$

$E = \dfrac{7.345\ V}{7}$ = 1.05 V at the equivalence point.

18.162 (a) $Zn(s) + 2\ Ag^+(aq) + H_2O(l) \rightarrow ZnO(s) + 2\ Ag(s) + 2\ H^+(aq)$

$\Delta H^\circ_{rxn} = \Delta H^\circ_f(ZnO) - [2\ \Delta H^\circ_f(Ag^+) + \Delta H^\circ_f(H_2O)]$

ΔH°_{rxn} = [(1 mol)(–350.5 kJ/mol)] – [(2 mol)(105.6 kJ/mol) + (1 mol)(–285.8 kJ/mol)]

ΔH°_{rxn} = –275.9 kJ

$\Delta S^\circ = [S^\circ(ZnO) + 2\ S^\circ(Ag)] - [S^\circ(Zn) + 2\ S^\circ(Ag^+) + S^\circ(H_2O)]$

ΔS° = [(1 mol)(43.7 J/(K· mol)) + (2 mol)(42.6 J/(K· mol))]
– [(1 mol)(41.6 J/(K· mol)) + (2 mol)(72.7 J/(K· mol)) + (1 mol)(69.9 J/(K· mol))

ΔS° = – 128.0 J/K

$\Delta G^\circ = \Delta H^\circ - T\Delta S^\circ$ = – 275.9 kJ – (298 K)(– 128.0 x 10^{-3} kJ/K) = –237.8 kJ

(b) 1 V = 1 J/C

$$\Delta G^\circ = -nFE^\circ \quad E^\circ = \frac{-\Delta G^\circ}{nF} = \frac{-(-237.8 \times 10^3 \text{ J})}{(2 \text{ mol } e^-)\left(\frac{96{,}500 \text{ C}}{1 \text{ mol } e^-}\right)} = 1.232 \text{ J/C} = 1.232 \text{ V}$$

$$E^\circ = \frac{0.0592 \text{ V}}{n} \log K; \quad \log K = \frac{nE^\circ}{0.0592 \text{ V}} = \frac{(2)(1.232 \text{ V})}{0.0592 \text{ V}} = 41.62$$

$$K = 10^{41.62} = 4 \times 10^{41}$$

(c) $E = E^\circ - \frac{0.0592 \text{ V}}{n} \log \frac{[H^+]^2}{[Ag^+]^2}$

The addition of NH_3 to the cathode compartment would result in the formation of the $Ag(NH_3)_2^+$ complex ion, which results in a decrease in Ag^+ concentration. The log term in the Nernst equation becomes larger and the cell voltage decreases.

On mixing equal volumes of two solutions, the concentrations of both solutions are cut in half.

	$Ag^+(aq)$	+	$2\ NH_3(aq)$	⇄	$Ag(NH_3)_2^+(aq)$
before reaction (M)	0.0500		2.00		0
assume 100% reaction	–0.0500		–2(0.0500)		+0.0500
after reaction (M)	0		1.90		0.0500
assume small back rxn	+x		+2x		–x
equil (M)	x		1.90 + 2x		0.0500 – x

$$K_f = 1.7 \times 10^7 = \frac{[Ag(NH_3)_2^+]}{[Ag^+][NH_3]^2} = \frac{(0.0500 - x)}{(x)(1.90 + 2x)^2} \approx \frac{0.0500}{(x)(1.90)^2}$$

Solve for x. $x = [Ag^+] = 8.15 \times 10^{-10}$ M

$$E = E^\circ - \frac{0.0592 \text{ V}}{n} \log \frac{[H^+]^2}{[Ag^+]^2} = 1.232 \text{ V} - \frac{0.0592 \text{ V}}{2} \log \frac{(1.00 \text{ M})^2}{(8.15 \times 10^{-10} \text{ M})^2} = 0.694 \text{ V}$$

(d) Calculate new initial concentrations because of dilution to 110.0 mL.

$$M_i \times V_i = M_f \times V_f; \quad M_f = [Cl^-] = \frac{M_i \times V_i}{V_f} = \frac{0.200 \text{ M} \times 10.0 \text{ mL}}{110.0 \text{ mL}} = 0.0182 \text{ M}$$

$$M_i \times V_i = M_f \times V_f; \quad M_f = [Ag^+] = \frac{M_i \times V_i}{V_f} = \frac{0.0500 \text{ M} \times 100.0 \text{ mL}}{110.0 \text{ mL}} = 0.0455 \text{ M}$$

$$M_i \times V_i = M_f \times V_f; \quad M_f = [NH_3] = \frac{M_i \times V_i}{V_f} = \frac{2.00 \text{ M} \times 100.0 \text{ mL}}{110.0 \text{ mL}} = 1.82 \text{ M}$$

Now calculate the $[Ag^+]$ as a result of the following equilibrium:

	$Ag^+(aq)$	+	$2\ NH_3(aq)$	⇄	$Ag(NH_3)_2^+(aq)$
before reaction (M)	0.0455		1.82		0
assume 100% reaction	–0.0455		–2(0.0455)		+0.0455
after reaction (M)	0		1.73		0.0455
assume small back rxn	x		+2x		–x
equil (M)	x		1.73 + 2x		0.0455 – x

$$K_f = 1.7 \times 10^7 = \frac{[Ag(NH_3)_2^+]}{[Ag^+][NH_3]^2} = \frac{(0.0455 - x)}{(x)(1.73 + 2x)^2} \approx \frac{0.0455}{(x)(1.73)^2}$$

Solve for x. $x = [Ag^+] = 8.94 \times 10^{-10}$ M

For AgCl, $K_{sp} = 1.8 \times 10^{-10}$

$IP = [Ag^+][Cl^-] = (8.94 \times 10^{-10}\text{ M})(0.0182\text{ M}) = 1.6 \times 10^{-11}$

$IP < K_{sp}$, AgCl will not precipitate.

Now calculate new initial concentrations because of dilution to 120.0 mL.

$$M_i \times V_i = M_f \times V_f;\quad M_f = [Br^-] = \frac{M_i \times V_i}{V_f} = \frac{0.200\text{ M} \times 10.0\text{ mL}}{120.0\text{ mL}} = 0.0167\text{ M}$$

$$M_i \times V_i = M_f \times V_f;\quad M_f = [Ag^+] = \frac{M_i \times V_i}{V_f} = \frac{0.0500\text{ M} \times 100.0\text{ mL}}{120.0\text{ mL}} = 0.0417\text{ M}$$

$$M_i \times V_i = M_f \times V_f;\quad M_f = [NH_3] = \frac{M_i \times V_i}{V_f} = \frac{2.00\text{ M} \times 100.0\text{ mL}}{120.0\text{ mL}} = 1.67\text{ M}$$

Now calculate the $[Ag^+]$ as a result of the following equilibrium:

	$Ag^+(aq)$ +	$2\ NH_3(aq)$ ⇄	$Ag(NH_3)_2^+(aq)$
before reaction (M)	0.0417	1.67	0
assume 100% reaction	–0.0417	–2(0.0417)	+0.0417
after reaction (M)	0	1.59	0.0417
assume small back rxn	+x	+2x	–x
equil (M)	x	1.59 + 2x	0.0417 – x

$$K_f = 1.7 \times 10^7 = \frac{[Ag(NH_3)_2^+]}{[Ag^+][NH_3]^2} = \frac{(0.0417 - x)}{(x)(1.59 + 2x)^2} \approx \frac{0.0417}{(x)(1.59)^2}$$

Solve for x. $x = [Ag^+] = 9.70 \times 10^{-10}$ M

For AgBr, $K_{sp} = 5.4 \times 10^{-13}$

$IP = [Ag^+][Br^-] = (9.70 \times 10^{-10}\text{ M})(0.0167\text{ M}) = 1.6 \times 10^{-11}$

$IP > K_{sp}$, AgBr will precipitate.

18.164 (a) Oxidation half reaction: $2\ [C_4H_{10}(g) + 13\ O^{2-}(s) \rightarrow 4\ CO_2(g) + 5\ H_2O(l) + 26\ e^-]$

Reduction half reaction: $13\ [O_2(g) + 4\ e^- \rightarrow 2\ O^{2-}(s)]$

Cell reaction: $2\ C_4H_{10}(g) + 13\ O_2(g) \rightarrow 8\ CO_2(g) + 10\ H_2O(l)$

(b) $\Delta H^\circ = [8\ \Delta H^\circ_f(CO_2) + 10\ \Delta H^\circ_f(H_2O)] - [2\ \Delta H^\circ_f(C_4H_{10})]$

$\Delta H^\circ = [(8\text{ mol})(-393.5\text{ kJ/mol}) + (10\text{ mol})(-285.8\text{ kJ/mol})]$
$\quad - [(2\text{ mol})(-126\text{ kJ/mol})] = -5754\text{ kJ}$

$\Delta S^\circ = [8\ S^\circ(CO_2) + 10\ S^\circ(H_2O)] - [2\ S^\circ(C_4H_{10}) + 13\ S^\circ(O_2)]$

$\Delta S^\circ = [(8\text{ mol})(213.6\text{ J/(K}\cdot\text{mol)}) + (10\text{ mol})(69.9\text{ J/(K}\cdot\text{mol)})]$
$\quad - [(2\text{ mol})(310\text{ J/(K}\cdot\text{mol)}) + (13\text{ mol})(205\text{ J/(K}\cdot\text{mol)})] = -877.2\text{ J/K}$

$\Delta G^\circ = \Delta H^\circ - T\Delta S^\circ = -5754\text{ kJ} - (298\text{ K})(-877.2 \times 10^{-3}\text{ kJ/K}) = -5493\text{ kJ}$

1 V = 1 J/C

$$\Delta G^\circ = -nFE^\circ;\quad E^\circ = -\frac{\Delta G^\circ}{nF} = -\frac{-5493 \times 10^3\text{ J}}{(52)(96{,}500\text{ C})} = 1.09\text{ J/C} = 1.09\text{ V}$$

$\Delta G^\circ = -RT \ln K$

$$\ln K = \frac{-\Delta G^\circ}{RT} = \frac{-(-5493\text{ kJ})}{(8.314 \times 10^{-3}\text{ kJ/K})(298\text{ K})} = 2217$$

$K = e^{2217} = 7 \times 10^{962}$

On raising the temperature, both K and E° will decrease because the reaction is exothermic ($\Delta H^\circ < 0$).

(c) C_4H_{10}, 58.12; 10.5 A = 10.5 C/s

$$\text{mass } C_4H_{10} = 10.5\text{ C/s} \times 8\text{ hr} \times \frac{60\text{ min}}{1\text{ hr}} \times \frac{60\text{ s}}{1\text{ min}} \times \frac{1\text{ mol e}^-}{96{,}500\text{ C}} \times \frac{2\text{ mol } C_4H_{10}}{52\text{ mol e}^-} \times \frac{58.12\text{ g } C_4H_{10}}{1\text{ mol } C_4H_{10}} = 7.00\text{ g } C_4H_{10}$$

$$n = 7.00\text{ g } C_4H_{10} \times \frac{1\text{ mol } C_4H_{10}}{58.12\text{ g } C_4H_{10}} = 0.120\text{ mol } C_4H_{10}$$

20 °C = 20 + 273 = 293 K

$$PV = nRT \qquad V = \frac{nRT}{P} = \frac{(0.120\text{ mol})\left(0.082\ 06\ \frac{\text{L}\cdot\text{atm}}{\text{K}\cdot\text{mol}}\right)(293\text{ K})}{\left(815\text{ mm Hg} \times \frac{1.00\text{ atm}}{760\text{ mm Hg}}\right)} = 2.69\text{ L}$$

18.166 (a) $4\,[Au(s) + 2\,CN^-(aq) \rightarrow Au(CN)_2^-(aq) + e^-]$ (oxidation half reaction)

$O_2(g) \rightarrow 2\,H_2O(l)$

$O_2(g) + 4\,H^+(aq) \rightarrow 2\,H_2O(l)$

$4\,e^- + O_2(g) + 4\,H^+(aq) \rightarrow 2\,H_2O(l)$ (reduction half reaction)

Combine the two half reactions.

$4\,Au(s) + 8\,CN^-(aq) + O_2(g) + 4\,H^+(aq) \rightarrow 4\,Au(CN)_2^-(aq) + 2\,H_2O(l)$

$4\,Au(s) + 8\,CN^-(aq) + O_2(g) + 4\,H^+(aq) + 4\,OH^-(aq)$
$\rightarrow 4\,Au(CN)_2^-(aq) + 2\,H_2O(l) + 4\,OH^-(aq)$

$4\,Au(s) + 8\,CN^-(aq) + O_2(g) + 4\,H_2O(l)$
$\rightarrow 4\,Au(CN)_2^-(aq) + 2\,H_2O(l) + 4\,OH^-(aq)$

$4\,Au(s) + 8\,CN^-(aq) + O_2(g) + 2\,H_2O(l) \rightarrow 4\,Au(CN)_2^-(aq) + 4\,OH^-(aq)$

(b) Add the following five reactions together. ΔG° is calculated below each reaction.

$4\,[Au^+(aq) + 2\,CN^-(aq) \rightarrow Au(CN)_2^-(aq)]$ $\quad K = (K_f)^4$

$\Delta G^\circ = -RT \ln K = -(8.314 \times 10^{-3}\text{ kJ/K})(298\text{ K}) \ln (6.2 \times 10^{38})^4 = -885.2\text{ kJ}$

$O_2(g) + 4\,H^+(aq) + 4\,e^- \rightarrow 2\,H_2O(l)$ $\quad E^\circ = 1.229\text{ V}$

$$\Delta G^\circ = -nFE^\circ = -(4\text{ mol e}^-)\left(\frac{96{,}500\text{ C}}{1\text{ mol e}^-}\right)(1.229\text{ V})\left(\frac{1\text{ J}}{1\text{ C}\cdot\text{V}}\right) = -474{,}394\text{ J} = -474.4\text{ kJ}$$

$4\,[H_2O(l) \rightleftharpoons H^+(aq) + OH^-(aq)]$ $\quad K = (K_w)^4$

$\Delta G^\circ = -RT \ln K = -(8.314 \times 10^{-3}\text{ kJ/K})(298\text{ K}) \ln (1.0 \times 10^{-14})^4 = +319.5\text{ kJ}$

$4\,[Au(s) \rightarrow Au^{3+}(aq) + 3\,e^-]$ $\qquad E^\circ = -1.498\ V$

$$\Delta G^\circ = -nFE^\circ = -(12\ mol\ e^-)\left(\frac{96{,}500\ C}{1\ mol\ e^-}\right)(-1.498\ V)\left(\frac{1\ J}{1\ C \cdot V}\right) = +1{,}734{,}684\ J = +1{,}734.7\ kJ$$

$4\,[Au^{3+}(aq) + 2\,e^- \rightarrow Au^+(aq)]$ $\qquad E^\circ = 1.401\ V$

$$\Delta G^\circ = -\,nFE^\circ = -(8\ mol\ e^-)\left(\frac{96{,}500\ C}{1\ mol\ e^-}\right)(1.401\ V)\left(\frac{1\ J}{1\ C \cdot V}\right) = -1{,}081{,}572\ J = -1{,}081.6\ kJ$$

Overall reaction:

$4\,Au(s) + 8\,CN^-(aq) + O_2(g) + 2\,H_2O(l) \rightarrow 4\,Au(CN)_2^-(aq) + 4\,OH^-(aq)$

$\Delta G^\circ = -\,885.2\ kJ - 474.4\ kJ + 319.5\ kJ + 1{,}734.7\ kJ - 1{,}081.6\ kJ = -387.0\ kJ$

19 Nuclear Chemistry

19.1 (a) In beta emission, the mass number is unchanged, and the atomic number increases by one. $^{106}_{44}\text{Ru} \rightarrow {}^{0}_{-1}\text{e} + {}^{106}_{45}\text{Rh}$

(b) In alpha emission, the mass number decreases by four, and the atomic number decreases by two. $^{189}_{83}\text{Bi} \rightarrow {}^{4}_{2}\text{He} + {}^{185}_{81}\text{Tl}$

(c) In electron capture, the mass number is unchanged, and the atomic number decreases by one. $^{204}_{84}\text{Po} + {}^{0}_{-1}\text{e} \rightarrow {}^{204}_{83}\text{Bi}$

19.2 $^{148}_{69}\text{Tm}$ decays to $^{148}_{68}\text{Er}$ by either positron emission or electron capture.

19.3 (a) ^{199}Au has a higher neutron/proton ratio and decays by beta emission. ^{173}Au has a lower neutron/proton ratio and decays by alpha emission.

(b) ^{196}Pb has a lower neutron/proton ratio and decays by positron emission. ^{206}Pb is nonradioactive.

19.4 $^{238}_{92}\text{U} \rightarrow {}^{238-(8\times4)-(6\times0)}_{92-(8\times2)-(6\times-1)}\text{X} = {}^{206}_{82}\text{Pb}$

19.5 $t_{1/2} = \dfrac{0.693}{k} = \dfrac{0.693}{1.08 \text{ x } 10^{-2}\text{ h}^{-1}} = 64.2 \text{ h}$

19.6 $k = \dfrac{0.693}{t_{1/2}} = \dfrac{0.693}{3.82 \text{ d}} = 0.181 \text{ d}^{-1}$

19.7 $\ln\left(\dfrac{N}{N_0}\right) = -0.693\left(\dfrac{t}{t_{1/2}}\right) = -0.693\left(\dfrac{16{,}230 \text{ y}}{5715 \text{ y}}\right) = -1.968$

$\dfrac{N}{N_0} = e^{-1.968} = 0.140;\quad \dfrac{N}{100\%} = 0.140;\quad N = 14.0\%$

19.8 Assume $N_0 = 100\%$ and $N = 89.2\%$ at $t = 5.00$ y

$$\ln\left(\frac{N}{N_0}\right) = (-0.693)\left(\frac{t}{t_{1/2}}\right); \qquad \frac{N}{N_0} = \frac{\text{Decay rate at time t}}{\text{Decay rate at time t} = 0}$$

$$\ln\left(\frac{89.2}{100}\right) = (-0.693)\left(\frac{5.00\text{ y}}{t_{1/2}}\right); \qquad t_{1/2} = 30.3\text{ y}$$

19.9

$$\ln\left(\frac{N}{N_0}\right) = (-0.693)\left(\frac{t}{t_{1/2}}\right); \qquad \frac{N}{N_0} = \frac{\text{Decay rate at time t}}{\text{Decay rate at t} = 0}$$

$$\ln\left(\frac{10{,}860}{16{,}800}\right) = (-0.693)\left(\frac{28.0\text{ d}}{t_{1/2}}\right); \qquad t_{1/2} = 44.5\text{ d}$$

19.10

$$\ln\left(\frac{N}{N_0}\right) = (-0.693)\left(\frac{t}{t_{1/2}}\right); \qquad \frac{N}{N_0} = \frac{\text{Decay rate at time t}}{\text{Decay rate at t} = 0}$$

$t_{1/2} = 44.5$ d

$$\ln\left(\frac{N}{16{,}800}\right) = (-0.693)\left(\frac{40.0\text{ d}}{44.5\text{ d}}\right)$$

$$\ln N - \ln(16{,}800) = (-0.693)\left(\frac{40.0\text{ d}}{44.5\text{ d}}\right)$$

$$\ln N = (-0.693)\left(\frac{40.0\text{ d}}{44.5\text{ d}}\right) + \ln(16{,}800) = 9.106$$

$N = e^{9.106} = 9011$ disintegrations/min

19.11 For ${}^{16}_{8}O$:

First, calculate the total mass of the nucleons (8 n + 8 p)

Mass of 8 neutrons = (8)(1.008 66) = 8.069 28
Mass of 8 protons = (8)(1.007 28) = 8.058 24
Mass of 8 n + 8 p = 16.127 52

Next, calculate the mass of a ^{16}O nucleus by subtracting the mass of 8 electrons from the mass of a ^{16}O atom.

Mass of ^{16}O atom = 15.994 91
–Mass of 8 electrons = $-(8)(5.486 \times 10^{-4})$ = –0.004 39
Mass of ^{16}O nucleus = 15.990 52

Then subtract the mass of the ^{16}O nucleus from the mass of the nucleons to find the mass defect:

Mass defect = mass of nucleons – mass of nucleus
= (16.127 52) – (15.990 52) = 0.137 00 u
Mass defect in grams = (0.137 00 u)(1.660 54 x 10^{-24} g/u) = 2.2749 x 10^{-25} g
Mass defect in g/mol = (2.2749 x 10^{-25} g)(6.022 x 10^{23} mol^{-1}) = 0.136 99 g/mol
Now, use the Einstein equation to convert the mass defect into the binding energy.
$\Delta E = \Delta mc^2 = (0.136\ 99\ \text{g/mol})(10^{-3}\ \text{kg/g})(3.00 \times 10^8\ \text{m/s})^2$
$\Delta E = 1.233 \times 10^{13}$ J/mol = 1.233×10^{10} kJ/mol

$$\Delta E = \frac{1.233 \times 10^{13}\ \text{J/mol}}{6.022 \times 10^{23}\ \text{nuclei/mol}} \times \frac{1\ \text{MeV}}{1.60 \times 10^{-13}\ \text{J}} \times \frac{1\ \text{nucleus}}{16\ \text{nucleons}} = 8.00\ \frac{\text{MeV}}{\text{nucleon}}$$

19.12 ^{6}Li atomic mass = (3 proton mass) + (3 neutron mass) + (3 electron mass) – (mass defect)
= (3 x 1.007 28) + (3 x 1.008 66) + (3 x 5.486 x 10^{-4}) – (0.034 37) = 6.0151 u
Mass defect in g/mol = 0.034 37 g/mol
Now, use the Einstein equation to convert the mass defect into the binding energy.
$\Delta E = \Delta mc^2 = (0.034\ 37\ \text{g/mol})(10^{-3}\ \text{kg/g})(3.00 \times 10^8\ \text{m/s})^2$
$\Delta E = 3.093 \times 10^{12}$ J/mol = 3.093×10^{9} kJ/mol

$$\Delta E = \frac{3.093 \times 10^{12}\ \text{J/mol}}{6.022 \times 10^{23}\ \text{nuclei/mol}} \times \frac{1\ \text{MeV}}{1.60 \times 10^{-13}\ \text{J}} \times \frac{1\ \text{nucleus}}{6\ \text{nucleons}} = 5.35\ \frac{\text{MeV}}{\text{nucleon}}$$

19.13 $$\Delta m = \frac{\Delta E}{c^2} = \frac{-820 \times 10^3\ \text{J}}{(3.00 \times 10^8\ \text{m/s})^2} = \frac{-820 \times 10^3\ \text{kg·m}^2/\text{s}^2}{(3.00 \times 10^8\ \text{m/s})^2} = -9.11 \times 10^{-12}\ \text{kg}$$

$$\Delta m = \frac{-9.11 \times 10^{-9}\ \text{g}}{2\ \text{mol NaCl}} = 4.56 \times 10^{-9}\ \text{g/mol}$$

19.14 $$\Delta m = \frac{\Delta E}{c^2} = \frac{-3.9 \times 10^{10}\ \text{J}}{(3.00 \times 10^8\ \text{m/s})^2} = \frac{-3.9 \times 10^{10}\ \text{kg·m}^2/\text{s}^2}{(3.00 \times 10^8\ \text{m/s})^2} = -4.3 \times 10^{-7}\ \text{kg}$$
$\Delta m = -4.3 \times 10^{-4}$ g

19.15 $^{1}_{0}n + ^{235}_{92}U \rightarrow ^{137}_{52}Te + ^{97}_{40}Zr + 2\ ^{1}_{0}n$

mass $^{235}_{92}U$	235.0439
mass $^{1}_{0}n$	1.008 66
–mass $^{137}_{52}Te$	–136.9254
–mass $^{97}_{40}Zr$	–96.9110
–mass 2 $^{1}_{0}n$	–(2)(1.008 66)
mass change	0.1988 u

$(0.1988 \text{ u})(1.660\ 54 \times 10^{-24} \text{ g/u})(6.022 \times 10^{23} \text{ mol}^{-1}) = 0.1988 \text{ g/mol}$
$\Delta E = \Delta mc^2 = (0.1988 \text{ g/mol})(10^{-3} \text{ kg/g})(3.00 \times 10^8 \text{ m/s})^2$
$\Delta E = 1.79 \times 10^{13} \text{ J/mol} = 1.79 \times 10^{10} \text{ kJ/mol}$

19.16 (a) $^{238}_{92}U \rightarrow {}^{232}_{90}Th + {}^{4}_{2}He$

(b)

mass ^{238}U	238.0508
mass ^{232}Th	–234.0436
mass ^{4}He	–4.0026
mass change	0.0046 u

$(0.0046 \text{ u/atom})(1.660\ 54 \times 10^{-24} \text{ g/u}) = 7.6 \times 10^{-27} \text{ g/atom}$

$(0.0046 \text{ u})(1.660\ 54 \times 10^{-24} \text{ g/u})(6.022 \times 10^{23} \text{ mol}^{-1}) = 0.0046 \text{ g/mol}$
$\Delta E = \Delta mc^2 = (0.0046 \text{ g/mol})(10^{-3} \text{ kg/g})(3.00 \times 10^8 \text{ m/s})^2$
$\Delta E = 4.1 \times 10^{11} \text{ J/mol} = 4.1 \times 10^8 \text{ kJ/mol}$
(c) Mass is lost and energy is released.

19.17 $^{1}_{1}H + {}^{2}_{1}H \rightarrow {}^{3}_{2}He$

mass ^{1}H	1.007 83
mass ^{2}H	2.014 10
–mass ^{3}He	–3.016 03
mass change	0.005 90 u

$(0.005\ 90 \text{ u})(1.660\ 54 \times 10^{-24} \text{ g/u})(6.022 \times 10^{23} \text{ mol}^{-1}) = 0.005\ 90 \text{ g/mol}$
$\Delta E = \Delta mc^2 = (0.005\ 90 \text{ g/mol})(10^{-3} \text{ kg/g})(3.00 \times 10^8 \text{ m/s})^2$
$\Delta E = 5.31 \times 10^{11} \text{ J/mol} = 5.31 \times 10^8 \text{ kJ/mol}$

19.18 $^{40}_{18}Ar + {}^{1}_{1}p \rightarrow {}^{40}_{19}K + {}^{1}_{0}n$

19.19 $^{238}_{92}U + {}^{12}_{6}C \rightarrow {}^{246}_{98}Cf + 4\,{}^{1}_{0}n$

19.20 $\ln\left(\frac{N}{N_0}\right) = (-0.693)\left(\frac{t}{t_{1/2}}\right)$; $\frac{N}{N_0} = \frac{\text{Decay rate at time t}}{\text{Decay rate at time t} = 0}$

$\ln\left(\frac{2.4}{15.3}\right) = (-0.693)\left(\frac{t}{5730 \text{ y}}\right)$; $t = 1.53 \times 10^4 \text{ y}$

19.21

	^{40}K	→	^{40}Ar
then (mmol)	N_0		0
change (mmol)	–0.95		+0.95
now (mmol)	N = 1.20		0.95

$N_0 - 0.95 = 1.20$ mmol therefore, $N_0 = 1.20 + 0.95 = 2.15$ mmol and N = 1.20 mmol

$$\ln\left(\frac{N}{N_0}\right) = (-0.693)\left(\frac{t}{t_{1/2}}\right)$$

$$\ln\left(\frac{1.20}{2.15}\right) = (-0.693)\left(\frac{t}{1.25 \times 10^9 \text{ y}}\right); \qquad t = 1.05 \times 10^9 \text{ y}$$

19.22 $^{235}_{92}U$

mass defect = (92 proton mass) + (143 neutron mass) + (92 electron mass) – $^{235}_{92}U$ mass

mass defect = (92 x 1.007 28) + (143 x 1.008 67) + (92 x 5.486 x 10^{-4}) – (235.043 929 9) = 1.916 11 u

(1.916 11 u/atom)(1.660 54 x 10^{-24} g/u) = 3.1818 x 10^{-24} g/atom

(1.916 11 u)(1.660 54 x 10^{-24} g/u)(6.022 x 10^{23} mol^{-1}) = 1.9161 g/mol

$\Delta E = \Delta mc^2 = (1.9161 \text{ g/mol})(10^{-3} \text{ kg/g})(3.00 \times 10^8 \text{ m/s})^2$

$\Delta E = 1.724 \times 10^{14}$ J/mol = 1.233×10^{11} kJ/mol

$$\Delta E = \frac{1.724 \times 10^{14} \text{ J/mol}}{6.022 \times 10^{23} \text{ nuclei/mol}} \times \frac{1 \text{ MeV}}{1.60 \times 10^{-13} \text{ J}} \times \frac{1 \text{ nucleus}}{235 \text{ nucleons}} = 7.62 \frac{\text{MeV}}{\text{nucleon}}$$

19.23

$^{235}_{92}U \rightarrow {}^{231}_{90}Th + {}^{4}_{2}He$

$^{231}_{90}Th \rightarrow {}^{231}_{91}Pa + {}^{0}_{-1}e$

$^{231}_{91}Pa \rightarrow {}^{227}_{89}Ac + {}^{4}_{2}He$

$^{227}_{89}Ac \rightarrow {}^{223}_{87}Fr + {}^{4}_{2}He$

$^{223}_{87}Fr \rightarrow {}^{219}_{85}At + {}^{4}_{2}He$

$^{219}_{85}At \rightarrow {}^{215}_{83}Bi + {}^{4}_{2}He$

$^{215}_{83}Bi \rightarrow {}^{215}_{84}Po + {}^{0}_{-1}e$

$^{215}_{84}Po \rightarrow {}^{211}_{82}Pb + {}^{4}_{2}He$

$$^{211}_{82}\text{Pb} \rightarrow {}^{211}_{83}\text{Bi} + {}^{0}_{-1}\text{e}$$
$$^{211}_{83}\text{Bi} \rightarrow {}^{207}_{81}\text{Tl} + {}^{4}_{2}\text{He}$$
$$^{207}_{81}\text{Tl} \rightarrow {}^{207}_{82}\text{Pb} + {}^{0}_{-1}\text{e}$$

19.24 $N_0 = 3.00$ and $N = 0.72$

$$\ln\left(\frac{N}{N_0}\right) = (-0.693)\left(\frac{t}{t_{1/2}}\right)$$

$$\ln\left(\frac{0.72}{3.00}\right) = (-0.693)\left(\frac{t}{7.03 \text{ x } 10^8 \text{ y}}\right); \qquad t = 1.45 \text{ x } 10^9 \text{ y}$$

19.25 $^{1}_{0}\text{n} + {}^{235}_{92}\text{U} \rightarrow {}^{140}_{56}\text{Ba} + {}^{93}_{36}\text{Kr} + 3\ {}^{1}_{0}\text{n}$

mass ^{1}n	1.00867
mass ^{235}U	235.0439
mass ^{140}Ba	−139.9106
mass ^{93}Kr	−92.9313
mass 3 ^{1}n	−3(1.00867)
mass change	0.1847

(0.1847 u)(1.660 54 x 10^{-24} g/u)(6.022 x 10^{23} mol^{-1}) = 0.1847 g/mol
$\Delta E = \Delta mc^2$ = (0.1847 g/mol)(10^{-3} kg/g)(3.00 x 10^8 m/s)^2
ΔE = 1.66 x 10^{13} J/mol = 1.66 x 10^{10} kJ/mol

19.26 2 billion years ago unusually rich uranium deposits, with the fissionable ^{235}U isotope at about 3% abundance, were flooded by groundwater, which acted as a moderator to slow the neutrons released by fission of ^{235}U, thereby allowing a nuclear chain reaction to take place. Today, however, the natural abundance of ^{235}U is only about 0.7% because of nuclear decay over the past few billion years. Because a chain-reaction is no longer self-sustaining at the present 0.7% level of ^{235}U, the conditions needed for natural reactors are no longer present on Earth.

Section Problems

Nuclear Reactions and Radioactivity (Sections 19.1 and 19.2)

19.28 Positron emission is the conversion of a proton in the nucleus into a neutron plus an ejected positron.
Electron capture is the process in which a proton in the nucleus captures an inner-shell electron and is thereby converted into a neutron.

19.30 In beta emission a neutron is converted to a proton and the atomic number increases.
In positron emission a proton is converted to a neutron and the atomic number decreases.

19.32 (a) ${}^{126}_{50}Sn \rightarrow {}^{0}_{-1}e + {}^{126}_{51}Sb$ (b) ${}^{210}_{88}Ra \rightarrow {}^{4}_{2}He + {}^{206}_{86}Rn$
(c) ${}^{77}_{37}Rb \rightarrow {}^{0}_{1}e + {}^{77}_{36}Kr$ (d) ${}^{76}_{36}Kr + {}^{0}_{-1}e \rightarrow {}^{76}_{35}Br$

19.34 The mass number decreases by four, and the atomic number decreases by two. This is characteristic of alpha emission. ${}^{214}_{90}Th \rightarrow {}^{210}_{88}Ra + {}^{4}_{2}He$

19.36 (a) ${}^{188}_{80}Hg \rightarrow {}^{188}_{79}Au + {}^{0}_{1}e$ (b) ${}^{218}_{85}At \rightarrow {}^{214}_{83}Bi + {}^{4}_{2}He$
(c) ${}^{234}_{90}Th \rightarrow {}^{234}_{91}Pa + {}^{0}_{-1}e$

19.38 (a) ${}^{162}_{75}Re \rightarrow {}^{158}_{73}Ta + {}^{4}_{2}He$ (b) ${}^{138}_{62}Sm + {}^{0}_{-1}e \rightarrow {}^{138}_{61}Pm$
(c) ${}^{188}_{74}W \rightarrow {}^{188}_{75}Re + {}^{0}_{-1}e$ (d) ${}^{165}_{73}Ta \rightarrow {}^{165}_{72}Hf + {}^{0}_{1}e$

19.40 ${}^{100}_{43}Tc \rightarrow {}^{0}_{1}e + {}^{100}_{42}Mo$ (positron emission)
${}^{100}_{43}Tc + {}^{0}_{-1}e \rightarrow {}^{100}_{42}Mo$ (electron capture)

Nuclear Stability (Section 19.3)

19.42 ${}^{160}W$ is neutron poor and decays by alpha emission. ${}^{185}W$ is neutron rich and decays by beta emission.

19.44 "Neutron rich" nuclides emit beta particles to decrease the number of neutrons and increase the number of protons in the nucleus.

19.46 $^{241}_{95}Am \rightarrow ^{237}_{93}Np + ^{4}_{2}He$

$^{237}_{93}Np \rightarrow ^{233}_{91}Pa + ^{4}_{2}He$

$^{233}_{91}Pa \rightarrow ^{233}_{92}U + ^{0}_{-1}e$

$^{233}_{92}U \rightarrow ^{229}_{90}Th + ^{4}_{2}He$

$^{229}_{90}Th \rightarrow ^{225}_{88}Ra + ^{4}_{2}He$

$^{225}_{88}Ra \rightarrow ^{225}_{89}Ac + ^{0}_{-1}e$

$^{225}_{89}Ac \rightarrow ^{221}_{87}Fr + ^{4}_{2}He$

$^{221}_{87}Fr \rightarrow ^{217}_{85}At + ^{4}_{2}He$

$^{217}_{85}At \rightarrow ^{213}_{83}Bi + ^{4}_{2}He$

$^{213}_{83}Bi \rightarrow ^{213}_{84}Po + ^{0}_{-1}e$

$^{213}_{84}Po \rightarrow ^{209}_{82}Pb + ^{4}_{2}He$

$^{209}_{82}Pb \rightarrow ^{209}_{83}Bi + ^{0}_{-1}e$

19.48 Each alpha emission decreases the mass number by four and the atomic number by two. Each beta emission increases the atomic number by one.

$^{232}_{90}Th \rightarrow ^{208}_{82}Pb$

$$\text{Number of } \alpha \text{ emissions} = \frac{\text{Th mass number} - \text{Pb mass number}}{4}$$

$$= \frac{232 - 208}{4} = 6\ \alpha \text{ emissions}$$

The atomic number decreases by 12 as a result of 6 alpha emissions. The resulting atomic number is (90 – 12) = 78.

Number of β emissions = Pb atomic number – 78 = 82 – 78 = 4 β emissions

Radioactive Decay Rates (Section 19.4)

19.50 $$k = \frac{0.693}{t_{1/2}} = \frac{0.693}{2.805\ d} = 0.247\ d^{-1}$$

19.52 $$t_{1/2} = \frac{0.693}{k} = \frac{0.693}{7.95 \times 10^{-3}\ d^{-1}} = 87.17d$$

$$\ln\left(\frac{N}{N_0}\right) = (-0.693)\left(\frac{t}{t_{1/2}}\right) = (-0.693)\left(\frac{185\text{ d}}{87.17\text{ d}}\right) = -1.4707$$

$$\frac{N}{N_0} = e^{-1.4707} = 0.2298; \qquad \frac{N}{100\%} = 0.2298; \qquad N = 23.0\%$$

19.54 $t_{1/2} = (102\text{ y})(365\text{ d/y})(24\text{ h/d})(3600\text{ s/h}) = 3.2167 \times 10^9\text{ s}$

$$k = \frac{0.693}{t_{1/2}} = \frac{0.693}{3.2167 \times 10^9\text{ s}} = 2.1544 \times 10^{-10}\text{ s}^{-1}$$

$$N = (1.0 \times 10^{-9}\text{ g})\left(\frac{1\text{ mol Po}}{209\text{ g Po}}\right)(6.022 \times 10^{23}\text{ atoms/mol}) = 2.881 \times 10^{12}\text{ atoms}$$

Decay rate $= kN = (2.1544 \times 10^{-10}\text{ s}^{-1})(2.881 \times 10^{12}\text{ atoms}) = 6.21 \times 10^2\text{ s}^{-1}$

621 α particles are emitted in 1.0 s.

19.56 Decay rate = kN

$$N = (1.0 \times 10^{-3}\text{ g})\left(\frac{1\text{ mol }^{79}\text{Se}}{79\text{ g}}\right)(6.022 \times 10^{23}\text{ atoms/mol}) = 7.6 \times 10^{18}\text{ atoms}$$

$$k = \frac{\text{Decay rate}}{N} = \frac{1.5 \times 10^5\text{/s}}{7.6 \times 10^{18}} = 2.0 \times 10^{-14}\text{ s}^{-1}$$

$$t_{1/2} = \frac{0.693}{k} = \frac{0.693}{2.0 \times 10^{-14}\text{ s}^{-1}} = 3.5 \times 10^{13}\text{ s}$$

$$t_{1/2} = (3.5 \times 10^{13}\text{ s})\left(\frac{1\text{ h}}{3600\text{ s}}\right)\left(\frac{1\text{ d}}{24\text{ h}}\right)\left(\frac{1\text{ y}}{365\text{ d}}\right) = 1.1 \times 10^6\text{ y}$$

19.58 $\ln\left(\frac{N}{N_0}\right) = (-0.693)\left(\frac{t}{t_{1/2}}\right); \quad \frac{N}{N_0} = \frac{\text{Decay rate at time t}}{\text{Decay rate at time t} = 0}$

$$\ln\left(\frac{6990}{8540}\right) = (-0.693)\left(\frac{10.0\text{ d}}{t_{1/2}}\right); \quad t_{1/2} = 34.6\text{ d}$$

19.60 (a) $1 \rightarrow 1/2 \rightarrow 1/4 \rightarrow 1/8$

After three half-lives, 1/8 of the strontium-90 will remain.

(b) $k = \frac{0.693}{t_{1/2}} = \frac{0.693}{29\text{ y}} = 0.0239\text{ y}^{-1} = 0.024\text{ y}^{-1}$

$$\text{(c)}\ t = \frac{\ln\frac{N}{N_o}}{-k} = \frac{\ln\frac{(Sr{-}90)_t}{(Sr{-}90)_o}}{-k} = \frac{\ln\frac{(0.01)}{(1)}}{-0.0239y^{-1}} = 193\ y$$

Energy Changes during Nuclear Reactions (Section 19.5)

19.62 The loss in mass that occurs when protons and neutrons combine to form a nucleus is called the mass defect. The lost mass is converted into the binding energy that is used to hold the nucleons together.

19.64 (a) For $^{52}_{26}Fe$:

First, calculate the total mass of the nucleons (26 n + 26 p)

Mass of 26 neutrons = (26)(1.008 66) = 26.225 16

Mass of 26 protons = (26)(1.007 28) = 26.189 28

Mass of 26 n + 26 p = 52.414 44

Next, calculate the mass of a ^{52}Fe nucleus by subtracting the mass of 26 electrons from the mass of a ^{52}Fe atom.

Mass of ^{52}Fe atom = 51.948 11

–Mass of 26 electrons = –(26)(5.486 x 10^{-4}) = –0.014 26

Mass of ^{52}Fe nucleus = 51.933 85

Then subtract the mass of the ^{52}Fe nucleus from the mass of the nucleons to find the mass defect:

Mass defect = mass of nucleons – mass of nucleus

= (52.414 44) – (51.933 85) = 0.480 59 u

Mass defect in g/mol:

(0.480 59 u)(1.660 54 x 10^{-24} g/u)(6.022 x 10^{23} mol^{-1}) = 0.480 58 g/mol

(b) For $^{92}_{42}Mo$:

First, calculate the total mass of the nucleons (50 n + 42 p)

Mass of 50 neutrons = (50)(1.008 66) = 50.433 00

Mass of 42 protons = (42)(1.007 28) = 42.305 76

Mass of 50 n + 42 p = 92.738 76

Next, calculate the mass of a ^{92}Mo nucleus by subtracting the mass of 42 electrons from the mass of a ^{92}Mo atom.

Mass of ^{92}Mo atom = 91.906 81

–Mass of 42 electrons = –(42)(5.486 x 10^{-4}) = –0.023 04

Mass of ^{92}Mo nucleus = 91.883 77

Then subtract the mass of the ^{92}Mo nucleus from the mass of the nucleons to find the mass defect:

Mass defect = mass of nucleons – mass of nucleus

= (92.738 76) – (91.883 77) = 0.854 97 u

Mass defect in g/mol:

$(0.854\ 99\ u)(1.660\ 54 \times 10^{-24}\ g/u)(6.022 \times 10^{23}\ mol^{-1}) = 0.854\ 99\ g/mol$

19.66 (a) For $^{58}_{28}Ni$:

First, calculate the total mass of the nucleons (30 n + 28 p)

Mass of 30 neutrons = (30)(1.008 66) = 30.259 80

Mass of 28 protons = (28)(1.007 28) = 28.203 84

Mass of 30 n + 28 p = 58.463 64

Next, calculate the mass of a ^{58}Ni nucleus by subtracting the mass of 28 electrons from the mass of a ^{58}Ni atom.

Mass of ^{58}Ni atom = 57.935 35

–Mass of 28 electrons = $-(28)(5.486 \times 10^{-4})$ = –0.015 36

Mass of ^{58}Ni nucleus = 57.919 99

Then subtract the mass of the ^{58}Ni nucleus from the mass of the nucleons to find the mass defect:

Mass defect = mass of nucleons – mass of nucleus

= (58.463 64) – (57.919 99) = 0.543 65 u

Mass defect in g/mol:

$(0.543\ 65\ u)(1.660\ 54 \times 10^{-24}\ g/u)(6.022 \times 10^{23}\ mol^{-1}) = 0.543\ 64\ g/mol$

Now, use the Einstein equation to convert the mass defect into the binding energy.

$\Delta E = \Delta mc^2 = (0.543\ 64\ g/mol)(10^{-3}\ kg/g)(3.00 \times 10^8\ m/s)^2$

$\Delta E = 4.893 \times 10^{13}\ J/mol = 4.893 \times 10^{10}\ kJ/mol$

$$\Delta E = \frac{4.893 \times 10^{13}\ J/mol}{6.022 \times 10^{23}\ nuclei/mol} \times \frac{1\ MeV}{1.60 \times 10^{-13}\ J} \times \frac{1\ nucleus}{58\ nucleons} = 8.76\ MeV/nucleon$$

(b) For $^{84}_{36}Kr$:

First, calculate the total mass of the nucleons (48 n + 36 p)

Mass of 48 neutrons = (48)(1.008 66) = 48.415 68

Mass of 36 protons = (36)(1.007 28) = 36.262 08

Mass of 48 n + 36 p = 84.677 76

Next, calculate the mass of a ^{84}Kr nucleus by subtracting the mass of 36 electrons from the mass of a ^{84}Kr atom.

Mass of ^{84}Kr atom = 83.911 51

–Mass of 36 electrons = $-(36)(5.486 \times 10^{-4})$ = –0.019 75

Mass of ^{84}Kr nucleus = 83.891 76

Then subtract the mass of the ^{84}Kr nucleus from the mass of the nucleons to find the mass defect:
Mass defect = mass of nucleons – mass of nucleus
= (84.677 76) – (83.891 76) = 0.786 00 u
Mass defect in g/mol:
$(0.786\ 00\ u)(1.660\ 54 \times 10^{-24}\ g/u)(6.022 \times 10^{23}\ mol^{-1}) = 0.785\ 98\ g/mol$
Now, use the Einstein equation to convert the mass defect into the binding energy.
$\Delta E = \Delta mc^2 = (0.785\ 98\ g/mol)(10^{-3}\ kg/g)(3.00 \times 10^8\ m/s)^2$
$\Delta E = 7.074 \times 10^{13}\ J/mol = 7.074 \times 10^{10}\ kJ/mol$

$$\Delta E = \frac{7.074 \times 10^{13}\ J/mol}{6.022 \times 10^{23}\ nuclei/mol} \times \frac{1\ MeV}{1.60 \times 10^{-13}\ J} \times \frac{1\ nucleus}{84\ nucleons} = 8.74\ MeV/nucleon$$

19.68 $^{174}_{77}Ir \rightarrow\ ^{170}_{75}Re +\ ^{4}_{2}He$

mass $^{174}_{77}Ir$	173.966 66
–mass $^{170}_{75}Re$	–169.958 04
–mass $^{4}_{2}He$	– 4.002 60
mass change	0.006 02 u

$(0.006\ 02\ u)(1.660\ 54 \times 10^{-24}\ g/u)(6.022 \times 10^{23}\ mol^{-1}) = 0.006\ 02\ g/mol$
$\Delta E = \Delta mc^2 = (0.006\ 02\ g/mol)(10^{-3}\ kg/g)(3.00 \times 10^8\ m/s)^2$
$\Delta E = 5.42 \times 10^{11}\ J/mol = 5.42 \times 10^8\ kJ/mol$

19.70 $$\Delta m = \frac{\Delta E}{c^2} = \frac{-92.2 \times 10^3\ J}{(3.00 \times 10^8\ m/s)^2} = \frac{-92.2 \times 10^3\ kg\cdot m^2/s^2}{(3.00 \times 10^8\ m/s)^2} = -1.02 \times 10^{-12}\ kg$$

$$\Delta m = \frac{-1.02 \times 10^{-9}\ g}{2\ mol\ NH_3} = 5.10 \times 10^{-10}\ g/mol$$

Fission and Fusion (Section 19.6)

19.72 (a) Nuclear fission is induced by bombarding a U-235 sample with beta particles.

19.74 $^{10}_{5}B +\ ^{1}_{0}n \rightarrow\ ^{7}_{3}Li +\ ^{4}_{2}He$

19.76 Fuel rods in a power plant cannot be used to make an atomic weapon unless the fuel rod is processed and significantly enriched in the fissionable U-235.

19.78 $2\ {}^{2}_{1}H \rightarrow {}^{3}_{2}He + {}^{1}_{0}n$

mass 2 ${}^{2}_{1}H$	2(2.0141)
–mass ${}^{3}_{2}He$	–3.0160
–mass ${}^{1}_{0}n$	–1.008 66
mass change	0.003 54 u

$(0.003\ 54\ u)(1.660\ 54 \times 10^{-24}\ g/u)(6.022 \times 10^{23}\ mol^{-1}) = 0.003\ 54\ g/mol$
$\Delta E = \Delta mc^2 = (0.003\ 54\ g/mol)(10^{-3}\ kg/g)(3.00 \times 10^8\ m/s)^2$
$\Delta E = 3.2 \times 10^{11}\ J/mol = 3.2 \times 10^8\ kJ/mol$

Nuclear Transmutation (Section 19.7)

19.80 (a) ${}^{109}_{47}Ag + {}^{4}_{2}He \rightarrow {}^{113}_{49}In$

(b) ${}^{10}_{5}B + {}^{4}_{2}He \rightarrow {}^{13}_{7}N + {}^{1}_{0}n$

19.82 ${}^{209}_{83}Bi + {}^{58}_{26}Fe \rightarrow {}^{266}_{109}Mt + {}^{1}_{0}n$

19.84 ${}^{238}_{92}U + {}^{2}_{1}H \rightarrow {}^{238}_{93}Np + 2\ {}^{1}_{0}n$

Detecting and Measuring Radioactivity (Section 19.8)

19.86 1 Sv = 1 Gy and 1 Gy = 1 J/kg
$5000\ \mu Sv = 5000 \times 10^{-6}\ Sv = 5000 \times 10^{-6}\ Gy = 5000 \times 10^{-6}\ J/kg$
joules absorbed = $(5000 \times 10^{-6}\ J/kg)(60\ kg) = 0.3\ J$

19.88 $4.0\ pCi = 4.0 \times 10^{-12}\ Ci$
$1\ Ci = 3.7 \times 10^{10}\ Bq = 3.7 \times 10^{10}$ disintegrations/s

(a) $4.0 \times 10^{-12}\ Ci \times \dfrac{3.7 \times 10^{10}\ \text{disintegrations/s}}{1\ Ci} \times \dfrac{60\ s}{1\ min} = 8.9$ disintegrations/min

(b) Rate $= \dfrac{8.9\ dis}{1\ min} \times \dfrac{60\ min}{1\ h} \times \dfrac{24\ h}{1\ d} = 1.3 \times 10^4$ dis/d

Rate $= 1.3 \times 10^4\ dis/d = kN = \left(\dfrac{0.693}{3.8\ d}\right)N$; solve for N

$N = 7.1 \times 10^4\ {}^{222}Rn$

19.90 1 Ci = 3.7×10^{10} disintegrations/s and 1.0 rad = 2.2×10^{11} disintegrations of ^{60}Co

$$1800 \text{ rad} \times \frac{2.2 \times 10^{11} \text{ disintegrations}}{1 \text{ rad}} = 3.96 \times 10^{14} \text{ disintegrations}$$

30 Ci = (30)(3.7×10^{10} disintegrations/s) = 1.11×10^{12} disintegrations/s

$$\frac{1.11 \times 10^{12} \text{ disintegrations/s}}{1 \text{ source}} \times 201 \text{ sources} = 2.23 \times 10^{14} \text{ disintegrations/s}$$

$$\text{time} = \frac{3.96 \times 10^{14} \text{ disintegrations}}{2.23 \times 10^{14} \text{ disintegrations/s}} = 1.8 \text{ s}$$

Some Applications of Nuclear Chemistry (Section 19.9)

19.92 $$\ln\left(\frac{N}{N_0}\right) = (-0.693)\left(\frac{t}{t_{1/2}}\right); \qquad \frac{N}{N_0} = \frac{\text{Decay rate at time t}}{\text{Decay rate at t} = 0}$$

$t_{1/2}$ = 5715 y

$$\ln\left(\frac{2.3}{15.3}\right) = (-0.693)\left(\frac{t}{5715 \text{ y}}\right)$$

age of bone = t = 1.6×10^4 y

19.94 U-238, $t_{1/2} = 4.47 \times 10^9$ yr

At time t, U-238 = 105 μmol and Pb-206 = 33 μmol

U-238 at time t_0 = (105 + 33) μmol = 138 μmol

$$\ln\left(\frac{N}{N_0}\right) = (-0.693)\left(\frac{t}{t_{1/2}}\right)$$

$$\ln\left(\frac{^{238}U_t}{^{238}U_0}\right) = \ln\left(\frac{105}{138}\right) = (-0.693)\left(\frac{t}{4.47 \times 10^9 \text{ yr}}\right)$$

age of rock = t = 1.8×10^9 yr

Chapter Problems

19.96 $$E = (1.50 \text{ MeV})\left(\frac{1.60 \times 10^{-13} \text{ J}}{1 \text{ MeV}}\right) = 2.40 \times 10^{-13} \text{ J}$$

$$\lambda = \frac{hc}{E} = \frac{(6.626 \times 10^{-34} \text{ J}\cdot\text{s})(3.00 \times 10^8 \text{ m/s})}{2.40 \times 10^{-13} \text{ J}} = 8.28 \times 10^{-13} \text{ m} = 0.000\,828 \text{ nm}$$

19.98 Mass of positron and electron

$= 2(9.109 \times 10^{-31}\ \text{kg})(6.022 \times 10^{23}\ \text{mol}^{-1}) = 1.097 \times 10^{-6}\ \text{kg/mol}$

$\Delta E = \Delta mc^2 = (1.097 \times 10^{-6}\ \text{kg/mol})(3.00 \times 10^8\ \text{m/s})^2$

$\Delta E = 9.87 \times 10^{10}\ \text{J/mol} = 9.87 \times 10^7\ \text{kJ/mol}$

19.100 For radioactive decay, $\ln \dfrac{N}{N_o} = -kt$

For ^{235}U, $k_1 = \dfrac{0.693}{t_{1/2}} = \dfrac{0.693}{7.04 \times 10^8\ \text{y}} = 9.84 \times 10^{-10}\ \text{y}^{-1}$

For ^{238}U, $k_2 = \dfrac{0.693}{t_{1/2}} = \dfrac{0.693}{4.47 \times 10^9\ \text{y}} = 1.55 \times 10^{-10}\ \text{y}^{-1}$

For ^{235}U, $\ln \dfrac{N_1}{N_{o1}} = -k_1t$ and $\ln \dfrac{N_1}{N_{o1}} + k_1t = 0$

For ^{238}U, $\ln \dfrac{N_2}{N_{o2}} = -k_2t$ and $\ln \dfrac{N_2}{N_{o2}} + k_2t = 0$

Set the two equations that are equal to zero equal to each other and solve for t.

$$\ln \frac{N_1}{N_{o1}} + k_1t = \ln \frac{N_2}{N_{o2}} + k_2t;$$

$$\ln \frac{N_1}{N_{o1}} - \ln \frac{N_2}{N_{o2}} = k_2t - k_1t = (k_2 - k_1)t$$

$$\ln \frac{\left(\frac{N_1}{N_{o1}}\right)}{\left(\frac{N_2}{N_{o2}}\right)} = (k_2 - k_1)t, \text{ now } N_{o1} = N_{o2}, \text{ so } \ln \frac{N_1}{N_2} = (k_2 - k_1)t$$

$\dfrac{N_1}{N_2} = 7.25 \times 10^{-3}$, so $\ln(7.25 \times 10^{-3}) = (1.55 \times 10^{-10}\ \text{y}^{-1} - 9.84 \times 10^{-10}\ \text{y}^{-1})t$

$$t = \frac{-4.93}{-8.29 \times 10^{-10}\ \text{y}^{-1}} = 5.9 \times 10^9\ \text{y}$$

The age of the elements is 5.9×10^9 y (6 billion years).

19.102 $^{232}_{90}Th \rightarrow {}^{208}_{82}Pb + 6\,{}^{4}_{2}He + 4\,{}^{0}_{-1}e$

Reactant: $^{232}_{90}Th$ nucleus = $^{232}_{90}Th$ atom – 90 e^-

Product: $^{208}_{82}Pb$ nucleus + (6)($^{4}_{2}He$ nucleus) + 4 e^-

= ($^{208}_{82}Pb$ atom – 82 e^-) + (6)($^{4}_{2}He$ atom – 2 e^-) + 4 e^-

= $^{208}_{82}Pb$ atom + (6)($^{4}_{2}He$ atom) – 90 e^-

Change: ($^{232}_{90}Th$ atom – 90 e^-) – [$^{208}_{82}Pb$ atom + (6)($^{4}_{2}He$ atom) – 90 e^-]

= $^{232}_{90}Th$ atom – [$^{208}_{82}Pb$ atom + (6)($^{4}_{2}He$ atom)] (electrons cancel)

Mass change = (232.038 054) – [(207.976 627) + (6)(4.002 603)]

= 0.045 809 u

(0.045 809 u)($1.660\ 54 \times 10^{-24}$ g/u)(6.022×10^{23} mol^{-1}) = 0.045 808 g/mol

$\Delta E = \Delta mc^2$ = (0.045 808 g/mol)(10^{-3} kg/g)(3.00×10^8 m/s)^2

$\Delta E = 4.12 \times 10^{12}$ J/mol = 4.12×10^9 kJ/mol

19.104 (a) For $^{50}_{24}Cr$:

First, calculate the total mass of the nucleons (26 n + 24 p)

Mass of 26 neutrons = (26)(1.008 66) = 26.225 16

Mass of 24 protons = (24)(1.007 28) = 24.174 72

Mass of 26 n + 24 p = 50.399 88

Next, calculate the mass of a ^{50}Cr nucleus by subtracting the mass of 24 electrons from the mass of a ^{50}Cr atom.

Mass of ^{50}Cr atom = 49.946 05

–Mass of 24 electrons = –(24)(5.486×10^{-4}) = –0.013 17

Mass of ^{50}Cr nucleus = 49.932 88

Then subtract the mass of the ^{50}Cr nucleus from the mass of the nucleons to find the mass defect:

Mass defect = mass of nucleons – mass of nucleus

= (50.399 88) – (49.932 88) = 0.467 00 u

Mass defect in g/mol:

(0.467 00 u)($1.660\ 54 \times 10^{-24}$ g/u)(6.022×10^{23} mol^{-1}) = 0.466 99 g/mol

Now, use the Einstein equation to convert the mass defect into the binding energy.

$\Delta E = \Delta mc^2$ = (0.466 99 g/mol)(10^{-3} kg/g)(3.00×10^8 m/s)^2

$\Delta E = 4.203 \times 10^{13}$ J/mol = 4.203×10^{10} kJ/mol

$$\Delta E = \frac{4.203 \times 10^{13}\ \text{J/mol}}{6.022 \times 10^{23}\ \text{nuclei/mol}} \times \frac{1\ \text{MeV}}{1.60 \times 10^{-13}\ \text{J}} \times \frac{1\ \text{nucleus}}{50\ \text{nucleons}} = 8.72\ \text{MeV/nucleon}$$

(b) For $^{64}_{30}Zn$:

First, calculate the total mass of the nucleons (34 n + 30 p)
Mass of 34 neutrons = (34)(1.008 66) = 34.294 44
Mass of 30 protons = (30)(1.007 28) = 30.218 40
Mass of 34 n + 30 p = 64.512 84

Next, calculate the mass of a ^{64}Zn nucleus by subtracting the mass of 30 electrons from the mass of a ^{64}Zn atom.
Mass of ^{64}Zn atom = 63.929 15
–Mass of 30 electrons = –(30)(5.486×10^{-4}) = –0.016 46
Mass of ^{64}Zn nucleus = 63.912 69

Then subtract the mass of the ^{64}Zn nucleus from the mass of the nucleons to find the mass defect:
Mass defect = mass of nucleons – mass of nucleus
= (64.512 84) – (63.912 69) = 0.600 15 u

Mass defect in g/mol:
(0.600 15 u)($1.660\,54 \times 10^{-24}$ g/u)(6.022×10^{23} mol^{-1}) = 0.600 14 g/mol

Now, use the Einstein equation to convert the mass defect into the binding energy.
$\Delta E = \Delta mc^2 = (0.600\,14\ \text{g/mol})(10^{-3}\ \text{kg/g})(3.00 \times 10^8\ \text{m/s})^2$
$\Delta E = 5.401 \times 10^{13}\ \text{J/mol} = 5.401 \times 10^{10}\ \text{kJ/mol}$

$$\Delta E = \frac{5.401 \times 10^{13}\ \text{J/mol}}{6.022 \times 10^{23}\ \text{nuclei/mol}} \times \frac{1\ \text{MeV}}{1.60 \times 10^{-13}\ \text{J}} \times \frac{1\ \text{nucleus}}{64\ \text{nucleons}} = 8.76\ \text{MeV/nucleon}$$

The ^{64}Zn is more stable because ΔE is larger.

19.106 $^{2}_{1}H + ^{3}_{2}He \rightarrow ^{4}_{2}He + ^{1}_{1}H$

mass $^{2}_{1}H$	2.0141
mass $^{3}_{2}He$	3.0160
–mass $^{4}_{2}He$	– 4.0026
–mass $^{1}_{1}H$	–1.0078
mass change	0.0197 u

(0.0197 u)($1.660\,54 \times 10^{-24}$ g/u)(6.022×10^{23} mol^{-1}) = 0.0197 g/mol
$\Delta E = \Delta mc^2 = (0.0197\ \text{g/mol})(10^{-3}\ \text{kg/g})(3.00 \times 10^8\ \text{m/s})^2$
$\Delta E = 1.77 \times 10^{12}\ \text{J/mol} = 1.77 \times 10^{9}\ \text{kJ/mol}$

19.108 $^{238}_{92}U + ^{1}_{0}n \rightarrow ^{239}_{94}Pu + 2\ ^{0}_{-1}e$

19.110 Each alpha emission decreases the mass number by four and the atomic number by two. Each beta emission increases the atomic number by one.

$^{237}_{93}Np \rightarrow ^{209}_{83}Bi$

$$\text{Number of } \alpha \text{ emissions} = \frac{\text{Np mass number} - \text{Bi mass number}}{4}$$

$$= \frac{237 - 209}{4} = 7\ \alpha \text{ emissions}$$

The atomic number decreases by 14 as a result of 7 alpha emissions. The resulting atomic number is (93 – 14) = 79.

Number of β emissions = Bi atomic number – 79 = 83 – 79 = 4 β emissions

19.112 (a) α emission: $^{226}_{89}Ac \rightarrow ^{222}_{87}Fr + ^{4}_{2}He$

β emission: $^{226}_{89}Ac \rightarrow ^{226}_{90}Th + ^{0}_{-1}e$

electron capture: $^{226}_{89}Ac + ^{0}_{-1}e \rightarrow ^{226}_{88}Ra$

(b) $t_{1/2} = \frac{0.693}{k} = \frac{0.693}{0.556\ d^{-1}} = 1.25\ d$

If 80% reacts, then 20% is left.

$$\ln\left(\frac{N}{N_0}\right) = (-0.693)\left(\frac{t}{t_{1/2}}\right)$$

$$\ln\left(\frac{20}{100}\right) = (-0.693)\left(\frac{t}{1.25\ d}\right)$$

$$t = \frac{\ln\left(\frac{20}{100}\right)(1.25\ d)}{(-0.693)} = 2.90\ d$$

Multiconcept Problems

19.114 $t_{1/2} = 138\ d = 138\ d \times \frac{1\ y}{365\ d} = 0.378\ y$

$$k = \frac{0.693}{t_{1/2}} = \frac{0.693}{0.378\ y} = 1.83\ y^{-1}$$

$$0.700\ mg \times \frac{1 \times 10^{-3}\ g}{1\ mg} = 7.00 \times 10^{-4}\ g$$

$$N_o = (7.00 \times 10^{-4}\ g)\left(\frac{1\ mol\ Po}{210\ g\ Po}\right)(6.022 \times 10^{23}\ atoms/mol) = 2.01 \times 10^{18}\ atoms$$

$$\ln\left(\frac{N}{N_o}\right) = -kt = -(1.83\ y^{-1})(1\ y) = -1.83$$

$$\frac{N}{N_o} = e^{-1.83} = 0.160$$

$N = 0.160\ N_o = (0.160)(2.01 \times 10^{18}\ atoms) = 0.322 \times 10^{18}$ atoms
atoms He = atoms Po decayed
atoms He = 2.01×10^{18} atoms – 0.322×10^{18} atoms = 1.688×10^{18} atoms

$$mol\ He = \frac{1.688 \times 10^{18}\ He\ atoms}{6.022 \times 10^{23}\ atoms/mol} = 2.80 \times 10^{-6}\ mol\ He$$

20 °C = 293 K

$$P = \frac{nRT}{V} = \frac{(2.80 \times 10^{-6}\ mol)\left(0.082\ 06\ \frac{L \cdot atm}{K \cdot mol}\right)(293\ K)}{0.2500\ L} = 2.69 \times 10^{-4}\ atm$$

$$P = 2.69 \times 10^{-4}\ atm \times \frac{760\ mm\ Hg}{1.00\ atm} = 2.04\ mm\ Hg$$

20 Transition Elements and Coordination Chemistry

20.1 (a) V, [Ar] $3d^3 4s^2$ (b) Co^{2+}, [Ar] $3d^7$
(c) Mn^{4+} in MnO_2, [Ar] $3d^3$ (d) Cu^{2+} in $CuCl_4^{2-}$, [Ar] $3d^9$

20.2 (a) Mn (b) Ni^{2+}
(c) Ag (d) Mo^{3+}

20.3 Z_{eff} increases from left to right across the first transition series.
(a) The transition metal with the lowest Z_{eff} (Ti) should be the strongest reducing agent because it is easier for Ti to lose its valence electrons. The transition metal with the highest Z_{eff} (Zn) should be the weakest reducing agent because it is more difficult for Zn to lose its valence electrons.
(b) The oxoanion with the highest Z_{eff} (FeO_4^{2-}) should be the strongest oxidizing agent because of the greater attraction for electrons. The oxoanion with the lowest Z_{eff} (VO_4^{3-}) should be the weakest oxidizing agent because of the lower attraction for electrons.

20.4 (a) $Cr_2O_7^{2-}$ (b) Cr^{3+} (c) Cr^{2+} (d) Fe^{2+} (e) Cu^{2+}

20.5 (a) $Cr(OH)_2$ (b) $Cr(OH)_4^-$ (c) CrO_4^{2-} (d) $Fe(OH)_2$ (e) $Fe(OH)_3$

20.6 $[Cr(NH_3)_2(SCN)_4]^-$

20.7 In $Na_4[Fe(CN)_6]$ each sodium is in the +1 oxidation state (+4 total); each cyanide (CN^-) has a −1 charge (−6 total). The compound is neutral; therefore, the oxidation state of the iron is +2.

20.8 (a)

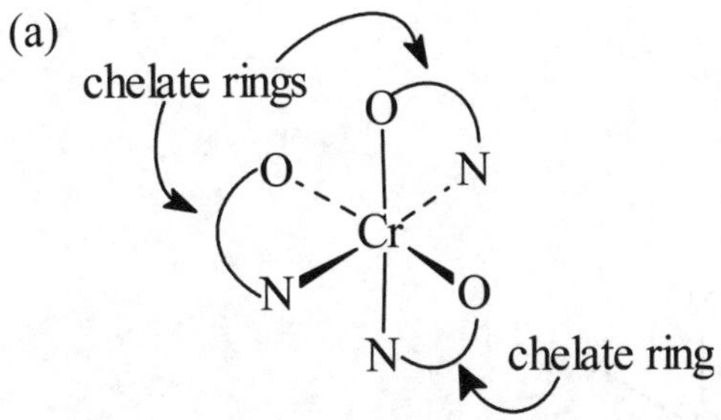

(b) Cr^{3+} is the Lewis acid. The glycinate ligand is the Lewis base. Nitrogen and oxygen are the ligand donor atoms. The chelate rings are identified in the drawing.
(c) The coordination number is 6. The coordination geometry is octahedral. The chromium is in the +3 oxidation state.

20.9 (a) tetraamminecopper(II) sulfate (b) sodium tetrahydroxochromate(III)
(c) triglycinatocobalt(III) (d) pentaaquaisothiocyanatoiron(III) ion

20.10 triamminetrichlorocobalt(III)

20.11 (a) $[Zn(NH_3)_4](NO_3)_2$ (b) $Ni(CO)_4$ (c) $K[Pt(NH_3)Cl_3]$ (d) $[Au(CN)_2]^-$

20.12 $[Pt(NH_3)_3Cl]Cl$, triamminechloroplatinum(II)chloride

20.13 Structures (1) and (4) are identical and are the cis isomer. Structures (2) and (3) are identical and are the trans isomer.

20.14 (1) and (2) are the same. (3) and (4) are the same. (1) and (2) are different from (3) and (4).

20.15 (a) Two diastereoisomers are possible.

NCS SCN Pt H_3N NH_3
cis

NCS NH_3 Pt H_3N SCN
trans

(b) No isomers are possible for a tetrahedral complex of the type MA_2B_2.
(c) Two diastereoisomers are possible.

NH_3 H_3N NH_3 Co O_2N NO_2 NO_2

NO_2 H_3N NH_3 Co O_2N NH_3 NO_2

20.16 (a) No isomers are possible for a complex of this type.

N N Pt Cl Cl

(b) Two diastereoisomers are possible.

Br N N Cr N N Br
trans

N N N Cr Br N Br
cis

20.17 (a) $[Fe(C_2O_4)_3]^{3-}$ can exist as enantiomers.

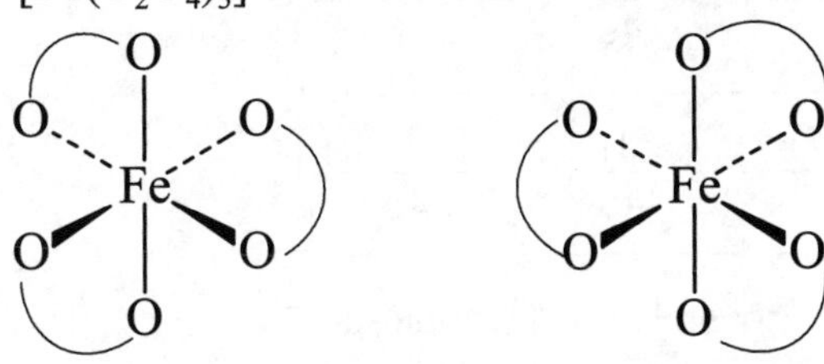

(b) $[Co(NH_3)_4en]^{3+}$ cannot exist as enantiomers.
(c) $[Co(NH_3)_2(en)_2]^{3+}$ can exist as enantiomers.

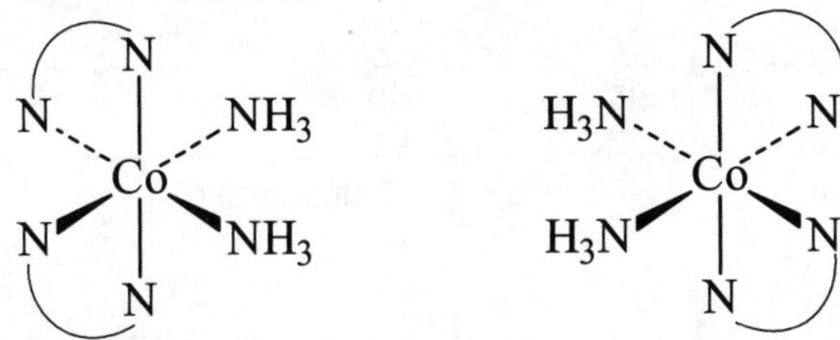

(d) $[Cr(H_2O)_4Cl_2]^+$ cannot exist as enantiomers.

20.18 (a) (2) and (3) are chiral and (1) and (4) are achiral.
(b) enantiomer of (2) enantiomer of (3)

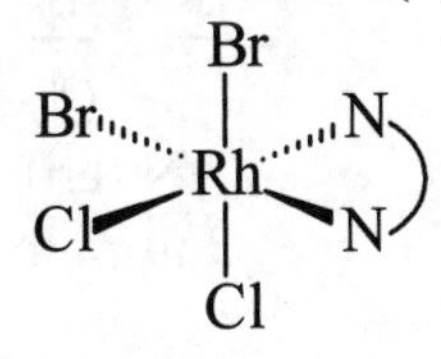

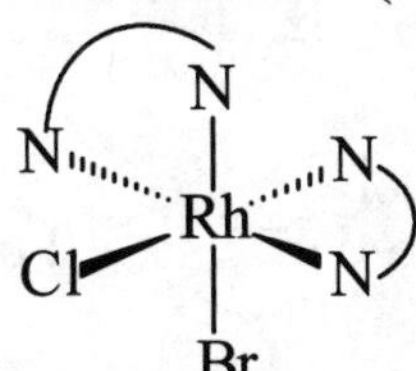

20.19 (a) The ion is absorbing in the red (625 nm), so the most likely color for the ion is blue.
(b) 625 nm = 625×10^{-9} m

$$E = h\frac{c}{\lambda} = (6.626 \times 10^{-34}\ J \cdot s)\left(\frac{3.00 \times 10^{8}\ m/s}{625 \times 10^{-9}\ m}\right) = 3.18 \times 10^{-19}\ J$$

20.20 (a)

V^{3+} [Ar] ↑ ↑ __ __ __ (3d) __ (4s) __ __ __ (4p)

$[VCl_4]^-$ [Ar] ↑ ↑ __ __ __ (3d) ↑↓ (4s) ↑↓ ↑↓ ↑↓ (4p)

sp^3 2 unpaired e^-

(b)

Pt^{2+} [Xe] ↑↓ ↑↓ ↑↓ ↑ ↑ (5d) __ (6s) __ __ __ (6p)

$[PtCl_4]^{2-}$ [Xe] ↑↓ ↑↓ ↑↓ ↑↓ ↑↓ (5d) ↑↓ (6s) ↑↓ ↑↓ __ (6p)

dsp^2 no unpaired e^-

20.21 (a) Fe^{3+} [Ar] ↑ ↑ ↑ ↑ ↑ (3d) _ (4s) _ _ _ (4p)

$[Fe(CN)_6]^{3-}$ [Ar] ↑↓ ↑↓ ↑ (3d) | ↑↓ ↑↓ (3d) ↑↓ (4s) ↑↓ ↑↓ ↑↓ (4p) |

d^2sp^3 1 unpaired e^-

(b) Co^{2+} [Ar] ↑↓ ↑↓ ↑ ↑ ↑ (3d) _ (4s) _ _ _ (4p)

$[Co(H_2O)_6]^{2+}$ [Ar] ↑↓ ↑↓ ↑ ↑ ↑ (3d) | ↑↓ (4s) ↑↓ ↑↓ ↑↓ (4p) ↑↓ ↑↓ (4d) | _ _ _ (4d)

sp^3d^2 3 unpaired e^-

20.22

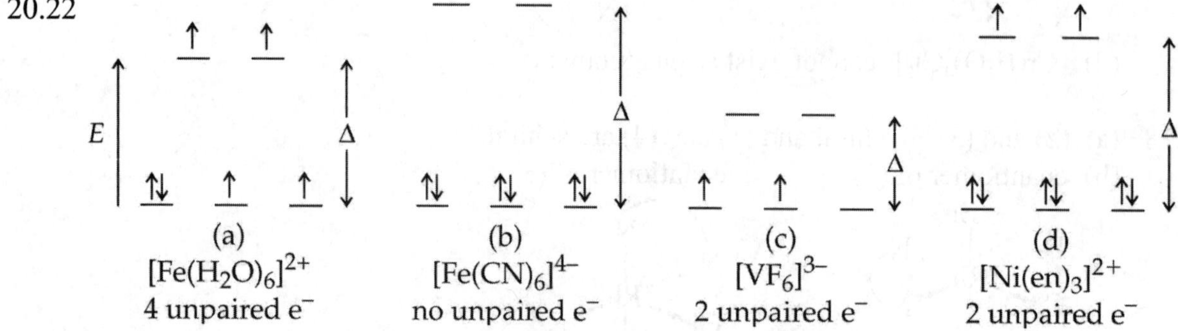

20.23 Cl^- is a weak field ligand and therefore $[FeCl_6]^{3-}$ is a high-spin complex with 5 unpaired electrons. CN^- is a strong field ligand and therefore $[Fe(CN)_6]^{3-}$ is a low-spin complex with 1 unpaired electron. $[FeCl_6]^{3-}$ is more paramagnetic because it has more unpaired electrons.

20.24 Both $[NiCl_4]^{2-}$ and $[Ni(CN)_4]^{2-}$ contain Ni^{2+} with a [Ar] $3d^8$ electron configuration.

(a) $[NiCl_4]^{2-}$ (tetrahedral)

↑↓ ↑ ↑
xy xz yz

↑↓ ↑↓
z^2 x^2-y^2

2 unpaired electrons

(b) $[Ni(CN)_4]^{2-}$ (square planar)

_
x^2-y^2

↑↓
xy

↑↓
z^2

↑↓ ↑↓
xz yz

no unpaired electrons

20.25 A diamagnetic four coordinate d^8 complex is most likely square planar.

20.26 (a) diamminedichloroplatinum(II)
(b) oxidation state = +2 and coordination number is 4
(c) Lewis acid is Pt^{2+} and Lewis bases are Cl^- and NH_3
(d) [Xe] $4f^{14}\ 5d^8$

20.27 $Pt(NH_3)_2Cl_2$

$\overline{\quad}$ x^2-y^2

↿⇂ xy

↿⇂ z^2

↿⇂ xz ↿⇂ yz

no unpaired electrons

20.28 (a) The chloride concentration is relatively high in blood plasma and, according to Le Chatelier's Principle, a high concentration of product shifts the equilibrium position toward the reactants. Inside the cell, the chloride concentration is lower, thus shifting the equilibrium positions toward the products.
(b) diammineaquachloroplatinum(II)
(c) According to the spectrochemical series, H_2O is a stronger field ligand than Cl^-. The crystal field splitting energy is larger, which corresponds to shorter wavelength of maximum absorption.

20.29 (a) +4
(b) diamminetetrachloroplatinum(IV)
(c) cis and trans isomers.

Cl, Cl, NH_3, Pt, Cl, NH_3, Cl

cis

NH_3, Cl, Cl, Pt, Cl, Cl, NH_3

trans

(d) Both the cis and trans isomers have symmetry planes and are achiral.

Chapter 20 – Transition Elements and Coordination Chemistry

Conceptual Problems

20.30

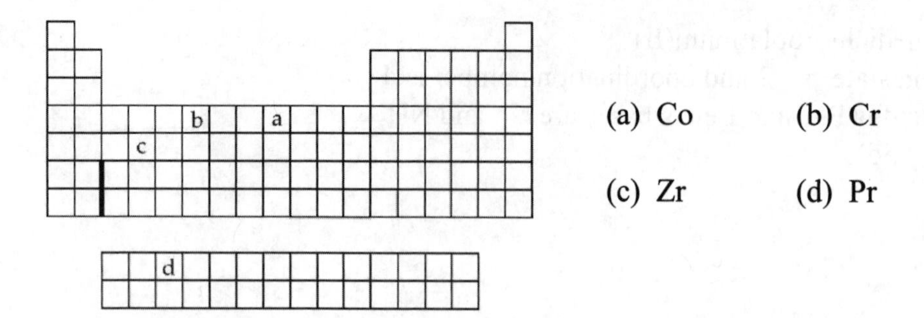

20.32 (a) The atomic radii decrease, at first markedly and then more gradually. Toward the end of the series, the radii increase again. The decrease in atomic radii is a result of an increase in Z_{eff}. The increase is due to electron-electron repulsions in doubly occupied d orbitals.
(b) The densities of the transition metals are inversely related to their atomic radii. The densities initially increase from left to right and then decrease toward the end of the series.
(c) Ionization energies generally increase from left to right across the series. The general trend correlates with an increase in Z_{eff} and a decrease in atomic radii.
(d) The standard oxidation potentials generally decrease from left to right across the first transition series. This correlates with the general trend in ionization energies.

20.34 (a) $\mathbf{NH_2}$–CH_2–CH_2–$\mathbf{NH_2}$ is a bidentate ligand. It can form a chelate ring using the atoms indicated in bold.
(b) CH_3–CH_2–CH_2–$\mathbf{NH_2}$ is a monodentate ligand.
(c) $\mathbf{NH_2}$–CH_2–CH_2–$\mathbf{NH}$–CH_2–$\mathbf{CO_2}^-$ is a tridentate ligand. It can form chelate rings using the atoms indicated in bold.
(d) $\mathbf{NH_2}$–CH_2–CH_2–NH_3^+ is a monodentate ligand. The first N can coordinate to a metal.

20.36 (a) $Na[Au(CN)_2]$
1 Na^+ 2 CN^-
The oxidation state of the Au is +1.
Coordination number = 2; Linear

$[CN—Au—NC]^-$

(b) $[Co(NH_3)_5Br]SO_4$
1 Br^- 1 SO_4^{2-} 5 NH_3 (no charge)
The oxidation state of the Co is +3.
Coordination number = 6; Octahedral

$[Co(NH_3)_5Br]^{2+}$: NH_3 (top), H_3N, NH_3, H_3N, NH_3 (equatorial), Br (bottom) around Co

(c) $Pt(en)Cl_2$
2 Cl^- en = $NH_2CH_2CH_2NH_2$ (no charge)
The oxidation state of the Pt is +2.
Coordination number = 4; Square planar

Cl Pt N
Cl N

(d) $(NH_4)_2[PtCl_2(C_2O_4)_2]$
2 NH_4^+ 2 Cl^- 2 $C_2O_4^{2-}$
The oxidation state of the Pt is +4.
Coordination number = 6; Octahedral

$[\ldots]^{2-}$ or $[\ldots]^{2-}$

20.38 (a) (1) chiral; (2) achiral; (3) chiral; (4) chiral
(b)

(1)   enantiomer of (1)

(3)   enantiomer of (3)

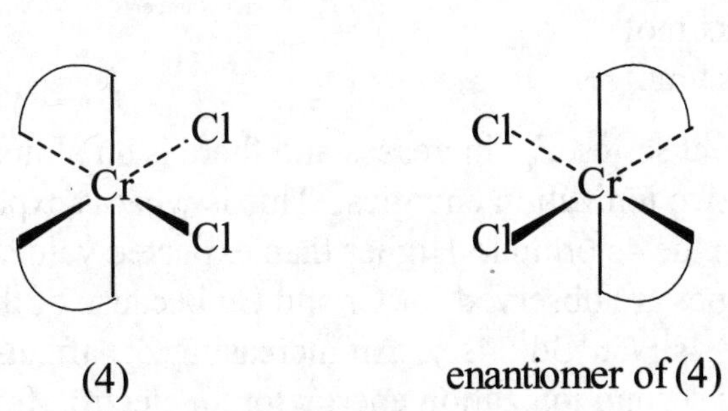

(4)   enantiomer of (4)

(c) (1) and (4) are enantiomers.

Section Problems

Electron Configurations and Properties of Transition Elements (Sections 20.1 and 20.2)

20.40 (a) Cr, [Ar] $3d^5 4s^1$ (b) Zr, [Kr] $4d^2 5s^2$ (c) Co^{2+}, [Ar] $3d^7$
(d) Fe^{3+}, [Ar] $3d^5$ (e) Mo^{3+}, [Kr] $4d^3$ (f) Cr(VI), [Ar] $3d^0$

20.42 (a) Cu^{2+}, [Ar] $3d^9$ ⥮ ⥮ ⥮ ⥮ ↿ (3d) 1 unpaired e^-

(b) Ti^{2+}, [Ar] $3d^2$ ↿ ↿ _ _ _ (3d) 2 unpaired e^-

(c) Zn^{2+}, [Ar] $3d^{10}$ ⥮ ⥮ ⥮ ⥮ ⥮ (3d) 0 unpaired e^-

(d) Cr^{3+}, [Ar] $3d^3$ ↿ ↿ ↿ _ _ (3d) 3 unpaired e^-

20.44 Ti is harder than K and Ca largely because the sharing of d, as well as s, electrons results in stronger metallic bonding.

20.46 (a) The decrease in radii with increasing atomic number is expected because the added d electrons only partially shield the added nuclear charge. As a result, Z_{eff} increases. With increasing Z_{eff}, the electrons are more strongly attracted to the nucleus, and atomic size decreases.
(b) The densities of the transition metals are inversely related to their atomic radii.

20.48 The smaller than expected sizes of the third-transition series atoms are associated with what is called the lanthanide contraction, the general decrease in atomic radii of the f-block lanthanide elements. The lanthanide contraction is due to the increase in Z_{eff} as the 4f subshell is filled.

20.50
Sc	(631 + 1235) = 1866 kJ/mol
Ti	(659 + 1310) = 1969 kJ/mol
V	(651 + 1410) = 2061 kJ/mol
Cr	(653 + 1591) = 2224 kJ/mol
Mn	(717 + 1509) = 2226 kJ/mol
Fe	(762 + 1562) = 2324 kJ/mol
Co	(760 + 1648) = 2408 kJ/mol
Ni	(737 + 1753) = 2490 kJ/mol
Cu	(745 + 1958) = 2703 kJ/mol
Zn	(906 + 1733) = 2639 kJ/mol

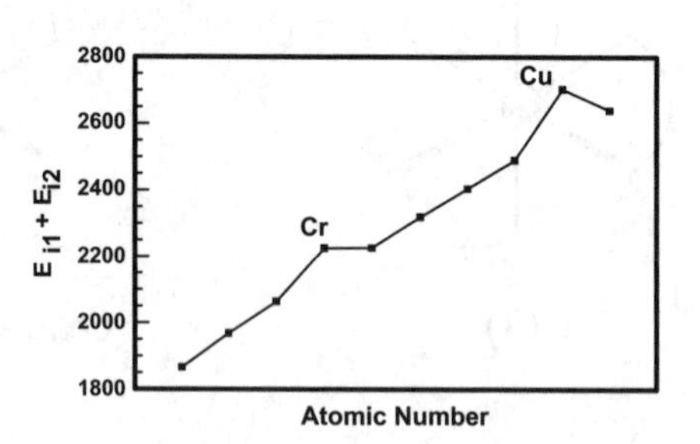

Across the first transition element series, Z_{eff} increases and there is an almost linear increase in the sum of the first two ionization energies. This is what is expected if the two electrons are removed from the 4s orbital. Higher than expected values for the sum of the first two ionization energies are observed for Cr and Cu because of their anomalous electron configurations (Cr $3d^5 4s^1$; Cu $3d^{10} 4s^1$). An increasing Z_{eff} affects 3d orbitals more than the 4s orbital and the second ionization energy for an electron from the 3d orbital is higher than expected.

20.52 Ti is more easily oxidized than is Zn because of a smaller Z_{eff}.

20.54 (a) $Cr(s) + 2\ H^+(aq) \rightarrow Cr^{2+}(aq) + H_2(g)$ (b) $Zn(s) + 2\ H^+(aq) \rightarrow Zn^{2+}(aq) + H_2(g)$
(c) N.R. (d) $Fe(s) + 2\ H^+(aq) \rightarrow Fe^{2+}(aq) + H_2(g)$

Oxidation States (Section 20.3)

20.56 (b) Mn (d) Cu

20.58 Sc(III), Ti(IV), V(V), Cr(VI), Mn(VII), Fe(VI), Co(III), Ni(II), Cu(II), Zn(II)

20.60 Cu^{2+} is a stronger oxidizing agent than Cr^{2+} because of a higher Z_{eff}.

20.62 A compound with vanadium in the +2 oxidation state is expected to be a reducing agent, because early transition metal atoms have a relatively low effective nuclear charge and are easily oxidized to higher oxidation states.

20.64 $Mn^{2+} < MnO_2 < MnO_4^-$ because of increasing oxidation state of Mn.

Chemistry of Selected Transition Elements (Section 20.4)

20.66 (a) $Cr_2O_3(s) + 2\ Al(s) \rightarrow 2\ Cr(s) + Al_2O_3(s)$
(b) $Cu_2S(l) + O_2(g) \rightarrow 2\ Cu(l) + SO_2(g)$

20.68 $Cr(OH)_3(s) + OH^-(aq) \rightarrow Cr(OH)_4^-(aq)$
The Cr in $Cr(OH)_4^-$ is in the +3 oxidation state. $Cr(OH)_4^-$ is deep green.

20.70 (c) $Cr(OH)_3$

20.72 (a) Add excess KOH(aq) and Fe^{3+} will precipitate as $Fe(OH)_3(s)$. $Na^+(aq)$ will remain in solution.
(b) Add excess NaOH(aq) and Fe^{3+} will precipitate as $Fe(OH)_3(s)$. $Cr(OH)_4^-(aq)$ will remain in solution.
(c) Add excess $NH_3(aq)$ and Fe^{3+} will precipitate as $Fe(OH)_3(s)$. $Cu(NH_3)_4^{2+}(aq)$ will remain in solution.

20.74 (a) $Cr_2O_7^{2-}(aq) + 6\ Fe^{2+}(aq) + 14\ H^+(aq) \rightarrow 2\ Cr^{3+}(aq) + 6\ Fe^{3+}(aq) + 7\ H_2O(l)$
(b) $4\ Fe^{2+}(aq) + O_2(g) + 4\ H^+(aq) \rightarrow 4\ Fe^{3+}(aq) + 2\ H_2O(l)$
(c) $Cu_2O(s) + 2\ H^+(aq) \rightarrow Cu(s) + Cu^{2+}(aq) + H_2O(l)$
(d) $Fe(s) + 2\ H^+(aq) \rightarrow Fe^{2+}(aq) + H_2(g)$

20.76 (a) $2\ CrO_4^{2-}(aq) + 2\ H_3O^+(aq) \rightarrow Cr_2O_7^{2-}(aq) + 3\ H_2O(l)$
(yellow) (orange)
(b) $[Fe(H_2O)_6]^{3+}(aq) + SCN^-(aq) \rightarrow [Fe(H_2O)_5(SCN)]^{2+}(aq) + H_2O(l)$
(red)

(c) $3\ Cu(s) + 2\ NO_3^-(aq) + 8\ H^+(aq) \rightarrow 3\ Cu^{2+}(aq) + 2\ NO(g) + 4\ H_2O(l)$
(blue)

(d) $Cr(OH)_3(s) + OH^-(aq) \rightarrow Cr(OH)_4^-(aq)$
$2\ Cr(OH)_4^-(aq) + 3\ HO_2^-(aq) \rightarrow 2\ CrO_4^{2-}(aq) + 5\ H_2O(l) + OH^-(aq)$
(yellow)

Coordination Compounds; Ligands (Sections 20.5 and 20.6)

20.78 (a) Ni^{2+} is the Lewis acid. Ethylenediamine is the Lewis base.
(b) Ethylenediamine is the ligand and the two N's are the donor atoms.
(c) $[Ni(en)_3]^{2+}$ is octahedral with a coordination number of 6.

20.80

	Coordination Number
(a) $[AgCl_2]^-$	2
(b) $[Cr(H_2O)_5Cl]^{2+}$	6
(c) $[Co(NCS)_4]^{2-}$	4
(d) $[ZrF_8]^{4-}$	8
(e) $[Fe(EDTA)(H_2O)]^-$	7

20.82 (a) $AgCl_2^-$
$2\ Cl^-$
The oxidation state of the Ag is +1.
(b) $[Cr(H_2O)_5Cl]^{2+}$
$4\ H_2O$ (no charge) $1\ Cl^-$
The oxidation state of the Cr is +3.
(c) $[Co(NCS)_4]^{2-}$
$4\ NCS^-$
The oxidation state of the Co is +2.
(d) $[ZrF_8]^{4-}$
$8\ F^-$
The oxidation state of the Zr is +4.
(e) $[Fe(EDTA)(H_2O)]^-$
H_2O (no charge) $EDTA^{4-}$
The oxidation state of the Fe is +3.

20.84 (a) $Co(NH_3)_3(NO_2)_3$
$3\ NH_3$ (no charge) $3\ NO_2^-$
The oxidation state of the Co is +3.
(b) $[Ag(NH_3)_2]NO_3$
$2\ NH_3$ (no charge) $1\ NO_3^-$
The oxidation state of the Ag is +1.
(c) $K_3[Cr(C_2O_4)_2Cl_2]$
$3\ K^+$ $2\ C_2O_4^{2-}$ $2\ Cl^-$
The oxidation state of the Cr is +3.

(d) $Cs[CuCl_2]$
1 Cs^+ 2 Cl^-
The oxidation state of the Cu is +1.

20.86 (a) $Ir(NH_3)_3Cl_3$ (b) $Ni(en)_2Br_2$ (c) $[Pt(en)_2(SCN)_2]^{2+}$

20.88

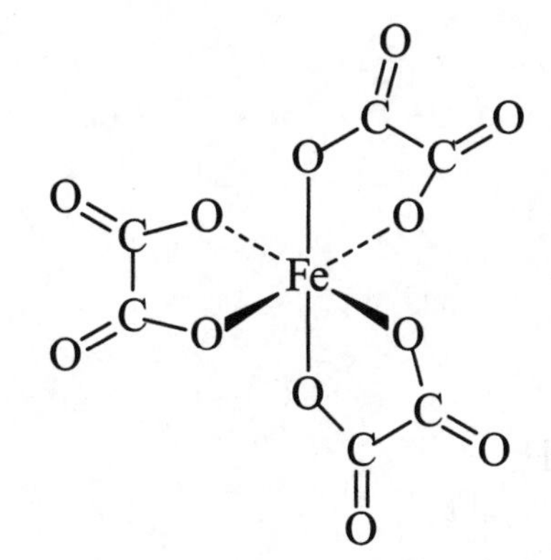

The iron is in the +3 oxidation state, and the coordination number is six. The geometry about the Fe is octahedral. The oxalate ligand is behaving as a bidentate chelating ligand. There are three chelate rings, one formed by each oxalate ligand.

Naming Coordination Compounds (Section 20.7)

20.90 (a) tetrachloromanganate(II) (b) hexaamminenickel(II)
(c) tricarbonatocobaltate(III) (d) bis(ethylenediamine)dithiocyanatoplatinum(IV)

20.92 (a) cesium tetrachloroferrate(III) (b) hexaaquavanadium(III) nitrate
(c) tetraamminedibromocobalt(III) bromide (d) diglycinatocopper(II)

20.94 (a) $[Pt(NH_3)_4]Cl_2$ (b) $Na_3[Fe(CN)_6]$
(c) $[Pt(en)_3](SO_4)_2$ (d) $Rh(NH_3)_3(SCN)_3$

Isomers (Sections 20.8 and 20.9)

20.96 (a) (1) $[Ru(NH_3)_5(NO_2)]Cl$, tetraamminenitroruthenium(II) chloride
(2) $[Ru(NH_3)_5(ONO)]Cl$, tetraamminenitritoruthenium(II) chloride
(3) $[Ru(NH_3)_5Cl]NO_2$, tetraamminechlororuthenium(II) nitrite
(b) (1) and (2) are linkage isomers.
(c) (1) and (2) are ionization isomers with (3).

20.98 (a) $[Cr(NH_3)_2Cl_4]^-$ can exist as cis and trans diastereoisomers.

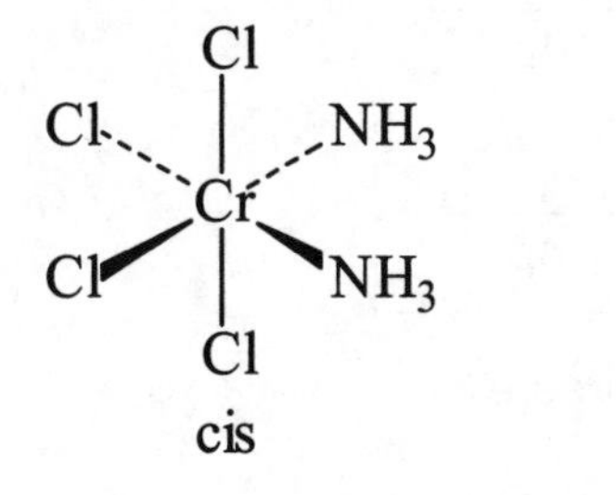

cis

NH3
Cl
Cl
Cr
Cl
Cl
NH3
trans

(b) $[Co(NH_3)_5Br]^{2+}$ cannot exist as diastereoisomers.
(c) $[FeCl_2(NCS)_2]^{2-}$ (tetrahedral) cannot exist as diastereoisomers.
(d) $[PtCl_2Br_2]^{2-}$ (square planar) can exist as cis and trans diastereoisomers.

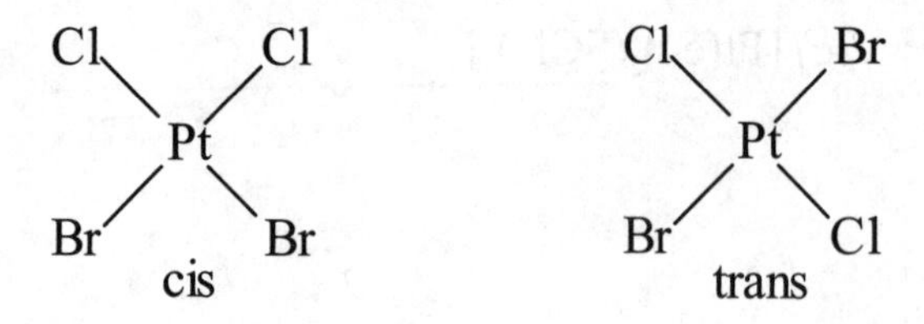

20.100 (c) cis-$[Cr(en)_2(H_2O)_2]^{3+}$ (d) $[Cr(C_2O_4)_3]^{3-}$

20.102

enantiomers

diastereoisomers

20.104 Plane-polarized light is light in which the electric vibrations of the light wave are restricted to a single plane. The following chromium complex can rotate the plane of plane-polarized light.

$[Cr(en)_3]^{3+}$

Color of Complexes; Valence Bond and Crystal Field Theories (Sections 20.10–20.12)

20.106 The measure of the amount of light absorbed by a substance is called the absorbance, and a graph of absorbance versus wavelength is called an absorption spectrum. If a complex absorbs at 455 nm, its color is orange (use the color wheel in Figure 20.26).

20.108 (a) $[Ti(H_2O)_6]^{3+}$

Ti^{3+} [Ar] ↑ _ _ _ _ (3d) _ (4s) _ _ _ (4p)

$[Ti(H_2O)_6]^{3+}$ [Ar] ↑ _ _ ↑↓ ↑↓ (3d) ↑↓ (4s) ↑↓ ↑↓ ↑↓ (4p)

d^2sp^3 1 unpaired e^-

(b) $[NiBr_4]^{2-}$

Ni^{2+}	[Ar]	↑↓ ↑↓ ↑↓ ↑ ↑ (3d)	__ (4s)	__ __ __ (4p)
$[NiBr_4]^{2-}$	[Ar]	↑↓ ↑↓ ↑↓ ↑ ↑ (3d)	↑↓ (4s)	↑↓ ↑↓ ↑↓ (4p)

sp^3 2 unpaired e^-

(c) $[Fe(CN)_6]^{3-}$ (low-spin)

Fe^{3+}	[Ar]	↑ ↑ ↑ ↑ ↑ (3d)	__ (4s)	__ __ __ (4p)
$[Fe(CN)_6]^{3-}$	[Ar]	↑↓ ↑↓ ↑ ↑↓ ↑↓ (3d)	↑↓ (4s)	↑↓ ↑↓ ↑↓ (4p)

d^2sp^3 1 unpaired e^-

(d) $[MnCl_6]^{3-}$ (high-spin)

Mn^{3+}	[Ar]	↑ ↑ ↑ ↑ __ (3d)	__ (4s)	__ __ __ (4p)	
$[MnCl_6]^{3-}$	[Ar]	↑ ↑ ↑ ↑ __ (3d)	↑↓ (4s)	↑↓ ↑↓ ↑↓ (4p)	↑↓ ↑↓ __ __ __ (4d)

sp^3d^2 4 unpaired e^-

20.110 (a) +3, M = Cr or Ni

(b)

$[Cr(OH)_4]^-$: [Ar] ↑ ↑ ↑ __ __ (3d) ↑↓ (4s) ↑↓ ↑↓ ↑↓ (4p)

Four sp^3 bonds to the ligands

$[Ni(OH)_4]^-$: [Ar] ↑↓ ↑↓ ↑ ↑ ↑ (3d) ↑↓ (4s) ↑↓ ↑↓ ↑↓ (4p)

Four sp^3 bonds to the ligands

(c) $[Cr(OH)_4]^-$

20.112 $[Ti(H_2O)_6]^{3+}$ Ti^{3+} $3d^1$

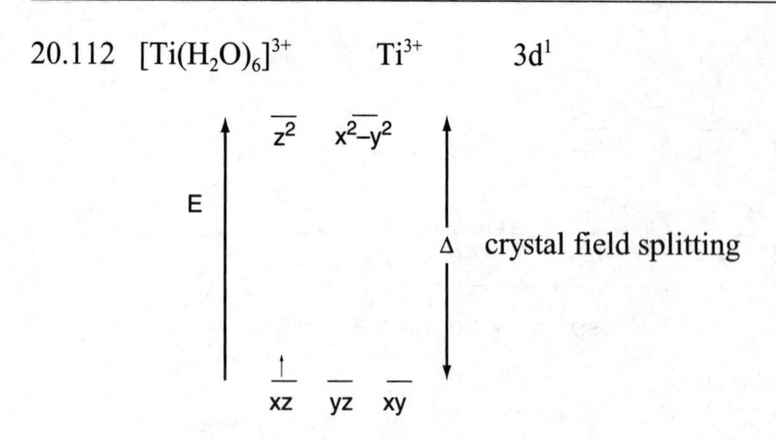

$[Ti(H_2O)_6]^{3+}$ is colored because it can absorb light in the visible region, exciting the electron to the higher-energy set of orbitals.

20.114 $\lambda = 544 \text{ nm} = 544 \times 10^{-9} \text{ m}$

$$\Delta = \frac{hc}{\lambda} = \frac{(6.626 \times 10^{-34} \text{ J} \cdot \text{s})(3.00 \times 10^{8} \text{ m/s})}{(544 \times 10^{-9} \text{ m})} = 3.65 \times 10^{-19} \text{ J}$$

$\Delta = (3.65 \times 10^{-19} \text{ J/ion})(6.022 \times 10^{23} \text{ ion/mol}) = 219{,}803 \text{ J/mol} = 220 \text{ kJ/mol}$

For $[Ti(H_2O)_6]^{3+}$, $\Delta = 240$ kJ/mol

Because $\Delta_{NCS^-} < \Delta_{H_2O}$ for the Ti complex, NCS^- is a weaker-field ligand than H_2O. If $[Ti(NCS)_6]^{3-}$ absorbs at 544 nm, its color should be red (use the color wheel in Figure 20.26).

20.116 (a) $[CrF_6]^{3-}$ (b) $[V(H_2O)_6]^{3+}$ (c) $[Fe(CN)_6]^{3-}$

(a) — —

↿ ↿ ↿

3 unpaired e^-

(b) — —

↿ ↿ —

2 unpaired e^-

(c) — —

⥮ ⥮ ↿

1 unpaired e^-

20.118 $Ni^{2+}(aq)$ $Zn^{2+}(aq)$

$Ni^{2+}(aq)$: ↿ ↿ / ⥮ ⥮ ⥮

$Zn^{2+}(aq)$: ⥮ ⥮ / ⥮ ⥮ ⥮

$Ni^{2+}(aq)$ is green because the Ni^{2+} ion can absorb light, which promotes electrons from the filled d orbitals to the higher energy half-filled d orbitals. $Zn^{2+}(aq)$ is colorless because the d orbitals are completely filled and no electrons can be promoted, so no light is absorbed.

20.120 Weak-field ligands produce a small Δ. Strong-field ligands produce a large Δ. For a metal complex with weak-field ligands, Δ < P, where P is the pairing energy, and it is easier to place an electron in either d_{z^2} or $d_{x^2-y^2}$ than to pair up electrons; high-spin complexes result. For a metal complex with strong-field ligands, Δ > P and it is easier to pair up electrons than to place them in either d_{z^2} or $d_{x^2-y^2}$; low-spin complexes result.

20.122

__ x^2-y^2

↑↓ xy

↑↓ z^2

↑↓ ↑↓ xz, yz

Square planar geometry is most common for metal ions with d^8 configurations because this configuration favors low-spin complexes in which all four lower energy d orbitals are filled, and the higher energy $d_{x^2-y^2}$ orbital is vacant.

Chapter Problems

20.124 (a) $[Mn(CN)_6]^{3-}$ Mn^{3+} [Ar] $3d^4$

CN^- is a strong-field ligand. The Mn^{3+} complex is low-spin.

__ __

↑↓ ↑ ↑ 2 unpaired e^-, paramagnetic

(b) $[Zn(NH_3)_4]^{2+}$ Zn^{2+} [Ar] $3d^{10}$

$[Zn(NH_3)_4]^{2+}$ is tetrahedral.

↑↓ ↑↓ ↑↓

↑↓ ↑↓ no unpaired e^-, diamagnetic

(c) $[Fe(CN)_6]^{4-}$ Fe^{2+} [Ar] $3d^6$

CN^- is a strong-field ligand. The Fe^{2+} complex is low-spin.

__ __

↑↓ ↑↓ ↑↓ no unpaired e^-, diamagnetic

(d) $[FeF_6]^{4-}$ Fe^{2+} [Ar] $3d^6$

F^- is a weak-field ligand. The Fe^{2+} complex is high-spin.

↑ ↑

↑↓ ↑ ↑ 4 unpaired e^-, paramagnetic

20.126 (a) $4\,[Co^{3+}(aq) + e^- \rightarrow Co^{2+}(aq)]$
$2\,H_2O(l) \rightarrow O_2(g) + 4\,H^+(aq) + 4\,e^-$
$4\,Co^{3+}(aq) + 2\,H_2O(l) \rightarrow 4\,Co^{2+}(aq) + O_2(g) + 4\,H^+(aq)$

(b) $4\,Cr^{2+}(aq) + O_2(g) + 4\,H^+(aq) \rightarrow 4\,Cr^{3+}(aq) + 2\,H_2O(l)$

(c) $3\,[Cu(s) \rightarrow Cu^{2+}(aq) + 2\,e^-]$
$Cr_2O_7^{2-}(aq) + 14\,H^+(aq) + 6\,e^- \rightarrow 2\,Cr^{3+}(aq) + 7\,H_2O(l)$
$3\,Cu(s) + Cr_2O_7^{2-}(aq) + 14\,H^+(aq) \rightarrow 3\,Cu^{2+}(aq) + 2\,Cr^{3+}(aq) + 7\,H_2O(l)$

(d) $2\,CrO_4^{2-}(aq) + 2\,H^+(aq) \rightarrow Cr_2O_7^{2-}(aq) + H_2O(l)$

20.128 $EDTA^{4-}$ in mayonnaise will complex any metal cations that are present in trace amounts. Free metal ions can catalyze the oxidation of oils, causing the mayonnaise to become rancid. The bidentate ligand $H_2NCH_2CO_2^-$ will not bind to metal ions as strongly as does the hexadentate $EDTA^{4-}$ and so would not be an effective substitute for $EDTA^{4-}$.

20.130 (a)

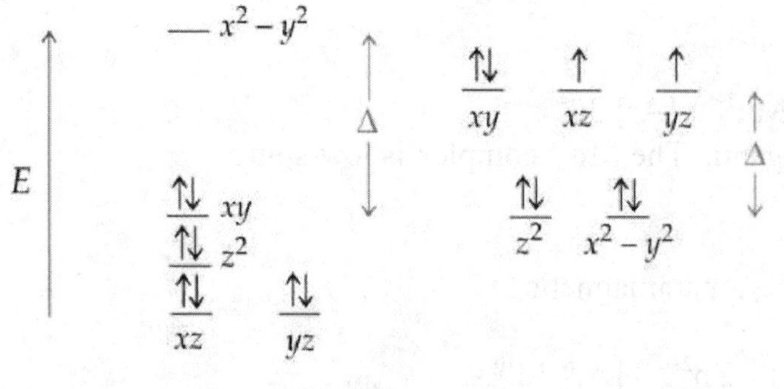

$x^2 - y^2$
↑↓ xy
↑↓ z^2
↑↓ xz ↑↓ yz
E, Δ

↑↓ xy ↑ xz ↑ yz
Δ
↑↓ z^2 ↑↓ $x^2 - y^2$

square planar nickel(II) tetrahedral nickel(II)

(b) $NiCl_2L_2$ is tetrahedral, $Ni(NCS)_2L_2$ is square planar.

L, NCS, Ni, SCN, L
L, NCS, Ni, L, NCS
L, Ni, Cl, L, Cl

(c) Square planar cis-$Ni(NCS)_2L_2$ and tetrahedral $NiCl_2L_2$ have a dipole moment.

20.132 Cr^{3+} is a $3d^3$ ion. Regardless of the crystal field splitting energy, the three electrons singly occupy the three lower energy d orbitals.

20.134 (a) $[Mn(H_2O)_6]^{2+}$ high-spin Mn^{2+}, $3d^5$

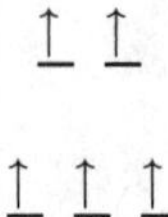

↑ ↑

↑ ↑ ↑

5 unpaired e^-

(b) $Pt(NH_3)_2Cl_2$ square-planar Pt^{2+}, $5d^8$

_

↿⇂

↿⇂

↿⇂ ↿⇂

no unpaired e^-

(c) $[FeO_4]^{2-}$ tetrahedral Fe(VI), $3d^2$

_ _ _

↿ ↿

2 unpaired e^-

(d) $[Ru(NH_3)_6]^{2+}$ low-spin Ru^{2+}, $4d^6$

_ _

↿⇂ ↿⇂ ↿⇂

no unpaired e^-

20.136 Linkage isomers:

$[NbCl_4(NCS)_2]^-$ (trans, both N-bonded NCS), $[NbCl_4(NCS)(SCN)]^-$ (trans), $[NbCl_4(SCN)_2]^-$ (trans, both S-bonded SCN)

Linkage isomers:

$[NbCl_4(NCS)_2]^-$ (cis, both N-bonded NCS), $[NbCl_4(NCS)(SCN)]^-$ (cis), $[NbCl_4(SCN)_2]^-$ (cis, both S-bonded SCN)

20.138 (a)

Co with ligands NO_2 (top), H_3N, NH_3, O_2N, NH_3, NO_2 (bottom); Co with ligands NO_2 (top), O_2N, NO_2, H_3N, NH_3, NH_3 (bottom)

(b) $[Co(NH_3)_4(NO_2)_2][Co(NH_3)_2(NO_2)_4]$

20.140 The nitro (–NO_2) complex is orange, which means it absorbs in the blue region (see color wheel) of the visible spectrum. The nitrito (–ONO) is red, which means it absorbs in the green region. The energy of the absorbed light is related to ligand field strength. Blue is higher energy than green, therefore nitro (–NO_2) is the stronger field ligand.

20.142 ML_2 (linear)

— z^2

— — xz, yz

— — $x^2–y^2$, xy

20.144 (a)

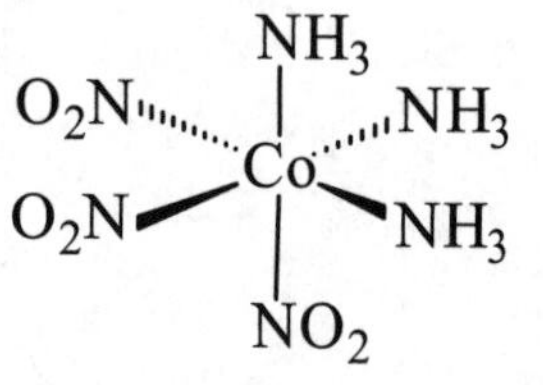

Isomer 1: Co with NH3 (top), NH3, NH3, O2N, O2N, NO2 (bottom)

Isomer 2: Co with NH3 (top), NH3, NO2, O2N, O2N, NH3 (bottom)

1 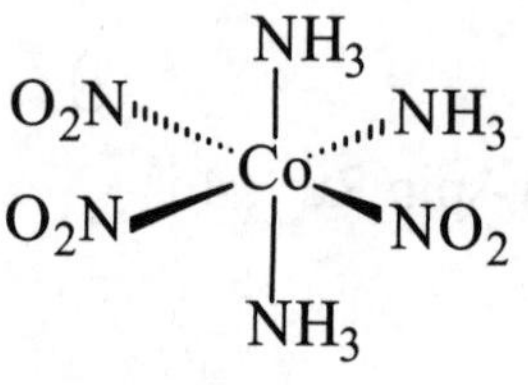 **2**

(b) Isomer **2** would give rise to the desired product because it has two trans NO_2 groups.

20.146 (a) $(NH_4)[Cr(H_2O)_6](SO_4)_2$, ammonium hexaaquachromium(III) sulfate

Cr^{3+} — —

↑ ↑ ↑

3 unpaired e^-

(b) $Mo(CO)_6$, hexacarbonylmolybdenum(0)

Mo^0 — —

↑↓ ↑↓ ↑↓

low-spin, no unpaired e^-

(c) $[Ni(NH_3)_4(H_2O)_2](NO_3)_2$, tetraamminediaquanickel(II) nitrate

Ni^{2+} ↑ ↑

↑↓ ↑↓ ↑↓

2 unpaired e^-

(d) $K_4[Os(CN)_6]$, potassium hexacyanoosmate(II)

Os^{2+} — —

↿⇂ ↿⇂ ↿⇂

low-spin, no unpaired e^-

(e) $[Pt(NH_3)_4](ClO_4)_2$, tetraammineplatinum(II) perchlorate

Pt^{2+} —

↿⇂

↿⇂

↿⇂ ↿⇂

low spin, no unpaired e^-

(f) $Na_2[Fe(CO)_4]$, sodium tetracarbonylferrate(–II)

Fe^{2-} ↿⇂ ↿⇂ ↿⇂

↿⇂ ↿⇂

no unpaired e^-

20.148 For transition metal complexes, observed colors and absorbed colors are generally complementary. Using the color wheel (Figure 20.26), the absorbed colors in the table are complementary colors to those observed.

	Observed Color	**Absorbed Color**	**Approximate λ (nm)**
$Cr(acac)_3$	red	green	530
$[Cr(H_2O)_6]^{3+}$	violet	yellow	580
$[CrCl_2(H_2O)_4]^+$	green	red	700
$[Cr(urea)_6]^{3+}$	green	red	700
$[Cr(NH_3)_6]^{3+}$	yellow	violet	420
$Cr(acetate)_3(H_2O)_3$	blue-violet	orange-yellow	600

The magnitude of Δ is comparable to the energy of the absorbed light from the low energy red end to the high energy violet end (ROYGBIV). The red of $[CrCl_2(H_2O)_4]^+$ is lower energy than the yellow of $[Cr(H_2O)_6]^{3+}$, so $Cl^- < H_2O$. Because $[CrCl_2(H_2O)_4]^+$ and $[Cr(urea)_6]^{3+}$ are both red, Δ for 6 urea's is approximately equal to Δ for 2 Cl^-'s and 4 H_2O's. Therefore, urea is between Cl^- and H_2O.

The spectrochemical series is: $Cl^- <$ urea $<$ acetate $< H_2O <$ acac $< NH_3$

Multiconcept Problems

20.150 (1) $Ni(H_2O)_6^{2+}(aq) + 6\,NH_3(aq) \rightleftarrows Ni(NH_3)_6^{2+}(aq) + 6\,H_2O(l)$ $K_f = 2.0 \times 10^8$

(2) $Ni(H_2O)_6^{2+}(aq) + 3\,en(aq) \rightleftarrows Ni(en)_3^{2+}(aq) + 6\,H_2O(l)$ $K_f = 4 \times 10^{17}$

(a) Reaction (2) should have the larger entropy change because three bidentate en ligands displace six water molecules.

(b) $\Delta G^\circ = \Delta H^\circ - T\Delta S^\circ$

Because ΔH°_1 and ΔH°_2 are almost the same, the difference in ΔG° is determined by the difference in ΔS°. Because ΔS°_2 is larger than ΔS°_1, ΔG°_2 is more negative than ΔG°_1 which is consistent with the greater stability of $Ni(en)_3^{2+}$.

(c) $\Delta H^\circ - T\Delta S^\circ = \Delta G^\circ = -RT \ln K_f$

$$\Delta H^\circ_1 - T\Delta S^\circ_1 - (\Delta H^\circ_2 - T\Delta S^\circ_2) = -RT \ln K_f(1) - [-RT \ln K_f(2)]$$

$$T\Delta S^\circ_2 - T\Delta S^\circ_1 = RT \ln K_f(2) - RT \ln K_f(1) = RT \ln \frac{K_f(2)}{K_f(1)}$$

$$\Delta S^\circ_2 - \Delta S^\circ_1 = R \ln \frac{K_f(2)}{K_f(1)} = [8.314\ J/(K \cdot mol)] \ln \frac{4 \times 10^{17}}{2.0 \times 10^8}$$

$$\Delta S^\circ_2 - \Delta S^\circ_1 = 178\ J/(K \cdot mol) \text{ or } 180\ J/(K \cdot mol)$$

20.152 (a) $Cr(s) + 2\,H^+(aq) \rightarrow Cr^{2+}(aq) + H_2(g)$

(b) $$\text{mol Cr} = 2.60\text{ g Cr} \times \frac{1\text{ mol Cr}}{52.00\text{ g Cr}} = 0.0500\text{ mol Cr}$$

mol H_2SO_4 = (0.050 00 L)(1.200 mol/L) = 0.060 00 mol H_2SO_4

The stoichiometry between Cr and H_2SO_4 is one to one, therefore Cr is the limiting reagent because of the smaller number of moles.

$$\text{mol } H_2 = 0.0500\text{ mol Cr} \times \frac{1\text{ mol } H_2}{1\text{ mol Cr}} = 0.0500\text{ mol } H_2$$

25 °C = 298 K

PV = nRT

$$V = \frac{nRT}{P} = \frac{(0.0500\text{ mol})\left(0.082\,06\ \frac{L \cdot atm}{K \cdot mol}\right)(298\text{ K})}{\left(735\text{ mm Hg} \times \frac{1.00\text{ atm}}{760\text{ mm Hg}}\right)} = 1.26\text{ L of } H_2$$

(c) 0.060 00 mol H_2SO_4 can provide 0.1200 mol H^+. 0.0500 mol Cr reacts with 2 x (0.0500 mol H^+) = 0.100 mol H^+. This leaves 0.0200 mol H^+ and 0.0600 mol SO_4^{2-}, which will give, after neutralization, 0.0200 mol HSO_4^- and 0.0400 mol SO_4^{2-}.

$[HSO_4^-]$ = 0.0200 mol/0.050 00 L = 0.400 M

$[SO_4^{2-}]$ = 0.0400 mol/0.050 00 L = 0.800 M

The pH of this solution can be determined from the following equilibrium:

	$HSO_4^-(aq)$ + $H_2O(l)$	⇄	$H_3O^+(aq)$	+ $SO_4^{2-}(aq)$
initial (M)	0.400		~0	0.800
change (M)	–x		+x	+x
equil (M)	0.400 – x		x	0.800 + x

$$K_{a2} = \frac{[H_3O^+][SO_4^{2-}]}{[HSO_4^-]} = 1.2 \text{ x } 10^{-2} = \frac{(x)(0.800 + x)}{0.400 - x}$$

$x^2 + 0.812x - 0.0048 = 0$

Use the quadratic formula to solve for x.

$$x = \frac{-(0.812) \pm \sqrt{(0.812)^2 - 4(1)(-0.0048)}}{2(1)} = \frac{-0.812 \pm 0.8237}{2}$$

x = 0.005 85 and –0.818

Of the two solutions for x, only the positive value of x has physical meaning, because x is the $[H_3O^+]$.

$[H_3O^+]$ = x = 0.005 85 M

pH = $-\log[H_3O^+] = -\log(0.005\ 85) = 2.23$

(d) Crystal field d-orbital energy-level diagram

E ↑

↑ z^2 — $x^2 - y^2$ Δ

↑ xz ↑ yz ↑ xy

Valence bond orbital diagram

$Cr(H_2O)_6^{2+}$ [Ar] ↑ ↑ ↑ ↑ __ (3d)   [↑↓ (4s)   ↑↓ ↑↓ ↑↓ (4p)   ↑↓ ↑↓] __ __ __ (4d)

sp^3d^2 4 unpaired e^-

(e) The addition of excess KCN converts $Cr(H_2O)_6^{2+}(aq)$ to $Cr(CN)_6^{4-}(aq)$. CN^- is a strong field ligand and increases Δ changing the chromium complex from high spin, with 4 unpaired electrons, to low spin, with only 2 unpaired electrons.

20.154 (a) Assume a 100.0 g sample of the chromium compound.

$$19.52 \text{ g Cr x } \frac{1 \text{ mol Cr}}{51.996 \text{ g Cr}} = 0.3754 \text{ mol Cr}$$

$$39.91 \text{ g Cl x } \frac{1 \text{ mol Cl}}{35.453 \text{ g Cl}} = 1.126 \text{ mol Cl}$$

$$40.57 \text{ g } H_2O \times \frac{1 \text{ mol } H_2O}{18.015 \text{ g } H_2O} = 2.252 \text{ mol } H_2O$$

$Cr_{0.3754}Cl_{1.126}(H_2O)_{2.252}$, divide each subscript by the smallest, 0.3754.

$Cr_{0.3754\,/\,0.3754}Cl_{1.126\,/\,0.3754}(H_2O)_{2.252\,/\,0.3754}$

$CrCl_3(H_2O)_6$

(b) $Cr(H_2O)_6Cl_3$, 266.45; AgCl, 143.32

For **A**: mol Cr complex = mol Cr = $0.225 \text{ g Cr complex} \times \frac{1 \text{ mol Cr complex}}{266.45 \text{ g Cr complex}} = 8.44 \times 10^{-4}$ mol Cr

mol Cl = mol AgCl = $0.363 \text{ g AgCl} \times \frac{1 \text{ mol AgCl}}{143.32 \text{ g AgCl}} = 2.53 \times 10^{-3}$ mol Cl

$$\frac{\text{mol Cl}}{\text{mol Cr}} = \frac{2.53 \times 10^{-3} \text{ mol Cl}}{8.44 \times 10^{-4} \text{ mol Cr}} = 3 \text{ Cl/Cr}$$

For **B**: mol Cr complex = mol Cr = $0.263 \text{ g Cr complex} \times \frac{1 \text{ mol Cr complex}}{266.45 \text{ g Cr complex}} = 9.87 \times 10^{-4}$ mol Cr

mol Cl = mol AgCl = $0.283 \text{ g AgCl} \times \frac{1 \text{ mol AgCl}}{143.32 \text{ g AgCl}} = 1.97 \times 10^{-3}$ mol Cl

$$\frac{\text{mol Cl}}{\text{mol Cr}} = \frac{1.97 \times 10^{-3} \text{ mol Cl}}{9.87 \times 10^{-4} \text{ mol Cr}} = 2 \text{ Cl/Cr}$$

For **C**: mol Cr complex = mol Cr = $0.358 \text{ g Cr complex} \times \frac{1 \text{ mol Cr complex}}{266.45 \text{ g Cr complex}} = 1.34 \times 10^{-3}$ mol Cr

mol Cl = mol AgCl = $0.193 \text{ g AgCl} \times \frac{1 \text{ mol AgCl}}{143.32 \text{ g AgCl}} = 1.34 \times 10^{-3}$ mol Cl

$$\frac{\text{mol Cl}}{\text{mol Cr}} = \frac{1.34 \times 10^{-3} \text{ mol Cl}}{1.34 \times 10^{-3} \text{ mol Cr}} = 1 \text{ Cl/Cr}$$

Because only the free Cl^- ions (those not bonded to the Cr^{3+}) give an immediate precipitate of AgCl, the probable structural formulas are:

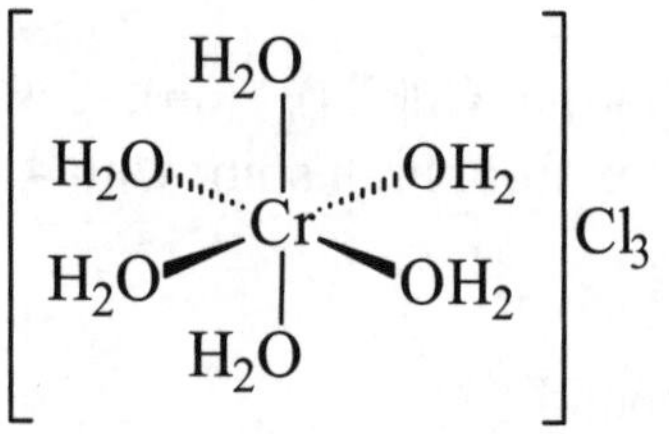

H_2O
H_2O OH_2
Cr Cl_3
H_2O OH_2
H_2O

A

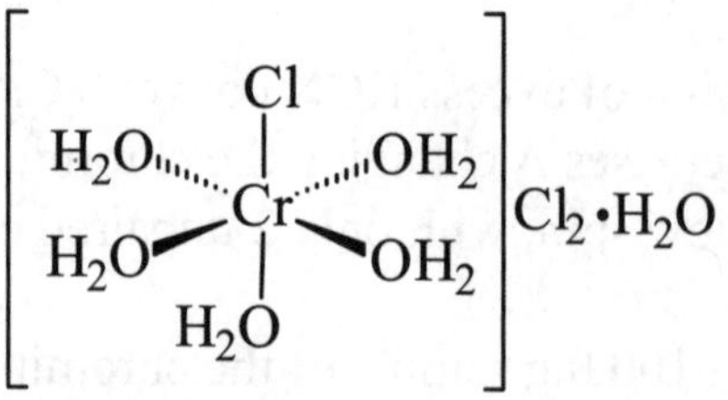

Cl
H_2O OH_2
Cr $Cl_2 \cdot H_2O$
H_2O OH_2
H_2O

B

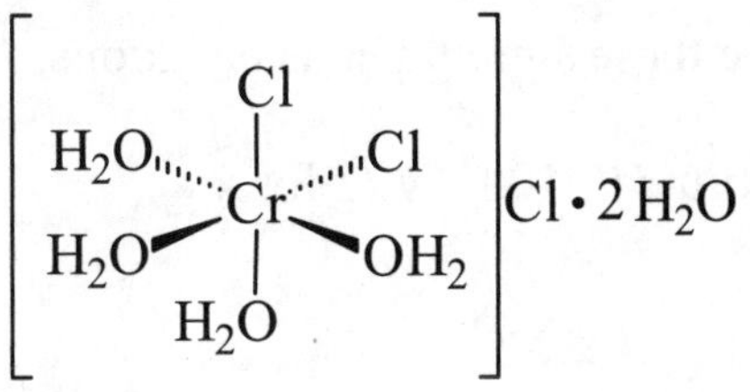

Cl
H_2O Cl
Cr $Cl \cdot 2H_2O$
H_2O OH_2
H_2O

C

Structure **C** can exist as either cis or trans diastereoisomers.

(c) H_2O is a stronger field ligand than Cl^-. Compound **A** is likely to be violet absorbing in the yellow. Compounds **B** and **C** have weaker field ligands and would appear blue or green absorbing in the orange or red, respectively.

(d) $\Delta T = K_f \cdot m \cdot i$
For **A**, i = 4; for **B**, i = 3; and for **C**, i = 2.
For **A**, $\Delta T = K_f \cdot m \cdot i = (1.86\ ^\circ C/m)(0.25\ m)(4) = 1.9\ ^\circ C$
freezing point $= 0\ ^\circ C - \Delta T = 0\ ^\circ C - 1.9\ ^\circ C = -1.9\ ^\circ C$
For **B**, $\Delta T = K_f \cdot m \cdot i = (1.86\ ^\circ C/m)(0.25\ m)(3) = 1.4\ ^\circ C$
freezing point $= 0\ ^\circ C - \Delta T = 0\ ^\circ C - 1.4\ ^\circ C = -1.4\ ^\circ C$
For **C**, $\Delta T = K_f \cdot m \cdot i = (1.86\ ^\circ C/m)(0.25\ m)(2) = 0.93\ ^\circ C$
freezing point $= 0\ ^\circ C - \Delta T = 0\ ^\circ C - 0.93\ ^\circ C = -0.93\ ^\circ C$

20.156 (a) $K = \dfrac{[Cr_2O_7^{2-}]}{[CrO_4^{2-}]^2[H^+]^2} = 1.00 \times 10^{14}$

$[Cr_2O_7^{2-}]/[CrO_4^{2-}]^2 = 1.00 \times 10^{14}\ [H^+]^2$
In neutral solution, $[H^+] = 1.0 \times 10^{-7}$ and $[Cr_2O_7^{2-}]/[CrO_4^{2-}]^2 = 1$, so $[Cr_2O_7^{2-}]$ and $[CrO_4^{2-}]$ are comparable.
In basic solution, $[H^+] < 1.0 \times 10^{-7}$ and $[Cr_2O_7^{2-}]/[CrO_4^{2-}]^2 < 1$, so $[CrO_4^{2-}]$ predominates.
In acidic solution, $[H^+] > 1.0 \times 10^{-7}$ and $[Cr_2O_7^{2-}]/[CrO_4^{2-}]^2 > 1$, so $[Cr_2O_7^{2-}]$ predominates.

(b) At pH = 4.000, the $[H^+] = 1.00 \times 10^{-4}$ M
Let $x = [Cr_2O_7^{2-}]$ and $y = [CrO_4^{2-}]$

$$\frac{[Cr_2O_7^{2-}]}{[CrO_4^{2-}]^2} = [H^+]^2(1.00 \times 10^{14})$$

$$\frac{[Cr_2O_7^{2-}]}{[CrO_4^{2-}]^2} = (1.00 \times 10^{-4})^2(1.00 \times 10^{14})$$

$$\frac{[Cr_2O_7^{2-}]}{[CrO_4^{2-}]^2} = 1.00 \times 10^{6} = \frac{x}{y^2}$$

Because there are 2 Cr atoms per $Cr_2O_7^{2-}$, the total Cr concentration is $2[Cr_2O_7^{2-}] + [CrO_4^{2-}]$, and therefore $2x + y = 0.100$.

$\frac{x}{y^2} = 1.00 \times 10^6$ and $2x + y = 0.100$ M; solve these simultaneous equations.

$x = (1.00 \times 10^6)y^2$ and $x = (0.100 - y)/2$; substitute $(0.100 - y)/2$ for x

$(0.100 - y)/2 = (1.00 \times 10^6)y^2$

$(2.00 \times 10^6)y^2 + y - 0.100 = 0$

Use the quadratic formula to solve for y.

$$y = \frac{-(1) \pm \sqrt{(1)^2 - 4(2.00 \times 10^6)(-0.100)}}{2(2.00 \times 10^6)} = \frac{(-1) \pm (894.2)}{4.00 \times 10^6}$$

$y = -2.24 \times 10^{-4}$ and $2.233 \times 10^{-4} = 2.23 \times 10^{-4}$

Of the two solutions for y, only the positive value of y has physical meaning because y is the $[CrO_4^{2-}]$.

$[CrO_4^{2-}] = 2.23 \times 10^{-4}$ M

$[Cr_2O_7^{2-}] = x = (1.00 \times 10^6)y^2 = (1.00 \times 10^6)(2.233 \times 10^{-4}\ M)^2 = 4.99 \times 10^{-2}$ M

(c) At pH = 2.000, the $[H^+] = 1.00 \times 10^{-2}$ M

Let $x = [Cr_2O_7^{2-}]$ and $y = [CrO_4^{2-}]$

$$\frac{[Cr_2O_7^{2-}]}{[CrO_4^{2-}]^2} = [H^+]^2(1.00 \times 10^{14})$$

$$\frac{[Cr_2O_7^{2-}]}{[CrO_4^{2-}]^2} = (1.00 \times 10^{-2})^2(1.00 \times 10^{14})$$

$$\frac{[Cr_2O_7^{2-}]}{[CrO_4^{2-}]^2} = 1.00 \times 10^{10} = \frac{x}{y^2}$$

Because there are 2 Cr atoms per $Cr_2O_7^{2-}$, the total Cr concentration is $2[Cr_2O_7^{2-}] + [CrO_4^{2-}]$, and therefore $2x + y = 0.100$.

$\frac{x}{y^2} = 1.00 \times 10^{10}$ and $2x + y = 0.100$ M; solve these simultaneous equations.

$x = (1.00 \times 10^{10})y^2$ and $x = (0.100 - y)/2$; substitute $(0.100 - y)/2$ for x

$(0.100 - y)/2 = (1.00 \times 10^{10})y^2$

$(2.00 \times 10^{10})y^2 + y - 0.100 = 0$

Use the quadratic formula to solve for y.

$$y = \frac{-(1) \pm \sqrt{(1)^2 - 4(2.00 \times 10^{10})(-0.100)}}{2(2.00 \times 10^{10})} = \frac{(-1) \pm (8.944 \times 10^4)}{4.00 \times 10^{10}}$$

$y = -2.24 \times 10^{-6}$ and $2.236 \times 10^{-6} = 2.24 \times 10^{-6}$

Of the two solutions for y, only the positive value of y has physical meaning because y is the $[CrO_4^{2-}]$.

$[CrO_4^{2-}] = 2.24 \times 10^{-6}$ M

$[Cr_2O_7^{2-}] = x = (1.00 \times 10^{10})y^2 = (1.00 \times 10^{10})(2.236 \times 10^{-6}\ M)^2 = 5.00 \times 10^{-2}$ M

21 Metals and Solid-State Materials

21.1 (a) $Cr_2O_3(s) + 2\ Al(s) \rightarrow 2\ Cr(s) + Al_2O_3(s)$
(b) $Cu_2S(s) + O_2(g) \rightarrow 2\ Cu(s) + SO_2(g)$
(c) $PbO(s) + C(s) \rightarrow Pb(s) + CO(g)$
(d) $2\ K^+(l) + 2\ Cl^-(l) \xrightarrow{\text{electrolysis}} 2\ K(l) + Cl_2(g)$

21.2 $CaO(s) + SiO_2(s) \rightarrow CaSiO_3(l)$ (slag)
The O^{2-} in CaO behaves as a Lewis base and SiO_2 is the Lewis acid. They react with each other in a Lewis acid-base reaction to yield $CaSiO_3$ (Ca^{2+} and SiO_3^{2-}).

21.3 The electron configuration for Hg is [Xe] $4f^{14}\ 5d^{10}\ 6s^2$. Assuming the 5d and 6s bands overlap, the composite band can accommodate 12 valence electrons per metal atom. Weak bonding and a low melting point are expected for Hg because both the bonding and antibonding MOs are occupied.

21.4 (a) The composite s-d band can accommodate 12 valence electrons per metal atom.
Hf [Xe] $6s^2\ 4f^{14}\ 5d^2$, 4 valence electrons (4 bonding, 0 antibonding)
The s-d band is 1/4 full, so Hf is picture (1).
Pt [Xe] $6s^2\ 4f^{14}\ 5d^8$, 10 valence electrons (6 bonding, 4 antibonding)
The s-d band is 5/6 full, so Pt is picture (2).
Re [Xe] $6s^2\ 4f^{14}\ 5d^5$, 7 valence electrons (6 bonding, 1 antibonding)
The s-d band is 7/12 full, so Re is picture (3).
(b) Re has an excess of 5 bonding electrons and it has the highest melting point and is the hardest of the three.
(c) Pt has an excess of only 2 bonding electrons and it has the lowest melting point and is the softest of the three.

21.5 Ge doped with As is an n-type semiconductor because As has an additional valence electron. The extra electrons are in the conduction band. The number of electrons in the conduction band of the doped Ge is much higher than for pure Ge, and the conductivity of the doped semiconductor is higher.

21.6 (a) (1), silicon; (2), white tin; (3), diamond; (4), silicon doped with aluminum
(b) (3) $<$ (1) $<$ (4) $<$ (2)
Diamond (3) is an insulator with a large band gap. Silicon (1) is a semiconductor with a band gap smaller than diamond. The conduction band is partially occupied with a few electrons and the valence band is partially empty. Silicon doped with aluminum (4) is a p-type semiconductor that has fewer electrons than needed for bonding and has vacancies (positive holes) in the valence band. White tin (2) has a partially filled s-p composite band and is a metallic conductor.

21.7 $E = 222 \text{ kJ/mol} \times \dfrac{1000 \text{ J}}{1 \text{ kJ}} \times \dfrac{1 \text{ mol}}{6.02 \times 10^{23}} = 3.69 \times 10^{-19} \text{ J}$

$$\nu = \frac{E}{h} = \frac{3.69 \times 10^{-19} \text{ J}}{6.626 \times 10^{-34} \text{ J}\cdot\text{s}} = 5.57 \times 10^{14} \text{ s}^{-1}$$

$$\lambda = \frac{c}{\nu} = \frac{3.00 \times 10^{8} \text{ m/s}}{5.57 \times 10^{14} \text{ s}^{-1}} = 5.39 \times 10^{-7} \text{ m} = 539 \times 10^{-9} \text{ m} = 539 \text{ nm}$$

21.8 $Si(OCH_3)_4 + 4\ H_2O \rightarrow Si(OH)_4 + 4\ HOCH_3$

21.9 $Ba[OCH(CH_3)_2]_2 + Ti[OCH(CH_3)_2]_4 + 6\ H_2O \rightarrow BaTi(OH)_6(s) + 6\ HOCH(CH_3)_2$

$$BaTi(OH)_6(s) \xrightarrow{\text{heat}} BaTiO_3(s) + 3\ H_2O(g)$$

21.10 (a) cobalt/tungsten carbide is a ceramic-metal composite.
(b) silicon carbide/zirconia is a ceramic-ceramic composite.
(c) boron nitride/epoxy is a ceramic-polymer composite.
(d) boron carbide/titanium is a ceramic-metal composite.

21.11 The color of the quantum dots depends on the wavelength of light they absorb, which is determined by band-gap energy. Different sizes of CdSe nanoparticles have different band-gap energies.

21.12 (a) 5.0 nm (b) 2.2 nm (c) 3.5 nm

21.13 CdSe diameter = 2 nm = 2×10^{-9} m and human hair diameter = 50 μm = 50×10^{-6} m

$\left(\dfrac{50 \times 10^{-6} \text{ m}}{2 \times 10^{-9} \text{ m}}\right)$ = 25,000 CdSe nanoparticles can fit across a human hair.

21.14 (a) Size (a) absorbs red light so it appears green, size (b) absorbs orange light so it appears blue, size (c) absorbs yellow light so it appears violet, and size (d) absorbs green light so it appears red.
(b) Particle sizes from smallest to largest are: $d < c < b < a$

21.15 The smaller the particle, the larger the band gap and the greater the shift in the color of the emitted light from the red to the violet. The yellow quantum dot is larger because yellow is closer to the red than is the blue.

Conceptual Problems

21.16 A – metal oxide; B – metal sulfide; C – metal carbonate; D – free metal

21.18 (a) (2), bonding MO's are filled.
(b) (3), bonding and antibonding MO's are filled.
(c) (3) < (1) < (2). Hardness increases with increasing MO bond order.

21.20 (a) (1) and (4) are semiconductors; (2) is a metal; (3) is an insulator
(b) (3) < (1) < (4) < (2). The conductivity increases with decreasing band gap.
(c) (1) and (4) increases; (2) decreases; (3) not much change.

21.22

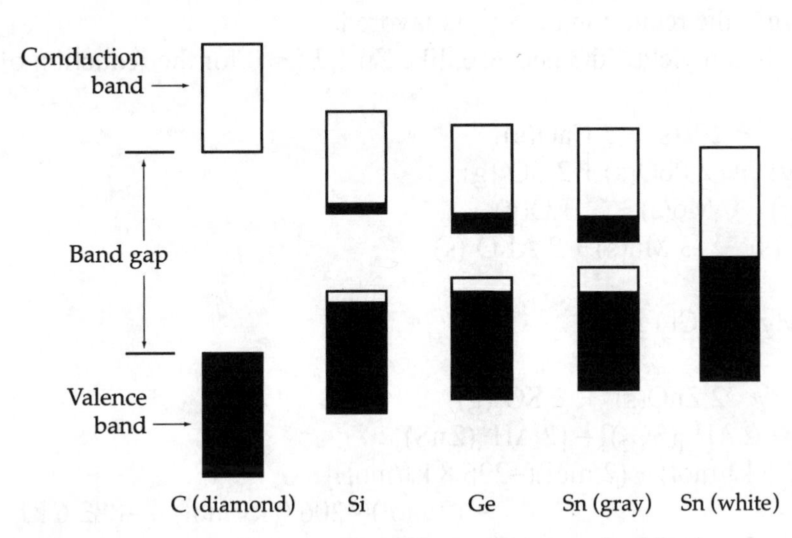

Section Problems
Sources of the Metallic Elements (Section 21.1)

21.24 TiO_2, MnO_2, and Fe_2O_3

21.26 (a) Cu is found in nature as a sulfide. (b) Zr is found in nature as an oxide.
(c) Pd is found in nature uncombined. (d) Bi is found in nature as a sulfide.

21.28 The less electronegative early transition metals tend to form ionic compounds by losing electrons to highly electronegative nonmetals such as oxygen. The more electronegative late transition metals tend to form compounds with more covalent character by bonding to the less electronegative nonmetals such as sulfur.

21.30 (a) Fe_2O_3, hematite (b) PbS, galena
(c) TiO_2, rutile (d) $CuFeS_2$, chalcopyrite

Metallurgy (Section 21.2)

21.32 The flotation process exploits the differences in the ability of water and oil to wet the surfaces of the mineral and the gangue. The gangue, which contains ionic silicates, is moistened by the polar water molecules and sinks to the bottom of the tank. The mineral particles, which contain the less polar metal sulfide, are coated by the oil and become attached to the soapy air bubbles created by the detergent. The metal sulfide particles are carried to the surface in the soapy froth, which is skimmed off at the top of the tank. This process would not work well for a metal oxide because it is too polar and will be wet by the water and sink with the gangue.

21.34 Hg^{2+} in HgS is reduced. S^{2-} in HgS is oxidized. O_2 is reduced.
Hg^{2+} in HgS and O_2 are oxidizing agents. S^{2-} in HgS is the reducing agent.

21.36 Because E° < 0 for Zn^{2+}, the reduction of Zn^{2+} is not favored.
Because E° > 0 for Hg^{2+}, the reduction of Hg^{2+} is favored.
The roasting of CdS should yield CdO because, like Zn^{2+}, E° < 0 for the reduction of Cd^{2+}.

21.38 (a) $V_2O_5(s) + 5\,Ca(s) \rightarrow 2\,V(s) + 5\,CaO(s)$
(b) $2\,PbS(s) + 3\,O_2(g) \rightarrow 2\,PbO(s) + 2\,SO_2(g)$
(c) $MoO_3(s) + 3\,H_2(g) \rightarrow Mo(s) + 3\,H_2O(g)$
(d) $3\,MnO_2(s) + 4\,Al(s) \rightarrow 3\,Mn(s) + 2\,Al_2O_3(s)$

(e) $MgCl_2(l) \xrightarrow{\text{electrolysis}} Mg(l) + Cl_2(g)$

21.40 $2\,ZnS(s) + 3\,O_2(g) \rightarrow 2\,ZnO(s) + 2\,SO_2(g)$

$\Delta H^\circ = [2\,\Delta H^\circ_f(ZnO) + 2\,\Delta H^\circ_f(SO_2)] - [2\,\Delta H^\circ_f(ZnS)]$

$\Delta H^\circ = [(2\text{ mol})(-350.5\text{ kJ/mol}) + (2\text{ mol})(-296.8\text{ kJ/mol})] - (2\text{ mol})(-206.0\text{ kJ/mol}) = -882.6\text{ kJ}$

$\Delta G^\circ = [2\,\Delta G^\circ_f(ZnO) + 2\,\Delta G^\circ_f(SO_2)] - [2\,\Delta G^\circ_f(ZnS)]$

$\Delta G^\circ = [(2\text{ mol})(-320.5\text{ kJ/mol}) + (2\text{ mol})(-300.2\text{ kJ/mol})] - (2\text{ mol})(-201.3\text{ kJ/mol}) = -838.8\text{ kJ}$

ΔH° and ΔG° are different because of the entropy change associated with the reaction. The minus sign for (ΔH° – ΔG°) indicates that the entropy is negative, which is consistent with a decrease in the number of moles of gas from 3 mol to 2 mol.

21.42 $FeCr_2O_4(s) + 4\,C(s) \rightarrow \underbrace{Fe(s) + 2\,Cr(s)}_{\text{ferrochrome}} + 4\,CO(g)$

(a) $FeCr_2O_4$, 223.84; Cr, 52.00; 236 kg = 236 x 10^3 g

$$\text{mass Cr} = 236 \times 10^3\text{ g} \times \frac{1\text{ mol }FeCr_2O_4}{223.84\text{ g}} \times \frac{2\text{ mol Cr}}{1\text{ mol }FeCr_2O_4} \times \frac{52.00\text{ g Cr}}{1\text{ mol Cr}} \times \frac{1.00\text{ kg}}{1000\text{ g}} = 110\text{ kg Cr}$$

(b) $$\text{mol CO} = 236 \times 10^3\text{ g} \times \frac{1\text{ mol }FeCr_2O_4}{223.84\text{ g}} \times \frac{4\text{ mol CO}}{1\text{ mol }FeCr_2O_4} = 4217.3\text{ mol CO}$$

$$PV = nRT;\quad V = \frac{nRT}{P} = \frac{(4217.3\text{ mol})\left(0.082\,06\ \frac{\text{L}\cdot\text{atm}}{\text{K}\cdot\text{mol}}\right)(298\text{ K})}{\left(740\text{ mm Hg} \times \frac{1.00\text{ atm}}{760\text{ mm Hg}}\right)} = 1.06 \times 10^5\text{ L CO}$$

21.44 $Ni^{2+}(aq) + 2\,e^- \rightarrow Ni(s)$; $1\ A = 1\ C/s$

$$\text{mass Ni} = 52.5\ \frac{C}{s} \times 8.00\ h \times \frac{3600\ s}{1\ h} \times \frac{1\ mol\ e^-}{96{,}500\ C} \times \frac{1\ mol\ Ni}{2\ mol\ e^-} \times \frac{58.69\ g\ Ni}{1\ mol\ Ni} \times \frac{1.00\ kg}{1000\ g}$$

mass Ni = 0.460 kg Ni

Iron and Steel (Section 21.3)

21.46 $Fe_2O_3(s) + 3\,CO(g) \rightarrow 2\,Fe(l) + 3\,CO_2(g)$
Fe_2O_3 is the oxidizing agent. CO is the reducing agent.

21.48 Slag is a by-product of iron production, consisting mainly of $CaSiO_3$. It is produced from the gangue in iron ore.

21.50 Molten iron from a blast furnace is exposed to a jet of pure oxygen gas for about 20 minutes. The impurities are oxidized to yield a molten slag that can be poured off.
$P_4(l) + 5\,O_2(g) \rightarrow P_4O_{10}(l)$
$6\,CaO(s) + P_4O_{10}(l) \rightarrow 2\,Ca_3(PO_4)_2(l)$ (slag)

$2\,Mn(l) + O_2(g) \rightarrow 2\,MnO(s)$
$MnO(s) + SiO_2(s) \rightarrow MnSiO_3(l)$ (slag)

21.52 $SiO_2(s) + 2\,C(s) \rightarrow Si(s) + 2\,CO(g)$
$Si(s) + O_2(g) \rightarrow SiO_2(s)$
$CaO(s) + SiO_2(s) \rightarrow CaSiO_3(l)$ (slag)

21.54
$3\,Fe_2O_3(s) + CO(g) \rightarrow 2\,Fe_3O_4(s) + CO_2(g)$ $\quad \Delta H^\circ = -46.4\ kJ$
$2 \times [Fe_3O_4(s) + CO(g) \rightarrow 3\,FeO(s) + CO_2(g)]$ $\quad \Delta H^\circ = 2(19.0\ kJ) = 38.0\ kJ$
$6 \times [FeO(s) + CO(g) \rightarrow Fe(s) + CO_2(g)]$ $\quad \Delta H^\circ = 6(-11.0\ kJ) = -66.0\ kJ$
$3\,Fe_2O_3(s) + 9\,CO(g) \rightarrow 6\,Fe(s) + 9\,CO_2(g)$ $\quad \Delta H^\circ = (-46.4 + 38.0 - 66.0) = -74.4\ kJ$
divide each coefficient by 3
$Fe_2O_3(s) + 3\,CO(g) \rightarrow 2\,Fe(s) + 3\,CO_2(g)$ $\quad \Delta H^\circ = -74.4\ kJ/3 = -24.8\ kJ$

21.56 No. In a blast furnace tungsten carbide (WC) would be formed.

Bonding in Metals (Section 21.4)

21.58

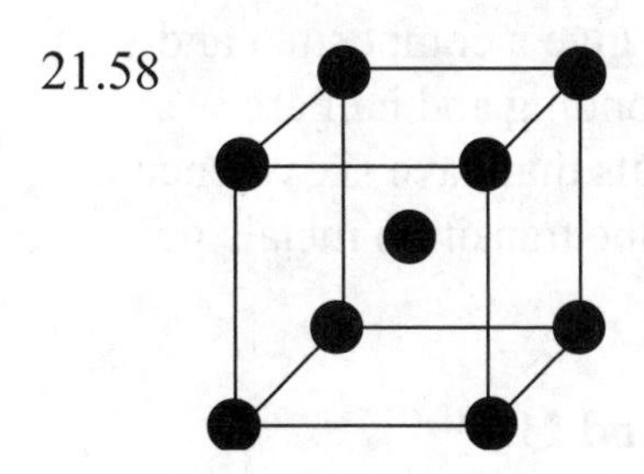

Each K has a single valence electron and has eight nearest neighbor K atoms. The valence electrons cannot be localized in an electron-pair bond between any particular pair of K atoms.

21.60 Malleability and ductility of metals follow from the fact that the delocalized bonding extends in all directions. When a metallic crystal is deformed, no localized bonds are broken. Instead, the electron sea simply adjusts to the new distribution of cations, and the energy of the deformed structure is similar to that of the original. Thus, the energy required to deform a metal is relatively small.

21.62 Ionic bonding is much stronger than metallic bonding.

21.64 The energy required to deform a transition metal like W is greater than that for Cs because W has more valence electrons and hence more electrostatic "glue".

21.66 The difference in energy between successive MOs in a metal decreases as the number of metal atoms increases so that the MOs merge into an almost continuous band of energy levels. Consequently, MO theory for metals is often called band theory.

21.68 The energy levels within a band occur in degenerate pairs; one set of energy levels applies to electrons moving to the right, and the other set applies to electrons moving to the left. In the absence of an electrical potential, the two sets of levels are equally populated. As a result there is no net electric current. In the presence of an electrical potential those electrons moving to the right are accelerated, those moving to the left are slowed down, and some change direction. Thus, the two sets of energy levels are now unequally populated. The number of electrons moving to the right is now greater than the number moving to the left, and so there is a net electric current.

21.70

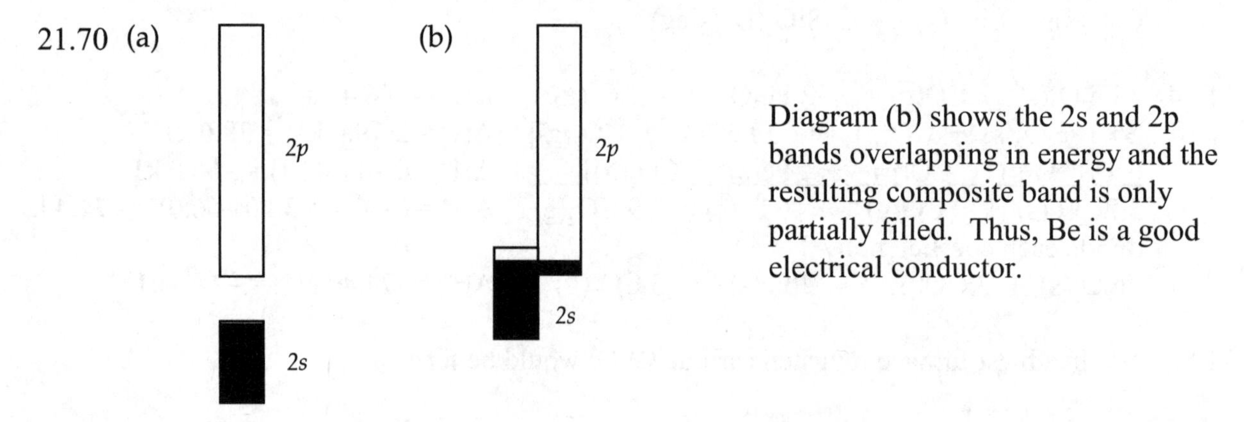

Diagram (b) shows the 2s and 2p bands overlapping in energy and the resulting composite band is only partially filled. Thus, Be is a good electrical conductor.

21.72 Transition metals have a d band that can overlap the s band to give a composite band consisting of six MOs per metal atom. Half of the MOs are bonding and half are antibonding, and thus one expects maximum bonding for metals that have six valence electrons per metal atom. Accordingly, the melting points of the transition metals go through a maximum at or near group 6B.

Semiconductors and Semiconductor Applications (Sections 21.5 and 21.6)

21.74 A semiconductor is a material that has an electrical conductivity intermediate between that of a metal and that of an insulator. Si, Ge, and Sn (gray) are semiconductors.

21.76 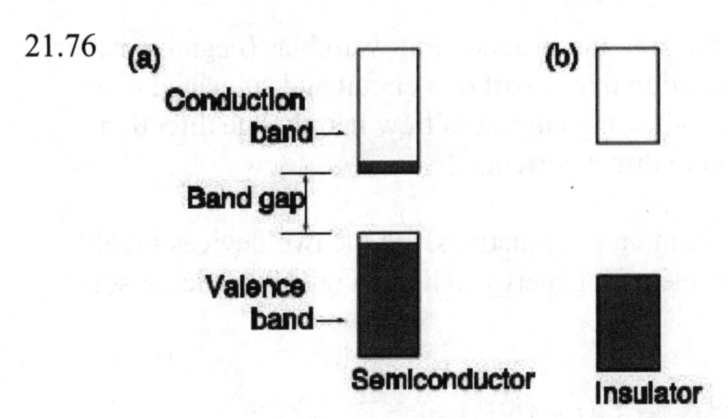

The MOs of a semiconductor are similar to those of an insulator, but the band gap in a semiconductor is smaller. As a result, a few electrons have enough energy to jump the gap and occupy the higher-energy, conduction band. The conduction band is thus partially filled, and the valence band is partially empty. When an electrical potential is applied to a semiconductor, it conducts a small amount of current because the potential can accelerate the electrons in the partially filled bands.

21.78 As the band gap increases, the number of electrons able to jump the gap and occupy the higher-energy conduction band decreases, and thus the conductivity decreases.

21.80 An n-type semiconductor is a semiconductor doped with a substance with more valence electrons than the semiconductor itself. Si doped with P is an example.

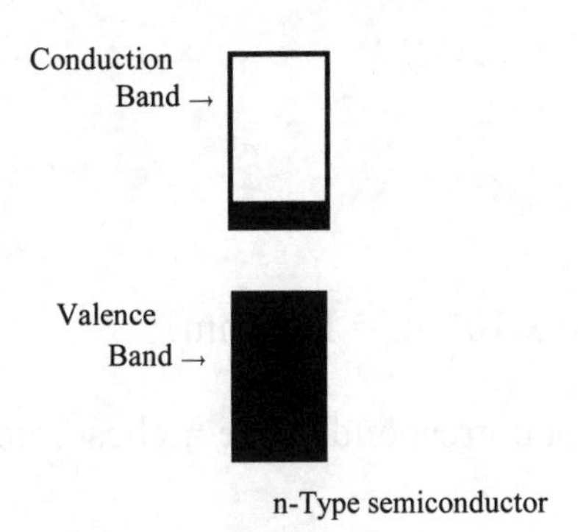

21.82 In the MO picture, the extra electrons occupy the conduction band. The number of electrons in the conduction band of the doped Ge is much greater than for pure Ge, and the conductivity of the doped semiconductor is correspondingly higher.

21.84 (a) p-type (In is electron deficient with respect to Si)
(b) n-type (Sb is electron rich with respect to Ge)
(c) n-type (As is electron rich with respect to gray Sn)

21.86 $Cd(CH_3)_2(g) + H_2Se(g) \rightarrow CdSe(s) + 2\ CH_4(g)$

21.88 Al_2O_3 < Ge < Ge doped with In < Fe < Cu

21.90 In a diode, current flows only when the junction is under a forward bias (negative battery terminal on the n-type side). A p-n junction that is part of a circuit and subjected to an alternating potential acts as a rectifier, allowing current to flow in only one direction, thereby converting alternating current to direct current.

21.92 Both an LED and a photovoltaic cell contain p-n junctions, but the two devices involve opposite processes. An LED converts electrical energy to light; a photovoltaic, or solar, cell converts light to electricity.

21.94 $E = 193\ \text{kJ/mol} \times \dfrac{1000\ \text{J}}{1\ \text{kJ}} \times \dfrac{1\ \text{mol}}{6.02 \times 10^{23}} = 3.21 \times 10^{-19}\ \text{J}$

$$\nu = \frac{E}{h} = \frac{3.21 \times 10^{-19}\ \text{J}}{6.626 \times 10^{-34}\ \text{J·s}} = 4.84 \times 10^{14}\ \text{s}^{-1}$$

$$\lambda = \frac{c}{\nu} = \frac{3.00 \times 10^{8}\ \text{m/s}}{4.84 \times 10^{14}\ \text{s}^{-1}} = 6.20 \times 10^{-7}\ \text{m} = 620 \times 10^{-9}\ \text{m} = 620\ \text{nm, orange light}$$

21.96 (a) InN has the smaller band-gap energy because In is larger than Ga.
(b) GaN would emit ultraviolet light and InN would emit red light.

21.98 $GaP_{0.50}As_{0.50} < GaP_{0.80}As_{0.20} < GaP_{1.00}As_{0.00}$

21.100 (a) $E = 107\ \text{kJ/mol} \times \dfrac{1000\ \text{J}}{1\ \text{kJ}} \times \dfrac{1\ \text{mol}}{6.02 \times 10^{23}} = 1.78 \times 10^{-19}\ \text{J}$

$$\nu = \frac{E}{h} = \frac{1.78 \times 10^{-19}\ \text{J}}{6.626 \times 10^{-34}\ \text{J·s}} = 2.68 \times 10^{14}\ \text{s}^{-1}$$

$$\lambda = \frac{c}{\nu} = \frac{3.00 \times 10^{8}\ \text{m/s}}{2.68 \times 10^{14}\ \text{s}^{-1}} = 1.12 \times 10^{-6}\ \text{m} = 1120 \times 10^{-9}\ \text{m} = 1120\ \text{nm}$$

(b) The wavelength is in the near IR and does not correspond to the highest intensity wavelength in the solar emission spectrum.

Superconductors (Section 21.7)

21.102 (1) A superconductor is able to levitate a magnet.
(2) In a superconductor, once an electric current is started, it flows indefinitely without loss of energy. A superconductor has no electrical resistance.

21.104 Some K^+ ions are surrounded octahedrally by six $C_{60}^{\ 3-}$ ions; others are surrounded tetrahedrally by four $C_{60}^{\ 3-}$ ions.

Ceramics and Composites (Sections 21.8 and 21.9)

21.106 Ceramics are inorganic, nonmetallic, nonmolecular solids, including both crystalline and amorphous materials. Ceramics have higher melting points, and they are stiffer, harder, and more resistant to wear and corrosion than are metals.

21.108 Ceramics have higher melting points, and they are stiffer, harder, and more wear resistant than metals because they have stronger bonding. They maintain much of their strength at high temperatures, where metals either melt or corrode because of oxidation.

21.110 The brittleness of ceramics is due to strong chemical bonding. In silicon nitride each Si atom is bonded to four N atoms and each N atom is bonded to three Si atoms. The strong, highly directional covalent bonds prevent the planes of atoms from sliding over one another when the solid is subjected to a stress. As a result, the solid cannot deform to relieve the stress. It maintains its shape up to a point, but then the bonds give way suddenly and the material fails catastrophically when the stress exceeds a certain threshold value. By contrast, metals are able to deform under stress because their planes of metal cations can slide easily in the electron sea.

21.112 Ceramic processing is the series of steps that leads from raw material to the finished ceramic object.

21.114 $Zr[OCH(CH_3)_2]_4 + 4\ H_2O \rightarrow Zr(OH)_4 + 4\ HOCH(CH_3)_2$

21.116 $(HO)_3Si–O–H + H–O–Si(OH)_3 \rightarrow (HO)_3Si–O–Si(OH)_3 + H_2O$
Further reactions of this sort give a three-dimensional network of Si–O–Si bridges. On heating, SiO_2 is obtained.

21.118 $2\ Ti(BH_4)_3(soln) \rightarrow 2\ TiB_2(s) + B_2H_6(g) + 9\ H_2(g)$

21.120 $3\ SiCl_4(g) + 4\ NH_3(g) \rightarrow Si_3N_4(s) + 12\ HCl(g)$

21.122 Graphite/epoxy composites are good materials for making tennis rackets and golf clubs because of their high strength-to-weight ratios.

Chapter Problems

21.124 $\Delta G = \Delta H - T\Delta S = -160.8\ kJ - (298\ K)(-0.410\ kJ/K) = -38.6\ kJ$
Because $\Delta G < 0$, the reaction is spontaneous at 25 °C (298 K).
Set $\Delta G = 0$ and solve for T to find the temperature at which the reaction becomes nonspontaneous.

$$0 = \Delta H - T\Delta S; \quad T = \frac{\Delta H}{\Delta S} = \frac{-160.8\ kJ}{-0.410\ kJ/K} = 392\ K = 119\ °C$$

21.126 $2\ Eu^{3+}(aq) + Zn(s) \rightarrow 2\ Eu^{2+}(aq) + Zn^{2+}(aq)$
$Eu^{2+}(aq) + SO_4^{2-}(aq) \rightarrow EuSO_4(s)$

21.128 The chemical composition of the alkaline earth minerals is that of metal sulfates and sulfites, MSO_4 and MSO_3.

21.130 Band theory better explains how the number of valence electrons affects properties such as melting point and hardness.

21.132 V [Ar] $3d^3\,4s^2$ Zn [Ar] $3d^{10}\,4s^2$
Transition metals have a d band that can overlap the s band to give a composite band consisting of six MOs per metal atom. Half of the MOs are bonding and half are antibonding. Strong bonding and a high enthalpy of vaporization are expected for V because almost all of the bonding MOs are occupied and all of the antibonding MOs are empty. Weak bonding and a low enthalpy of vaporization are expected for Zn because both the bonding and the antibonding MOs are occupied.

21.134 With a band gap of 130 kJ/mol, GaAs is a semiconductor. Because Ge lies between Ga and As in the periodic table, GaAs is isoelectronic with Ge.

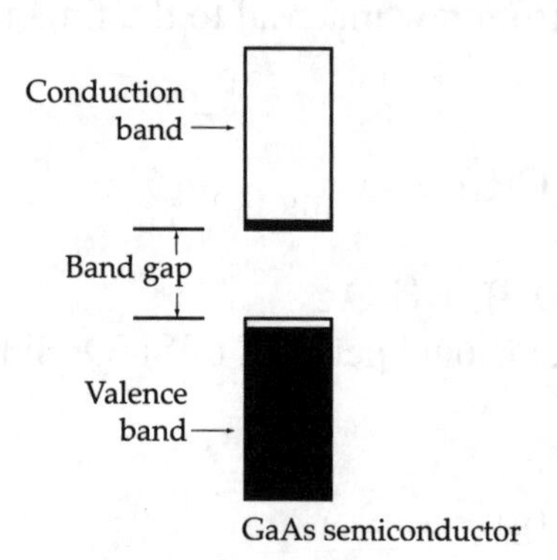

21.136 $YBa_2Cu_3O_7$, 666.20; $Cu(OCH_2CH_3)_2$, 153.67
$Y(OCH_2CH_3)_3$, 224.09; $Ba(OCH_2CH_3)_2$, 227.45

$$\text{mol } Cu(OCH_2CH_3)_2 = 75.4\text{ g} \times \frac{1\text{ mol}}{153.67\text{ g}} = 0.4907\text{ mol } Cu(OCH_2CH_3)_2$$

$$\text{mass } Y(OCH_2CH_3)_3 = 0.4907\text{ mol } Cu(OCH_2CH_3)_2 \times$$
$$\frac{1\text{ mol } Y(OCH_2CH_3)_3}{3\text{ mol } Cu(OCH_2CH_3)_2} \times \frac{224.09\text{ g } Y(OCH_2CH_3)_3}{1\text{ mol } Y(OCH_2CH_3)_3} = 36.7\text{ g } Y(OCH_2CH_3)_3$$

$$\text{mass } Ba(OCH_2CH_3)_2 = 0.4907\text{ mol } Cu(OCH_2CH_3)_2 \times$$
$$\frac{2\text{ mol } Ba(OCH_2CH_3)_2}{3\text{ mol } Cu(OCH_2CH_3)_2} \times \frac{227.45\text{ g } Ba(OCH_2CH_3)_2}{1\text{ mol } Ba(OCH_2CH_3)_2} = 74.4\text{ g } Ba(OCH_2CH_3)_2$$

mass $YBa_2Cu_3O_7$ = 0.4907 mol $Cu(OCH_2CH_3)_2$ x

$$\frac{1 \text{ mol } YBa_2Cu_3O_7}{3 \text{ mol } Cu(OCH_2CH_3)_2} \times \frac{666.20 \text{ g } YBa_2Cu_3O_7}{1 \text{ mol } YBa_2Cu_3O_7} = 109 \text{ g } YBa_2Cu_3O_7$$

21.138 (a) $6\ Al(OCH_2CH_3)_3 + 2\ Si(OCH_2CH_3)_4 + 26\ H_2O \rightarrow$
$6\ Al(OH)_3(s) + 2\ Si(OH)_4(s) + 26\ HOCH_2CH_3$
sol

(b) H_2O is eliminated from the sol through a series of reactions linking the sol particles together through a three-dimensional network of O bridges to form the gel.
$(HO)_2Al–O–H + H–O–Si(OH)_3 \rightarrow (HO)_2Al–O–Si(OH)_3 + H_2O$

(c) The remaining H_2O and solvent are removed from the gel by heating to produce the ceramic, $3\ Al_2O_3 \cdot 2\ SiO_2$.

21.142 (a)

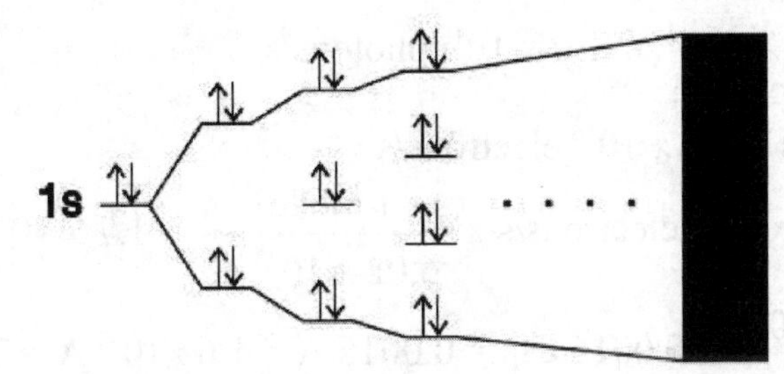

This material is an insulator because all MOs are filled, preventing the movement of electrons.

(b) 

Neutral hydrogen atoms have only 1 valence electron, compared with 2 in H^-. Partially empty antibonding MOs will allow the movement of electrons, so the doped material will be a conductor.

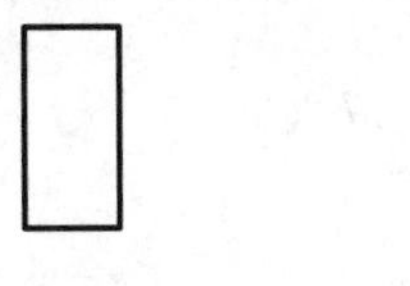

(c) The missing electrons in the doped material create "holes" that are positive charge carriers. This type of doped material is a p-type semiconductor.

21.144 (a) Because nitrogen has one more valence electron than carbon, nitrogen-doped diamond would be an n-type semiconductor.

(b)

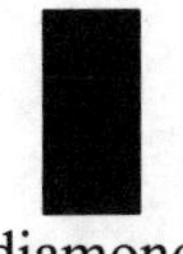

diamond (insulator) | nitrogen-doped diamond (n-type semiconductor)

(c) 425 nm = 425 x 10^{-9} m

$$E = \frac{hc}{\lambda} = (6.626 \times 10^{-34}\ J \cdot s)\left(\frac{3.00 \times 10^{8}\ m/s}{425 \times 10^{-9}\ m}\right)\left(\frac{1\ kJ}{1000\ J}\right)(6.022 \times 10^{23}\ /mol)$$

E = 282 kJ/mol

Multiconcept Problems

21.146 (a) Eu^{2+}, [Xe] $4f^7$ (b) BM = $\sqrt{n(n+2)} = \sqrt{7(7+2)}$ = 7.94 BM

21.148 660 nm = 660 x 10^{-9} m and 3.0 mW = 3.0 x 10^{-3} W = 3.0 x 10^{-3} J/s

$$E = h\frac{c}{\lambda} = (6.626 \times 10^{-34}\ J \cdot s)\left(\frac{3.00 \times 10^{8}\ m/s}{660 \times 10^{-9}\ m}\right) = 3.0 \times 10^{-19}\ J/photon$$

$$\#\ of\ photons/s = \frac{3.0 \times 10^{-3}\ J/s}{3.0 \times 10^{-19}\ J/photon} = 1.0 \times 10^{16}\ photons/s$$

# of electrons/s = # of photons/s = 1.0 x 10^{16} electrons/s

$$\#\ of\ moles\ of\ electrons/s = 1.0 \times 10^{16}\ electrons/s \times \frac{1\ mol\ e^-}{6.02 \times 10^{23}\ e^-} = 1.7 \times 10^{-8}\ mol\ e^-/s$$

$$A = 1.7 \times 10^{-8}\ mol\ e^-/s \times \frac{96{,}500\ C}{1\ mol\ e^-} = 0.0016\ C/s = 0.0016\ A = 1.6 \times 10^{-3}\ A = 1.6\ mA$$

21.150 $SiO_2(s) + 2\ C(s) \rightarrow Si(s) + 2\ CO(g)$

(a) $\Delta H^\circ = 2\ \Delta H^\circ_f(CO) - \Delta H^\circ_f(SiO_2)$

ΔH° = (2 mol)(–110.5 kJ/mol) – (1 mol)(–910.7 kJ/mol) = 689.7 kJ

$\Delta S^\circ = [S^\circ(Si) + 2\ S^\circ(CO)] - [S^\circ(SiO_2) + 2\ S^\circ(C)]$

ΔS° = [(1 mol)(18.8 J/(K· mol)) + (2 mol)(197.6 J/(K· mol))]
– [(1 mol)(41.5 J/(K · mol)) + (2 mol)(5.7 J/(K · mol))]

ΔS° = 361.1 J/K = 361.1 x 10^{-3} kJ/K

$\Delta G^\circ = \Delta H^\circ - T\Delta S^\circ$ = 689.7 kJ – (298.15 K)(361.1 x 10^{-3} kJ/K) = 582.0 kJ

(b) The reaction is endothermic because $\Delta H^\circ > 0$.

(c) The number of moles of gas increases from 0 to 2 mol, therefore, $\Delta S^\circ > 0$.

(d) Because $\Delta G^\circ > 0$, the reaction is nonspontaneous at 25 °C and 1 atm pressure of CO.

(e) To determine the crossover temperature, set $\Delta G^\circ = 0$ and solve for T.

$\Delta G^\circ = 0 = \Delta H^\circ - T\Delta S^\circ$

$$\Delta H^\circ = T\Delta S^\circ;\ T = \frac{\Delta H^\circ}{\Delta S^\circ} = \frac{689.7\ kJ}{361.1 \times 10^{-3}\ kJ/K} = 1910\ K = 1637\ ^\circ C$$

21.152 $C(s) + CO_2(g) \rightarrow 2\ CO(g)$

(a) CO_2, 44.01

$$mol\ CO_2 = 100.0\ g\ CO_2 \times \frac{1\ mol\ CO_2}{44.01\ g\ CO_2} = 2.272\ mol\ CO_2$$

$\Delta H^\circ = [2\ \Delta H^\circ_f(CO)] - \Delta H^\circ_f(CO_2)$

ΔH° = (2 mol)(–110.5 kJ/mol) – (1 mol)(–393.5 kJ/mol) = 172.5 kJ

$\Delta S^\circ = [2\ S^\circ(CO)] - [S^\circ(C) + S^\circ(CO_2)]$
$\Delta S^\circ = (2\ mol)(197.6\ J/(K \cdot mol)) - [(1\ mol)(5.7\ J/(K \cdot mol)) + (1\ mol)(213.6\ J/(K \cdot mol))]$
$\Delta S^\circ = 175.9\ J/K = 175.9 \times 10^{-3}\ kJ/K$
at 500 °C (773 K):
$\Delta G^\circ = \Delta H^\circ - T\Delta S^\circ = 172.5\ kJ - (773\ K)(175.9 \times 10^{-3}\ kJ/K) = 36.5\ kJ$

$$\ln K_p = \frac{-\Delta G^\circ}{RT} = \frac{-36.5\ kJ/mol}{[8.314 \times 10^{-3}\ kJ/(K \cdot mol)](773\ K)} = -5.68$$

$K_p = e^{-5.68} = 3.4 \times 10^{-3}$

$$P_{CO_2} = \frac{nRT}{V} = \frac{(2.272\ mol)\left(0.082\ 06\ \frac{L \cdot atm}{K \cdot mol}\right)(773\ K)}{50.00\ L} = 2.88\ atm$$

	$C(s)$ + $CO_2(g)$	⇄ $2\ CO(g)$
initial (atm)	2.88	0
change (atm)	$-x$	$+2x$
equil (atm)	$2.88 - x$	$2x$

$$K_p = \frac{(P_{CO})^2}{P_{CO_2}} = 3.4 \times 10^{-3} = \frac{(2x)^2}{2.88 - x}$$

$4x^2 + (3.4 \times 10^{-3})x - 9.79 \times 10^{-3} = 0$
Use the quadratic formula to solve for x.

$$x = \frac{-(3.4 \times 10^{-3}) \pm \sqrt{(3.4 \times 10^{-3})^2 - (4)(4)(-9.79 \times 10^{-3})}}{2(4)} = \frac{(-3.4 \times 10^{-3}) \pm (0.396)}{8}$$

$x = 0.049\ 08$ and $-0.049\ 93$
Of the two solutions for x, only the positive value of x has physical meaning because 2x is the partial pressure of CO.
$P_{CO_2} = 2.88 - x = 2.88 - 0.049\ 08 = 2.831\ atm$
$P_{CO} = 2x = 2(0.049\ 08) = 0.0982\ atm$
$P_{total} = P_{CO_2} + P_{CO} = 2.831 + 0.0982 = 2.93\ atm$

$$[CO] = \frac{n}{V} = \frac{P}{RT} = \frac{(0.0982\ atm)}{\left(0.082\ 06\ \frac{L \cdot atm}{K \cdot mol}\right)(773\ K)} = 1.5 \times 10^{-3}\ M$$

$$[CO_2] = \frac{n}{V} = \frac{P}{RT} = \frac{(2.831\ atm)}{\left(0.082\ 06\ \frac{L \cdot atm}{K \cdot mol}\right)(773\ K)} = 4.46 \times 10^{-2}\ M$$

(b) at 1000 °C (1273 K):
$\Delta G^\circ = \Delta H^\circ - T\Delta S^\circ = 172.5\ kJ - (1273\ K)(175.9 \times 10^{-3}\ kJ/K) = -51.4\ kJ$

$$\ln K_p = \frac{-\Delta G^\circ}{RT} = \frac{-(-51.4\ kJ/mol)}{[8.314 \times 10^{-3}\ kJ/(K \cdot mol)](1273\ K)} = 4.86$$

$K_p = e^{4.86} = 1.3 \times 10^2$

$$P_{CO_2} = \frac{nRT}{V} = \frac{(2.272\ \text{mol})\left(0.082\ 06\ \frac{\text{L}\cdot\text{atm}}{\text{K}\cdot\text{mol}}\right)(1273\ \text{K})}{50.00\ \text{L}} = 4.75\ \text{atm}$$

	C(s) + CO_2(g)	⇄ 2 CO(g)
initial (atm)	4.75	0
change (atm)	–x	+2x
equil (atm)	4.75 – x	2x

$$K_p = \frac{(P_{CO})^2}{P_{CO_2}} = 1.3 \times 10^2 = \frac{(2x)^2}{4.75 - x}$$

$4x^2 + (1.3 \times 10^2)x - 617.5 = 0$

Use the quadratic formula to solve for x.

$$x = \frac{-(1.3 \times 10^2) \pm \sqrt{(1.3 \times 10^2)^2 - (4)(4)(-617.5)}}{2(4)} = \frac{(-1.3 \times 10^2) \pm (163.6)}{8}$$

x = 4.200 and –36.70

Of the two solutions for x, only the positive value of x has physical meaning because 2x is the partial pressure of CO.

$P_{CO_2} = 4.75 - x = 4.75 - 4.200 = 0.55$ atm

$P_{CO} = 2x = 2(4.200) = 8.40$ atm

$P_{total} = P_{CO_2} + P_{CO} = 0.55 + 8.40 = 8.95$ atm

$$[CO] = \frac{n}{V} = \frac{P}{RT} = \frac{(8.40\ \text{atm})}{\left(0.082\ 06\ \frac{\text{L}\cdot\text{atm}}{\text{K}\cdot\text{mol}}\right)(1273\ \text{K})} = 8.04 \times 10^{-2}\ \text{M}$$

$$[CO_2] = \frac{n}{V} = \frac{P}{RT} = \frac{(0.55\ \text{atm})}{\left(0.082\ 06\ \frac{\text{L}\cdot\text{atm}}{\text{K}\cdot\text{mol}}\right)(1273\ \text{K})} = 5.3 \times 10^{-3}\ \text{M}$$

(c) $\Delta G^\circ = \Delta H^\circ - T\Delta S^\circ$; The equilibrium shifts to the right with increasing temperature because ΔH° is positive (endothermic reaction) and ΔS° is positive. Therefore, ΔG° is more negative at higher temperatures.

21.154 (a)

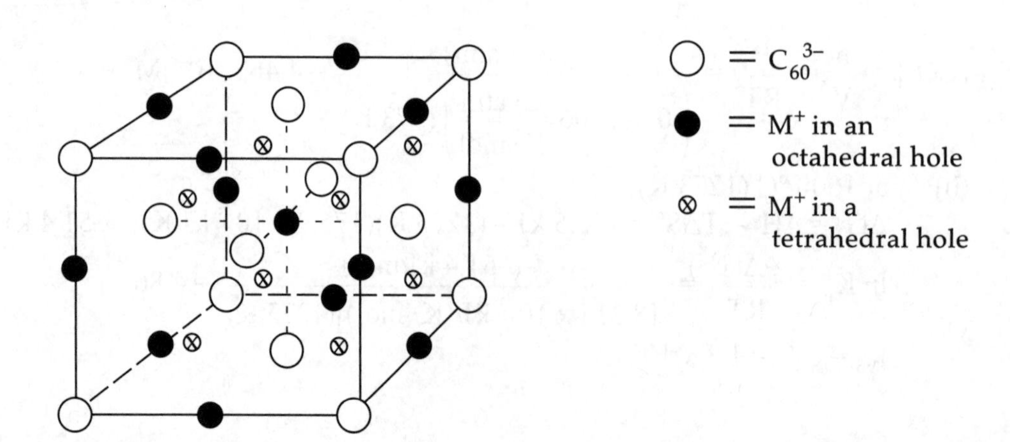

(b) There are 4 C_{60}^{3-} ions, 4 octahedral holes, and 8 tetrahedral holes per unit cell.

(c) Octahedral holes: (1/2,1/2,1/2), (1/2,0,0), (0,1/2,0), (0,0,1/2)
Tetrahedral holes: (1/4,1/4,1/4), (3/4,1/4,1/4), (1/4,3/4,1/4), (3/4,3/4,1/4), (1/4,1/4,3/4), (3/4,1/4,3/4), (1/4,3/4,3/4), (3/4,3/4,3/4)

(d) Let the unit cell edge = a.
The face diagonal is equal to 4R = 4(500 pm) = 2000 pm
$a^2 + a^2 = (2000)^2$; $2a^2 = 4 \times 10^6$; $a^2 = 2 \times 10^6$; $a = \sqrt{2 \times 10^6} = 1414$ pm
$a = 2R(C_{60}^{3-}) + 2R(\text{octahedral hole}) = 1414$ pm

$$R(\text{octahedral hole}) = \frac{1414\text{ pm} - 2R(C_{60}^{3-})}{2} = \frac{1414\text{ pm} - 2(500\text{ pm})}{2} = 207\text{ pm}$$

The tetrahedron that defines the tetrahedral hole can be thought of as being found inside a cube with edge = a/2 = 707 pm. This cube is located in one corner of the unit cell.
The face diagonal of this cube = $2R(C_{60}^{3-}) = 1000$ pm.
The body diagonal of this cube = $\sqrt{707^2 + 1000^2} = 1225$ pm
Body diagonal = $2R(C_{60}^{3-}) + 2R(\text{tetrahedral hole}) = 1225$ pm

$$R(\text{tetrahedral hole}) = \frac{1225\text{ pm} - 2R(C_{60}^{3-})}{2} = \frac{1225\text{ pm} - 2(500\text{ pm})}{2} = 112\text{ pm}$$

(e) Na^+ will fit into the octahedral and tetrahedral holes without expanding the C_{60}^{3-} framework. K^+ and Rb^+ will fit into the octahedral holes without expanding the C_{60}^{3-} framework but will fit into the tetrahedral holes only if the C_{60}^{3-} framework is expanded.

22 The Main-Group Elements

22.1 (a) B is above Al in group 3A, and therefore B is more nonmetallic than Al.
(b) Ge and Br are in the same row of the periodic table, but Br (group 7A) is to the right of Ge (group 4A). Therefore, Br is more nonmetallic.
(c) Se (group 6A) is more nonmetallic than In because it is above and to the right of In (group 3A).
(d) Cl (group 7A) is more nonmetallic than Te because it is above and to the right of Te (group 6A).

22.2 Element A

22.3 (a) HNO_3 H_3PO_4

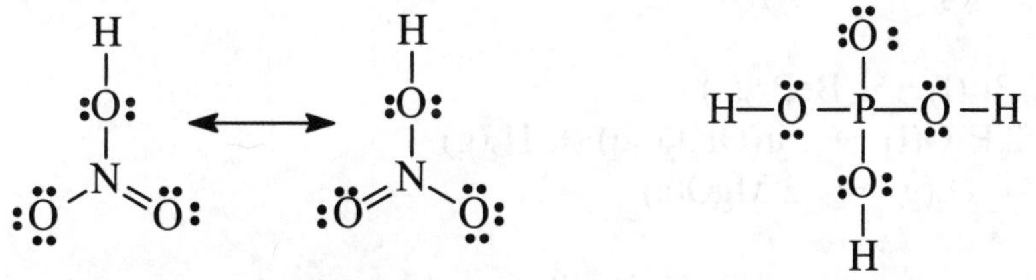

Nitrogen can form very strong pπ - pπ bonds. Phosphorus forms weaker pπ - pπ bonds, so it tends to form more single bonds.
(b) The larger S atom can accommodate six bond pairs in its valence shell, but the smaller O atom is limited to two bond pairs and two lone pairs.

22.4 Carbon forms strong π bonds with oxygen. Silicon does not form strong π bonds with oxygen, and what results are chains of alternating silicon and oxygen singly bonded to each other.

22.5 (a) SiH_4, covalent (b) KH, ionic (c) H_2Se, covalent

22.6 (a) $SrH_2(s) + 2\ H_2O(l) \rightarrow 2\ H_2(g) + Sr^{2+}(aq) + 2\ OH^-(aq)$
(b) $KH(s) + H_2O(l) \rightarrow H_2(g) + K^+(aq) + OH^-(aq)$

22.7 (a) A, KH; B, MgH_2; C, H_2O; D, HCl
(b) HCl
(c) $KH(s) + H_2O(l) \rightarrow H_2(g) + K^+(aq) + OH^-(aq)$
$MgH_2(s) + 2\ H_2O(l) \rightarrow 2\ H_2(g) + Mg^{2+}(aq) + 2\ OH^-(aq)$
(d) HCl reacts with water to give an acidic solution. KH and MgH_2 react with water to give a basic solution.

22.8 (a) (1) ZrH_x, interstitial (2) PH_3, covalent (3) HBr, covalent (4) LiH, ionic
(b) (1) and (4) are likely to be solids at 25 °C. (2) and (3) are likely to be gases at 25 °C. Covalent hydrides, like (2) and (3), form discrete molecules and have only relatively weak intermolecular forces, resulting in gases. (4) is an ionic metal hydride with strong ion-ion forces holding the 3-dimensional lattice together in the solid state. (1) is an interstitial hydride with the metal atoms in a solid crystal lattice and H's occupying holes.
(c) $LiH(s) + H_2O(l) \rightarrow H_2(g) + Li^+(aq) + OH^-(aq)$

22.9 (a) A, NaH; B, PdH_x; C, H_2S; D, HI
(b) NaH (ionic); PdH_x (interstitial); H_2S and HI (covalent)
(c) H_2S and HI (molecular); NaH and PdH_x (3-dimensional crystal)
(d) NaH: Na +1, H –1
H_2S: S –2, H +1
HI: I –1, H +1

22.10 (a) O^{2-} (b) O_2^{2-} (c) O_2^-

22.11 (a) $2\ Cs(s) + 2\ H_2O(l) \rightarrow 2\ Cs^+(aq) + 2\ OH^-(aq) + H_2(g)$
(b) $Rb(s) + O_2(g) \rightarrow RbO_2(s)$

22.12 (a) $Be(s) + Br_2(l) \rightarrow BeBr_2(s)$
(b) $Sr(s) + 2\ H_2O(l) \rightarrow Sr(OH)_2(aq) + H_2(g)$
(c) $2\ Mg(s) + O_2(g) \rightarrow 2\ MgO(s)$

22.13 An ethane-like structure is unlikely for diborane because it would require 14 valence electrons and diborane only has 12. The result is two three-center, two-electron bonds between the borons and the bridging hydrogen atoms.

22.14 (a) Each C atom is sp^2 hybridized with trigonal planar geometry. (b) The unhybridized p orbitals on each carbon atom are perpendicular to the hexagonal arrangement of carbon atoms on the sheet. The p orbitals have sideways overlap and electrons are mobile in the extended bonding system.

22.15 :C≡O: :Ö=C=Ö: [carbonate ion: C bonded to :Ö: (single), :O (double), and :Ö: (single)] 2-

Carbon monoxide will have the strongest carbon-oxygen bond because it is a triple bond.

22.16 $Hb–O_2 + CO \rightleftharpoons Hb–CO + O_2$
Mild cases of carbon monoxide poisoning can be treated with O_2. Le Châtelier's principle says that adding a product (O_2) will cause the reaction to proceed in the reverse direction, back to $Hb–O_2$.

22.17 (a) $Si_8O_{24}^{16-}$ (b) $Si_2O_5^{2-}$

22.18

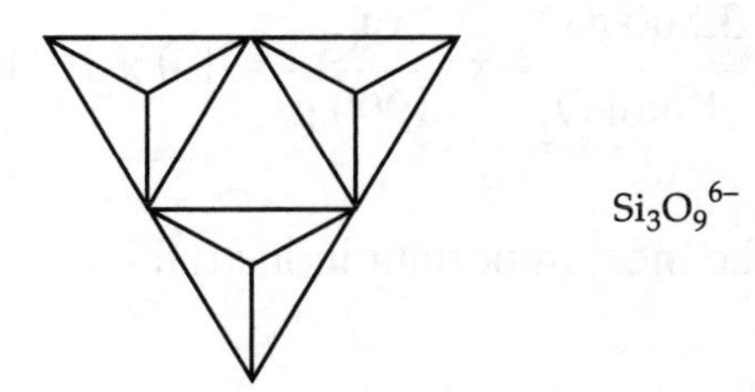

$Si_3O_9^{6-}$

22.19 (a) –3 (b) –2 (c) +1 (d) +4

22.20

$$\overset{-1}{:\ddot{N}}=\overset{+1}{N}=\overset{0}{\ddot{O}}: \longleftrightarrow \overset{0}{:N}\equiv\overset{+1}{N}-\overset{-1}{:\ddot{O}:} \longleftrightarrow \overset{-2}{:\ddot{N}}-\overset{+1}{N}\equiv\overset{+1}{O}:$$

The middle resonance structure makes the greatest contribution to the resonance hybrid because formal charges are minimized and the negative formal charge resides on the most electronegative element, oxygen.

22.21 NO_2 because reactants are favored for an exothermic reaction at high temperatures.

22.22 A is Li; B is Ga; C is C
(a) Li_2O, Ga_2O_3, CO_2
(b) Li_2O is the most ionic. CO_2 is the most covalent.
(c) CO_2 is the most acidic. Li_2O is the most basic.
(d) Ga_2O_3 is amphoteric and can react with both $H^+(aq)$ and $OH^-(aq)$.

22.23 (a) $Li_2O(s) + H_2O(l) \rightarrow 2\ Li^+(aq) + 2\ OH^-(aq)$
(b) $SO_3(l) + H_2O(l) \rightarrow H^+(aq) + HSO_4^-(aq)$
(c) $Cr_2O_3(s) + 6\ H^+(aq) \rightarrow 2\ Cr^{3+}(aq) + 3\ H_2O(l)$
(d) $Cr_2O_3(s) + 2\ OH^-(aq) + 3\ H_2O(l) \rightarrow 2\ Cr(OH)_4^-(aq)$

22.24 (a) SO_3^{2-}, HSO_3^-, SO_4^{2-}, HSO_4^- (b) HSO_4^- (c) SO_3^{2-} (d) HSO_4^-

22.25 (a) $H-\ddot{\underset{\cdot\cdot}{S}}-H$, bent.
(b) $:\ddot{\underset{\cdot\cdot}{O}}-\ddot{S}=\ddot{O}: \longleftrightarrow :\ddot{O}=\ddot{S}-\ddot{\underset{\cdot\cdot}{O}}:$, bent, S is sp^2 hybridized.
(c) Resonance structures of SO_3:
$:\ddot{\underset{\cdot\cdot}{O}}-S(-\ddot{O}:)=\ddot{O}: \longleftrightarrow :\ddot{O}=S(-\ddot{O}:)-\ddot{\underset{\cdot\cdot}{O}}: \longleftrightarrow :\ddot{\underset{\cdot\cdot}{O}}-S(=\ddot{O}:)-\ddot{\underset{\cdot\cdot}{O}}:$
trigonal planar, S is sp^2 hybridized.

22.26 $H_2(g) + 1/2\ O_2(g) \rightarrow H_2O(g)$ $\Delta H° = -242$ kJ

$$\text{mol } H_2 = 1.45 \times 10^6\ L \times \frac{0.088\ kg}{1\ L} \times \frac{1000\ g}{1\ kg} \times \frac{1\ mol\ H_2}{2.016\ g\ H_2} = 6.33 \times 10^7\ mol\ H_2$$

$$q = 6.33 \times 10^7 \text{ mol } H_2 \times \frac{242 \text{ kJ}}{1 \text{ mol } H_2} = 1.5 \times 10^{10} \text{ kJ}$$

$$\text{mass } O_2 = 6.33 \times 10^7 \text{ mol } H_2 \times \frac{0.5 \text{ mol } O_2}{1 \text{ mol } H_2} \times \frac{32.00 \text{ g } O_2}{1 \text{ mol } O_2} \times \frac{1 \text{ kg}}{1000 \text{ g}} = 1.0 \times 10^6 \text{ kg } O_2$$

22.27 The steam-hydrocarbon reforming process is the most important industrial preparation of hydrogen.

$$CH_4(g) + H_2O(g) \xrightarrow[\text{Ni catalyst}]{1100\ ^\circ C} CO(g) + 3\ H_2(g)$$

$$CO(g) + H_2O(g) \xrightarrow{400\ ^\circ C} CO_2(g) + H_2(g)$$

$$CO_2(g) + 2\ OH^-(aq) \rightarrow CO_3^{2-}(aq) + H_2O(l)$$

22.28 (a) $H_2O(g) + C(s) \xrightarrow{1000\ ^\circ C} CO(g) + H_2(g)$

(b) $C_3H_8(g) + 3\ H_2O(g) \rightarrow 7\ H_2(g) + 3\ CO(g)$

22.29 Hydrogen can be stored as a solid in the form of solid interstitial hydrides or in the recently discovered tube-shaped molecules called carbon nanotubes.

22.30 Assume 12.0 g of Pd with a volume of 1.0 cm^3.

$V_{H_2} = 935 \text{ cm}^3 = 935 \text{ mL} = 0.935 \text{ L}$

$$PV = nRT; \quad n_{H_2} = \frac{PV}{RT} = \frac{(1.00 \text{ atm})(0.935 \text{ L})}{\left(0.082\ 06 \frac{\text{L} \cdot \text{atm}}{\text{K} \cdot \text{mol}}\right)(273 \text{ K})} = 0.0417 \text{ mol } H_2$$

$n_H = 2 n_{H_2} = 0.0834 \text{ mol H}$

$$12.0 \text{ g Pd} \times \frac{1 \text{ mol Pd}}{106.42 \text{ g Pd}} = 0.113 \text{ mol Pd}$$

$Pd_{0.113}H_{0.0834}$

$Pd_{0.113/0.113}H_{0.0834/0.113}$

$PdH_{0.74}$

g H = (0.0834 mol H)(1.008 g/mol) = 0.0841 g H

$$d_H = 0.0841 \text{ g/cm}^3; \quad M_H = \frac{0.0834 \text{ mol}}{0.001 \text{ L}} = 83.4 \text{ M}$$

22.31 (a) TiH_2, 49.88; Assume 1.0 cm^3 of TiH_2, which has a mass of 3.75 g.

$$3.75 \text{ g } TiH_2 \times \frac{1 \text{ mol } TiH_2}{49.88 \text{ g } TiH_2} = 0.0752 \text{ mol } TiH_2$$

$$0.0752 \text{ mol } TiH_2 \times \frac{2 \text{ mol H}}{1 \text{ mol } TiH_2} = 0.150 \text{ mol H}$$

$0.150 \text{ mol H} \times \dfrac{1.008 \text{ g H}}{1 \text{ mol H}} = 0.151 \text{ g H}$

$d_H = 0.15 \text{ g/cm}^3$; the density of H in TiH_2 is about 2.1 times the density of liquid H_2.

(b)

$$PV = nRT; \quad V = \frac{nRT}{P} = \frac{\left(0.15 \text{ g} \times \dfrac{1 \text{ mol}}{2.016 \text{ g}}\right)\left(0.082\ 06 \dfrac{\text{L} \cdot \text{atm}}{\text{K} \cdot \text{mol}}\right)(273 \text{ K})}{1.00 \text{ atm}} = 1.7 \text{ L } H_2$$

$1.7 \text{ L} = 1.7 \times 10^3 \text{ mL} = 1.7 \times 10^3 \text{ cm}^3$

22.32 (a) NH_3 is the Lewis base and BH_3 is the Lewis acid.
(b) Nitrogen and boron both have sp^3 hybrid orbitals with bond angels of close to 109.5°.

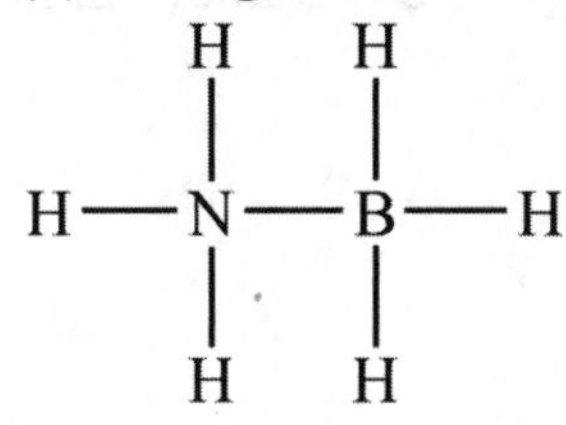

Conceptual Problems

22.34

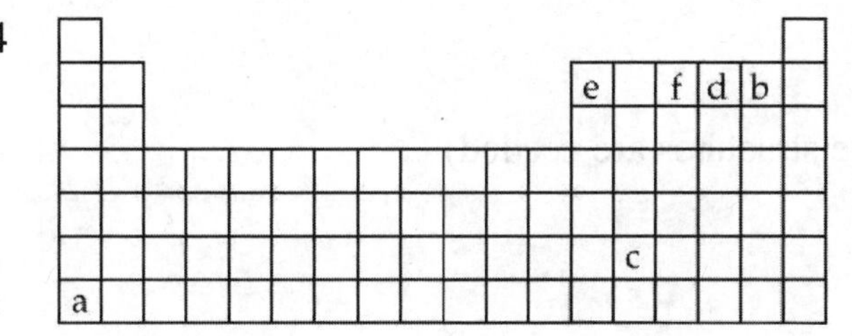

22.36 (a) (1) covalent (2) ionic (3) covalent (4) interstitial
(b) (1) H, +1; other element, –3 (2) H, –1; other element, +1
(3) H, +1; other element, –2

22.38 (a) The ionic hydride (4) has the highest melting point.
(b) (1), (2), and (3) are covalent hydrides. (1) and (2) can hydrogen bond, (3) cannot. Consequently, (3) has the lowest boiling point.
(c) (1), water, and (4), the ionic hydride react together to form $H_2(g)$.

22.40 (a) (1) –2, +2; (2) –2, +1; (3) –2, +5
(b) (1) three-dimensional; (2) molecular; (3) molecular
(c) (1) solid; (2) gas or liquid; (3) gas or liquid
(d) (2) hydrogen; (3) nitrogen

22.42 (1) is CO_2. The molecule is linear because C has two charge clouds. There are two C=O double bonds.

(2) is SO_2. The molecule is bent because S has three charge clouds. There is one S–O single bond and one S=O double bond. SO_2 has two resonance structures so each S–O bond appears to be a bond and a half.
CO_2 has the stronger bonds.

22.44 (a) N_2, O_2, F_2, P_4 (tetrahedral), S_8 (crown-shaped ring), Cl_2
(b) :N≡N: :Ö=Ö: :F̤̈–F̤̈: :C̤̈l—C̤̈l:

:P—P: (tetrahedral P_4) S S S S S S S S (crown-shaped S_8 ring)

(c) The smaller N and O can form strong π bonds, whereas P and S cannot. In both F_2 and Cl_2, the atoms are joined by a single bond.

22.46 (a) CO_2, Cl_2O_7, SO_3, N_2O_5
(b) :Ö=C=Ö:

:Ö–Cl(O)(O)–Ö–Cl(O)(O)–Ö:

:Ö=S(O)(O)   (resonance structures are needed)

:O=N(O)–Ö–N(O)=O: (resonance structures are needed)

Section Problems
General Properties and Periodic Trends (Sections 22.1 and 22.2)

22.48 (a) Cl (group 7A) is to the right of S (group 6A) in the same row of the periodic table. Cl has the higher ionization energy.
(b) Si is above Ge in group 4A. Si has the higher ionization energy.
(c) O (group 6A) is above and to the right of In (group 3A) in the periodic table. O has the higher ionization energy.

22.50 (a) Al is below B in group 3A. Al has the larger atomic radius.
(b) P (group 5A) is to the left of S (group 6A) in the same row of the periodic table. P has the larger atomic radius.
(c) Pb (group 4A) is below and to the left of Br (group 7A) in the periodic table. Pb has the larger atomic radius.

22.52 (a) I (group 7A) is to the right of Te (group 6A) in the same row of the periodic table. I has the higher electronegativity.
(b) N is above P in group 5A. N has the higher electronegativity.
(c) F (group 7A) is above and to the right of In (group 3A) in the periodic table. F has the higher electronegativity.

22.54 (a) Sn is below Si in group 4A. Sn has more metallic character.
(b) Ge (group 4A) is to the left of Se (group 6A) in the same row of the periodic table. Ge has more metallic character.
(c) Bi (group 5A) is below and to the left of I (group 7A) in the periodic table. Bi has more metallic character.

22.56 In each case the more ionic compound is the one formed between a metal and nonmetal.
(a) CaH_2 (b) Ga_2O_3 (c) KCl (d) $AlCl_3$

22.58 Molecular (a) B_2H_6 (c) SO_3 (d) $GeCl_4$
Extended three-dimensional structure (b) $KAlSi_3O_8$

22.60 (a) Sn (b) Cl (c) Sn (d) Se (e) B

22.62 The smaller B atom can bond to a maximum of four nearest neighbors, whereas the larger Al atom can accommodate more than four nearest neighbors.

22.64 In O_2 a π bond is formed by 2p orbitals on each O. S does not form strong π bonds with its 3p orbitals, which leads to the S_8 ring structure with single bonds.

Group 1A: Hydrogen (Sections 22.3)

22.66 (a) $Zn(s) + 2\,H^+(aq) \rightarrow H_2(g) + Zn^{2+}(aq)$
(b) at 1000 °C, $H_2O(g) + C(s) \rightarrow CO(g) + H_2(g)$
(c) at 1100 °C with a Ni catalyst, $H_2O(g) + CH_4(g) \rightarrow CO(g) + 3\,H_2(g)$
(d) There are a number of possibilities. (b) and (c) above are two; electrolysis is another:
$2\,H_2O(l) \rightarrow 2\,H_2(g) + O_2(g)$

22.68 $CaH_2(s) + 2\,H_2O(l) \rightarrow 2\,H_2(g) + Ca^{2+}(aq) + 2\,OH^-(aq)$
CaH_2, 42.09; 25 °C = 298 K

$$PV = nRT; \quad n_{H_2} = \frac{PV}{RT} = \frac{(1.00\text{ atm})(2.0 \times 10^5\text{ L})}{\left(0.082\ 06\,\frac{\text{L}\cdot\text{atm}}{\text{K}\cdot\text{mol}}\right)(298\text{ K})} = 8.18 \times 10^3\text{ mol H}_2$$

$$8.18 \times 10^3\text{ mol H}_2 \times \frac{1\text{ mol CaH}_2}{2\text{ mol H}_2} \times \frac{42.09\text{ g CaH}_2}{1\text{ mol CaH}_2} \times \frac{1\text{ kg}}{1000\text{ g}} = 1.7 \times 10^2\text{ kg CaH}_2$$

22.70 (a) MgH_2, H^- (b) PH_3, covalent (c) KH, H^- (d) HBr, covalent

22.72 H_2S – covalent hydride, gas, weak acid in H_2O
NaH – ionic hydride, solid (salt like), reacts with H_2O to produce H_2
PdH_x – metallic (interstitial) hydride, solid, stores hydrogen

22.74 (a) CH_4, covalent bonding (b) NaH, ionic bonding

22.76 (a) H—$\ddot{\underset{\cdot\cdot}{Se}}$—H , bent (b) H—$\ddot{As}$—H (with third H below As), trigonal pyramidal

(c) SiH_4 (Si bonded to four H), tetrahedral

22.78 A nonstoichiometric compound is a compound whose atomic composition cannot be expressed as a ratio of small whole numbers. An example is PdH_x. The lack of stoichiometry results from the hydrogen occupying holes in the solid state structure.

Group 1A and 2A: Alkali and Alkaline Earth Metals (Sections 22.4 and 22.5)

22.80 Predicted for Fr: melting point ≈ 23 °C; boiling point ≈ 650 °C; density ≈ 2 g/cm^3; atomic radius ≈ 275 pm

22.82 (a) $2\,K(s) + 2\,H_2O(l) \rightarrow 2\,K^+(aq) + 2\,OH^-(aq) + H_2(g)$
(b) $2\,K(s) + Br_2(l) \rightarrow 2\,KBr(s)$
(c) $K(s) + O_2(g) \rightarrow KO_2(s)$

22.84 $2\,Mg(s) + O_2(g) \rightarrow 2\,MgO(s)$
$MgO(s) + H_2O(l) \rightarrow Mg(OH)_2(aq)$

22.86
anode $Mg^{2+}(l) + 2\,e^- \rightarrow Mg(l)$
cathode $2\,Cl^-(l) \rightarrow Cl_2(g) + 2\,e^-$
overall $Mg^{2+}(l) + 2\,Cl^-(l) \rightarrow Mg(l) + Cl_2(g)$

Group 3A: Boron (Section 22.6)

22.88 (a) Al (b) Tl (c) B

22.90 +3 for B, Al, Ga and In; +1 for Tl

22.92 Boron is a hard semiconductor with a high melting point. Boron forms only molecular compounds and does not form an aqueous B^{3+} ion. $B(OH)_3$ is an acid.

22.94 (a) An electron deficient molecule is a molecule that doesn't have enough electrons to form a two-center, two-electron bond between each pair of bonded atoms. B_2H_6 is an electron deficient molecule.

(b) A three-center, two-electron bond has three atoms bonded together using just two electrons. The B–H–B bridging bond in B_2H_6 is a three-center, two-electron bond.

Group 4A: Carbon and Silicon (Sections 22.7 and 22.8)

22.96 (a) Pb (b) C (c) Si (d) C

22.98 (a) $GeBr_4$, tetrahedral; Ge is sp^3 hybridized.
(b) CO_2, linear; C is sp hybridized.
(c) CO_3^{2-}, trigonal planar; C is sp^2 hybridized.
(d) $SnCl_3^-$, trigonal pyramidal; Sn is sp^3 hybridized.

22.100 Diamond is a very hard, high melting solid. It is an electrical insulator.
Diamond has a covalent network structure in which each C atom uses sp^3 hybrid orbitals to form a tetrahedral array of σ bonds. The interlocking, three-dimensional network of strong bonds makes diamond the hardest known substance with the highest melting point for an element. Because the valence electrons are localized in the σ bonds, diamond is an electrical insulator.

22.102 Graphene is a two-dimensional array of hexagonally arranged carbon atoms just one atom thick, essentially one layer of graphite. Graphene is extremely strong and flexible, and is a superb conductor of electricity.

22.104 CO bonds to hemoglobin and prevents it from carrying O_2. CN^- bonds to cytochrome oxidase and interferes with the electron transfer associated with oxidative phosphorylation.

22.106 Silicon and germanium are semimetals, and tin and lead are metals. Silicon is a hard, gray, semiconducting solid that melts at 1414 °C. It crystallizes in a diamondlike structure but does not form a graphitelike allotrope because of the relatively poor overlap of silicon p orbitals. Germanium is a relatively high-melting, brittle semiconductor that has the same crystal structure as diamond and silicon. Tin exists in two allotropic forms: the usual silvery white metallic form called white tin and a brittle, semiconducting form with the diamond structure called gray tin. Both white tin and lead are soft, malleable, low-melting metals. Only the metallic form occurs for lead.

22.108

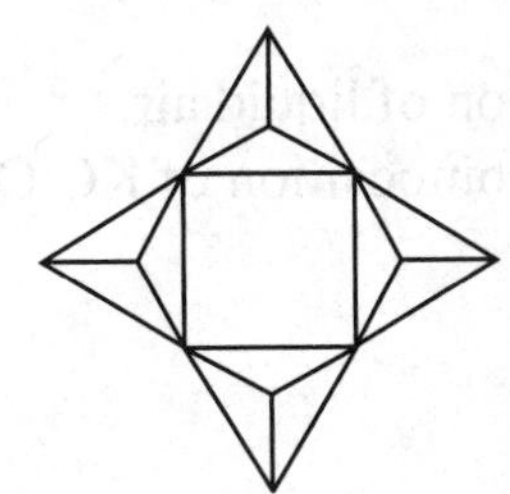

$Si_4O_{12}^{8-}$

Group 5A: Nitrogen and Phosphorus (Sections 22.9 and 22.10)

22.110 (a) P (b) Sb and Bi (c) N (d) Bi

22.112 (a) N_2O, +1 (b) N_2H_4, −2 (c) Ca_3P_2, −3
(d) H_3PO_3, +3 (e) H_3AsO_4, +5

22.114 **:N≡N:**
N_2 is unreactive because of the large amount of energy necessary to break the N≡N triple bond.

22.116 White phosphorus consists of tetrahedral P_4 molecules with 60° bond angles.

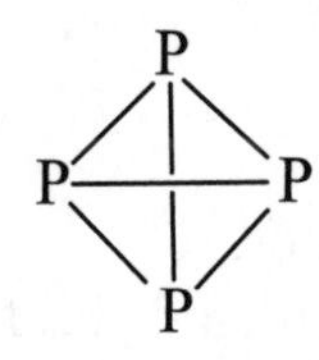

Red phosphorus is polymeric.
White phosphorus is reactive due to the considerable strain in the P_4 molecule.

22.118 (a) The structure for phosphorous acid is

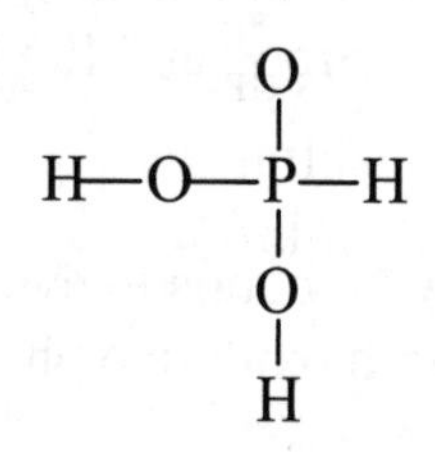

Only the two hydrogens bonded to oxygen are acidic.
(b) Nitrogen forms strong π bonds, and in N_2 the nitrogen atoms are triple bonded to each other. Phosphorus does not form strong pπ - pπ bonds, and so the P atoms are single bonded to each other in P_4.

Group 6A: Oxygen and Sulfur (Sections 20.11 and 20.12)

22.120 (a) O (b) Te (c) Po (d) O

22.122 (a) O_2 is obtained in industry by the fractional distillation of liquid air.
(b) In the laboratory, O_2 is prepared by the thermal decomposition of $KClO_3(s)$.

$$2\ KClO_3(s) \xrightarrow[MnO_2]{heat} 2\ KCl(s) + 3\ O_2(g)$$

22.124 (a) $4\ Li(s) + O_2(g) \rightarrow 2\ Li_2O(s)$ (b) $P_4(s) + 5\ O_2(g) \rightarrow P_4O_{10}(s)$
(c) $4\ Al(s) + 3\ O_2(g) \rightarrow 2\ Al_2O_3(s)$ (d) $Si(s) + O_2(g) \rightarrow SiO_2(s)$

22.126 **:Ö::Ö:** The electron dot structure shows an O=O double bond. It also shows all electrons paired. This is not consistent with the fact that O_2 is paramagnetic.

22.128 An element that forms an acidic oxide is more likely to form a covalent hydride. C and N are examples.

22.130 Li_2O < BeO < B_2O_3 < CO_2 < N_2O_5 (see Figure 22.16)

22.132 N_2O_5 < Al_2O_3 < K_2O < Cs_2O (see Figure 22.16)

22.134 (a) CrO_3 (higher Cr oxidation state) (b) N_2O_5 (higher N oxidation state)
(c) SO_3 (higher S oxidation state)

22.136 (a) $Cl_2O_7(l) + H_2O(l) \rightarrow 2\,H^+(aq) + 2\,ClO_4^-(aq)$
(b) $K_2O(s) + H_2O(l) \rightarrow 2\,K^+(aq) + 2\,OH^-(aq)$
(c) $SO_3(l) + H_2O(l) \rightarrow H^+(aq) + HSO_4^-(aq)$

22.138 (a) $ZnO(s) + 2\,H^+(aq) \rightarrow Zn^{2+}(aq) + H_2O(l)$
(b) $ZnO(s) + 2\,OH^-(aq) + H_2O(l) \rightarrow Zn(OH)_4^{2-}(aq)$

22.140 (a) rhombic sulfur – yellow crystalline solid (mp 113 °C) that contains crown-shaped S_8 rings.
(b) monoclinic sulfur – an allotrope of sulfur in which the S_8 rings pack differently in the crystal.
(c) plastic sulfur – when sulfur is cooled rapidly, the sulfur forms disordered, tangled chains, yielding an amorphous, rubbery material called plastic sulfur.
(d) Liquid sulfur between 160 and 195 °C becomes dark reddish-brown and very viscous forming long polymer chains (S_n, n > 200,000).

22.142 (a) $Zn(s) + 2\,H_3O^+(aq) \rightarrow Zn^{2+}(aq) + H_2(g) + 2\,H_2O(l)$
(b) $BaSO_3(s) + 2\,H_3O^+(aq) \rightarrow H_2SO_3(aq) + Ba^{2+}(aq) + 2\,H_2O(l)$
(c) $Cu(s) + 2\,H_2SO_4(l) \rightarrow Cu^{2+}(aq) + SO_4^{2-}(aq) + SO_2(g) + 2\,H_2O(l)$
(d) $H_2S(aq) + I_2(aq) \rightarrow S(s) + 2\,H^+(aq) + 2\,I^-(aq)$

22.144 (a) Acid strength increases as the number of O atoms increases.
(b) In comparison with S, O is much too electronegative to form compounds of O in the +4 oxidation state. Also, an S atom is large enough to accommodate four bond pairs and a lone pair in its valence shell, but an O atom is too small to do so.
(c) Each S is sp^3 hybridized with two lone pairs of electrons. The bond angles are therefore 109.5°. A planar ring would require bond angles of 135°.

Group 7A and 8A: Halogen and Noble Gases (Sections 22.13 and 22.14)

22.146 (a) At is in Group 7A. The trend going down the group is gas → liquid → solid. At, being at the bottom of the group, should be a solid.
(b) At is likely to react with Na just like the other halogens, yielding NaAt.

22.148 $MnO_2(s) + 2\,Br^-(aq) + 4\,H^+(aq) \rightarrow Mn^{2+}(aq) + 2\,H_2O(l) + Br_2(aq)$

22.150 (a) Assume a 100.0 g sample. From the percent composition data, a 100.0 g sample contains 25.25 g Ti, and 74.75 g Cl.

$$25.25 \text{ g Ti} \times \frac{1 \text{ mol Ti}}{47.87 \text{ g Ti}} = 0.5275 \text{ mol Ti}$$

$$74.75 \text{ g Cl} \times \frac{1 \text{ mol Cl}}{35.45 \text{ g Cl}} = 2.109 \text{ mol Cl}$$

$Ti_{0.5275}Cl_{2.109}$; divide each subscript by the smaller, 0.5275.
$Ti_{0.5275/0.5275}Cl_{2.109/0.5275}$
The empirical and molecular formula is $TiCl_4$, titanium tetrachloride.
(b) $Ti(s) + 2\ Cl_2(g) \rightarrow TiCl_4(g)$
(c) $TiCl_4(l) + 2\ Mg(s) \rightarrow Ti(s) + 2\ MgCl_2(s)$

22.152 (a) $HBrO_3$, +5 (b) HIO, +1

22.154 (a) HIO_3 [Lewis structure: :Ö—Ï—Ö—H with :Ö: below I] trigonal pyramidal

(b) ClO_2^- [Lewis structure: [:Ö—C̈l—Ö:]⁻] bent

(c) HOCl H—Ö—C̈l: bent

(d) IO_6^{5-} [Lewis structure: I bonded to six :Ö: atoms, bracketed, 5−] octahedral

22.156 Oxygen atoms are highly electronegative. Increasing the number of oxygen atoms increases the polarity of the O–H bond and increases the acid strength.

Chapter Problems

22.158 (a) $Si_3O_{10}^{8-}$
(b) The charge on the anion is 8–. Because the Ca^{2+} to Cu^{2+} ratio is 1:1, there must be 2 Ca^{2+} and 2 Cu^{2+} ions in the formula for the mineral. There are also 2 waters. The formula of the mineral is: $Ca_2Cu_2Si_3O_{10} \cdot 2\ H_2O$

22.160 $I_2O_5(aq) + H_2O(l) \rightarrow 2\ HIO_3(aq)$; HIO_3 is iodic acid.

22.162 (a) Ga (b) In (c) Pb
Metals are better electrical conductors than nonmetals (S and P) or semimetals (B).

22.164 C, Si, Ge and Sn have allotropes with the diamond structure.
Sn and Pb have metallic allotropes.
C (nonmetal), Si (semimetal), Ge (semimetal), Sn (semimetal and metal), Pb (metal)

22.166 (a) In diamond each C is covalently bonded to four additional C atoms in a rigid three-dimensional network solid. Graphite is a two-dimensional covalent network solid of carbon sheets that can slide over each other. Both are high melting because melting requires the breaking of C–C bonds.

22.168 Cl—S—S—Cl

22.170 NH_3, $K_b = 1.8 \times 10^{-5}$; N_2H_4, $K_b = 8.9 \times 10^{-7}$; NH_2OH, $K_b = 9.1 \times 10^{-9}$
The strongest base, NH_3, will react to the greatest extent with HNO_2.

22.172 (a) C as diamond
(b) $Cl_2(g) + H_2O(l) \rightarrow HOCl(aq) + H^+(aq) + Cl^-(aq)$
(c) NO (d) NO_2 (e) BF_3 (f) Al_2O_3 (g) Si (h) HNO_3
(i) C as diamond, graphite, and fullerene.

22.174 Carbon is a versatile element that can form millions of very stable compounds with elements such as N, O, and H. Biomolecules contain chains and rings with many C–C bonds. Si–Si bonds are much less stable and chains of Si atoms are uncommon. In addition, carbon can form very stable pπ-pπ multiple bonds. On the other hand, the chemistry of silicon (which cannot form stable pπ-pπ bonds) is dominated by structures based on the SiO_4^{4-} anion.

Multiconcept Problems

22.176 (a)

```
      O
      |
  HO—P—H
      |
      OH
```

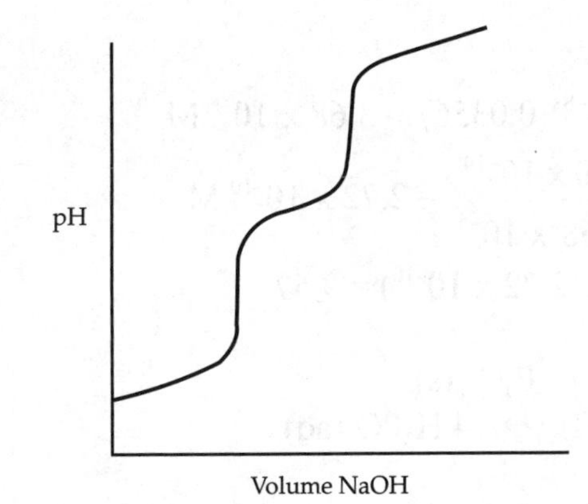


```
      OH
      |
  HO—P:
      |
      OH
```

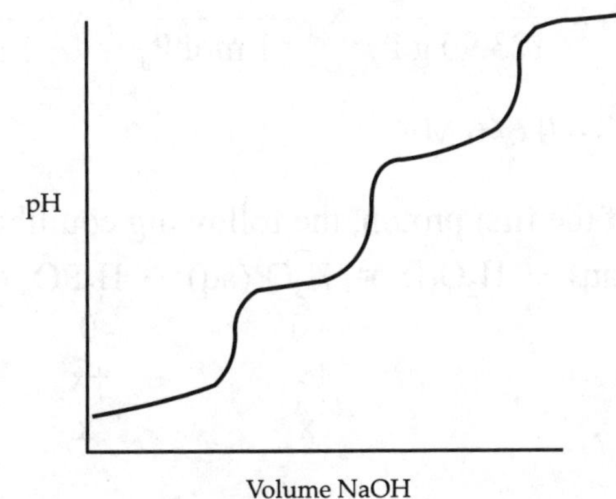

(b) $K_{a1} = 1.0 \times 10^{-2}$; $pK_{a1} = -\log(1.0 \times 10^{-2}) = 2.00$
$K_{a2} = 2.6 \times 10^{-7}$; $pK_{a2} = -\log(2.6 \times 10^{-7}) = 6.59$

At the first equivalence point, $pH = \dfrac{pK_{a1} + pK_{a2}}{2} = \dfrac{2.00 + 6.59}{2} = 4.29$

mmol HPO_3^{2-} = (30.00 mL)(0.1240 mmol/mL) = 3.72 mmol HPO_3^{2-}
volume NaOH to reach second equivalence point

$$= 3.72 \text{ mmol } HPO_3^{2-} \times \frac{2 \text{ mmol NaOH}}{1 \text{ mmol } HPO_3^{2-}} \times \frac{1.00 \text{ mL}}{0.1000 \text{ mmol NaOH}} = 74.40 \text{ mL}$$

At the second equivalence point only Na_2HPO_3, a basic salt, is in solution.

$$[HPO_3^{2-}] = \frac{3.72 \text{ mmol}}{30.00 \text{ mL} + 74.40 \text{ mL}} = 0.0356 \text{ mmol/mL} = 0.0356 \text{ M}$$

$$K_b = \frac{K_w}{K_{a2}} = \frac{1.0 \times 10^{-14}}{2.6 \times 10^{-7}} = 3.8 \times 10^{-8}$$

	$HPO_3^{2-}(aq)$ + $H_2O(l)$	⇄ $H_2PO_3^{2-}(aq)$	+ $OH^-(aq)$
initial (M)	0.0356	0	~0
change (M)	–x	+x	+x
equil (M)	0.0356 – x	x	x

$$K_b = \frac{[H_2PO_3^-][OH^-]}{[HPO_3^{2-}]} = 3.8 \times 10^{-8} = \frac{(x)(x)}{0.0356 - x} \approx \frac{x^2}{0.0356}$$

Solve for x.

$$x = [OH^-] = \sqrt{(3.8 \times 10^{-8})(0.0356)} = 3.68 \times 10^{-5} \text{ M}$$

$$[H_3O^+] = \frac{K_w}{[OH^-]} = \frac{1.0 \times 10^{-14}}{3.68 \times 10^{-5}} = 2.72 \times 10^{-10} \text{ M}$$

$pH = -\log[H_3O^+] = -\log(2.72 \times 10^{-10}) = 9.57$

22.178 (a) $P_4(s) + 5\,O_2(g) \rightarrow P_4O_{10}(s)$
$P_4O_{10}(s) + 6\,H_2O(l) \rightarrow 4\,H_3PO_4(aq)$
(b) P_4, 123.90

$$\text{mol } H_3PO_4 = 5.00 \text{ g } P_4 \times \frac{1 \text{ mol } P_4}{123.90 \text{ g } P_4} \times \frac{1 \text{ mol } P_4O_{10}}{1 \text{ mol } P_4} \times \frac{4 \text{ mol } H_3PO_4}{1 \text{ mol } P_4O_{10}} = 0.1614 \text{ mol}$$

$$[H_3PO_4] = \frac{0.1614 \text{ mol}}{0.2500 \text{ L}} = 0.646 \text{ M}$$

For the dissociation of the first proton, the following equilibrium must be considered:

	$H_3PO_4(aq)$ + $H_2O(l)$	⇄ $H_3O^+(aq)$	+ $H_2PO_4^-(aq)$
initial (M)	0.646	~0	0
change (M)	–x	+x	+x
equil (M)	0.646 – x	x	x

$$K_{a1} = \frac{[H_3O^+][H_2PO_4^-]}{[H_3PO_4]} = 7.5 \times 10^{-3} = \frac{x^2}{0.646 - x}$$

$x^2 + (7.5 \times 10^{-3})x - (4.84 \times 10^{-3}) = 0$

Solve for x using the quadratic formula.

$$x = \frac{-(7.5 \times 10^{-3}) \pm \sqrt{(7.5 \times 10^{-3})^2 - (4)(1)(-4.84 \times 10^{-3})}}{2(1)} = \frac{(-7.5 \times 10^{-3}) \pm 0.139}{2}$$

x = 0.0658 and –0.0733; Of the two solutions for x, only the positive value of x has physical meaning, because x is the $[H_3O^+]$.

$x = 0.0658\text{ M} = [H_2PO_4^-] = [H_3O^+]$

Only the dissociation of the first proton contributes a significant amount of H_3O^+.

$pH = -\log[H_3O^+] = -\log(0.0658) = 1.18$

(c) $3\,Ca^{2+}(aq) + 2\,H_3PO_4(aq) \rightarrow Ca_3(PO_4)_2(s) + 6\,H^+(aq)$

$Ca_3(PO_4)_2$, 310.18

$$\text{mass } Ca_3(PO_4)_2 = 0.1614 \text{ mol } H_3PO_4 \times \frac{1 \text{ mol } Ca_3(PO_4)_2}{2 \text{ mol } H_3PO_4} \times \frac{310.18 \text{ g } Ca_3(PO_4)_2}{1 \text{ mol } Ca_3(PO_4)_2} = 25.0 \text{ g}$$

(d) $Zn(s) + 2\,H^+(aq) \rightarrow H_2(g) + Zn^{2+}(aq)$; the gas is H_2.

$$\text{mol } H_2 = 0.1614 \text{ mol } H_3PO_4 \times \frac{6 \text{ mol } H^+}{2 \text{ mol } H_3PO_4} \times \frac{1 \text{ mol } H_2}{2 \text{ mol } H^+} = 0.242 \text{ mol } H_2$$

PV = nRT; 20 °C = 293 K

$$V = \frac{nRT}{P} = \frac{(0.242 \text{ mol})\left(0.082\,06 \frac{L \cdot atm}{K \cdot mol}\right)(293 \text{ K})}{\left(742 \text{ mm Hg} \times \frac{1.00 \text{ atm}}{760 \text{ mm Hg}}\right)} = 5.96 \text{ L}$$

23 Organic and Biological Chemistry

23.1 (a) 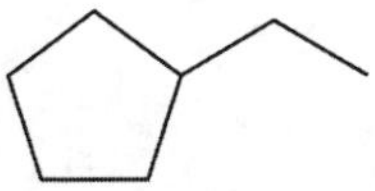 (b)

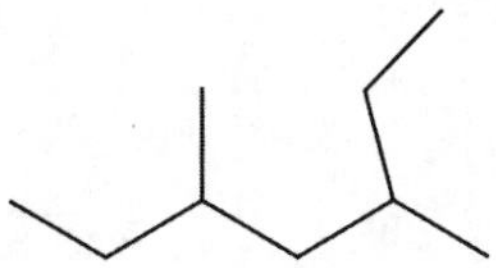

23.2 (a) C_8H_{16} (b) C_7H_{16}

23.3 Structures (a) and (c) are identical. They both contain a chain of six carbons with two $-CH_3$ branches at the fourth carbon and one $-CH_3$ branch at the second carbon. Structure (b) is different, having a chain of seven carbons.

23.4 $CH_3CH_2CH_2CH_2CH_2CH_3$

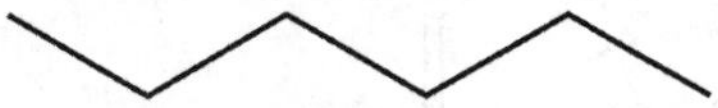

$CH_3CH(CH_3)CH_2CH_2CH_3$

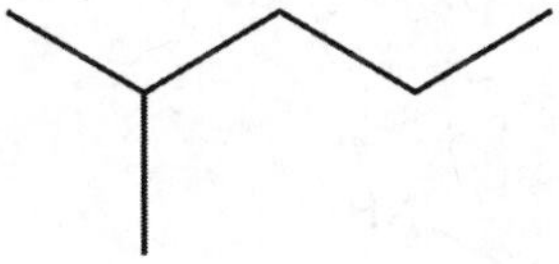

$CH_3CH_2CH(CH_3)CH_2CH_3$

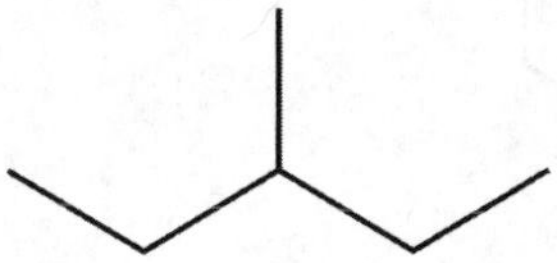

$CH_3C(CH_3)_2CH_2CH_3$

$CH_3CH(CH_3)CH(CH_3)CH_3$

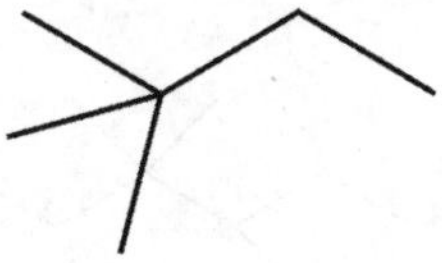

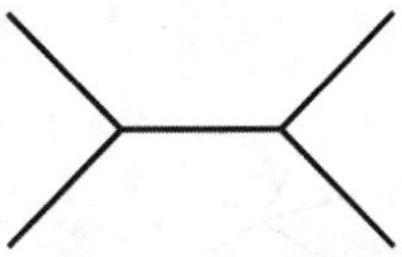

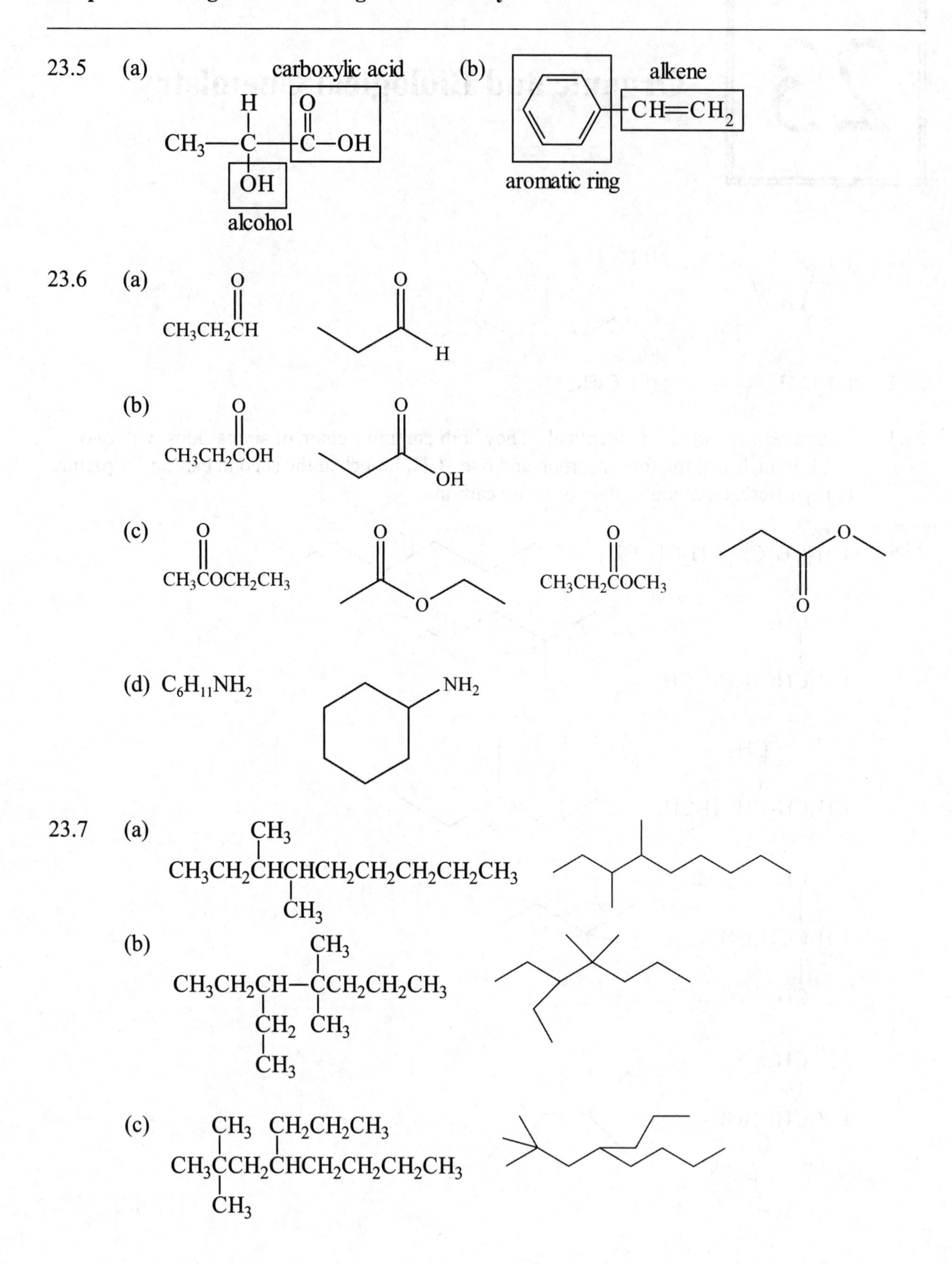
23.5
(a)
carboxylic acid
H
O
CH3—C—C—OH
OH
alcohol
(b)
alkene
CH=CH2
aromatic ring
23.6
(a)
O
CH3CH2CH
O
H
(b)
O
CH3CH2COH
O
OH
(c)
O
CH3COCH2CH3
O
O
O
CH3CH2COCH3
O
O
(d) C6H11NH2
NH2
23.7
(a)
CH3
CH3CH2CHCHCH2CH2CH2CH2CH3
CH3
(b)
CH3
CH3CH2CH—CCH2CH2CH3
CH2 CH3
CH3
(c)
CH3 CH2CH2CH3
CH3CCH2CHCH2CH2CH2CH3
CH3

(d) 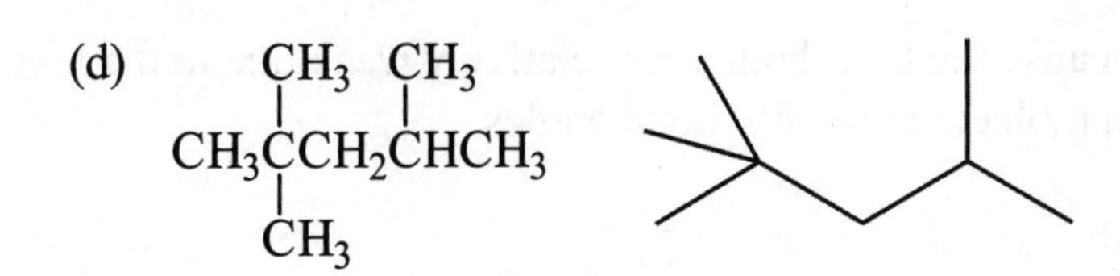

23.8 pentane

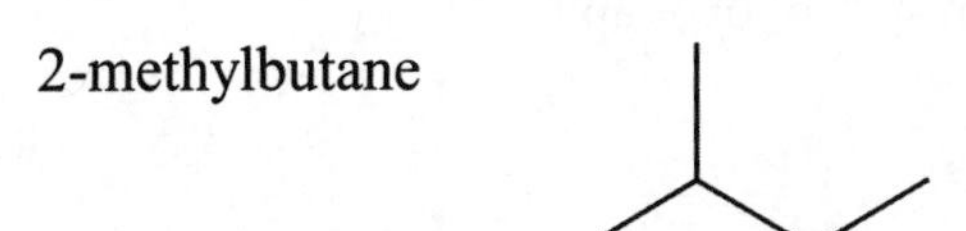

2,2-dimethypropane

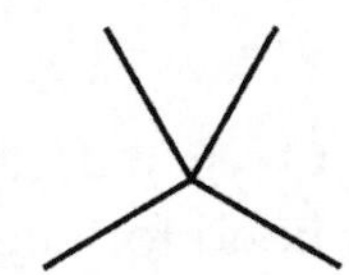

23.9 (a) 3-methyl-1-butene

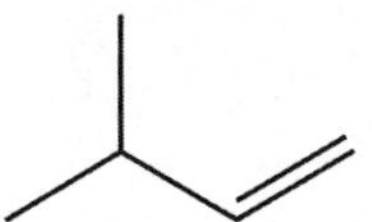

(b) 4-methyl-3-heptene

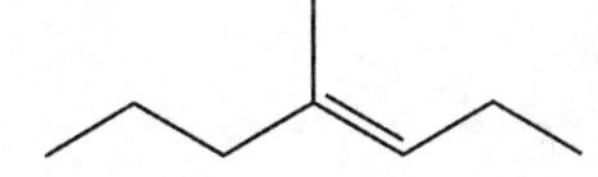

(c) 3-ethyl-1-hexyne

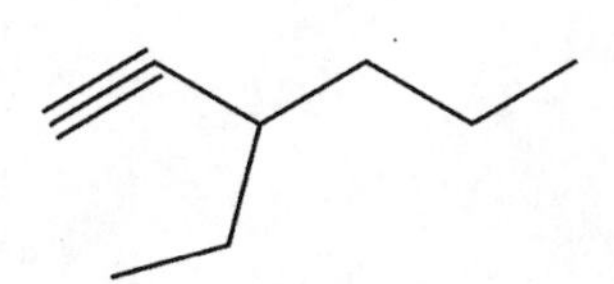

23.10 (a) 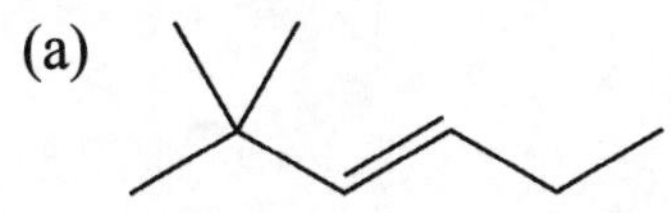

(b)

23.11 (a) aldopentose (b) ketotriose (c) aldotetrose

23.12

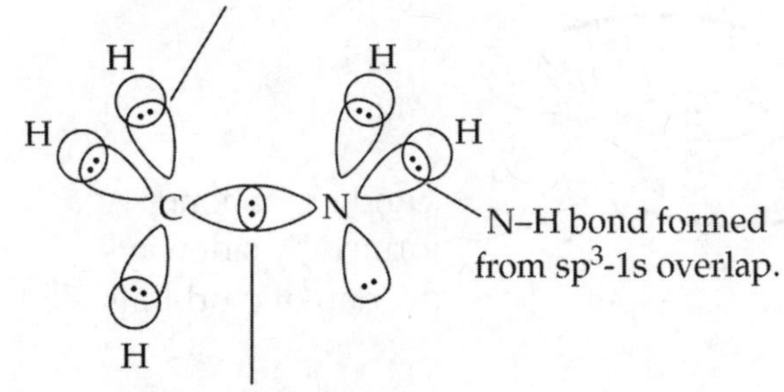

23.13 The molecule is unstable because the C–C bonds are relatively weak due to the less effective orbital overlap in a molecule with 60° bond angles.

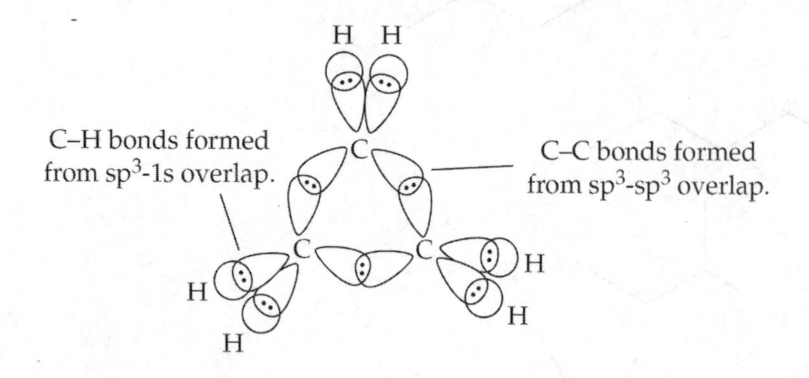

23.14 (a)

Carbon–Oxygen double bond consists of one σ bond formed by head on overlap of sp^2 orbitals ...

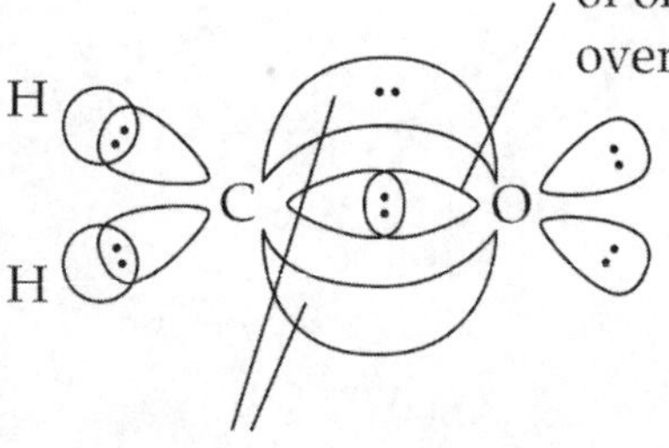

... and sideways overlap of p orbitals

(b) No (c) Yes

23.15 (a) $H_2C=C=\ddot{O}\!:$

(b) Carbon–Carbon double bond consists of one σ bond formed by head on overlap of sp^2 and sp orbitals...

Carbon–Oxygen double bond consists of one σ bond formed by head on overlap of sp and sp^2 orbitals...

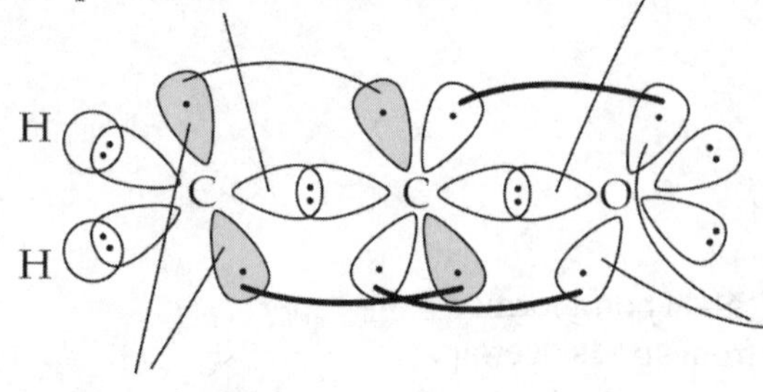

...and one π bond formed by sideways overlap of p orbitals

...and one π bond formed by sideways overlap of p orbitals

23.16 (a) Does not exhibit cis-trans isomerism.
(b) The cis isomer was shown in the problem.

The trans isomer is shown here:

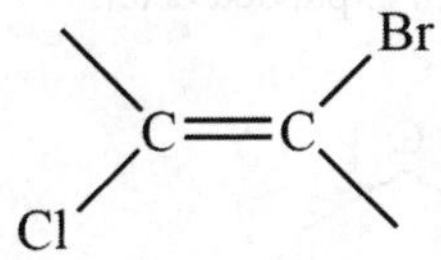

Br
C=C
Cl

(c) Does not exhibit cis-trans isomerism.

23.17 (a) Fumaric acid is a trans isomer.
(b) Maleic acid

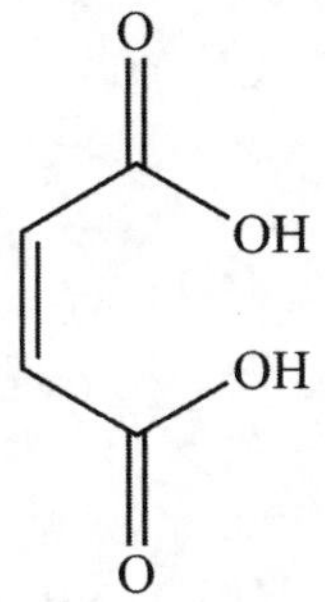

O
OH
OH
O

(c) No, because succinic acid does not have a carbon-carbon double bond.

23.18 (a) trans (b) cis

23.19 (a) 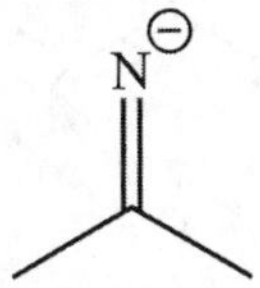 N⊖ (b)  ⊕ (c) 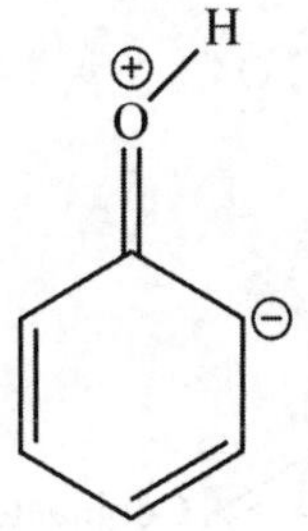 H ⊕O ⊖

23.20 (a) 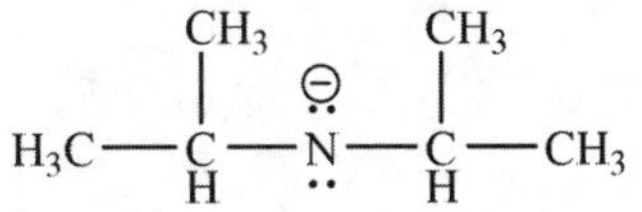 CH3 CH3 H3C—C—N—C—CH3 H H (b) ⊖ O ⊕ N O ⊖ C ⊕ H2C CH H2C CH2 C H2 (c) H C⊖ HC CH HC—CH

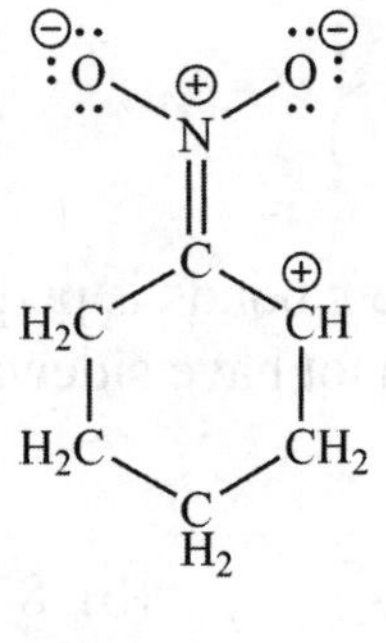

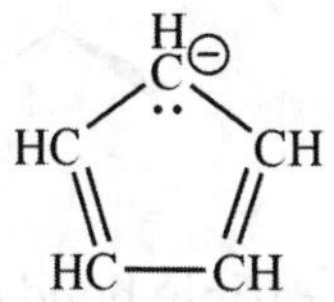

23.21 (a) 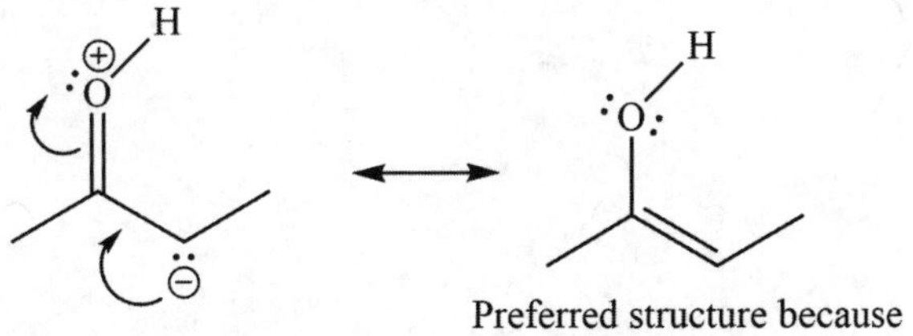 H ⊕O ⊖ ⟷ H O

Preferred structure because formal charges are zero

(b) 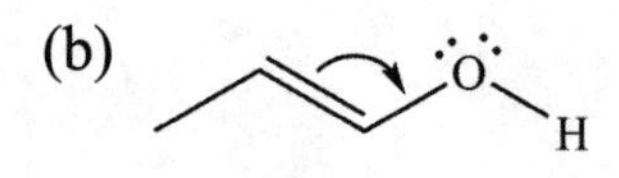

Moving electrons as indicated would result in an incorrect electron-dot structure because oxygen would have an expanded octet.

(c)

Both resonance structures are equivalent

(d)

Preferred resonance structure because negative formal charge is on more electronegative oxygen

23.22 (a) (b)

23.23

Not conjugated

Conjugated

23.24 In the triple bond one of the π bonds is perpendicular to the other π bonds in the molecule. Therefore, it cannot have sideways overlap with the p orbitals that make up the other π bonds.

23.25 (a) (b) 8

23.26

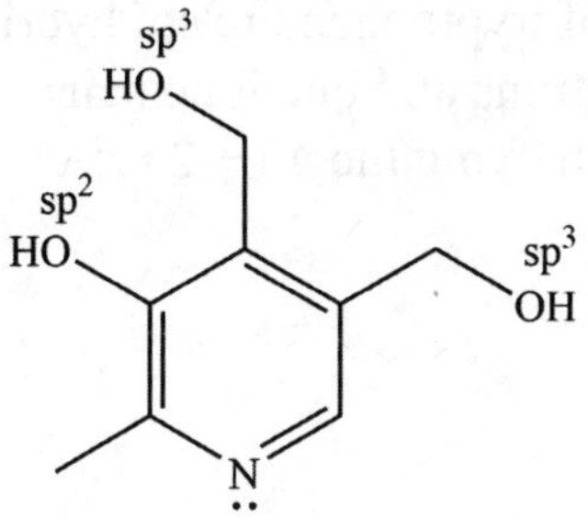

sp^3 HO

sp^2 HO

sp^3 OH

N

23.27

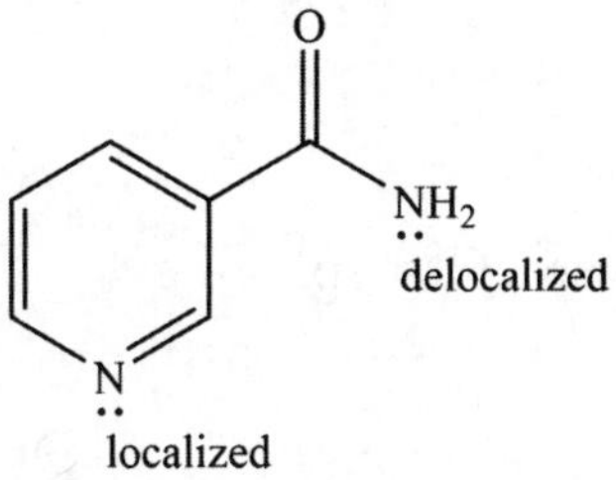

O

NH_2 delocalized

N localized

The localized lone pair of electrons is in an sp^2 hybrid orbital that is in a plane perpendicular to the orbitals used for the delocalized electrons in the aromatic ring.

23.28 Val-Cys

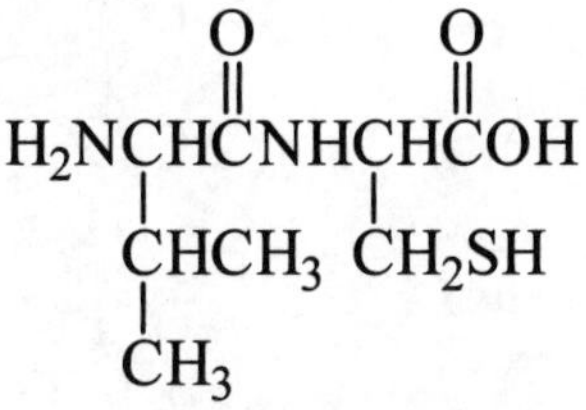

$H_2NCH(CH(CH_3)CH_3)C(=O)NHCH(CH_2SH)C(=O)OH$

Cys-Val

$H_2NCH(CH_2SH)C(=O)NHCH(CH(CH_3)CH_3)C(=O)OH$

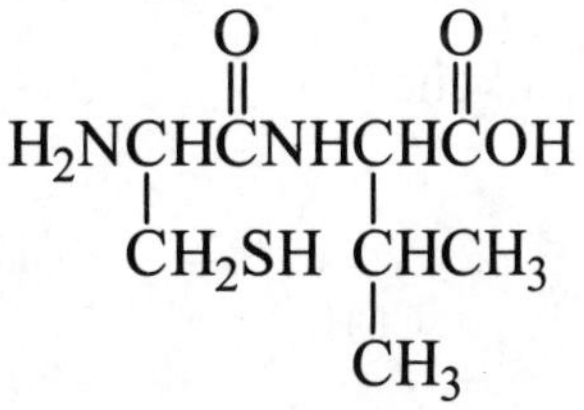

23.29 (a) 4 (b) Phe-nonpolar, Asn-polar, Trp-nonpolar, Ala-nonpolar

23.30 Proline is not aromatic because it does not have a ring of sp^2 hybridized atoms. Tyrosine is aromatic because it has a six-member ring of sp^2 hybridized carbon atoms that are part of a conjugated system. There are 6π electrons and satisfying the $4n + 2$ rule for aromaticity.

23.31 (a)

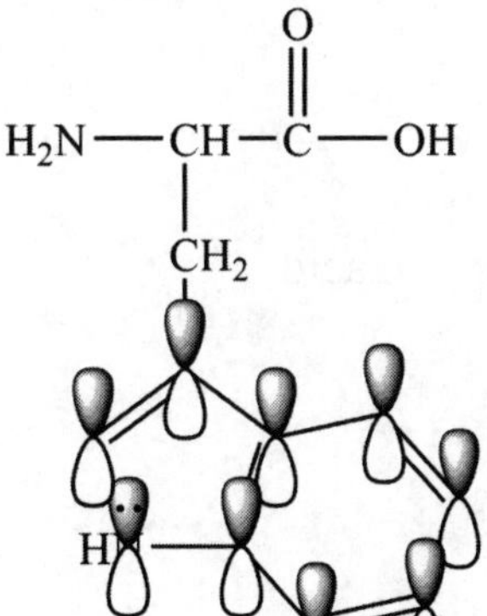

$H_2N—CH—C(=O)—OH$

CH_2

H N

(b) All atoms in the two-ring system of the side chain of tryptophan are sp^2 hybridized and conjugated. Four double bonds contribute 8 π electrons and one lone pair contributes 2 π electrons for a total of 10 π electrons, satisfying the 4n + 2 rule.

23.32

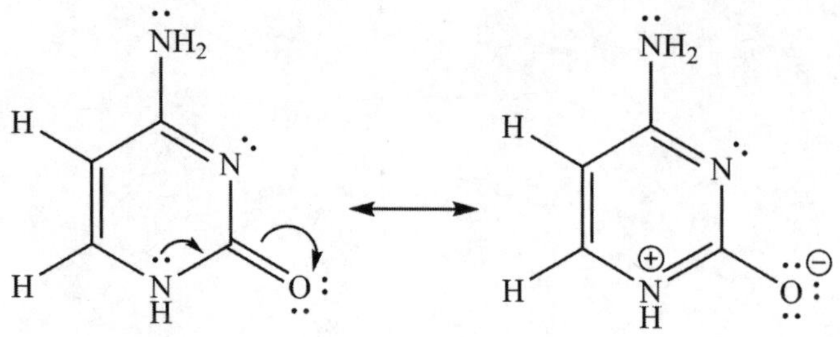

NH_2 H H N N H O ⟷ NH_2 H H N N H O

23.33 (a) ester, alcohol, alkene (b) O = sp^2, $C_a = sp^3$, $C_b = sp^2$, $C_c = sp^3$

(c) HO HO O O HO OH ⟷ HO HO O O HO OH

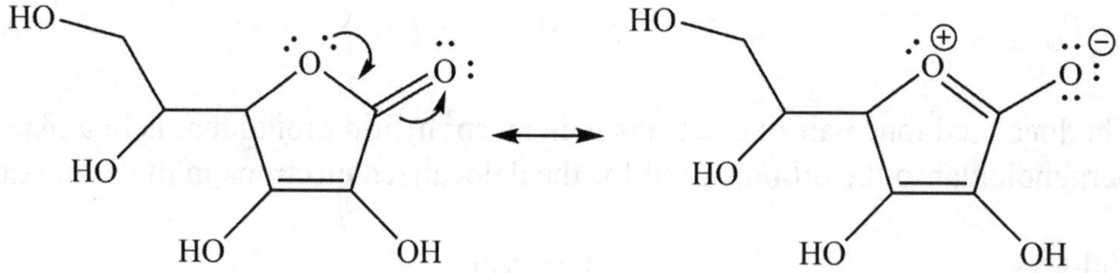

23.34 (a) carbonyl, amine, alkene

(b) O CH_3 H_3C N N O N N CH_3

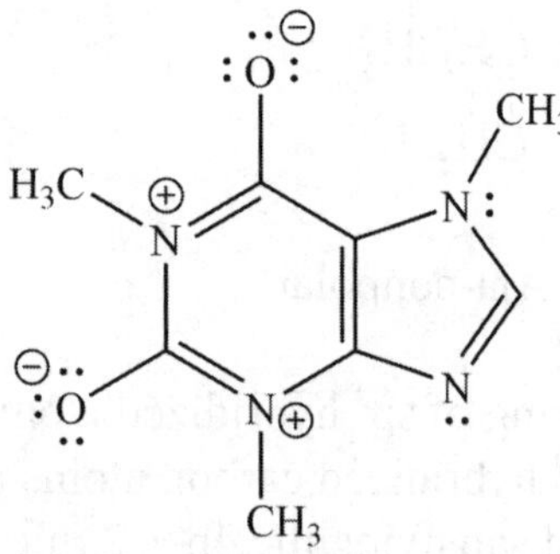

(c) sp^2

(d) 10 electrons in the conjugated system so it meets the 4n + 2 rule. Caffeine is an aromatic compound.

Conceptual Problems

23.36 (a) alkene, ketone, ether (b) alkene, amine, carboxylic acid

23.38 (a) serine (b) methionine

23.40 Ser-Val

Section Problems

Isomers, Functional Groups, and Naming (Sections 23.1–23.3)

23.42 In a straight-chain alkane, all the carbons are connected in a row. In a branched-chain alkane, there are branching connections of carbons along the carbon chain.

23.44 $CH_3CH_2CH_2CH_2CH_2CH_2CH_3$

23.46 (a) No, because they contain different numbers of carbons and hydrogens.
(b) They are isomers of each other.
(c) No, they are identical.

23.48 A functional group is a part of a larger molecule and is composed of an atom or group of atoms that has a characteristic chemical behavior. They are important because their chemistry controls the chemistry in molecules that contain them.

23.50 (a) $CH_3CH_2\overset{\overset{O}{\|}}{C}CH_2CH_3$ (b) $CH_3CH_2CH_2\overset{\overset{O}{\|}}{C}OCH_2CH_3$ (c) $NH_2CH_2\overset{\overset{O}{\|}}{C}OH$

23.52 $CH_3CH_2CH_2OH$ $CH_3\overset{\overset{OH}{|}}{C}HCH_3$ $CH_3CH_2OCH_3$

23.54 ester, aromatic ring, and amine

23.56 (a) 4-ethyl-3-methyloctane
(b) 4-isopropyl-2-methylheptane
(c) 2,2,6-trimethylheptane
(d) 4-ethyl-4-methyloctane

23.58 (a)

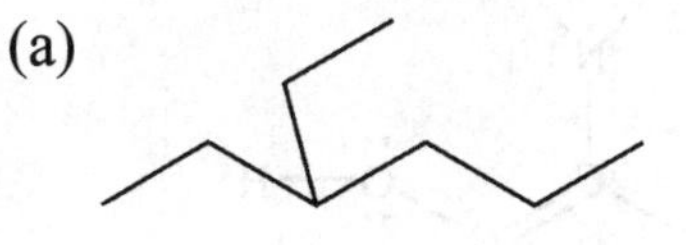

(b)

(c)

(d)

23.60 $CH_2{=}C{=}CHCH_2CH_3$ $CH_2{=}CHCH{=}CHCH_3$ $CH_2{=}CHCH_2CH{=}CH_2$

$CH_3CH{=}C{=}CHCH_3$ $CH_2{=}\overset{\overset{CH_3}{|}}{C}CH{=}CH_2$ $CH_2{=}C{=}\overset{\overset{CH_3}{|}}{C}CH_3$

23.62 (a) 4-methyl-2-pentene
(b) 3-methyl-1-pentene
(c) 1,2-dichlorobenzene, or o-dichlorobenzene
(d) 2-methyl-2-butene
(e) 7-methyl-3-octyne

23.64 (a) C_3H_4 (b) C_4H_8 (c) C_5H_6

23.66 (a)

```
     H   H
     |   |     ..
H—C—C—O—H
     |   |     ..
     H   H
```

C_2H_6O

(b)

```
     H   H
     |   |     ..
H—C—C—N—H
     |   |     |
     H   H    H
```

C_2H_7N

(c)

```
            H
            |
   H       N       H
    \    / ..  \    /
H—C          C—H
H—C          C—H
    /    \       /  \
   H       C       H
          /  \
         H    H
```

$C_5H_{11}N$

(d)

```
     H   H        H
     |   |    ..   |
H—C—C—O—C—H
     |   |    ..   |
     H   H        H
```

C_3H_8O

(e)

```
     H   H   H
     |   |   |    ..
H—C—C—C—F:
     |   |   |    ..
     H   H   H
```

C_3H_7F

(f)

```
          ..
     H   O:  H
     |   ||   |
H—C—C—C—H
     |         |
     H        H
```

C_3H_6O

23.68 (a)

```
          ..
          O:
          ||        ..
          C         O—H
        /    \    /   ..
H—C          C
H /             ||
  H             O:
                ..
```

$C_3H_4O_3$

(b)

```
       H    H
         \ /
          N:
          |          ..
 H      C         O—H
   \  /   \\   /  ..
    C          C
    ||          |
    C          C
   /  \     //  \
 H      C         H
          |
          H
```

C_6H_7NO

(c)

```
      H   H
 H    |    |   H
   \  |    |  /
    C—C
    |      |
  :O:    :O:
    |      |
    C—C
  /   |    |  \
 H   H   H   H
```

$C_4H_8O_2$

Carbohydrates (Section 23.4)

23.70 An aldose contains the aldehyde functional group while a ketose contains the ketone functional group.

23.72

$$\begin{array}{c} \overset{\displaystyle O}{\|} \\ HOCH_2\underset{\displaystyle \underset{\displaystyle OH}{|}}{C}H\,C\,CH_2OH \end{array}$$

23.74 (a) constitutional isomers (b) anomers

Valence Bond Theory, Cis-Trans Isomers (Section 23.5)

23.76 (a) All C–H bond angles are ~109.5°.

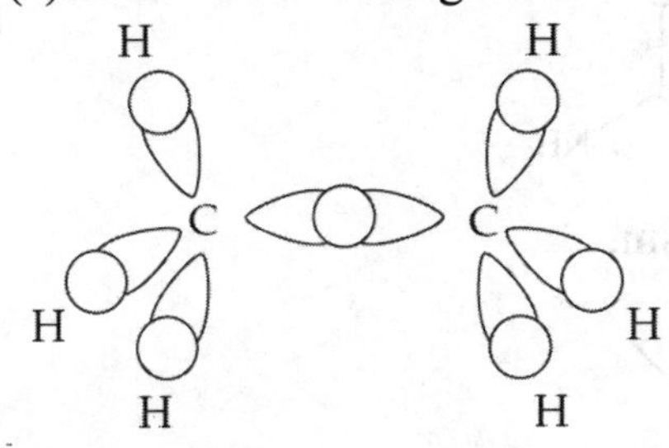

(b) Two C–H bond angles are ~120° and one is 90°.

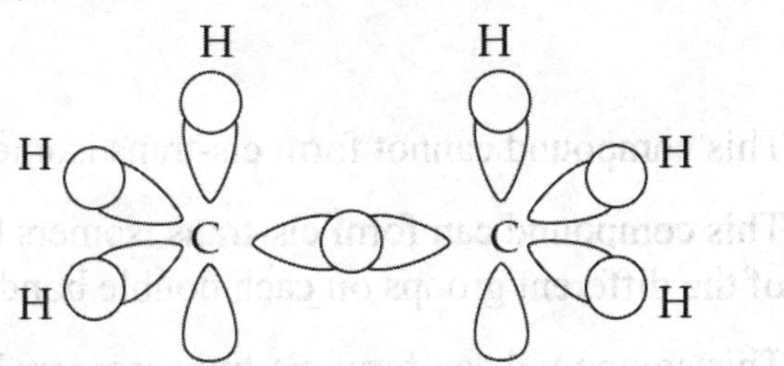

(c) In structure (b), the 90° bond angle introduces a larger repulsion and lower stability. Structure (a) is more favorable.

23.78

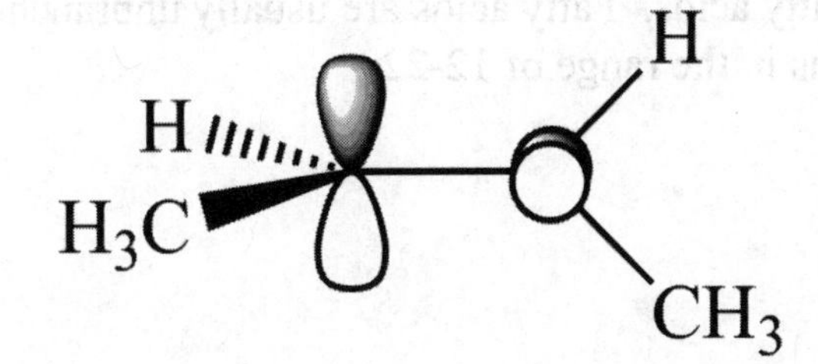

23.80

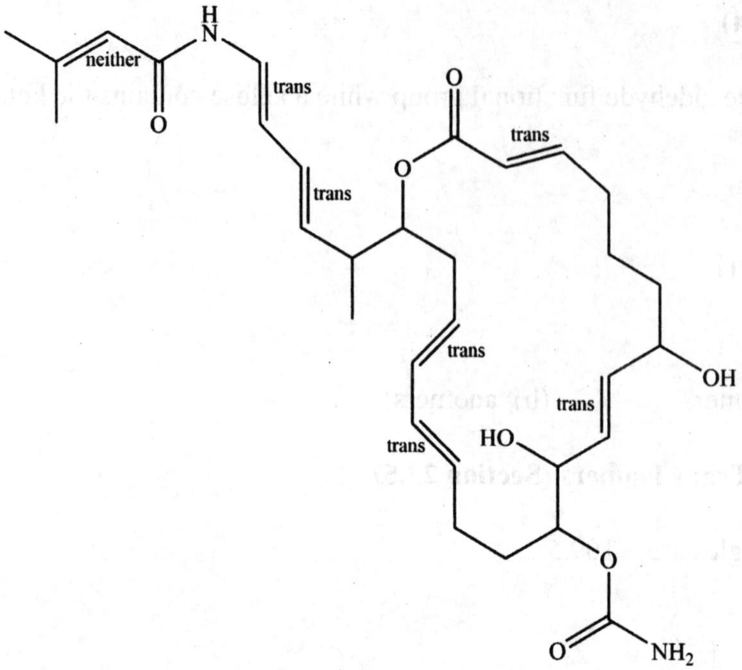

neither
trans
trans
trans
trans
trans
trans

HO
OH
NH_2

23.82 Compounds (b) and (c) exhibit cis–trans isomerism.

(b) 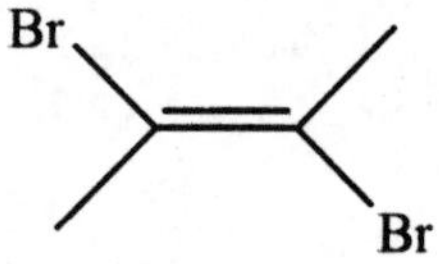 Br, Br — trans

(c) 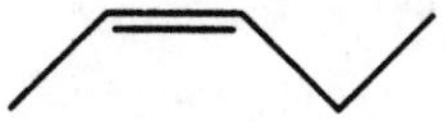 cis

23.84 (a) $CH_2{=}CHCH_2CH_2CH_2CH_3$ This compound cannot form cis-trans isomers.

(b) $CH_3CH{=}CHCH_2CH_2CH_3$ This compound can form cis-trans isomers because of the different groups on each double bond C.

(c) $CH_3CH_2CH{=}CHCH_2CH_3$ This compound can form cis-trans isomers because of the different groups on each double bond C.

Lipids (Section 23.6)

23.86 Long-chain carboxylic acids are called fatty acids. Fatty acids are usually unbranched and have an even number of carbon atoms in the range of 12-22.

23.88

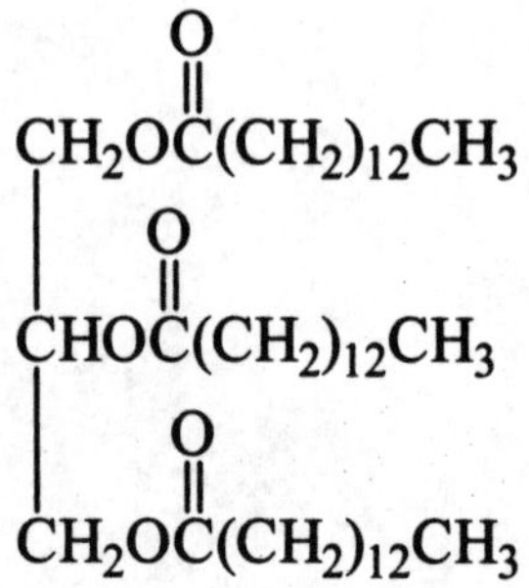

$CH_2OC(=O)(CH_2)_{12}CH_3$

$CHOC(=O)(CH_2)_{12}CH_3$

$CH_2OC(=O)(CH_2)_{12}CH_3$

23.90

$$CH_3(CH_2)_{14}\overset{\overset{\displaystyle O}{\|}}{C}O(CH_2)_{15}CH_3$$

23.92 (a) 18:2 (ω-6) (b) 16:0

23.94 Only (c) is an unsaturated fatty acid. It has double bonds in a cis configuration. This geometry substantially disrupts intermolecular forces, leading to oils (liquids) at room temperature.

23.96 (a) true (b) true (c) false (d) true

23.98 None are possible. (a) has only 15 carbons. (b) is fully hydrogenated. (c) is not hydrogenated.

Formal Charge and Resonance (Section 23.7)

23.100 (a)

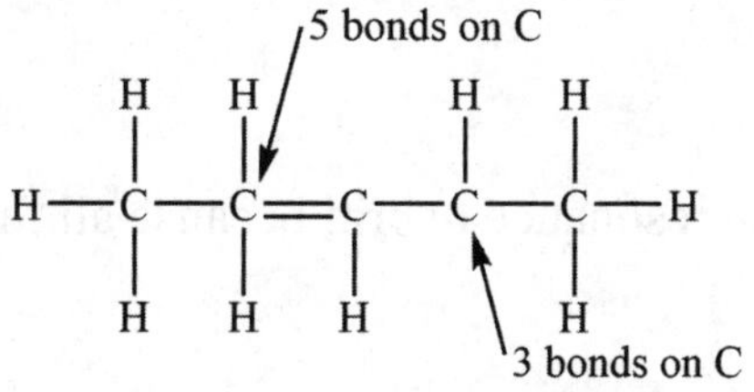

5 bonds on C

3 bonds on C

This structure is not valid.

(b)

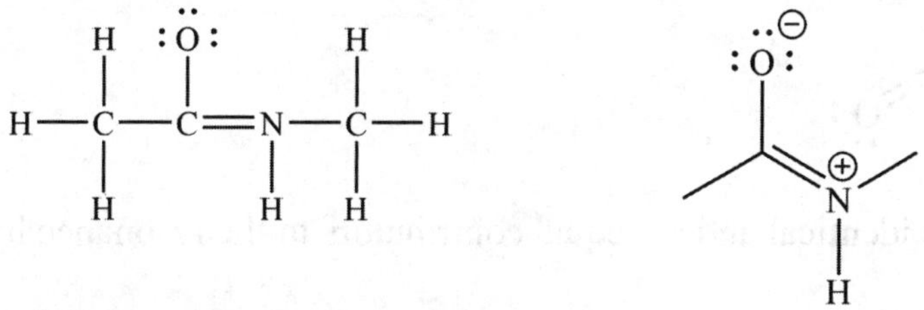

(c) incomplete octet

5 bonds on N & C

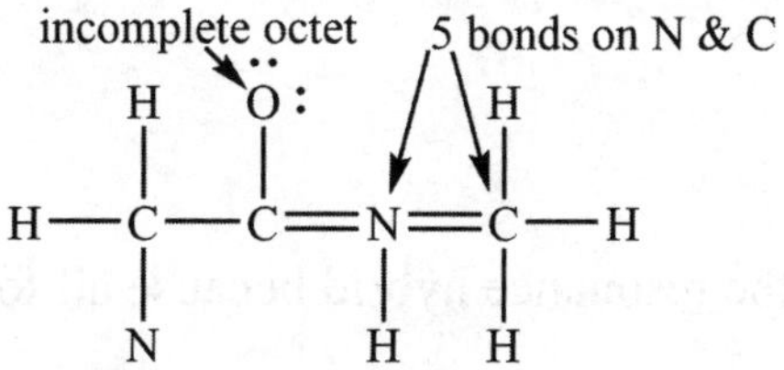

This structure is not valid.

23.102 (a)

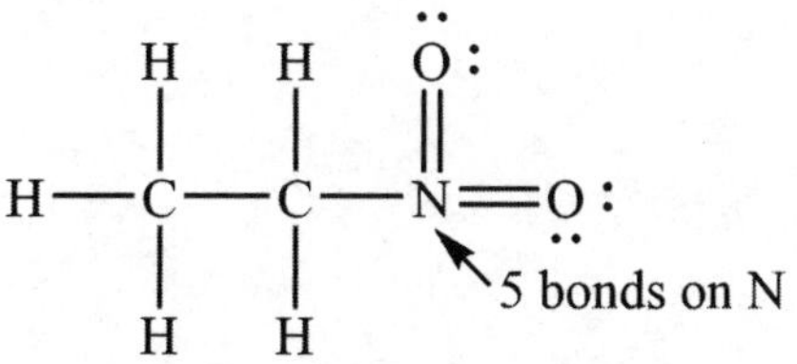

5 bonds on N

This structure is not valid.

(b)

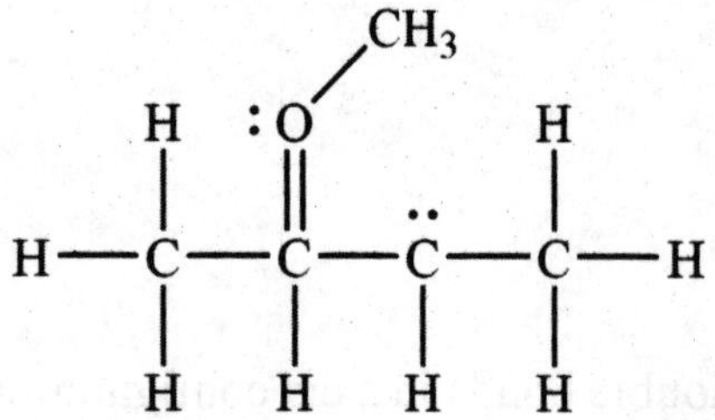

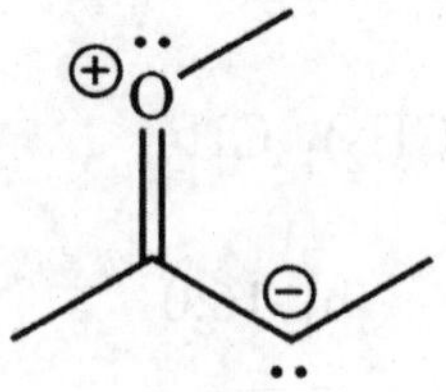

(c)

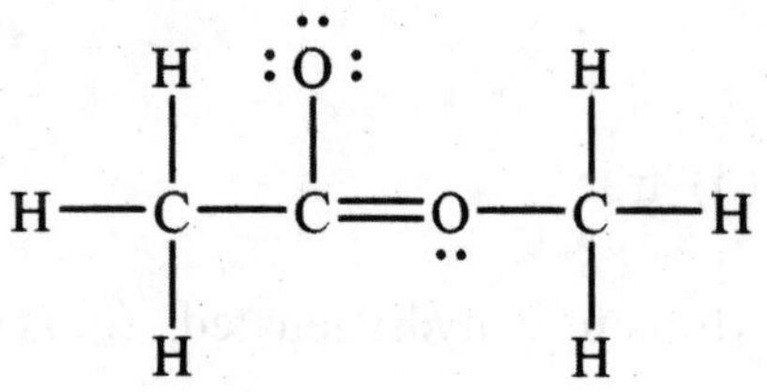

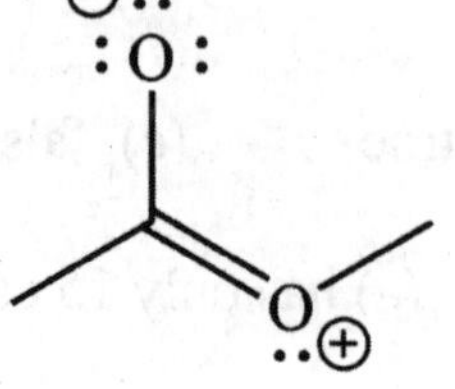

23.104 (a)

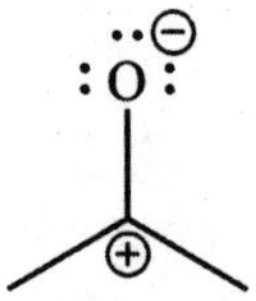

The original structure contributes more to the resonance hybrid because all formal charges are zero.

(b)

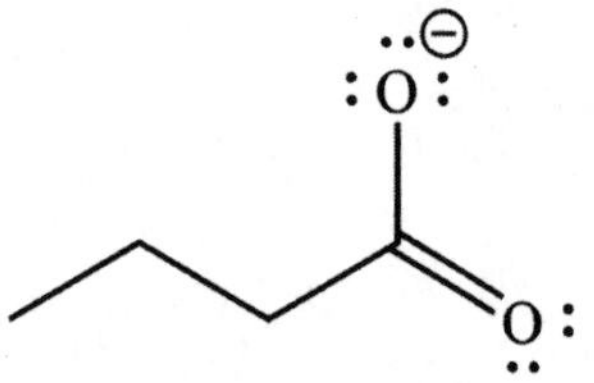

The two structures are identical and are equal contributors to the resonance hybrid.

23.106 (a)

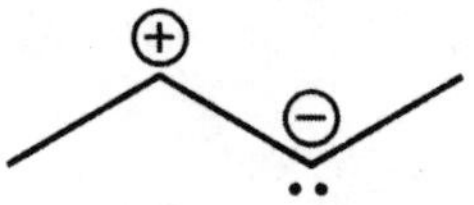

The original structure contributes more to the resonance hybrid because all formal charges are zero.

(b)

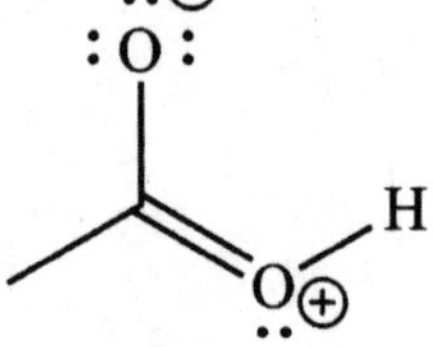

The original structure contributes more to the resonance hybrid because all formal charges are zero.

23.108 They are resonance structures.

23.110 (a)

(b)

Conjugation Systems (Section 23.8)

23.112 (a)

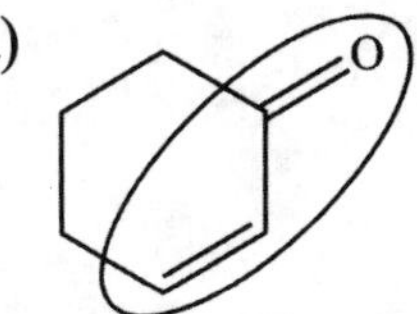

(b)

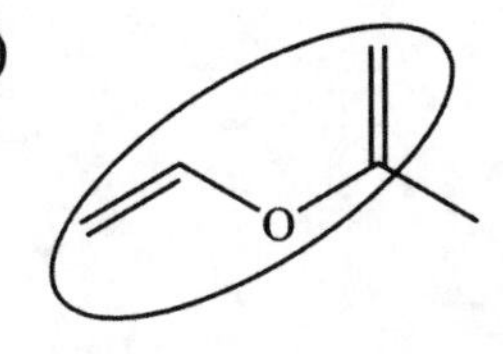

(c)

23.114

23.116

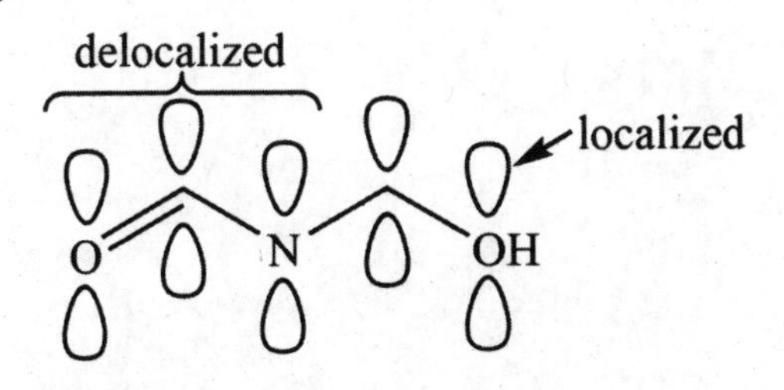

Proteins (Section 23.9)

23.118 (a) serine (b) threonine (c) proline (d) phenylalanine (e) cysteine

23.120 Val-Ser-Phe-Met-Thr-Ala

23.122 Met-Ile-Lys, Met-Lys-Ile, Ile-Met-Lys, Ile-Lys-Met, Lys-Met-Ile, Lys-Ile-Met

Aromaticity and Molecular Orbital Theory (Section 23.10)

23.124 Only (b) is aromatic.

23.126 There are 6 π-electrons, 2 each in the 2 π bonds and 2 in the p-orbital of the NH nitrogen.

Nucleic Acids (Section 23.11)

23.128 Just as proteins are polymers made of amino acid units, nucleic acids are polymers made up of nucleotide units linked together to form a long chain. Each nucleotide contains a phosphate group, an aldopentose sugar, and an amine base.

23.130

-O, P, O, O—CH_2, O, base

O, O, P, -O, O—CH_2, O, base

O

23.132 Original: T–A–C–C–G–A
Complement: A–T–G–G–C–T

23.134

$:\ddot{O}:^{\ominus}$, $\overset{\oplus}{N}H$, $\overset{\oplus}{N}H$, $:\ddot{O}:^{\ominus}$

Chapter Problems

23.136 (a)
$$\begin{array}{l}\quad CH_3\\ \quad\ |\\ CH_3CHCH_2CH_2CH_2CH_2CH_3\end{array}$$

(b)
$$\begin{array}{l}\quad CH_3\quad\ \ CH_2CH_3\\ \quad\ |\qquad\quad |\\ CH_3CHCH_2CHCH_2CH_3\end{array}$$

(c) $CH_3CH_2CH(CH_3)\text{–}C(CH_2CH_3)(CH_3)CH_2CH_2CH_2CH_3$

(d) $CH_3CH(CH_3)CH_2C(CH_3)_2CH_2CH_2CH_3$

(e) CH_3, CH_3 (1,1-dimethylcyclopentane)

(f) $CH_3CH_2CH(CH_3)CH(CH(CH_3)CH_3)CH_2CH_2CH_3$

23.138 $CH_3(CH_2)_{18}C(=O)O(CH_2)_{31}CH_3$

23.140 There are alkene and alkyne functional groups.

23.142 $C_{11}H_{12}N_2O_2$

H, N, H_2N, O, HO

23.144 (a) ketone (b) aldehyde (c) ketone (d) amide (e) ester

23.146 The bicyclic ring system is aromatic with 10 π electrons that satisfies the 4n +2 rule.

23.148 (a) a fat

$CH_2OC(=O)(CH_2)_{16}CH_3$

$CHOC(=O)(CH_2)_{16}CH_3$

$CH_2OC(=O)(CH_2)_{14}CH_3$

(b) a vegetable oil

$$
\begin{array}{l}
CH_2O\overset{O}{\overset{\|}{C}}(CH_2)_7CH{=}CH(CH_2)_7CH_3 \\
|\\
CHO\overset{O}{\overset{\|}{C}}(CH_2)_7CH{=}CH(CH_2)_7CH_3 \\
|\\
CH_2O\overset{O}{\overset{\|}{C}}(CH_2)_7CH{=}CHCH_2CH{=}CH(CH_2)_4CH_3
\end{array}
$$

(c) an aldotetrose

$$HOCH_2\overset{OH}{\overset{|}{C}}H\underset{OH}{\underset{|}{C}}H\overset{O}{\overset{\|}{C}}H$$

23.150

$$CH_3(CH_2)_{16}\overset{O}{\overset{\|}{C}}O(CH_2)_{21}CH_3$$

Multiconcept Problems

23.152 (a) Calculate the empirical formula. Assume a 100.0 g sample of fumaric acid.

$$41.4 \text{ g C} \times \frac{1 \text{ mol C}}{12.01 \text{ g C}} = 3.45 \text{ mol C}$$

$$3.5 \text{ g H} \times \frac{1 \text{ mol H}}{1.008 \text{ g H}} = 3.47 \text{ mol H}$$

$$55.1 \text{ g O} \times \frac{1 \text{ mol O}}{16.00 \text{ g O}} = 3.44 \text{ mol O}$$

Because the mol amounts for the three elements are essentially the same, the empirical formula is CHO (29).

(b) Calculate the molar mass from the osmotic pressure.

$$\Pi = MRT; \quad M = \frac{\Pi}{RT} = \frac{\left(240.3 \text{ mm Hg} \times \dfrac{1.00 \text{ atm}}{760 \text{ mm Hg}}\right)}{\left(0.082\ 06 \dfrac{\text{L}\cdot\text{atm}}{\text{K}\cdot\text{mol}}\right)(298 \text{ K})} = 0.0129 \text{ M}$$

$$(0.1000 \text{ L})(0.0129 \text{ mol/L}) = 1.29 \times 10^{-3} \text{ mol fumaric acid}$$

$$\text{fumaric acid molar mass} = \frac{0.1500 \text{ g}}{1.29 \times 10^{-3} \text{ mol}} = 116 \text{ g/mol}$$

molecular mass = 116

(c) Determine the molecular formula. $\dfrac{\text{molar mass}}{\text{empirical formula mass}} = \dfrac{116}{29} = 4$

molecular formula = $C_{(1 \times 4)}H_{(1 \times 4)}O_{(1 \times 4)} = C_4H_4O_4$

From the titration, the number of carboxylic acid groups can be determined.

$$\text{mol } C_4H_4O_4 = 0.573 \text{ g} \times \frac{1 \text{ mol } C_4H_4O_4}{116 \text{ g}} = 0.004\ 94 \text{ mol } C_4H_4O_4$$

mol NaOH used = (0.0941 L)(0.105 mol/L) = 0.0099 mol NaOH

$$\frac{\text{mol NaOH}}{\text{mol } C_4H_4O_4} = \frac{0.0099 \text{ mol}}{0.004\ 94 \text{ mol}} = 2$$

Because 2 mol of NaOH are required to titrate 1 mol $C_4H_4O_4$, $C_4H_4O_4$ is a diprotic acid. Because $C_4H_4O_4$ gives an addition product with HCl and a reduction product with H_2, it contains a double bond.

HO—C(=O)—CH=CH—C(=O)—OH (cis)   HO—C(=O)—CH=CH—C(=O)—OH (trans) $H_2C{=}C(\overset{O}{\overset{\|}{C}}{-}OH)_2$

(d) The correct structure is HO—C(=O)—CH=CH—C(=O)—OH (trans, H atoms on opposite sides of the double bond)

23.154 (a) $-CO_2H$ is the more acidic group because it has the smaller pK_a (larger K_a).

(b) At pH = 4.00 only HA and H_2A^+ are present in appreciable amounts.

$$pH = pK_a + \log\frac{[\text{Base}]}{[\text{Acid}]}$$

$$4.00 = 2.34 + \log\frac{[HA]}{[H_2A^+]};\quad 4.00 - 2.34 = \log\frac{[HA]}{[H_2A^+]};\quad 1.66 = \log\frac{[HA]}{[H_2A^+]}$$

$$\frac{[HA]}{[H_2A^+]} = 10^{1.66} = 45.7$$

Let [HA] = 100 – x and $[H_2A^+]$ = x

$$\frac{100 - x}{x} = 45.7;\ 100 - x = 45.7x;\ 100 = 46.7x;\ x = \frac{100}{46.7} = 2.1$$

HA = 100 – x = 98% and H_2A^+ = x = 2.1%

(c) At pH = 8.50 only HA and A^- are present in appreciable amounts.

$$pH = pK_a + \log\frac{[\text{Base}]}{[\text{Acid}]}$$

$$8.50 = 9.69 + \log\frac{[A^-]}{[HA]};\quad 8.50 - 9.69 = \log\frac{[A^-]}{[HA]};\quad -1.19 = \log\frac{[A^-]}{[HA]}$$

$$\frac{[A^-]}{[HA]} = 10^{-1.19} = 0.0646$$

Let [HA] = 100 – x and $[A^-]$ = x

$$\frac{x}{100 - x} = 0.0646;\ 0.0646(100 - x) = x;\ 6.46 - 0.0646x = x;\ 6.46 = 1.0646x$$

$$x = \frac{6.46}{1.0646} = 6.1; \qquad HA = 100 - x = 94\% \text{ and } A^- = x = 6.1\%$$

(d) The maximum amount of HA is found at a pH that is midway between the two pK_a values, 2.34 and 9.69. That pH = 6.01.